Lecture Notes in Networks and Systems 799

The series "Lecture Notes in Networks and Systems" publishes the latest developments in Networks and Systems—quickly, informally and with high quality. Original research reported in proceedings and post-proceedings represents the core of LNNS.

Volumes published in LNNS embrace all aspects and subfields of, as well as new challenges in, Networks and Systems.

The series contains proceedings and edited volumes in systems and networks, spanning the areas of Cyber-Physical Systems, Autonomous Systems, Sensor Networks, Control Systems, Energy Systems, Automotive Systems, Biological Systems, Vehicular Networking and Connected Vehicles, Aerospace Systems, Automation, Manufacturing, Smart Grids, Nonlinear Systems, Power Systems, Robotics, Social Systems, Economic Systems and other. Of particular value to both the contributors and the readership are the short publication timeframe and the worldwide distribution and exposure which enable both a wide and rapid dissemination of research output.

The series covers the theory, applications, and perspectives on the state of the art and future developments relevant to systems and networks, decision making, control, complex processes and related areas, as embedded in the fields of interdisciplinary and applied sciences, engineering, computer science, physics, economics, social, and life sciences, as well as the paradigms and methodologies behind them.

Indexed by SCOPUS, INSPEC, WTI Frankfurt eG, zbMATH, SCImago.

All books published in the series are submitted for consideration in Web of Science.

For proposals from Asia please contact Aninda Bose (aninda.bose@springer.com).

Alvaro Rocha · Hojjat Adeli ·
Gintautas Dzemyda · Fernando Moreira ·
Valentina Colla
Editors

Information Systems and Technologies

WorldCIST 2023, Volume 1

Editors
Alvaro Rocha
ISEG
Universidade de Lisboa
Lisbon, Cávado, Portugal

Hojjat Adeli
College of Engineering
The Ohio State University
Columbus, OH, USA

Gintautas Dzemyda
Institute of Data Science and Digital Technologies
Vilnius University
Vilnius, Lithuania

Fernando Moreira
DCT
Universidade Portucalense
Porto, Portugal

Valentina Colla
TeCIP Institute
Scuola Superiore Sant'Anna
Pisa, Italy

ISSN 2367-3370 ISSN 2367-3389 (electronic)
Lecture Notes in Networks and Systems
ISBN 978-3-031-45641-1 ISBN 978-3-031-45642-8 (eBook)
https://doi.org/10.1007/978-3-031-45642-8

This Springer imprint is published by the registered company Springer Nature Switzerland AG
The registered company address is: Gewerbestrasse 11, 6330 Cham, Switzerland

Preface

This book contains a selection of papers accepted for presentation and discussion at the 2023 World Conference on Information Systems and Technologies (WorldCIST'23). This conference had the scientific support of the Sant'Anna School of Advanced Studies, Pisa, University of Calabria, Information and Technology Management Association (ITMA), IEEE Systems, Man, and Cybernetics Society (IEEE SMC), Iberian Association for Information Systems and Technologies (AISTI), and Global Institute for IT Management (GIIM). It took place in Pisa city, Italy, 4–6 April 2023.

The World Conference on Information Systems and Technologies (WorldCIST) is a global forum for researchers and practitioners to present and discuss recent results and innovations, current trends, professional experiences and challenges of modern Information Systems and Technologies research, technological development, and applications. One of its main aims is to strengthen the drive toward a holistic symbiosis between academy, society, and industry. WorldCIST'23 was built on the successes of: WorldCIST'13 held at Olhão, Algarve, Portugal; WorldCIST'14 held at Funchal, Madeira, Portugal; WorldCIST'15 held at São Miguel, Azores, Portugal; WorldCIST'16 held at Recife, Pernambuco, Brazil; WorldCIST'17 held at Porto Santo, Madeira, Portugal; WorldCIST'18 held at Naples, Italy; WorldCIST'19 held at La Toja, Spain; WorldCIST'20 held at Budva, Montenegro; WorldCIST'21 held at Terceira Island, Portugal; and WorldCIST'22, which took place online at Budva, Montenegro.

The Program Committee of WorldCIST'23 was composed of a multidisciplinary group of 339 experts and those who are intimately concerned with Information Systems and Technologies. They have had the responsibility for evaluating, in a 'blind review' process, and the papers received for each of the main themes proposed for the Conference were: A) Information and Knowledge Management; B) Organizational Models and Information Systems; C) Software and Systems Modeling; D) Software Systems, Architectures, Applications, and Tools; E) Multimedia Systems and Applications; F) Computer Networks, Mobility, and Pervasive Systems; G) Intelligent and Decision Support Systems; H) Big Data Analytics and Applications; I) Human-Computer Interaction; J) Ethics, Computers & Security; K) Health Informatics; L) Information Technologies in Education; M) Information Technologies in Radiocommunications; and N) Technologies for Biomedical Applications.

The conference also included workshop sessions taking place in parallel with the conference ones. Workshop sessions covered themes such as: Novel Computational Paradigms, Methods, and Approaches in Bioinformatics; Artificial Intelligence for Technology Transfer; Blockchain and Distributed Ledger Technology (DLT) in Business; Enabling Software Engineering Practices Via Latest Development's Trends; Information Systems and Technologies for the Steel Sector; Information Systems and Technologies for Digital Cultural Heritage and Tourism; Recent Advances in Deep Learning Methods and Evolutionary Computing for Health Care; Data Mining and Machine Learning in Smart Cities; Digital Marketing and Communication, Technologies, and Applications;

Digital Transformation and Artificial Intelligence; and Open Learning and Inclusive Education Through Information and Communication Technology.

WorldCIST'23 and its workshops received about 400 contributions from 53 countries around the world. The papers accepted for oral presentation and discussion at the conference are published by Springer (this book) in four volumes and will be submitted for indexing by WoS, Scopus, EI-Compendex, DBLP, and/or Google Scholar, among others. Extended versions of selected best papers will be published in special or regular issues of leading and relevant journals, mainly JCR/SCI/SSCI and Scopus/EI-Compendex indexed journals.

We acknowledge all of those that contributed to the staging of WorldCIST'23 (authors, committees, workshop organizers, and sponsors). We deeply appreciate their involvement and support that was crucial for the success of WorldCIST'23.

April 2023

Alvaro Rocha
Hojjat Adeli
Gintautas Dzemyda
Fernando Moreira
Valentina Colla

Organization

Honorary Chair

Hojjat Adeli	The Ohio State University, USA

General Chair

Álvaro Rocha	ISEG, University of Lisbon, Portugal

Co-chairs

Gintautas Dzemyda	Vilnius University, Lithuania
Sandra Costanzo	University of Calabria, Italy

Workshops Chair

Fernando Moreira	Portucalense University, Portugal

Local Organizing Committee

Valentina Cola (Chair)	Scuola Superiore Sant'Anna—TeCIP Institute, Italy
Marco Vannucci	Scuola Superiore Sant'Anna—TeCIP Institute, Italy
Vincenzo Iannino	Scuola Superiore Sant'Anna—TeCIP Institute, Italy
Stefano Dettori	Scuola Superiore Sant'Anna—TeCIP Institute, Italy

Advisory Committee

Ana Maria Correia (Chair)	University of Sheffield, UK
Brandon Randolph-Seng	Texas A&M University, USA

Program Committee

Ana Carla Amaro	Universidade de Aveiro, Portugal
Ana Isabel Martins	University of Aveiro, Portugal
Anabela Tereso	University of Minho, Portugal
Anabela Gomes	University of Coimbra, Portugal
Anacleto Correia	CINAV, Portugal
Andrew Brosnan	University College Cork, Ireland
Andjela Draganic	University of Montenegro, Montenegro
Aneta Polewko-Klim	University of Białystok, Institute of Informatics, Poland
Aneta Poniszewska-Maranda	Lodz University of Technology, Poland
Angeles Quezada	Instituto Tecnologico de Tijuana, Mexico
Anis Tissaoui	University of Jendouba, Tunisia
Ankur Singh Bist	KIET, India
Ann Svensson	University West, Sweden
Anna Gawrońska	Poznański Instytut Technologiczny, Poland
Antoni Oliver	University of the Balearic Islands, Spain
Antonio Jiménez-Martín	Universidad Politécnica de Madrid, Spain
Aroon Abbu	Bell and Howell, USA
Arslan Enikeev	Kazan Federal University, Russia
Beatriz Berrios Aguayo	University of Jaen, Spain
Benedita Malheiro	Polytechnic of Porto, ISEP, Portugal
Bertil Marques	Polytechnic of Porto, ISEP, Portugal
Boris Shishkov	ULSIT/IMI-BAS/IICREST, Bulgaria
Borja Bordel	Universidad Politécnica de Madrid, Spain
Branko Perisic	Faculty of Technical Sciences, Serbia
Carla Pinto	Polytechnic of Porto, ISEP, Portugal
Carlos Balsa	Polythecnic of Bragança, Portugal
Carlos Rompante Cunha	Polytechnic of Bragança, Portugal
Catarina Reis	Polytechnic of Leiria, Portugal
Célio Gonçalo Marques	Polythenic of Tomar, Portugal
Cengiz Acarturk	Middle East Technical University, Turkey
Cesar Collazos	Universidad del Cauca, Colombia
Christine Gruber	K1-MET, Austria
Christophe Guyeux	Universite de Bourgogne Franche Comté, France
Christophe Soares	University Fernando Pessoa, Portugal
Christos Bouras	University of Patras, Greece
Christos Chrysoulas	London South Bank University, UK
Christos Chrysoulas	Edinburgh Napier University, UK
Ciro Martins	University of Aveiro, Portugal
Claudio Sapateiro	Polytechnic of Setúbal, Portugal
Cosmin Striletchi	Technical University of Cluj-Napoca, Romania
Costin Badica	University of Craiova, Romania

Gian Piero Zarri	University Paris-Sorbonne, France
Giovanni Buonanno	University of Calabria, Italy
Gonçalo Paiva Dias	University of Aveiro, Portugal
Goreti Marreiros	ISEP/GECAD, Portugal
Graciela Lara López	University of Guadalajara, Mexico
Habiba Drias	University of Science and Technology Houari Boumediene, Algeria
Hafed Zarzour	University of Souk Ahras, Algeria
Haji Gul	City University of Science and Information Technology, Pakistan
Hakima Benali Mellah	Cerist, Algeria
Hamid Alasadi	Basra University, Iraq
Hatem Ben Sta	University of Tunis at El Manar, Tunisia
Hector Fernando Gomez Alvarado	Universidad Tecnica de Ambato, Ecuador
Hector Menendez	King's College London, UK
Hélder Gomes	University of Aveiro, Portugal
Helia Guerra	University of the Azores, Portugal
Henrique da Mota Silveira	University of Campinas (UNICAMP), Brazil
Henrique S. Mamede	University Aberta, Portugal
Henrique Vicente	University of Évora, Portugal
Hicham Gueddah	University Mohammed V in Rabat, Morocco
Hing Kai Chan	University of Nottingham Ningbo China, China
Igor Aguilar Alonso	Universidad Nacional Tecnológica de Lima Sur, Peru
Inês Domingues	University of Coimbra, Portugal
Isabel Lopes	Polytechnic of Bragança, Portugal
Isabel Pedrosa	Coimbra Business School - ISCAC, Portugal
Isaías Martins	University of Leon, Spain
Issam Moghrabi	Gulf University for Science and Technology, Kuwait
Ivan Armuelles Voinov	University of Panama, Panama
Ivan Dunđer	University of Zabreb, Croatia
Ivone Amorim	University of Porto, Portugal
Jaime Diaz	University of La Frontera, Chile
Jan Egger	IKIM, Germany
Jan Kubicek	Technical University of Ostrava, Czech Republic
Jeimi Cano	Universidad de los Andes, Colombia
Jesús Gallardo Casero	University of Zaragoza, Spain
Jezreel Mejia	CIMAT, Unidad Zacatecas, Mexico
Jikai Li	The College of New Jersey, USA
Jinzhi Lu	KTH-Royal Institute of Technology, Sweden
Joao Carlos Silva	IPCA, Portugal

Miguel Melo	INESC TEC, Portugal
Mihai Lungu	University of Craiova, Romania
Mircea Georgescu	Al. I. Cuza University of Iasi, Romania
Mirna Muñoz	Centro de Investigación en Matemáticas A.C., Mexico
Mohamed Hosni	ENSIAS, Morocco
Monica Leba	University of Petrosani, Romania
Nadesda Abbas	UBO, Chile
Narjes Benameur	Laboratory of Biophysics and Medical Technologies of Tunis, Tunisia
Natalia Grafeeva	Saint Petersburg University, Russia
Natalia Miloslavskaya	National Research Nuclear University MEPhI, Russia
Naveed Ahmed	University of Sharjah, United Arab Emirates
Neeraj Gupta	KIET group of institutions Ghaziabad, India
Nelson Rocha	University of Aveiro, Portugal
Nikola S. Nikolov	University of Limerick, Ireland
Nicolas de Araujo Moreira	Federal University of Ceara, Brazil
Nikolai Prokopyev	Kazan Federal University, Russia
Niranjan S. K.	JSS Science and Technology University, India
Noemi Emanuela Cazzaniga	Politecnico di Milano, Italy
Noureddine Kerzazi	Polytechnique Montréal, Canada
Nuno Melão	Polytechnic of Viseu, Portugal
Nuno Octávio Fernandes	Polytechnic of Castelo Branco, Portugal
Nuno Pombo	University of Beira Interior, Portugal
Olga Kurasova	Vilnius University, Lithuania
Olimpiu Stoicuta	University of Petrosani, Romania
Patricia Zachman	Universidad Nacional del Chaco Austral, Argentina
Paula Serdeira Azevedo	University of Algarve, Portugal
Paula Dias	Polytechnic of Guarda, Portugal
Paulo Alejandro Quezada Sarmiento	University of the Basque Country, Spain
Paulo Maio	Polytechnic of Porto, ISEP, Portugal
Paulvanna Nayaki Marimuthu	Kuwait University, Kuwait
Paweł Karczmarek	The John Paul II Catholic University of Lublin, Poland
Pedro Rangel Henriques	University of Minho, Portugal
Pedro Sobral	University Fernando Pessoa, Portugal
Pedro Sousa	University of Minho, Portugal
Philipp Jordan	University of Hawaii at Manoa, USA
Piotr Kulczycki	Systems Research Institute, Polish Academy of Sciences, Poland

Contents

Human-Computer Interaction

Health Informatics

Computer Networks, Mobility and Pervasive Systems

Physarum-Inspired Enterprise Network Redesign

Sami J. Habib(✉) and Paulvanna N. Marimuthu

Computer Engineering Department, Kuwait University, Safat , P. O. Box 5969, 13060 Kuwait City, Kuwait

{sami.habib,paulvanna.m}@ku.edu.kw

Abstract. We have developed a Physarum-inspired redesign algorithm to redesign the existing enterprise network (EN) with an objective function to minimize the traffic through the backbone. In contrast to the classical Physarum model, which forms an initial network by exploring its surroundings, our redesign algorithm is developed under the constraint that the topology of EN is given. The nodes and edges of EN are analogous to the food sources and connecting tubes of the Physarum network, whereas the traffic demand on the edges of EN is reversed to match the tubular length. Our redesign algorithm exploits the foraging behavior of Physarum to generate the possible maximum-flow paths by selecting the combination of each node as a source against the remaining as a destination. A factor called affinity factor is defined to select the highly associated node(s) of each source by traversing the generated paths from each source. A clustering algorithm is added to group the associated nodes together, and the procedure is repeated for all nodes. The experimental results on a weighted undirected network revealed that the redesign algorithm manages to reduce EN backbone traffic by a maximum of 78% from the initial single-clustered EN by optimally distributing the nodes within the clusters.

Keywords: Affinity factor · Backbone traffic · Clustering · Network redesign · Physarum foraging

1 Introduction

The enterprise network has been utilized for many commercial and social applications as it is the basis of the global internet. The internal behavior is expected to handle dynamic workloads; thus, EN infrastructure should evolve incrementally to cope with the increasing workloads. We view an existing EN as a two-level structure, where the backbone network forms the higher-level and serves as a traffic concentration point and possible connection to the global internet; the lower-level network has a set of clients added over a timespan and the traffic between any two nodes is routed through the backbone network. In our prior works on network redesign, we developed a number of custom-made tools to reduce the traffic at the backbone, where the redesign schemes generated alternate topologies by relocating the clients randomly within the clusters; the

A. Rocha et al. (Eds.): WorldCIST 2023, LNNS 799, pp. 3–13, 2024.
https://doi.org/10.1007/978-3-031-45642-8_1

search space is explored using bio-inspired algorithms, such as Genetic Algorithm [1, 2], Ant-colony Optimization [3], Molecular Assembly [4] and Tabu Search [5]. In this paper, we used Physarum's shortest path finding abilities to generate a set of traffic-balanced clusters, so as to reduce the traffic on the backbone.

Physarum Polycephalum, a unicellular amoeboid forages for food sources through a network of tubes that assemble and disassemble in response to the availability of nutrients in the food sources [6]. When nutrients are abundant in the searched food sources, the tubes are short and thick, and when nutrients are scarce, the tubes are long and narrow in diameter. The patterns of connectivity, known as the protoplasmic tube network, evolve to strike a balance between distance and quantity of nutrients. This intelligent behavior aids Physarum in addressing issues with decision-making [7, 8], shortest path finding [9], and spatial optimization [10, 11]. Nakagaki et al. [12] first demonstrated that Physarum could manage the path planning challenges, wherein the computer simulations converge to the shortest path for any input graph. He adopted the poiseuille flow to model the fluid flow through the tubuloid structure of Physarum. Bonifaci et al. [13] have provided analytical proof for this convergence model. Subsequently, there were many improvement to the path finding algorithm. In this work, we have followed the work in [14] and we have considered the discretization of Physarum's dynamic behavior to solve the EN redesign problem.

In this paper, our main objective function is to generate clusters of highly communicating nodes, so as to reduce the inter-cluster traffic through the backbone. We view EN as weighted undirected graph analogous to the Physarum network [15], and the weights on the edges represent the traffic flow between two nodes. Since the weights on the edges represent the strength to associate the nodes, the weights are inverted to match them to the tube length in Physarum shortest path finding model. We have developed the redesign algorithm under the constraint that the topology of EN is given. A set of possible shortest paths is generated using the classical Physarum Solver [16] by selecting a node in the network as a source node at a time and the remaining nodes as destinations. A new factor called affinity factor is defined to find the highly associated node with each source and it is defined as a ratio of the frequency of the node appearance in the generated paths to the total number of paths for the selected source. We have introduced a function that uses the affinity factor of nodes to combine compatible pairs in forming clusters, with the output being inter-cluster traffic routed through the backbone. The function is repeatedly called to check all the nodes in the network. The experimental results of a weighted undirected network with 10 nodes resulted in two clusters, and it showed that the traffic flow through the backbone is reduced by 78% than the existing single clustered network.

2 Related Work

Our extensive literature survey revealed that the research on Physarum computing has become more popular in recent years after the maze-solving experiment carried out by Nagasaki et al. [12]. Subsequent research works confirmed that the cognitive-skills of Physarum is used to solve wide variety of problems, such as shortest-path generation [17], building real-world transportation networks [18, 19], and Steiner-tree problems [20, 21].

The Physarum model is well-studied from the computational point of view from the first mathematical modeling introduced by Tero et al. [22], where he described the Physarum protoplasm–nutrient transportation behavior using Hagen-Poiseuille law. The Physarum-model is refined by Adamatsky [23] using reaction-diffusion mechanism, and Zhang and Yan [24] developed a general mathematical model to solve network optimization problems for finding shortest path in directed and in undirected networks. A recent research Physarum dealt with a new stochastic optimizer [25], which is based on the oscillatory behavior of Physarum.

Our extended study on Physarum skill based network clustering problem revealed that there are very few works dealt with community detection [26], where the authors developed Physarum-inspired initial candidate solution for refining the solutions developed using Ant Colony Optimization and Markov-clustering algorithms; moreover, the authors employed different parameters, such as frequency of communication with its neighbors, and degree of connectivity to detect the community structure in social networks.

In our humble research endeavor, we have explored many algorithms for distributed system redesign including the network topology and data management. Here, we defined a factor called affinity factor to study the node associations with a focus to generate clusters from the generated paths.

3 Modeling Enterprise Network Redesign

We have considered EN as a weighted graph G (V, E, W), where V is the set of client nodes, E is the set of edges connecting the clients, and W is the set of weights on the edges. The redesign problem is viewed as a node consolidation problem with an objective function to minimize the inter-cluster traffic through the backbone as stated in Eq. (1). Here, the sum of the weights on the edges $w(e_{ij})$ across the clusters represents the backbone traffic; the term δ_{ij} is the Boolean variable, and it is equal to 1 if the edge is present between the clients (i,j). We have added a few constraints to govern the objective function. Constraint (2) is added to ensure that the weight on the edges should be positive or zero. Constraint (3) is added to limit the maximum number of clusters to $\frac{|N|}{2}$ and Constraint (4) confirms that the sum of nodes within the generated clusters should be equal to N.

$$\min \phi = \sum_{\substack{e_{ij} \in E \\ i \in c_l \& j \in c_k \\ i \neq j \& l \neq k}} w(e_{ij}) * \delta_{ij} \tag{1}$$

Subject to:

$$w(e_{ij}) \geq 0, \ \forall e_{ij} \in E \tag{2}$$

$$2 \leq cl_i \leq \frac{|N|}{2}, i = 1, 2, 3, \ldots M \tag{3}$$

$$\sum_{1}^{N/2} |cl| = N \tag{4}$$

3.1 Modeling Physarum Model for Network Redesign

The foragers searching for food sources usually employ information on food density, quality, etc. to locate the resources, and then they may decide how long to spend on exploiting the resources. In Physarum's' lifecycle, it searches for food sources within an area in its exploratory phase and it tries to connect the explored food sources to the central through the shortest path during its exploitation stage. In this work, we modified the classical Physarum solver algorithm proposed by Tero et al. [22] with the constraint that EN topology is given rather than the network formed by Physarum in its exploratory phase. We view EN as an N-clustered network analogous to the initial network formed by Physarum, where the food sources form the vertices and the veins that connect them form the tubular structures. The traffic flow on the edges of EN network form the weights and the weights are inverted to match the tube length of Physarum's network. In our work, we did not consider the physical distance and the actual location of the nodes while generating the clusters; instead, we focused on their associations as a constraint to facilitate a degree of freedom in choosing a topology according to the requirements of the application. The flow path generated by the modified algorithm reflects the path with highest demand on the edges between the nodes on the path generated between the selected source and destination. The developed model checks all combinations of source and sink; if N is the total number of nodes in EN, then, the path finding algorithm is repeated for N_{C_2} times.

The flow rate of the tubes in Physarum varies dynamically and it is directly proportional to the pressure variations in the tube, which in turn varies with the amount of nutrients at the food sources. The traffic demand $\left(T_{ij}\right)$ through an edge e_{ij} is associated with the flux Q_{ij} through the tube between the food source and the central node in Physarum and it is represented using Eq. (5), which is formulated according to [16]; the food source is the node $n_i \in V$ and the backbone is analogous to central node. The term D_{ij} is the diameter of the tube and D_{ij} matrix is defined with values (0 1] initially. The parameter V is the set of nodes distributed within EN. If there is no edge exist between any two nodes, then the term $t_{ij} = 0$ and the traffic flow through backbone is also zero, as analogous to the vanishing of tubular structure if the flux flow is zero. The term $D_{ij} * t_{ij}$ gives the variation of diameter.

$$T_{ij} = \frac{D_{ij}}{W_{ij}} * (P_i - P_j), \, where \, w_{ij} = \frac{1}{t_{ij}} \tag{5}$$

The pressure is calculated using Eq. (6) according to [16].

$$\sum_i \frac{D_{ij}^h}{W_{ij}} \left|P_i - P_j\right| = \begin{cases} I_0 \; I \; is \; a \; source \\ -\, I_0 \; I \; is \; a \; destin \\ 0 \; otherwise \end{cases} \tag{6}$$

The tube diameter updating is done using Eq. (7) according to [16]. Here h is the time step and the term Δ is the mesh size. The diameter evolves using Eq. (7).

$$\frac{D_{ij}^{h+1} - D_{ij}^h}{\Delta} = \left|T_{ij}^h\right| - D_{ij}^{h+1} \tag{7}$$

We introduce a new factor, called affinity factor as stated in Eq. (8), and it is defined as the number of times the node n_i is present in the set of paths P generated by a source to reach the destinations. Here, the term s represents the source node and the term δ_{ip} is a Boolean term that represents the presence of node $n_i \in N$ in path $p_j \in P$.

$$A(n_i) = \frac{\sum_{i=1}^{N-1} n_i * \delta_{ip}}{|P|}, \forall p_i \in P \; and \; i \neq s \tag{8}$$

The clusters are formed by combining the source s with the node n_i that is having highest affinity factor in the generated path set P, as demonstrated in Eq. (9), where the suffix k represents the cluster number k. The generated clusters are further consolidated by checking the affinity of nodes outside the clusters.

$$c_k = \{\{s\} \cup \{n_i\} | A(n_i) = \max\} \tag{9}$$

4 Proposed Method

The developed EN redesign algorithm is demonstrated in *Algorithm 1*, where we have modified the classical Physarum algorithm proposed by Tero et al. [22] to suit our problem domain. The generation of maximum flow path is described in lines 2 to 7. Since the classical Physarum algorithm was designed for a single source and sink, the algorithm is applied many times to generate a set of paths for each source while simultaneously designating the other nodes as sinks. The combination is selected with replacement, since the given EN is an undirected network graph. The construction of the final set of clusters is described in lines 8 to 16 along with the computation of the affinity factor and cluster formulation. The initial cluster generation and the redesign operation of generated clusters check for compliance with the given constraints in Eqs. (2) to (4) while forming the clusters.

Redesign_Algorithm()

1. ***Begin***
2. *G= Generate_undirected_weighted_EN (V, E, W)*
3. ***For** each node n_i in V as source*
4. ***For** node n_k in V except n_i*
5. *Path [n_i, n_k] =**Call Physarum_path_generation** (G, n_i, n_k)*
6. ***End for***
7. ***End for***
8. *Initialize cluster={ }*
9. ***For** node n_i in V as source s_i*
10. *A(i) = **max** (Estimate **node_affinity_pairs** (s_i, $\mathbf{Path}(s_i)$) using Equation (8))*
11. *cluster = **init_cluster** (s_i, A(i), cluster)*
12. ***End for***
13. *Compute the inter-cluster traffic (cluster)*
14. *output clusters*
15. ***End***

Algorithm 1: Physarum based EN redesign algorithm.

The modified Physarum path generation classical algorithm is described in *Algorithm 2*. As a preprocessing step, the network G is generated with the inverted traffic demand as the weight on the edge. The diameter D_{ij} matrix has a range (0 1]; with the assumption of flux flow in Physarum matched with the traffic demand at each node, the traffic flow matrix and diameter of the edges are updated and the algorithm is used find the maximum flow path. The algorithm terminates when it reaches the maximum step size *M*. The cluster formation is illustrated using init_cluster algorithm, as demonstrated in *Algorithm 3*. The algorithm checks the presence of either the source node or its affinity node in the existing cluster and it forms a new cluster if both does not exists; if either of the node exists, the other node is added to the same cluster. The algorithm breaks if both the nodes are present.

Physarum_path_generation (G, source, destination)

1. *Begin*
2. *Initialize D_0*
3. *Max_iteration = M*
4. *For i from 1 to M*
5. *Choose n_i as Physarum central node*
6. *Choose n_k as the food source*
7. *Calculate the pressure p_i based on Equation (6)*
8. *Calculate the traffic flow Ti based on Equation (5)*
9. *Update D using Equation (7)*
10. *End For*
11. *Output generated_path(n_i to n_k)*
12. *End*

Algorithm 2: Physarum based path generation algorithm.

1. *init_cluster (s_i, A(i), cluster)*
2. *Begin*
3. *Check the presence of s_i or A(i) in* ***cluster*** *;*
4. *If (either si or A(i))*
5. *add non-existing node to the cluster of other*
6. *Elsif (both_not_present in* ***cluster****)*
7. *New_cluster* = $node\{si\} \cup \{node(A(i))\}$
8. *else break*
9. *Update* ***cluster***
10. *End*

Algorithm 3: Generation of final clusters.

5 Results and Discussion

We have coded the redesign_algorithm in a Python programming environment, where we utilized '*networkx*' package to simulate the existing EN graph. We considered a small enterprise network comprised of 10 nodes with traffic demand between the nodes ranging from 4 Mbit/s to 14 Mbit/s, and it is represented as a sparse weighted undirected graph, as shown in Fig. 1. As a preprocessing step, the traffic demand on the edges of the input EN is inverted to match the distance in Physarum. The diameter D_{ij} matrix is filled with a value of 0.5 except for the diagonal elements as stated in Eq. (10). The algorithm first generated a set of paths for each node using the modified Physarum algorithm, and a generated path between source node 1 and destination node 9 is shown in Fig. 2.

$$D = \begin{bmatrix} 0 & \cdots & 0.5 \\ \vdots & \ddots & \vdots \\ 0.5 & \cdots & 0 \end{bmatrix} \tag{10}$$

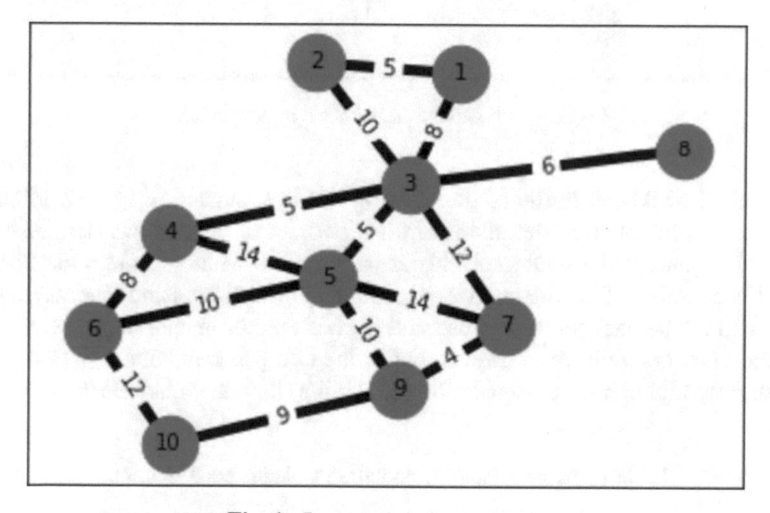

Fig. 1. Input enterprise network.

Subsequently, the affinity factors of each node towards its source that is present in the generated set of paths are estimated; a sample calculation describing the source node, the nodes present in the generated paths, the associated affinity factors and the set of clusters are tabulated in Table 1. The affinity factor has a value between 0 and 1and it is maximal when the selected node appears in all of the generated paths. The second row of Table 1 depicts the particular case where the source node is already a member of a cluster and the affinity node is included in the existing cluster.

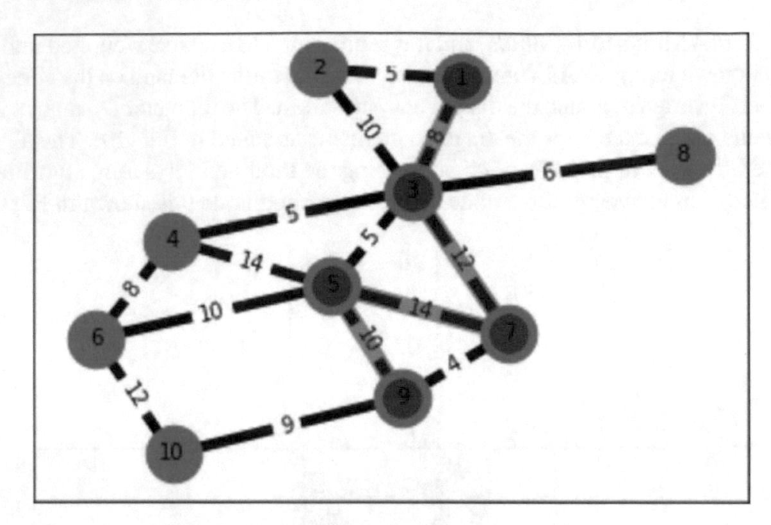

Fig. 2. Path generated between node 1 and 9.

The initial backbone traffic of the existing EN is estimated to be 132 Mbit/s. The final network generated by the algorithm has two set of clusters, as demonstrated in Fig. 3 which generated a final inter-cluster traffic of 29 Mbit/s through the backbone. The backbone traffic of the final network is found to be optimum and thus, saving a total of 78% load on the backbone. We observed from the experimental results that the node distribution is found uniform; however, out of the two generated clusters, one is loaded with a slightly higher intra-cluster traffic (63 Mbit/s) than the other cluster (41 Mbit/s).

Table 1. Affinity factor Calculation and clusters formation.

s.no	Source node	No of paths	Nodes in the path	Affinity factors	Clusters
1	2	8	3	1	2,3
2	3	7	7 8 4	0.71 0.14 0.14	2,3,7
3	4	6	5 6 3	0.5 0.33 0.17	4,5
4	8	2	3	1	2,3,7,8

We are continuing our research towards improving our algorithm to be suitable for large size networks and also improving the optimization of the cluster formation by considering further information within each node.

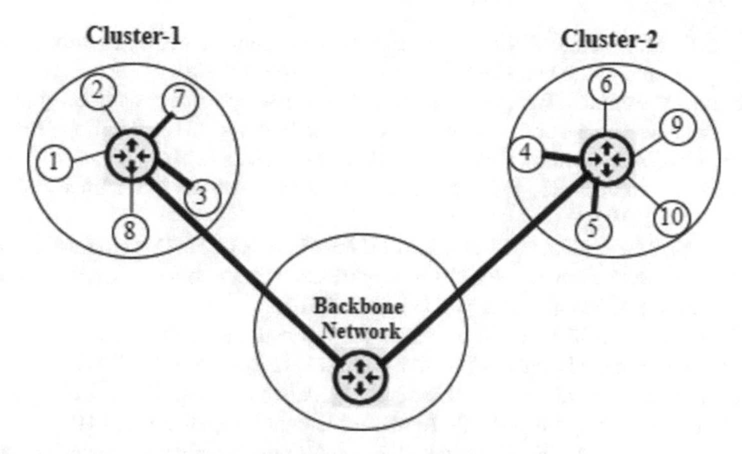

Fig. 3. Final network.

6 Conclusions

We have developed a Physarum-inspired redesign algorithm to redesign the existing enterprise network (EN) in order to reduce the inter-cluster communications routed through the backbone network. The classical Physarum model is modified under the constraint that the topology of EN is given and the traffic demand along the edge is reversed to match the tube length of the Physarum network. We have introduced a factor called the affinity factor to select highly associated node(s) of each source by traversing the generated paths of each source. A cluster formation algorithm is added to cluster the associated nodes together, and the procedure is repeated for all nodes in EN. The experimental results on a weighted EN graph revealed that the redesign algorithm manages to reduce the backbone traffic by a maximum of 78% through the optimal distribution of nodes within the clusters.

We are continuing our research by varying the EN network size and combining the developed algorithm with the classical bio-inspired algorithms to improve EN performance.

Acknowledgement. This work was supported by Kuwait University under a research grant no. EO08/19.

References

1. Habib, S.J., Marimuthu, P.N., Taha, M.: Network consolidation through soft computing. In the Proceedings of International Symposium on Methodologies for Intelligent Systems, Prague, September 14–17 (2009)
2. Habib, S.J., Marimuthu, P.N., Zaeri, N.: Carbon-aware enterprise network through redesign. Comput. J. **58**(2), 234–245 (2015)
3. Habib, S.J., Marimuthu, P.N.: A bio-inspired tool for managing resilience in enterprise networks with embedded intelligent formulation. Expert. Syst. **35**(1), 1–14 (2018)

4. Habib, S.J., Marimuthu, P.N.: Self-organization in ambient networks through molecular assembly. J. Ambient. Intell. Humaniz. Comput. **2**, 165–173 (2011)
5. Habib, S.J., Marimuthu, P.N., Hussain, T.H.: Enterprise network sustainability through bio-inspired scheme. In: The Proceedings of International Conference on Bio-Inspired Computing - Theories and Applications, October 16–19, Wuhan, China (2014)
6. Nakagaki, T., Yamada, H., Hara, M.: Smart network solutions in an amoeboid organism. Biophys. Chem. **107**(1), 1–5 (2004)
7. Iwayama, K., Zhu, L., Hirata, Y., Aono, M., Hara, M., Aihara, K.: Decision-making Ability of physarum polycephalum enhanced by its co-ordinated spatio-temporal oscillatory dynamics. Bioinspiration and Biomimetics, vol. 11, no. 3 (2016)
8. Gao, C., et al.: Does being multi-headed make you better at solving problems? A survey of physarum-based models and computations. Phys. Life Rev. **29**, 1–26 (2019)
9. Zhang, X., Zhang, Y., Zhang, Z., Mahadevan, S., Adamatzky, A., Deng, Y.: Rapid physarum algorithm for shortest path problem. Appl. Soft Comput. **23**, 19–26 (2014)
10. Dhawale, D., Kamboj, V.K., Anand, P.: An effective solution to numerical and multi-disciplinary design optimization problems using chaotic slime mold algorithm. Engineering with Computers, Springer, online, pp.1–39 (2021)
11. Zhu, L., Aono, M., Kim, S.J., Hara, M.: Amoeba-based computing for traveling salesman problem: long-term correlations between spatially separated individual cells of physarum polycephalum. Biosystems **112**(1), 1–10 (2013)
12. Nakagaki, T., Yamada, H., Tóth, Á.: Maze-solving by an amoeboid organism. Nature **407**(5964), 439–442 (2000)
13. Bonifaci, V.: Physarum can compute shortest paths: a short proof. Inf. Process. Lett. **113**(1–2), 4–7 (2013)
14. Tero, A., Kobayashi, R., Nakagaki, T.: Physarum Solver: A Biologically Inspired Method of Road-Network Navigation. Phys. A Stat. Mech. Appl. **363**(1), 115–119 (2006)
15. Baumgarten, W., Ueda, T., Hauser, M.J.: Plasmodial vein networks of the slime mold physarum polycephalum form regular graphs. Physicsl Rev. E: Stat. Nonlinear, Soft Matter Phys. **82**(4), 1–6 (2010)
16. Sun, Y.: Physarum-inspired Network Optimization: A Review. Computing Research Repository-Computer Science, Emerging Technologies, eprint-vol.1712.02910 (2017)
17. Nakagaki, T., et al.: Minimum-risk path finding by an adaptive Amoebal network. Phys. Rev. Lett. **99**(6), 068104 (2007)
18. Watanabe, S., Tero, A., Takamatsu, A., Nakagaki, T.: Traffic optimization in railroad networks using an algorithm mimicking an amoeba-like organism, physarum plasmodium. BioSystems **105**(3), 225–232 (2011)
19. Liu, Y., Gao, C., Zhang, Z.: Simulating transport networks with a physarum foraging model. IEEE Access **7**, 23725–23739 (2019)
20. Sun, Y., Hameed, P.N., Verspoor, K., Halgamuge, S.: A physarum inspired prize-collecting Steiner tree approach to identify subnetworks for drug repositioning. BMC Syst. Biol. **10**(S5), 25–38 (2016)
21. Hsu, S., Massolo, F.I.S., Schaposnik, L.P.: A physarum-inspired approach to the euclidean steiner tree problem. Sci. Rep. **12**(14536) (2022)
22. Tero, A., Kobayashi, R., Nakagaki, T.: A Mathematical model for adaptive transport network in path finding by true slime mold. J. Theor. Biol. **244**(4), 553–564 (2006)
23. Adamatzky, A.: Physarum machines: encapsulating reaction-diffusion to compute spanning tree. Naturwissenschaften **94**(12), 975–980 (2007)
24. Zhang, X., and Yan, C.: Physarum-Inspired Solutions to Network Optimization Problems, Shortest Path Solvers – from Software to Wetware, Springer Professional, Bristol, UK, pp. 329–363 (2018)

25. Li, S., Chen, H., Wang, M., Heidari, A.A., Mirjalili, S.: Slime mould algorithm: a new method for stochastic optimization. Futur. Gener. Comput. Syst. **111**, 300–323 (2020)
26. Gao, C., Liang, M., Li, X., Zhang, Z., Wang, Z., Zhou, Z.: Network community detection based on the physarum-inspired computational framework. IEEE/ACM Trans Comput. Biol.Bioinformatics **15**(6), 1916–1928 (2018)

Improving LoRaWAN RSSI-Based Localization in Harsh Environments: The Harbor Use Case

Azin Moradbeikie[1,2], Ahmad Keshavarz[1], Habib Rostami[1], Sara Paiva[2], and Sérgio Ivan Lopes[2,3,4](✉)

[1] PGU—Persian Gulf University, Bandar Bushehr, Iran
{amoradbeiki,a.keshavarz,habib}@pgu.ac.ir

[2] ADiT-Lab—Instituto Politécnico de Viana do Castelo, Rua Escola Industrial e Comercial NunÁlvares, 4900-347 Viana do Castelo, Portugal
spaiva@estg.ipvc.pt

[3] CiTin—Centro de Interface Tecnológico Industrial, 4970-786 Arcos de Valdevez, Portugal
sil@estg.ipvc.pt

[4] IT—Instituto de Telecomunicações, 3810-193 Aveiro, Portugal

Abstract. Recently, LoRaWAN communications have become a widely used technology within IoT ecosystems due to their long-range coverage, low-cost, and native RSSI-based location capabilities. However, RSSI-based localization has low accuracy due to interference in propagation, such as the multipath and fading phenomena, becoming more critical in harsh and dynamic environments like airports or harbors. A harbor has a wide area with a combination of distinct landscapes (sea, river, urban areas, etc.), and distinct infrastructures (buildings, large steel structures, etc.). In this paper, we evaluate and present a harbor assets localization system that uses a LoRaWAN-based multi-slope path-loss modeling approach. For this purpose, a harbor scale LoRaWAN testbed, composed of three gateways (GWs) and a mobile end node, has been deployed and used to characterize RSSI-based multi-slope path-loss modeling under realistic conditions. Experimental data have been collected over three days in different dynamic scenarios and used for ranging and location estimation comparison using distinct methods, i.e., RSSI-based and fingerprinting. Based on the achieved results, by a correct partitioning of the test environment based on its specific environmental conditions, a decrease between 4 and 10 dBm in path-loss estimation error can be achieved. This path-loss estimation error decrements provide 50% improvement in distance estimation accuracy.

Keywords: IIoT · LoRaWAN · RSSI · Fingerprinting · Localization

1 Introduction

Due to the steady development of cities and industries in recent years, the demand for goods transportation for industrial and manufacturing as part of

A. Rocha et al. (Eds.): WorldCIST 2023, LNNS 799, pp. 14–25, 2024.
https://doi.org/10.1007/978-3-031-45642-8_2

supply chains has increased. Harbors act as the primary player in supporting and managing the global import and export of goods. According to the Trans-European Transport Network (TEN-T) report [1], 74% of the goods imported and exported and 37% of exchanges within the European Union transit through harbors. Furthermore, Asia dominates the global maritime trade arena, with 41% of goods loaded and 62% of goods unloaded. The increase of the maritime trade area drives the harbor to continue to rely on a sustainable expansion of promising technologies [2], such as Industrial Internet of Things (IIoT), Big Data, Cloud Computing, and communication technologies, and thus move towards a smart harbor, and thus promoting process optimization and sustainability.

Due to the large area of ports and continuous long-time operation requirements of assets, the adopted communication network should provide long-range coverage with low power consumption to provide an energy-efficient solution. These requirements make newly emerged LPWAN (Low Power Wide Area Networks) technologies — such as Sigfox, LoRa, and NB-IoT—an appropriate candidate for communications within the smart harbor. Between these technologies, LoRaWAN features (long range, easy deployment, increased battery life, cost efficiency, and reduced latency) [3] make it a potential solution for secure end-node communication in the IoT ecosystem [4], and thus provide cost-effective Location-Enabled IoT (LE-IoT). In addition, a smart harbor needs robust and reliable localization and tracking methods to provide efficient harbor management by supporting decision-making [5,6].

Received Signal Strength Indicator (RSSI), Time of Arrival (ToA), and Angle of Arrival (AoA) are the three main methods for signal feature-based localization. ToA and AoA methods need costly hardware, which makes them expensive solutions for harbor because of the need for extensive and high number deployment. On the other hand, RSSI-based localization has become the most cost-effective option for end-node localization, however, it faces several relevant challenges in the harbor. In the RSSI localization method, the attenuation of RF signals is used for distance estimation based on a prior known path-loss model. The estimated distance is then used in the location estimation of the end node. Since LoRaWAN technology enables long-distance communication, an end node can be in near or far and in different directions of a gateway (GW). So, it will experience different power attenuation due to distinct environmental situations and different types of land surfaces and weather conditions (such as temperature and humidity) [7]. Moreover, high environmental dynamics (led by moving vessels, quay cranes, terminal tractors, containers, etc.) make it hard to determine a static path-loss model for the harbor. For example, when a vessel arrives at a harbor, containers are unloaded from it and moved to the yard of the harbor (temporary storage area for containers). Unloading containers to the yard is similar to the sudden creation of a building in the harbor. This sudden change leads to changes in signal path loss. In addition, a multitude of metallic components and surfaces and the fading effect resulting from the sea proximity, make harbors harsh environments for using RSSI-based localization methods. Furthermore, as the localization of containers in the harbor is an important requirement, the adopted method for location estimation should be low power consumption. Encouraged by the aforementioned challenges, this paper, at first, describes a LoRaWAN architecture for harbor assets locating and provides an evaluation of the environment dynamic effect on the distance estimation error. Next, a detailed characterizing of path-loss measurements in various roads of harbor caused by distinct environmental situations (such as Line-of-Sight and Non-Line-of-Sight effects and different types of land surfaces) is presented to prove the importance of multi-slope path-loss model consideration for accuracy improvement.

The remainder of the paper is organized as follows: Sect. 2 reviews related work, Sect. 3 describes the deployed LoRaWAN testbed architecture used in the experimental procedure, in Sect. 4 the results are presented, and lastly, in Sect. 5 conclusions are put forward.

2 Related Work

There is a growth in research to provide an end node localization method as a substitute for Global Navigation Satellite Systems (GNSS) because of their higher power consumption and higher operational cost. In February 2015, the LoRaWAN Alliance released a Long-Range Wide Area Network (LoRaWAN) geolocation white paper. LoRaWAN uses the LoRa Chirp Spread Spectrum (CSS) modulation method that provides long-range coverage [8]. Several existing works provide measurements of the LoRa technology performance as a communication protocol (including coverage, capacity, delay, and throughput) in various applications [10,11,13,14]. In [10], the authors measure the coverage performance of LoRaWAN in an urban area. Their results show that the amount of successfully delivered packets exceeds 80% for up to 5 km distances. Authors in [11] introduce setups of the performance measurements to analyze the scalability of the LoRaWAN. They realized that more than 60% of the packets can be received from a distance of up to 30 km on water.

Industrial environments are one of the most challenging cases for LoRa usage. In [13], the authors evaluate the performance of a LoRaWAN network in industrial scenarios. In the proposed model, different IIoT end nodes communicate with a central controller to provide monitoring and sensing information. For this purpose, they use the NS-3 *lorawan* module as a simulator for their evaluation. Authors in [16] demonstrate the feasibility of a LoRaWAN to be used for data collecting in marine environments. In the proposed model, the transmitting device is placed in the middle of the sea and the gateway is placed ashore. They set up an operating network and proved that the best SF to be exploited for the whole system is 7 since it ensured limited packet losses. Authors in [17] propose a real-time monitoring infrastructure for the remote, real-time control of offshore sea farms based on LoRaWAN by using Fixed Nodes and Mobile Sinks. However, authors in [18] show that the floating LPWAN suffers significant performance degradation, compared to the static terrestrial deployments. They present a novel channel access method for floating LPWAN. In [19], present a new method for communicating and auto-adapting to the altering requirements and typical conditions of a marine environment based on LoRaWAN protocol to routinely transfer data between the open sea and the land. In the all mentioned papers, the authors adopt LoRaWAN just as a communication network and not for localization and tracking purposes.

Localization and tracking are one of the interesting uses of LoRaWAN. In [9], the feasibility of LoRaWAN adoption for a GNSS-less localization has been proved. In [15], the authors present a boat tracking and monitoring system based on LoRa. Their obtained results showed the validity of LoRa aiming at ships tracking in port. In [20], the authors evaluated the scalability of LoRa devices in the network of the LoRa radio technology for geolocation and tracking using ns-3 for a harbor use case. They mention that, if 500 nodes are deployed in an area spanning a radius of 2500 m, the probability of successful transmission is greater than 85%. They do not provide a real case of study, but they stated to derive the appropriate LoRaWAN implementation in harbor application, its parameters (like SF value, number of nodes, update rate, and coverage area) have to be determined accordingly. Despite the growing research in LoRa-based networks and localization methods, there is still a need for further development and evaluation of LoRaWAN in industries with harsh environments. In this paper, we set up a real implementation of LoRaWAN for harbor asset tracking in an industry with a harsh environment.

3 LoRaWAN Experimental Testbed

In this section, we present the components and the architecture of the LoRaWAN testbed, which consists of three deployed LoRaWAN gateways (one on the rooftop of the harbor central building and two more on the Telecommunication tower) and one end node that is moved across the Bushehr harbor. As the localization accuracy in the up part of the harbor is most important for the company, the location of GWs are chosen in such a way as to provide triangulation of GWs for the up part of the harbor. The distance of GW1 from GW2 and GW3 is equal to 790 m and 920 m, respectively. The topology of the LoRaWAN testbed and its surrounding environment is illustrated in Fig. 1.

Each gateway is equipped with an MCU and an SX1301 digital baseband transceiver and has been installed on antenna towers with heights between 30 and 35 m (Fig. 1). The LoRaWAN end node is implemented with an MCU, a transceiver SX1276, and a GPS unit mounted on the roof of a car with a height equal to 1.5 m (Fig. 1). While the car is moving in the harbor, the LoRaWAN end nodes will transmit packets to the gateways at intervals of 9 s. A packet includes GPS coordinates, timestamps, and battery charge information. The corresponding SNR and RSSI of the end node are also transmitted to the gateways.

All the packets are transmitted with a spreading factor, bandwidth, coding rate, and channel equal to 7, 125 kHz, 4/5, and 868 kHz respectively. All the data were collected in the harbor area from Feb 6, 2022, to Feb 15, 2022. We logged over 2500 records sensed by at least two gateways in total. The collected is depicted with red points in Fig. 2.

4 Evaluation Results

Harsh and high dynamics of the harbor environment leads to vast changes in PL parameters and measured RSSI subsequently. This issue caused high variance in measured RSSI for the same location which makes RSSI-based localization methods so challenging. For showing this problem, in this paper, experiments took place in two different situations (busy and solitude days in the harbor). The variance of measured RSSI based on distance from GW1 on these different days is shown in Fig. 3.

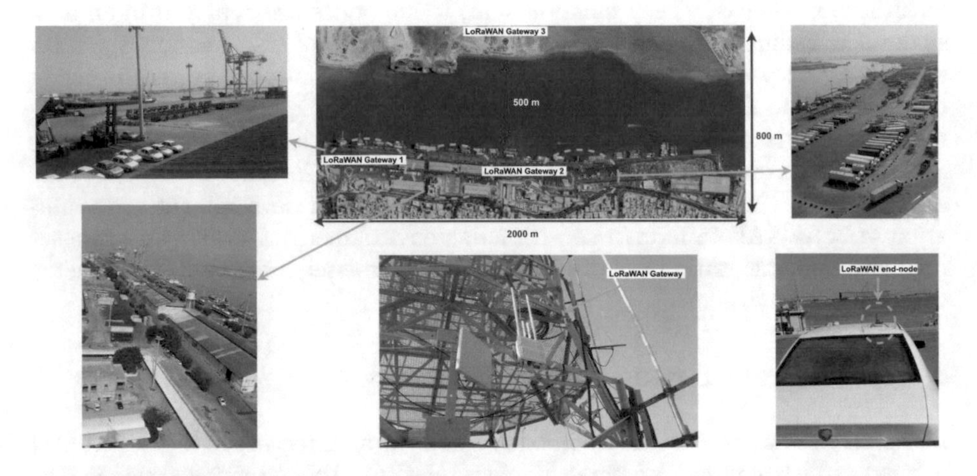

Fig. 1. Harbor LoRaWAN testbed with Gateways with an end node identified.

Fig. 2. Spatial data collected with the LoRaWAN testbed.

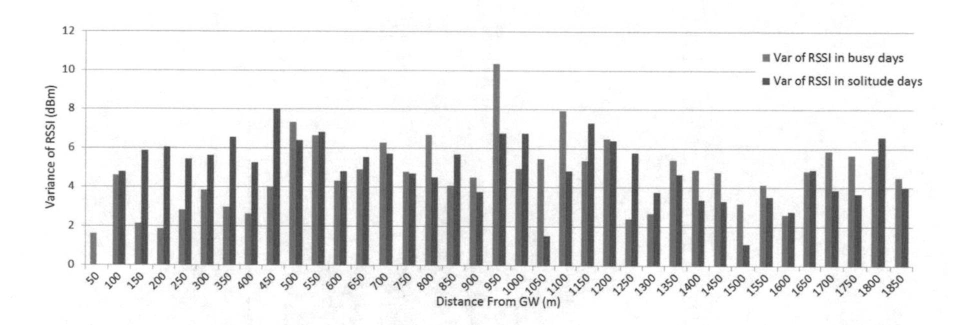

Fig. 3. The variance of measured RSSI-based on distance from GW1 in busy and solitude days.

On a busy day, a vessel anchored in a specific terminal leads to increase truck traffic in the yard between GW1 and GW2 to unload containers from the vessel. These dynamics force the environment to experience several radio propagation phenomena, e.g., multipath, fading and interference, impacting the variance of the received and measured RSSI values, showing distinct patterns for busy and solitude days. Several RSSI measurements for the same distance from a GW, on different days, increase the overall localization error. This variance can also be seen for distances of less than 500 m, cf. Fig. 3. Moreover, there are also similar changes in the RSSI variance measured by GW2 which indicates that these propagation-related phenomena occur all around the harbor.

In the following of this section, the accuracy of distance estimation by using a single PL model is presented and the improvement scale of estimated distance by using a multi-slope PL model is reported. Then, location estimation accuracy by using the fingerprint method is presented.

4.1 Distance Estimation by Using Single PL Method

To compute the distance estimation accuracy in the harbor, first, we used the measured RSSI to calculate the path-loss by using Equation (1).

$$PL = RSSI + SNR + P_{tx} + G_{tx} \tag{1}$$

where SNR represents the signal-to-noise ratio (dB), P_{tx} is the transmission power, and G_{tx} corresponds to the gain of the transmitter. The measured PL (dBm) data can be conveniently modeled using the first-order fit to compute the path loss model as Eq. (2).

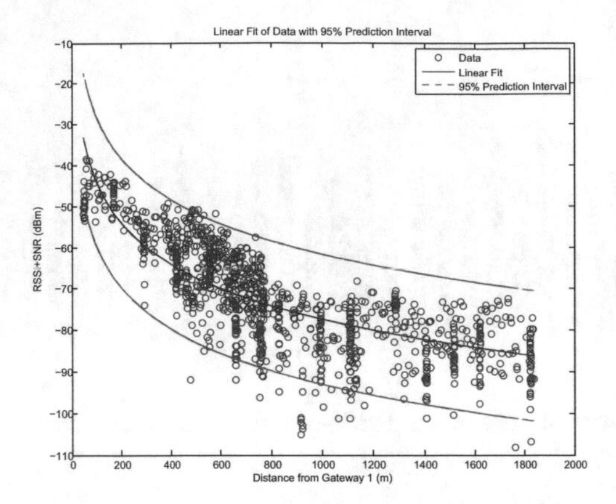

Fig. 4. Measured and expected path-loss in a Busy day.

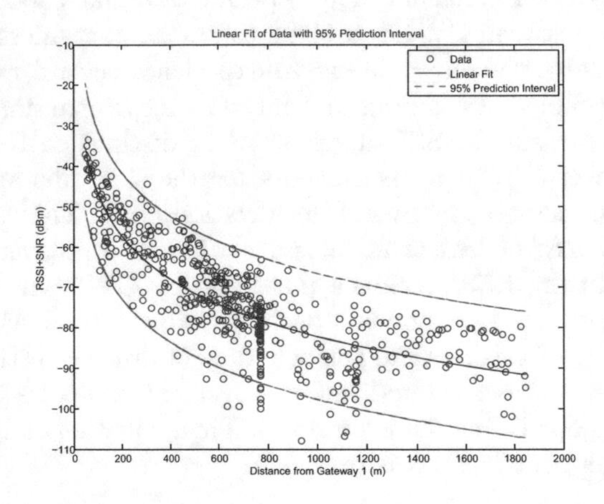

Fig. 5. Measured and expected path-loss in a Solitude day.

$$PL = A + 10nLog(d) \tag{2}$$

where n is the path loss exponent, A is the path loss value at a reference distance equal to one meter from the receiver, and d is the distance between an end node and the LoRaWAN gateway [12]. Figure 4 and 5 show the measured path loss (marked with blue dots) and the expected path loss of GW1 (solid red curve) for a busy and solitude day in the harbor, respectively. The PL diversity leads to various path loss exponents on busy and solitude days. There is the same situation for GW2. The calculated path loss exponents and average distance estimation error of three GWs are listed in Table 1.

4.2 Distance Estimation by Using Multi-slope PL Method

As mentioned, end nodes at the same distance, but in different directions of a gateway will experience different power attenuation due to distinct environmen-

Table 1. Path-loss characteristics obtained experimentally.

Gateway ID	Measurement Scenario	n	A	Average Distance Estimation Error
GW1	Busy day	3.329	22.58	805.5 m
	Solitude day	5.557	24.51	678.8 m
GW2	Busy day	0.9815	40.08	412.0 m
	Solitude day	1.4271	31.09	462.9 m
GW3	Busy and Solitude day	6.4	102	761.3 m

tal conditions (Line-of-Sight and Non-Line-of-Sight effect) and different types of land surfaces.

On this basis, it is important to provide an optimum environment partitioning and compute the path-loss model for each partition. To evaluate the effect of distinct environmental conditions on the path-loss, the collected data for a specific part of the harbor (Fig. 6, (a)) is split into two distinct sets (Fig. 6, (b) and (c)).

The selected data in Fig. 6, (b) and (c) has the same distance from GW1. But, they have various environmental situations (Fig. 6, (b) has less than 10 m distance from sea and Fig. 6, (c) surrounded by high buildings that lead to the loss of the Line-of-Sight effect in some parts). Figure 7 shows the measured path loss (marked with blue dots) and the expected path loss (solid red curve) for specified parts in Fig. 6, respectively. As it can be seen in Fig. 7, by dividing the environment into two parts, the data samples in Fig. 6, (a) are split into two parts such that it leads to a decrease in the difference between estimated and measured path loss. This results in a decrease in the average distance estimation error and improves distance estimation accuracy. The two slope path-loss model can be presented as Eq. 3.

$$PL = A_i + 10n_i Log(d_i), \quad i = 1, ..., k \cap d_i \subseteq Part_i \tag{3}$$

where $i = 1, 2, ..., k$ is equal to the number of considered parts of the divided environment. The calculated two slope path loss exponents for the split parts are listed in Table 2. Based on the result, for distance estimation improvement, it is important to consider the multi-slope path-loss model.

Table 2. Two slope path-loss characteristics obtained experimentally.

#	Near to the Sea		Near to the Buildings	
	n	A	n	A
GW1	5.221	83.128	2.602	0.84
GW2	1.886	17.675	1.954	21.94
GW3	1.946	117	2.637	116

4.3 Location Estimation by Using Fingerprint Method

The accuracy and usability of the fingerprinting algorithm implementation for large areas of the harbor for localization are presented in this subsection. We estimated the locations of the end node by using the fingerprint method on each busy and solitude day separately. For this purpose, the environment is split into 5×5 m squares. The location of the end node is estimated based on the constructed radio map of the harbor on the same day. The average location estimation error is equal to 346 m. The most important drawback of the fingerprint-based localization method is its requirement for frequent signal map updating. As the harbor has a highly dynamic environment, the requirement for signal map updating becomes a huge problem that makes it an inappropriate method.

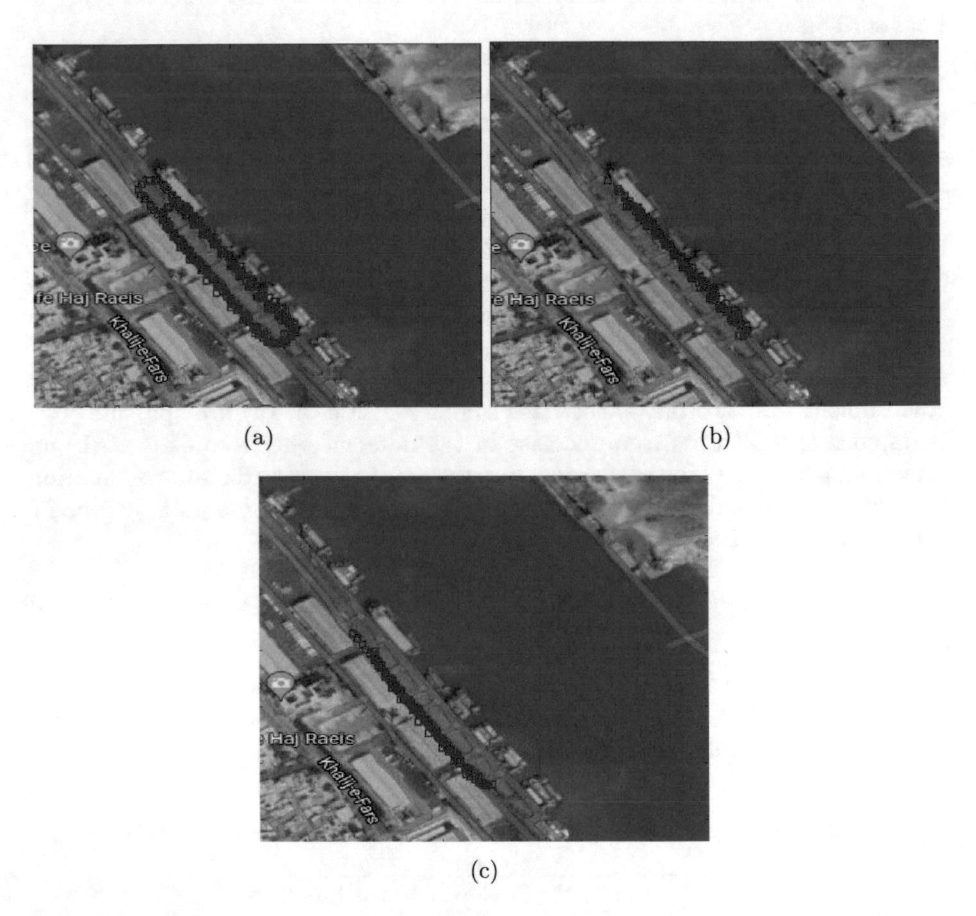

Fig. 6. Selected part of the harbor for distinct environmental situations effect evaluation: a) all data points; b) less than 10 m distance from the sea; and c) surrounded by high buildings and losing the Line-of-Sight effect in some parts.

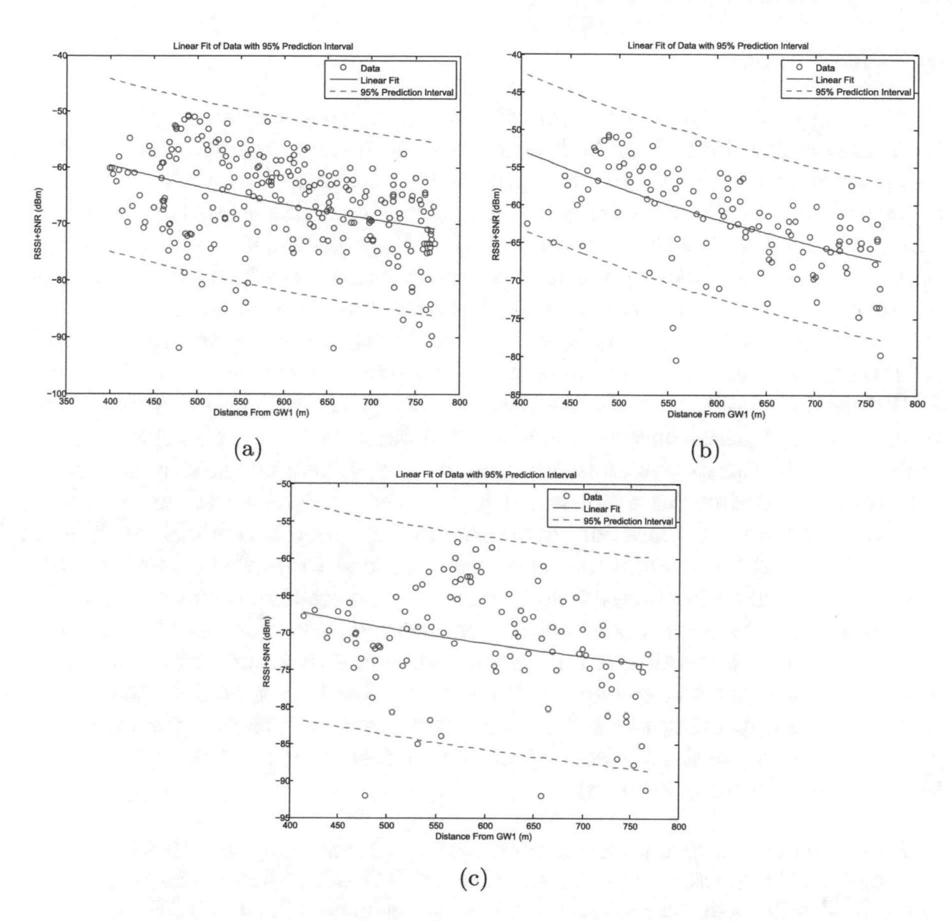

Fig. 7. Selected part of the harbor for distinct environmental situations effect evaluation: a) all data points; b) less than 10 m distance from the sea; and c) surrounded by high buildings and losing the Line-of-Sight effect in some parts.

According to the stated contents, based on the highly dynamic environment of the harbor, RSSI-based localization is a more efficient method. There are many RSSI-based localization methods. An overview of localization methods using RSSI in public LoRaWan is presented in Table 3. But these methods are executed in normal environments and it is necessary to customize these methods to provide an acceptable localization accuracy in highly dynamic environments.

Table 3. An overview of localization methods using RSSI

Method	ANN [21]	ANN [22]	Linear Ridge [23]
Localization accuracy	340 m	381 m	784 m

5 Conclusion

In this paper, the implementation of a LoRaWAN-based localization system for a harbor has been described, and the experimental characterization of its harsh environment. Based on the authors' knowledge, it is the first paper that reviewed the high dynamics effect of the harbor on RSSI-based localization by using LoRaWAN. Results have shown that fingerprint-based localization has better performance (346 m) which is not acceptable for the harbor environment. Furthermore, its accuracy depends highly on the completeness of generated radio map information for the target area. In addition, updating the radio map is an important issue that decreases its performance. On the other hand, with RSSI-based methods, the path-loss parameters regarding different gateways have been estimated based on their location, and their estimated distance accuracy increases for higher distances. In addition, it is dependent on the dynamics of the environment. To improve RSSI-based localization, it is essential to provide an optimum environment partitioning and multi-slope path-loss model for different parts. Based on the results, this approach can lead to between 4 and 10 dBm improvements in estimated path-loss accuracy that leads to distance estimation improvement between 100 and 300 m. On average, this provides 50% improvement in distance estimation accuracy. In addition, a dynamic scene analysis for low and high distances can improve the estimated path-loss model. Future work will focus on improving the LoRaWAN positioning accuracy in the harbor by considering a sequential model of continuously measured RSSI data, to improve the estimated location accuracy.

Acknowledgment. This work has been partially supported by the TECH (NORTE-01-0145-FEDER-000043) project, under the PORTUGAL 2020 Partnership Agreement, funded through the European Regional Development Fund (ERDF).

References

1. Scipioni, S.: European Ports: Potential Gateways to International Trade (2020)
2. Fraga-Lamas, P., Lopes, S.I., Fernández-Caramés, T.M.: Green IoT and Edge AI as Key Technological Enablers for a Sustainable Digital Transition towards a Smart Circular Economy: An Industry 5.0 Use Case. Sensors 2021, 21, 5745. https://doi.org/10.3390/s21175745
3. Moradbeikie, A., Keshavarz, A., Rostami, H., Paiva, S., Lopes, S.I.: GNSS-free outdoor localization techniques for resource-constrained IoT architectures: a literature review. Appl. Sci. **11**, 10793 (2021). https://doi.org/10.3390/app112210793
4. Torres, N., Pinto, P., Lopes, S.I.: Security Vulnerabilities in LPWANs-an attack vector analysis for the IoT ecosystem. Appl. Sci. **11**, 3176 (2021). https://doi.org/10.3390/app11073176
5. Moradbeikie, A. Keshavarz, A., Rostami, H., Paiva, S., Lopes, S.I.: LoRaWAN as a cost-effective technology for assets localization in smart harbors, IEEE Smart Cities Newsletter, October (2021). https://smartcities.ieee.org/newsletter/october-2021
6. Pasandi, H.B., et al.: Low-cost traffic sensing system based on LoRaWAN for urban areas, In Proceedings of the 1st International Workshop on Emerging Topics in Wireless, pp. 6–11 (2022)

7. Moradbeikie, A., Keshavarz, A., Rostami, H., Paiva, S., Lopes, S.I.: Improvement of RSSI-based LoRaWAN Localization using Edge-AI, Edge-IoT 2021 - 2nd EAI International Conference on Intelligent Edge Processing in the IoT Era, Virtual, November 24–26 2021 (2021)
8. Liando, J.C., Gamage, A., Tengourtius, A.W., Li, M.: Known and unknown facts of LoRa: experiences from a large-scale measurement study. ACM Trans. Sensor Networks (TOSN) **15**, 1–35 (2019)
9. Fargas, B.C., Petersen, M.N.: GPS-free geolocation using LoRa in low-power WANs. In: 2017 global internet of things summit (Giots), pp. 1–6. IEEE (2017)
10. Petajajarvi, J., Mikhaylov, K., Roivainen, A., Hanninen, T., Pettissalo, M.: On the coverage of LPWANs: range evaluation and channel attenuation model for LoRa technology. In: 2015 14th International Conference on ITS Telecommunications (ITST), pp. 55–59. IEEE (2015)
11. Petäjäjärvi, J., Mikhaylov, K., Pettissalo, M., Janhunen, J., Iinatti, J.: Performance of a low-power wide-area network based on LoRa technology: doppler robustness, scalability, and coverage. In: International Journal of Distributed Sensor Networks, vol. 13 (2017)
12. Pahlavan, K., Levesque, A.H.: Wireless Information Networks. Wiley, New York (2005)
13. Magrin, D., Capuzzo, M., Zanella, A., Vangelista, L., Zorzi, M.: Performance analysis of LoRaWAN in industrial scenarios. IEEE Trans. Industr. Inf. **17**, 6241–6250 (2020)
14. Sanchez-Iborra, R., Sanchez-Gomez, J., Ballesta-Viñas, J., Cano, M.D., Skarmeta, A.F.: Performance evaluation of LoRa considering scenario conditions. Sensors **18**, 772 (2018)
15. Sanchez-Iborra, R., Liaño, G.I., Simoes, C., Couñago, E., Skarmeta, A.F.: Tracking and monitoring system based on LoRa technology for lightweight boats. Electronics **8**(1) (2019). https://doi.org/10.3390/electronics8010015
16. Parri, L., Parrino, S., Peruzzi, G., Pozzebon, A.: Low Power Wide Area Networks (LPWAN) at sea: performance analysis of offshore data transmission by means of LoRaWAN connectivity for marine monitoring applications. Sensors **19**, 3239 (2019). https://doi.org/10.3390/s19143239
17. Parri, L., Parrino, S., Peruzzi, G., Pozzebon, A.: A LoRaWAN network infrastructure for the remote monitoring of offshore sea farms. In: IEEE International Instrumentation and Measurement Technology Conference (I2MTC), pp. 1–6 (2020). https://doi.org/10.1109/I2MTC43012.2020.9128370
18. Wang, Y., Zheng, X., Liu, L., Ma, H.: PolarTracker: attitude-aware channel access for floating low power wide area networks. In: IEEE INFOCOM 2021-IEEE Conference on Computer Communications, pp. 1–10. IEEE (2021)
19. Pensieri, S., et al.: Evaluating LoRaWAN connectivity in a marine scenario. J. Marine Sci. Eng. **9**(11), 1218 (2021)
20. Priyanta, I.F., Golatowski, F., Schulz, T., Timmermann, D.: Evaluation of LoRa technology for vehicle and asset tracking in smart harbors. In: IECON 2019-45th Annual Conference of the IEEE Industrial Electronics Society, vol. 1, pp. 4221–4228. IEEE (2019)
21. Daramouskas, I., Vaggelis, K., Michael, P.: Using neural networks for RSSI location estimation in LoRa networks. In: 2019 10th International Conference on Information, Intelligence, Systems and Applications (IISA), pp. 1–7. IEEE (2019)
22. Nguyen, T.A.: Lora localisation in cities with neural networks (2019)
23. Janssen, T., Berkvens, R., Weyn, M.: Benchmarking RSS-based localization algorithms with LoRaWAN. Internet Things **11**, 100235 (2020)

Impact of Traffic Sampling on LRD Estimation

João Mendes[1], Solange Rito Lima[1], Paulo Carvalho[1], and João Marco C. Silva[2](✉)

[1] Centro Algoritmi, Universidade do Minho, Braga, Portugal
a71862@alunos.uminho.pt, {solange,pmc}@di.uminho.pt
[2] INESC TEC, Universidade do Minho, Braga, Portugal
joao.marco@inesctec.pt

Abstract. Network traffic sampling is an effective method for understanding the behavior and dynamics of a network, being essential to assist network planning and management. Tasks such as controlling Service Level Agreements or Quality of Service, as well as planning the capacity and the safety of a network can benefit from traffic sampling advantages.

The main objective of this paper is focused on evaluating the impact of sampling network traffic on: (i) achieving a low-overhead estimation of the network state and (ii) assessing the statistical properties that sampled network traffic presents regarding the eventual persistence of Long-Range Dependence (LRD). For that, different Hurst parameter estimators have been used. Facing the impact of LRD on network congestion and traffic engineering, this work will help clarify the suitability of distinct sampling techniques in accurate network analysis.

Keywords: Traffic Sampling · Self-Similarity · Long Range Dependence

1 Introduction

In today's networks, due to the huge traffic volumes and speeds, it is increasingly harder and more expensive to understand the behaviour of a network through the analysis of the total number of packets. Consequently, the use of traffic sampling, which consists in electing a small number of packets for analysis, becomes essential. Even though sampling techniques reduce the amount of collected data drastically, sampling impacts on the performance of network equipment, and leads usually to a reduction in the estimation accuracy of network status and the statistical properties of traffic.

Since the identification of self-similarity and Long-Range Dependence (LRD) as properties of Internet traffic [1,9,11], representing a paradigm shift from Poisson and memory-less processes, there have been extensive research to estimate LRD in distinct network environments [10]. The presence of LRD in network traffic reveals the existence of statistically significant correlations across large time scales [7], meaning that traffic burtiness effects may be time-persistent.

A. Rocha et al. (Eds.): WorldCIST 2023, LNNS 799, pp. 26–36, 2024.
https://doi.org/10.1007/978-3-031-45642-8_3

Facing the impact of self-similarity and LRD on traffic engineering, namely on establishing proper network configurations and network congestion control, where bursts inside bursts may lead to unpredictable queuing delay and loss, understanding the impact of sampling on LRD estimation becomes crucial. However, till present, no focus has been given to verifying if LRD property is maintained when traffic sampling is applied, and how sampling techniques affect the subsequent LRD estimation algorithms.

In this paper, the evaluation of the impact of selective analysis of network traffic will be explored at two levels: (i) in the overhead reduction when estimating network status; (ii) in the statistical properties of traffic, such as self-similarity and LRD, when comparing sampled to the original network traffic. In more detail, this involves: studying the different sampling techniques and the properties of network traffic; applying different sampling techniques to real-world datasets using the sampling framework proposed in [12]; and (iii) comparing the different traffic sampling techniques in terms of volume of data and accuracy in estimating the statistical properties of traffic. This will allow a better understanding of the behaviour and impact of sampling on network traffic analysis.

This paper is organized as follows: Sect. 2 provides background concepts and related work on traffic sampling and LRD; Sect. 3 details the methodology and tools used in the data processing and LRD estimation phase; Sect. 4 debates the evaluation results and main findings; Sect. 5 summarizes the conclusions and future work.

2 Background

This section provides a short overview of traffic sampling techniques and fractal properties of Internet traffic, such as self-similarity and LRD, and corresponding estimation methods.

2.1 Sampling

A way to lighten the burden of handling massive Internet traffic volumes is resorting to traffic sampling techniques to obtain a subset of the overall traffic for subsequent analysis. This subset should be yet representative of the whole traffic behaviour, allowing tasks such as traffic classification and protocol modeling, using substantially smaller data volumes. There are different sampling techniques and algorithms, with varying rates of data collected, computational cost and the ability of accurately representing the network behaviour. Due to its relevance, we briefly overview them next.

Sampling Techniques Overview - Sampling techniques can be broadly classified according to their: *Selection Scheme*, which determines the selection pattern of packets to be collected; *Selection Trigger*, which defines the type of trigger, which determines the start and the end of a sampling instance; and *Granularity*, which determines if sampling is carried out on a packet or flow basis (see Fig. 1).

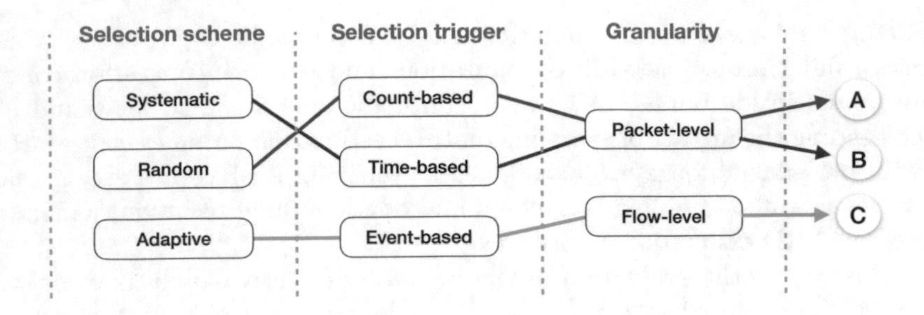

Fig. 1. Sampling techniques classification

As regards the selection scheme, the pattern to be collected can be ruled by a deterministic, random or adaptive function. While a probabilistic approach (random) can be used to overcome possible biasing of deterministic choice, adaptive approaches are able to adjust the pattern dynamically, e.g., based on ongoing traffic measurements. As regards the selection trigger, sampling can be prompted based on a counter, a timer or an event.

The sampling techniques within class A (see Fig. 1) are common in most sampling tools, e.g., Cisco NetFlow and sFlow. According to some authors, time-based triggers are less robust than count-based when applied to traffic characterization [4,12], being affected by the burstiness of network traffic. However, class B techniques may be suitable for applications requiring the analysis of consecutive packets, such as IDS [12] or for multi-channel wireless traffic characterization [13]. Event-based sampling techniques (class C), relying on content-dependent schemes, are oriented for monitoring a specific service or component, adapting the sampling frequency in response to some observed traffic event [5].

2.2 Properties of Internet Traffic

Self-similarity and scaling phenomena have dominated backbone Internet traffic analysis for the past two decades. However, self-similarity and LRD analysis is hindered by the difficulty of identifying dependence and estimating its parameters unambiguously [7].

Self-Similarity and Long Range Dependence - Internet traffic may exhibit intriguing temporal correlation properties and long memory [2], visible in plots of packet or byte traffic arrivals which are similar and bursty in different time scales. This behaviour is known as self-similarity or scale-invariance may hold true for different time granularity or scales, from a few hundred milliseconds up to hundreds of seconds. This means that a dilated portion of a traffic trace exhibits similar statistical characteristics than the whole. "Dilating" is applied on both amplitude and time axes of the sample path, according to a scaling parameter H called Hurst parameter. On the other hand, LRD is a long-memory property observed in large time scales.

A random process $X(t)$ (e.g., cumulative packet arrivals in time interval $[0,t]$) is said to be self-similar, with Hurst parameter $H > 0, H \in R$, if $X(t) =_d a^{-H}X(at)$, $\forall a > 0$, and $=_d$ denoting equality of all finite-order distributions [2].

The class of self-similar processes is usually restricted to that of self-similar with stationary increments (SSSI) processes, which are the "integral" of a stationary process [2]. The parameter H characterizes self-similar processes, but it is often used to label the LRD increments of SSSI processes. In this work, the expression "Hurst parameter of a LRD process" (characterized by γ) denotes actually the parameter $H = (\gamma + 1)/2$ of its integral SSSI parent process. The predominant way to quantify LRD is through the Hurst parameter, which is a scalar [7], and cannot be calculated definitively, only estimated. There are two categories of Hurst parameter estimators: those operating in the time domain, and those operating in frequency or wavelet domain.

2.3 Methods for Estimating the Parameters H and γ

Variance-Time Plot. The variance-time plot method [2,11] studies the variance σ^2(m) of the aggregated sequence obtained by dividing the time-series $\{x_j\}$ into non-overlapping blocks of length m and averaging them, i.e. $X_k^{(m)} = \frac{1}{m}\sum_{i=(k-1)m+1}^{km} x_i$. for k=1,2,...,N/m [2]. The variance of the aggregated sequence can be estimated as the sample variance,

$$\sigma^2(m) = \frac{1}{N/m}\sum_{k=1}^{N/m}\left[x_k^{(m)}\right]^2 - \left[\frac{1}{N/m}\sum_{k=1}^{N/m} X_k^{(m)}\right]^2 \tag{1}$$

For large N/m and m, this variance obeys the power law $\sigma^2(m) \sim m^{\gamma-1}\sigma_x^2$. Thus, by linear regression on a log-log plot of $\sigma^2(m)$ as a function of m, the parameters γ and H can be estimated.

Rescaled Adjusted Range Statistic (R/S). The R/S statistic, one of the main time-domain methods for LRD estimation [6,14], is defined as:

$$\frac{R(n)}{S(n)} = \frac{1}{S(n)}\left\{\max_{0\le l\le n}\left[Y(l) - \frac{l}{n}Y(n)\right] - \min_{0\le l\le n}\left[Y(l) - \frac{l}{n}Y(n)\right]\right\} \tag{2}$$

Assuming LRD of $\{x_j\}$ such as $S_Y(f) \sim c_2|f|^{-r}$ for $f \to 0, 0 < r < 1$, when $n \to \infty$, its expected value is given by $E\left[\frac{R(n)}{S(n)}\right] \sim C_H n^H$, where C_R is a positive constant. To estimate the Hurst parameter using the statistic R/S, initially, the input sequence $\{x_j\}$ is divided in k blocks. Then, for each *lag-n*, *R/S* is computed at up to k starting points, taken evenly in different blocks. By linear regression on a log-log plot of R(n)/S(n) as a function of n, the parameters γ and H can be estimated.

Periodogram. This method plots the logarithm of the spectral density of the time series in comparison with the logarithm of the frequency. A function $I(\lambda)$ can be defined over a time series $X(t)$, as $I(\lambda) = \frac{1}{2\pi N}|\sum_{j=1}^{N} X_j e^{ij\lambda}|^2$, where λ is a frequency, N is the number of terms in the series, and X_j is the data under analysis. As $I(\lambda)$ is an estimator of the spectral density, a series with LRD should have a periodogram which is proportional to $|\lambda|^{1-2H}$ close to the origin. Therefore, a regression of the logarithm of the periodogram on the logarithm of λ should give a coefficient of $1 - 2H$. The slope provides an estimate of H.

3 Experimental Environment

This section describes the methodology and tools used to set up the testbed for LRD estimation over sampled datasets. For this purpose, both the sampling framework, the tools used to transform traffic traces into proper input data for the sampling framework and the LRD estimation algorithms are described next.

3.1 Sampling Framework and Datasets

The sampling framework used to obtain sampled data over large-scale public datasets was previously implemented and tested [12]. This framework includes a full range of sampling techniques as described in Sect. 2.1. The framework allows to apply these techniques to both online and offline traffic sampling in the form of pcap files. An integrated way of performing sampling, and the one used for obtaining sampled data sets, is using a Python script that applies all of the different sampling techniques previously referred to with pre-defined parameters, such as the sampling size and the interval between samples.

Regarding the network traffic used as input, this work resorts to public datasets from Optical Carrier OC48 and OC192 links (2.48Gbit/s and 9.95Gbit/s, respectively), gathered from an ISP infrastructure in the American west coast and maintained by CAIDA[1]

3.2 Data Preprocessing

CAIDA datasets undergone a preprocessing phase before serving as inputs both for the sampling framework and for LRD analysis. This preliminary phase involved the use of Mergecap tool for merging multiple CAIDA files, ordering data by timestamp, and Wireshark as traffic analyser.

Even though Wireshark has the possibility to turn a .pcap file into a .csv file with all data required for LRD estimation algorithms, namely the number of packets/bytes for each equally spaced time interval, it has limitations regarding the number of lines supported in the .csv file. To overcome this limitation, Editcap tool (with flag -i) was used to split the files composing the original dataset into smaller ones. In practice, Editcap reads part of or all captured packets from

[1] Available at: https://publicdata.caida.org/datasets/passive/.

an *infile*, optionally converts them in various ways, and writes the resulting packet capture to an *outfile* (or outfiles), by default, in .pcapng format. Each outfile is created with an infix _nnnnn[_YYYYmmddHHMMSS], comprasing the serial number of the output file and the timestamp of its first packet.

The TShark protocol analyser was also used to generate statistics, such as the packets and/or number of bytes from the input file, and to collect packets/bytes statistics per time interval, with resolutions from seconds (s) to microseconds (μs). The following code snippet shows how to obtain a specific statistic, the number of bytes per 0.1 s interval, from all .pcap files in the current folder.

```
def tshark100ms():
    startdir='.'
    for root, dirs, files in os.walk(startdir):
        for file in files:
            if file.endswith('.pcap'):
                fName = os.path.join(root,file)
                packet = subprocess.run([f'tshark -r "{fName}"
                -qz io,stat,0.1 > {fName}0_1.txt'],
                capture_output=True,text=True,
                shell=True).stdout
                cmd = f"sed -i '/=======================/,$!d'
                {fName}0_1.txt"
```

Using the previous tools, it is possible to extract the number of packets/bytes per time interval from numerous text files, maintaining the chronological order in the resulting file. In particular, the granularity of 0.1 s, 0.01 s and 0.001 s were used. Each file will serve as input for the statistics and LRD estimation algorithms. Instead of using the raw number of bytes, the data was normalized using the natural logarithmic function (*ln*). The final text file containing the number of bytes per time interval was obtained after applying the Python *math.log* function. This file will serve later as input for the Selfis tool.

3.3 Selfis - LRD Estimation

The evaluation results were obtained using the self-similarity analysis software tool SELFIS [8], capable of applying the Hurst estimation algorithms, described in Sect. 2.2, to an input file containing the time-series. Basic statistics, the graphical display of the power spectrum function and the autocorrelation function can also be obtained.

4 Results and Discussion

The impact of traffic sampling on LRD estimation is assessed using two previously collected traces, a three hours long OC48 and two hours long OC192 publicly available, as described in Sect. 3.

LRD is comparatively analyzed resorting to three different time-domain Hurst parameter estimators, i.e., the *aggregated variance*, the *rescaled range* (R/S), and the *periodogram*. Each estimator is applied to three different time series corresponding to the amount of data from the two traces in the time windows of 0.1 s, 0.01 s and 0.001 s. Notice the variation in the observation time window directly impacts on the communication overhead since smaller windows imply data being transmitted to the estimator more frequently. The same process is applied to the resulting traces from applying five sampling techniques, namely: (i) SystC 1/100 - systematic count-based scheme with 1-out-of-100 packet selected for the sampled trace; (ii) RandC 1/100 - uniform probabilistic count-based scheme with 1-out-of-100 packet selected for the sampled trace; (iii) SystT 100/500 - systematic time-based scheme with the sampling frequency of 100ms-out-of-500ms; (iv) LP 100/200 - adaptive linear prediction scheme with a fixed sample size of 100ms and the initial interval between samples of 200ms; and (v) MuST 200/500 - multiadaptive scheme with the initial sample size of 200ms and the initial interval between samples of 500ms.

For the count-based schemes, the frequencies were selected based on [3], while the time-based frequencies were selected following the results in [12].

Considering that the main objective of using traffic sampling is to reduce the amount of data acquired and processed in multiple monitoring tasks, Fig. 2 shows the data reduction achieved when considering the sampling schemes in use with different frequencies. Although the reduction ratio can be previously defined for count-based schemes, for time-based approaches, the bursty nature of the traffic leads to a variable volume of selected packets in the sample. Even though, regardless of the sampling strategy adopted, Fig. 2 shows that the volume of data was reduced to $\approx 20\%$ (worst case) when compared with the original (i.e., unsampled) OC192 trace.

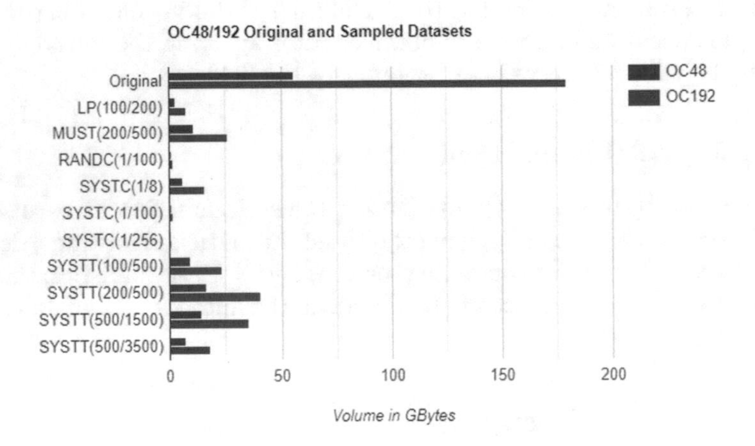

Fig. 2. Volume of data reduction through sampling.

Figure 3 illustrates the visual correlation of the number of packets (OC48) for two different time scales. Note that increasing the observation time-window does not lead to a smoother plot, which ratifies the self-similar behaviour described in Sect. 2.2. Although not represented here, a similar behaviour is observed for the OC192 trace.

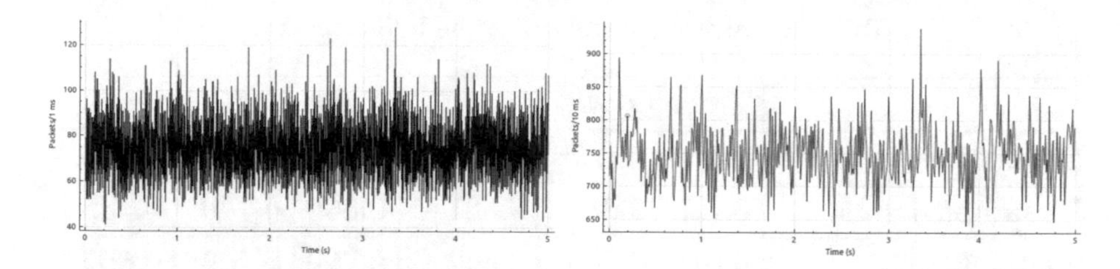

Fig. 3. Visual observation of self-similarity - total number of packets for a 0.001 s (left) and 0.01 s (right) time scale.

When comparing the sampling impact of estimating the Hurst parameter through the selected time-domain estimators, Tables 1-3 provide interesting outcomes. Firstly, for the aggregate variance (see Table 1), count-based techniques present a better relationship between the accuracy of LRD estimation and the amount of data required. However, as network measurements are typically multipurpose, the time-based multiadaptive scheme (i.e., MuST) may represent a good option. For instance, as discussed in [12], this technique provides better results for flow-based traffic classification.

Table 1. Hurst parameter using the aggregate variance estimator

	Trace OC48			Trace OC192		
Sampling scheme	**0.1 s**	**0.01 s**	**0.001 s**	**0.1 s**	**0.01 s**	**0.001 s**
Unsampled	0.857	0.923	0.908	0.784	0.828	0.835
SystC 1/100	0.815	0.835	0.727	0.832	0.818	0.804
RandC 1/100	0.814	0.837	0.740	0.793	0.819	0.804
SystT 100/500	0.429	0.611	0.776	0.822	0.720	0.714
LP 100/200	0.649	0.671	0.771	0.560	0.584	0.760
MuST 200/500	0.635	0.833	0.836	0.729	0.770	0.801

Figure 4 presents the visual representation of the R/S estimator for the unsampled OC48 trace aggregated bytes in 0.1 s time-window. When applying this estimator to the sampled traffic (see Table 2), similar results for the aggregate variance are observed, including an accuracy degradation of the MuST technique for OC48 trace when aggregated over 0.1 s time intervals. In this case, a comparable result is achieved for the OC192 trace for the same time aggregation. Here, the other adaptive sampling technique (LP) also provides valuable results for the OC48 trace, which ratifies the promising suitability of this type of sampling for multipurpose measurement tasks.

Table 2. Hurst parameter using the R/S estimator

	Trace OC48			Trace OC192		
Sampling scheme	**0.1 s**	**0.01 s**	**0.001 s**	**0.1 s**	**0.01 s**	**0.001 s**
Unsampled	0.793	0.658	0.621	0.800	0.740	0.732
SystC 1/100	0.617	0.565	0.542	0.790	0.703	0.650
RandC 1/100	0.613	0.564	0.542	0.793	0.703	0.650
SystT 100/500	0.473	0.412	0.514	0.686	0.297	0.458
LP 100/200	0.747	0.409	0.585	0.598	0.474	0.658
MuST 200/500	0.281	0.614	0.626	0.445	0.650	0.774

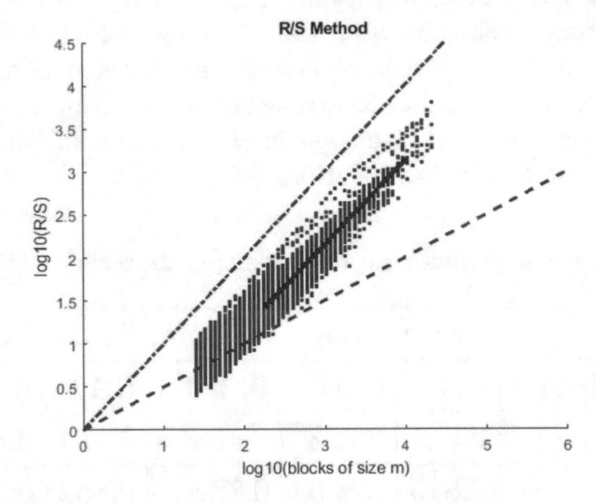

Fig. 4. Visual observation of R/S estimator for the OC48 trace.

The comparative accuracy of time-based and count-based schemes is similar when resorting to the periodogram estimator. In this case, time-based techniques outperform count-based ones for particular time windows. In this sense, this estimator might be the most suitable to be combined with different measurement tasks while providing realistic estimations for LRD-based analysis.

Although the 0.001 s observation time-window might increase the communication overhead related to measurement tasks, globally, this dimension presents a better accuracy regardless of the trace analysed. This conclusion is based on the sum of the estimation error for all compared sampling techniques (Table 3).

Table 3. Hurst exponent for the periodogram estimator

	Trace OC48			Trace OC192		
Sampling scheme	**0.1 s**	**0.01 s**	**0.001 s**	**0.1 s**	**0.01 s**	**0.001 s**
Unsampled	1.053	0.663	0.589	0.857	0.676	0.755
SystC 1/100	0.731	0.543	0.505	0.742	0.658	0.648
RandC 1/100	0.822	0.539	0.507	0.867	0.661	0.649
SystT 100/500	0.540	0.371	0.514	1.035	0.435	0.652
LP 100/200	0.775	0.485	0.565	0.650	0.553	0.716
MuST 200/500	0.728	0.669	0.582	0.575	0.662	0.762

5 Conclusions

The volume of data traversing modern networks demands that measurement-based tasks are able to provide accurate results without processing all the underlying packets. Traffic sampling is the primary strategy used to achieve this balance. However, different sampling techniques provide distinct accuracy results for different monitoring tasks. Thereby, this work presented a comparative analysis of multiple sampling techniques applied to estimating temporal correlation and long memory behaviour of real backbone traffic. The results show that although varying in accuracy, sampling processes preserve the statistical traffic properties related to self-similarity while reducing the analyzed volume of data by 80%, in the worst case. Globally, higher accuracy was observed from count-based techniques, which also select the lower amount of data for samples. However, these results should be put into the context of multipurpose network measurements. For instance, if sampled traffic is also used for flow-based classification or intrusion detection, time-based techniques could provide better overall accuracy. Therefore, future work will consider different monitoring tasks and explore tuning adaptive techniques toward Hurst parameter estimation.

Acknowledgements. This work is financed by National Funds through the Portuguese funding agency, FCT - Fundação para a Ciôncia e a Tecnologia, within project LA/P/0063/2020, and by FCT, within the R&D Units Project Scope: UIDB/00319/2020.

References

1. Beran, J., Sherman, R., Taqqu, M.S., Willinger, W.: Long-range dependence in variable-bit-rate video traffic. IEEE Trans. Commun. **43** (1995)
2. Bregni, S.: Compared accuracy evaluation of estimators of traffic long-range dependence. In: 2014 IEEE Latin-America Conference on Communications (2014)

3. Carela-Español, V., Barlet-Ros, P., Aparicio, A., Solé-Pareta, J.: Analysis of the impact of sampling on NetFlow traffic classification. Comput. Netw. **55** (2011)
4. Claffy, K.C., Polyzos, G.C., Braun, H.W.: Application of sampling methodologies to network traffic characterization. SIGCOMM **23** (1993)
5. Du, Y., Huang, H., Sun, Y.E., Chen, S., Gao, G.: Self-adaptive sampling for network traffic measurement. In: IEEE Conference on Computer Communications (INFOCOM) (2021)
6. Hurst, H.E., Black, R.P., Simaika, Y.M.: Long-term storage : an experimental study. Constable London (1965)
7. Karagiannis, T., Molle, M., Faloutsos, M.: Long-range dependence ten years of internet traffic modeling. IEEE Internet Comput. **8** (2004)
8. Karagiannis, T.: SELFIS: a short tutorial (2002)
9. Leland, W.E., Taqqu, M.S., Willinger, W., Wilson, D.V.: On the self-similar nature of ethernet traffic. In: Conference Proceedings on Communications Architectures, Protocols and Applications, SIGCOMM (1993)
10. Montanari, A., Taqqu, M., Teverovsky, V.: Estimating long-range dependence in the presence of periodicity: an empirical study. Math. Comput. Modelling (1999)
11. Paxson, V., Floyd, S.: Wide area traffic: the failure of poisson modeling. IEEE/ACM Trans. Netw. **3** (1995)
12. Silva, J.M.C., Carvalho, P., Lima, S.R.: A modular traffic sampling architecture: bringing versatility and efficiency to massive traffic analysis. J. Netw. Syst. Manage. **25** (2017)
13. Tamma, B., Bs, M., Rao, R.: Time-based sampling strategies for multi-channel wireless traffic characterization in tactical cognitive networks (2008)
14. Taqqut, M., Teverovsky, V., Willinger, W.: Estimators for long-range dependence: an empirical study. Fractals **03** (1995)

Features Extraction and Structure Similarities Measurement of Complex Networks

Haji Gul[1], Feras Al-Obeidat[2], Munir Majdalawieh[2], Adnan Amin[1], and Fernando Moreira[3(✉)]

[1] Center for Excellence in Information Technology, Institute of Management Sciences, Lahore, Pakistan
adnan.amin@imsciences.edu.pk

[2] College of Technological Innovation, Zayed University, Abu Dhabi, United Arab Emirates
{feras.al-obeidat,munir.majdalawieh}@zu.ac.ae

[3] REMIT, IJP, Universidade Portucalense, 4200-072 Porto, Portugal
fmoreira@uportu.pt

Abstract. Various models have been proposed to shed light on the evolution mechanisms of real-world complex networks (e.g., Facebook, Twitter, etc.) that can be expressed in terms of graph similarity. Generally, state-of-the-art research has assumed that complex networks in the real world are jointly driven by (i) multiplex features rather than a single pure mechanism, and (ii) a focus on either local or global features of complex networks. Nonetheless, the extent to which these characteristics interact to influence network evolution is not entirely clear. This study introduces an approach for calculating graph similarity based on a variety of graph features, including graph cliques, entropy, spectrum, Eigenvector centrality, cluster coefficient, and cosine similarity. Initially, each network structure was closely analyzed, and multiple features were extracted and embedded in a vector for the aim of similarity measurement. The experiments demonstrate that the proposed approach outperforms other graph similarity methods. Additionally, we find that the approach based on cosine similarity performs significantly better in terms of accurate estimations (i.e., 0.81 percent) of overall complex networks, compared to the Shortest Path Kernel (SPK) at 0.69 percent and the Weisfeiler Lehman Kernel (WLK) at 0.67 percent.

Keywords: Complex Network Analysis · Graph Features Extraction · Complex Network Similarity · Graph Classification

1 Introduction

Network science is widely used in the real world to describe the characteristics of complex systems. When a complex system is represented as a graph, nodes are typically used to represent objects, while edges represent the relationships between objects. From various fields, scientists represent complex systems as

A. Rocha et al. (Eds.): WorldCIST 2023, LNNS 799, pp. 37–47, 2024.
https://doi.org/10.1007/978-3-031-45642-8_4

graphs from a variety of perspectives, including protein-protein interaction [1], social networks [2], author collaboration, terrorist activities, traffic, animal and human contact, and football complex networks, to name only a few. A graph represents the interaction or collaboration among the components of a complex network system. Where components can be expressed by nodes and the interaction between them by links. For example, a protein, a terrorist, a football player, and a Facebook user are nodes, and the interactions between proteins, football players, or Facebook users are the links or edges [3]. Complex network graphs provide a strong basis to analyze complex systems, which is very helpful in many aspects, such as resolving and predicting scientific future discoveries of complex networks [4]. Estrada and Ernesto [5] analyzed the transitivity and normal distance in electrical, biological, and social real-world data and discovered that they all function similarly. According to [6], the authors found numerous topological characteristics in the structure of a community, depending on whether the considered data is connected to social, medical, terrorist, correspondence, or router network systems.

The majority of existing graph similarity kernel-based methods concentrate only on either local or global complex network properties. These kernels decompose graphs into their constituent structural components and add up the pairwise similarities between them. There are kernels that compare graphs based on subtree [7], random-walk [8], small-sub-graph [9], shortest path [10], and cycles [11]. Each complex network topological attribute is critical in resolving the network's most pressing issues. To solve the link prediction problem and enhance eleven, a solution based on common neighbor and clustering coefficient complex network features has been recently introduced [12]. Similarly, a method based on entropy is created to identify the variation of network topology [13]. Zahiri et al. extended the Girvan-Newman community detection algorithm by utilizing centrality. Graph energy [14] also has an important role in generating the scene of a graph. Cerqueti et al. introduced a method based on clustering coefficient identification of risk assessment [15]. Additionally, a hybrid method based on entropy and centrality of complex network features has been introduced to identify the organizations of terrorists [16].

In the existing literature, the researchers assumed a very small number of complex networks and, more importantly, a very limited number of features. Furthermore, the focus on a few features could not be able to provide efficient information about the various complex network systems. Laura et al. [17], introduced an algorithm that works only on graph local neighbor features. The shortest graph kernel [18] works on the global feature path of a graph. Similarly, the Weisfeiler-Lehman sub-tree kernel [19] counts pairs of similar sub-tree structures in pairs of graphs and checks the number of iterations. It also fails to check the similarity between smaller and larger graphs. However, local feature-based techniques utilized for larger graphs may suffer from their local environment and fail to perform consistently, as several interesting characteristics of these complex networks may not be captured in local and global structures. Therefore, there is a need to develop a method that covers the gap between the existing methods.

In this paper, we present a model to compute complex network structure similarity that is based on multiple advanced local and global features and employs cosine similarity.

2 Review on Similarity Measurements

In this section, we presented a brief systematic literature review of the complex network similarity measures, and we concluded the section by identifying gaps in the target domain.

Similarities Measurements for Real-World Complex Network: Researchers Barabasi and Albert [20], Watts and Strogatz [21] developed scale-free and small-world complex network models in 1999. Recent research has largely focused on network reconstruction and analysis, with some studies also focusing on complex network systems similarity measurement and classification. A variety of approaches have been presented to analyze and compute the similarities between real-world complex systems.

According to the study [22], a well-known way is to utilize a graph kernel to measure the similarity of various complex networks. Recently, a few innovative strategies have risen to the top of prior approaches proposed by scholars [23]. These strategies use a deep learning system to learn information-driven graph representations. The Weisfeiler-Lehman (WL) computation is summarized in [24] by figuring out how to encode only the necessary highlights from a node neighborhood during each cycle.

The work in Mnih et al. [25], on the other hand, took into account how to guide CNN to the most important articles for the visual analysis response task. Although topological attentional handling has been applied effectively to numerous issues, the majority of current work lies in the fields of language handling or computer vision. Interestingly, according to [24], a method uses neural convolution architecture to measure a portion of the information network.

Kashima [26], clustered the vectors using distance methods after mapping networks to feature vectors. In order to construct the feature vectors, which are mined using search engines on the Web. Similarly, Prulj captures a network's topology using graphlets [27]. Graphlets capture the local structure properties by capturing all potential sub-graphs of a small number of nodes. By keeping track of the frequencies of all the different sized graphlets, the graphlet frequency distribution can be used to compare networks [28]. This isn't feasible for huge graphs because it needs a precise count of each graphlet. It's been employed in aerial picture comparison [29], scene classification [30], and learning on synthetic networks [31]. DeltaCon [32], a scalable technique for same-sized networks based on node influence, was proposed by Koutra et al. It compares two networks with an identical node set by computing the influence of each node on the nodes of the other network. For each network, these values are recorded as matrices, and the difference between the matrices is then calculated to produce an affinity score that measures similarity.

Some studies have also started to turn to the use of structural considerations in real-world networks to compute similarities. Choi et al. [33] proposed a model in which topological features focus on clinical ontology. The work by Costa, L. da F., et al. [34], included extensive coverage of network characterization. The characterization, assessment, segmentation, modelling, and validation of complex networks all require measurements of their structure. Several additional, powerful measures have been presented, which were first limited to simple properties such as vertex degree, clustering coefficient, and shortest path length.

2.1 Graph Similarity Measurement Algorithms

Some state-of-the-art graph similarity algorithms are listed below. These are also evaluated over eleven complex network data-sets.

Shortest Path Kernel (SPK) [10]: The transformation of the original graphs into shortest-path graphs is the first and most important step in the shortest-path kernel. The input graph's nodes are replicated in the shortest-paths graph. In addition to the input graph, there is an edge linking all nodes connected by a walk. The shortest distance between starting node v_i and destination node v_j is labelled on every edge between these two nodes. Mathematically, the shortest path kernel can be expressed as,

$$SPK(G_1, G_2) = \sum_{e_1 \epsilon E_1} \sum_{e_2 \epsilon E_2} k_{walk}(e_1, e_2), \tag{1}$$

On edge walks of length 1, $k_w^{(1)}alk(e_i, e_j)$ is a positive semi-definite kernel. It compares the lengths of the shortest paths that correspond to edges e_i and e_j, as well as the labels of their destination vertices.

Weisfeiler Lehman kernel (WLK) [19]: This method labeling the correlative in G and G', which means that nodes in G and G' will obtain identical new labels if their multi-set labels are identical. As a result, one iteration of Weisfeiler-Lehman relabeling can be thought of as a function $r(V, E, li) = (V, E, l_{i+1})$ that alters all graphs similarly. It's interesting to note that r is dependent on the graphs we're looking at. The mathematical formula is given below,

$$k_{WL}^{(h)}(G, G') = k(G_1, G_1') + k(G_2, G_2') + ...k(G_h, G_h') \tag{2}$$

On the other hand, the proposed study significantly covers the gap identified in the literature, specifically handling the clinical Ontology that works on coordination with non-cyclic and single-domain in a real-world complex system. Furthermore, we have also analyzed and utilized multiple other crucial graph properties. Another study used topological management to solve the problem of vertex portrayal learning [35], which aims to learn how to embed a vertex in a network graph. Additionally, the proposed work also extracts a variety of graph features and measures the similarity of complex networks based on these properties. Finally, we compared and evaluated these graphs using cosine similarity.

3 Materials and Methods

Initially, a 0/1 adjacency matrix was produced, with 0 indicating that there is no link between the pair of vertices in the arrangement and 1 indicating that there is a relationship between the pair of vertices. If there is a connection between two vertices, it is numerically expressed by $e = (x, y)$. Where x is the starting and y is the targeting vertex. The term "*link*" refers to anything that connects two points. It might be a relationship between two Facebook users/vertices, communication between two karate club members, the interaction between pairs of dolphins, or a terrorist joint attack other possibilities.

Features Extractions: Complex networks have a variety of structural features. The required formulas to extract complex network features are Maximum Degree, total nodes, total edges, the sum of degrees, average degree, algebraic connectivity, sum of average neighbors, average path length, no of cliques, cluster coefficient, average closeness, diameter, distance distribution, edge betweenness, eigenvector centrality, graph energy, radius, spectrum, entropy, and graph density.

Proposed Framework: The complex graph consists of multiple features related to the path, degree, neighbors, cycles, triangles, etc. Each feature of a complex network graph has its own mathematical representation for discovering it. Section 3, expresses the mathematical formula representation and a brief explanation of these aspects. We have analyzed in detail these complex networks given in Sect. 3 and reported numerous features given in Table 1. Based on these features, the complex networks are compared and the similarity among them is computed using cosine similarity which has not been used yet for graph comparisons, see Table 2 Cosine similarity is a popular method used to compute

Table 1. Features extracted from various complex networks.

Features	Karate	Dolphin	Sampson	Con USA	Zebra	Kangaroo	Human Con	Terrorist	Football	Highland	Iceland
Max Degree	17	12	28	8	14	15	31	29	19	10	24
Total Nodes	33	62	18	49	26	14	41	63	35	15	75
Total Edges	78	159	189	107	111	91	336	243	118	58	114
Sum of Length Degree	34	62	18	49	27	17	43	64	35	6	75
Sum of Degree	156	318	378	214	222	182	672	486	236	116	228
Average Degree	4.5882	5.129	21	4.3673	8.2222	10.7059	15.6279	7.5938	6.7429	7.25	3.04
Algebraic Connectivity	0.4685	0.173	9.6164	0.098	0	0.9713	4.0517	0.3304	0.5738	2.3353	0.0871
Sum of Avg Neighbour	326.7471	431.1208	261.4933	235.4405	243.3484	207.998	806.4569	790.1343	362.1553	122.3468	864.2063
Avg Path Length	2.4082	3.357	1.6078	4.1633	1.8417	1.3603	1.6711	2.691	2.1227	1.5417	3.1996
No of Clique	5	5	11	3	10	10	9	11	6	5	4
Cluster Coefficients	0.2557	0.3088	0.8538	0.4062	0.8448	0.8406	0.5639	0.561	0.3293	0.5271	0.1573
Avg Closeness	0.4394	0.3123	0.6638	0.2559	0	0.8099	0.6216	0.3914	0.5002	0.6983	0.3271
Diameter	5	8	3	11	Inf	3	3	6	5	3	6
Distance Distribution	1	1	1	1	0.7379	1	1	1	1	1	1
Edge Betweeness	5.38E+03	1.94E+04	4.87E+03	1.02E+04	4.64E+03	2.76E+03	2.61E+04	2.22E+04	8.17E+03	2.0315E+03	1.7165E+04
Eigenvector centrality	−4.978	5.628	4.1637	5.6226	−3.9252	3.8975	6.1102	5.5225	−4.9301	−3.8692	5.9877
Graph Energy	48.3032	107.0244	64.7188	85.3129	48.6507	32.6507	112.3721	115.5351	64.1984	32.2907	74.89
Graph Radius	3	5	2	6	Inf	2	2	3	3	2	3
Graph Spectrum	156	318	378	214	222	182	672	486	236	116	228
Entropy	1.9805	2.3517	2.187	1.8137	1.5886	2.1501	2.8849	2.6706	2.4339	1	1.554
Link Density	0.139	0.0841	1.2353	0.091	0.3162	0.6691	0.3721	0.1205	0.1983	0.4833	0.0411

the similarity between two entities. Additionally, all these have been analyzed, and embedded each graph features in vector for cosine. The graphical view and explanation of the proposed model is included in Fig. 1.

The cosine similarity technique measures the proportion of likeness between two non-zero vectors of an innermost item space. It is characterized by the rising cosine of the point between them, which is equivalent to the innermost result of similar vectors standardized to all have length 1. The 0° of cosine value is 1, and it is under 1 for any point in the span radiance $(0, \pi]$.

Following that, it is a judgment of direction rather than extent: two vectors with similar directions have a cosine similitude of 1, two vectors arranged at 90° comparative with one another have a comparability of 0, and two vectors completely opposite each other have a closeness of (0, -1), independent of their size. The cosine method is especially utilized in optimistic planetary systems, where the consequence is approximately constrained to [0, 1].

Each real-world network attribute is put away in an independent vector, and bit by bit, every one of the vectors contrasts with each other by discovering the similitude. Finally, we reported the similarities among these networks in Table 2 The experiment code is available through the link [1].

$$CS(x, y) = \frac{x.y}{||x||X||y||} \tag{3}$$

The methodology (given below) was utilized for the experiments.

$$CS(x, y) = \frac{\sum_{a=1}^{N} x_a X y_a}{\sqrt{\sum_{a=1}^{N} x^2 X} \sqrt{\sum_{a=1}^{N} y^2}} \tag{4}$$

DataSets: All the data-sets (undirected and unweighted) are preprocessed, and irrelevant information is removed. These networks are expressed by a graph to accomplish various activities. Total eleven data-sets are analyzed and the majority of them are obtained from[2],[3],[4]. The features have been extracted are: karate, dolphin, Sampson, contiguous USA, zebra, human contact, kangaroo, train bombing, highland, and Iceland.

Evaluation Setup: The assessment criteria are briefly outlined in this section. In this study, eleven complex network datasets from diverse disciplines were investigated. Preprocessing eliminates unnecessary information from these datasets, such as loops and weights. Then there's a features vector that contains all of the identified features of each dataset. Finally, cosine similarity is used to evaluate these feature vectors.

---

[1] https://github.com/hajigul/Graph_S.
[2] https://networkrepository.com/.
[3] https://snap.stanford.edu/data/.
[4] http://konect.cc/networks/.

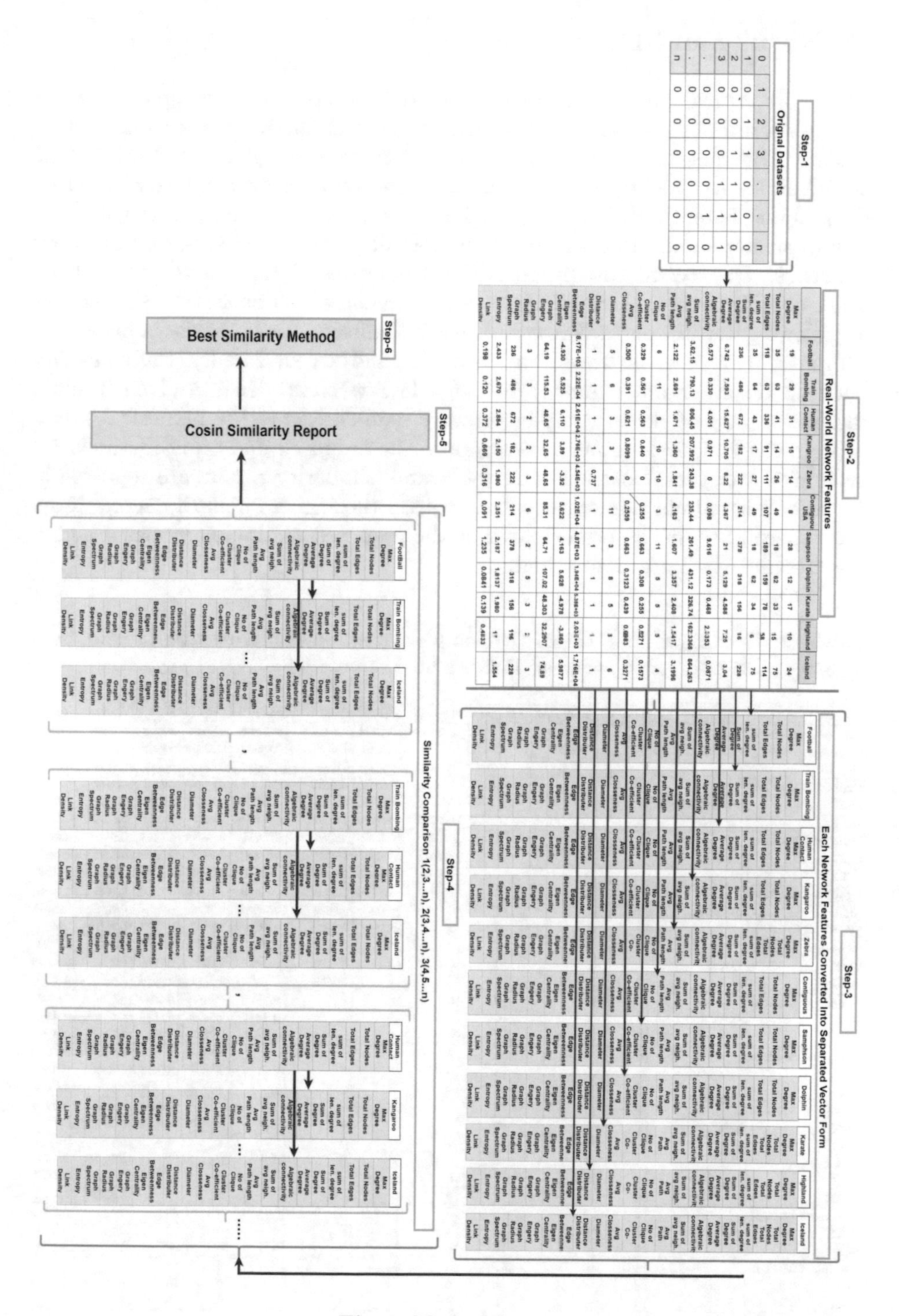

Fig. 1. Methodology

4 Results and Discussion

The most critical phase is to investigate the structure of a complex network and use a distinct mathematical formula to extract the resultant value of each feature. Different network features are determined and given in Table 1. The cosine similarity method has been used to find an efficient similarity based on these multiple features. The upper triangle is given in Table 2, where the first row expresses similarities between the football complex network and all other networks. The second row presents the similarities of train bombing with all other complex networks. Similarities of human contact networks with other networks are given in the third row of Table 2. Similarly, the similarities have been computed and given in rows four, five, six, and seven. Finally, the similarity between dolphins and karate is included in row eight. In the given Table 2, there are totals of eleven complex network data-sates and the numerical values express the similarities among each other. The proposed and other methods are evaluated over these data-sets and the overall similarity results are reported in Table 2. Where bold numerical values express the higher similarity score. From the experiments, it has been clear that the proposed methodology has a higher graph similarity score than others.

Table 2. Overall, in this Table, the complex network similarity results have been reported. The best result is denoted with a bold value.

Methods	Data-Sates	Football	Train Bombing	Human	Kangaroo	Zebra	Conti USA	Sampson	Dolphin	Karate	Highland	Iceland
Proposed Method		**1**	**0.7723**	0.6776	**0.9163**	**0.7761**	**0.9579**	**0.9143**	**0.9565**	**0.963**	**0.9793**	0.9201
SPK	**Football**	**1**	0.5841	**0.6925**	0.7382	0.6356	0.7876	0.7908	0.8688	0.8938	0.9087	**0.9223**
WLK		**1**	0.7203	0.5321	0.8008	0.7313	0.8098	0.8385	0.5869	0.7391	0.8185	0.8369
Proposed Method		0	**1**	**0.9724**	0.5697	0.5279	**0.6472**	**0.8014**	**0.8803**	0.6532	**0.9552**	**0.9278**
SPK	**Train Bombing**	0	**1**	0.9565	0.6324	0.7309	0.5812	0.6307	0.7403	0.6136	0.6874	0.7236
WLK		0	**1**	0.5187	**0.7635**	**0.7589**	0.6285	0.4102	0.7087	**0.6768**	0.7108	0.7281
Proposed Method		0	0	**1**	0.4975	0.4422	0.5689	**0.7621**	**0.7996**	0.5493	**0.9701**	**0.8576**
SPK	**Human**	0	0	**1**	**0.7015**	0.4178	0.5327	0.6784	0.6947	0.5312	0.6987	0.7452
WLK		0	0	**1**	0.5896	**0.6323**	**0.5864**	0.6899	0.7314	**0.6325**	0.6784	0.5689
Proposed Method		0	0	0	**1**	**0.7466**	**0.9714**	**0.8242**	**0.8166**	**0.9418**	0.7512	0.8465
SPK	**Kangaroo**	0	0	0	**1**	0.5336	0.5699	0.7301	0.7325	0.7125	0.7254	**0.8701**
WLK		0	0	0	**1**	0.5122	0.8999	0.7758	0.6987	0.4827	**0.7965**	0.6981
Proposed Method		0	0	0	0	**1**	**0.8037**	0.6477	**0.7272**	**0.7946**	0.6955	0.8023
SPK	**Zebra**	0	0	0	0	**1**	0.4177	**0.8377**	0.6247	0.7741	**0.7745**	**0.8288**
WLK		0	0	0	0	**1**	0.7562	0.552777	0.5821	0.6978	0.6587	0.6875
Proposed Method		0	0	0	0	0	**1**	**0.8863**	**0.8949**	**0.9466**	0.8637	0.8435
SPK	**Conti USA**	0	0	0	0	0	**1**	**0.8875**	0.7321	0.8874	**0.8759**	**0.8563**
WLK		0	0	0	0	0	**1**	0.5874	0.7841	0.7952	0.7569	0.7784
Proposed Method		0	0	0	0	0	0	**1**	**0.9479**	**0.793**	**0.9671**	**0.7248**
SPK	**Sampson**	0	0	0	0	0	0	**1**	0.4009	0.6879	7725	0.7021
WLK		0	0	0	0	0	0	**1**	0.7994	0.5278	0.6751	0.6914
Proposed Method		0	0	0	0	0	0	0	**1**	**0.8746**	**0.8995**	**0.8946**
SPK	**Dolphin**	0	0	0	0	0	0	0	**1**	0.8146	0.7848	0.8125
WLK		0	0	0	0	0	0	0	**1**	0.7046	0.6745	0.7562
Proposed Method		0	0	0	0	0	0	0	0	**1**	**0.9448**	**0.9598**
SPK	**Karate**	0	0	0	0	0	0	0	0	**1**	0.7896	0.6587
WLK		0	0	0	0	0	0	0	0	**1**	0.6258	0.7895
Proposed Method		0	0	0	0	0	0	0	0	0	**1**	**0.8367**
SPK	**HT**	0	0	0	0	0	0	0	0	0	**1**	0.7854
WLK		0	0	0	0	0	0	0	0	0	**1**	0.6871
Proposed Method		0	0	0	0	0	0	0	0	0	0	**1**
SPK	**Iceland**	0	0	0	0	0	0	0	0	0	0	**1**
WLK		0	0	0	0	0	0	0	0	0	0	**1**

5 Conclusions

In the network science analysis and similarities of real-world complex networks pose a crucial problem. We empirically analyzed various networks in this paper and extracted many important features embedded in vectors. The cosine similarity method was then used to find the similarity of these networks' structures. The similarity is calculated based on the network features. Some of the most discriminant measures are cliques, connected components, edge betweenness, link density, network entropy, spectrum, network radius, network energy, Eigenvector centrality, distribution distance, diameter, closeness, cluster coefficient, algebraic connectivity, and average degree. It could be extended in the future to work with large complex networks from different domains.

References

1. Mohamed, S.K., Nováček, V., Nounu, A.: Discovering protein drug targets using knowledge graph embeddings. Bioinformatics **36**(2), 603–610 (2020)
2. Alamsyah, A., Rahardjo, B., et al.: Social network analysis taxonomy based on graph representation. arXiv preprint arXiv:2102.08888 (2021)
3. Gul, H., Amin, A., Adnan, A., Huang, K.: A systematic analysis of link prediction in complex network. IEEE Access **9**, 20531–20541 (2021)
4. Gul, H., Amin, A., Nasir, F., Ahmad, S.J., Wasim, M.: Link prediction using double degree equation with mutual and popular nodes. In: Rocha, Á., Adeli, H., Dzemyda, G., Moreira, F., Ramalho Correia, A.M. (eds.) WorldCIST 2021. AISC, vol. 1368, pp. 328–337. Springer, Cham (2021). https://doi.org/10.1007/978-3-030-72654-6_32
5. Estrada, E.: The Structure of Complex Networks: Theory and Applications. Oxford University Press, Oxford (2012)
6. Lancichinetti, A., Kivelä, M., Saramäki, J., Fortunato, S.: Characterizing the community structure of complex networks. PLoS ONE **5**(8), e11976 (2010)
7. Mahé, P., Vert, J.-P.: Graph kernels based on tree patterns for molecules. Mach. Learn. **75**(1), 3–35 (2009)
8. Sugiyama, M., Borgwardt, K.: Halting in random walk kernels. Adv. Neural. Inf. Process. Syst. **28**, 1639–1647 (2015)
9. Kriege, N., Mutzel, P.: Subgraph matching kernels for attributed graphs. arXiv preprint arXiv:1206.6483 (2012)
10. Borgwardt, K.M., Kriegel, H.-P.: Shortest-path kernels on graphs. In: Fifth IEEE International Conference on Data Mining (ICDM'05), p. 8. IEEE (2005)
11. Pinto, S., et al.: Rapid rewiring of arcuate nucleus feeding circuits by leptin. Science **304**(5667), 110–115 (2004)
12. Ghorbanzadeh, H., Sheikhahmadi, A., Jalili, M., Sulaimany, S.: A hybrid method of link prediction in directed graphs. Expert Syst. Appl. **165**, 113896 (2021)
13. Liu, J., Zhou, S.: A new measure of network robustness: network cluster entropy. In: Lin, L., Liu, Y., Lee, C.-W. (eds.) SocialSec 2021. CCIS, vol. 1495, pp. 175–191. Springer, Singapore (2021). https://doi.org/10.1007/978-981-16-7913-1_13
14. Suhail, M., et al.: Energy-based learning for scene graph generation. In: Proceedings of the IEEE/CVF Conference on Computer Vision and Pattern Recognition, pp. 13936–13945 (2021)

15. Cerqueti, R., Clemente, G.P., Grassi, R.: Systemic risk assessment through high order clustering coefficient. Ann. Oper. Res. **299**(1), 1165–1187 (2021)
16. Spyropoulos, A.Z., Bratsas, C., Makris, G.C., Ioannidis, E., Tsiantos, V., Antoniou, I.: Entropy and network centralities as intelligent tools for the investigation of terrorist organizations. Entropy **23**(10), 1334 (2021)
17. Zager, L.A., Verghese, G.C.: Graph similarity scoring and matching. Appl. Math. Lett. **21**(1), 86–94 (2008)
18. Yanardag, P., Vishwanathan, S.: Deep graph kernels. In: Proceedings of the 21th ACM SIGKDD International Conference on Knowledge Discovery and Data Mining, pp. 1365–1374 (2015)
19. Shervashidze, N., Schweitzer, P., Van Leeuwen, E.J., Mehlhorn, K., Borgwardt, K.M.: Weisfeiler-lehman graph kernels. J. Mach. Learn. Res. **12**(9) (2011)
20. Barabási, A.-L., Albert, R.: Emergence of scaling in random networks. Science **286**(5439), 509–512 (1999)
21. Watts, D.J., Strogatz, S.H.: Collective dynamics of 'small-world' networks. Nature **393**(6684), 440–442 (1998)
22. Nikolentzos, G., Meladianos, P., Vazirgiannis, M.: Matching node embeddings for graph similarity. In: Proceedings of the AAAI Conference on Artificial Intelligence, vol. 31 (2017)
23. Duvenaud, D., et al.: Convolutional networks on graphs for learning molecular fingerprints. arXiv preprint arXiv:1509.09292 (2015)
24. Niepert, M., Ahmed, M., Kutzkov, K.: Learning convolutional neural networks for graphs. In: International Conference on Machine Learning, pp. 2014–2023. PMLR (2016)
25. Mnih, V., Heess, N., Graves, A., Kavukcuoglu, K.: Recurrent models of visual attention. arXiv preprint arXiv:1406.6247 (2014)
26. Kashima, H., Inokuchi, A.: Kernels for graph classification. In: ICDM Workshop on Active Mining (2002)
27. Pržulj, N.: Biological network comparison using graphlet degree distribution. Bioinformatics **23**(2), e177–e183 (2007)
28. Rahman, M., Bhuiyan, M.A., Rahman, M., Al Hasan, M.: GUISE: a uniform sampler for constructing frequency histogram of graphlets. Knowl. Inf. Syst. **38**(3), 511–536 (2014)
29. Zhang, L., Han, Y., Yang, Y., Song, M., Yan, S., Tian, Q.: Discovering discriminative graphlets for aerial image categories recognition. IEEE Trans. Image Process. **22**(12), 5071–5084 (2013)
30. Zhang, L., Bian, W., Song, M., Tao, D., Liu, X.: Integrating local features into discriminative graphlets for scene classification. In: Lu, B.-L., Zhang, L., Kwok, J. (eds.) ICONIP 2011. LNCS, vol. 7064, pp. 657–666. Springer, Heidelberg (2011). https://doi.org/10.1007/978-3-642-24965-5_74
31. Janssen, J., Hurshman, M., Kalyaniwalla, N.: Model selection for social networks using graphlets. Internet Math. **8**(4), 338–363 (2012)
32. Koutra, D., Shah, N., Vogelstein, J.T., Gallagher, B., Faloutsos, C.: DeltaCon: principled massive-graph similarity function with attribution. ACM Trans. Knowl. Discov. Data (TKDD) **10**(3), 1–43 (2016)
33. Gul, H., Amin, A., Nasir, F., Ahmad, S.J., Wasim, M.: Link prediction using double degree equation with mutual and popular nodes. In: Rocha, Á., Adeli, H., Dzemyda, G., Moreira, F., Ramalho Correia, A.M. (eds.) WorldCIST 2021. AISC, vol. 1368, pp. 328–337. Springer, Cham (2021). https://doi.org/10.1007/978-3-030-72654-6_32

34. Costa, L.D.F., Rodrigues, F.A., Travieso, G., Villas Boas, P.R.: Characterization of complex networks: a survey of measurements. Adv. Phys. **56**(1), 167–242 (2007)
35. Chen, K., Wang, J., Chen, L.-C., Gao, H., Xu, W., Nevatia, R.: ABC-CNN: an attention based convolutional neural network for visual question answering. arXiv preprint arXiv:1511.05960 (2015)

Big Data Analytics and Applications

Technology Use by Nigerian Smallholder Farmers and the Significant Mediating Factors

Enobong Akpan-Etuk(✉)

Instutute of Management, University of Bolton, Bolton, England
enobongakpanetuk@gmail.com

Abstract. The willingness of smallholder farmers in Nigeria to adopt new agricultural technologies has been low, despite the technological advancement in agriculture. Although technology is seen as the main route out of the traditional methods of food production and poverty, the rate of adoption of agricultural technology remains low among farmers in Nigeria. Factors affecting the adoption of agricultural technologies include the acquisition of information, characteristics of the technology, the education of farmers, social capital, farm size, household size, finance, self-control and hope. This paper explored the literature on the influence of the factors that determine the adoption of technology by smallholder farmers. And proposes that any exogenous influences and offer of support must recognize smallholder farmers' personal resources when promoting technology adoption.

Keywords: Smallholder · Technology · Adoption · Exogenous

1 Introduction

In developing countries such as Nigeria, the rural population consist of smallholder farmers (Lowder et al. 2016; Lerberghe 2008) who cultivate few acres of land for agriculture and making little or no use of technologies (Morton 2007; Min et al.; 2017; Altieri and Koohafkan 2008). Despite their low yields of crops, smallholder farmers still produce the major share of food supply in most of the developing countries. Therefore, the agricultural productivity of smallholder farmers has made a tremendous impact in alleviating poverty and improving food security (Samberg et al. 2016; Parry, 2007). Thus, terms such as poverty reduction, productivity enhancement, environmental mitigation or protection, adaptation to climate change, modernization and agricultural technology developments have been the main focus of many agricultural policy projects and programs (Tal and Bukchin 2017; Loevinsohn et al. 2013). These developments have been defined as the improved methods, means, physical practices and techniques that enhances agricultural outputs (Jan et al. 2009). Such agricultural developments offer improved management regimes such as water and irrigation management, pest and weed management, soil fertility management (Loevinsohn et al. 2013).

New agricultural technologies improve agricultural processes thereby helping in reducing poverty, protecting the environment, enhancing productivity, and enabling farmers to adapt to climate change (Tal and Buckchin 2017; Loevinsohn et al. 2013).

A. Rocha et al. (Eds.): WorldCIST 2023, LNNS 799, pp. 51–64, 2024.
https://doi.org/10.1007/978-3-031-45642-8_5

Technology use enhances the growth of agricultural outputs and improves the methods, means, practices and physical techniques of food production (Jain et al. 2009). Introduced technologies are also important in boosting soil quality, crop production and reduces nutrition and food insecurity (Sennuga and Fadiji 2020). Current agricultural technologies improve on-farm food production and post-production, aids in the delivery of healthy and safe agricultural products, and considers environmental, social and economic sustainability (Sannuga et al. 2020; FAO 2010). New agricultural technologies assist smallholder farmers in crop management and protection, land management, fertilizer management, pest management, water and irrigation management, conservation agriculture and land management (Sannuga et al. 2020; FAO 2010). Therefore, increasing agricultural productivity of smallholder farmers through technology is likely to have a huge impact on increasing food security and alleviating poverty.

Research has shown that new agricultural technological advances improve productivity of smallholder farmers (Tessa and Kurukulasuriya 2010, Porter et al. 2014; Biagini et al. 2014). Yet, there are limiting factors to the use of new technologies by smallholder farmers (Buckchin and Kerret 2018).

This paper, therefore, explores the reasons most smallholder farmers fail to use new agricultural technologies, and to examine the underlying factors. The next section reviews the literature on technology use in agriculture.

2 Literature Review

2.1 Agricultural Technologies

Loevinsohn et al. (2013) defines technology as a method and means of producing goods and services which includes physical techniques and different methods of organization. The authors also explain that technology is classified as 'new' when it has just been introduced to a group of farmers or a new place. Technology is the information or knowledge that enables some tasks to be performed, enables some services to be rendered effectively, and aids the productions of certain goods or products (Lavison 2013). The use of technology improves certain situations, changes the status quo to more desirable outcomes, simplifies work, saves time, and improves labour (Bonabana-Wabbi 2002).

Agricultural technologies include several kinds of enhanced practices and techniques that increases agricultural output (Jain et al. 2009). These are the technologies used in water and soil management, irrigation, weed control and crop breeding. Agricultural technologies used for protecting soil degradation are referred to as conservation agriculture (Prasai et al. 2018; Hobbs et al. 2007). These technologies are used to enrich the soil, conduct soil analysis, determine soil characteristics, composition and nutrient level, increase biodiversity and the natural biological processes below and above the ground surface, protects the soil from degradation and erosion (Tiwari et al. 2018; Karki and Shrestha, 2014). Agricultural technologies involve complex management processes and systems that delivers information to agricultural users and farmers (Tebaldi and Gobjila 2018). These technologies are diverse, and they are designed mainly for use in the agricultural sector, for example, the farming robotic. There are other digital technologies or solutions that are used in the agricultural sector and other sectors of the economy e.g. drones.

Agricultural technologies use integrated and novel technologies that are designed to enable farmers increase food production, reduce waste and improve food safety (United Nations Global Impact 2017; Sarker et al. 2020). Examples of some of the digital solutions in agriculture includes Information and Communication Technology (Walter et al. 2017, El Bilali and Allahyari, 2018; Aker 2011, food traceability technology (World Economic Forum (WEF) 2019; Bosona and Gebresenbet 2013; Astill et al. 2019), big data and internet-of-things (IoT) (Sarker et al. 2020; Garrett 2017). Digital technology in agriculture has also been used to address the efficient use of water in agriculture (Nhamo et al. 2020, Hunt and Daughtry 2018), precise and variable inputs applications (Koutsos and Menexes 2019; Iost Filho et al. 2019; Finger et al. 2019), the use of aerial imaging technology to monitor crops (Iost Filho et al. 2019), the use of wireless sensors to monitor the weather (Hunt and Daughtry, 2018) the use of quick response (QR) and blockchain systems to trace the contamination of foods along the food supply chain (Tian 2016; Rotz et al. 2019a; Kamilaris et al. 2019; Deichmann et al. 2016; Ayaz et al. 2019) and many more. There are also socio-economic benefits of using digital agricultural technologies in agriculture. This increases the income of farmers and increases their job profile. On the other hand, there social concerns around the impact of digital agricultural technologies. According to Rotz et al. (2019b), agricultural technologies can increase inequality in the agricultural sector mainly between smallholder farmers and agri-food corporations.

2.2 The Social, Economic and Environmental Impact of Digital Agricultural Technology

Digital agricultural technologies can be used to connect farmers to the market. With technologies such as digital platforms, farmers have the ability to connect and sell their produce directly to consumers, retailers and restaurants, thereby mitigating post-harvest waste (FAO 2017b). According to Deichmann et al. (2016), information exchange through social networks can act as a tool for warning farmers against emerging disaster that could affect their crops or impact their farms.

Adopting digital agriculture could also lead to a range of environmental benefits. Technologies such as digital simulations, RFID and GPS are used to design distribution routes for food trucks, therefore reducing spoilage and waste (Tian 2016; Springael et al. 2018; Padilla et al. 2018). The use of the blockchain system in agriculture can improve the consumer awareness about sustainability and environmental characteristics of food production (Kamilaris et al. 2019). According to Handford et al. (2014), digital designs e.g. nanotechnology could contribute to precision agriculture, and thus, more use of agrochemicals and feed additives that leads to the reduction of agricultural waste.

The advancement of technology such as vertical and cellular agriculture has created opportunity to increase the efficiency of resources (Helliwell and Burton, 2021). Vertical agriculture, mostly in urban areas have the potential to reduce or prevent waste and maximize economic efficiency by reducing harvesting and delivery time to the market (Babbitt et al. 2019; Avgoustaki and Xydis 2020).

2.3 Technology Adoption

Technology adoption refers to the adoption of a new technology or innovation by individuals or a group of people. The adoption of a new innovation or idea is a mental process that happens in 5 stages:

a- Awareness stage: In the awareness stage, the individual or farmer is aware of the idea but lacks the information about the new technology.
b- Interest stage: This is a stage where the individual or farmer has information about the technology, makes more enquiry on how the new technology works, and how affordable it is.
c- Mental stage: At this stage, the individual or farmer has obtained all the information about the new agricultural technology and knows what it is all about.
d- Trial stage: The individual obtains the new agricultural technology and tries it to know how valuable it is.
e- Adoption stage: Is when the individual or farmer starts using the technology (Sennuga and Oyewole, 2020; Cheteni et al. 2014; Rogers 2013).

The adoption of new agricultural technologies has been associated with improved food production, increased earnings and employment opportunities, lower food prices and reduction in poverty (Sennuga et al. 2020, Kasirye 2010). Contrarily, non-adoption of new agricultural technologies leads to poverty, low crop yields, socio-economic stagnation and deprivation (Jain et al. 2009).

2.4 The Factors that Determine the Adoption of Technology by Smallholder Farmers

There are different dynamics that determines whether a farmer will decide on using technology. The choice to use technology may be as a result of certain conditions which includes, individual decisions: the cost of adopting the technology, the farmer's uncertain beliefs about technology, educational level, and the kind of technology itself (Hall and Khan, 2002). Understanding the factors that influence a farmer's decision to use technology is essential for the disseminators and generators of the technologies, and researchers tracking the use of the technologies for improvements (Hall and Khan 2002).

Traditionally, researchers believed that a farmer's adoption behaviour to technology relates to personal endowment and characteristics, institutional constraints, lack of information about the technology, uncertainty, and risk (Koppel 1994; Rogers 2003; Uaiene 2009; Kohli and Sigh, 1997, Foster and Rosenzweig 1996; Feder et al. 1985). Yet, others have included learning, social networks, and personal resources such as self-control and hope in the factors that determine the adoption of technology by smallholder farmers in Africa.

1- Acquisition of Information

Chawinga et al. (2018a, b) explained that the low rate of adoption of technology by smallholder farmers is due to the lack of information and training concerning the technology to be used. Accessing the information about these technologies is important in improving food production (Phiri et al. 2018a; Mwangi and Kariuki 2015), reducing uncertainty about the technology, and could replace a farmers' subjective opinion about the technology with an objective assessment of the technology (Mwangi and Kariuki, 2015). Therefore, disseminating information to farmers about the importance of the technology and the different techniques that could be used in agricultural production, could help in the acceptance of the technology (Phiri et al. 2018b; Obayelu et al. 2007).

However, having access to the information about agricultural technologies does not guarantee that those technologies will be adopted by farmers, as they may be evaluated and perceived subjectively (Uaiene et al. 2009; Prager and Posthumus 2010; Murage et al. 2015). While the information about the new technology may be beneficial to the farmers, they may not trust the information source or be motivated to learn about its recommendations (Jin et al, 2015; Fisher 2013; Harris 2018). The farmers may suspect that the information may have environmental, political, or financial implications (Buck and Alwang, 2011). Thus, trusting the source of the information plays a significant role in adopting the new agricultural technologies and methods (Schewe and Stuart 2017; Jin et al. 2015; Fisher 2013; Buck and Alwang, 2011). The smallholder farmers' trust in the information source is very important and plays the main role in adopting the new agricultural technology and methods (Schewe and Stuart, 2017; Jin et al. 2015; Fisher 2013; Buck and Alwang 2011).

Therefore, receiving the information about the new agricultural technologies and trusting that information is vital in the adoption of the technology.

2- Characteristics of the Technology

The kind of technology under consideration will determine if the farmer will adopt it or not. Trialability can also be the main determinant of the adoption of a particular technology (Doss 2003). Farmers who get the opportunity to try out a particular technology are more likely to adopt that technology. This is because they would have experienced the importance of the technology during the trial period. According to Mignouna et al. (2011), farmers who perceive technologies as being important and meets their needs in food production are likely to adopt it. Karugia et al. (2004) also stress that farmers should be involved in the evaluation of a new agricultural technology before it is introduced to them. It is therefore important to ensure that farmers and beneficiaries of new technologies are fully involved in the initial phase, but not in the later stages, like normal practice.

3- Education of the Farmer

Education plays an important role, and a major factor in farmer's decision to adopt new agricultural technologies. The level of the farmer's education increases his or her ability to obtain and process the information that are relevant to the new agricultural technology (Namara et al. 2013; Lavison 2013; Mignouna et al. 2011). According to Ajewole (2010) and Okunlola et al. 2011), the level of education of farmers has a significant and positive influence on their adoption of a new technology. This is because a higher level of education influences a person's thoughts and attitudes, making them more rationale, open and able to analyze information and realize the benefits of technology (Waller et al. 1988). The educational level of the farmers will increase their ability to adopt new agricultural technologies (Mignounal et al. 2011). Education enables the farmers to read labels and manuals, follow directions and instructions on how to operate machines and use tools that are required for farming. According to Croppensted et al. (2003), highly educated farmers are more likely to adopt new agricultural technologies.

They are more knowledgeable about the benefits of new technologies and have a better mental attitude towards the acceptance of new technologies (Sennuga et al. 2020; Caswell et al. 2011; Mwangi and Kariuki 2015; Deressa et al. 2011).

4- Social Capital

Another significant influencing factor is social capital. Social capital helps farmers obtain technical information through communication, technical learning and other channels. Therefore, improving farmers' awareness of new agricultural technologies and promotes their enthusiasm to adopt those technologies. According to Peng and Lin (2022), social capital affects the relationships of individuals. The relationship changes and network of farmers affects their individual decision making. A social network formed through close interaction and relationships among individuals encourages information sharing. Smallholder farmers access information through different social networks and this helps them reduce the cost of learning and enables them to make decisions on adopting new technologies (Han et al. 2022). According to Husen et al (2017), due to the limited level of education of some smallholder farmers, they will listen to information about technologies from individuals in their social network. Farmer will tend to listen to other agricultural technology adaptors in their social network before adopting those technologies. Being a member of a social group likely enhances social capital and promotes information exchange and trust (Mignouna et al. 2011). Farmers learn about new agricultural technologies and benefit from each other when they belong to the same social group (Uaiene et al. 2009). According to Akankwasa (2010), farmers who have social network connections, and belong to community organizations, engage in social learning about new technologies, which raises the likelihood of adopting technologies.

5- Farm Size

According to Akudugu et al. (2012), farm size affects the adoption of new agricultural technologies. Farmers that engage in large-scale farming are likely to adopt new agriculture than farmers who engage in small-scale farming (Kasenge 1998). Several researches have also suggested that farmers with large farms are more likely to adopt new agricultural technologies (Fuglie and Kascak 2003; Parvan 2011; Moser and Barrett 2008).

6- Household Size

Research has shown that smallholder farmers from a large household are more likely to adopt new agricultural technologies compared to those from a smaller household. The household size indicates the measure of labour that is available in the family, with large families having access to more labour (Mwangi and Kariuki 2015). Also, when introducing new agricultural technologies to the farm, farmers with large households have the capacity to reduce labour constraints in the farm (Mignouna et al. 2011). This show that smallholder farmers with large households generates more income from the use of their family labour and new agricultural technologies (Mwangi and Kariuki 2015).

7- Farming Experience

Research has shown that farming experience is a major factor in the adoption of new agricultural technology by smallholder farmers. Smallholder farmers who have been farming for many years have the skills, technical know-how and knowledge, and this contributes to their ability to make decision about the adoption of new agricultural technology (Yishak 2005; Melaku 2005; Petros 2010). The more the experience the smallholder farmer has in cultivating crops, the more they are aware of the technology needed in the farm (Matata et al. 2010). Where the smallholder farmers have limited farming experience, the rate of technology adoption was low (Araya and Mohammad 2014).

8- Finance

Finance is classified as the financial intervention from institutions such as banks, non-governmental and government agencies. Finance is identified as one of the factors that influence technology adoption by smallholder farmers. Various researches have reported the correlation between access to finance and technology adoption among smallholder farmers, due to the high cost of the new agricultural technologies (Mudhara et al. 2003; Makate et al. 2019; Habtemariam et al. 2019; Diagne 2009). Finance empowers smallholder farmers to purchase new agricultural technologies (Fisher and Carr 2015). Smallholder farmers also use finance to acquire farmland, hire labour and get the required training and education needed to operate the new agricultural technology. The affordability, accessibility and availability of finance was found to influence the adoption of technology by smallholder farmers (Akrofi et al. 2019; Awotide et al. 2015).

9- Self-control

According to DeWall et al. (2008), self-control and decision making are closely linked. Decision making is mostly reliant on having the ability to make sophisticated judgement and reason logically based on diverse information (Vohs and Baumeister 2011). Making difficult decisions such as adopting new technology, leads to mental depletion and this could also lead to making default choices. Some experiment in law and medicine support the claim that people make intuitive and low-effort decisions when they are depleted (Vohs and Baumeister 2011; Gallagher et al. 2011; Danziger et al. 2011). However, some studies that was reviewed by (Vohs and Baumeister 2011) proposed that self-control could counter depletion effects. Therefore, higher self-control could lead to non-default decisions, by making people overcome the challenges in decision making. The smallholder farmer's decision to adopt new technology would be a non-default decision as this involves changing the status quo. Higher level of self-control may prompt people to make more deliberate and careful choices rather than sticking to their familiar routine (Rosenbaum 1980; De Ridder and Lensvelt-Mulders 2018). It is hypothesized that a smallholder farmer with a higher level of self-control will more likely adopt new agricultural technology. Research has shown that different interventions can be used to boost self-control (Ronen and Rosenbaum 2010; Pressley 1079; Muraven et al. 1999; Duckworth 2011).

10- Hope

Hope is connected to the adoption of technology by smallholder farmers. According to Snyder's hope theory (Snyder 2000), goal-oriented, cognitive hope is made up of three components: (a) the goals that people wish should happen to them; (b) having the ability to come up with ways to achieve goals; (c) agency thinking: the ability to put both pathways to use. The study of Bukchin and Kerret (2018), showed the connection between hope and technology adoption by smallholder farmers. The theoretical analysis was later proved in the field about the important link between hope, problem-solving and action-taking in uncertain circumstances. The findings in the research showed that the farmers that are filled with hope were more likely to visualize the desired effects of technology, and they tend to expect the technology to improve their lives. This makes them adopt the technology. Moreover, the smallholder farmers that are filled with hope are more likely to overcome their apprehension about the risk involved in adopting the new technology. Research has shown that different interventions can be used to boost hope (Vilaythong et al. 2009; Sydney 2000; Feldman and Dreher 2012; Daugherty et al. 2018).

3 Conclusions

This paper has revealed that the adoption of new agricultural technologies by smallholder farmers depends on certain factors which includes among others: the acquisition of information, characteristics of the technology, education of farmers, social capital, farm size, household size, finance, farming experience and having self-control and hope. Given the importance of information, hope and self-control in technology adoption, I would propose that non-governmental organizations, government agencies, agricultural trainers and developmental organizations recognize smallholder's farmers' personal resources when promoting technology adoption. Apart from offering information on financial means and farming techniques, training programs should be provided to farmers to boost their self-control skills and hope. The promotion and use of new agricultural technologies are essential for increasing food production, food security, alleviating poverty especially for smallholder farmers in Nigeria.

References

Ajewole, C.: Farmer's response to adoption of commercially available organic fertilizers in Oyo State Nigeria African. J. Agricult. Res. **5**(18), 2497–2503 (2010)

Aker, J.C.: Dial "A" for agriculture: a review of information and communication technologies for agricultural extension in developing countries. Agric. Econ. **42**(6), 631–647 (2011)

Akrofi, N.A., Sarpong, D.B., Somuah, H.A.S., OseiOwusu, Y.: Paying for privately installed irrigation services in Northern Ghana: the case of the smallholder Bhungroo irrigation technology. Agric. Water Manag. **216**, 284–293 (2019)

Akudugu, M., Guo, E., Dadzie, S.: Adoption of modern agricultural production technologies by farm households in Ghana: what factors influence their decisions? J. Biol. Agric. Healthcare **2**(3) (2012)

Altieri, M.A., Koohafkan, P.: Enduring Farms: Climate Change, Smallholders and Traditional Farming Communities, vol. 6 Third World Network (TWN) Penang (2008)

Araya, S., Mohammed, H.: Adoption of improved local wheat seed production systems in Meskan and Sodo districts of Ethiopia. Seed Technol. **36**(2), 151–160 (2014)

Astill, J., et al.: Transparency in food supply chains: a review of enabling technology solutions. Trends Food Sci. Technol. **91**, 240–247 (2019)

Awotide, B., Abdoulaye, T., Alene, A., Manyong, V.M.: Impact of access to credit on agricultural productivity: evidence from smallholder cassava farmers in Nigeria. In: 2015 Conference of International Association of Agricultural Economists (IAAE), Milan, Italy. pp. 1–34 (2015)

Ayaz, M., Ammad-Uddin, M., Sharif, Z., Mansour, A., Aggoune, E.-H.M.: Internet- of-Things (IoT)-based smart agriculture: toward making the elds talk. IEEE Access **7**, 129551–129583 (2019)

Biagini, B., Kuhl, L., Gallagher, K.S., Ortiz, C.: Technology transfer for adaptation. Nat. Clim. Chang. **4**(9), 828 (2014)

Bonabana-Wabbi, J.: Assessing Factors Affecting Adoption of Agricultural Technologies: The Case of Integrated Pest Management (IPM) in Kumi District, M.Sc. thesis Eastern Uganda (2002)

Bosona, T., Gebresenbet, G.: Food traceability as an integral part of logistics management in food and agricultural supply chain. Food Contr. **33**(1), 32–48 (2013)

Buck, S., Alwang, J.: Agricultural extension, trust, and learning: results from economic experiments in Ecuador. Agric. Econ. **42**(6), 685–699 (2011)

Bukchin, S., Kerret, D.: Food for hope: the role of personal resources in farmers' adoption of green technology. Sustainability **10**(5), 1615 (2018)

Bukchin, S., Kerret, D., (2019). Goal-oriented Hope and Sustainable Technology Adoption by Smallholder Farmers. Manuscript submitted for publication

Caswell, M., Fuglie, K., Ingram, C., Jans, S., Kascak, C.: Adoption of Agricultural Production Technologies: Lessons Learned from the U.S. Department of Agriculture Area Studies Project, Agriculture Economic Report-792, U.S. Department of Agriculture (USDA), Washington, D.C., USA (2001)

Cheteni, P., Mushunje, A., Taruvinga, A.: Barriers and incentives to potential adoption of biofuels crops by smallholder farmers in the Eastern Cape Province, South Africa. Munich Personal RePEc Archive (MPRA) Paper No. 59029. Crop Protect. 76, pp. 83–91 (2014)

Croppenstedt, A., Demeke, M., Meschi, M.: Technology Adoption in the Presence of constraints: the case of fertilizer demand in Ethiopia. Rev. Dev. Econ. **1**(7), 58–70 (2003)

Danziger, S., Levav, J., Avnaim-Pesso, L.: Extraneous factors in judicial decisions. Proc. Natl. Acad. Sci. **108**(17), 6889–6892 (2011)

Daugherty, D.A., Runyan, J.D., Steenbergh, T.A., Fratzke, B.J., Fry, B.N., Westra, E.: Smartphone delivery of a hope intervention: another way to flourish. PLoS ONE **13**(6), e0197930 (2018)

De Ridder, D.T., Lensvelt-Mulders, G.: Taking stock of self-control: a meta-analysis of how trait self-control relates to a wide range of behaviors self- regulation and self-control. Routledge, pp. 221–274 (2018)

Deichmann, U., Goyal, A., Mishra, D.: Will digital technologies transform agriculture in developing countries? Agric. Econ. **47**(S1), 21–33 (2016)

Deressa, T.T., Hassan, R.M., Ringler, C., Tekie, A., Mahmud, Y.: Determinants of farmers' choice of adaptation methods to climate change in the Nile Basin of Ethiopia. Glob. Environ. Chang. **19**, 248–255 (2011)

DeWall, C.N., Baumeister, R.F., Masicampo, E.: Evidence that logical reasoning depends on conscious processing. Conscious. Cognit. **17**(3), 628–645 (2008)

Diagne, A.: Technological change in smallholder agriculture: Bridging the adoption gap by understanding its source. UC Berkeley: Center for Effective Global Action (2009)

Doss, C.R.: Understanding Farm Level Technology Adoption: Lessons Learned from CIMMYT's Microsurveys in Eastern Africa. CIMMYT Economics Working Paper 03–07. Mexico, D.F.: CIMMYT (2003)

Duckworth, A.L.: The significance of self-control. Proc. Natl. Acad. Sci. **108**(7), 2639–2640 (2011)

El Bilali, H., Allahyari, M.S.: Transition towards sustainability in agriculture and food systems: role of information and communication technologies. Inf. Process. Agric. **5**(4), 456–464 (2018)

FAO: Human Energy Requirement, Food and Nutrition Technical Report 1. Food and Agriculture Organization of the United Nations, Rome (2010)

Feder, G., Just, R.E., Zilberman, D.: Adoption of agricultural innovations in developing countries: a survey. Econ. Dev. Cult. Change **33**(2), 255–298 (1985)

Feldman, D.B., Dreher, D.E.: Can hope be changed in 90 minutes? Testing the efficacy of a single-session goal-pursuit intervention for college students. J. Happiness Stud. **13**(4), 745–759 (2012)

Finger, R., Swinton, S.M., El Benni, N., Walter, A.: Precision farming at the nexus of agricultural production and the environment. Annu. Rev. Resour. Econ. **11**, 313–335 (2019)

Fisher, M., Carr, E.R.: The influence of gendered roles and responsibilities on the adoption of technologies that mitigate drought risk: the case of drought-tolerant maize seed in eastern Uganda. Glob. Environ. Chang. **35**, 82–92 (2015)

Fisher, R.: 'A gentleman's handshake': the role of social capital and trust in transforming information into useable knowledge. J. Rural Stud. **31**, 13–22 (2013)

Frazier, P.A., Tix, A.P., Barron, K.E.: Testing moderator and mediator effects in counseling psychology research. J. Couns. Psychol. **51**(1), 115 (2004)

Food and Agricultural Organization (FAO): Save Food for a Better Climate: Converting the Food Loss and Waste Challenge into Climate Action (2017b). http://www.fao.org/save-food/news-and-multimedia/news/news-details/en/c/1062697/. Accessed 6 Dec 2022

Fuglie, K.O., Kascak, C.A.: Adoption and diffusion of natural-resourceconserving agricultural technology. Rev. Agric. Econ. **23**(2), 386–403 (2001)

Gallagher, P., Fleeson, W., Hoyle, R.H.: A self-regulatory mechanism for personality trait stability: contra-trait effort. Soc. Psychol. Person. Sci. **2**(4), 335–342 (2011)

Garrett, R.: Plug in to future food connections: the Internet of Things (IoT) will better connect the food and beverage industry to reduce food loss and make food safer to eat. (Cover story). Food Logist. 184, 16 (2017)

Habtemariam, L.T., Mgeni, C.P., Mutabazi, K.D., Sieber, S.: The farm income and food security implications of adopting fertilizer micro-dosing and tiedridge technologies under semi-arid environments in central Tanzania. J. Arid Environ. **166**, 60–67 (2019)

Hall, B.H., Khan. B.: Adoption of New Technology. New Economy handbook (2002)

Han, M., Liu, R., Ma, H., Zhong, K., Wang, J., Xu, Y.: The impact of social capital on farmers' willingness to adopt new agricultural technologies: empirical evidence from China. Agriculture **12**, 1368 (2022)

Handford, C.E., Dean, M., Spence, M., Henchion, M., Elliott, C.T., Campbell, K.: Awareness and attitudes towards the emerging use of nanotechnology in the agri- food sector. Food Contr. **57**, 24 (2015)

Harris, S.: Agricultural Information Needs and Food Access in the Stann Creek District of Belize (2018)

Hunt, E.R., Daughtry, C.S.T.: What good are unmanned aircraft systems for agricultural remote sensing and precision agriculture? Int. J. Rem. Sens. **39**(15–16), 5345–5376 (2018)

Husen, N.A., Loos, T.K., Siddig, K.: Social capital and agricultural technology adoption among Ethiopian farmers. Am. J. Rural. Dev. **5**, 65–72 (2017)

Iost Filho, F.H., Heldens, W.B., Kong, Z., de Lange, E.S.: Drones: innovative technology for use in precision pest management. J. Econ. Entomol. **113**(1), 1–25 (2019)

Jain, R., Arora, A., Raju, S.: A novel adoption index of selected agricultural technologies: linkages with infrastructure and productivity. Agric. Econ. Res. Rev. **22**, 109–120 (2009)

Jain, R., Arora, A., Raju, S.: A novel adoption index of selected agricultural technologies: linkages with infrastructure and productivity. Agric. Econ. Res. Rev. **22**(1) (2009)

Jin, S., Bluemling, B., Mol, A.P.: Information, trust and pesticide overuse: interactions between retailers and cotton farmers in China. NJAS - Wageningen J. Life Sci. **72**, 23–32 (2015)

Kamilaris, A., Fonts, A., Prenafeta-Bold´υ, F.X.: The rise of blockchain technology in agriculture and food supply chains. Trends Food Sci. Technol. **91**, 640–652 (2019)

Karki, T., Shrestha, J.: Conservation agriculture: significance, challenges and opportunities in Nepal. Adv. Plant. Agric. Res. **1**(5), 00029 (2014)

Karugia, S., Baltenweck, I., Waithaka, M., Miano, M., Nyikal, R., Romney, D.: Perception of technology and its impact on technology uptake: the case of fodder legume in central Kenya highlands. In: The Role of Social Scientists Proceedings of the Inaugural Symposium, 6 to 8 December 2004, Grand Regency Hotel, Nairobi, Kenya (2004)

Kasenge V.: Socio-economic factors influencing the level of soil management technologies on fragile land. In: Shayo-Ngowi, A.J., Ley, G., Rwehumbiza, F.B.R. (eds.) Proceedings of the 16th Conference of Soil Science Society of East Africa, 13th-19th, Tanga, Tanzania, vol. 8, pp. 102–112 (1998)

Kasirye: Constraints to Agriculyural Technology adoption in Uganda: Evidence from the 2005/06–2009/10 Uganda National Panel Survey. AgEcon- Research in Agricultural & Applied Economics (2013)

Koppel, B.M. (ed.): Induced Innovation Theory and International Agricultural Development: A Reassessment the Johns Hopkins University Press, Baltimore (1994)

Koutsos, T., Menexes, G.: Economic, agronomic, and environmental bene ts from the adoption of precision agriculture technologies: a systematic review. Int. J. Agric. Environ. Inf. Syst. **10**(1), 40–56 (2019)

Lavison, R.: Factors Influencing the Adoption of Organic Fertilizers in Vegetable Production in Accra, Msc Thesis, Accra Ghana (2013)

Lerberghe, W.V.: The World Health Report 2008: Primary Health Care: Now More than Ever (2008)

Loevinsohn, M., Sumberg, J., Diagne, A., Whitfield, S.: Under what circumstances and conditions does adoption of technology result in increased agricultural productivity? Syst. Rev (2013)

Lowder, S.K., Skoet, J., Raney, T.: The number, size, and distribution of farms, smallholder farms, and family farms worldwide. World Dev. **87**, 16–29 (2016)

Makate, C., Makate, M., Mango, N., Siziba, S.: Increasing resilience of smallholder farmers to climate change through multiple adoption of proven climate-smart agriculture innovations. Lessons from Southern Africa. J. Environ. Manage. **231**, 858–868 (2019)

Matata, P., Ajayi, O.O., Oduol, P., Agumya, A.: Socio-economic factors influencing adoption of improved fallow practices among smallholder farmers in Western Tanzania. Afr. J. Agric. Res. **5**, 818–823 (2010)

Melaku G.: Adoption and Profitability of Kenyan top bar hive beekeeping technology: A study in Ambasel woreda of Ethiopia. Unpublished MSc thesis, Alemaya University, Alemaya, Ethiopia (2005)

Mignounal, D.B., Manyong, V.M., Mutabazi, K.D.S., Senkondo, E.M.: Determinants of adopting imazapyrresistant maize for Striga control in Western Kenya: a factors influencing adoption of improved agricultural technologies (IATs) among smallholder farmers in Kaduna State Nigeria double-hurdle approach. J. Dev. Agric. Econ. **11**(3), 572–586 (2011)

Min, S., Huang, J., Bai, J., Waibel, H.: Adoption of intercropping among smallholder rubber farmers in Xishuangbanna. China. Int. J. Agric. Sustain. **15**(3), 223–237 (2017)

Min, S., Huang, J., Bai, J., Waibel, H.: Adoption of intercropping among small- holder rubber farmers in Xishuangbanna. China. Int. J. Agric. Sustain. **15**(3), 223–237 (2017)

Morton, J.F.: The impact of climate change on smallholder and subsistence agriculture. Proc. Natl. Acad. Sci. 104 (50), 19680–19685 (2007). Murage, A., Pittchar, J., Midega, C., Onyango, C., Khan, Z., 2015. Gender specific S

Morton, J.F.: The impact of climate change on smallholder and subsistence agriculture. Proc. Natl. Acad. Sci. **104**(50), 19680–19685 (2007)

Moser, C., Barrett, C.B.: The disappointing adoption dynamics of a yield increasing, low external input technology: the case of SRI in Madagascar. Agric. Syst. **76**(3), 1085–1100 (2003)

Mudhara, M., Hilderbrand, P.E., Nair, P.: Potential for adoption of Sesbania sesban improved fallows in Zimbabwe: a linear programming-based case study of small-scale farmers. Agrofor. Syst. **59**, 307–315 (2003)

Murage, J., Pittchar, C., Midega, C., Onyango, Z.K.: Gender specific perceptions and adoption of the climate-smart push–pull technology in eastern Africa (2015)

Muraven, M., Baumeister, R.F., Tice, D.M.: Longitudinal improvement of self-regulation through practice: building self-control strength through repeated exercise. J. Soc. Psychol. **139**(4), 446–457 (1999)

Mwangi, M., Kariuki, S.: Factors determining adoption of new agricultural technology by smallholder farmers in developing countries. J. Econ. Sust. Dev. **5**(6), 208–216 (2015)

Namara, E., Weligamage, P., Barker, R.: Prospects for adopting system of rice intensification in Sri Lanka: A socioeconomic assessment. Research Report 75. Colombo, Sri Lanka: International Water Management Institute (2003)

Nhamo, L., et al.: Prospects of improving agricultural and water productivity through unmanned aerial vehicles. Agriculture **10**(7), 256 (2020)

Obayelu, A., Ajayi, O., Oluwalana, E., Ogunmola, O.: What does literature say about the determinants of adoption of agricultural technologies by smallholders farmers? (2017). Ong, A.D., Van Dulmen, M.H., 2006. Oxford Handbook of Methods in Positive Psychology. Oxford University Press

Okunlola, O., Oludare, O., Akinwalere, B.: Adoption of new technologies by fish farmers in Akure Ondo state, Nigeria. J. Agric. Technol. **7**(6), 1539–1548 (2011)

Padilla, M.P.B., Canabal, P.A.N., Pereira, J.M., Rian˜o, H.E.H.: Vehicle routing problem for the minimization of perishable food damage considering road conditions. Logist. Res. **11**, 2 (2018)

Parry, M.L.: Climate Change 2007-impacts, Adaptation and Vulnerability: Working Group II Contribution to the Fourth Assessment Report of the IPCC, vol. 4 Cambridge University Press (2007)

Parvan, A.: Agricultural technology adoption: issues for consideration when scaling-up. Cornell Pol. Rev. **1**(1), 1–12 (2011)

Peng, L.L.; Lin, L.: Social organization, social capital and social integration of floating population: an empirical study. J. Nanjing Agric. Univ. (Soc. Sci. Ed.), **22**, 43–52 (2022)

Petros, T.: Adoption of Conservation Tillage Technologies in Metema Woreda, North Gondar Zone, Ethiopia. An M.Sc thesis Submitted to School of Graduate Studies of Haramaya University (2010)

Phiri, A., Chipeta, G.T., Chawinga, W.D.: Information Behaviour of Rural Smallholder Farmers in Some Selected Developing Countries: A Literature Review. Information Development, 0266666918804861 (2018a)

Phiri, A., Chipeta, G.T., Chawinga, W.D.: Information needs and barriers of rural smallholder farmers in developing countries: a case study of rural smallholder farmers in Malawi. Information Development, 0266666918755222 (2018b)

Porter, J.R., Xie, L., Challinor, A.J., Cochrane, K., Howden, S.M.: Chapter 7: Food Security and Food Production Systems. Cambridge University Press. Prager, K., Posthumus, H., 2010. Socio-economic factors influencing farmers' adoption of soil conservation practices in Europe. Hum. Dimens. Soil Water Conserv. 12, 1–21 (2014)

Prager, K., Posthumus, H.: Socio-economic factors influencing farmers' adoption of soil conservation practices in Europe. Hum. Dimens. Soil Water Conserv. **12**, 1–21 (2010)

Prasai, H.K., Sah, S.K., Gautam, A.K., Regmi, A.P.: Conservation agriculture responses to productivity and profitability of mungbean under maize based cropping system in far western Nepal. J. Pure Appl. Algebr. **3**(1), 63–82 (2018)

Pressley, M.: Increasing children's self-control through cognitive interventions. Rev. Educ. Res. **49**(2), 319–370 (1979)

Rogers, E.M.: Diffusion of Innovations, 5th edn. The Free Press, New York (2003)

Ronen, T., Rosenbaum, M.: Developing learned resourcefulness in adolescents to help them reduce their aggressive behavior: preliminary findings. Res. Soc. Work. Pract. **20**(4), 410–426 (2010)

Rosenbaum, M.: A schedule for assessing self-control behaviors: preliminary findings. Behav. Ther. **11**(1), 109–121 (1980)

Rotz, S., et al.: The politics of digital agricultural technologies: a preliminary review. Sociol. Rural. **59**(2), 203–229 (2019)

Rotz, S., et al.: Automated pastures and the digital divide: how agricultural technologies are shaping labour and rural communities. J. Rural. Stud. **68**, 112–122 (2019)

Samberg, L.H., Gerber, J.S., Ramankutty, N., Herrero, N., West, P.C.: Subnational distribution of average farm size and smallholder contributions to global food production. Environ. Res. Lett. **11**(12), 124010 (2016)

Sarker, M.N.I., Islam, M.S., Murmu, H., Rozario, E.: Role of big data on digital farming. Int. J. Sci. Technol. Res. **9**(4) (2020)

Schewe, R.L., Stuart, D.: Why don't they just change? Contract farming, informational influence, and barriers to agricultural climate change mitigation. Rural. Sociol. **82**(2), 226–262 (2017)

Sennuga, S.O., Baines, R.N., Conway, J.S., Angba, C.W.: Awareness and adoption of improved agricultural technologies among smallholder farmers in relation to the adopted villages programme: the case study of Northern Nigeria. Int. J. Biol. Agric. Healthcare **10**(6), 34–49 (2020)

Sennuga, S.O., Conway, J.S., Sennuga, M.A.: Impact of information and communication technologies (ICTs) on agricultural productivity among smallholder farmers: evidence from Sub-Saharan African communities. Int. J. Agric. Extens. Rural Dev. Stud. **7**(1), 27–43 (2020)

Sennuga, S.O., Fadiji, T.O.: Effectiveness of traditional extension models among rural dwellers in Sub-Saharan African Communities. Int. J. Adv. Res. **8**(4), 401–415 (2020)

Sennuga, S.O., Oyewole, S.O.: Exploring the effectiveness of agricultural technologies training among smallholder farmers in Sub-Saharan African communities. Eur. J. Train. Dev. Stud. **7**(4), 1–15 (2020)

Snyder, C.R., et al.: The will and the ways: development and validation of an individual differences measure of hope. J. Personal. Soc. Psychol. **60**(4), 570 (1991)

Springael, J., Paternoster, A., Braet, J.: Reducing postharvest losses of apples: optimal transport routing (while minimizing total costs). Comput. Electron. Agric. **146**, 136–144 (2018)

Tal, A., Bukchin, S. (eds.): Execut. Sum. Proceed March. Terlau, W., Hirsch, D., Blanke, M., (2019). Smallholder Farmers as a Backbone for the Implementation of the Sustainable Development Goals. Sustainable Development

Tebaldi, E., Gobjila, A.: Romania: systematic country diagnostic. Background Note - Agriculture (2018). http://documents1.worldbank.org/curated/en/698251530897736576/pdf/128044-SCD-PUBLIC-P160439-RomaniaSCD BackgroundNoteAgriculture.pdf. Accessed 15 Dec 2022

Tessa, B., Kurukulasuriya, P.: Technologies for climate change adaptation: emerging lessons from developing countries supported by UNDP. J. Int. Aff. 17–31 (2010)

Tian, F.: An agri-food supply chain traceability system for china based on RFID & blockchain technology. In: Paper presented at the 2016 13th international conference on service systems and service management (ICSSSM), pp. 1–6. IEEE (2016)

Tiwari, T., Jackson, T., Chatterjee, K.: Scaling Out Conservation Agriculture Based Sustainable Intensification Approaches for Smallholder Farmers: Regional Policy Brief for Agricultural Policy Makers (2018)

Uaiene, R., Arndt, C., Masters, W.: Determinants of Agricultural Technology Adoption in Mozambique. Discussion papers No. 67E (2009)

United Nations Global Impact: Digital Agriculture: Feeding the Future (2017). http://breakthrough.unglobalcompact.org/disruptive-technologies/digital-agriculture/. Accessed 12 Dec 2022

Vilaythong, A.P., Arnau, R.C., Rosen, D.H., Mascaro, N.: Humor and hope: can humor increase hope? Humor **16**(1), 79–90 (2003)

Vohs, K.D., Baumeister, R.F.: Handbook of Self-Regulation: Research, Theory, and Applications. Guilford Publications (2011)
Waller, B., Hoy, W., Henderson, L., Stinner, B., Welty, C.: Matching innovation with potential users: a case study of potato IPM practices. Agric. Ecosyst. Environ. **70**, 203–215 (1998)
Walter, A., Finger, R., Huber, R., Buchmann, N.: Opinion: smart farming is key to developing sustainable agriculture. Proc. Natl. Acad. Sci. Unit. States Am. **114**(24), 6148–6150 (2017)
World Economic Forum (WEF): Innovation with a Purpose: Improving Traceability in Food Value Chains through Technology Innovations (2019). http://www.3.weforum.org/docs/WEF_Traceability_in_food_value_chains_Digital.pdf. Accessed 15 Dec 2022
Gecho, Y.: Determinants of Adoption of Improved Maize Technology in Damot Gale Woreda, Wolaita, Ethiopia. MSc thesis. Alemaya University (2005)

Balancing Plug-In for Stream-Based Classification

Francisco de Arriba-Pérez[1], Silvia García-Méndez[1], Fátima Leal[2], Benedita Malheiro[3,4](✉), and Juan Carlos Burguillo-Rial[1]

[1] Information Technologies Group, atlanTTic, University of Vigo, Vigo, Spain
{farriba,sgarcia}@gti.uvigo.es, J.C.Burguillo@uvigo.es
[2] REMIT, Universidade Portucalense, Porto, Portugal
fatimal@upt.pt
[3] INESC TEC, Porto, Portugal
mbm@isep.ipp.pt
[4] ISEP, Polytechnic Institute of Porto, Porto, Portugal

Abstract. The latest technological advances drive the emergence of countless real-time data streams fed by users, sensors, and devices. These data sources can be mined with the help of predictive and classification techniques to support decision-making in fields like e-commerce, industry or health. In particular, stream-based classification is widely used to categorise incoming samples on the fly. However, the distribution of samples per class is often imbalanced, affecting the performance and fairness of machine learning models. To overcome this drawback, this paper proposes Bplug, a balancing plug-in for stream-based classification, to minimise the bias introduced by data imbalance. First, the plug-in determines the class imbalance degree and then synthesises data statistically through non-parametric kernel density estimation. The experiments, performed with real data from Wikivoyage and Metro of Porto, show that Bplug maintains inter-feature correlation and improves classification accuracy. Moreover, it works both online and offline.

Keywords: Data bias · fairness · imbalanced data sets · machine learning algorithm · stream classification

1 Introduction

Artificial Intelligence (AI) has been growing exponentially over the last decade. New intelligent systems have emerged in multiple domains to support humans in their personal and professional routines. Specifically, machine learning (ML) models can solve predictive and classification problems, *e.g.*, predict preferences or classify symptoms. Classification involves predicting a class label for a given observation. However, since the distribution of samples is frequently imbalanced, the learning algorithms become biased towards the majority group [12]. This affects mostly the predictive performance of the minority class. Learning from imbalanced data sets and data streams is among the leading challenges in the domain. When processing data streams, ML models must cope with data properties evolving over time, and disproportionate class distribution [7].

A. Rocha et al. (Eds.): WorldCIST 2023, LNNS 799, pp. 65–74, 2024.
https://doi.org/10.1007/978-3-031-45642-8_6

To address this problem, this paper contributes with a balancing plug-in for stream-based classification, which synthesises new samples on the fly. Bplug is composed of three sub-modules: (*i*) class distribution analysis which verifies the imbalance degree of the data; (*ii*) parameter setting to maintain the inter-feature correlation among synthetic and original samples; and (*iii*) sample generation based on non-parametric Kernel Density Estimation (KDE). The proposed solution works both with data streams (online processing) and data sets (offline or batch processing). It generates new samples of the minority class based on a data stream sliding window (online) or the full data set (offline).

The experiments were performed with two real data sets obtained from Wikivoyage and Metro of Porto. The results show that Bplug effectively maintains the correlation among variables and their distribution function. The rest of this paper is organised as follows. Section 2 provides the literature review. Section 3 describes the proposed method. Section 4 reports the experimental results. Finally, Sect. 5 summarises and discusses the outcomes.

2 Related Work

The classification relies on a wide range of algorithms, such as decision trees, neural networks, Bayesian networks, nearest neighbours or support vector machines. However, imbalanced class distribution is a challenging scenario for most ML classification algorithms [18]. Learning from imbalanced data, where the number of observations is not equally distributed among classes, causes bias and unfairness. This unfairness, caused by data under-representation, discriminates between minority and majority classes.

To minimise the bias introduced by imbalanced data, the literature provides several offline data-level sampling methods for the data pre-processing stage. They aim to balance the data set, guarantee that all targets have the same number of samples [4], and optimise classifier performance [2]. The literature presents two standard sampling methods for class balancing: (*i*) Random Over-Sampling (ROS) that duplicates observations in the minority class, modifying the imbalance degree to any level [10]; and (*ii*) Random Under-Sampling (RUS) that removes observations from the majority class, balancing the original data. A well-known ROS approach is the Synthetic Minority Over-sampling Technique (SMOTE) [5]. SMOTE relies on the k-Nearest Neighbours (KNN) algorithm to synthesise minority class samples. In turn, RUS explores KNN [1] and clustering [13]. Moreover, over and undersampling methods can be combined through ensemble techniques like boosting, *e.g.*, RUSBoost [17] and SMOTEBoost [6], or bagging, *e.g.*, overBagging and underBagging [8].

In the online context, incoming samples may arrive quickly and in large volumes. Consequently, stream-based learning requires fast incremental model updating and tends to forget older training instances. Nguyen *et al.* (2011) argue that, in these circumstances, incremental model updating can further amplify the bias towards the majority class [15]. Nonetheless, scant research has been conducted to address the problem of data imbalance in stream-based scenarios.

In this regard, Korycki *et al.* (2020) propose a method to change imbalanced ratios among multiple classes in evolving data streams [11]. It uses active learning combined with stream-based oversampling, based on information about current class ratios and classifier errors in each class, to create new meaningful instances.

Concerning open-source packages, there are some solutions based on statistical approaches. Synthia, an open-source Python package, models and parameterises data using empirical and parametric methods [14]. Synthetic Data Vault (SDV) synthesizes data using generative models of relational databases [16]. SDV creates a complete database model by intersecting related database tables and computing their statistics. However, both packages implement offline processing. Contrary to existing open-source solutions, this contribution balances class instances in both batch and stream-based scenarios. Bplug synthesises data based on the current data imbalance degree, maintaining inter-feature correlation.

3 Proposed Method

The synthetic data generation module comprises three sub-modules: (*i*) class distribution analysis; (*ii*) parameter setting; and (*iii*) sample generation. Figure 1 illustrates the method proposed for both stream and batch ML operation.

Class distribution analysis determines the sample distribution per class by organising the input data samples into c class buckets. These buckets are used as input data to train the synthetic data generation module.

Parameter optimisation aims to maintain the inter-feature correlation and encompasses:

1. Detection of the feature data types present in the experimental input data to avoid the generation of incorrect feature values (*e.g.*, decimal values on integer features).
2. Determination of the feature value ranges. Since all features may include and be affected by outliers, only the values between the 10th percentile and 90th percentile are considered.
3. Computing of average feature values, based on the identified feature ranges, as the difference between the maximum and minimum values divided by the number of samples in the interval.
4. Selection of the minimum nonzero feature values for the minimum bandwidth of feature kernel distribution functions.
5. Selection of the best Gaussian and top-hat functions based on the log-likelihood metric and a 5-fold cross-validation [3].

Synthetic data generation relies on non-parametric KDE [19], where kernel distribution functions and bandwidths are configuration parameters.

- **Online synthesis** performs class distribution analysis, parameter optimisation and sample generation for the minority classes. It uses an n-sized sliding window, where n corresponds to the number of original samples (O) required to overcome cold start[1]. The synthetic data generation starts

[1] If $O < 20\,000$, it is advisable that $n = O/2$, otherwise $n = O/4$.

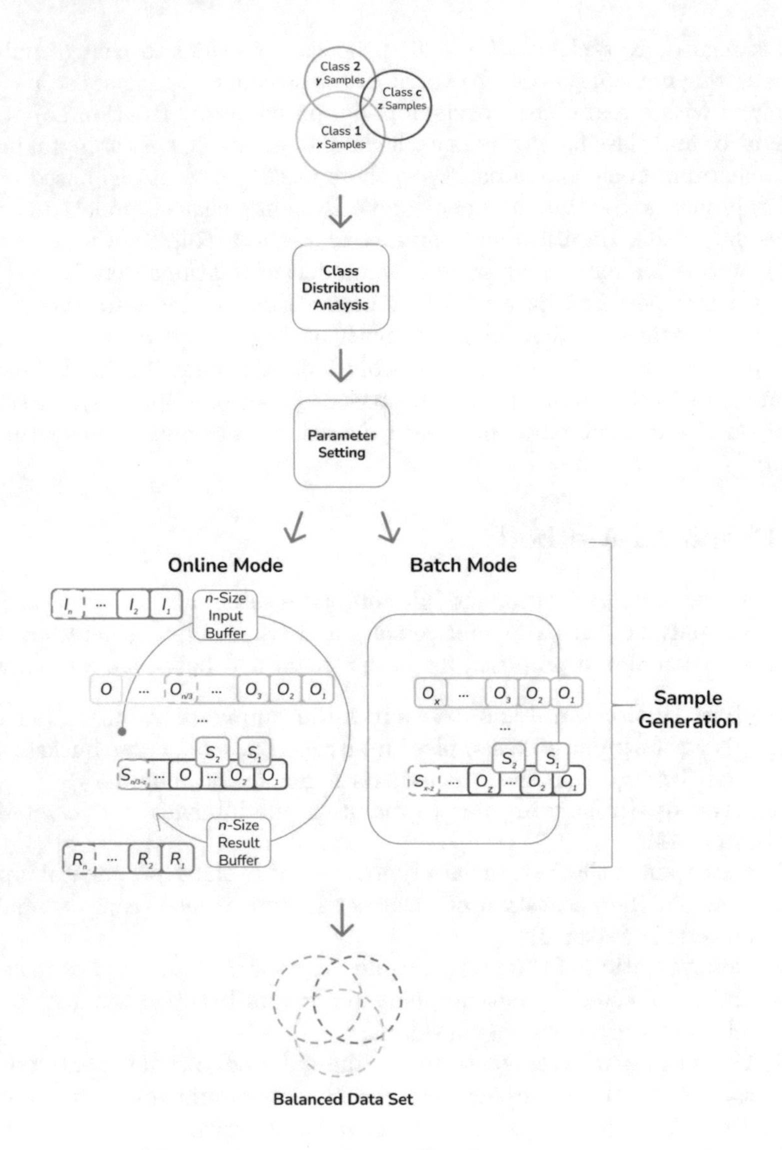

Fig. 1. Synthetic data generation scheme.

when the n-size sliding window is full. Being c the number of classes, then, the number of samples per class can be represented as $\frac{n}{c}$. The classes with a larger number of samples are randomly down-sampled, and the remaining original samples are randomly combined with the synthetic (S) samples. The merged samples are saved in a buffer of size n. This process is continuously repeated with the next sliding window.

- **Offline synthesis** performs class distribution analysis, parameter optimisation and sample generation for the minority classes using a window of the size of the data set.

This hybrid method combines the downsampling of the majority classes with oversampling of the minority classes.

4 Experiments and Results

This section details the Bplug data balancing experimental set up, covering implementation, data sets and performance results, and closes with a comprehensive comparison with competing data generators. The comparison between the experimental results, described in terms of distribution functions and inter-feature correlations, and the results from the literature enable a full analysis of the proposed solution. All experiments were performed on a server with the following hardware specifications: (*i*) Operating System: Ubuntu 18.04.2 LTS 64 bits; (*ii*) Processor: IntelCore i9-9900K 3.60 GHz; (*iii*) RAM: 32 GB DDR4; (*iv*) Disk: 500 GB (7200 rpm SATA) + 256 GB SSD. The experiments analyse the statistical profile of the original and synthetic data in two distinct application domains: a crowdsourced data set (Wikivoyage) and a predictive maintenance data set (MetroPT). The best distribution function was selected using `GridSearchCV`[2] from scikit-learn Python library. Moreover, the adopted kernel density estimation corresponds to the scikit-learn Python library implementation[3].

4.1 Crowdsourced Data Set Results

The experimental data set[4] is composed of 40 000 reviews (39 602 made by humans and 398 by bots) from Wikivoyage[5] [9]. The online experiment employs a 10 000-sample sliding window. The resulting data stream comprises 40 000 entries (20 000 samples made by humans and 20 000 samples made by bots, respectively). Figure 2 compares the imbalanced and balanced data with the feature distribution functions and inter-feature correlations. Particularly, Fig. 2(a) and 2(b) show that the corresponding feature distribution functions are identical, except for the now balanced target feature `isbot_union`.

4.2 Transportation Data Set Results

The MetroPT data set[8] was provided by Metro of Porto, the urban metro public transport service in Porto, Portugal. It is intended to be used for predictive maintenance and contains 15 000 samples (5001 corresponding to transport failures and 9999 to non-transport failures).

---

[2] Available at https://scikit-learn.org/stable/modules/generated/sklearn.model_selection.GridSearchCV.html, December 2022.

[3] Available at https://scikit-learn.org/stable/modules/generated/sklearn.neighbors.KernelDensity.html, December 2022.

[4] Available from the corresponding author on reasonable request.

[5] Available at https://www.wikivoyage.org, December 2022.

The online experiment adopts a 7500-sample sliding window. The resulting data includes 15 000 entries (7500 error samples and 7500 non-error samples). Figure 3 compares the imbalanced and balanced data through feature distribution functions and inter-feature correlations. Once again, the corresponding feature distribution functions are identical for all features except for the now balanced target feature `isbot_union` (see Fig. 3(a) and 3(b)).

In this experiment, the quality of the synthetic data is sustained over the slightly changed feature distribution functions and inter-feature correlations between the imbalanced and balanced data.

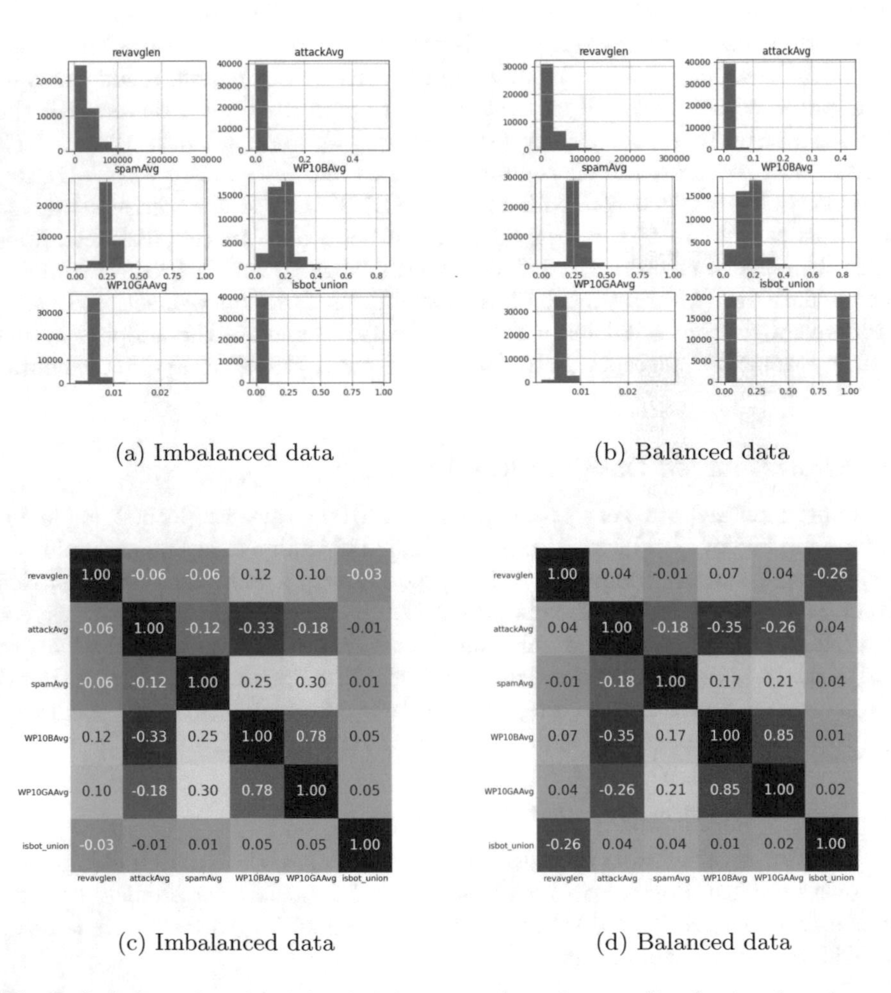

(a) Imbalanced data (b) Balanced data

(c) Imbalanced data (d) Balanced data

Fig. 2. Imbalanced and balanced Wikivoyage data: feature distribution functions and inter-feature correlation.

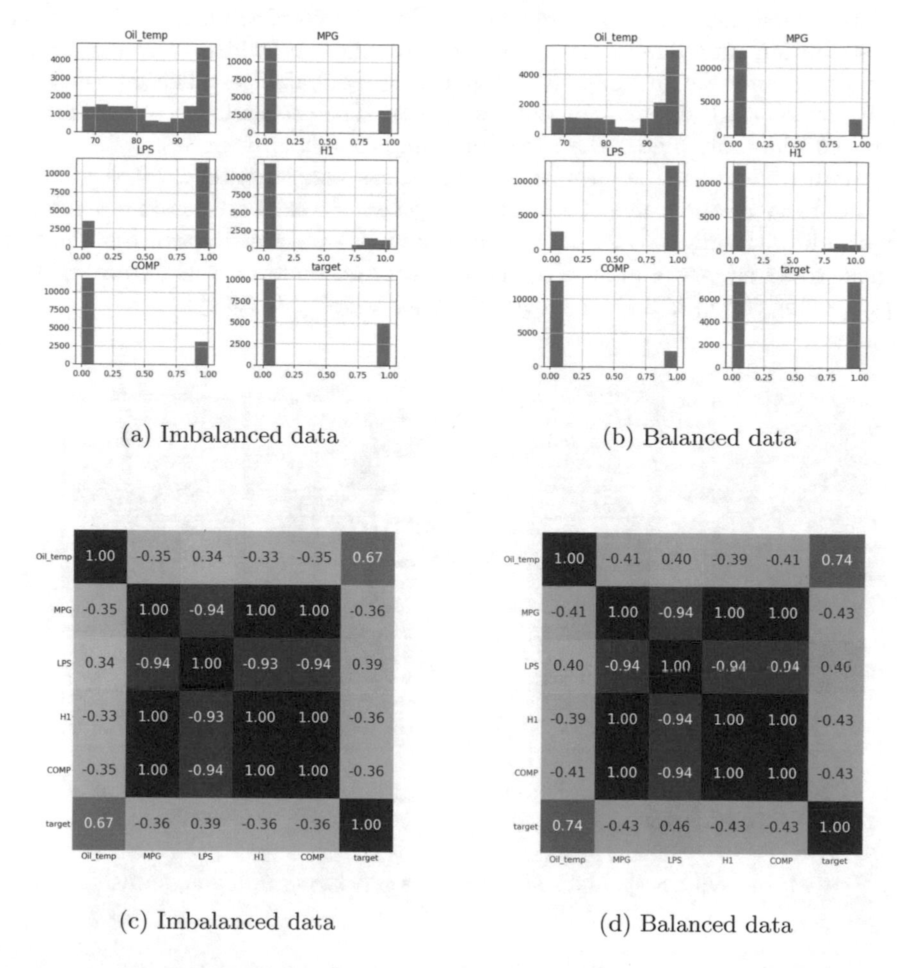

(a) Imbalanced data

(b) Balanced data

(c) Imbalanced data

(d) Balanced data

Fig. 3. Imbalanced and balanced MetroPT data: feature distribution functions and inter-feature correlation.

4.3 Comparison with Synthetic Data Generators

The proposed plug-in was compared against the synthetic data open-source packages found in the literature: (*i*) SDV; and (*ii*) Synthia. This comparison was performed with both data sets and packages offline since neither Synthia nor SDV supports the online operation. The results per package refer to a single data set because package behaviour remains identical regardless of the data set.

Synthetic Data Vault produced, in the case of Wikivoyage data set, another imbalanced data set with 29 131 human and 10 869 bot contributions. Figure 4 compares the feature distribution functions of Bplug and SDV. Specifically, Fig. 4(a) and 4(b) show Bplug's superior ability to mimic the original feature distribution functions compared to SDV. In this sense, SDV fails to detect the distribution functions of most features and the bandwidth of features with small dynamic ranges, namely in the case of probabilities. The inter-feature correlation analysis is not possible with SDV. The processing time (1811.04 s) was considerably higher than that of Bplug (0.54 s).

Synthia failed, in the case of MetroPT data set, to detect: (*i*) the target variable type and range (it assigns negative decimal values to an integer variable with only two feasible values); and (*ii*) the distribution functions of most features (see Fig. 5). The inter-feature correlation is maintained. The processing time (0.13 s) was lower than with Bplug (0.33 s).

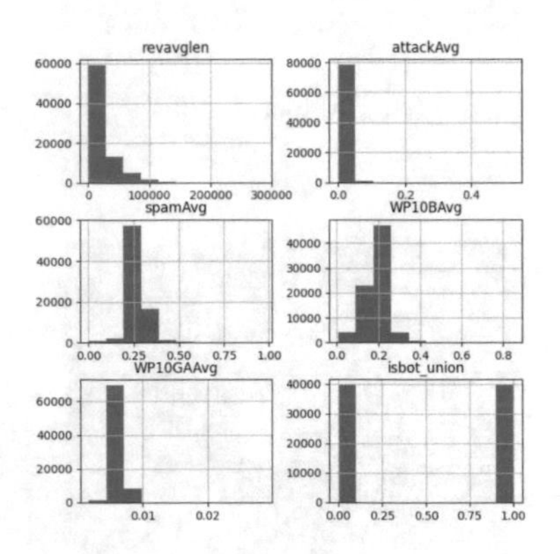

(a) Distribution functions (Bplug in batch mode)

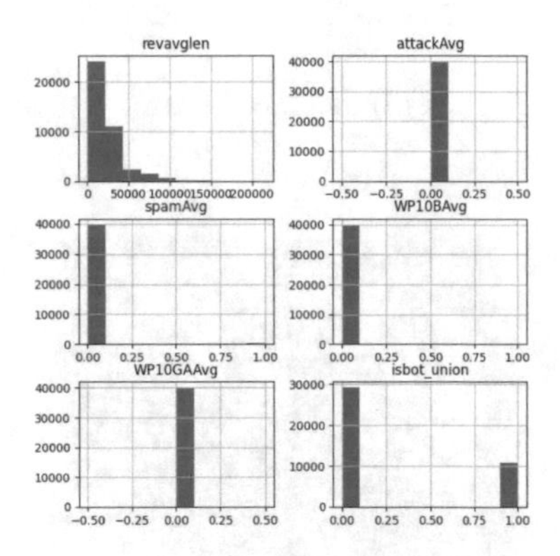

(b) Distribution functions (SDV)

Fig. 4. Wikivoyage data set: comparison between Bplug and SDV.

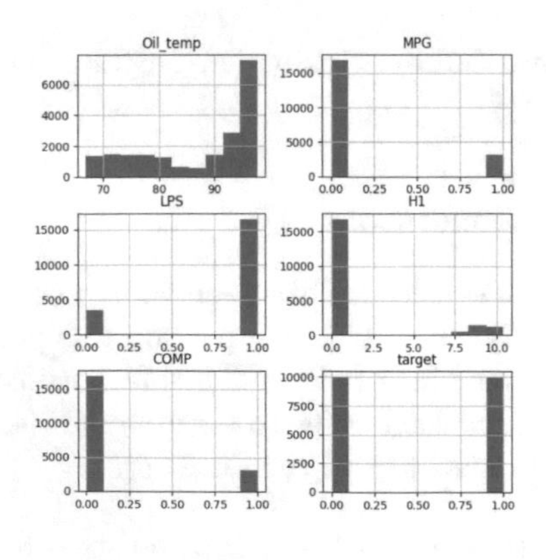

(a) Distribution functions (Bplug in batch mode)

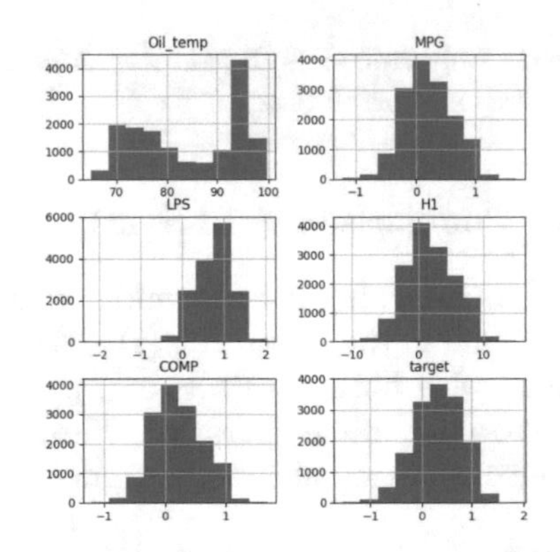

(b) Distribution functions (Synthia)

Fig. 5. MetroPT data set: comparison between Bplug and Synthia.

5 Conclusions

Classification is a popular problem explored by ML. To ensure the good performance of the ML classifiers, experimental samples must be equally distributed among target classes. In addition, class imbalance biases the models towards the majority group. To minimise this problem, this paper proposes Bplug, a pre-processing plug-in to balance data before the training stage. Moreover, the proposed solution operates not only with offline classification but also with stream-based classification, where data properties may change over time.

The proposed plug-in integrates: (*i*) class distribution analysis to verify the imbalance degree; (*ii*) parameter setting to maintain the inter-feature correlation between synthetic and original samples; and (*iii*) sample generation. This solution contributes to improving the performance and fairness of ML classification models. Bplug was tested and evaluated with the Wikivoyage and the MetroPT data sets. The experiments show that inter-feature correlation is maintained after class balancing. Moreover, the plug-in was compared with existing open-source packages for synthetic data generation, *i.e.*, SDV and Synthia, which were unable to detect the distribution functions of most features. In particular, Synthia does not detect the target variables and their corresponding type. Contrary to these solutions, Bplug works both in online and offline modes. To sum up, this paper describes a balancing plug-in for stream-based classification. As future work, the plan is to explore alternative bias mitigation techniques, analyse the effect of the majority class downsampling in the classification and determine the optimal window size, all in streaming scenarios.

Acknowledgements. This work was partially supported by: (*i*) Xunta de Galicia grant ED481B-2021-118 and ED481B-2022-093, Spain; and (*ii*) Portuguese National Funds through the FCT - Fundação para a Ciência e a Tecnologia (Portuguese Foundation for Science and Technology) as part of project UIDB/50014/2020 (DOI: 10.54499/UIDP/50014/2020 — https://doi.org/10.54499/UIDP/50014/2020).

References

1. Abu Alfeilat, H.A., et al.: Effects of distance measure choice on K-nearest neighbor classifier performance: a review. Big Data **7**(4), 221–248 (2019)
2. Batista, G.E.A.P.A., Prati, R.C., Monard, M.C.: A study of the behavior of several methods for balancing machine learning training data. ACM SIGKDD Explorations Newsletter **6**, 20–29 (2004)
3. Berrar, D.: Cross-validation. In: Encyclopedia of Bioinformatics and Computational Biology, pp. 542–545. Elsevier (2019)
4. Branco, P., Torgo, L., Ribeiro, R.P.: Pre-processing approaches for imbalanced distributions in regression. Neurocomputing **343**, 76–99 (2019)
5. Chawla, N.V., Bowyer, K.W., Hall, L.O., Kegelmeyer, W.P.: Smote: synthetic minority over-sampling technique. J. Artif. Intell. Res. **16**, 321–357 (2002)
6. Chawla, N.V., Lazarevic, A., Hall, L.O., Bowyer, K.W.: Smoteboost: improving prediction of the minority class in boosting. In: Proceedings of European Conference on Principles of Data Mining and Knowledge Discovery, vol. 2838, pp. 107–119. Springer (2003)

7. Fernández, A., García, S., Galar, M., Prati, R.C., Krawczyk, B., Herrera, F.: Learning from imbalanced data streams. In: Learning from Imbalanced Data Sets, pp. 279–303. Springer (2018)
8. Galar, M., Fernandez, A., Barrenechea, E., Bustince, H., Herrera, F.: A review on ensembles for the class imbalance problem: Bagging-, boosting-, and hybrid-based approaches. IEEE Trans. Syst. Man Cybern. Part C (Appl. Rev.) **42**, 463–484 (2012)
9. García-Méndez, S., et al.: Simulation, modelling and classification of wiki contributors: spotting the good, the bad, and the ugly. Simul. Model. Pract. Theory **120**, 102616 (2022)
10. He, H., Garcia, E.A.: Learning from imbalanced data. IEEE Trans. Knowl. Data Eng. **21**(9), 1263–1284 (2009)
11. Korycki, L., Krawczyk, B.: Online oversampling for sparsely labeled imbalanced and non-stationary data streams. In: Proceedings of 2020 International Joint Conference on Neural Networks (IJCNN), pp. 1–8. IEEE (2020)
12. Krawczyk, B.: Learning from imbalanced data: open challenges and future directions. Progress Artif. Intell. **5**(4), 221–232 (2016)
13. Lin, W.C., Tsai, C.F., Hu, Y.H., Jhang, J.S.: Clustering-based undersampling in class-imbalanced data. Inf. Sci. **409–410**, 17–26 (2017)
14. Meyer, D., Nagler, T.: Synthia: multidimensional synthetic data generation in python. J. Open Source Softw. **6**, 2863 (2021)
15. Nguyen, H.M., Cooper, E.W., Kamei, K.: Online learning from imbalanced data streams. In: Proceedings of 2011 International Conference of Soft Computing and Pattern Recognition (SoCPaR), pp. 347–352. IEEE (2011)
16. Patki, N., Wedge, R., Veeramachaneni, K.: The synthetic data vault. In: Proceedings of 2016 IEEE International Conference on Data Science and Advanced Analytics (DSAA), pp. 399–410. IEEE (2016)
17. Seiffert, C., Khoshgoftaar, T.M., Hulse, J.V., Napolitano, A.: Rusboost: a hybrid approach to alleviating class imbalance. IEEE Trans. Syst. Man Cybern. - Part A: Syst. Hum. **40**, 185–197 (2010)
18. Sun, Y., Wong, A.K., Kamel, M.S.: Classification of imbalanced data: a review. Int. J. Pattern Recognit Artif Intell. **23**, 687–719 (2009)
19. Węglarczyk, S.: Kernel density estimation and its application. ITM Web of Conferences **23**, 1–8 (2018)

Explainable Classification of Wiki Streams

Silvia García-Méndez[1], Fátima Leal[2], Francisco de Arriba-Pérez[1], Benedita Malheiro[3,4](✉), and Juan Carlos Burguillo-Rial[1]

[1] Information Technologies Group, atlanTTic, University of Vigo, Vigo, Spain
{sgarcia,farriba}@gti.uvigo.es, J.C.Burguillo@uvigo.es
[2] REMIT, Universidade Portucalense, Porto, Portugal
fatimal@upt.pt
[3] INESC TEC, Porto, Portugal
mbm@isep.ipp.pt
[4] ISEP, Polytechnic Institute of Porto, Porto, Portugal

Abstract. Web 2.0 platforms, like wikis and social networks, rely on crowdsourced data and, as such, are prone to data manipulation by ill-intended contributors. This research proposes the transparent identification of wiki manipulators through the classification of contributors as benevolent or malevolent humans or bots, together with the explanation of the attributed class labels. The system comprises: (*i*) stream-based data pre-processing; (*ii*) incremental profiling; and (*iii*) online classification, evaluation and explanation. Particularly, the system profiles contributors and contributions by combining features directly collected with content- and side-based engineered features. The experimental results obtained with a real data set collected from Wikivoyage – a popular travel wiki – attained a 98.52% classification accuracy and 91.34% macro F-measure. In the end, this work seeks to address data reliability to prevent information detrimental and manipulation.

Keywords: Classification · data modelling · intelligent decision support system · natural language processing · stream processing

1 Introduction

Online behaviour has dramatically changed thanks to ubiquitous Internet access, the popularity of mobile devices and the proliferation of crowdsourcing platforms. Such platforms offer content to users based on data voluntarily shared by other users. This data-based model expects users to be unbiased, truthful and trust each other. In reality, the idealised operation is affected by ill-intended contributors who provide biased, misleading or wrong information on purpose. As a whole, the user crowd, which may include humans and non-humans (bots), generates a continuous stream of contributions known as edits or revisions. Bots and humans can be well- and ill-intended, generate content, extract content and execute actions [18].

A. Rocha et al. (Eds.): WorldCIST 2023, LNNS 799, pp. 75–84, 2024.
https://doi.org/10.1007/978-3-031-45642-8_7

Wikivoyage[1], a well-known and worldwide travel wiki constitutes an illustrative case. This wiki provides information related to popular destinations, comprising data about highlights, gastronomy, cultural activities, *etc.* The information is organised in pages, also known as articles, corresponding to locations, regions, countries or continents. Similarly, TripAdvisor[2] offers travel-related reviews on destinations, accommodation, food, transports, *etc.* The information is also organised on location-based pages. In both platforms, pages are collaboratively edited by the user crowd.

Artificial Intelligence (AI) techniques are a natural choice to mine crowdsourced data streams effectively. Moreover, the latest AI research trends focus on the design of accountable, responsible and transparent algorithms [4]. As such, a judicious combination of AI techniques can identify and explain on the fly why contributions are malevolent or benevolent, helping to signal unwelcome content within crowdsourcing platforms.

The present research proposes a transparent stream-based solution to classify contributors into benevolent or malevolent humans or bots in real-time. Considering the state of the art, namely our previous approach [6], the current work models contributors and contributions based not only on side but also content-related features and explains the attributed class labels, all on the fly. Experimental results on a real data set gathered from Wikivoyage endorse this profiling refinement. Ultimately, the proposed solution enables the early detection and isolation of malevolent contributors and their contributions within crowdsourcing platforms.

The rest of the paper is organised as follows. Section 2 reviews key-related work on the modelling and classification of wiki contributors. Section 3 describes this novel proposal. Section 4 presents the experimental results. Finally, Sect. 5 concludes the article.

2 Related Work

The popularity and openness of the crowdsourcing model constitute both a strength and a threat, namely, for social media and collaborative wiki pages [5]. For some years, the research community has been battling misinformation [21]. In this line, several works address the latter concern from an offline perspective. In early research, Adler *et al.* (2011) [1] propose WikiTrust, a vandalism detection solution for Wikipedia[3] which exploits spatial-temporal metadata together with engineered side-based features for classification. Choi *et al.* (2016) [3] analyse malicious and legitimate campaigns in crowdsourcing platforms to extract representative features (context- and content-related features) and compare their solution with baselines obtained from the literature. Similarly, Yamak *et al.* (2016) present a sockpuppet[4] detection system for Wikipedia [20] where

[1] Available at https://www.wikivoyage.org, December 2022.

[2] Available at www.tripadvisor.com, December 2022.

[3] Available at https://en.wikipedia.org, December 2022.

[4] Fake account created by ill-intended users to disseminate biased content.

users are profiled based on side features extracted from a data set of blocked Wikipedia users and their contributions.

In 2017, Wikimedia deployed the Objective Revision Evaluation Service[5] (ORES). It explores side-based features like the number of sections and references to predict the quality of revisions (damaging, good faith, *etc.*) and articles (spam, attacked, vandalised, *etc.*) in real-time. Green & Spezzano (2017) [7] combine user-editing features with ORES predictions. Sarabadani *et al.* (2017) [16] advance a vandalism detection system based on content-related and user-related features. Velayutham *et al.* (2017) [19] use non-textual side-based features for bot detection in Twitter[6]. Zheng, Albano *et al.* (2019) [22] implement a semi-supervised model (combine rules and manually annotated data) for Wikipedia using, once again, side-based features. Moreover, Zheng, Yuan *et al.* (2019) [23] explore side user behaviour features, train a Long Short-Term Memory (LSTM) model with benevolent contributors and build a Generative Adversarial Network to detect ill-intended users.

Concerning the online identification of wiki manipulators, there is scant research. The sole exceptions are the works of Heindorf *et al.* (2016, 2019) [8,9] and our own previous research [6]. Heindorf *et al.* (2016, 2019) present an online vandalism detector that classifies new revisions instantaneously, exploring side features associated with user behaviour. In our previous work, we simulate, model and classify wiki contributors with the help of side-based engineered features from original and simulated data, ensuring classes remain balanced.

Even though algorithmic transparency is essential to promote user trust in AI solutions and enhance user experience [10,14], the majority of the reviewed solutions rely on opaque techniques. The exceptions are the vandalism detectors proposed by Liu and Lu (2019) [13] and Subramanian *et al.* (2019) [17], which explain outcomes through visual analytical graphics.

Table 1 compares the current proposal with the most related research mentioned in terms of engineered features, classification goal, offline/stream-based processing, and if they provide explanations. The current online method profiles and labels contributors and contributions mostly using content- and side-based engineered features. Moreover, it explains class labels through graphs and textual information.

Table 1. Comparison with related work from the literature.

Related work	Profiling (engineered features)	Classification		Processing	Explaining
		Contributor	Contribution		
Heindorf *et al.* (2016, 2019) [8,9]	Side-based		✓	Online	
Liu and Lu (2019) [13]	Side-based	✓		Offline	Visual
Subramanian *et al.* (2019) [17]	Side-based	✓		Offline	Visual
García-Méndez *et al.* (2022) [6]	Side-based	✓	✓	Online	
Current proposal	Content-based Side-based	✓	✓	Online	Textual Visual

[5] Available at https://ores.wikimedia.org, December 2022.

[6] Available at https://twitter.com, December 2022.

3 Proposed Method

The proposed solution comprises: (*i*) stream-based data processing with feature engineering and feature selection; (*ii*) incremental profiling; (*iii*) online classification; (*iv*) evaluation using standard classification metrics; and (*v*) classification explainability. Figure 1 describes this scheme.

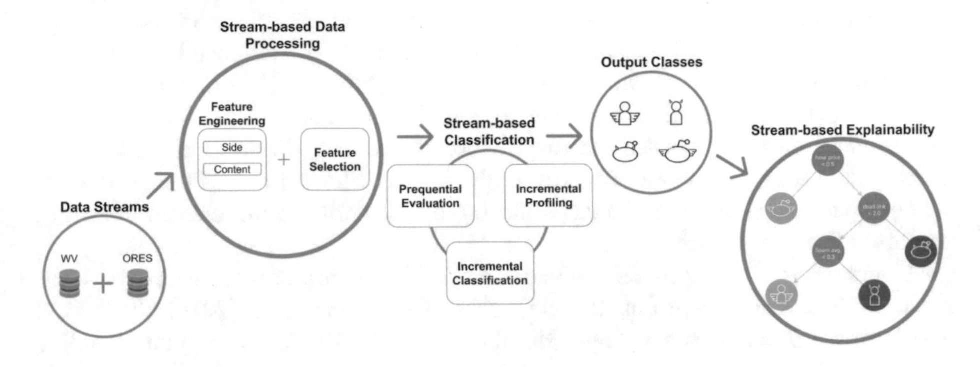

Fig. 1. Online classification of wiki contributors and contributions.

Stream-based data processing is essential to ensure the consistency of the feature engineering process and also take full advantage of the ML models. Particularly, this process is composed of two steps: (*i*) feature engineering and (*ii*) feature selection.

- **Feature engineering** manipulates the collected data to create new features. Table 2 details the features engineered from the experimental data set. The distribution is as follows: 4403 benevolent and 602 malevolent humans, and 311 benevolent and 49 malevolent bots.
- **Feature selection** intends to select the best features of the data set to ensure the best classification results with the smallest feature set. For that purpose, a wrapper-type feature selection algorithm is applied.

Incremental profiling relies on the features incrementally computed to create the contributor profiles. Table 3 details those features which contemplate side contributor-related features, the ones obtained from ORES and content-derived features.

Online classification employs popular stream-based ML models (both single and ensemble) from the literature, inspired by their competitive performance in similar use cases [2,9,15]: (*i*) Naive Bayes (NB); (*ii*) Decision Tree (DT); (*iii*) Random Forest (RF); and (*iv*) Boosting Classifier (BC).

Evaluation protocol relies on accuracy and F-measure both in macro-averaging and micro-averaging [11,12] to evaluate the performance and effectiveness of the system. The latter two approaches are essential to computing the evaluation metrics globally and per class due to the imbalanced nature of the experimental data set.

Table 2. Features considered for both contributor and contribution profiling.

Number	Feature name	Source	Type
1	Review ID	Wikivoyage	
2	Editor ID	Wikivoyage	
3	Article name	Wikivoyage	
4	Size of the revision	Wikivoyage	
5	Edit damaging & good faith probability	ORES	
6	Edit item quality (A/B/C/D/E) probability	ORES	
7	New article quality (OK/attack/spam/vandalism) probability	ORES	
8	Article quality (wp10: B/C/FA/GA/start/stub) probability	ORES	
9	Characters inserted in the edit	Engineered	Content-based
10	Word n-grams in the edit	Engineered	Content-based
11	Number of characters inserted in the edit	Engineered	Side-based
12	Number of links in the edit	Engineered	Side-based
13	Number of repeated links in the edit	Engineered	Side-based
14	Number of edits per article	Engineered	Side-based
15	Number of edits per week	Engineered	Side-based
16	Number of articles revised per week	Engineered	Side-based

Table 3. Incremental features used for both contributor and contribution classification.

Number	Feature name
1	Average number of edits per article
2	Average number of edits per week
3	Average number of articles edited per week
4	Average ORES edit damaging & good faith probability
5	Average ORES edit item quality (A/B/C/D/E) probability
6	Average ORES new article quality (OK/attack/spam/vandalism) probability
7	Average ORES article quality (wp10: B/C/FA/GA/start/stub) probability
8	Average size of the edit
9	Average number of links in the edit
10	Average number of repeated links in the edit
11	Average number of characters inserted in the edit
12	Cumulative word n-grams of the edit

Explainability enable to understand the results. The proposed solution uses interpretable ML models to perform both contributor and contribution classification. This work employs both graph-based and natural language (using templates) approaches to provide the end users with enough level of transparency to understand the generated predictions.

4 Experiments and Results

The experiments intend to evaluate the proposed explainable method using stream processing. Stream-based ML models were incrementally trained and evaluated using `EvaluatePrequential`[7]. The experiments were conducted on a server with the following hardware specifications: (*i*) Operating System: Ubuntu 18.04.2 LTS 64 bits; (*ii*) Processor: IntelCore i9-10900K 2.80 GHz; (*iii*) RAM: 96 GB DDR4; and (*iv*) Disk: 480 GB NVME + 500 GB SSD.

Data set[8] was gathered using MediaWiki[9] between 14th January 2020 and 21st June 2020. Specifically, it is composed of 5365 reviews made by 369 contributors about 1824 pages. To extract the content of the reviews, the system issues GET requests to the Wikivoyage API[10] with the *compare* method. Then, the textual data is processed with the spaCy tool[11] with the `en_core_web_lg` model[12] to perform lemmatization and remove stopwords[13]. Finally, word *n*-grams are computed using the incremental text content from the revisions using `CountVectorizer`[14] Python library using as configuration parameters: `max_df_in=0.7`, `min_df_in=0.001`, `wordgram_range_in=(1,4)`.

Feature selection performs a recursive search over configurable parameter ranges using Recursive Feature Elimination (RFE[15]) and Linear Support Vector Classifier (LinearSVC[16]) with the following configuration parameters: `penalty =l1`, `dual=False`, `step=0.05` and `n_jobs=-1`. As a result, the

Table 4. Classification results.

Classifier	Accuracy	*F*-measure					Time (s)
		Macro	BH	MH	BB	MB	
NB	87.84	64.31	93.05	33.41	66.23	64.53	3.03
DT	94.90	75.60	97.44	80.48	86.97	37.50	31.63
RF	**98.52**	**91.34**	**100.00**	**94.59**	88.94	**81.82**	30.76
BC	97.57	84.79	98.99	91.66	**90.31**	58.19	921.25

[7] Available at https://scikit-multiflow.readthedocs.io/en/stable/api/generated/skmultiflow.evaluation.EvaluatePrequential.html, December 2022.
[8] Available from the corresponding author on reasonable request.
[9] Available at www.pypi.org/project/mediawiki-utilities, December 2022.
[10] Available at https://en.wikivoyage.org/w/api.php, December 2022.
[11] Available at https://spacy.io, December 2022.
[12] Available at https://spacy.io/models/en#en_core_web_lg, December 2022.
[13] Available at https://gist.github.com/sebleier/554280, December 2022.
[14] Available at https://scikit-learn.org/stable/modules/generated/sklearn.feature_extraction.text.CountVectorizer.html, December 2022.
[15] Available at www.scikit-learn.org/stable/modules/generated/sklearn.feature_selection.RFE.html, December 2022.
[16] Available at www.scikit-learn.org/stable/modules/generated/sklearn.svm.LinearSVC.html, December 2022.

selected features are: 2, 3, 5 (A, D, E), 6 (spam, vandalism) in Table 3, and certain word n-grams.

Online Classification employs multiple classification ML models: (*i*) NB[17]; (*ii*) DT[18]; (*iii*) RF[19]; and (*iv*) BC[20]. Table 4 shows the results for benevolent human (BH), malevolent human (MH), benevolent bot (BB) and malevolent bot (MB) categories. The first classifier, NB, allows to establish a preliminary baseline with acceptable accuracy and macro F-measure values, 87.84 % and 64.31 %, respectively. However, the micro F-measure for the malevolent human contributors is still rather low. The same applies to the DT and the time-consuming BC model for the case of malevolent bot contributors. Consequently, the best results are attained by the RF model with accuracy, and macro and micro F-measure values all above 80 %.

Explainability is provided graphically and textually. The proposed solution uses the `get_model_description` method[21] from scikit-multiflow to traverse the decision path learned by the DT model. The visual description is obtained with the `dtreeviz` library[22]. Figure 2 displays, for a given sample, the decision path (arrows) and the corresponding natural language explanations.

Literature comparison highlights the contributions of the proposed method. The methods compared are offline (except [6]) and opaque. The remaining systems do not provide directly comparable metrics for discussion [8,9,13,16,17]. Regarding contributor classification, Yamak *et al.* (2016) [20] reported accuracy and macro F-measure values of 99.80 % (1.28 and 8.46 percent points higher, respectively). The solution by Green & Spezzano (2017) [7] attained 82.10 % accuracy (16.42 percent points lower). The accuracy obtained by Velayutham *et al.* (2017) [19] is 83.00 % (15.52 percent points lower). Last but not least, Zheng, Albano *et al.* (2019) [22] reported a macro F-measure value of 85.00 % (6.34 percent points lower). The higher performance of Yamak *et al.* (2016) [20] probably results from exploiting a Wikipedia data set containing exclusively blocked users and their contributions. The lower performance of the remaining works likely derives from profiling users solely based on side features. Considering the classification of contributions, the system by Adler *et al.* (2011) [1] and by Choi *et al.* (2016) [3] attained an accuracy value of 99 % (0.48 percent points higher) and 99.2 % (0.68 percent points higher). The identical performance of these solutions and the current proposal sustain the relevance of content-based features in the classification of wiki streams.

---

[17] Available at https://scikit-multiflow.readthedocs.io/en/latest/api/generated/skmultiflow.bayes.NaiveBayes.html, December 2022.

[18] Available at https://scikit-multiflow.readthedocs.io/en/stable/api/generated/skmultiflow.trees.ExtremelyFastDecisionTreeClassifier.html, December 2022.

[19] Available at https://scikit-multiflow.readthedocs.io/en/stable/api/generated/skmultiflow.meta.AdaptiveRandomForestClassifier.html#skmultiflow.meta.AdaptiveRandomForestClassifier, December 2022.

[20] Available at https://scikit-multiflow.readthedocs.io/en/stable/api/generated/skmultiflow.meta.OnlineBoostingClassifier.html, December 2022.

[21] Available at https://scikit-multiflow.readthedocs.io/en/stable/api/generated/skmultiflow.trees.HoeffdingTreeClassifier.html, December 2022.

[22] Available at https://pypi.org/project/dtreeviz, December 2022.

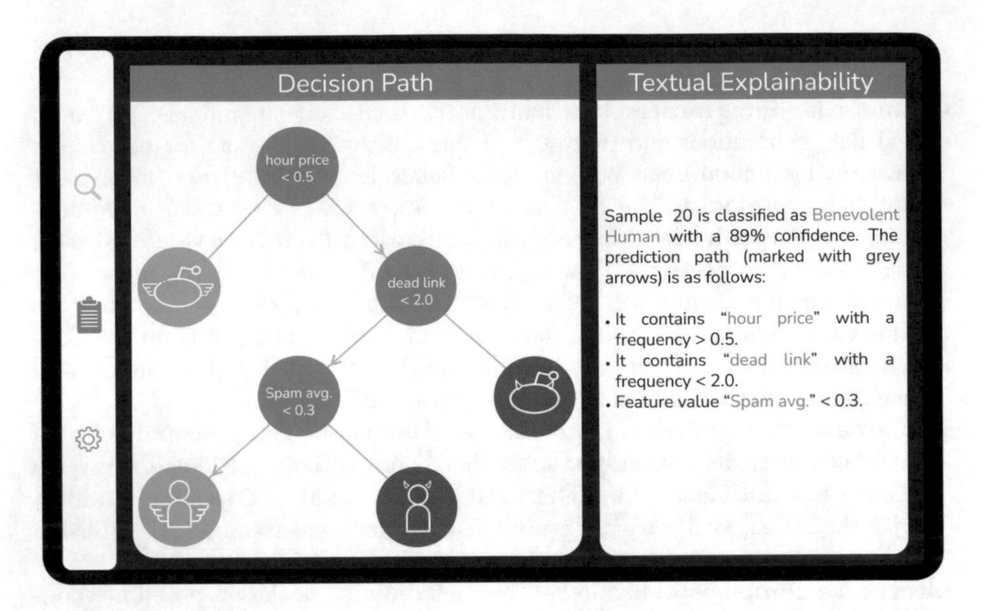

Fig. 2. Visual and natural language description of the classification.

Zheng, Yuan *et al.* (2019) [23] addressed both contributor and contribution classification, obtaining 89.73 % accuracy (8.79 percent points lower) and 90.10 % macro F-measure (1.24 percent points lower). In our previous work [6], we processed online the data streams to distinguish between benevolent and malevolent as well as human and non-human contributors on the fly with a classification accuracy of up to 91.44 % (7.08 percent points lower). The aforementioned solutions present lower performance and are opaque.
Summing up, the proposed stream-based solution generally reports superior performance in the classification of both contributors and contributions versus the literature. In addition, the method is transparent, using textual and visual explanations.

5 Conclusions

The proliferation of platforms that exploit and depend on data produced by the crowd raises data reliability concerns. The gravity of the problem becomes evident when it comes to decision making, *e.g.*, in online purchasing, users tend to trust more peer opinions than expert advice.

The present research presents a real-time transparent solution to classify wiki users as benevolent or malevolent human or non-human directly from a live wiki data stream. User profiling combines content- and side-based engineered features with Wikivoyage and ORES data. Moreover, it provides the end user with textual and visual explanations of class labels extracted from the learned decision tree. As such, the main contributions of this work include the enrichment of contributor and contribution profiles with content- and side-based features, and the transparent classification of contributors and contributions through textual and visual explanations of the assigned class labels, all in real-time.

The experimental results obtained with a real data set gathered from Wikivoyage show that this extended profiling approach boosts classification performance. Specifically, it attains 98.52% accuracy and 91.34% macro F-measure.

This work addresses data reliability by identifying data manipulation and manipulators in real-time. In future work, the plan is to isolate malevolent contributors and contributions and generate synthetic data to balance the experimental data set.

Acknowledgements. This work was partially supported by: (*i*) Xunta de Galicia grant ED481B-2021-118 and ED481B-2022-093, Spain; and (*ii*) Portuguese National Funds through the FCT - Fundação para a Ciência e a Tecnologia (Portuguese Foundation for Science and Technology) as part of project UIDB/50014/2020 (DOI: 10.54499/UIDP/50014/2020 — https://doi.org/10.54499/UIDP/50014/2020).

References

1. Adler, B.T., de Alfaro, L., Mola-Velasco, S.M., Rosso, P., West, A.G.: Wikipedia vandalism detection: combining natural language, metadata, and reputation features. In: Gelbukh, A. (ed.) CICLing 2011. LNCS, vol. 6609, pp. 277–288. Springer, Heidelberg (2011). https://doi.org/10.1007/978-3-642-19437-5_23
2. Amaral, G., Piscopo, A., Kaffee, L.A., Rodrigues, O., Simperl, E.: Assessing the quality of sources in Wikidata across languages: a hybrid approach. J. Data Inf. Quality **13**(4), 1–35 (2021)
3. Choi, H., Lee, K., Webb, S.: Detecting malicious campaigns in crowdsourcing platforms. In: Proceedings of the IEEE/ACM International Conference on Advances in Social Networks Analysis and Mining, pp. 197–202. IEEE (2016)
4. Dignum, V.: Responsibility and artificial intelligence. The oxford handbook of ethics of AI 4698, 215 (2020)
5. Egger, R., Gula, I., Walcher, D. (eds.): Open tourism: Open innovation, crowdsourcing and co-creation challenging the tourism industry, Tourism on the Verge. Springer (2016)
6. García-Méndez, S., et al.: Simulation, modelling and classification of wiki contributors: spotting the good, the bad, and the ugly. Simul. Model. Pract. Theory **120**, 102616 (2022)
7. Green, T., Spezzano, F.: Spam users identification in Wikipedia via editing behavior. In: Proceedings of the International AAAI Conference on Web and Social Media, pp. 532–535. AAAI (2017)
8. Heindorf, S., Potthast, M., Stein, B., Engels, G.: Vandalism Detection in Wikidata. In: Proceedings of the ACM International Conference on Information and Knowledge Management, pp. 327–336. ACM (2016)
9. Heindorf, S., Scholten, Y., Engels, G., Potthast, M.: Debiasing vandalism detection models at Wikidata. In: Proceedings of the ACM Web Conference, pp. 670–680. ACM (2019)

10. Leal, F., García-Méndez, S., Malheiro, B., Burguillo, J.C.: Explanation plug-in for stream-based collaborative filtering. In: Proceedings of the World Conference on Information Systems and Technologies, pp. 42–51. Springer (2022)
11. Liu, T., Chen, Z., Zhang, B., Ma, W.Y., Wu, G.: Improving text classification using local latent semantic indexing. In: Proceedings of the IEEE International Conference on Data Mining, pp. 162–169. IEEE (2004)
12. Liu, Y., Loh, H.T., Sun, A.: Imbalanced text classification: a term weighting approach. Expert Syst. Appl. **36**(1), 690–701 (2009)
13. Liu, Z., Lu, A.: Explainable visualization for interactive exploration of CNN on Wikipedia vandal detection. In: Proceedings of the IEEE International Conference on Big Data, pp. 2354–2363. IEEE (2019)
14. Naiseh, M., Jiang, N., Ma, J., Ali, R.: Personalising explainable recommendations: literature and conceptualisation. In: Proceedings of the World Conference on Information Systems and Technologies, pp. 518–533. Springer (2020)
15. Salutari, F., Hora, D.D., Dubuc, G., Rossi, D.: Analyzing Wikipedia users' perceived quality of experience: a large-scale study. IEEE Trans. Netw. Serv. Manage. **17**(2), 1082–1095 (2020)
16. Sarabadani, A., Halfaker, A., Taraborelli, D.: Building automated vandalism detection tools for Wikidata. In: Proceedings of the International Conference on World Wide Web Companion, pp. 1647–1654. ACM (2017)
17. Subramanian, S.S., Pushparaj, P., Liu, Z., Lu, A.: Explainable visualization of collaborative vandal behaviors in Wikipedia. In: Proceedings of the IEEE Symposium on Visualization for Cyber Security, pp. 1–5. IEEE (2019)
18. Tsvetkova, M., García-Gavilanes, R., Floridi, L., Yasser, T.: Even good bots fight: the case of Wikipedia. PLoS ONE **12**(2), e0171774 (2017)
19. Velayutham, T., Tiwari, P.K.: Bot identification: helping analysts for right data in Twitter. In: Proceedings of the International Conference on Advances in Computing, Communication, & Automation, pp. 1–5. IEEE (2017)
20. Yamak, Z., Saunier, J., Vercouter, L.: Detection of multiple identity manipulation in collaborative projects. In: Proceedings of the International Conference Companion on World Wide Web, pp. 955–960. ACM (2016)
21. Zhang, X., Ghorbani, A.A.: An overview of online fake news: characterization, detection, and discussion. Inf. Process. Manage. **57**(2), 102025 (2020)
22. Zheng, L.N., Albano, C.M., Vora, N.M., Mai, F., Nickerson, J.V.: The roles bots play in Wikipedia. In: Proceedings of the ACM on Human-Computer Interaction, vol. 3, pp. 1–20. ACM (2019)
23. Zheng, P., Yuan, S., Wu, X., Li, J., Lu, A.: One-class adversarial nets for fraud detection. In: Proceedings of the AAAI Conference on Artificial Intelligence, pp. 1286–1293. AAAI Press (2019)

Reconstruction of Meteorological Records with PCA-Based Analog Ensemble Methods

Murilo M. Breve[1,2], Carlos Balsa[1,2], and José Rufino[1,2](✉)

[1] Research Centre in Digitalization and Intelligent Robotics (CeDRI), Instituto Politécnico de Bragança, Campus de Santa Apolónia, 5300-253 Bragança, Portugal
{murilo.breve,balsa,rufino}@ipb.pt

[2] Laboratório para a Sustentabilidade e Tecnologia em Regiões de Montanha (SusTEC), Instituto Politécnico de Bragança, Campus de Santa Apolónia, 5300-253 Bragança, Portugal

Abstract. The Analog Ensemble (AnEn) method has been used to reconstruct missing data in time series with base on other correlated time series with full data. As the AnEn method benefits from the use of large volumes of data, there is a great interest in improving its efficiency. In this paper, the Principal Component Analysis (PCA) technique is combined with the classical AnEn method and a K-means cluster-based variant, within the context of reconstructing missing meteorological data at a particular station using information from neighboring stations. This combination allows to reduce the dimension of the number of predictor time series, while ensuring better accuracy and higher computational performance than the AnEn methods: it reduces prediction errors by up to 30% and achieves a computational speedup of up to 2x.

Keywords: Meteorological data reconstruction · Analogue ensemble · K-means clustering · Principal component analysis · MATLAB · R

1 Introduction

Information about past weather states is crucial to many scientific domains and practical applications. In the renewable energy field, for instance, it is vital to know the historical weather data and meteorological patterns, in order to estimate the productive potential of a given site, before making substantial financial investments [9]. However, full meteorological data may not always be available or may be absent altogether. In this scenario, data reconstruction techniques come into play. These should be numerically accurate and computationally efficient.

A well-known approach for meteorological data reconstruction is the Analog Ensemble (AnEn) method. Initially, it was used as a post-processing technique, to improve the accuracy of deterministic numerical forecast models [13]: past observations that are similar to the forecast are used to enhance the accuracy of

A. Rocha et al. (Eds.): WorldCIST 2023, LNNS 799, pp. 85–96, 2024.
https://doi.org/10.1007/978-3-031-45642-8_8

the forecast. The AnEn method can also be used directly for weather forecasting [6,18]. More recently [5], AnEn was used to reconstruct data of a meteorological variable by means of data from other variables at the same site, or based on data from the same or other variables from neighbor locations.

Compared to other machine learning methods, implementing the AnEn approach is considered relatively simple [1]. Concerning prediction assertiveness, a comparison [12] between AnEn and a Convolutional Neural Network (CNN), as post-processing methods of a Weather Research and Forecasting (WRF) model, showed that both methods improved equally the prediction accuracy. Similarly, in a homogeneous comparison [16] of the same methods, used as Empirical-statistical downscaling techniques, AnEn outperformed the CNN.

Large training datasets (historical observations from which missing values are derived) are advantageous for the AnEn methods: the more data is available, the easier it is to capture the variation tendency of the variable(s) to be reconstructed [7,16]. At the same time, more data entails more processing time. Hence, there is a lot of interest in improving the computational efficiency of these methods, while preserving (and ideally improving) their numerical accuracy. To this end, several variants of the AnEn methods have been investigated.

ClustAnEn (Cluster-based AnEn) is a variant of the AnEn method, based on K-means clustering, that is particularly efficient from a computational standpoint [3,4]. In this variant, there is a prior grouping of all feasible analogs, which allows selecting the analogs only by their group, instead of searching all possible analogs one by one. In addition to significantly reduce computational costs, this variant does not reduce the accuracy of the reconstructions.

Another approach that can be used to leverage the AnEn method is its combination with the Principal Components Analysis (PCA) technique. PCA is commonly used in multivariate statistics to minimize the datasets size while maintaining the most important information. Recently, PCA proved to be effective in the context of the reconstruction of meteorological data when combined with the AnEn method, originating the PCAnEn hybrid method [2].

This work consolidates previous investigations on the PCAnEn method and expands it further by applying the same rationale (integration with the PCA technique) to the ClustAnEn method (which results in the new PCClustAnEn variant). Thus, data coming from multiple predictor stations is reduced to one or two principal components (*PC*s) that are then used (instead of the original variables records) when applying the original AnEn method or its ClustAnEn variant, to reconstruct the missing data. The PCAnEn and PCClustAnEn techniques are also compared in numerical accuracy and computational efficiency.

The rest of the paper is organized as follows. Section 2 describes the dataset used and points out the correlations between meteorological variables and stations. Sections 3 and 4 introduce the PCA technique and combine it with the AnEn methods. Section 5 applies the new methods to reconstructs meteorological variables. Section 6 lays out final considerations and future work directions.

2 Meteorological Dataset

The data used in this paper for the reconstruction experiments originates from the US government National Data Buoy Center (NDBC) [14]. NBDC operates a network of data gathering buoys and coastal stations dispersed across various regions of the globe. This work uses data from various stations placed in the south of the Chesapeake Bay and surroundings (see Fig. 1). The predicted station is WDSV2 (name in red), and the predictor stations (name in white) are located within a 30 km radius from it (note: in the remaining of the paper, the suffix "V2" is omitted for simplicity). These stations are either at (buoys) or close to (coastal stations) sea level, and share similar climatological conditions.

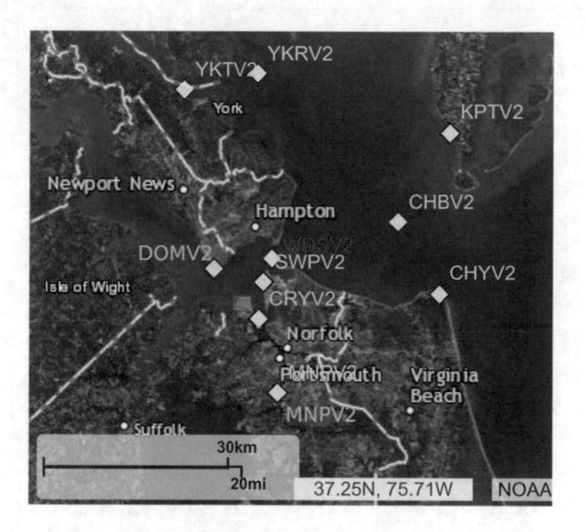

Fig. 1. Geolocation of the meteorological stations [14].

Several measurements or variables are taken at each station. These variables differ, depending on if the station is in a buoy or not. For this work, a common set of variables, available at all stations, was considered: atmospheric pressure ($PRES$) [mbar], air temperature ($ATMP$) [°C], wind speed ($WSPD$) [m/s] and peak gust speed (GST) [m/s]. Depending on the variable, the measurements are taken every 6 min or correspond to averages over a 6 min period.

Some basic properties of these variables may be seen in Table 1, namely the global average value in the dataset and the availability, for each station, between 2010 until the end of the year 2019. In this study, only variables with at least 85% of availability (in bold) were used. Moreover, because some stations are unable to comply with this degree of availability, the combination of stations used for each variable may be different.

To decide which variables will later be combined in the experiments, a preliminary study must be performed on the correlation between variables and stations. This is important because when variables are sufficiently correlated, the PCA method may be used to retain more data in fewer dimensions.

Table 1. Meteorological dataset characterization.

Station	**WSPD**		**GST**		**PRES**		**ATMP**	
	Mean	Avail.(%)	Mean	Avail.(%)	Mean	Avail.(%)	Mean	Avail.(%)
WDS	5.7	**97.5**	6.6	**97.5**	1017.4	**93.6**	16.5	**87.9**
YKR	5.9	**98.0**	6.9	**98.0**	1017.4	**98.6**	15.9	**98.5**
YKT	4.3	**97.7**	5.4	**97.7**	1017.3	**98.4**	16.0	**98.2**
MNP	2.6	**96.4**	4.1	**96.5**	1017.5	**97.9**	16.8	**97.7**
CHY	5.4	**95.5**	6.9	**95.5**	1017.0	31.1	16.1	**97.0**
DOM	3.9	**97.5**	5.3	**97.5**	1017.8	**98.3**	16.1	**98.2**
KPT	4.7	**97.4**	6.0	**97.5**	NA	0	NA	0
SWP	NA	0	NA	0	1017.7	**96.1**	NA	0
CRY	4.1	82.5	15.6	80.5	1017.6	82.8	16.5	34.3

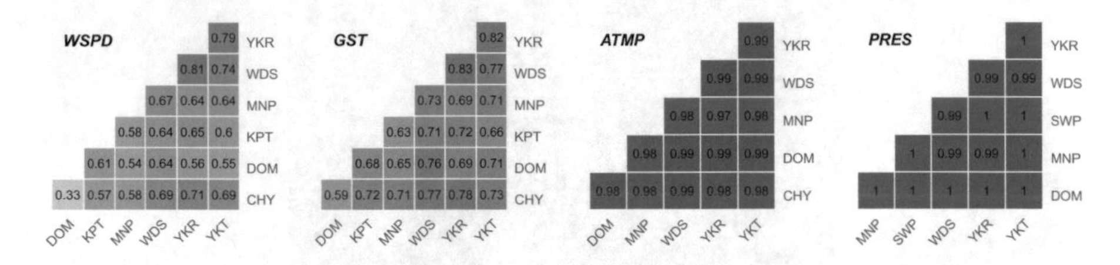

Fig. 2. Correlation between stations for each variable.

The correlations between the stations for each variable, relative to the same time period of Table 1, are shown in Fig. 2. In general, all stations are somehow correlated across all variables, with the $ATMP$ and $PRES$ variables showing the strongest correlations (1 or near 1).

Figure 3 presents the correlations between different meteorological variables within the same station. Due to the percentage of data available in the KPT and CHY stations, only records of two ($WSPD$ and GST) and three ($WSPD$, GST and $ATMP$) variables, respectively, are used for these stations. In the other stations, the records of all variables are available (see Table 1). The correlation between $WSPD$ and GST is strong across all stations. Between $ATMP$ and $PRES$, there is a minor inverse correlation. The other variables at different stations showed a weak correlation. Thus, only $WSPD$ and GST are used together in the experiments, once they are the only strongly correlated variables.

3 Determination of the Principal Components

Following the study on the correlation between variables and stations, the PCA approach may then be applied to reduce the dimensionality of the datasets.

Firstly, the dimensions with most data dispersion are identified. This enables to identify the principal components (PCs) that best distinguish the dataset

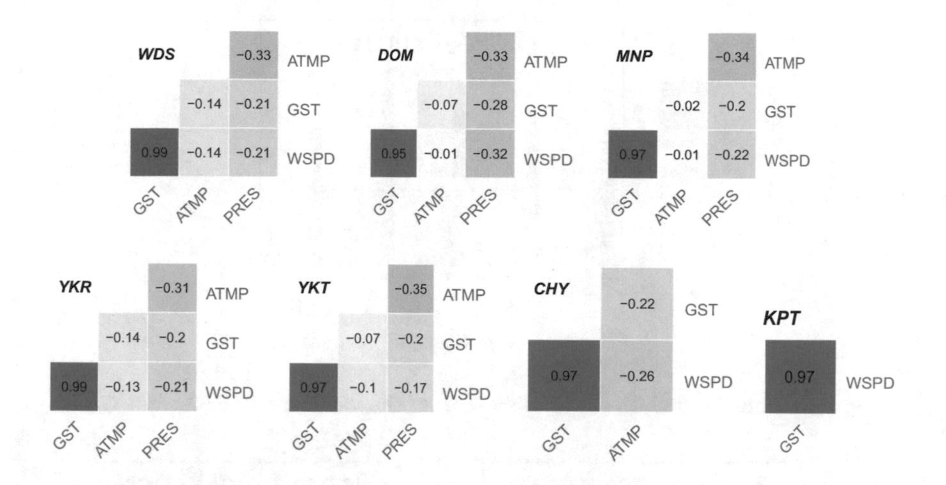

Fig. 3. Correlation between variables at each station.

under study. Consider that the dataset corresponding to the multiple predictor stations is represented by the data matrix $H \in \mathbb{R}^{m \times n}$, where each column h_i, with $i = 1, \ldots, n$, includes the scaled and normalized records of a single variable. Then, the thin singular value decomposition of H gives $H = U\Sigma V^T$, where $U \in \mathbb{R}^{m \times n}$, $\Sigma \in \mathbb{R}^{n \times n}$ and $V \in \mathbb{R}^{m \times n}$ (for details see [10]). The diagonal matrix Σ contains the singular values σ_i of H, for $i = 1, \ldots, n$, where $\sigma_1 > \sigma_2 > \ldots > \sigma_n$. The right singular vectors v_i are the *principal components directions* of H.

The vector $z_1 = Hv_1$ has the largest sample variance (σ_1^2/m) amongst all linear combinations of the columns of H, and so z_1 is the first principal component (PC_1). The second principal component (PC_2) is $z_2 = Hv_2$, once v_2 corresponds to the second largest variance (σ_2^2/m). The remaining principal components are defined similarly. The new variables are linear combinations of the columns of H, i.e., they are linear combinations of the normalized original variables $h_1, h_2, \ldots, h_n$, given by

$$z_j = v_{1j}h_1 + v_{2j}h_2 + \ldots + v_{nj}h_n \quad \text{for} \quad j = 1, 2, \ldots, n \tag{1}$$

where the coefficients v_{ij} (called *loadings*), with $i = 1, 2, \ldots, n$, are the elements of the vector v_j. The value of a coefficient is proportional to how significant a particular variable is in the principal component. It is expected that a few of the first principal components accurately reflect the original dataset, since they are likely to account for a significant proportion of the overall variation [17].

Figures 4 and 5 show the standard deviations of each PC for different amounts of input stations (# Stations). The standard deviation threshold of 1 is shown by a dotted line. PCs with standard deviation values above this line have more variance and, consequently, more information than the original normalized variables, whose standard deviation is equal to 1. Note that the variables from the WDS station were not included in the original variables because WDS was used only as the predicted station.

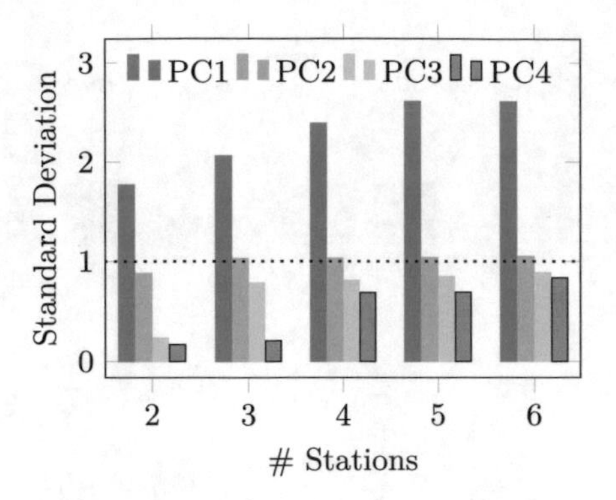

Fig. 4. Standard deviation of the *PC*s from the variables *WSPD* and *GST*.

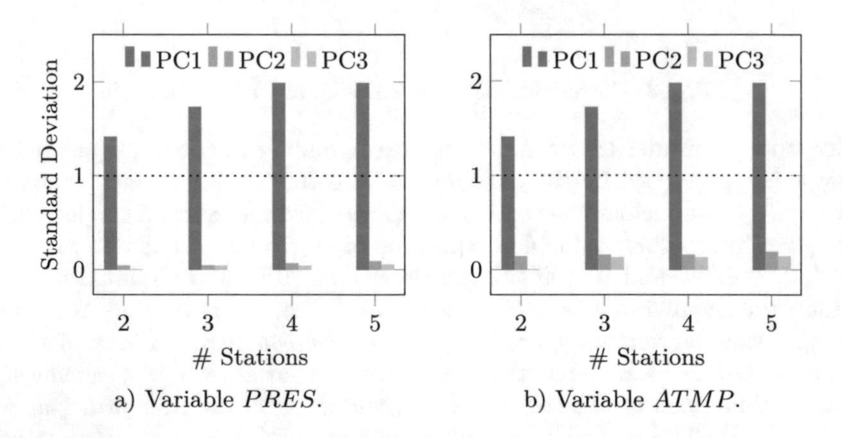

a) Variable *PRES*. b) Variable *ATMP*.

Fig. 5. Standard deviation of the *PC*s from the variables *PRES* and *ATMP*.

In Fig. 4 the Principal Component Analysis is conducted from a data matrix that includes two meteorological variables, *WSPD* and *GST*, from distinct stations. Except for the 2-station arrangement, both *PC*1 and *PC*2 exhibited values of standard deviation higher than one. These *PC*s with standard deviation greater than one were selected for the training phase.

The standard deviations of the *PC*s computed from the *PRES* and *ATMP* variables are shown in Fig. 5. It is crucial to highlight that, unlike the variables *WSPD* and *GST* (see Fig. 4), the *PRES* and *ATMP* variables were examined individually because they do not correlate sufficiently (see Fig. 3). Both variables showed the same pattern of standard deviation values, indicating that the *PC*1 was enough to capture most information of the data in all configurations of stations. Furthermore, in contrast to the *WSPD* and *GST* analysis, more information was contained in the *PC*1, since the relative concentration of standard deviation in this *PC* was significantly larger.

4 Analog Ensemble Combined with PCA

The AnEn method is able to reconstruct missing data in a time series. In this work, the time series builds on meteorological data and the reconstruction at a *predicted* station is carried out using data from neighbor *predictor* stations.

The process starts by identifying, in each predictor station, the predictor value for the same moment in time for which the predicted value must be reconstructed. Then, past historical values are found, in the predictor stations dataset, that are similar to the predictor value. These past values are called *analogs*. In the next step, the analogs are matched in time with corresponding observations (also historical measurements) of the predicted station. Finally, the missing value is predicted (reconstructed) by averaging the matching observations. The reconstruction error can then be evaluated if the real observed value for that instant is indeed available (as it is the case in this work).

The previous simplified description considers single numerical values for the predictor and analogs. In fact, these are vectors of $2k+1$ values (measurements), recorded at successive instants of the same time window, and $k > 0$ is an integer representing the width of each half-window (past and future) around the central instant. Therefore, analogs are established based on the similarity of vectors, and not single values, which enables the selection of analogs based on similar weather trends rather than single similar values [5]. The mapping of analogs (vectors) into observations (single values) is then based on the analogs central values.

In addition, when using multiple predictor stations, the analogs identified for each station may be required to overlap in time (dependent approach), or not (independent approach). In the current work, it was used the first alternative.

When the PCA technique is combined with the AnEn method (PCAnEn), the principal components (*PC*s) generated from the datasets of the predictor stations are used instead of the original datasets. This allows to use data from a larger number of stations, without increasing the computational effort.

The same idea may be applied to the K-means variant of the AnEn method. This variant enables to reduce the number of operations needed to determine the analogs of a given predictor. This is accomplished by replacing the comparison with all possible analogs, by a comparison with the clusters produced using the K-means clustering method. More precisely, the comparison is made with the centroids of the clusters, whose number is much less than the number of possible analog vectors (see [4] for details). Thus, similarly to what happens in the PCAnEn method, the PCClustAnEn method involves replacing the original predictor datasets with corresponding *PC*s prior to the clustering.

5 Experimental Evaluation

Both the PCAnEn and PCClustAnEn methods were put to the test for the reconstruction in the WDS station of the four variables selected for this study ($WSPD$, GST, $ATMP$ and $PRES$), every 6m, from 10m to 6pm period, during the full year of 2019 (prediction period). The remaining stations (within a 30 km radius around WDS, and with more than 85% of data available), were used as predictor stations, considering the training period of 2011 to 2018.

The reconstruction was performed by two separate implementations of the methods, one in R [15] and another in MATLAB [11]. This allowed for the mutual verification of the numerical results and provided an opportunity to compare the implementations performance-wise. The computing system used to execute the methods was a KVM-based virtual machine (with 16 virtual cores of a Intel Xeon W-2195 CPU, 64 GB of RAM and 256 GB of SSD) hosted on the CeDRI cluster, running Linux Ubuntu 20.04 LTS, R 4.2.2 and MATLAB R2021a.

Besides testing the PCAnEn and PClustAnEn methods, the corresponding non-PCA variants (AnEn and ClustAnEn) applied to the original datasets were also tested. This way, the specific impact of the PCA technique may also be accessed. To make the comparison fair, AnEn and ClustAnEn were tested using as predictors the same two stations used to test the PCAnEn and PClustAnEn with a 2-station configuration (#Stations = 2). This means that the variable is predicted from the same variable located in the two closest stations, thus ensuring the most favourable configuration to the AnEn and ClustAnEn methods.

The accuracy of the predicted/reconstructed values is assessed by comparing them to the exact values recorded at station WDS during the prediction period. The comparison is done by means of the Root Mean Square (RMSE) error that measures, simultaneously, the systematic and the random error [8].

Table 2. RMSE of the reconstruction with different methods.

Method	# Stations	**MATLAB**				**R**			
		WSPD	*GST*	*ATMP*	*PRES*	*WSPD*	*GST*	*ATMP*	*PRES*
PCAnEn	2	1.65	1.95	1.01	0.51	1.65	1.97	1.01	0.51
	3	1.32	1.52	0.84	0.48	1.32	1.52	0.84	0.48
	4	1.27	1.46	0.78	**0.45**	1.27	1.45	0.78	**0.45**
	5	**1.19**	**1.36**	**0.71**	0.61	**1.19**	**1.36**	**0.71**	0.61
	6	1.24	1.42	—	—	1.24	1.44	—	—
AnEn	2	1.68	1.77	0.86	0.59	1.67	1.76	0.86	0.59
PCClustAnEn	2	1.65	1.95	1.01	0.52	1.65	1.95	1.01	0.52
	3	1.32	1.50	0.84	0.48	1.32	1.51	0.84	0.48
	4	1.27	1.45	0.78	**0.45**	1.28	1.45	0.78	**0.45**
	5	**1.20**	**1.35**	**0.72**	0.61	**1.20**	**1.36**	**0.72**	0.60
	6	1.27	1.40	—	—	1.25	1.42	—	—
ClustAnEn	2	1.69	1.73	0.88	0.60	1.69	1.74	0.87	0.56

Table 2 presents the RMSE values for all tests performed. For each test, the number of *PC*s used was 1 or 2, after the values of the respective standard deviation, as explained in Sect. 3. Between PCAnEn and PCClustAnEn, there were no noteworthy changes in accuracy. The 5-station setup demonstrated a lower RMSE than the non-PCA approaches for the majority of variables. The higher errors are obtained with the 2-station configurations, in which case there's no sensible advantage in using the PCA variants over the non-PCA ones. The reductions in error rates from the PCA implementations ranged from ≈18% to ≈30%, for the best setting of each variable, compared to the non-PCA methods. These considerations apply to both implementations (R and MATLAB).

Regarding the computational performance, Fig. 6 shows the processing times of the MATLAB (M) and R (R) codes, with different amounts of stations, for the reconstruction of the *WSPD* variable, using all the CPU cores (16) available in the test bed computational system. The *WSPD* variable was chosen for the performance evaluation because a) it is available for more stations (recall Table 2), and b) it requires 2 *PC*s to represent the original variables when using 3 or more stations. The same is also valid for the *GST* variable, whose processing times are either the same (PCA-based approaches) or similar (other approaches).

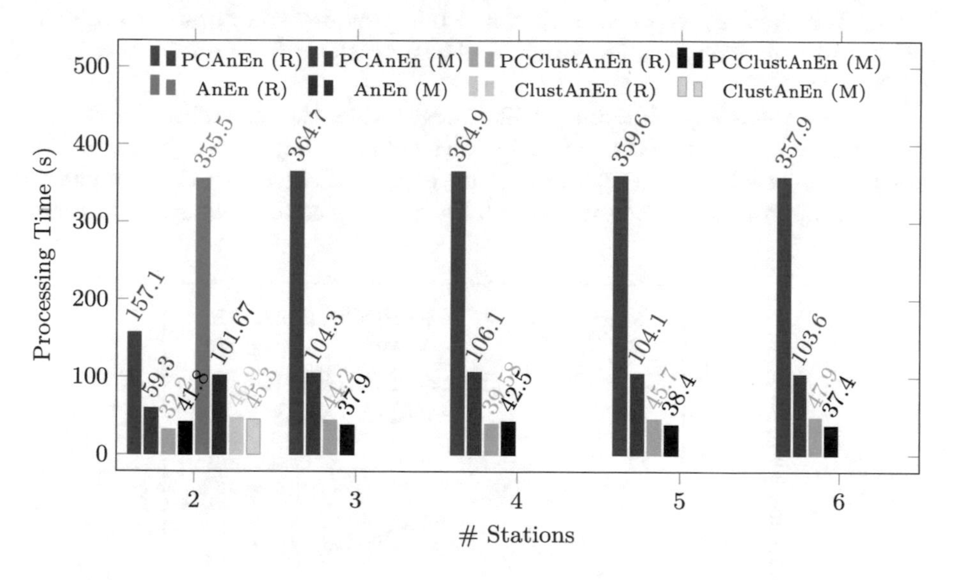

Fig. 6. Reconstruction time of *WSPD* with 16 cores (2 to 6 stations).

When using clustering (ClustAnEn and PCClustAnEn), the reconstruction times are the lowest, and the variations are small for different numbers of stations, whether using PCA or not; also, for the only scenario where it makes sense to use the non-PCA approaches (2 stations), using clustering alone (ClustAnEn) is slower than combining it with PCA (PCClustAnEn).

Without clustering (AnEn and PCAnEn), the processing times are noticeably higher. When applying PCA (PCAnEN), the highest times are obtained with 3 or more stations (using 2 *PC*s), and they are similar; these times roughly double the time with 2 stations (using 1 PC); thus, without clustering, the number of *PC*s used has a noticeable influence (direct proportionality) on the processing times. For the 2-stations scenario, not using PCA (AnEn) doubles the processing times compared to using PCA (PCAnEn), being equivalent to using PCA with more than 2 stations, once it is also using two time series.

Lowering the processing times is important, but it shouldn't be at the expense of higher reconstruction errors. Ideally, the reconstruction should be faster and also more accurate. The smallest RMSE errors for the $WSPD$ variable are obtained with PCA-based methods using 5 stations, whether clustering is used (PCClustAnEn) or not (PCAnEn) – recall Table 2. However, clustering ensures much lower processing times, with a speedup between $\gtrapprox$2,7 (MATLAB code) and $\gtrapprox$7,8 (R code). Comparing the processing times of PCClustAnEn with 5 stations, with the ones of ClustAnEn with 2 stations (the best provided by not using PCA) yields almost none speedup (46,9/45,7 = 1,03 and 45,3/38,4 = 1,18); however, the RMSE error of PCClustAnEn with 5 stations is only 1,2/1,69≈70% of the error of ClustAnEn with 2 stations, thus favouring the first approach.

The impact on performance of using or not the PCA method is perceivable in the 2-stations scenario. Here, using PCA provides speedups ranging from 2,26 to 1,08, for comparable methods (AnEn vs PCAnEn, and ClustAnEn vs PCClustAnEn).

Another advantage of adding PCA emerges when two variables, like $WPSD$ and GST, are used together in the analysis. Once they share the same time series, PCA-based methods can predict both variables in a single run, unlike the non-PCA approaches, which would require two runs of the reconstruction code.

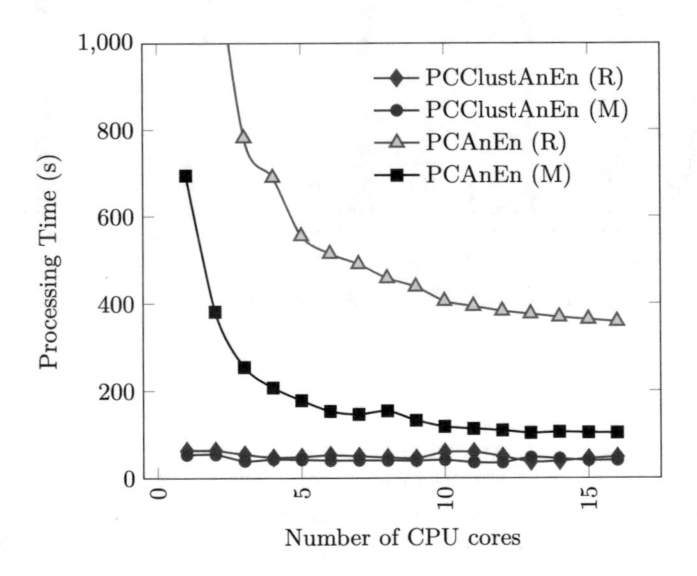

Fig. 7. Reconstruction time of $WSPD$ with 1 to 16 cores (6 stations).

The MATLAB code was found consistently faster than the R code. This is visible in Fig. 6 for 16 cores, and can also be seen in Fig. 7 for a variable number of cores. However, under PCClustAnEn the differences were minor, meaning both implementations are equally efficient when applying the K-means clustering. More important, PCClustAnEn required much less processing times, in all configurations, compared to PCAnEn. Also, PCClustAnEn mostly doesn't benefit from the extra cores, in opposition to the PCAnEn method, where the search for analogues is the biggest code hotspot and is easily parallelizable.

It should also be stressed that both R and MATLAB were used with default configurations, without any extra performance tuning to optimize their behavior.

6 Conclusion

This paper describes the combination of Analog Ensemble (AnEn) methods with Principal Component Analysis (PCA), allowing for data from several stations to be reduced to a smaller number of time series, corresponding to the Principal Components (*PC*s). These are then used, instead of the original variables records, to reconstruct data missing in the records of a meteorological site.

In our experiments, the PCA technique improved the assertiveness of prediction without compromising the computational performance, since it is possible to increase the number of stations without increasing the quantity of input time series. It was also shown that the efficacy of PCA is heavily influenced by the correlation between the time series of several predictors, as higher correlation allows for a high proportion of information/variance in the first components.

Furthermore, two different implementations of the methods studied were used and compared, one in MATLAB and other in R. This allowed to double-check the numerical results and gain insight on the potential performance impact of choosing either implementation. The scalability of both codes was also studied, in a medium-scale multicore system. The performance evaluation showed the superiority of the AnEn methods where PCA is combined with clustering.

Future applications of the same methodology with larger datasets are planned, to assess more accurately the effects of the quantity, correlation and closeness of predictor stations. The R code will also be tuned to improve its performance.

Acknowledgements. This work was supported by national funds through FCT/MCTES (PIDDAC): CeDRI, UIDB/05757/2020 (DOI: 10.54499/UIDB/05757/2020) and UIDP/05757/2020 (DOI: 10.54499/UIDB/05757/2020); and SusTEC, LA/P/0007/2020 (DOI: 10.54499/LA/P/0007/2020).

References

1. Alessandrini, S.: Predicting rare events of solar power production with the analog ensemble. Solar Energy **231**, 72–77 (2022). https://doi.org/10.1016/j.solener.2021.11.033. https://www.sciencedirect.com/science/article/pii/S0038092X21009920
2. Balsa, C., Breve, M.M., André, B., Rodrigues, C.V., Rufino, J.: PCAnEn - Hindcasting with Analogue Ensembles of Principal Components. In: Garcia, M.V., Gordón-Gallegos, C. (eds.) CSEI 2022. LNNS, vol. 678, pp. 169–183. Springer, Cham (2022). https://doi.org/10.1007/978-3-031-30592-4_13

3. Balsa, C., Rodrigues, C.V., Araújo, L., Rufino, J.: Hindcasting with cluster-based analogues. In: Guarda, T., Portela, F., Santos, M.F. (eds.) ARTIIS 2021. CCIS, vol. 1485, pp. 346–360. Springer, Cham (2021). https://doi.org/10.1007/978-3-030-90241-4_27
4. Balsa, C., Rodrigues, C.V., Araújo, L., Rufino, J.: Cluster-based analogue ensembles for hindcasting with multistations. Computation **10**(6), 91 (2022). https://doi.org/10.3390/computation10060091
5. Balsa, C., Rodrigues, C.V., Lopes, I., Rufino, J.: Using analog ensembles with alternative metrics for hindcasting with multistations. ParadigmPlus **1**(2), 1–17 (2020). https://journals.itiud.org/index.php/paradigmplus/article/view/11
6. Birkelund, Y., Alessandrini, S., Byrkjedal, Ø., Monache, L.D.: Wind power prediction in complex terrain using analog ensembles. J. Phys. Conf. Ser. **1102**(1), 012,008 (2018). https://doi.org/10.1088/1742-6596/1102/1/012008. https://dx.doi.org/10.1088/1742-6596/1102/1/012008
7. Castellano, C.M., DeGaetano, A.T.: Downscaling extreme precipitation from cmip5 simulations using historical analogs. J. Appl. Meteorol. Climatol. **56**(9), 2421 – 2439 (2017). https://doi.org/10.1175/JAMC-D-16-0250.1. https://journals.ametsoc.org/view/journals/apme/56/9/jamc-d-16-0250.1.xml
8. Chai, T., Draxler, R.R.: Root mean square error (RMSE) or mean absolute error (MAE)? – Arguments against avoiding RMSE in the literature. Geosci. Model Dev. **7**(3), 1247–1250 (2014). https://doi.org/10.5194/gmd-7-1247-2014
9. Davò, F., Alessandrini, S., Sperati, S., Monache, L.D., Airoldi, D., Vespucci, M.T.: Post-processing techniques and principal component analysis for regional wind power and solar irradiance forecasting. Sol. Energy **134**, 327–338 (2016). https://doi.org/10.1016/j.solener.2016.04.049
10. Eldén, L.: Matrix methods in data mining and pattern recognition. SIAM, Philadelphia, PA, USA (2007)
11. MATLAB: version 9.10.0.1602886 (R2021a). The MathWorks Inc., Natick, Massachusetts (2021)
12. Meech, S., Alessandrini, S., Chapman, W., Delle Monache, L.: Post-processing rainfall in a high-resolution simulation of the 1994 piedmont flood. Bulletin of Atmospheric Science and Technology **1**(3), 373–385 (2020). https://doi.org/10.1007/s42865-020-00028-z
13. Monache, L.D., Eckel, F.A., Rife, D.L., Nagarajan, B., Searight, K.: Probabilistic weather prediction with an analog ensemble. Mon. Weather Rev. **141**(10), 3498–3516 (2013). https://doi.org/10.1175/mwr-d-12-00281.1
14. National Weather Service: National Data Buoy Center. https://www.ndbc.noaa.gov
15. R Core Team: R: A Language and Environment for Statistical Computing. R Foundation for Statistical Computing, Vienna, Austria (2022). https://www.R-project.org/
16. Rozoff, C.M., Alessandrini, S.: A comparison between analog ensemble and convolutional neural network empirical-statistical downscaling techniques for reconstructing high-resolution near-surface wind. Energies **15**(5) (2022). https://doi.org/10.3390/en15051718. https://www.mdpi.com/1996-1073/15/5/1718
17. Spence, L., Insel, A., Friedberg, S.: Elementary Linear Algebra: A matrix Approach. Pearson Education Limited (2013)
18. Zhang, X., Li, Y., Lu, S., Hamann, H.F., Hodge, B.M., Lehman, B.: A solar time based analog ensemble method for regional solar power forecasting. IEEE Trans.on Sustain. Energy **10**(1), 268–279 (2019). https://doi.org/10.1109/TSTE.2018.2832634

Analysis of the Characteristics of Different Peer-To-Peer Risky Loans

Bih-Huang Jin[1](✉), Yung-Ming Li[2], and Kuan-Te Ho[2]

[1] Department of Business Administration, Tunghai University, Taichung, Taiwan
bihuang@thu.edu.tw

[2] Institute of Information Management, National Chiao Tung University, Hsinchu, Taiwan

Abstract. This study analyzes data from Lending Club 2011 January to 2016 January. We use survival analysis and proportional hazards model to find what loan characteristics will have lower default rate in high-risk group and low risk group. The grouping way is through the lending Club credit score. Our study provides a way to analyze loan characteristics to reduce information asymmetry and default rate. To earn higher interest and take the principal back have always been the biggest issue in the financial world. Our research gives the advices through survival analysis with empirical data. The results show the repayment characteristics of the high-risk group and the low-risk group is similar. Except for the following four characteristics, 'mortgage', 'education', 'home improvement', and 'medical' are opposite.

Keywords: FinTech · P2P Lending · Loan Risk · Data Analysis

1 Introduction

With the rapid development of peer-to-peer lending, the relevant platform has sprung up, such as Lending Club, Zopa, Prosper, SoFi and a non-profit platform, Kiva. What difference between traditional bank and peer to peer lending platform has made the P2P lending platform so popular? First, Traditional banks are willing to take on the lower risk, so traditional banks cannot face a broader risk of the borrower, and on the contrary, investors therefore cannot earn high return. Second, slower loan issuance speed also caused traditional bank declined; in general, P2P borrowing platform only need only 7 days, while the traditional banks may need 20 days. Third, in most P2P lending platforms, the assessment of the borrower is based on the predefined rules of automated handling, and the borrower takes the corresponding borrowing rate as long as the information is filled in. However, the traditional bank's multiple layers of examination and approval are not as transparent as a P2P platform. Forth, traditional bank operation strategies are more conservative. By the contrast, risk borrower can get loans in the P2P platform because the lender can choose different risk appetite; the lender with risk preference can choose the high-risk borrower.

However, in order to pursue convenience and wider range of risks, more dangerous problems have to be faced. Because the lack of collateral on peer-to-peer lending, how

A. Rocha et al. (Eds.): WorldCIST 2023, LNNS 799, pp. 97–105, 2024.
https://doi.org/10.1007/978-3-031-45642-8_9

to determine credit risk has become a very important issue. Which of the personal information messages provided by borrower is truly important? What are the difference characteristics between the high risk and the low-risk borrower when both of them will repay the loan? This study explores these problems by empirical analysis of the Lending Club's data.

At the first beginning of this study, we found out an interesting phenomenon in more than 277 thousand data from Lending Club, an online peer to peer lending platform. In the big amount of data, 49.9 thousand data were fully repaid but which belonged to high-risk groups. Among these data, more than 47 thousand loans were fully repayment and early repayment before due, which represents the borrower with high-risk credit score will still repay on time or even in advance and completely. Usually, a lender's investment on a borrower with good credit score often can only earn lower pay. While the high investment return is generally considered to bear a higher risk. But the data shows that even the poor credit score there are still many borrowers who are punctual repayment. This phenomenon tells us that it is important and valuable to find out who will fully repay from a poorly rated credit score.

Credit rating in Lending Club can be divided into 7 levels from A to G; grade A for the credit risk is good, followed by lower to grade G for the higher credit risk. The loan interest rate of borrower is in accordance with the grade of credit rating. Each level has its corresponding borrowing rate, loan interest rates can be regarded as risk compensation to lender, so the higher the risk, the higher the borrowing rate. In the peer-to-peer lending, a lender invests in loan that meets its own risk preferences and gets paid accordingly. However, as an investor, anyone wants to get higher payoff. Hence, a lender would like to know how to earn higher interest rate from a high risk but safe loan. Prior scholars have tried to construct various regression models to predict loan default rates, or to investigate the characteristics of default loans, thus helping investors to choose loan or borrower with lower default probability [1–4]. But in the past, scholars haven't divided borrowers into a high-risk group and a low-risk group to discuss and explore characteristics of the high-risk borrower that does not default, or the differences of characteristics between the high-risk borrower and the low-risk borrower.

Therefore, this study divides borrowers into a high-risk group and a low-risk group, and through empirical analysis to observe the different characteristics among high-risk borrower and low risk borrower. Investors then can apply the detections of this study to acquire high interest rates from high-risk borrowers and receive full repayment. This research can help investors to earn higher and do not need to take on great risks.

Research Questions

RQ1: Are high-risk borrower group and low-risk borrower group different in terms of loan default characteristics?

RQ2: What are high-risk borrower group's default characteristics? What are low risk borrower group's default characteristics?

2 Related Literature

The success of P2P loans is the success of repayment, but not the success of borrow. It is that the repayment complete or not to decide the default rate. However, how can a lender know a borrower is going to default or not? Because there is information asymmetry existing between a lender and a borrower, the only way for a lender to determine whether borrower will default or not is through credit score.

Therefore, past scholars tried many methods, such as decision tree (DT), neural network (NNs), SVM, etc.; they hoped to predict borrower default rate from past historical data. Jin and Zhu [2] used data mining to predict default risk of peer-to-peer (P2P) lending loan. Emekter et al. [1] used binary logistic regression and cox proportional hazard to evaluating credit risk and loan performance. An alternative concept was presented, estimating loan profitability by IRR profit scoring systems [4]. Besides statistic methods, M. F. Lin et al. [5] approved a loan borrow or lend from friends could lower the default rates. The discovery of a strong social network relationship is an important factor in determining the success of borrowing and reducing the risk of default. Freedman and Jin [6] studied the importance of social relations in determining loan grants. The results show that borrowers with social relationships are more likely to get loan funds and get lower interest rates. Some scholars discuss the relationship between P2P lending and crowd wisdom. When the credit information is very limited, the borrower to seek the wisdom of the crowd.

These methods above, whether it is the wisdom of crowd, social networks, or data analysis to building regression model, are designed to reduce the information asymmetry between lender and borrower in peer-to-peer lending. Information asymmetry is a core problem of online P2P lending. A lender usually wants to get as much valid information about borrower as possible, in the other side, a borrower might try to hide their weaknesses in order to get the interest rate as low as possible. In order to help the lender to make right decisions based on valid information, the P2P lending platform forces its borrower to provide financial information and personal data that are verified by an external agency. These are called characteristics of these P2P borrowing determinants because they have a significant impact on the success of the borrower's loan and required interest rate [7]. Jin and Zhu [2] focused on feature selection and uses random forest method and importance analysis to propose the selected social and economic factors in addition to the bank commonly used ones. This helps lenders to understand the determinants of default risk.

3 Research Methodologies

Survival analysis is used to observe the probability and influence variable of a specific event in a group during a period of time [8]. In the loan data of this study, the starting time of all samples can be clearly grasped. Some samples, however, have not yet been fully repaid or default at the end of the observation period (January, 2016). These groups of samples are called right censored data. Our research focused on fully repayment and default data, so we deleted uncertainty data.

Survival analyses based on different allocation assumptions have relatively different models. Such as Weibull distribution survival model, Log-logistic survival model, Log-normal survival model, proportional hazards model, among which the proportion of risk model is the most used. This study also used the proportion of risk model as an empirical basis for this model.

Proportional Hazards Model (PHM) was proposed by Cox in 1972 and is most widely used in survival analysis in the statistical area [8]. Relative to other models, such as Logit Model, Probit Model, Life Table, Accelerated Failure Time Model, the proportional hazard model is less limited to obey a probability assignment for a data or residual term. On the other hand, the proportional hazard model, in addition to being able to cover variables that do not change its value by time. It is also suitable for dealing with time-dependent covariates or time-varying covariates that change their value over time.

FICO score is the most common and convincing credit score in the US which is created by the Fair Isaac Corporation. The lenders use the borrower's FICO score and details of the borrower's credit report to assess the credit risk and determine whether to expand the credit. The national average FICO Score is 695. The highest FICO score is 850 and the lowest is score 300. The calculating method of the FICO score comes from the Fair Isaac Corporation which applies different weightings to components of credit such as payment history, accounts owed, length of credit history, new credit etc.

Our research divided borrowers in to two groups, high-risk group and low-risk group. The rule of group dividing is according to the credit rate of Lending Club. Lending Club's credit rate is referenced by FICO score.

Our data is from Lending Club 2011 January to 2016 January. Lending club is the biggest peer to peer lending plat form in the US. In terms of market share, Lending Club accounted for about 75% of the US P2P market, much bigger than the earliest established platform in US, Prosper. Prosper has only accounted for about 4%. Lending Club has become a monopoly of the US P2P lending industry giants. There are seven grades to distinguish different risk borrowers, from A to G. The loan level indicates the degree of risk and return.

4 Data Analysis and Results

The definition of variables is according to lending club dictionary and Table 1 shows the variables of our data. Grade is intended to distinguish between high-risk and low-risk groups, A-C is a low-risk group, and D-G is a high-risk group. Among the variables of measuring loan characteristics, 14 different loan purposes are included, from debt consolidation to home improvement or loans to start up a small business. In variable of borrower characteristics, including annual income of borrower, inquiries last 6 months, housing situation, revolving line utilization rate and number of open accounts etc. Housing situation can be divided into own, rent, mortgage and other. Months since last delinquency means the number of months since the borrower's last delinquency. However, our study separates this variable into four levels, recently, a few years, long time ago and no delinquency. Three ratios are included in borrower indebtedness, that relate to loan amount to annual income, annual instalment to income, debt to income. These are very important indicators of personal borrowing. These indicators will be calculated based on the borrower's information and the amount of loan.

Tables 2 and 3 shows the survival analysis results, by means of 29 Cox regressions, one for each explanatory variable. The Table provides the regression coefficients, standard errors, risk ratios and significance of p-values. A positive regression coefficient β for an explanatory variable means that the risk is higher. A negative regression coefficient β for an explanatory variable means that the risk is lower. For discrete variables,

Table 1. Variables used in the study

Variables	Definition
Borrower Assessment	
Grade	Lending Club categorizes borrowers into seven different loan grades from A down to G, A-grade being in the safest
Interest Rate	Interest rate on the loan
Loan Characteristics	
Loan Purpose	14 loan purposes: wedding, credit card, car loan, major purchase, home improvement, debt consolidation, house, vacation, medical, moving, renewable energy, educational, small business, and other
Loan Amount	The listed amount of the loan applied for by the borrower
Borrower Characteristics	
Annual Income	The annual income provided by the borrower during registration
Housing Situation	Own, rent, mortgage and other
Inquiries Last 6 Months	The number of inquiries by creditors during the past 6 months
Public Records	Number of derogatory public records
Revolving Utilization	Revolving line utilization rate, or the amount of credit the borrower is using relative to all available revolving credit
Open Accounts	The number of open credit lines in the borrower's credit file
Months Since Last Delinquency	The number of months since the borrower's last delinquency. Separate it into four period, recently, a few years, long time ago and no delinquency.
Borrower Indebtedness	
Loan Amount to Annual Income	Loan amount to annual income
Annual Instalment to Income	The annual payment owed by the borrower divided by the annual income provided by the borrower during registration
Debt to Income	Borrower's debt to income ratio. Monthly payments on the total debt obligations, excluding mortgage, divided by self-reported monthly income.

Reference: Lending Club Dictionary

we use an example to explain hazard ratio. Take ‘car loan’ for example, the risk of loans for ‘car loan’ is 1.301 times higher than the risk of loans for ‘no car loan’ in low-risk group. And the risk of loans for ‘rent’ is 0.846 times lower than the risk of loans for ‘no rent’ in low-risk group. For continuous variable, hazard ratio can be interpreted as the predicted change in the risk for a unit increase in the explanatory variable. But the risk ratio for increasing n units is 〚exp(β)〛 ^n. The Table reveals important practical findings for lenders.

In low-risk group, according to Table 2, the risk ratio of the wedding variable is 1.813, which means that low risk borrower with the lending purpose is wedding has 1.813 times default risk of the non-wedding borrower in condition of other variables fixed. Meanwhile, the default risk for borrower’s housing situation is mortgage is 1.041 times higher than non-rent. In this case, low risk group has an opposite result comparing to high-risk group. In addition to continuous variables, it is also a similar concept. Take interest rate for example, for each additional unit on interest rate will increase 1.097 times in the default risk. In the case of low-risk borrower, the default risk increased to 3.486 times for each additional unit of annual instance to annual income.

To sum up in low-risk group, for discrete group, our study found with these variable ‘rent’, ‘other’, ‘home improvement’, ‘medical’, ‘small business’, the risk of default will lower. And with these variables, ‘mortgage’, ‘own’, ‘debt consolidation’, ‘credit card’, ‘car loan’, ‘education’, ‘house’, ‘major purchase’, ‘vacation’, ‘wedding’, ‘a few years’, ‘long time ago’, ‘no delinquency’ the default risk will be higher than without these characteristics. It will be no different with or without the variable 'recently', 'vacation' in last delinquency of the default risk in low-risk group.

Table 2. Cox regression analysis for survival time of low-risk loans

Predictors	Parameter estimates(β)	Standard error	Hazard ratio exp(β)	P-value
Discrete Variable				
Housing Situation				
rent	-0.167	0.196	0.846	0.396
mortgage	0.041	0.196	1.041	0.836
own	-0.044	0.073	0.745	0.549
other	-0.112	0.197	0.894	0.569
Table 6 Cox regression analysis for survival time of low risk loans (continue)				
Loan purpose				
Debt consolidation	0.147	0.069	1.158	0.032 **
Credit card	0.184	0.069	1.202	0.008 ***
Car loan	0.263	0.090	1.301	0.003 ***
Education	0.064	0.166	1.066	0.700
Home improvement	-0.037	0.073	0.964	0.616
House	0.233	0.114	1.262	0.041 **
Major purchase	0.182	0.080	1.200	0.023 **
Medical	-0.093	0.089	0.911	0.295
Small business	-0.263	0.082	0.768	0.001 ***
Vacation	-0.007	0.102	0.993	0.944
Wedding	0.595	0.113	1.813	0.000 ***
Last delinquency				
Recently	0.000	0.000	1.000	0.000 ***
A few years	0.287	0.024	1.332	0.000 ***
Long time ago	0.352	0.023	1.421	0.000 ***
No delinquency	0.393	0.021	1.481	0.000 ***
Continous Variables				
Borrower Assessment				
Interest rate	0.092	0.004	1.097	0.000 ***
Loan Characteristics				
Loan amount	0.000	0.000	1.000	0.000 ***
Borrower Characteristics				
Annual income	0.000	0.000	1.000	0.026 **
Inquiries last 6 months	0.046	0.006	1.047	0.000 ***
Public records	0.155	0.012	1.168	0.000 ***
Revolving utilization	0.004	0.000	1.004	0.000 ***
Open accounts	0.008	0.001	1.008	0.000 ***
Borrower Indebtedness				
Loan Amount to Annual Income	1.808	0.855	6.097	0.035 **
Annual Instalment to Annual Income	1.249	2.128	3.486	0.557
Debt to Income	0.023	0.001	1.023	0.000 ***

Under two-tailed test, ***1% significant level, **5% significant level, *10% significant level

In Table 3, exp(β) presents as hazard ratio. First, for the dummy variable, there are two types of variable values. Just like a discrete variable in lending purpose-wedding, the only two possible value of wedding is "1" or "0". "1" means lending purpose is wedding and "0" means lending purpose is not wedding. The hazard ratio is expressed as a multiple relationship between default risk of wedding borrowers and non-wedding borrowers. For example, in Table 3, the risk ratio of the wedding variable is 1.897, which means that high risk borrower with the lending purpose is wedding has 1.897 times default risk of the non-wedding borrower in condition of other variables fixed. Likewise, the default risk for borrower's housing situation is mortgage is 0.852 times less than non-mortgage. We can infer housing situation is mortgage is a good loan characteristic in high-risk group. In addition to continuous variables, it is also a similar concept. Take interest rate for example, for each additional unit on interest rate will increase 1.055 times in the default risk. In the case of high-risk borrower, the default risk is reduced to 0.886 times for each additional unit of annual instance to annual income.

To sum up in high-risk group, for discrete group, our study found that these variables 'rent', 'mortgage', 'other', 'education', 'small business', the risk of default will lower. And with these variables, 'own', 'debt consolidation', 'credit card', 'car loan', 'home improvement', 'house', 'major purchase', 'medical', 'medical', 'vacation', 'wedding', 'a few years', 'long time ago', 'no delinquency', the default risk will be higher than without these characteristics. It will be no different with or without the variable 'recently' in last delinquency of the default risk in high-risk group.

Table 3. Cox regression analysis for survival time of high-risk loans

Predictors	Parameter estimates(β)	Standard error	Hazard ratio exp(β)	P-value
Discrete Variable				
Housing Situation				
rent	-0.329	0.259	0.720	0.204
mortgage	-0.161	0.259	0.852	0.535
own	-0.260	0.260	0.771	0.317
other	0.106	0.062	1.112	0.087 *
Loan purpose				
Debt consolidation	0.184	0.059	1.202	0.002 ***
Credit card	0.304	0.063	1.356	0.000 ***
Car loan	0.282	0.112	1.325	0.012 **
Education	-0.062	0.185	0.940	0.738
Home improvement	0.136	0.071	1.146	0.056 *
House	0.136	0.102	1.146	0.183
Major purchase	0.054	0.082	1.056	0.507
Medical	0.058	0.081	1.060	0.471
Small business	-0.169	0.069	0.844	0.015 **
Vacation	0.165	0.096	1.179	0.085 *
Wedding	0.640	0.118	1.897	0.000 ***
Last delinquency				
Recently	0.000	0.000	1.000	0.000 ***
A few years	0.080	0.032	1.083	0.013 **
Long time ago	0.182	0.032	1.200	0.000 ***
No delinquency	0.059	0.029	1.061	0.043 *
Continuous Variables				
Borrower Assessment				
Interest rate	0.053	0.005	1.055	0.000 ***
Loan Characteristics				
Loan amount	0.000	0.000	1.000	0.000 ***
Borrower Characteristics				
Annual income	0.000	0.000	1.000	0.000 ***
Inquiries last 6 months	0.035	0.006	1.036	0.000 ***
Public records	0.064	0.016	1.066	0.000 ***
Revolving utilization	0.000	0.000	1.000	0.302
Open accounts	0.000	0.002	1.000	0.834
Borrower Indebtedness				
Loan Amount to Annual Income	2.134	1.038	8.446	0.040 *
Annual Instalment to Annual Income	-0.121	2.383	0.886	0.959
Debt to Income	0.024	0.001	1.024	0.000 ***

Under two-tailed test, ***1% significant level, **5% significant level, *10% significant level.

Figure 1 shows the survival curves of high-risk groups and low risk groups. As the Tables 2 and 3 shows, except of the loan characteristic 'mortgage', 'education', 'home improvement', and 'medical', other trend of characteristics between high-risk group and low risk group are the same. However, these four loan characteristics are on the opposite. The default rate will be lower if the loan with 'home improvement' and 'medical' and without 'mortgage' and 'educational' in low-risk group. The default rate will be lower if the loan with 'mortgage' and 'educational' and without 'home improvement' and 'medical' in low-risk group.

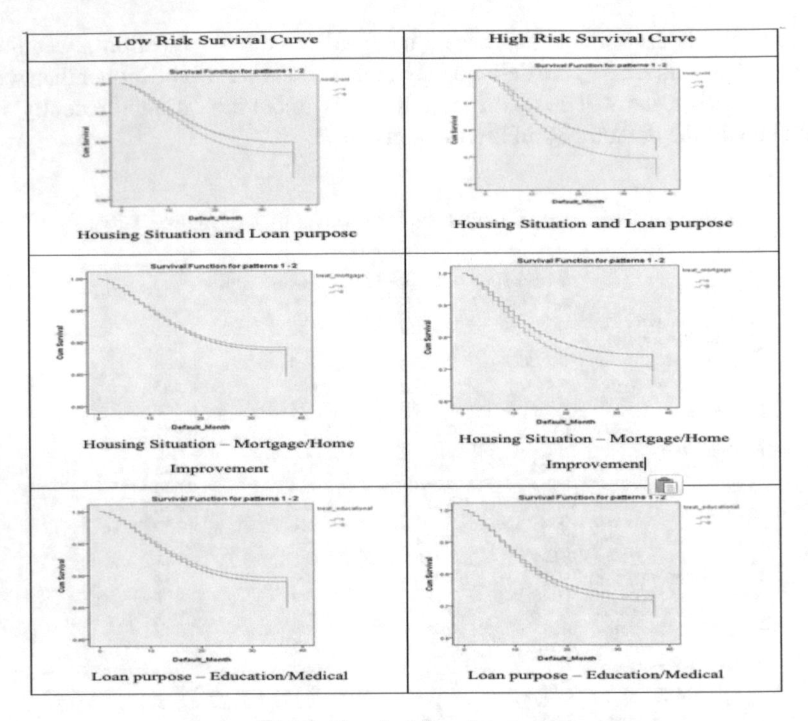

Fig. 1. Survival function curve

5 Conclusion

The transaction cost of a P2P loan company may take less than traditional financial institutions because its business model is simpler. They do not capture the deposit, they do not comply with the strict bank supervision, they do not remain idle balances; they just make contact with the borrower and the lender. In addition, most of the process is automated to complete through the online platform. The operational cost is the most important factor to explain the bank's interest space. P2P lending platform, like other online business use technology as a strength. This strength can lead to improved efficiency which is a very important factor in the lend and borrow of a loan in the market. In the P2P lending business model, credit risk is distinguished by individuals, lenders take risks to other people to borrow money. The problem of asymmetric information is huge. Therefore, the quality of the information provided on P2P lending site is very important. This information can be provided by third parties, such as external credit rating or it can be extracted from the platform in itself, such as loans allocated to each level.

This study provides a way to analyze loan characteristics to reduce information asymmetry. To earn higher interest and take the principal back has always been the biggest issue in the financial world. Our research gives the advices to lenders through survival analysis with empirical data. The result shows high risk group with loan characteristics of 'rent', 'mortgage', 'other', 'education', 'small businesses will have lower default rate. And if loan characteristics without 'own', 'debt consolidation', 'credit card', 'car loan', 'home improvement', 'house', 'major purchase', 'medical', 'medical', 'vacation', 'wedding', 'a few years', 'long time ago', 'no delinquency' will also have lower default rate.

References

1. Emekter, R., Tu, Y., Jirasakuldech, B., Lu, M.: Evaluating credit risk and loan performance in online peer-to-peer (P2P) lending. Appl. Econ. **47**(1), 54–70 (2015). https://doi.org/10.1080/00036846.2014.962222
2. Jin, Y., Zhu, Y.: A data-driven approach to predict default risk of loan for online peer-to-peer (P2P) lending. In: Proceedings - 2015 5th International Conference on Communication Systems and Network Technologies, CSNT 2015, pp. 609–613 (2015). https://doi.org/10.1109/CSNT.2015.25
3. Malik, M., Thomas, L.C.: Modelling credit risk of portfolio of consumer loans. J. Oper. Res. Soc. **61**(3), 411–420 (2010). https://doi.org/10.1057/jors.2009.123
4. Serrano-Cinca, C., Gutiérrez-Nieto, B., López-Palacios, L.: Determinants of default in P2P lending. PLoS ONE **10**(10), e0139427 (2015). https://doi.org/10.1371/journal.pone.0139427
5. Lin, M., Prabhala, N.R., Viswanathan, S.: Judging borrowers by the company they keep: friendship networks and information asymmetry in online peer-to-peer lending. Manage. Sci. **59**(1), 17–35 (2013). https://doi.org/10.1287/mnsc.1120.1560
6. Freedman, S., Jin, G.Z.: The information value of online social networks: lessons from peer-to-peer lending. Int. J. Ind. Organ. **51**, 185–222 (2017)
7. Bachmann, A., et al.: Online peer-to-peer lending-a literature review. J. Internet Bank. Commer. **16**(2), 1 (2011)
8. Cox, D.R.: Models and life-tables regression. J. R. Stat. Soc. Ser. B **34**(2), 187–220 (1972)

Physics-Informed Autoencoders with Intrinsic Differential Equations for Anomaly Detection in Industrial Processes

Marcus J. Neuer[1,2](✉), Andreas Wolff[3], and Nils Hallmanns[3]

[1] RWTH Aachen, Templergraben 55, 52062 Aachen, Germany
Marcus.Neuer@rwth-aachen.de
[2] innoRIID GmbH., Merowinger Platz 1, 40225 Duesseldorf, Germany
[3] Betriebsforschungsinstitut (BFI), Sohnstr. 69, 40699 Duesseldorf, Germany

Keywords: machine learning · differential equation · auto differentiation · anomaly detection · physics-informed

1 Introduction

Anomaly detection plays a central role in monitoring industrial processes. The common understanding is that the time series of a variable contains visible traces of anomalies - which can be selected and validated by suited algorithms. But the anomaly may not be incorporated in the functional dependency with t. In some cases, the actual anomaly could be detectable in the noise behaviour, in the standard deviation or even the moments of the stochastic process that is associated with the measurement variable.

The field of anomaly detection - sometimes also outlier detection - has therefore a longstanding scientific tradition. Linear techniques like the principal component analysis (PCA) have been used by Shyu et al. [11] in 2006 for detecting exotic failures. Sensor faults states were investigated by Kameswari et al. in [3], focusing on use cases in process industry. In this context, autoregressive neural network structures like autoencoders have been studied a while for detecting anomalies: In [1], Chen et al. present an autoencoder ensemble for outlier detection. Neuer et al. showed in [6,7] how autoencoder eigenspace analysis can help to quantify uncertainty of the anomaly prediction.

As autoencoders are essentially neural networks that are trained in an unsupervised fashion - they force the neural network to provide an identity mapping from input data to output data, propagating the data through a bottleneck - there are no labels necessary for training. From a technical perspective this is practical, as the concise form of an anomaly may not be known or has so far not been discovered.

In a celebrated review work, Karniadakis et al. [4] discussed several methods for incorporating mathematical relations into the body of learning algorithms, called physics-informed machine learning. Physics-informed neural networks, a

A. Rocha et al. (Eds.): WorldCIST 2023, LNNS 799, pp. 106–112, 2024.
https://doi.org/10.1007/978-3-031-45642-8_10

specific form of these approaches utilize prior knowledge about the application scenario to enhance their prediction capabilities. This can be as simple as calculating the optimal preparatory transformation or as sophisticated, as integrating a differential equation while training the network.

The present work shows an anomaly detection with an autoencoder that was enhanced by physics knowledge, here the abstract differential equation for including cooling behaviour. In different stages, we extend our initial concept by a continuous wavelet pre-transformation of the input temperature curves.

2 Physics-Informed Machine Learning

2.1 Transformations

Mostly, physics-information is provided to a network via transformations of the input data. Often, this step is called data enrichment. A common example is the inclusion of a Fourier transformation right to oscillatory processes. Note, normally deep learning networks can easily train out the optimum transformation on their own. But the question remains why we should not provide all knowledge we have about the process to the machine learning code? Why keep information hidden, that could have been beneficial for the training?

And in fact one central result of papers on physics-informed learning, like e.g. stated in Raissi et al. [8], is that the additional information reduces the amount of data necessary for successful training of the predictor. What is the reason for this? The neural network has to learn the optimum transformation of the data together with the actual prediction capability. If we provide the best possible representation of our data, more training capacity can be spend for the important prediction - an idea that can also be found in Hinton et al. fundamental works on this topic [2].

The right transformation was also focused by Renard et al. in [9,10]. In this approach the authors search for a fragment of the time series - the shapelet -, which represents the series the best. This shapelet is indeed a true part of the time series itself.

In our present case, we will use a transformation into the continuous wavelet space to identify a fragment of the resulting wavelet space,

$$\mathrm{CWT}[x(t)](a,b) = \frac{1}{|a|^{\frac{1}{2}}} \int_{-\infty}^{\infty} x(t)\psi\left(\frac{t-b}{a}\right) dt, \tag{1}$$

when applying a wavelet of Ricker-type,

$$\psi(t) = \frac{2}{\sqrt{3\sigma}\pi^{\frac{1}{4}}} \left[1 - \left(\frac{t}{\sigma}\right)\right] \exp\left(-\frac{t^2}{2\sigma^2}\right), \tag{2}$$

which is practically a second derivative of a gaussian kernel. Note that in (1), a and b appear as parameters, spanning the wavelet space. Figure 1 shows an illustration of the shape of this wavelet and how it differs when varying the width a.

Identifying a suited section of the wavelet space is comparable to the procedure shown in [7], where a dedicated set of eigenvalues and eigenvectors from the PCA could be found to separate the anomalous scenarios from the normal operating conditions.

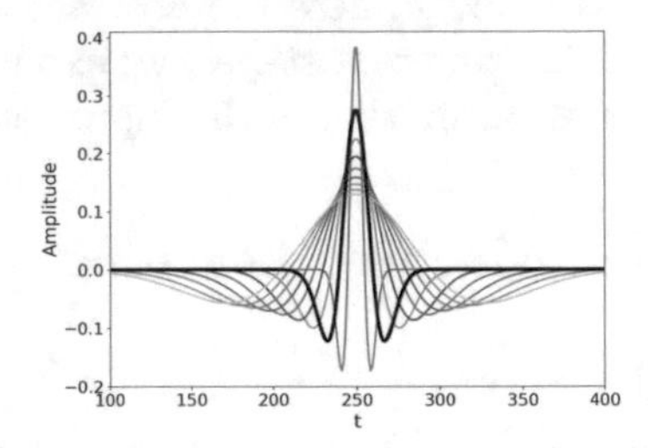

Fig. 1. Ricker wavelet illustration, varying the width of the wavelet.

2.2 Integrating Ordinary Differential Equations

Lagaris et al. showed already in their paper from 1998 [5], how a neural network can be used to integrate ordinary and partial differential equations. The main notion of this paper is, that a differential equation of the form

$$x' = f(x, t) \tag{3}$$

can be approximated by a neural network $\mathcal{N}$ by

$$\mathcal{N}(t) \approx x(t). \tag{4}$$

Consequently, this leads to

$$\mathcal{N}'(t) \approx f(x, t) \tag{5}$$

which allows us to incorporate the differential equation as such in the loss function of the network,

$$L = \sqrt{\sum_i \left(\frac{d\mathcal{N}(t_i)}{dt} - f(x, t) \right)^2}. \tag{6}$$

Lagaris stated also in [5] that the substitution

$$g(t) = x_0 + t\mathcal{N}(t) \tag{7}$$

will ensure that the intial conditions of the ODE are properly met (see $g(t = 0)$ to verify this) and that the modified loss function can be written as

$$L = \sqrt{\sum_i \left(\frac{dg(t_i)}{dt} - f(x,t) \right)^2}, \tag{8}$$

due to (6). The advantage of this approach is that the neural network training via gradient descent automatically optimises (8) and finally solves the differential equation numerically. For details regarding partial differential equations, please refer to [8] or [4] who provide more details on more involved applications.

In modern machine learning frameworks like Tensorflow, the alteration of the loss function can be done conveniently. The `tf.GradientTape` functionality allows us to implement the underlying so-called automatic differentiation directly in our autoencoder source code.

3 Autoencoder Architecture for Anomaly Detection

Figure 2 shows the chosen autoencoder structure. The input quantities T_0 and t are necessary for the automatic differentiation process in Tensorflow. In the latent space, namely to one of the latent neurons, we apply the differential equation. A decoding stage can be coupled to the latent layer depending on the training step. As activation function, we use leaky-relu, learning rate is $l = 0.0001$ and as optimisers we apply Adam and stochastic gradient descent (SGD) as provided by the Tensorflow toolchain.

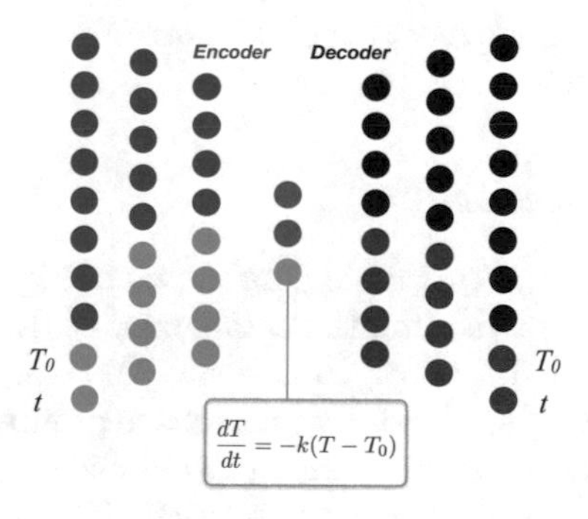

Fig. 2. Autoencoder structure with an automatic differentiation enforcing Newtons cooling law in the latent layer.

The training is cascaded like a split-step algorithm. First the differential equation part is trained, using only the lower coloured neurons of the encoder stage and the lowest neuron in the latent layer. Then an intermediate training will use the full encoding and decoding stage (overwriting parts of previously trained equation). This procedure is iteratively repeated until both gradient descents have converged.

Note that the convergence for the automatic differentiation is substantially more time consuming than the normal autoencoder training.

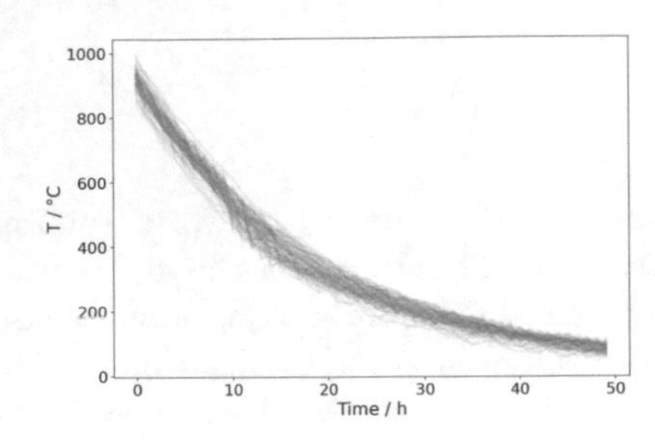

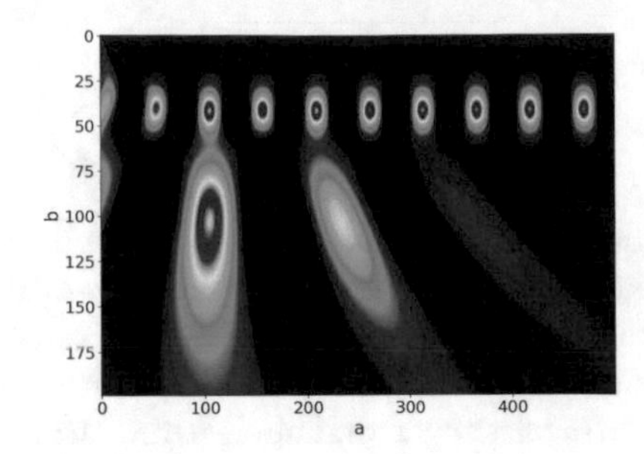

Fig. 3. Anonymised and renormalised data from cooling down a series of 480 steel coils (left). Result of the continuous wavelet transform (left), yielding the whole wavelet space including relevant structures to distinghuish different groups.

4 Results

4.1 Test Data

We apply the above described method to real world data from a steel production, namely to the cool down of coils. These coils exhibit normal behaviour and multiple classes of anomalies. 480 data sets with cool downs have been collected for investigation. Just from visually checking this data in Fig. 3 the anomalies cannot be detected. Neither can the data be grouped or classified. In the data set, there are 4 different groups of curves, all corresponding to process behaviour that we want to distinguish.

4.2 Wavelet Preprocessing

The results of the wavelet preprocessing are shown in Fig. 3. A complete image of the wavelet space is shown, yielding the complexity of the Ricker transformation kernel. In the right hand side of Fig. 3 an example cross-section of this wavelet space is plotted, now marking the different existing groups of temperature curves in individual colors. The wavelet transformation shows a better separability of the blue and red cool down functions. Nevertheless, a full separation is not visibible.

4.3 Results Due to the Inclusion of the Differential Equation

Upon inclusion of the differential equation, we evaluate the autoencoder on a test subset of data (that was not used for training). To visualise the quality of the separation we use an aggressively small architecture with 5 latent neurons. Figure 4 shows the neuron answers to the test cases, coloring the groups of different curves in blue, yellow, green and red. Note that both red and yellow are anomalies. Separation in latent neuron 1, LN1, is nearly perfect. But also 2 can be used to distinguish between green/red and yellow/blue.

Fig. 4. Result of the autoencoder with wavelet enrichment and differential equation training. For a latent space larger or equal to the dimension of 5, there is always one neuron that succeeds in fully separating the four clusters.

5 Summary

We presented the application of an autoencoder that was enhanced with two physics-informed techniques: a) modification of the input data by wavelet transformation and b) inclusion of an differential equation in the latent neuron layer. The autoencoder can be used to distinguish between four groups of cooldown curves, allowing to detect. anomalous temperature behaviour in the cooling process. The data was based on real-world measurements, that have been anonymised and renormalised. The application of the anomaly detection also works on the unmodified data sets.

References

1. Chen, J., Sathe, S., Aggarwal, C., Turaga, D.: Outlier detection with autoencoder ensembles. In: Proceedings of the 2017 SIAM International Conference on Data Mining (2017)
2. Hinton, G.E.: Learning multiple layers of representation. Trends Cogn. Sci. **11**, 428–434 (2007)
3. Kameswari, U.S., Babu, I.R.: Sensor data analysis and anomaly detection using predictive analytics for process industries. In: 2015 IEEE Workshop on Computational Intelligence: Theories, Applications and Future Directions (WCI), pp. 1–8 (2015)
4. Karniadakis, G.E., Kevrekidis, I.G., Lu, L., Perdikaris, P., Wang, S., Yang, L.: Physics-informed machine learning. Nat. Rev. Phys. **3**, 422–440 (2021)
5. Lagaris, I.E., Likas, A., Fotiadis, D.I.: Artificial neural networks for solving ordinary partial differential equations. IEEE Trans. Neural Networks **9**(5), 987–1000 (1998)
6. Neuer, M.J.: Quantifying uncertainty in physics-informed variational autoencoders for anomaly detection. In: Colla, V., Pietrosanti, C. (eds.) ESTEP 2020. AISC, vol. 1338, pp. 28–38. Springer, Cham (2021). https://doi.org/10.1007/978-3-030-69367-1_3
7. Neuer, M.J., Quick, A., George, T., Link, N.: Anomaly and causality analysis in process data streams using machine learning with specialized eigenspace topologies. In: European Steel Technology and Application Days (METEC & 4th ESTAD) (2019)

8. Raissi, M., Perdikaris, P., Karniadakis, G.E.: Physics-informed neural networks: a deep learning framework for solving forward and inverse problems involving nonlinear partial differential equations. J. Comput. Phys. **378**, 686–707 (2019)
9. Renard, X., Rifqi, M., Erray, W., Detyniecki, M.: Random-shapelet: an algorithm for fast shapelet discovery. In: IEEE International Conference on Data Science and Advanced Analytics (DSAA) (2015)
10. Renard, X., Rifqi, M., Fricout, G., Detyniecki, M.: East representation: fast discovery of discriminant temporal patterns from time series. In: ECML/PKDD Workshop on Advanced Analytics and Learning on Temporal Data (2016)
11. Shyu, M.L., Chen, S.C., Sarinnapakorn, K., Chang, L.: Principal component-based anomaly detection scheme. Stud. Comput. Intell. **9**, 19 (2006)

Big Data as a Tool for Analyzing Academic Performance in Education

Manuel Ayala-Chauvin[1], Boris Chucuri-Real[2], Pedro Escudero-Villa[1], and Jorge Buele[2(✉)]

[1] Centro de Investigaciones de Ciencias Humanas y de La Educación (CICHE), Universidad Indoamérica, Ambato 180103, Ecuador
{mayala,pedroescudero}@uti.edu.ec

[2] Carrera de Ingeniería Industrial, Facultad de Ingeniería Industria y Producción, Universidad Indoamérica, Ambato 180103, Ecuador
bchucuri@indoamerica.edu.ec, jorgebuele@uti.edu.ec

Abstract. Educational processes are constantly evolving and need upgrading according to the needs of the students. Every day an immense amount of data is generated that could be used to understand children's behavior. This research proposes using three machine learning algorithms to evaluate academic performance. After debugging and organizing the information, the respective analysis is carried out. Data from eight academic cycles (2014–2021) of an elementary school are used to train the models. The algorithms used were Random Trees, Logistic Regression, and Support Vector Machines, with an accuracy of 93.48%, 96.86%, and 97.1%, respectively. This last algorithm was used to predict the grades of a new group of students, highlighting that most students will have acceptable grades and none with a grade lower than 7/10. Thus, it can be corroborated that the daily stored data of an elementary school is sufficient to predict the academic performance of its students using computational algorithms.

Keywords: Big Data · Academic Performance · Education · Support Vector Machines

1 Introduction

The last decades have seen significant changes in teaching methods and processes that have challenged the adaptability of education [1]. The traditionalist method used in many academic institutions lacked technological tools directly influence the students. In Latin America, education was based on factors related to the performance and experience of the teacher and not on a defined methodology and support tools [2]. In third-world countries, the incorporation of technological tools is taking place slowly, and the scarcity of economic resources is one of the main factors [3]. Despite these shortcomings, adopting the online modality during the COVID-19 emergency was possible, demonstrating its usefulness in emergency contexts [4]. To enhance the teaching-learning processes, it is necessary to integrate them with robotics, machine learning, and immersive environments [5, 6], including the analysis of the information generated periodically through big

A. Rocha et al. (Eds.): WorldCIST 2023, LNNS 799, pp. 113–122, 2024.
https://doi.org/10.1007/978-3-031-45642-8_11

data. According to Gartner, using Big Data to handle a large volume of data is an innovative and cost-effective way of information processing [7]. The research results have consolidated it as a beneficial technology for decision-making and process automation in the last decade [7].

Big Data has promoted the development of new proposals within the field of education, enhancing educational management with efficient use of data. In Ecuador, the analysis of this information is a recent topic since the literature review showed studies from 2017. Urena-Torres analyzes the information on undergraduate enrollments at the Universidad Técnica Particular de Loja (UTPL) [8]. In contrast, Baldeon Egas evaluate students' qualifications at the Universidad Tecnológica Israel, the study modalities, and the trends in the future [9]. Tejedor, S analyzes journalism courses at six universities in Ecuador, Colombia, and Spain to propose curricular changes [10], and Villegas-Ch evaluates the student satisfaction from an Ecuadorian university proposing the information use for future decision-making [11]. Finally, the most recent work uses machine learning algorithms to evaluate the variables influencing mathematics performance in senior high school students [12].

Most of the research in Ecuador is carried out in higher-level institutions since secondary-level institutions do not store information. Therefore, there is a waste of information that does not generate knowledge. Nevertheless, it is a widespread practice, especially in developing countries that need standards and guidelines to manage this information [2]. Many Big Data tools allow for predicting student behavior based on information obtained in the past [12]. This is used for the approach of new strategies to optimize current processes. Therefore, the purpose of this study was to analyze the academic performance of a secondary educational unit for which information was processed using data filtering techniques. A predictive model was proposed using Python software to evaluate academic performance through predictions. As a working hypothesis using Big Data will allow the identification of academic performance, being a precedent for future research.

This paper contains four sections, including the introduction in Sect. 1. Section 2 describes the methods and materials and Sect. 3 the results. The conclusions are shown in Sect. 4.

2 Methods and Materials

This research seeks to solve a practical problem that has been identified in the field of education through the exploration of current studies reported in Ecuador. In this context, it was necessary to directly approach an educational institution to obtain information on academic performance. The data used in this article were collected at the "Isabel la Católica" Basic Education School, located in Píllaro town in the province of Tungurahua in Ecuador. We work with the registers from the total population of students. The information of 2076 students in 8 different academic cycles were collected and filtered.

2.1 Obtaining and Processing Information

All physically and digitally documented information that may be useful (grades and academic averages by cycles, socioeconomic records, and others) were requested in writing from the institution's management. Free access to the internal database was also requested. The information was digitalized and stored in calculus sheets. During the information processing, classification of it was made according to the average score, classroom identification, level, academic cycle, and subject [13]. On the other hand, it was possible to have access to the files of each student where their personal and family information was detailed. Integrating the data, it was possible to determine a suitable organization (after data purification) for the implementation of the predictive model with the Python tool. This can be seen in Fig. 1.

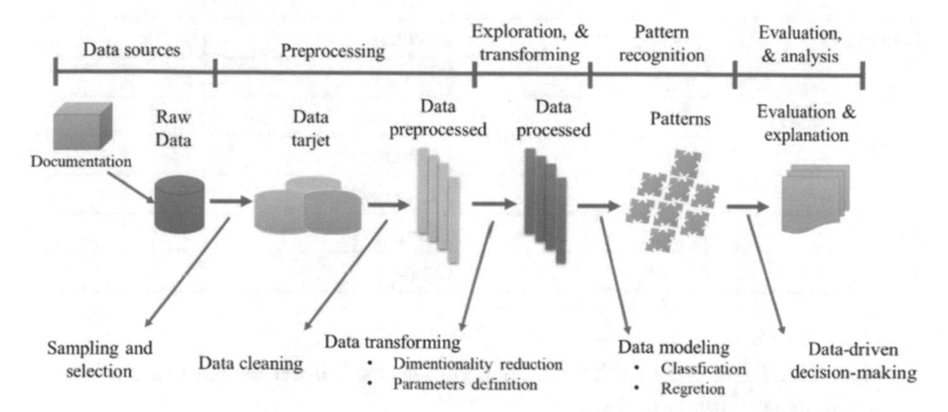

Fig. 1. Data collection and analysis process.

2.2 Identification and Organization of Data

Data from 3 student cuts were taken, where the data from the initial periods are found in promotions; 2013–2014, 2014–2015, 2015–2016, and end in promotions; 2018–2019, 2019–2020, and 2020–2021 respectively. Table 1 describes the coding used for the treatment of socio-demographic information. The sector of the city where the student lives, the family economic level, the type of housing, and their sex were considered.

All this information is organized within a matrix, in which only the grades in each subject and the previously assigned codes of the sociodemographic variables were placed. The start and end periods assigned by the variables AI and AF, respectively were also placed, and the degree of study (GE). Python version 3.10.5 software was used for the information processing; this open-source tool is commonly used to solve problems in a simple and efficient way due to its versatility. Python has a close relationship with scikit-learn or machine learning module that is easily complemented by its interface. All this facilitates its use for the development of statistical and data analysis algorithms. For the visualization of the code Jupyter was used; an application that facilitates the interaction of work linked to projects related to analysis, statistics, and technological innovation. It

Table 1. Coding of the variables.

Group	Abreviation	Subgrup	Code
Adress	DIR	North	0
		South	1
		East	2
		West	3
Economy	ECO	Regular	0
		Good	1
		Very good	2
		Excellent	3
Housing	VIVI	Owned	0
		Rented	1
		Borrowed	2
		Antichresis	3
		With loans	4
Gender	SEX	Female	0
		Male	1

has a user-friendly interface and contains complete work libraries that make it ideal for the development of this proposal.

2.3 Data Analysis

In Python, a preparation stage was carried out that begins with the analysis of the behavior of the data for each student through diagrams and histograms. Figure 2 shows the data of the participants of a school year as a practical example of what was obtained in each variable. The histogram shows the mathematics grades for the 2018 period where is displayed the range of grades obtained by students before the COVID pandemic.

Likewise, the behavior of the data was studied with the realization of various diagrams by subject based on the students' grades during the academic periods 2014 - 2021. Figure 3 shows the mathematics grades for these periods.

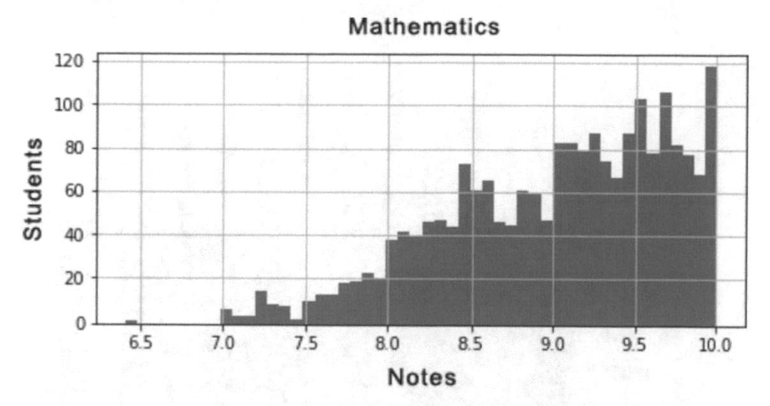

Fig. 2. Histogram of math scores

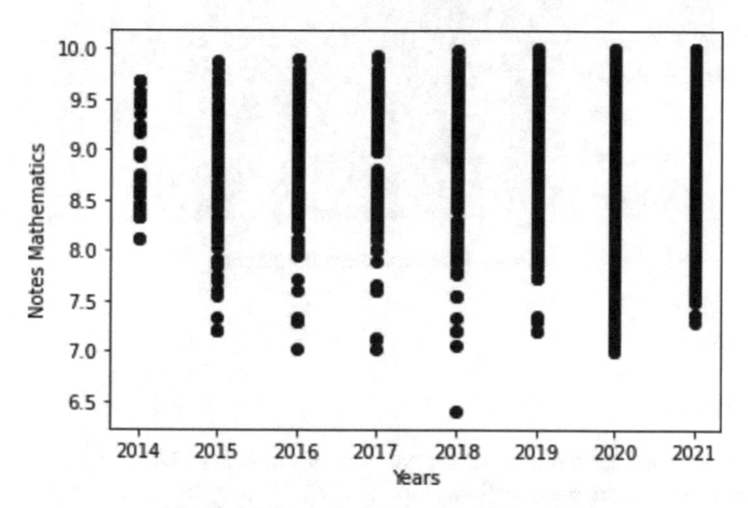

Fig. 3. Mathematics grades for 2014 - 2021

2.4 Correlation Between Attributes

Using the Pearson statistic, the correlation between variables was determined using parameters such as the covariance between students. The relationship between sociodemographic variables such as gender, address, economy, and housing were low or almost nil. Between the qualifications of the various subjects, we found a positive correlation; between natural and social sciences was quite strong. In Fig. 4 the respective correlation diagram is presented, showing the relationships using colors.

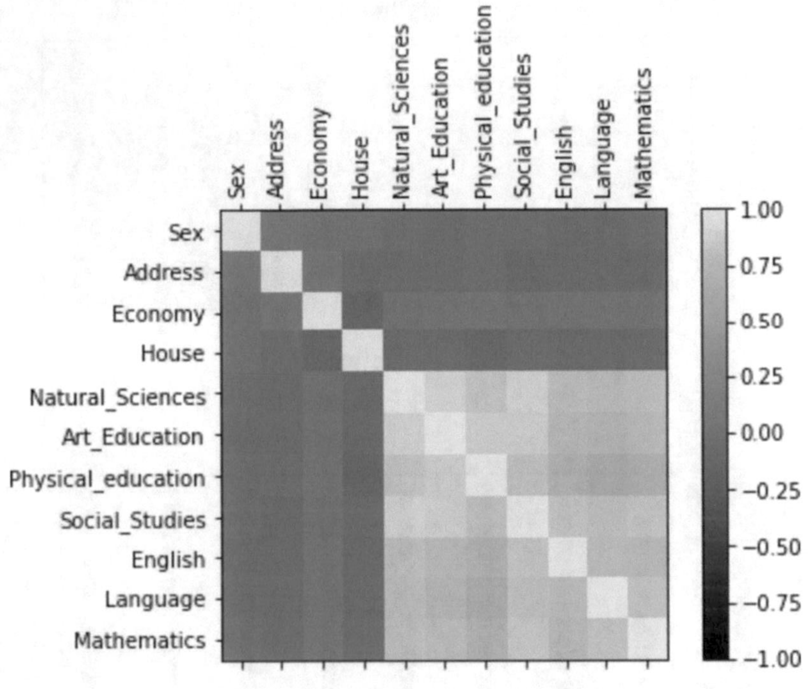

Fig. 4. Correlation matrix diagram

3 Results

3.1 Classifiers

Based on the information obtained, support vector machines (SVM), random trees, and logistic regression were used as classifiers. In each of them, the percentage of precision and error in the prediction of academic performance was analyzed. For the training of the classifiers, 99.8% of the total data was used, obtaining a confusion matrix for each classifier. In addition, classification parameters have been established in each column, considering academic performance. The defined ranges are: "regular" from 7 to 7.99 points, "good" for 8 - 8.99, "very good" for 9 - 9.99, and "excellent" for those with 10 points.

Support Vector Machines. The algorithm was trained with 99.8% of the data, and with the remaining data, the respective prediction was made, resulting in the confusion matrix presented in Fig. 5. As a result, there were 147 data with a prediction of "good" academic performance and 4 data considered "very good" that were wrong and should be "good" (false negatives). Additionally, based on the result of the confusion matrix yields an accuracy of 97.1% in training data prediction.

Random Trees. The confusion matrix generated in Fig. 6 shows an accuracy of 96.86%. It was obtained that 139 students predict being within a "good" academic performance group, and 4 are false positives.

```
Confusion_matrix
[[148   0   2   0]
 [  0   1   4   0]
 [  3   2 235   0]
 [  1   0   0  18]]
```

```
Accuracy
0.9710144927536232
```

Fig. 5. Confusion matrix with Support Vector Machines for academic performance.

```
Confusion_matrix
[[144   0   4   2]
 [  0   0   5   0]
 [  8   0 232   0]
 [  8   0   0  11]]
```

```
Accuracy
0.9347826086956522
```

Fig. 6. Confusion matrix with Random trees of academic performance.

Logistic Regression. The confusion matrix generated in Fig. 7 shows an accuracy of 95.89% and an error of 4.35%. It was obtained that 139 students predict being within a "good" academic performance group, and 4 are false positives.

```
Confusion_matrix
[[149   0   1   0]
 [  0   0   5   0]
 [  5   1 234   0]
 [  1   0   0  18]]
```

```
Accuracy
0.9685990338164251
```

Fig. 7. Confusion matrix with Logistic regression for academic performance.

The findings of this study show that thanks to the use of tools related to big data, results related to academic performance can be predicted, improving part of educational management, and optimizing resources in this process. On the other hand, when analyzing the three predictive models proposed, it was possible to understand that the use of the model with Support Vector Machines presents greater precision in the prediction with a much lower percentage of error (Fig. 8a). For this reason, a prediction with test data was proposed considering new data in a group of 32 students in order to evaluate the proposed model (Fig. 8b).

Finally, obtaining the prediction results for the academic performance, each parameter was established in percentage values for the academic performance, values illustrated in Fig. 9.

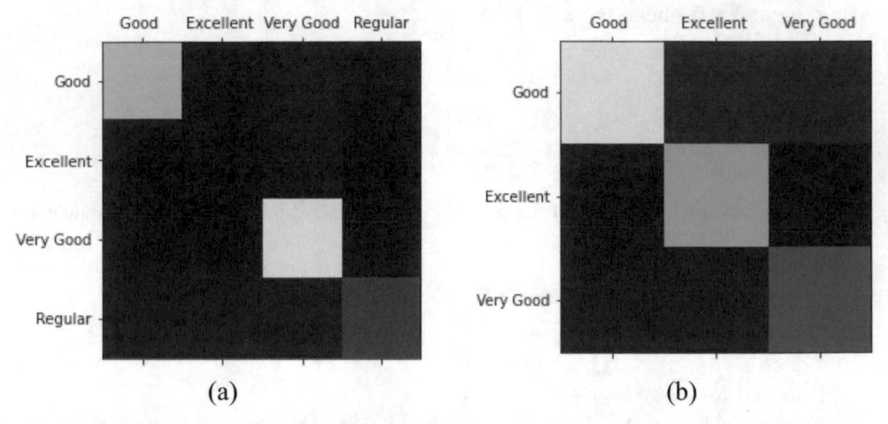

Fig. 8. Support Vector Machines: (a) Confusion Diagram. (b) Confusion matrix after training on academic performance.

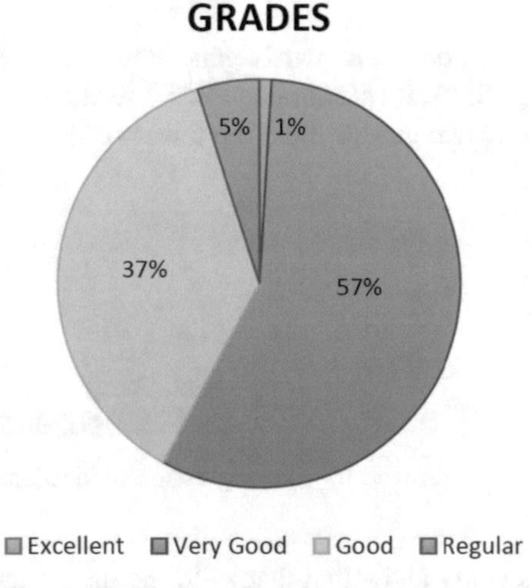

Fig. 9. Prediction of academic performance.

4 Conclusions

Ecuador's academic system is based on outdated learning models and processes that may need to be revised for students. The ineffectiveness of government entities is reflected in poor academic performance and low grades in regional evaluations. To this must be added school dropout and poverty in society, making the relation-ship between teacher and student more expensive. As mentioned in [14], education is linked to technology, generating essential changes to improve the student experience. In Ecuador, important advances are already being made in data mining and big data to analyze academic performance better. Based on this, the predictive model presented precise results that show students' behavior and allow making decisions in the long term faster and more efficiently. As was seen in the literature review, in Ecuador, most of the research has been conducted on higher education students, where there is a more significant amount of information. In our study, we worked with data from an elementary school, fostering the basis for a process of technological evolution and adaptability. The analysis significantly improved the school's database because it used the information, they did not consider necessary. The prediction's quantitative analysis of academic performance will provide a broader picture to the school's administration and faculty.

This analysis will reduce the use of physical and energy resources, avoiding local environmental degradation and contributing to environmental protection. It allows for optimizing teachers' time, who will no longer need to hold several meetings to make decisions. In addition, the use of infrastructure and energy consumption within the institution is reduced by face-to-face or virtual meetings. The use of physical documentation is also reduced, which contributes to saving resources and protecting the environment. With this predictive model, it is possible to generate a plan to reinforce students' knowledge with low academic performance. In addition, they can be grouped according to their needs to create group tutorials to optimize the teaching-learning processes.

The main limitation was that the socioeconomic information of the students was limited. In future work, we plan to use teacher training and experience data to find new patterns related to academic performance and improve the model's prediction.

Acknowledgments. Universidad Indoamérica for its support and financing under the "Big data analysis and its impact on society, education and industry" project.

References

1. Ruiz, T.R., Nunez, G.: Characterization of enrollment in higher education in Ecuador. In: 2022 IEEE Sixth Ecuador Technical Chapters Meeting (ETCM), pp. 1–4. Institute of Electrical and Electronics Engineers (IEEE), Quito (2022). https://doi.org/10.1109/etcm56276.2022.9935738
2. Suárez Monzón, N., Gómez Suárez, V., Lara Paredes, D.G.: Is my opinion important in evaluating lecturers? Students' perceptions of student evaluations of teaching (SET) and their relationship to SET scores. Educ. Res. Eval. **27**, 117–140 (2022). https://doi.org/10.1080/13803611.2021.2022318
3. Monzón, N.S., Jadán-Guerrero, J., Mesa, M.L.C., Andrade, M.V.A.: Digital transformation of education: technology strengthens creative methodological productions in Master's programs in education. In: Russo, D., Ahram, T., Karwowski, W., Di Bucchianico, G., Taiar, R. (eds.) Intelligent Human Systems Integration 2021: Proceedings of the 4th International Conference on Intelligent Human Systems Integration (IHSI 2021): Integrating People and Intelligent Systems, February 22-24, 2021, Palermo, Italy, pp. 663–668. Springer, Cham (2021). https://doi.org/10.1007/978-3-030-68017-6_98
4. Arias-Flores, H., Guadalupe-Lanas, J., Pérez-Vega, D., Artola-Jarrín, V., Cruz-Cárdenas, J.: Emotional state of teachers and university administrative staff in the return to face-to-face mode. Behav. Sci. **12**(11), 420 (2022). https://doi.org/10.3390/bs12110420
5. Buele, J., López, V.M., Varela-Aldás, J., Soria, A., Palacios-Navarro, G.: Virtual environment application that complements the treatment of dyslexia (VEATD) in children. In: Rocha, Á., Ferrás, C., Marin, C.E.M., García, V.H.M. (eds.) Information Technology and Systems: Proceedings of ICITS 2020, pp. 330–339. Springer, Cham (2020). https://doi.org/10.1007/978-3-030-40690-5_33

6. Ortiz, J.S., Palacios-Navarro, G., Andaluz, V.H., Guevara, B.S.: Virtual reality-based framework to simulate control algorithms for robotic assistance and rehabilitation tasks through a standing wheelchair. Sensors. **21**, 5083 (2021). https://doi.org/10.3390/s21155083
7. Gärtner, B., Hiebl, M.R.W.: Issues with big data. In: Quinn, M., Strauss, E. (eds.) The Routledge Companion to Accounting Information Systems, pp. 161–172. Routledge (2017). https://doi.org/10.4324/9781315647210-13
8. Urena-Torres, J.P., Tenesaca-Luna, G.A., Arciniegas, M.B.M.: Analysis and processing of academic data from a higher institution with tools for big data. In: Iberian Conference on Information Systems and Technologies, CISTI. IEEE Computer Society, Lisbon (2017). https://doi.org/10.23919/CISTI.2017.7975822
9. Egas, P.F.B., Saltos, M.A.G., Toasa, R.: Application of data mining and data visualization in strategic management data at Israel technological university of ecuador. In: Botto-Tobar, M., León-Acurio, J., Cadena, A.D., Díaz, P.M. (eds.) Advances in Emerging Trends and Technologies: Volume 1, pp. 419–431. Springer, Cham (2020). https://doi.org/10.1007/978-3-030-32022-5_39
10. Tejedor, S., Ventin, A., Martinez, F., Tusa, F.: Emerging lines in the teaching of university communication the inclusion of data journalism in universities in Spain, Colombia and Ecuador. In: Iberian Conference on Information Systems and Technologies, CISTI. IEEE Computer Society, Seville (2020). https://doi.org/10.23919/CISTI49556.2020.9141013
11. Villegas-Ch, W., Roman-Cañizares, M., Jaramillo-Alcázar, A., Palacios-Pacheco, X.: Data analysis as a tool for the application of adaptive learning in a university environment. Appl. Sci. **10**, 1–19 (2020). https://doi.org/10.3390/app10207016
12. Espinosa-Pinos, C.A., Ayala-Chauvín, I., Buele, J.: Predicting academic performance in mathematics using machine learning algorithms. In: Valencia-García, R., Bucaram-Leverone, M., Del Cioppo-Morstadt, J., Vera-Lucio, N., Jácome-Murillo, E. (eds.) Technologies and Innovation: 8th International Conference, CITI 2022, Guayaquil, Ecuador, November 14–17, 2022, Proceedings, pp. 15–29. Springer, Cham (2022). https://doi.org/10.1007/978-3-031-19961-5_2
13. García-Magariño, I., Bedia, M.G., Palacios-Navarro, G.: FAMAP: a framework for developing m-health apps. In: Rocha, Á., Adeli, H., Reis, L.P., Costanzo, S. (eds.) Trends and Advances in Information Systems and Technologies: Volume 1, pp. 850–859. Springer, Cham (2018). https://doi.org/10.1007/978-3-319-77703-0_83
14. Núñez-Naranjo, A.F., Ayala-Chauvin, M., Riba-Sanmartí, G.: Prediction of University dropout using machine learning. In: Rocha, Á., Ferrás, C., López-López, P.C., Guarda, T. (eds.) Information Technology and Systems: ICITS 2021, Volume 1, pp. 396–406. Springer, Cham (2021). https://doi.org/10.1007/978-3-030-68285-9_38

Ethics, Computers and Security

Phishing and Human Error. Beyond Training and Simulations

Jeimy J. Cano M.(✉)

Facultad de Derecho, Universidad de los Andes, Cra 1E No.18A-10, Bogotá, Colombia
jjcano@yahoo.com

Abstract. Phishing is catalogued in the literature as one of the most common and at the same time most effective attacks at the time of realizing an adverse event in an organization. It is an action based on intelligence, deception and distraction that focuses on individuals to motivate them to take actions that end up with the compromise of their sensitive data. In this sense, companies have focused their efforts to prevent this type of attacks based on training and simulations, understanding phishing as a result contrary to what is expected by the corporation, which has ended up being ineffective. Consequently, this article proposes the understanding of phishing as a consequence, where the context and everything that involves the individual is what is relevant to explain the unexpected behavior, following the theory of human error that, without releasing the person from responsibility for their actions, understands and highlights aspects that go beyond the regular and standard training, to motivate a vigilant and more resistant to deception posture.

Keywords: Phishing · Human Error · Training · Context

1 Introduction

There are many international studies and reports that insist and detail that 95% of security failures or breaches are due to "human errors" [1]. "Errors" that are located in behaviors or actions of people in particular scenarios and specific contexts, which the reports do not comment or detail in this regard. In this sense, the term "human errors" only provides information or basic statistics about something that an individual performed that was not in accordance with what was expected by the organization and whose consequence was configured in a breach or materialization of a risk in a company.

In this sense, if we want to go deeper into this statistic, it is necessary to explore a different path to what is currently known and find in the essence of the word "error" a possible explanation for the behavior of a person at a particular time. While traditional strategies for dealing with deception in the digital context (where phishing is located) insist on training and raising people's awareness, little has been explored about what surrounds the person and the understanding of why previously established controls fail to prevent the materialization of deception [2].

Understanding the "error" not as an outcome, but as a consequence, establishes the basis of the strategy to better explore phishing, that reliable deception that invites or

A. Rocha et al. (Eds.): WorldCIST 2023, LNNS 799, pp. 125–133, 2024.
https://doi.org/10.1007/978-3-031-45642-8_12

motivates a person to disclose personal, sensitive or financial information, which ends up exposing the individual to the effects of a malicious code that can generate irreparable damage to their information or that of their company [3].

If we understand phishing as a consequence of a series of previous events that occur around the person, it is viable to investigate different alternatives on its treatment, since it is not only the final action that ends in the breach, but the whole previous process that takes place and the failures of the controls so that the deception can be effectively carried out. In short, to place the analysis not in the final effects that are taken, but in all the previous intelligence that the adversary advances and the exploitability that is derived by placing the deception in a reliable and known area for individuals.

Phishing is the most efficient, attractive and agile attack vector an attacker has, if he manages to pull off a reliable deception (combined with a credible distraction) to get through the critical judgment and reasonable doubts that a trained and educated person has to resist this type of strategies developed by attackers [4].

Therefore, this article aims to propose unconventional alternatives to making people more resistant to the reliable deception proposed by adversaries, without prejudice to the fact that sooner or later the aggressor will succeed and it will be necessary to act in a coherent, coordinated and committed way to reduce the impact and collateral effects that a situation like this can cause in the organization. To this end, this document begins with an understanding of phishing, then analyzes why we are prone to failure, to focus on the explanation of phishing from the theory of human error and finally to offer some conclusions.

2 What is Phishing?

A review of the extensive literature available on the subject reveals some general agreements on its definition. The concept is associated with the practice of fishing, where an individual tries to place bait on a hook that is attractive and motivating so that the fish "bites", hooks itself, and then catches it. In this sense, this work of the fisherman, which requires patience, time and opportunity to achieve its objective, is replicated in the digital context to effectively carry out a deception.

Phishing is based on three fundamental elements: intelligence, deception and distraction. Intelligence as a product resulting from the evaluation, integration, analysis and interpretation of the information gathered by the adversary about his target [5]. It is the informed framework that the attacker develops in order to have the necessary data to shape his attack in the most effective way possible. This exercise can take the aggressor months or sometimes years, depending on the motivations and objectives that have been outlined.

With the concrete result of the intelligence, the deception is configured, the strategy that allows to simulate a credible and properly documented aspect of the victim's reality that is sufficiently persuasive and valid to motivate a concrete action. It is the moment of the design parallel to everyday life that places the person in a reliable and safe zone to act and establish a relationship with the aggressor agent, without the latter being aware of it. The invisibility cloak is achieved thanks to efficient prior intelligence work.

Once the artifice is created, we proceed to develop the distraction, that moment of loss of attention and limited critical judgment of the person, created by the reliability of the environment and previous conditions elaborated around the context of falsehood, to advance in the development of the deception strategy that leads to the compromise of the information of the person or groups of people targeted by this exercise.

In this sense, a phishing attack is a planned and detailed process that seeks to persuade and conquer the victim to perform a specific behavior that generally leads the individual to hand over personal, financial or sensitive data to an unidentified third party. This process has a series of basic steps that are presented below [6]:

- Intelligence: Once the attacker selects the victim or target group, he establishes the strategy to collect reliable information about the person or group of people who are of interest, or who are in the imagination of the potential victims, so that when the deception is carried out, it becomes a natural part of their context and the actions to be taken.
- Fabricating the story: Establishing the strategy to capture the victim's attention based on the previous intelligence available. This exercise requires precision and timing to best situate the deception in the imagination of the person or group of people in order to generate the required momentum and action through a known and relevant discourse for the dynamics of potential victims.
- Email design: The illusion is constructed based on the appearance of the email (e.g., email address structure, subject header and content), to give the appearance of legitimate communication by using institution logos, terminology and message tone to create authenticity in the victim's mind.
- Deployment pivot: Establish the reliable generator domain and the most appropriate time to send the designed message to the email accounts selected to carry out the deception. This step must take care of technical details in the distribution to overcome the possible technological controls available in the organizations.
- Spoofed website: Once the user selects/clicks on the hyperlink included in the email message, he/she is directed to a spoofed website that also looks authentic and legitimate in its design, and subsequently the victims' personal data is unsuspectingly captured.

Once this process is carefully deployed and planned, the attacker, like the fisherman, proceeds to hope that his deception strategy will be successful and bear the expected fruits. The information gathered from the victims will be the loot that the attacker will then use to advance to the next stage of his plans. He may sell the information obtained, create new scams or extortions, or act as an intermediary for other criminals to carry out more elaborate and costly attacks such as ransomware.

Phishers have a different motivational agenda than other adversaries. These criminal profiles generally carry out their actions by [7]:

- Financial benefits: *phishers* can use stolen banking credentials for financial gain.
- Identity concealment: *phishers* sell identities to others who may be criminals looking for ways to hide their identities and activities.
- Fame and notoriety: *phishers* may target victims for peer recognition.

In general, *phishers* develop four types of attacks, using at least five specific methods detailed below in Tables 1 and 2.

Table 1. Phishing attacks

Attacks	Description
Spear phishing	Targeted to specific users of the organization
Whaling	Targeted to senior executives (CEO, CFE, CRO, etc.)
Non-selective phishing	Distribution of phishing emails to as many employees as possible, hoping to catch at least one
Business Email Compromise (BEC)	A phishing attack in which the email appears to come from the company, usually from management or someone in authority

Source [8]:

Table 2. Phishing methods

Methods	Description
Impersonation	The attacker poses as a trusted sender by spoofing a domain or email address
Filter evasion	The attacker uses images instead of text in the body of the email, so anti-phishing filters are unable to detect suspicious text
URL Malicious	The email includes a link that leads to a malicious website that may contain malware or is used to capture credentials
Embedded malware	The email may contain a file or attachment containing malware
Malicious pop-up window	The attacker places a pop-up window on top of a legitimate web page that prompts the user to enter his credentials

Source [8]:

3 Human Error Theory. Why Are We Prone to Error?

Error is a natural condition of the human being. The individual remains all the time in gray work and in the process of learning, which requires a permanent improvement from the exercise of their daily relationships. In this sense, people's behavior is situated from the perception of their environment, the relationships they build, the social dynamics and their personal inclinations, which configure a personal development that crosses their human logics and the decisions they make in a social scenario.

Beyond this transcendent reading, science has been studying for many years the issue of "error" as a fundamental element to make the distinction of safety in the industrial context. That is, in the scenario of the operational discipline, of the performance of people in factory environments or production lines where different events can lead an individual to generate accidents or injuries that end up affecting their welfare and the operating model of the companies [9]. It is no coincidence that in these environments "zero accidents" is spoken of as a mantra that is constantly repeated to remind that a person's life is the most important thing and therefore, it is necessary to "take care of me, to take care of you".

In this scenario, the literature establishes at least four categories of human error. These categories emphasize human actions and their challenges, rather than the consequences derived from them, since it is natural for the individual to make mistakes and it would be very easy to assign responsibility without further analysis. The categories are [9]:

- Error - People don't know what to do, or they think they do, and they are wrong.
- Violation - The person knows what to do, but chooses not to do it.
- Mismatch - The activity does not match the person's ability, or the person knows what needs to be done, and intends to do it, but the activity is beyond his or her physical or mental ability.
- Inattention - The person knows what to do, intend to do it and is capable of doing it, but does not do it or does it incorrectly.

When looking at these categories, it can be seen that training or coaching is not always the right answer, since there are different flavors of "error" that require an in-depth review of the person and his context, to understand what led to the behavior and not just make him feel bad, or point out that he performed an action that ended in an undesired event with implications for both the person and the organization.

When a person configures an "error", he/she is going through an internal conceptual structure that starts with knowledge, is tuned with will, mobilized with desire and nourished by the dynamics of the context where he/she works, to create an amalgam of feelings and actions that are unique in time, mode and place that end in a condition not expected or different from what was expected by the one who evaluates the action in the light of a standard, good practice or expected result [10].

Following Reason's reflections [11], people can be exposed to at least two types of conditions that lead to unforeseen situations in the development of their activities: active errors and latent conditions. While active errors are presented by direct exposure to the activity or action with immediate consequences, latent conditions are associated with activities that are distanced in time and space from the operative action, which generates consequences that remain latent in the system until an adverse event is triggered. Therefore, the more distant the individuals are from the direct activity, the greater their potential for damage and the possible generation of domino effects depending on how coupled and interconnected their components are.

4 Phishing. A Human Error in the Digital Context

Deception in the digital context is the most natural tendency and the strategy most used by adversaries to achieve their objectives. To the extent that deception is credible and reliable, it can advance their purposes and create the instability and uncertainty that characterizes their actions. Deception in the digital context is increasingly more elaborate and credible as access to information by adversaries is abundant and open, without prejudice to the partners they may have to validate and complement the data they have available at the time [12].

If we understand phishing as a consequence of a series of considerations and elements around the individual, it is possible to consider different strategies to make the individual more resistant to this phenomenon [13]. If a reading of phishing is made from the categories of error previously detailed (see Table 3), some lines of analysis can be observed that can suggest treatment alternatives that connect the action of the individual and the posture of support and accompaniment that is required, rather than one that only advises and insists on one or another behavior.

Table 3. Phishing as human error

Category	Phishing	Foundation for success
Error	Person thinks he/she knows the scope of deception and believes he/she knows how to identify and act in such a situation	Overconfidence / Poor critical judgment
Violation	It is a thoughtful and conscious individual decision that makes the difference in the final action	Failure of technology mechanism / Context conditions
Mismatch	Limited ability of the individual to recognize the different types and evolution of deception	Recognition of known patterns / General knowledge of deception
Slips and inattention	A digital reality in the midst of social dynamics that creates distractions and deceptions for individuals	Distractions / fatigue / stress

Source: Own elaboration

If phishing is understood from the category of error, it warns a person who has received information, but not training on the subject, who thinks he knows the scope of deception and possibly believes he knows how to identify and act in a situation like this. This scenario generates a feeling of excessive confidence that weakens the person's critical judgment, creating a zone of cognitive opacity that is exploited by the attacker's reliable intelligence to reliably elaborate the deception. Faced with this issue, the challenge is to always keep those who claim to "know and be prepared" out of the comfort zone in order to generate permanent breaks in their daily lives that allow them to remain vigilant and active in order to respond.

This initial condition of error is configured by a series of latent conditions that are created around the individual, which, being distant in time and space, create a zone of cognitive invisibility in the individual's performance that leads to a feeling of confidence that ends in an adverse event, in this case the materialization of the previously designed deception.

If phishing is seen as a violation, a meditated and conscious individual decision is what makes the difference in the final action. In this line, the intentionality of the individual is what sets the tone of the analysis, and it is there where technological tools can validate the behavior of a person and establishes what contextual conditions are those that enable and trigger an action against the previous knowledge that one has. The aforementioned, in social and cognitive perspective, is known to have a harmful effect on their physical and mental integrity, with adverse consequences that end up with calls for attention or increased supervision. Therefore, the strategy is to simplify the action to be executed, increase supervision and accompany the individual from their particular reality and dynamics.

In this reading as a violation, the issue of active error is noticed, translated into a direct exposure to the adverse event, in which the behavior that is executed has direct consequences on the individual, which leads to a subsequent reflection on what has happened and the recognition of the action taken whose effects are not only for him, but for all those with whom he is in relation and with the company.

If phishing is understood as a mismatch, a different reality of treatment is configured given the need to expand and deepen the individual's ability to recognize the different types and evolution of deception. As previously indicated, the accelerated transformation of phishing, such as "voice phishing", highlights the limited knowledge and little skill developed by an individual to date, which necessarily implies an update of the techniques of deception and the high exploitability of this reality. The vulnerability is not in the lack of training, nor in the cognitive capacity of the person, but in the creation of permanent learning patterns that lead him to question frequently what he does not recognize as known or reliable. To deal with phishing in this context is to enable learning windows that test and challenge individuals' prior knowledge so that they develop patterns of action that conform not only to what they know, but to know and modified contexts that may be unusual.

This type of error based on mismatch is again connected with latent conditions, now in the training and lack of updating of the person in the face of the evolution of deception. To the extent that people distance themselves from this reality, they do not continue their training and create gray zones in their behavior patterns that, although they respond to previous conditions of knowledge, do not adjust to the current challenges of attackers and their ability to produce more sophisticated deceptions.

If we review phishing as a lack of attention, we recognize a digital reality and social dynamics that creates any number of distractions for individuals, where in the same instant there may be different activities in progress that permanently change the focus of concentration. This context, added to the natural pressures of the work activity and the need to achieve results, and if we add to this, personal and family concerns, an ideal scenario is configured for the intelligence performed by the adversary to bear fruit with the deception he has designed. To deal with phishing from this perspective, it is necessary to reduce moments of stress, redesign the activity to have fewer distractions and enable disconnection practices that allow the individual to regain control, taking him out of the inertia of the business dynamics.

Inattention is directly associated with active error. It is the right moment of the concrete action where the planned activity is performed outside the parameters due to different circumstances. Lack of concentration, lack of care and many times distractions create the natural scenarios for people to end up clicking on the link that arrives by mail, knowing and knowing in advance the characteristics of the deception and the harmful effects of this action on mail messages clearly with phishing characteristics.

As can be seen, phishing treated as a human error offers a differentiated treatment palette, which leads security and control managers to study the specific reality of individuals to develop a differentiated exploitability profile [14], which allows taking specific actions that make people less susceptible to this deception, and why not, create a credible and concrete deterrent effect that leads adversaries to change or update their techniques which must necessarily adjust the security posture and the development of the organizational culture of information security.

5 Conclusions

To date, complete frameworks called "anti-phishing" have been reported in the literature that establish a treatment strategy covering technological controls, organizational aspects and human factors, which rely on traditional positions that end in the application of standards and good practices that understand the unexpected behavior of the person as an outcome and not as a consequence [6].

As phishing techniques move towards less perceptible, more simulated and covert attacks, the "known recipes" begin to lose their effectiveness as they are located in those known and basic elements of digital deception that a person is expected to recognize [15]. Therefore, by understanding phishing as a consequence, it is possible to explore alternatives from the categories of error, which generate patterns of behavior and learning, which are installed in the cognitive reflection of the individual, not as a call to execution by insistence, but to the vigilant posture by experience, which is none other than that which is accompanied by three realities:

- Sooner or later you are going to fall for the deception.
- There will be negative consequences that will need to be addressed and managed.
- Confusion will be present to distract and act erratically.

Faced with these three realities, security and control professionals should change their position to one of accompaniment and guidance, which seeks to reduce or limit the negative consequences of the behavior carried out, and above all to be present and supportive in the face of the uncertainty that is unleashed on the person, that does not stigmatize or re-victimize the person, but helps him to create a zone of stability and confidence that brings him back to reality and makes him part of business resilience that quickly absorbs the adverse event, learns from it and bounces back quickly to continue its operation [16].

Phishing, more than a threat against security or corporate cybersecurity, is a rational and mental threat that takes advantage of cognitive vulnerabilities, which are located in social vulnerabilities, inherent to the dynamics of the context and which are enabled and enhanced by technological vulnerabilities. In this sense, the capacity for influence, distraction and deception by the adversary will always be open to as many possibilities as there are variants in the convergence of these three vulnerabilities [17].

So, the response to this constantly evolving threat is not always training or simulations, which reveal personal weakness or exposure that create direct or indirect finger-pointing (which may end up with effects contrary to those expected) for those who end up "clicking", but a differentiated treatment posture and vision that makes a difference when it is possible to place some category of error in a specific situation or dynamics of the organization.

References

1. WEF: The Global Risk Report 2022. World Economic Forum (2022). https://www3.weforum.org/docs/WEF_The_Global_Risks_Report_2022.pdf
2. Reason, J.: Human error: models and management. BMJ **320**(7237), 768–70 (2000). https://doi.org/10.1136/bmj.320.7237.768. https://www.ncbi.nlm.nih.gov/pmc/articles/PMC1117770/
3. Alshammari, Z.: Phishing attacks cybersecurity. ISSA Journal (2022)
4. Alkhalil, Z., Hewage, C., Nawaf, L., Khan, I.: Phishing attacks: a recent comprehensive study and a new anatomy. Front. Comput. Sci. **3**, 563060 (2021). https://doi.org/10.3389/fcomp.2021.563060
5. Jiménez, F.: Handbook of Intelligence and Counterintelligence. Third Edition. Seville, Spain: CISDE Editorial (2019)
6. Frauenstein, E.D., von Solms, R.: An enterprise anti-phishing framework. In: Dodge, R.C., Futcher, L. (eds) Information Assurance and Security Education and Training. WISE WISE 2013 2013 2011 2009. IFIP Advances in Information and Communication Technology, vol 406. Springer, Berlin, Heidelberg (2013). https://doi.org/10.1007/978-3-642-39377-8_22
7. Yu, W.D., Nargundkar, S., Tiruthani, N.: A phishing vulnerability analysis of web based systems. In: Proceedings of the 13th IEEE Symposium on Computers and Communications (ISCC 2008), pp. 326–331. IEEE, Marrakech, Morocco (2008). https://doi.org/10.1109/ISCC.2008.4625681
8. Blankenship, J., O'Malley, C.: Best Practices: Phishing Prevention. Protect Against Email-Borne Threats With Forrester's Layered Approach. Forrester Research (2019). https://www.proofpoint.com/au/resources/threat-reports/forrester-best-practices-phishing-prevention
9. Kletz, T.: Learning from Accidents in Industry. Gulf Professional Publishing, Oxford, UK (2001)
10. De la Torre, S.: Learning from mistakes. El tratamiento didáctico de los errores como estrategia de innovación. Buenos Aires, Argentina: Editorial Magisterio del Río de la Plata. (2004)
11. Reason, J.: Managing the Risks of Organizational Accidents. Ashgate, Aldershot, England (1997)
12. Steves, M., Greene, K., Theofanos, M.: Categorizing human phishing difficulty: a phish scale. J. Cybersecurity. **6**(1), tyaa009 (2020). https://doi.org/10.1093/cybsec/tyaa009
13. Edmondson, A.: Strategies for learning from failure. Harvard Business Review (2011). https://hbr.org/2011/04/strategies-for-learning-from-failure
14. Stalling, W.: Effective cybersecurity. A Guide to Using Best Practices and Standards. USA: Addisson Wesley. (2019)
15. Carpenter, P., Roer, K.: The security culture playbook. An Executive Guide to Reducing Risk and Developing Your Human Defense Layer. Wiley, Hoboken, NJ. USA (2022)
16. Brumfield, C.: A medical model for reducing cybersecurity risk behavior. CSO Computer-World (2022). https://cso.computerworld.es/tendencias/un-modelo-medico-para-reducir-el-comportamiento-de-riesgo-en-ciberseguridad
17. Barojan, D.: Building digital resilience ahead of elections and beyond. In: Jayakumar, S., Ang, B., Anwar, N.D. (eds.) Disinformation and Fake News, pp. 61–73. Springer, Singapore (2021). https://doi.org/10.1007/978-981-15-5876-4_5

Towards a Cybersecurity Awareness Plan for Casinos: An Initial Review for Internal Workers

Jaime Díaz[1], Rodrigo Guzmán[2], Jeferson Arango-López[3], Jorge Hochstetter[1], Gabriel M. Ramirez V.[4], and Fernando Moreira[5](✉)

[1] Depto. Cs. de la Computación e Informática, Universidad de La Frontera, Temuco, Chile
jaimeignacio.diaz@ufrontera.cl
[2] Consultora Tecnológica RGIT, Temuco, Chile
[3] Depto. de Sistemas e Informática, Universidad de Caldas, Manizales, Colombia
[4] Facultad de Ingenierías, Universidad de Medellín, Medellín, Colombia
[5] REMIT, IJP, Universidade Portucalense and IEETA, Universidade de Aveiro, Aveiro, Portugal
fmoreira@upt.pt

Abstract. Public and private institutions have invested in IT to increase their information security. Along with investments, the human factor is dominant. In that sense, countries have also implemented their own Computer Security Incident Response Teams (CSIRTs), whose main objective is to minimize and control the damage in case of a security breach. In the case of the Chilean government, with its CSIRT, they propose new guidelines for IT standards related to cybersecurity in the country's gaming casinos. This incorporation includes creating internal policies, procedures, protocols, and procurement. The objective of this article is to design a model for creating a cybersecurity awareness and education campaign based on the recommendations of the National Institute of Standards and Technology (NIST) and ISO 27001. The methodology consists of the evaluation of these alternatives and the declaration of 5 preliminary stages. On this occasion, we evaluated the first of them, evaluating all the internal workers of the company to form the subsequent initiatives.

Keywords: cybersecurity awareness · policy development · electronic commerce · education

1 Introduction

In the last decade, both public and private organizations and institutions have invested in IT to increase their information security [1, 2]. These investments must be accompanied by cultural changes within the entities, which are essential for the investments and efforts to be implemented and maintained over time [3–5].

The preponderant role played by the human factor in computer systems and cybersecurity issues, in general, is enormous. The responsibility is shared, given the need to adopt good practices and the power of information [3, 6, 7]. In this regard, there are

A. Rocha et al. (Eds.): WorldCIST 2023, LNNS 799, pp. 134–143, 2024.
https://doi.org/10.1007/978-3-031-45642-8_13

currently Computer Security Incident Response Teams (CSIRTs) whose main objective is to minimize and control damage in the event of a cyberattack [3]. They also perform the functions of advising, responding, and recovering normal operations, as well as preventing future incidents from occurring. To achieve this, it acts as a coordinator of all areas, individuals, and processes involved in an incident [3, 6, 7].

Chile's CSIRT, formalized in 2019 as a Department within the government structure, specifically in the Undersecretariat of the Interior and Public Security, has established more than 70 cybersecurity agreements with organizations across the country [8, 9]. In 2021 the CSIRT of Chile entered into a collaboration agreement on Cybersecurity with the Superintendence of Gaming Casinos (SCJ), the regulatory body in Chile of Gaming Casinos; as a result of this collaboration, new standards and regulations for information management and implementation of Cybersecurity in the country's gaming casinos are published. The SCJ decides to reactivate and give the necessary urgency to the Cybersecurity project, focusing mainly on the regulatory body's requirements. Given this requirement, it is necessary to implement a cybersecurity awareness plan around education, good practices, and responsibility in managing digital technologies to disseminate and publicize the various measures implemented within the company [3]. Therefore, the SCJ, after knowing the current scenario on cybersecurity issues within the casinos and being aware of the importance of cybersecurity, suggests that it is necessary to design an awareness and education campaign that allows a correct information flow from top management to the hosts of the company. Thus, delivering the necessary tools for digital security and hygiene in the global context of the operation. This initiative is general and could be replicated without inconvenience in other institutions.

The project includes creating internal policies, procedures, and protocols that give formality and structure to the actions carried out or to be carried out in the area of information security and cybersecurity. It also includes purchasing equipment and software to provide an additional layer of protection to the servers of the different casino systems. This paper proposes to cover this area. The objective of this article is to initiate a discussion on the design of a model to create a cybersecurity awareness and education campaign based on the recommendations of the National Institute of Standard and Technology (NIST) Special Publications (SP) 800-16 and SP 800-50 and the existing general cybersecurity plan for casinos [10, 11]. Our contribution consists of presenting an awareness model and a pre-assessment of internal employees for casinos. We seek to identify initial knowledge, which will initiate subsequent intervention work.

2 Background and Related Concepts

2.1 Cybersecurity Culture

Cybersecurity is defined as a set of guidelines, policies, and tools that aim to create a climate of digital trust by protecting the assets of organizations and individuals with ICT support. ISACA defines *cybersecurity* as "protecting information assets, addressing threats to information processed, stored, and transported by interconnected information systems." [12]. The new trends in mobility and connectivity present significant new challenges. The technology, in constant growth and sophistication, goes hand in hand with security incidents and breaches that specialized criminals can potentially exploit [13–15].

Cybersecurity professionals and users who operate information systems must be informed and flexible in identifying, managing, and reacting to new threats such as APTs (Advanced Persistent Threats) [16]. They must exhibit a high degree of situational awareness. This awareness takes time, as it is usually developed through experience within a specific organization. Each organization has its own distinctive culture. This means conditions vary significantly from one organization to another.

Despite the possibilities for personal and professional development that cyberspace offers us, cyber threat occupies a prominent place among the risks and threats to internal and external security. An effective response to the threat is only possible with the involvement of all an organization's sectors through a cybersecurity culture. Consequently, in the first instance, it is necessary to reflect on the basic premises for a solid cybersecurity culture to take root in the organization [17].

2.2 ISO 27001

ISO 27001 is an international standard issued by the International Organization for Standardization (ISO) and describes how to manage information security in an enterprise [18]. This standard can be implemented in any organization, for-profit or not-for-profit, private or public, small or large. The standard provides a methodology for implementing information security management in an organization. ISO 27001 is based on the PDCA quality management theory (also known as the Deming cycle: Plan-Do-Check-Act).

2.3 NIST

The National Institute of Standards and Technology (NIST), an agency under the U.S. Department of Commerce, aims to promote innovation and industrial competition through advances in standards and technology to enhance economic stability and quality of life [19]. It developed the NIST Cybersecurity Framework that uses a common language to guide companies of all sizes to manage and reduce cybersecurity risks and protect their information. This framework does not provide new cybersecurity functions or categories but compiles best practices (ISO, ITU, CIS, among others) and groups them according to affinity. It uses business drivers to guide cybersecurity activities and considers cyber risks as part of the organization's risk management processes. The framework consists of three parts: the basic framework, the framework profile, and the implementation levels.

Our current approach is based on the NITS special publications described below.

NIST SP 800-50 Training and Awareness Program. NIST Special Publication 800-50 creates an IT training and awareness program. It supports the specific requirements outlined in the Federal Information Security Management Act (FISMA) of 2002 and the U.S. Office of Management and Budget (OMB) Circular A-130 [10].

The publication identifies the four critical steps in the lifecycle of an IT security awareness and training program: (i) Awareness and Training Program Design; (ii) Awareness and Training Material Development; (iii) Program Implementation; and (iv) Post-Implementation.

NIST SP 800-16 Role-Based Training Model. This publication describes information technology and cybersecurity role-based training for Federal Departments and Agencies and Organizations (Federal Organizations) [11]. SP 800-16 is based on the premise that learning is continuous. Specifically, learning in this context begins with awareness, evolves into training, and evolves into education [11]. It defines the information security learning required when an individual assumes different roles within an organization and different responsibilities for information systems. This paper uses the model to identify the knowledge, skills, and abilities an individual needs to perform the information security responsibilities specific to each of his or her roles in the organization.

3 Methodology

The first part of this project contemplates the development of a model to create a cybersecurity awareness campaign that can be replicated in any existing business unit within the gaming casinos. NIST Special Publications SP 800-16 and SP 800-50 [10, 11], will be applied based on the project currently designed by the company under the ISO 27001 framework [18].

The first three phases of the model proposed the use of SP 800-16 in charge of providing context and identification of the company, defining roles, and identifying activities to be performed in the campaign. The subsequent three phases of the model use SP 800-50 for the design of the concrete material to be applied, implementation of the campaign, and finally, the set of evaluations. Figure 1 explains the relationship between the proposed stages.

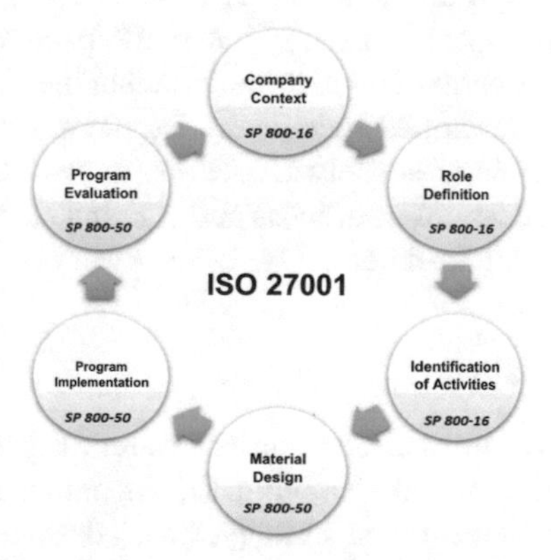

Fig. 1. Proposed awareness model

3.1 Company Context

This phase is the initial stage of the model. We seek to identify the company or organization in the following areas: (i) Knowledge and existence of information security and cybersecurity policies and procedures; (ii) Identification of employees, functions, and educational level; (iii) Identify Systems and Owners; (iv) Level of the organization (local, national or international); (v) Regulatory framework or current legislation impacting the organization; (vi) Campaign centralization model; (vii) Define the objective of the campaign; and (viii) Pre-assessment through a knowledge test.

3.2 Role Definition

In this phase, we will define the roles on which we will focus our campaign. Roles may vary according to the need and context of the company, its vision, and its campaign objective. Many organizations may reduce their roles due to a lack of personnel and limited resources.

3.3 Identification of Key Activities

The third phase of the model consists of assigning activities and knowledge that each role must have. We will rely on Annex B of SP 800-16 [11], *"Knowledge and skills catalog."* We will create our customized matrix of roles and activities that should be in knowledge according to our objectives. We will identify the support activities and the necessary resources to carry out our campaign successfully.

3.4 Material Design

At this stage, we must already consider "What behavior do we want to reinforce" (awareness) and "What skill do we want the audience to learn and apply" (training)? In both cases, the focus should be on specific material that participants should integrate into their work. Attendees will pay attention and incorporate what they see or hear in a session if they feel the material was developed specifically for them. Any presentation that feels canned, impersonal, and generic is shelved like any annual or traditional compliance induction or training. A successful awareness and training campaign can be effective if the material is exciting and up-to-date.

3.5 Program Implementation

We can only implement an information security awareness program or campaign only after proper implementation of the following: (i) An initial assessment of identified needs; (ii) A developed implementation strategy; (iii) A designed customized awareness and training materials; and (iv) The plan's implementation will focus its resources on covering the broad spectrum of the organization, from top management to operational positions.

3.6 Program Evaluation

The program we have designed and implemented can quickly become obsolete if we pay insufficient attention to technological advances, IT infrastructure, organizational changes, and current threats. We must be aware of this potential problem and incorporate mechanisms into our strategy to ensure that the program remains relevant and meets the overall objectives.

Formal evaluation and feedback are critical components of any information security awareness and training program. Continuous improvement can only occur with a good idea of how our designed program performs. In addition, we must prepare the feedback mechanism to absorb and evaluate compliance with the initial objectives.

4 A Preliminary Test

For the design of the pre-assessment, an internal communication team was created with the participation of a journalist, graphic designer, Corporate Systems Manager, and the Information Security Project Manager.

This initial test was applied through the Office 365 App Forms, given the ease of implementation and user control management, and the Company has licenses for all hosts. Working meetings were held to design and confirm the set of questions. The following were analyzed: Recurring incidents in the IT support platform, brainstorming on particular situations, and the INCIBE awareness KIT recommendations [20]. We defined that the initial test would have 50 questions (see Table 1) grouped into seven areas of interest in cybersecurity.

Table 1. Initial Test Question Areas

Scope	Questions
WIFI Networks	7
e-Commerce	6
Web browsing	6
Malware Awareness	6
Information Privacy	5
Use of Credentials	15
Email	5

Each question is worth 1 point if answered correctly and 0 points if answered incorrectly. Table 2 shows the final grade for each evaluation.

Table 3 shows the set of questions for one of the worst-performing dimensions.

In order to evaluate if the knowledge of the hosts is in accordance with this initial evaluation, the following weighting table is elaborated, which was discussed and agreed with the Dreams S.A. Information Security Committee (Table 4).

All workers currently active in operation are invited to participate in this initial test via email. As a result of the COVID-19 pandemic, only a tiny group of them are active due to the total closure of operations of the Gaming Casinos in Chile by direct instruction of the superintendence (SCJ).

Table 2. Scoring definitions

	Definition
TPE	Total Evaluation Points
TPC	Total Correct Questions
Results	(TPC * 100)/TPE

Table 3. Pre-test questions: e-commerce section

e-Commerce
1.- The use of instant money transfer companies should be used for (...)
2.- We are interested in a discounted product within a well-known e-commerce platform. After sending a message to the seller using the web messaging system, the seller asks us to (...)
3.- Many online shopping platforms offer us the option of saving our credit card details. When should I accept it?
4.- User comments are an indicator of the reliability of a website, as feedback from other buyers can give us beneficial information about product quality or reliability (...)
5.- We must periodically review the movements of our bank account if we make purchases on online shopping platforms
6.- Which of the following payment methods is the safest? (...)

Table 4. Weighting table according to evaluation results

	from	to
Over expected knowledge	>=85	100
Expected knowledge	>=70	<85
Knowledge below expected	>=50	<70
Knowledge well below expected	>=0	<50

5 Results

Participation in the initial test was 88.2% of all people who were active with their employment contracts at that date. We used Qlik Sense for the analysis and presentation of the results. We concluded that the overall average of participants was 84.8% knowledge of cybersecurity matters, giving an average score of 42.41 out of 50. In addition: (i) 81.3% of the participants were men; (ii) The average age was 42.7 years; and (iii) The management departments with the highest participation were: Integral Security, Games, and Administration and Finance.

From the information related to the cybersecurity questions, we can determine that: (i) The overall average evaluation was 84.8%, equivalent to 42.41 points out of 50;

(ii) The lowest scoring category was e-commerce with 70.8%, followed by wireless networks with 80.7%, use of credentials with 81.2%, and knowledge of malware with 87.1%; (iii) The best-evaluated categories were those related to e-mail use, with 96.3%, and information security, with 95.5%; (iv) The host with the lowest evaluation scored 62%; (v) Only one worker obtained the highest score and belonged to the IT area; (vi) Overall, IT Management obtained an average of 92%, equivalent to 46 points out of 50; and (vii) The management that passed the desired threshold of 90% were Systems with 92% and Operations with 91.3%. Table 5 shows the five topics with the lowest scores.

Table 5. Lowest scores

	%
Recognition of secure passwords	26%
Recognition of secure payment methods	32%
Application of Windows updates	36%
Security in Wifi-home networks	40%
Use of free applications	57%

Table 6 shows thc results of the survey according to the suggested classification.

Table 6. Results according to designed classification

	Surveyed	%
Over expected knowledge	12	16%
Expected knowledge	58	77,33%
Knowledge below expected	5	6,67%
Knowledge well below expected	0	0%
Total	**75**	100%

It should be noted that the workers participating in the initial test all have supervisory, administrative, or management roles. In addition, they have formal higher education. Also, they are more prepared on issues of digital jobs within the company.

6 Conclusions and Discussion

According to the Atlantic council's 2019 cybersecurity report [21], 43% of data breaches in 2018 were carried out through social attacks, and only 57% of affected companies reported having allocated resources to cybersecurity issues. This scenario shows how essential cybersecurity topics are for IT specialists and the educational and awareness processes that must be applied to the entire company. A robust IT security program can only be implemented with significant attention to training users on security policies, procedures, and techniques. In addition, those who manage the IT infrastructure must have the necessary skills to perform their assigned tasks effectively.

We expect that this first approach will help us to (i) Be a real contribution to reducing the digital gap of workers in gaming casinos; (ii) Strengthen the weakest link in the chain in cybersecurity areas, providing the necessary tools in the face of a possible cybersecurity incident; (iii) Implement internal protocols, policies, and measures related to cybersecurity; (iv) Create a digital awareness campaign based on roles according to the functions and participation in the processes of the different companies; and (v) Have a model that can be replicated in other business units or companies linked to gaming casinos in Chile.

Regarding the limitations of this iteration, we have only worked on Chilean casino gaming regulations. This scenario could complicate if the regulation varies significantly in other regions. However, within Latin America, Chile continues to pioneer this type of initiative, and we seek to generate a replication process to obtain positive results in subsequent iterations.

The results of the model's effectiveness are based on the correct execution of the implementation phases of the original awareness plan implemented by Dreams S.A. for the following periods. Our future work will be to continue with the phases of the proposed model, evaluate knowledge and then enter into a continuous improvement process.

Acknowledgments. This work was partially supported by the Universidad de La Frontera, Temuco, Chile, through the DIUFRO Project under Grant DI21-0016.

References

1. Dhillon, G., Smith, K., Dissanayaka, I.: Information systems security research agenda: exploring the gap between research and practice. J. Strateg. Inf. Syst. **30**, 101693 (2021). https://doi.org/10.1016/j.jsis.2021.101693
2. Shukla, A., Katt, B., Nweke, L.O., Yeng, P.K., Weldehawaryat, G.K.: System security assurance: a systematic literature review. Comput. Sci. Rev. **45**, 100496 (2022). https://doi.org/10.1016/j.cosrev.2022.100496
3. AlDaajeh, S., Saleous, H., Alrabaee, S., Barka, E., Breitinger, F., Raymond Choo, K.-K.: The role of national cybersecurity strategies on the improvement of cybersecurity education. Comput. Secur. **119**, 102754 (2022). https://doi.org/10.1016/j.cose.2022.102754
4. Mishra, A., Alzoubi, Y.I., Anwar, M.J., Gill, A.Q.: Attributes impacting cybersecurity policy development: an evidence from seven nations. Comput. Secur. **120**, 102820 (2022). https://doi.org/10.1016/j.cose.2022.102820
5. Alanazi, M., Freeman, M., Tootell, H.: Exploring the factors that influence the cybersecurity behaviors of young adults. Comput. Human Behav. **136**, 107376 (2022). https://doi.org/10.1016/j.chb.2022.107376
6. García, A.A.: Ciberseguridad: ¿por qué es importante para todos? SigloVeintiuno Editores (2019)
7. Gale, M., Bongiovanni, I., Slapnicar, S.: Governing cybersecurity from the boardroom: challenges, drivers, and ways ahead. Comput. Secur. **121**, 102840 (2022). https://doi.org/10.1016/j.cose.2022.102840

8. Circular de ciberseguridad – SCJ. https://www.scj.cl/marco-normativo/normativas-en-consulta/circular-de-ciberseguridad. Accessed 17 Oct 2022
9. Superintendencia de Casinos de Juego publica normativa en ciberseguridad junto al CSIRT de Gobierno. https://www.csirt.gob.cl/noticias/superintendencia-de-casinos-de-juego-publica-normativa-en-ciberseguridad-junto-al-csirt-de-gobierno/. Accessed 07 Oct 2022
10. Wilson, M., Hash, J.: Building an Information Technology Security Awareness and Training Program. National Institute of Standards and Technology, Gaithersburg, MD (2003). https://doi.org/10.6028/nist.sp.800-50
11. Toth, P., Klein, P.: A Role-Based Model for Federal Information Technology/Cybersecurity Training, 3rd draft. National Institute of Standards and Technology (2014)
12. Isaca: Guía de Estudio de Fundamentos de la Ciberseguridad, 3a Edición
13. Gargiulo, C., Sgambati, S.: Active mobility in historical centres: towards an accessible and competitive city. Transp. Res. Procedia **60**, 552–559 (2022). https://doi.org/10.1016/j.trpro.2021.12.071
14. Chevalier, A., Charlemagne, M.: When connectivity makes safer routes to school: conclusions from aggregate data on child transportation in Shanghai. Transp. Res. Interdisc. Perspect. **8**, 100267 (2020). https://doi.org/10.1016/j.trip.2020.100267
15. Cheng, Y.-H., Chen, S.-Y.: Perceived accessibility, mobility, and connectivity of public transportation systems. Transp. Res. Part A: Policy Pract. **77**, 386–403 (2015). https://doi.org/10.1016/j.tra.2015.05.003
16. Advanced persistent threat – Glossary. https://csrc.nist.gov/glossary/term/advanced_persistent_threat. Accessed 13 Oct 2022
17. Cano, R.: Ciberseguridad y ciberdefensa. Retos y perspectivasen un mundo digital/Cybersecurity and cyberdefense. Challenges and perspectives in a digital world. RISTI (Revista Iberica de Sistemas e Tecnologias
18. Calder, A.: Information Security Based on ISO 27001/ISO 27002. Van Haren (2009)
19. National Institute of Standards and Technology | NIST. https://www.nist.gov/. Accessed 13 Oct 2022
20. Kit de concienciación. https://www.incibe.es/protege-tu-empresa/kit-concienciacion. Accessed 13 Oct 2022
21. Forsey, A.: Annual Report 2019/2020: Shaping the Global Future Together. https://www.atlanticcouncil.org/in-depth-research-reports/report/annual-report-2019-2020-shaping-the-global-future-together/. Accessed 17 Oct 2022

Ethical Framework for the Software Development Process: A Systematic Mapping Study

Lucrecia Llerena[1(✉)], Henry Perez[1], John Plazarte[1], John W. Castro[2], and Nancy Rodríguez[1]

[1] Quevedo State Technical University, Quevedo, Ecuador
{lllerena,henry.perez2017,john.plazarte2017, nrodriguez}@uteq.edu.ec

[2] Departamento de Ingeniería Informática y Ciencias de la Computación, Universidad de Atacama, Copiapó, Chile
john.castro@uda.cl

Abstract. Ethics is one of the main elements that maintain the organization within the software development process in companies. Therefore, it is necessary to apply codes of ethics to meet this requirement. This research aims to propose an ethical framework that facilitates understanding and possible implementations of ethical issues during the software development process. For this purpose, a systematic mapping study (SMS) was performed. After performing the SMS, we did not find a study that integrates an ethical framework into the software development process. Therefore, we propose an ethical integration framework for the software development process because it is considered imperative that every developer understands ethical principles and can apply them during the software development for responsible decision-making. We conclude, first, that software companies do not follow specific ethical guidelines, resulting in unethical outcomes in developing their software products. Secondly, implementing a code of ethics will depend on how each software developer knows, and knows how to act in a way that is consistent with ethical values.

Keywords: Ethics · Ethical Conduct · Framework · Software · Software Company

1 Introduction

The difficulties in software development are overwhelming [1] because some companies do not have ethical standard that every employee must follow [2]. Concerning the above, there is a possibility that any engineer or employee of the company may leak private information during software development [3]. Although there are no specific tools for ethical and sustainable production, the integration of its concept has not yet reached many companies worldwide [4]. However, software companies nowadays face changes in their negotiation policies since, to maintain a competitive advantage with companies

A. Rocha et al. (Eds.): WorldCIST 2023, LNNS 799, pp. 144–154, 2024.
https://doi.org/10.1007/978-3-031-45642-8_14

of more significant importance, they must follow well-defined guidelines on ethics [5]. This is because not having ethics integrated can be overwhelming, causing the company to affect its image. Currently, in companies such as Google, Meta, and Amazon, it is not known what is done with their users' data since these companies do not follow the terms of ethics and morality completely, and this could mean a collateral impact with their corporate image [6]. The data these companies have stored on their servers can be filtered without the owner's user noticing, but this does not mean that companies are infringing. The user has to stop using their services if the company does not allow to improve its policies and the way of working with its engineering departments [7]. This is important to implement an ethical framework that helps each client feel safe and confident with the work being done (e.g., software development) and that the data transmitted by them is ethically protected [6]. In this way, every employee in the development company can recognize, interpret and act on ethical decisions.

A Systematic Literature Review (SLR) aims to search for and identify relevant material related to a given topic. In contrast, a Systematic Mapping Study (SMS) provides a more general overview of the research conducted on a topic [8]. Due to the above, we carried out an SMS to find literature on integrating a code of ethics in the software development process. Among the primary studies found, on the one hand, few works with proposals for ethical frameworks for the area of artificial intelligence (AI) are reported. On the other hand, studies on implementing an ethical framework for software development are null. This indicates that ethical guidelines have not been implemented in software development companies. Consequently, these companies do not follow a development methodology supported by ethics. Therefore, we have combined information from an AI code of ethics with software development to propose an ethical framework to improve this process. Our work significantly contributes to the field of ethics and, in general, to software development teams and educational institutions. It is essential to propose an ethical framework to avoid unethical activities in professional performance. To the best of our knowledge, no framework provides the ethical issues arising in each software life cycle phase. In addition, the proposal of an ethical framework is helpful in the practice of software engineering, so the following contributions are presented in this paper: (i) Define an ethical framework for educational institutions or companies that require a software development team; (ii) generalize the ethical framework for any stage of software development (e.g., in the stages of the waterfall model, this methodology is considered the basis of many development methodologies); (iii) to apply the ethical framework in the software product environments and in the behavior of the professionals that exercise the software development activity.

Paper Organization. Section 2 describes some concepts of ethics focused on technological development. Sect. 3 describes the research method performed. In Sect. 4, the proposal for implementing a framework in the software development process is made, and the framework is described. Section 5 discusses the results obtained and the estimated impact of implementing it in the development process. In addition, this section describes the limitations of the study. Finally, Sect. 6 describes the findings and future research.

2 Related Work

In this section, related work identified during the research is briefly described. Boyd [2] designs a framework for observing how information sets contribute to ethical commitment, associating it with ethical sensitivity. The use of this ethical framework is efficient for learning algorithm training domains. Therefore, some of its ethical guidelines can be adapted for software development. Almahmoud et al. [5] describe how the computers use already trained machine learning models. The concern, however, is to determine the quality of the model considering ethical, engineering, operational, and legal implications. This may ethically affect software development because the information captured corresponds to users who use a given product, and this data may be leaked. This study does not report the codes of ethics that can be used in software companies. Boyd and Shilton [9] offer a framework for ethical engagement by matching ethical elements with ethical sensitivity. This ethical engagement framework makes theoretical and methodological contributions to the study of how ethics is operationalized in design teams by providing valuable language for studies of collaboration and communication around ethics. Vakkuri et al. [10] propose an ethical framework for developing AI. ECCOLA is a sprint-to-sprint process that facilitates ethical thinking during AI development and provides questions for the organization of various ethical issues in AI systems. In ECCOLA there are 21 cards divided into eight topics. These topics are those of AI ethics. ECCOLA is intended to be used throughout the design and development process. Also, Vakkuri et al. [11] describe how companies use new technologies and high-level tools to manage the ethics of AI to help industry organizations create more ethical AI systems. The authors surveyed 211 software companies to find out their current ethical status; as a result, applying the survey provides insight into how a software company can or could apply ethical practices within its development department. Thus, using an ethical framework in computer knowledge results in a social impact through innovative practices [12]. Islam [13] proposes a framework for integrating ethical values as requirements in the specifications of the software engineering process and ensuring the development of IA applications. This framework builds on using a traditional software lifecycle for developing IA applications by specifying the requirements management phase and implementing its ethical framework. Implementing ethical practices is essential for developing IA applications to address ethical considerations throughout the software development lifecycle in any learning area. In related work, some ethical frameworks have been found that focus on specific fields but not specifically on software development. However, these frameworks do not consider all phases of software development. Therefore, it is essential to know the possible ethical integrations that could be applied in all stages of software development for decision-making in a high-risk situation, especially in areas of high technological dependency.

3 Research Method

Kitchenham et al. [8] proposes four activities for the SMS: (i) define the research questions, (ii) plan the review that consists of defining the tasks to be carried out, (iii) execute the review based on selecting primary studies, evaluating the quality of the studies, extracting relevant data, and synthesizing the data, (iv) preparing the review report. At the beginning of the research, a search was carried out to identify other literature reviews related to an ethical framework for the software development process. As a result of this search, we did not find a literature review regarding the topic of interest. For this reason, we made an SMS following the guidelines of Kitchenham et al. [8]. Our research aims to answer the following research questions (RQ): (**RQ1**) What ethical frameworks have been implemented in software development companies? (**RQ2**) How should codes of ethics be applied in software development companies? (**RQ3**) What is the impact of applied codes of ethics on the software development process? To ensure that relevant information was obtained, an SMS was performed in four databases (DB): ScienceDirect, Scopus, ACM Digital Library, and IEEE Xplore, selecting the year 2017 as the start date and the year 2022 as the end date. We use updated research with a maximum of 6 years of publication. We started with the identification of keywords based on the research questions. These keywords are Ethics, ethical conduct, software company, development companies, development industries, framework, and software. Based on the keywords, the following search string was constructed: *("ethical conduct" OR ethics) AND ("software company" OR "development companies" OR "development industries") AND (framework OR software).* The criteria used to select the primary studies are summarized below: (i) **Inclusión**: The studies found must be directly related to the use of a code of ethics in software development companies AND the works must be written in English, AND their publication date must be equal to or greater than 2017; (ii) **Exclusion**: The studies do not report aspect related to the implementation of a code of ethics in software development companies OR works that make proposals for the integration of codes of ethics but do not validate them in real cases.

The primary strategy for study selection is explained below. First, the search was performed once the search string and fields for each DB had been defined. The set of studies resulting from the search was referred to as "Studies found." The studies were reviewed by examining the title, keywords, and abstract (the latter two, when available). Those studies that met the inclusion criteria were included in the "Preselected Studies" group, and duplicate studies were removed between each DB (i.e., between different search terms in the same DB). Then duplicates were removed between all DB. After, the selection criteria were again applied to the full text of the studies belonging to the "Preselected Studies" group. Finally, the group obtained from these studies has been called "Primary Studies" (PS) (see: https://n9.cl/1d8fo). The results of applying the filters in the selection process for each DB can be seen in Table 1.

Therefore, to better understand the SMS's structuring, a PRISMA flowchart (see Fig. 1) was structured to allow the reader to quickly understand the basic procedures and study selection throughout the review process.

Table 1. Search results.

Database	Studies found	Pre-selected studies	PS
IEEE Xplore	5	3	0
ScienceDirect	430	5	5
ACM	84	6	5
Scopus	11	1	1
Total	**422**	**16**	**11**

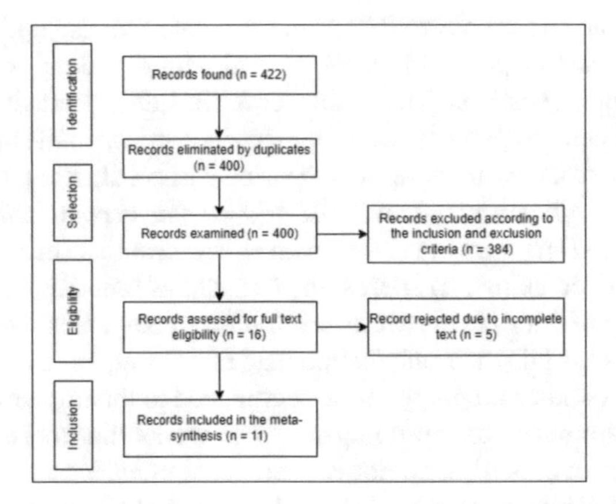

Fig. 1. PRISMA flowchart

4 Proposed Solution

As a solution, the creation of an ethical framework to be applied in the software development process is proposed. For this purpose, the main ethical issues in software companies and the development process must be analyzed. The framework sets the ethical principles for both the product and the profession. The principles of ethics in the product emphasize the software development process, and the principles of ethics in the profession emphasize the actions of workers in software companies. We have selected the ECCOLA ethical framework [10] as our proposal's basis since it is close to establishing a general model of ethical guidelines in software development. Taking as a reference this framework for IA development [10], we propose the Ethical Framework for the Process of Software Development (EFPSD). The difference between the ECCOLA framework and our EFPSD proposal is that the former is designed for exclusive application in IA. While EFPSD is a proposal to be implemented in the software development process, it can also be considered as a resource to be used in the educational environment (e.g., developing ethical scenarios, ethics workshops, and learning from their work environment). EFPSD is divided into two parts. One part is designated for the product developed (product) and the other for the work conducted by each individual in his area (profession). Each of these parts contains subparts that will be specific topics of ethics for the different stages of development. Table 2 describes the aspects that must be considered during the software development process. On the one hand, most of these aspects were inspired by the primary studies identified in the SMS. On the other hand, some aspects were described based on our work experience in different organizations and on the knowledge received during our training. The aspects considered are products and profession. On the one hand, a product refers to an idea or customer request developed by software teams. These teams do not have a way to apply ethics to their developed products; for this, a set of codes of ethics was elaborated in the proposed framework (See Table 2). On the other hand, a profession refers to a person's activities in software teams. These activities are not always performed ethically because there is no way to restrict the actions exercised. For this reason, a set of codes of ethics was elaborated in the proposed framework (see Table 3). EFPSD involves complying with internal processes for each software development life cycle stage. It is proposed to structure the EFPSD for each stage of the cascade model [14] since it is one of the most used models during software development. However, it is the old model and is still the most seen by all companies, institutions, and universities. This model has been selected because it is the most used in software development companies due to its ease of use and the management of deliverables for each stage is very simple. The following are the aspects of EFPSD that should be applied at each stage of the cascade model. These aspects will be grouped according to the product and profession components. According to these two components, each case is analyzed to classify the aspects and ethical codes depending on the stage of the software development process. Each aspect of EFPSD will depend on the companies/institutions that require it in their software development processes. However, this framework is intended for different software development methodologies.

Communication. The software engineer and the customer meet to gather software requirements [14]. In this way, engineers must comply with the EFPSD aspects: Product (quality and reliability) and profession (knowledge, support, and environment), and consider that the product (software) to be developed should be of higher quality and that its price should be commensurate with the size of its development.

Planning. This phase defines resources, time, and other information relevant to the project [14]. Some examples of these essential resources and information are requirements analysis, data tracking, etc. Aspects of the EFPSD that should be applied in the planning stage are product (goal and documentation) and profession (responsibility). Therefore, at this stage, every software engineer must consider the software product's main objectives, its documentation, and the responsibility to perform the tasks correctly.

Modeling. It is related to software analysis and design [14]. The aspects of EFPSD to be considered are product (objective, standard, and documentation) and profession (knowledge, support, and responsibility). So, every software engineer must consider the software product objectives, standards, documentation and knowledge, support, and responsibility, to achieve good software modeling.

Construction. It refers to software code and testing. The aspects of the EFPSD to be applied in this phase are product (standard, objectives, documentation, and maintenance) and profession (responsibility and illegal acts). In this way, the construction of the software can be done without problems in the work environment.

Deployment. It refers to the delivery of the project to the client [14]. The aspects of the EFPSD to be applied in the deployment phase are product (quality, objectives, standard, documentation, reliability, maintenance) and profession (environment, knowledge, support, responsibility, and illegal acts). In order to ensure a satisfactory delivery and to promote the approval of the software by the customer.

Table 2. Product code of ethics.

Aspects	Ethical codes	PS
Quality	a) Ensure high-quality software b) The price must be representative of the size and quality of the software	[PS1] [PS10]
Objective	a) Meet the main objectives of the project b) Do not deviate from the objectives of the project c) Not to infringe on clients' decisions regarding their objectives	[PS4] [PS7]
Standard	a) Respect the standards and laws set forth by the client, organization, or state b) Follow a generalization standard of User Interface and business logic	[PS2] [PS3] [PS11]
Documentation	a) Avoid the alteration and distribution of private information b) All information will be documented and distributed to the company or person in charge of the system c) Comply with the terms and conditions implemented by the company	[PS2] [PS4]
Reliability	a) Respect the data stored in the database b) Avoid data leakage to different institutions c) Do not use user data for research unless users have agreed to the terms	
Maintenance	a) Document updates and protect current or editable data b) Perform maintenance without intentionally introducing more defects c) Do not leave errors during the maintenance of software for profit	[PS5] [PS6]

Table 3. Profession code of ethics.

Aspects	Ethical codes	PS
Environment	a) Have an organized environment in the company b) Avoid labor conflicts c) Respect the other professionals and staff of the company d) Carry out the activities destined during the working day e) Do not infringe on the tasks of other professionals	[PS4] [PS9]

(*continued*)

Table 3. (*continued*)

Aspects	Ethical codes	PS
Knowledge	a) Encourage the learning of Junior developers with the help of Senior developers, in the software development company they are working on b) Do not direct other professionals on the wrong learning path c) Do not discriminate for lack of knowledge in technical aspects	[PS10]
Support	a) Support co-workers in problems they are unable to solve b) Provide knowledge and recommendations to co-workers during the software development process	[PS7]
Responsibility	a) Perform tasks properly b) Deliver the tasks in the corresponding time c) Be held accountable for mistakes made during development d) Do not steal software functionalities	[PS2] [PS10] [PS11]
Illegal acts	a) Ensure that the development process is within the company; if the company validates the process outside, it must follow the safety regulations b) Avoid disseminating company information	[PS3] [PS5] [PS8] [PS9]

5 Discussion of Results and Limitations

Figure 2 synthesizes the results using two bubbles scatter plots. The graph represents the number of studies published per year, according to publication type (journal or conference). Thus, the bubbles are located at the intersections between the two axes and their size is proportional to the number of publications for each combination of values. As seen on the left side of Fig. 2, 58% of research papers are published in journals. While on the right side of Fig. 2, interest in the study of the codes of ethics of the ACM in Software Engineering has decreased considerably since 2017, and this interest will grow in 2020. Next, each research question will be answered.

RQ1: What ethical frameworks have been implemented in software development companies? A collection of studies on ethical principles based on technologies that are applied in AI [3] was found [13]. It is essential to mention that this collection of principles is not oriented to software development, but this does not mean that codes of ethics are unnecessary and that companies do not apply them because of scarce information about an ethical framework. There is an ethical framework that is specifically aimed at AI technology development, among the most prominent of which is ECCOLA [10]. ECCOLA is a sprint-to-sprint process facilitating ethical thinking during the development of AI [10]. This framework provides several questions from different ethical issues to evaluate. Although ECCOLA is only used during the design and development process of AI, it served as the basis for the proposal of an ethical framework in the software development process, considering some aspects, such as the product and the professional side.

RQ2: How should codes of ethics be applied in software development companies? There is no specific way to apply codes of ethics in these companies. However, the

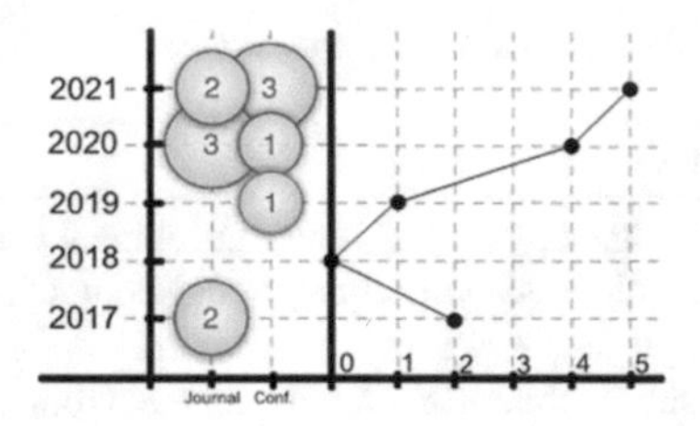

Fig. 2. Mapping for the primary study distribution. Source: The authors

ECCOLA [3] ethics application methodology can be followed as a basis to improve efficiency in teamwork so that the implementation of an ethical framework focused on software development can not only be applied in companies but is also intended as a learning guide in the systems area for to students and educational institutions [9].

RQ3: What is the impact of applied codes of ethics on the software development process? The code of ethics impact is most noticeable within software development companies because they use users to test their projects [1, 11]. Codes of ethics help protect the user from disclosing their data without their consent. In addition, consequently, to the work behavior that some people usually have, it is essential to cover this area in the proposal of an ethical framework. Implementing an ethical framework for the product under development is favorable because it improves its quality and reduces the development time of the software product [7]. We know that the lack of morale affects the work environment of software development companies and produces poor performance in collaborative work [15].

The research is qualitative and has some limitations concerning the validity of its contributions. In terms of construct validity, as well as internal validity, we did not find any problems. The external validity of the SMS is threatened by including only studies written in English. On the other hand, regarding reliability, it is considered that PSS authors may make errors of judgment when analyzing outstanding publications, so publications of great importance may have been overlooked. Throughout the systematic search, certain aspects have been found that may jeopardize the validity of the study. The main threat to validity is to commit bias in the selection process of primary studies. To reduce this bias, we followed the guidelines suggested by Kitchenham et al. [8]. Some considerations were made to ensure the study selection's validity. First, we searched the four most relevant DB, ScienceDirect, Scopus, ACM, and IEEE Xplore. Therefore, we obtained the most relevant studies from various journals and conferences on the topic of interest. However, if additional DB (e.g., SpringerLink) had been included, new results and additional information could have been obtained. Regarding access to content, researchers successfully accessed the full text of all studies that met the inclusion/exclusion criteria, so no studies lacked the full text. Another threat to validity is the application of the criteria for selecting and analyzing abstracts of the studies found. To minimize subjectivity, two research team members conducted the selection process in parallel, and the selected studies were agreed upon in a group meeting. Regarding the validity of the data, the analysis was performed on a sample of 11 primary studies. Data synthesis and extraction were used in these studies to look for possible implementations of codes of ethics in the software development process.

6 Conclusions and Future Work

This work aims to provide an ethical framework to support software companies for the better development of their products. For this purpose, an SMS has been carried out to structure an ethical framework considering all the stages of the software development process. As a result of the SMS, we found no specific ethical framework oriented to the software development process. However, the closest we found was an IA ethical framework called ECCOLA. This framework aims to help companies create more ethical IA tools. Therefore, to support ethical issues arising during the software development process, the ECCOLA framework has been considered as the basis for our proposal. To support ethical issues in the software development process, we propose EFPSD, an ethical framework that involves aspects related to the product and the profession in the software development area. In this way, it is not only the product developed ethically but also the activities performed by the professionals, allowing them to be more organized and, therefore, to obtain a product of excellence. In addition, EFPSD includes different aspects that occur in different scenarios during software product development. In the scenarios, the ethical codes that should be considered in the different stages of the software development process have been described (see Table 2). The cascade methodology has been used as an example of implementing these ethical codes in each phase. With the research performed, it is determined that, although there are studies that propose ethical guidelines related to IA, no reported guidelines are applied to software development. Likewise, this study does not show how to apply the ethical principles of the EFPSD. Therefore, it is essential to use this ethical framework for software development and to detail its application in the process, considering each stage of the software life cycle (case study), with the hope that this framework will be applied in existing software developments. Finally, to standardize the ethical aspects of EFPSD, the cascade methodology was taken as a basis. In this way, companies can replicate our research with their different development methodologies. In future work, we propose the implementation of EFPSD in the software development of an Ecuadorian company. With this, we can analyze our proposal's impact and identify which aspects need to be corrected.

References

1. Nabbosa, V., Kaar, C.: Societal and ethical issues of digitalization. In: 2020 International Conference on Big Data in Management (ICBDM'20), pp. 118–124, Manchester, UK (2020)
2. Boyd, K.L.: Datasheets for datasets help ML engineers notice and understand ethical issues in training data. ACM Hum. Comput. Interact. **5**(CSCW2), 1–27 (2021)
3. Heidelberg, M., Kelman, A., Hopkins, J., Allen, M.E.: The evolution of data ethics in clinical research and drug development. Ethics Med. Publ. Health **14** (2020)
4. Aydemir, F.B., Dalpiaz, F.: Poster: ethics-aware software engineering. In: ICSE-Companion'18, pp. 228–229, Gothenburg, Sweden (2018)

5. Almahmoud, J., Deline, R., Drucker, S.M.: How teams communicate about the quality of ML models: a case study at an international technology company. In: ACM Hum. Comput. Interact. **5**(GROUP), 1–24, article 222 (2021)
6. Härkönen, H., Naskali, J., Kimppa, K.: Hub companies shaping the future: the ethicality and corporate social responsibility of platform economy giants. In: 2nd ACM SIGSOFT International Workshop on Software-Intensive Business: Start-Ups, Platforms, and Ecosystems (IWSiB'19), pp. 48–53, Tallinn, Estonia (2019)
7. Lowry, P.B., Zhang, J., Wu, T.: Nature or nurture? A meta-analysis of the factors that maximize the prediction of digital piracy by using social cognitive theory as a framework. Comput. Human. Behav. **68**, 104–120 (2017)
8. Kitchenham, B.A., Budgen, D., Brereton, P.: Evidence-Based Software Engineering and Systematic Reviews, 1st edn. Chapman & Hall/CRC, Boca Raton (2015)
9. Boyd, K.L., Shilton, K.: Adapting ethical sensitivity as a construct to study technology design teams. ACM Hum. Comput. Interact. **5**(GROUP), 1–29 (2021)
10. Vakkuri, V., Kemell, K.-K., Jantunen, M., Halme, E., Abrahamsson, P.: ECCOLA—a method for implementing ethically aligned AI systems. J. Syst. Softw. **182**, 111067 (2021)
11. Vakkuri, V., Kemell, K.-K., Kultanen, J., Abrahamsson, P.: The current state of industrial practice in artificial intelligence ethics. IEEE Softw. **37**(4), 50–57 (2020)
12. Kumar, S., Kremer-Herman, N.: Integrating ethics across computing: an experience report of three computing courses engaging ethics and societal impact through roleplaying, case studies, and service learning. In: 2019 IEEE Frontiers in Education Conference (FIE'19), pp. 1–5, Covington, KY, USA (2019)
13. Islam, Z.U.: Software engineering methods for responsible artificial intelligence. In: 20th International Joint Conference on Autonomous Agents and MultiAgent Systems (AAMAS'21), pp. 1802–1803, Virtual Event, UK (2021)
14. Pressman, R.S.: Ingeniería del Software-Un EnfoquePráctico, 5th edn. McGraw-Hill, New York (2010)
15. Brinkman, B., Flick, C., Gotterbarn, D., Miller, K., Vazansky, K., Wolf, M.J.: Dynamic technology challenges static codes of ethics. ACM SIGCAS Comput. Soc. **47**(3), 7–24 (2017)

Social Engineering as the Art of Deception in Cyber-Attacks: A Mapping Review

Javier Guaña-Moya[1(✉)] and Diego Ávila-Pesantez[2]

[1] Pontificia Universidad Católica del Ecuador, Quito, Ecuador
eguana953@puce.edu.ec, eguana953@pucesa.edu.ec
[2] Escuela Superior Politécnica de Chimborazo, Riobamba, Ecuador

Abstract. Social engineering is the art of getting users to compromise information systems, presenting itself as a novel deception strategy. Social engineers target individuals with access to information to manipulate them into disclosing sensitive data or even carrying out malicious attacks through influence and persuasion. These types of attacks have become the most significant cybersecurity threats. This bibliographic review aims to analyze social engineering due to its impact on the productivity and efficiency of organizations, using deception and knowledge of human vulnerabilities to establish gaps in information security and gain access to secure data. We conclude that social engineering acts based on human exposure through attacks in four clearly defined phases. These attacks can be classified, first, considering the target as humans and computers, and finally, according to how the attack is made, as social, technical, and physical.

Keywords: Social engineering · cybersecurity · attacks · victims

1 Introduction

The 21st century has been characterized by the accelerated movement of business, media, social interaction, and education toward Internet platforms. As a result, the amount and importance of information flowing through the digital environment have increased exponentially [1]. This reality has caused an increase in criminal activity in cyberspace, which has materialized as data leakage attacks, malware, ransomware, and phishing, among others. As a result, in recent years, traditional modes of attack, such as hacking, have proven slightly less successful [2]. However, alternative areas of vulnerabilities have been exposed, among which social engineering stands out, which has gained more attention due to the way it is used in all modes of cyber-attacks [3].

Over the years, there have been several attempts to provide a concrete definition of the term social engineering attacks in the literature, each of which is slightly different but has the same general meaning [4]. Concerning individuals, it has been described as "Human Hacking," that is, the art of tricking people into revealing personal credentials that are then used to gain access to networks or accounts, taking advantage of human vulnerabilities using social interaction, influence, and persuasion to achieve cybersecurity violations [5]. At the business level, it is defined as a violation of organizational

A. Rocha et al. (Eds.): WorldCIST 2023, LNNS 799, pp. 155–163, 2024.
https://doi.org/10.1007/978-3-031-45642-8_15

security through interaction with people to deceive them and force them to break routine security procedures, compromising information systems. Therefore, technical protection measures are often ineffective against these attacks, and people generally believe they are good at detecting them. However, research indicates that users perform very poorly in detecting lies and deception [6].

Social engineering attacks traditionally follow a 4-step template: target investigation, form a relationship with the subject, exploit the connection and formulate an attack, and finally exit without a trace [7]. In this way, investigating this topic has gone from searching the dumpster to extracting information from the target on social media platforms [8]. With the development of Machine Learning, the second and third steps have been automated with social bot tracking and interaction with targets on social networks. The wide range of technologies and methodologies used for social engineering attacks, and the speed with which they evolve, make the development of software solutions such as spam and bot detectors a game of cat and mouse [9].

This bibliographic review aims to analyze the social engineering techniques that affect the productivity and efficiency of organizations. These techniques, using deception and knowledge of human vulnerabilities, establish gaps in information security and gain access to secure data.

2 Methodology

The current investigation was developed based on the systematic literature review regulations established by Kitchenham [10] to obtain information related to the research questions that arise from its development. This methodology shows the main stages: A) Review planning; B) Carrying out the review; C) Analysis of results.

A. Review Planning: This study analyzes social engineering as a severe problem in the cybersecurity field due to its impact on the production and effectiveness of organizational processes. Usually, social engineering uses fraud and human vulnerabilities to implant information security gaps and make sensitive data available that must be highly secure. For the development of the topic, we propose the following research questions:

- Q1: What are the bases of social engineering?
- Q2: What are the most common social engineering attacks, and what are the phases?
- Q3: How are social engineering attacks classified according to purpose?
- Q4: What are the current strategies to prevent social engineering attacks?

Digital databases, such as ACM Digital Library, IEEE eXplorer, Science Direct Elsevier, Scopus, and Springer Link, were used, which deal with social engineering and technology issues. Academic journals and technical publications between 2015 and 2022 were considered among the sources of information. The search strategy was based on aspects related to the research questions, using the following keywords (in English) as parameters: (a) social engineering, (b) persuasion, (c) human vulnerability to ("information security" OR "cybersecurity" OR "attack"). In addition, to refine the selection, the following criteria were applied (see Table 1).

Table 1. Selection criteria

Inclusion Criteria	Exclusion Criteria
Articles that address the definition of social engineering	Information posted on general websites
Documents outlining the forms and types of social engineering attacks	Documents with irrelevant contributions
Articles with information about defense methods developed to prevent social engineering attacks	Blog information

B. Carrying out the review: The articles were selected based on this phase's search strings and selection criteria. The titles, content, and conclusions were reviewed in each one, which allowed determining their contribution to the questions posed. As a result of the search, 163 documents were identified. Only 38 documents that met the established criteria were selected.
C. Analysis of results: The foundation of social engineering was determined by answering Q1: What are the foundations of social engineering?

Social engineering is based on the basic principle that the user is the weakest link in the information chain, considering that no system in the world not depends to a lesser or greater degree on a human being [11]. Therefore, social engineering is a universal and platform-independent vulnerability [12]. Cybersecurity experts often state that a truly secure computer is offline, to which social engineers reply that there will always be an opportunity to convince someone to turn it on [13].

On the other hand, at a more fundamental level, important discoveries have been made in social psychology about the principles of persuasion. Mainly the work of Cialdini [14], an expert in the area, indicates that persuasion is frequently used as a reference in contributions to social engineering research, considering that, despite being more focused on persuasion within marketing, the fundamental principles are crucial for anyone who needs to understand how deception works. Considering that social engineering tries to deceive decision-makers, which is similar to the stratagems used thousands of years ago, it can be said that it has existed in many forms throughout history and will continue to exist [15]. For this reason, removing social engineering violations is virtually impossible. Some authors point out that security awareness training does not reduce this vulnerability to zero probably [16].

The persistent cybersecurity threat caused by social engineering has roots in exploited human vulnerabilities rather than computer system security flaws. These facts are because computer vulnerabilities can be patched to perfection while existing human vulnerabilities universally accompany individuals throughout life and only disappear when the human race does [5, 17].

Breda et al. [18] point out that a talented practitioner of this discipline understands and perceives the patterns of social interaction to manipulate the psychological aspects of the human mind. When manipulation is achieved, the attacker can execute an efficient security attack without investing heavily in breaking security techniques. However, a social engineer educated in computer science can also complement the attack with technological means to achieve malicious intentions [19].

Q2: What are the most common social engineering attacks, and what are the phases?

Today, social engineering attacks represent the most significant threat facing cybersecurity. Some authors point out that it can be detected but have yet to be stopped [20]. Social engineers prey on victims to obtain sensitive information, which can be used for specific purposes or sold on the black market and dark web [21]. According to Mouton et al. [22], social engineering attacks differ despite having a typical pattern with four similar phases. Similarly, Abass [23] points out that most pirates use these steps to start the attack, considering that this order provides a roadmap to approach the target safely and without suspicion. These phases are:

1. Collect information about the target: In this phase, also called information gathering, the attacker selects victims based on specific requirements. Over the past decade, some of the biggest security threats have come from the use of social media. The rapid growth of these technologies allows millions of users to post daily on Facebook, Twitter, and many other networks [24]. That type of published information may seem like it is optional. Still, it is the key to launching a successful attack since it includes photos, likes, dislikes, personal information, location, friend information, and business information [7]. The danger of making a large amount of information available is that a curious attacker can piece together these sources and get a clear picture of an individual or company. Additionally, once this information is published, it is practically impossible to remove it from a social network due to the high possibility that it will be forwarded to others and posted again [25]. The truth is that with this information in hand, the attacker can use social engineering to make a convincing impersonation of that person or win a business using privileged information [22].
2. Develop a relationship with the target: At this stage, the attacker gains the victims' trust through direct contact or email communication. Trust development can be achieved through insider information, usually presenting oneself as an older group member, based on the human nature of trusting others until they prove themselves untrustworthy [26]. Suppose someone claims to be a specific person. This claim is usually accepted in that case, so a skilled hacker will often try to exploit this weakness before spending time and effort on other methods of cracking passwords or gaining access to systems [22].
3. Exploit the available information and execute the attack: This stage is called the gaming phase. At this stage, the attacker emotionally influences the victims to provide the attacker with sensitive information or make security mistakes. For this reason, this phase focuses on maintaining the compliance momentum built in the second step without raising suspicion, using manipulative tactics to get the target into a desired emotional state appropriate to the plan [23]. The attacker studies the victim's emotional state and then uses it for his own benefit. This exploitation can occur by disclosing seemingly insignificant information or granting or transferring access to the attacker. In this phase, the actual attack begins without alerting the victim that she is being attacked [27]. Significant examples of successful exploitation include: Personal Information – holding the door open or allowing the attacker to enter the

premises; Provide social proof by introducing the social engineer to the rest of the company's staff; Expose trade secrets in discussion with a suspected partner.
4. Traceless output: The attacker leaves without any evidence, leaving the victim feeling that they did something good for someone else, which allows for possible future interactions [28].

Q3: How are social engineering attacks classified according to purpose?

Social engineering attacks can be classified into two categories: human and computer. In human-based attacks, the attacker executes the attack in person by interacting with the target to collect the desired information. However, in this way, the attacker can only influence a limited number of victims [29]. While software-based attacks are carried out using devices such as computers or mobile phones to obtain information from the targets, reaching many victims in a few seconds [30]. Social engineering toolkit is one of the hacking attacks used for spear phishing emails [31]. Social-engineering attacks can also be classified into three categories, depending on how the attack is carried out: social, technical, and physical attacks. Socially-based attacks are carried out through relationships with victims to play on psychology and emotion. These attacks are the most dangerous and successful since they involve human interactions, the best examples being baiting and spear phishing [32].

Technical-based attacks are carried out over the Internet using social networks and online service websites. These attacks collect desired information such as passwords, credit card details, and security questions [33]. Finally, physical attacks refer to physical actions performed by the attacker to collect information about the target, such as searching for valuable documents in dumpsters [34]. Social engineering attacks can combine all of these aspects described. Examples of social engineering attacks include phishing, phishing on tech support calls, shoulder surfing, dumpster diving, theft of important documents, diversion theft, fake software, harassment, quid pro quo, pretexting, tracking, pop-ups, robocalls, ransomware, online, social engineering, reverse social engineering, and phone social engineering [7, 35].

Social engineering attacks can be classified into various categories based on multiple perspectives. Thus, they can be classified into two types according to the entity involved: humans or software. They can be classified into three categories depending on how the attack is carried out: social, technical, and physical. Therefore, through the analysis of the different existing classifications of social engineering attacks, these attacks can also be classified into two broad categories: direct and indirect. The first type of attack uses direct contact between the attacker and the victim to carry out the attack. They refer to attacks carried out through physical, visual, or voice interactions. However, they may also require the attacker's presence in the victim's work area to execute the attack. Examples of these attacks are physical access, shoulder surfing, dumpster diving, telephone social engineering, pretexts, identity theft in technical support calls, and theft of important documents [34, 36, 37]. Attacks classified in the indirect category do not require the attacker's presence to launch an attack. These attacks include phishing, fake software, pop-ups, ransomware, SMSishing, online social engineering, and reverse social engineering [7].

Q4: What are the strategies to prevent social engineering attacks?

Companies are investing more and more money and resources to establish effective strategies against social engineering attacks. However, existing detection methods have fundamental limitations, and countermeasures are ineffective in dealing with the increasing number of such attacks [38]. On the one hand, human-based techniques are limited by the subjective decisions of humans. On the other hand, technology-based approaches can also be limited since technological vulnerabilities can be exploited. In addition, this type of attack evolves day after day, and the attackers are getting smarter and stronger [39]. Therefore, there is a great need for more effective detection techniques and countermeasures to detect and minimize their impact [40]. Because humans are a security challenge on any network, developing employee training programs is essential. More specifically, direct this education at the early levels of school education, considering that training students at an early age can minimize the number of victims in the future. For this reason, each country's investment in cybernetics education must be increased [31].

On the other hand, according to Sandoval [13], some of the practical strategies for this type of defense are: A) For no reason, disclose sensitive information with strangers or in public places, such as social networks, advertisements, and web pages, between others; B) Suppose that suspect that there is a possibility that someone is trying to commit a deception. In that case, it is essential to demand identification and attempt to reverse the situation, trying to obtain as much information about the suspect as possible without alerting him; C) Implement a set of security policies that minimize risky actions and, very importantly, communicate them to all company users; D) Carry out physical security controls to reduce the inherent danger to people; E) Routinely perform audits and pen tests using social engineering to detect security holes of this nature; F) Develop information security awareness programs.

3 Discussion

As reviewed, social engineering is based on two clearly defined aspects. In the first place is human vulnerability, considering that the individuals or users who intervene in the information systems represent the weakest element of the process, regardless of the technological platform used [11, 12]. Through strategies of deception and persuasion, social engineers manipulate people, knowing full well that there is always the opportunity to control people's actions to achieve their goal of violating information security [31]. Second, there are errors and security flaws in computer systems. However, even though they can be corrected or patched as a preventive measure against a computer attack, they will always be at risk of human vulnerabilities, as pointed out by Dhiman et al. [17].

Regarding the stages of social engineering attacks, authors such as Mouton et al. [22] and Abass [23] agree that they are divided into four phases, all applicable to victims: 1) Collection of the most significant amount of information from the target, mainly through the use of social networks [24]; 2) Development of a relationship with the target, which can be achieved by direct contact or by electronic messaging, to create trust [26]; 3) Taking advantage of the information collected and executing the attack to get the target to unconsciously adapt their actions to the plan established by the social engineer [23]; 4) Leave without leaving traces, in such a way that the door is left open for future attacks [28].

On the other hand, social engineering attacks can be classified as human [29] and computerized [30, 31]. When considering how attacks are carried out, they can be technical [33], physical [34], and social. According to Patil and Devale [32], social attacks are the ones with the highest risk because they involve the psychological and emotional aspects of the victims. Despite the constant development of software with various strategies for information security, the human factor will always be the main challenge when designing and applying detection and evasion mechanisms for social engineering attacks [38, 40]. Therefore, the primary strategy to avoid them must be based on the training and education of individuals from homes, educational institutes, and organizations, providing knowledge and tools that prevent and avoid the risks to which computer systems are permanently subjected.

4 Conclusions

The vertiginous evolution of humanity has been presented mainly by the increase in knowledge accessible to all, which has been promoted and facilitated to a great extent by the development of the information age, together with the growing use of the Internet, which has established an absolute reliance on the World Wide Web (WWW) in virtually all human activities. However, the digital environment as an infrastructure favors the increase in cybercrime based on the needs of society to become an increasingly protected environment.

Considering that social engineering acts based on basic parameters such as human vulnerability and failures in information systems, this has allowed cybersecurity to develop to increase the sophistication of processes constantly. However, people today are more exposed because cybercrime is applied by threat actors, who do not necessarily have substantial technical knowledge about information systems but focus on exploiting human weaknesses very well through elaborate deception. These people are known as social engineers, and it has been demonstrated that the human component is at the center of the infection chain in most cyber-attacks.

Likewise, social engineering is constantly increasing both in sophistication and ruthless efficiency, achieving the development of four very well-structured phases, such as the collection of information about the target; development of the relationship with the target; exploitation of the available information and execution of the attack and, finally, exit without leaving any evidence.

On the other hand, this development allows social engineering attacks to be divided into two large blocks corresponding to the objective and the way they are carried out. The first block is classified as human and computer, while the second block is classified as social, technical, and physical.

Consequently, it can be predicted that social engineering will be the most predominant attack vector within cyber security and will be the subject of increasingly in-depth studies as it evolves. All this is to recommend, educate and apply good practices and prevention mechanisms for individuals and organizations, considering that the human factor represents the most significant risk within the security of information systems.

References

1. Dwivedi, Y.K., et al.: Setting the future of digital and social media marketing research: perspectives and research propositions. Int. J. Inf. Manag. **59**, 102168 (2021). https://doi.org/10.1016/j.ijinfomgt.2020.102168
2. Bolton, R.N., et al.: Customer experience challenges: bringing together digital, physical and social realms. J. Serv. Manag. **29**(5), 776–808 (2018)
3. Verizon Business Ready: 2019 Data Breach Investigations Report. (2019). Accedido: 8 de julio de 2022. [En línea]. Disponible en: https://www.phishingbox.com/downloads/Verizon-Data-Breach-Investigations-Report-DBIR-2019.pdf
4. Klimburg-Witjes, N., Wentland, A.: Hacking humans? Social engineering and the construction of the "deficient user" in cybersecurity discourses. Sci. Technol. Hum. Values **46**(6), 1316–1339 (2021)
5. Wang, Z., Sun, L., Zhu, H.: Defining social engineering in cybersecurity. IEEE Access **8**, 85094–85115 (2020). https://doi.org/10.1109/ACCESS.2020.2992807
6. Pîrnău, M.: Considerations on preventing social engineering over the internet, p. 12
7. Salahdine, F., Kaabouch, N.: Social engineering attacks: a survey. Future Internet **11**(4), Art. n.o 4 (2019). https://doi.org/10.3390/fi11040089
8. Algarni, A., Xu, Y., Chan, T.: An empirical study on the susceptibility to social engineering in social networking sites: the case of Facebook. Eur. J. Inf. Syst. **26**(6), 661–687 (2017). https://doi.org/10.1057/s41303-017-0057-y
9. Venkatesha, S., Reddy, K.R., Chandavarkar, B.R.: Social engineering attacks during the COVID-19 pandemic. SN Comput. Sci. **2**(2), 78 (2021)
10. Kitchenham, B.: Procedures for Performing Systematic Reviews, vol. 33. Keele University, Keele, UK (2004). [En línea]. Disponible en: https://www.researchgate.net/publication/228756057_Procedures_for_Performing_Systematic_Reviews
11. Fan, W., Lwakatare, K., Rong, R.: Social engineering: I-E based model of human weakness for attack and defense investigations. Int. J. Comput. Netw. Inf. Secur. **09**, 1–11 (2017)
12. Wang, Z., Zhu, H., Sun, L.: Social engineering in cybersecurity: effect mechanisms, human vulnerabilities and attack methods. IEEE Access **9**, 11895–11910 (2021). https://doi.org/10.1109/ACCESS.2021.3051633
13. Sandoval, E.: Ingeniería Social: Corrompiendo la mente humana. Defensa Digital **10**, 23–28 (2011)
14. Cialdini, R.B.: Influence: Science and Practice the Comic. Writers of the Round Table Press (2012)
15. Hadnagy, C.: Ingeniería social. El arte del hacking personal. Anaya Multimedia (2011)
16. van Mourik, D.-J.: Targeted attacks and the human vulnerability How to assess susceptibility to targeted cyber attacks exploiting human vulnerabilities. February 2017, Accedido: 9 de julio de 2022. [Enlínea]. Disponible en: https://hdl.handle.net/1887/64557
17. Dhiman, P., Wajid, S.A., Quraishi, F.F.: A comprehensive study of social engineering - the art of mind hacking. Int. J. Sci. Res. Comput. Sci. Eng. Inf. Technol. **2**(6), 543–548 (2017)
18. Breda, F., Barbosa, H., Morais, T.: Social engineering and cyber security, pp. 4204–4211 (2017). https://doi.org/10.21125/inted.2017.1008
19. Mouton, F., Malan, M.M., Leenen, L., Venter, H.S.: Social engineering attack framework. In: 2014 Information Security for South Africa, Johannesburg, South Africa, ago. 2014, pp. 1–9 (2014). https://doi.org/10.1109/ISSA.2014.6950510
20. Libicki, M.: Could the issue of DPRK hacking benefit from benign neglect? Georgetown J. Int. Aff. **19**, 83–89 (2018)
21. Alkhalil, Z., Hewage, C., Nawaf, L., Khan, I.: Phishing attacks: a recent comprehensive study and a new anatomy. Front. Comput. Sci. **3** (2021). Accedido: 13 de agosto de 2022. [En línea]. Disponible en: https://www.frontiersin.org/articles/10.3389/fcomp.2021.563060

22. Mouton, F., Leenen, L., Venter, H.S.: Social engineering attack examples, templates and scenarios. Comput. Secur. **59**, 186–209 (2016)
23. Abass, I.A.M.: Social engineering threat and defense: a literature survey. J. Inf. Secur. **09**(04), Art. No. 04 (2018)
24. Sushama, C., Kumar, M., Neelima, P.: Privacy and security issues in the future: a social media. Mater. Today Proc. (2021). https://doi.org/10.1016/j.matpr.2020.11.105
25. Lohani, S.: Social Engineering: Hacking into Humans». Rochester, NY, 5 de febrero de 2019. Accedido: 15 de agosto de 2022. [En línea]. Disponible en: https://papers.ssrn.com/abstract=3329391
26. Jang-Jaccard, J., Nepal, S.: A survey of emerging threats in cybersecurity. J. Comput. Syst. Sci. **80**(5), 973–993 (2014)
27. Chizari, H., Zulkurnain, A., Hamidy, A., Husain, A.: Social engineering attack mitigation. Int. J. Math. Comput. Sci. 188–198 (2015)
28. Nyirak, A.: Social Engineering Tools. Security Through Education (2017). Accedido: 9 de julio de 2022. Disponible en: https://www.social-engineer.org/framework/se-tools/
29. Airehrour, D., Nair, N., Madanian, S.: Social engineering attacks and countermeasures in the New Zealand banking system: advancing a user-reflective mitigation model. Information **9**, 110 (2018). https://doi.org/10.3390/info9050110
30. Thomas, C.: Computer security threats. In: IntechOpen (2020). https://doi.org/10.5772/intechopen.93041
31. Guaña-Moya, J., Chiluisa-Chiluisa, M.A., del Carmen Jaramillo-Flores, P., Naranjo-Villota, D., Mora-Zambrano, E.R., Larrea-Torres, L.G.: Phishing attacks and how to prevent them. In: 2022 17th Iberian Conference on Information Systems and Technologies (CISTI), pp. 1–6. IEEE, June 2022
32. Patil, P., Devale, P.R.: A literature survey of phishing attack technique, **5**(4), 3 (2016)
33. Kalniņš, R., Puriņš, J., Alksnis, G.: security evaluation of wireless network access points. Appl. Comput. Syst. **21**(1), 38–45 (2017)
34. Pokrovskaia, N.N., Snisarenko, S.O.: Social engineering and digital technologies for the security of the social capital' development. In: 2017 International Conference «Quality Management, Transport and Information Security, Information Technologies» (IT&QM&IS), September 2017, pp. 16–18 (2017). https://doi.org/10.1109/ITMQIS.2017.8085750
35. Yasin, A., Fatima, R., Liu, L., Yasin, A., Wang, J.: Contemplating social engineering studies and attack scenarios: a review study. Secur. Priv. **2**, e73 (2019)
36. Aroyo, A.M., Rea, F., Sandini, G., Sciutti, A.: Trust and social engineering in human robot interaction: will a robot make you disclose sensitive information, conform to its recommendations or gamble?. IEEE Robot. Autom. Lett. **3**(4), 3701–3708 (2018). https://doi.org/10.1109/LRA.2018.2856272
37. Beckers, K., Pape, S.: A serious game for eliciting social engineering security requirements. In: 2016 IEEE 24th International Requirements Engineering Conference (RE), September 2016, pp. 16–25 (2016). https://doi.org/10.1109/RE.2016.39
38. Cullen, A., Armitage, L.: The social engineering attack spiral (SEAS). In: 2016 International Conference on Cyber Security and Protection of Digital Services (Cyber Security), June 2016, pp. 1–6 (2016). https://doi.org/10.1109/CyberSecPODS.2016.7502347
39. Aldawood, H., Skinner, G.: Contemporary cyber security social engineering solutions, measures, policies, tools and applications: a critical appraisal. p. 15 (2019)
40. Heartfield, R., Loukas, G.: A taxonomy of attacks and a survey of defence mechanisms for semantic social engineering attacks. ACM Comput. Surv. **48**(3), 37:1–37:39 (2015). https://doi.org/10.1145/2835375

Several Online Pharmacies Leak Sensitive Health Data to Third Parties

Robin Carlsson[1], Sampsa Rauti[1(✉)], Sini Mickelsson[1], Tuomas Mäkilä[1], Timi Heino[1], Elina Pirjatanniemi[2], and Ville Leppänen[1]

[1] University of Turku, Turku, Finland
{crcarl,sjprau,sini.mickelsson,tdhein,ville.leppanen}@utu.fi
[2] Abo Akademi, Turku, Finland
elina.pirjatanniemi@abo.fi

Abstract. As the demand for digital services keeps growing, online pharmacies have become a very important part of essential digital services. When sensitive personal data such as medicine orders are processed, privacy issues become increasingly important. In this paper, we take a look at personal data delivered to third parties in 20 Finnish online pharmacies. More specifically, we study whether the data on prescription medicine orders is leaked out to third-party analytics services. Our findings reveal that 14 (70%) of studied pharmacies send out information about the customers intending to order prescription medicines to third parties, and 7 (35%) of the pharmacies even leak the data about the specific prescription medicine a specific customer is ordering. We also discuss implications of the data leaks and give suggestions on how this alarming state of affairs can be alleviated.

Keywords: Online pharmacies · Web privacy · Network traffic analysis · Data concerning health

1 Introduction

Digital technologies are a powerful tool for delivering essential services when customers have challenges in using services onsite. Especially many vulnerable people such as the elderly, those with serious health conditions, and sometimes people living in rural areas can benefit from digital services [17]. As connectivity has improved and services can reach a greater portion of the population, legislators have also understood the importance of digital service delivery. In Finland, for instance, the Act on the Provision of Digital Services (306/2019) was passed in 2019 in order to improve the accessibility, quality and security of digital services and allow everyone to use them equally. The COVID-19 pandemic has only made the need for online services more pronounced and accelerated the digital transformation [2].

As the demand for digital services rises, online privacy issues also become increasingly important. One very important example of essential digital services

A. Rocha et al. (Eds.): WorldCIST 2023, LNNS 799, pp. 164–175, 2024.
https://doi.org/10.1007/978-3-031-45642-8_16

are online pharmacies, which we are focusing on in this paper. Sensitive information such as prescription medicine orders are processed in these services, making strict data privacy vitally important. The traditional brick-and-mortar pharmacies are obliged to keep confidential the information regarding the customer's medical conditions, unless there is a lawful basis to turn the information over to third parties such as the customer's explicit consent or a legal obligation. The pharmacies are presumed only to use the information on the customer's medication for the purpose of achieving good patient care and to comply with their mandatory obligations. It is also important to set up the pharmacy facilities so that customers can do business privately and confidentially.

In the current digital world, however, the same principles of confidentiality and privacy do not always appear to be followed so well in practice. Third-party analytics services and tracking mechanisms are widespread across the web [13,19], even in essential web services and on websites maintained by public sector bodies [6,18]. In several essential online services, sensitive personal data can be unintentionally delivered to third parties if the developers and data protection officers have not been careful when designing websites. With this in mind, the current study conducts a network traffic analysis for 20 Finnish online pharmacies in order to find out whether personal data and sensitive information is being leaked to third-party analytics services.

The contributions of this paper are as follows. Along with Zheutlin et al.'s study on data-tracking in online pharmacies [20], to the best of our knowledge, the current study is one of the first studies to address third-party tracking in online pharmacies. In their brief article, Zheutlin et al. discuss the prevalence of data tracking among online pharmacies but do not go into the detail about what kind of data is shared to the third-party analytics services. The current study not only researches whether tracking happens and is widespread, but also provides a technical analysis on personal data shared with third parties and investigates whether sensitive information is being delivered to analytics services in several online pharmacies. In our technical analysis, we focus on the use case in which a customer searches and places an order for a prescription medicine. Specifically, we want to find out whether the intent to order a specific prescription medicine is leaked out to third parties. In doing so, the authors also revealed serious data leaks among several Finnish online pharmacies. We have reported these privacy issues to the appropriate authorities, and one of the studied pharmacies has already made significant changes to its website and privacy practices. Finally, the current study discusses the implications of our results for web development and online pharmacies in general.

The rest of the paper is organized as follows. Section 2 reviews the related work. Section 3 introduces the setting of the study, the used data set and methodology. Section 4 discusses the results of the study, providing an analysis on personal data the studied online pharmacies leak to third parties. Section 5 discusses the implications of our key findings and provides some guidelines to improve the privacy of online pharmacies. Finally, Sect. 6 concludes the paper.

2 Related Work

Zheutlin et al. [20] study the prevalence of data tracking among online pharmacies. Among accredited digital pharmacies in the US, analytics services were found to be common, with over two-thirds of the studied pharmacies using at least two third-party services to capture user data. The most prevalent data-tracking services were Facebook and Google Analytics. While the authors do note that the results raises concerns about how health data – such as information on prescription medication – is shared, they do not present any more details about the nature of the data delivered to the third-party analytics services. Therefore, our current study can be seen as a continuation of their work, although the set of studied pharmacies is different.

In a recent study, Huo et al. [7] studied potential health data leaks and analytics on 459 online patient portals and 4 telehealth websites. Their findings indicate that 14% of patient portals include Google Analytics. Also, 9 websites contained services disclosing sensitive health data, such as medications and laboratory results, to third parties. It is quite obvious that It is quite obvious that many health care operators and maintainers of health related websites have not been aware of these problems or lacked the technological skills to detect and fix them. Indeed, in several medical studies (see e.g. [8,15]), analytics services are still being used as a part of health related web platforms, most likely without fully realizing the risks involved. The results of Huo et al., just like the findings of our current study, highlight a lack of a privacy-by-design approach, and put emphasis on the importance of educating website developers and data protection officers about this issue.

In a similar vein, Zheutlin et al. studied data-tracking on government, non-profit, and commercial health-related websites [21]. The results indicate that it is relatively common for health-related websites to provide information to third-parties, the average number of analytics services on the studied websites was 2.11. Unfortunately, in many cases, finding and displaying health information online is not a private action. This problem does not only concern health-related websites but many other critical services as well. Recent studies have shown that even many public sector bodies can leak sensitive information to third parties from their websites [6].

Finally, it is also important to have the appropriate tools and methods for studying network traffic and compliance issues on websites. The users need to know if websites really act according to their data privacy choices. Martinez et al. [10] present new algorithms and a measure to assess user tracking compliance and the confidence in the analytics services. From a technical viewpoint, it is necessary to be able to accurately identify personal data from internet traffic when researching what kind of information is delivered to analytics services. To this end, Liu et al. [9] develop a new method of discovering various types of personal data carried within network traffic.

3 Study Setting and Methodology

From the list of legal Finnish online pharmacies[1] maintained by the Finnish Medicines Agency, Fimea, 20 online pharmacies were chosen to be studied. If a chosen online pharmacy did not sell prescription medicines at all, the pharmacy was discarded from the set of selected pharmacies and a new one was chosen at random. In this study, we chose not to refer to the pharmacies by their real names, but instead call them Pharmacies 1–20.

The online pharmacies were tested by first navigating to the store page, clearing the browser cache and cookies and then reloading the page. From reloading onwards, all the network traffic associated with use of the pharmacy was captured using Google Chrome's DevTools. The cache was disabled while recording and the captured traffic was preserved as a HTTP Archive file. When arriving at the pharmacy website, all cookies were consented to upon request.

While the exact navigation on websites varies between the studied online pharmacies, the experiment was always continued until the intent to order a prescription medicine was clear – for example, a button for ordering a specific medicine was pressed or the medicine was successfully added as part of the order. The goal was to find out whether the intent to order a specific prescription medicine was delivered to third-party analytics services. To place a final order, chatting with a pharmacist is required in Finnish online pharmacies. We ended our experiment before this phase.

The 20 online pharmacies tested all provided a search bar, and the name of a prescription medicine was entered into the search. If the product came up in the search (i.e. prescription medicines were searchable items), its product page was opened. Finally, if the product page led to a separate page for ordering the product, that page was also accessed to see whether sensitive information about the user's intent to purchase this prescription medicine was sent to analytics services.

If the prescription medicine was found in the search and the product page was accessed, but no separate page for ordering the product was available, we registered and logged into the online pharmacy to initiate an order. After registering, the ordering process was continued until contact with a pharmacy assistant was required.

If prescription medicines were not available in the search, but the pharmacy offered a link or button for beginning the ordering process directly, this option was used instead. An order was initiated, and the correct product was then selected to be part of the order. The ordering process was interrupted before requesting that a pharmacy assistant contact the user.

Moreover, after running the experiments described above, the privacy policies of the chosen online pharmacies were analyzed. Each privacy policy document was read to find out whether anything about sending data on ordered prescription medicines to third parties was mentioned. The studied privacy policies were analyzed by two researchers and any disagreements were discussed until agreement was reached.

---

[1] https://www.fimea.fi/apteekit/verkkopalvelutoiminta/lailliset_apteekin_verkkopalvelut.

Finally, we briefly summarize what is meant by the term *personal data*. In this paper, the term is given the same meaning as in the EU General Data Protection Regulation (GDPR)[2]. Pursuant to Article 4(1) of the GDPR "personal data" refers to any information relating to an identified or identifiable person, in other words, data based on which an individual can be identified directly or indirectly. According to the GDPR a person can be identifiable specifically based on a reference to an identifier such as name, location data, an online identifier or to one or more factors specific to the physical, physiological, genetic, mental, economic, cultural or social identity of the said person. In general, the scope of the term is considered to be broad (see e.g. [1,3]; [4, p. 113], [12, p. 41–42]).

4 Results: Analysis on Personal Data Sent to Third Parties

Table 1 shows the third-party analytics services that were found on the pharmacy websites while performing our experiments and recording network traffic. We can see that only 6 pharmacies did not have any analytics on the pages we tested. Pharmacies 1 and 2 had 4 different third-party services, which can be considered a large number as we only followed a path consisting of a few different pages. When there were any analytics services on the studied website, Google was almost always present, except in two cases where Pingdom, a Swedish website monitoring service was used. Other frequently appearing third-party services were Facebook and Giosg. The latter is not an analytics service per se but rather a company providing live chat services.

The data sent to third parties contains items such as IP addresses, device and user specific identifiers, User-Agent headers with information on operating system and browser, and other technical pieces of information such as screen size. When it comes to identifying the user, an important piece of data is often the device's IP address, which is delivered along with every web request.

According to the preamble to the GDPR, in determining whether a person is identifiable, all the means reasonably likely to be used to identify the person directly or indirectly should be taken into consideration. This includes all objective facts, such as the costs and the amount of time required for identification as well as the available technology and technological developments.[3] Pursuant to the case Breyer of the Court of Justice of the European Union, IP addresses can be considered as personal data even if identifying of a person requires acquiring additional information from a third party [1].

In accordance with recital 26 of the GDPR, data protection provisions do not apply to anonymized information. Google Analytics, for example, can be

[2] Regulation (EU) 2016/679 of the European Parliament and of the Council of 27 April 2016 on the protection of natural persons with regard to the processing of personal data and on the free movement of such data, and repealing Directive 95/46/EU (General Data Protection Regulation). Official Journal of the European Union, L 119, 4 May 2016.

[3] Recital 26 of the preamble to the GDPR.

configured to anonymize data by removing the last octet of the user's IP address. It is questionable, however, whether this anonymization renders data anonymous "in such a manner that the data subject is not or no longer identifiable". This is because even if the IP address is partially anonymized, lots of other technical information is sent along it, making identification of the user possible, especially for big data collectors such as Google. Taking all this into consideration, it can be argued that in the case of the studied online pharmacies, a significant risk exists that the data could be linked to the person visiting the website.

Table 1. Pharmacies and detected third-party analytics services.

	Third-party analytics services
Pharmacy 1	Facebook, Giosg, Google, New Relic
Pharmacy 2	Crazy Egg, Facebook, Giosg, Google
Pharmacy 3	Google
Pharmacy 4	Facebook, Google
Pharmacy 5	
Pharmacy 6	
Pharmacy 7	
Pharmacy 8	
Pharmacy 9	Google
Pharmacy 10	
Pharmacy 11	Google, Pingdom
Pharmacy 12	Facebook, Google, Pingdom
Pharmacy 13	Pingdom
Pharmacy 14	Pingdom
Pharmacy 15	Google, Pingdom
Pharmacy 16	Google, Pingdom
Pharmacy 17	Google
Pharmacy 18	Giosgm Google
Pharmacy 19	Google, Hotjar
Pharmacy 20	

However, the most interesting and sensitive information sent out to third parties concerns the customer (identified by an IP address or a device identifier) visiting the product page of a specific prescription medicine, or visiting the prescription medicine order page. The former of these indicates interest in a specific medicine and the latter indicates that the customer intends to place an order. These alone are personal data items that should not be sent to third parties. When the customer's visit on a product page of a specific prescription medicine can be connected to an order by the same customer (indicating an intention to order a specific medicine) and this data is sent to a third party, a strong argument can be made that sensitive health related data is being leaked.

The aforementioned connection between a specific medicine and an order can be strong or weak. The strong connection is formed when the previously visited page (referer) is sent to the third party when the customer is on the prescription medicine order page. This way, the third party can immediately see that the order has been initiated from a product page of a specific medicine, which indicates the intent of ordering the prescription medicine in question. Often the medicine name was even contained in the product page's URL address. The weaker connection is created when the previously visited page is not directly sent to a third party, but both the product page of a medicine and the order page include analytics from the same third party. The third party can now see from timestamps that these pages have been visited consecutively and deduce that an order for a specific medicine is likely being placed. In both of these cases, the third party can say

Table 2. The data items related to viewing and ordering prescription medicines delivered to analytics services.

	Website had analytics	Data on specific medicine being viewed is sent to 3rd party	Data on intent to make order is sent to 3rd party	Intent to order can be connected to specific medicine
Pharmacy 1	X	X	X	X
Pharmacy 2	X	X	X	X
Pharmacy 3	X	X	X	X
Pharmacy 4	X	X	X	X
Pharmacy 5				
Pharmacy 6				
Pharmacy 7				
Pharmacy 8				
Pharmacy 9	X	X	X	X
Pharmacy 10				
Pharmacy 11	X		X	
Pharmacy 12	X		X	
Pharmacy 13	X		X	
Pharmacy 14	X		X	
Pharmacy 15	X		X	
Pharmacy 16	X		X	
Pharmacy 17	X		X	
Pharmacy 18	X	X	X	X
Pharmacy 19	X	X	X	X
Pharmacy 20				

with great certainty that the customer intends to order a specific prescription medicine, although the weaker connection requires more analysis.

Table 2 shows what kind of information about the user's actions is delivered to the analytics providers from each studied pharmacy website. The first column indicates whether the online pharmacy in question had any third-party analytics. The second column designates whether the information about the user viewing a specific medicine's product page is sent to the analytics providers. The third column indicates whether the intent to make an order was leaked. Finally, the fourth column shows whether the intent to order can be linked to a specific medicine.

We can see that in 7 cases, the intent to order a specific medicine was leaked to one or more third-party analytics providers. Altogether, the intent to make an order was leaked in 14 cases, although 7 of these leaks did not have information on the ordered medicine. Although the sample of 20 online pharmacies is not a very large dataset, it is definitely an alarming observation that 14 pharmacies were leaking information about sensitive medicine orders and 7 of these pharmacies revealed the exact prescription medicine a specific user intended to order to third parties.

The different colors in Table 2 indicate different platforms used to build the pharmacy websites. The first platform, marked in red, is a platform used for many different web stores, not only online pharmacies. It has an easy option to integrate Google Analytics to the web store, for example. However, half of the pharmacies built with Platform 1 have chosen not use any analytics. Platform 2, shown in green, is a Finnish platform specifically built for online pharmacies. Nevertheless, analytics are used on every website built on this platform. It noteworthy, however, that on websites built with Platform 2, information on a specific medicine was never sent to analytics services. Finally, Platform 3, in blue, is also a Finnish solution for online pharmacies, emphasizing both use of analytics and security in its advertising. Judging from Pharmacies 17–19, these two goals seem to be in conflict when it comes to practice.

Table 3 shows the information privacy policies contained about delivering sensitive health related data to third parties. The table contains only those 16 online pharmacies that were found to have analytics on their websites. We can see that 10 out of 16 pharmacy websites denied sending any data about medicines or products users have displayed or ordered, although our network traffic analysis clearly proves this happens. Three studied pharmacies admitted that this information can be shared with third parties, although they did not explicitly state that information about intended prescription medicine orders is being sent out. The used language was more subtle, stating that the collected personal data, among many other personal data items, includes information on ordered products. In another section of the privacy policy, it was then stated that personal data can be shared with third parties. One of the privacy policies (Pharmacy 17) did not clearly indicate whether personal data is given to third parties. It is also worth noting that Pharmacy 1 explicitly stated in its cookie consent banner that information about prescriptions or medication is not collected.

Table 3. The information privacy policies contained about delivering sensitive medical data to third parties. The table only contains pharmacies that had analytics on their websites.

	Mention about medicine orders being delivered to 3rd parties	Mention about medicine orders NOT being sent to 3rd parties
Pharmacy 1		X
Pharmacy 2		X
Pharmacy 3	X	
Pharmacy 4	X	
Pharmacy 9	X	
Pharmacy 11		X
Pharmacy 12		X
Pharmacy 13		X
Pharmacy 14		X
Pharmacy 15		X
Pharmacy 16		X
Pharmacy 17		
Pharmacy 18		X
Pharmacy 19		X

It is clear that pharmacies did not adequately inform the users about the fact that sensitive health related data is turned over to third parties. It was also evident in many cases that a privacy policy document had been directly copied from another online pharmacy without sufficiently paying attention to their contents and applicability to the online pharmacy in question. Several privacy policies – or at least large sections of the documents – were identical with each other, even sharing the same spelling mistakes on several occasions.

5 Discussion

Our results have showed that the current state of data privacy in Finnish online pharmacies gives reason for great concern. Sensitive information such as data revealing a person's intention to order prescription medicine should be granted special protection. Also, in many cases this data can be further used to deduce what diseases a person is suffering from. This is especially the case when several medicines are ordered or when the third-party analytics service has an opportunity to observe numerous orders over time, revealing details on a person's medical history. It is clear this information should not be disclosed to third parties. It is also possible that analytics services end up collecting information of an intention to purchase medicine which does not result in ordering the medicine in question due to intervention from the pharmacist. Thus, the collected data can also lead into misinterpretations about the person's health status.

In most cases, the health related data and the third-party analytics companies receiving the data were not mentioned in privacy policy documents. Even if the collection and sharing of data would be transparent, in this context transferring data concerning intention to order prescription medicines to a third party can be considered highly unethical and unnecessary [16].

In this study, we only covered a subset of all Finnish online pharmacies. Judging from the numbers of online pharmacies with serious privacy problems in this subset, however, it is likely that there are dozens of more pharmacy websites that leak health related data to third parties. Therefore, it is safe to say that the problem is much bigger, both in Finland and in other countries. While it is impossible to say whether the analytics providers really store and use the health related data they receive or whether it is discarded, it is unacceptable that data is sent out to begin with. It is important to note, however, that the leaked data about intended prescription medicine orders only goes to analytics service providers who do not necessarily have an incentive to use it further, and the data likely does not end up in open data market. Making use of the data would probably also require some manual work and additional knowledge about how the specific online pharmacy is implemented in many cases.

The software platforms developers use to build online pharmacy websites are a significant factor contributing to the privacy problem. These platforms often readily offer the option to effortlessly deploy and turn on third-party analytics services, which makes it easy to enable analytics without fully appreciating the consequences. While letting health related data leak to third parties may often be unintentional, in the case of online pharmacies the implications can be very serious. It may not immediately occur to developers that in terms of privacy, an online pharmacy should not be treated like an average web store. Software developers and data protection officers should pay more attention to what kind of data flows out from their websites. This is easy to accomplish with a similar setup as the one used in this study. Such data flow analysis should always be an integral part of the web development and testing process. The privacy practises of the used platforms should be carefully assessed and analytics should not be used on pages which reveal vulnerable aspects of users. If analytics are deemed necessary, the data should be stored locally by the pharmacy (e.g. using open source analytics solutions such as Matomo [5]), not delivered to a third party.

It is also obvious from our results that online pharmacies have failed to write clear and truthful privacy policies, which is unfortunately a common problem in today's web services [11]. The analyzed privacy policy documents do not provide appropriate information about processing activities. Following a small number of standardized templates when composing privacy policy documents could make them easier to produce and understand [14].

Finally, it is worth noting that our study also aims to have a societal impact by making the results available for online pharmacies so that data leaks can be fixed and unnecessary analytics services are removed from critical pages. As a result of this study, Pharmacy 1 has already removed analytics services from its website, revamped its privacy policy, and reconsidered its privacy practices. We have also reported our findings on other online pharmacies, which will hopefully also help to improve privacy on their websites.

6 Conclusions

We have presented a study of data leaks on Finnish pharmacy websites. We found that out of 20 studied online pharmacies, 14 pharmacies were leaking information about prescription medicine orders, and 7 of these pharmacies revealed to third parties the exact prescription medicine a specific user intended to order. Although the sample is relatively small, the result that 70% of the studied pharmacies leak health related personal data is highly concerning.

Our findings warrant more research with a wider data set, and we are already in the process of extending this study to cover all Finnish online pharmacies. In the future, online pharmacies in other countries should also be studied in the same manner. Data leaks could also be further studied by experimenting with different consent choices on pharmacy websites.

We also hope that these results are a wake-up call for software developers and data protection officers involved in maintaining essential services that involve sensitive data. It is vitally important for service operators to understand their accountability for protecting customer's privacy in areas where they are particularly vulnerable, including raised awareness and control over the used online platforms and the related design choices. At the same time, users should be clearly informed of what personal data is processed and which parties process it. When it comes to pharmacy websites, the use of any external analytics service, let alone several of them, is difficult to justify. Customers should be able to trust online pharmacies just as much as they trust traditional brick-and-mortar pharmacies.

Acknowledgements. This research has been funded by Academy of Finland project 327397, IDA – Intimacy in Data-Driven Culture.

References

1. Case C-582/14, Patrick Breyer v. Bundesrepublik Deutschland [2016] ECLI:EU:C:2016:779, paragraph 49
2. Almeida, F., Santos, J.D., Monteiro, J.A.: The challenges and opportunities in the digitalization of companies in a post-COVID-19 world. IEEE Eng. Manag. Rev. **48**(3), 97–103 (2020)
3. Article 29 Data Protection Working Party: Opinion 4/2007 on the concept of personal data. adopted on 20th june. wp 136, p. 4
4. Bygrave, L., Tosoni, L.: Article 4(1). personal data. In: Kuner, C., Bygrave, L., Docksey, C., Drechsler, L. (eds.) The EU General Data Protection Regulation: A Commentary. Oxford University Press, Oxford, United Kingdom (2020)

5. Gamalielsson, J., et al.: Towards open government through open source software for web analytics: The case of Matomo. JeDEM-eJournal of eDemocracy Open Gov. **13**(2), 133–153 (2021)
6. Heino, T., Carlsson, R., Rauti, S., Leppänen, V.: Assessing discrepancies between network traffic and privacy policies of public sector web services. In: Proceedings of the 17th International Conference on Availability, Reliability and Security, pp. 1–6 (2022)
7. Huo, M., Bland, M., Levchenko, K.: All eyes on me: inside third party trackers' exfiltration of phi from healthcare providers' online systems. In: Proceedings of the 21st Workshop on Privacy in the Electronic Society, pp. 197–211 (2022)
8. Linardon, J., Rosato, J., Messer, M.: Break binge eating: reach, engagement, and user profile of an internet-based psychoeducational and self-help platform for eating disorders. Int. J. Eat. Disord. **53**(10), 1719–1728 (2020)
9. Liu, Y., Song, H.H., Bermudez, I., Mislove, A., Baldi, M., Tongaonkar, A.: Identifying personal information in internet traffic. In: Proceedings of the 2015 ACM on Conference on Online Social Networks. COSN '15, pp. 59-70. Association for Computing Machinery, New York, NY, USA (2015). https://doi.org/10.1145/2817946.2817947
10. Martínez, D., Calle, E., Jové, A., Pérez-Solà, C.: Web-tracking compliance: websites' level of confidence in the use of information-gathering technologies. Comput. Secur. **122**, 102873 (2022)
11. Mulder, T.: Health apps, their privacy policies and the GDPR. Eur. J. Law Technol. (2019)
12. Purtova, N.: The law of everything. Board concept of personal data and future of EU data protection law. Innov. Technol. **10**(1), 40–81 (2018)
13. Quintel, D., Wilson, R.: Analytics and privacy. Inf. Technol. Libr. **39**(3) (2020)
14. Rowan, M., Dehlinger, J.: A privacy policy comparison of health and fitness related mobile applications. Procedia Comput. Sci. **37**, 348–355 (2014)
15. Santin, O., McShane, T., Hudson, P., Prue, G.: Using a six-step co-design model to develop and test a peer-led web-based resource (PLWR) to support informal carers of cancer patients. Psychooncology **28**(3), 518–524 (2019)
16. Schwartz, P.M.: Privacy, ethics, and analytics. IEEE Secur. Priv. **9**(3), 66–69 (2011)
17. Somenahalli, S., Shipton, M.: Examining the distribution of the elderly and accessibility to essential services. Procedia. Soc. Behav. Sci. **104**, 942–951 (2013)
18. Thompson, N., Ravindran, R., Nicosia, S.: Government data does not mean data governance: lessons learned from a public sector application audit. Gov. Inf. Q. **32**(3), 316–322 (2015)
19. Wambach, T., Bräunlich, K.: The evolution of third-party web tracking. In: Camp, O., Furnell, S., Mori, P. (eds.) ICISSP 2016. CCIS, vol. 691, pp. 130–147. Springer, Cham (2017). https://doi.org/10.1007/978-3-319-54433-5_8
20. Zheutlin, A.R., Niforatos, J.D., Sussman, J.B.: Data-tracking among digital pharmacies. Ann. Pharmacotherapy, 10600280211061757 (2022)
21. Zheutlin, A.R., Niforatos, J.D., Sussman, J.B.: Data-tracking on government, non-profit, and commercial health-related websites. J. Gen. Intern. Med. **37**(5), 1315–1317 (2022)

Verifying the Situation of Cybersecurity in Portugal's Municipalities

Teófilo Branco Júnior(✉) and Isabel Celeste Fonseca

University of Minho, Braga, Portugal
teofilotb@hotmail.com

Abstract. This study presents the results of the application of a survey designed to verify if cybersecurity requirements are being applied in municipal administrations. To check how these information security measures are implemented, we developed a questionnaire containing the leading cyber security checkpoints related to local IT administration. These checkpoints were elaborated based on recommendations from institutions specialized in cybersecurity, scientific literature, and current legislation. The conformity or not of the questions formulated in this survey aims to provide a panorama of the current situation about information security. Furthermore, the analysis of the survey answers by the Municipality made it possible to reveal the stage they are at concerning the practices desired to implement secure and effective cybersecurity in the operation of their information systems. Therefore, we can prepare specific reports based on the Municipality' answers. Thus, the results indicated which cybersecurity gaps need to resolve, practices without compliance with the legislation in force, and how the Municipality can mitigate the security risks involved.

Keywords: Cybersecurity · Cybersecurity Law · Cybersecurity Strategies · Data Privacy · e-Government

1 Introduction

A Capgemini's study, with IDC and Instituto Politecnico di Milano, on the level of e-Government services in Europe, was published in eGovernment Benchmark 2021[1]. The study, which evaluated more than seven thousand web pages from 36 European countries, concluded that more than eight in ten of the assessed public services (81%) are now available online. On the other hand, there was also a sharp increase in cyber-attack cases since 2019 [2]. Moreover, cyber-attacks have also targeted critical national infrastructures such as healthcare and services.

There is a diversity of targets of cyber threats, ranging from public institutions to ordinary citizens. The implications can be catastrophic because, among these institutions, there can be hospitals, schools, and even nuclear power plants [3].

According to [4, 2], like traditional crime, cybercrime is often described by the crime triangle involving three factors: a victim, motive, and opportunity. The victim is the target, the reason is the aspect that leads the criminal to commit the attack, and the opportunity

A. Rocha et al. (Eds.): WorldCIST 2023, LNNS 799, pp. 176–186, 2024.
https://doi.org/10.1007/978-3-031-45642-8_17

is the vulnerability for the crime to be achieved. This study presents the results of the application of a survey designed to verify if cybersecurity requirements are applied in municipal administrations in northern Portugal covered in the project "Smart Cities and Law, E.Governance and Rights: Contributing to the definition and implementation of a Global Strategy for Smart Cities, Ref.: NORTE-01–0145-FEDER-000063".

The present work aims to provide alternatives to improve information security management to avoid the occurrence of cyber-attacks, loss of critical data, and exposure of confidential data and mitigate risks of continuity of services provided in Municipalities covered by the project.

This paper presents the cybersecurity studies that guided this survey's development, its application in a few Municipalities, and the results obtained.

1.1 Research Objectives

The objective of this study and others that complement it is to provide insight into the desired conditions for preparing the Municipality for the change to smart cities. From the maturing understanding of e-Governance, it will be possible to design a global plan for the digital transition of local governments [5]. About the cybersecurity discipline, the goal was to assess the level of readiness of the Municipality to adopt smart cities securely.

In this regard, identifying checkpoints related to cybersecurity principles is an important indicator. Through this research, it is possible to identify risks, raise the awareness of local administrations, propose recommendations, and assist in developing projects to correct existing gaps. In addition, the analyses can justify obtaining the resources needed to implement the required measures to mitigate cyber-attack risks.

1.2 Methodology

We used a qualitative approach with inductive logic, where generalizations are applied to describe the topic of cybersecurity in organizations [6].

The research explores IT management's perception of the current cyber security situation in local connectivity infrastructure and IT resource management.

We employed research techniques such as literature review [7], interviews with IT administrators [8], and PhDs to develop a survey to identify lacks related to the principles and practices regarding cybersecurity. With the application and analysis of this survey, we believe it will be possible to verify whether current methods are appropriate for compliance with cybersecurity principles and requirements or whether there is a need to implement measures to mitigate existing risks.

2 Background

The solution proposed in this study explores concepts about cybersecurity to improve and optimize the administrative management of local governments. In this regard, we identify some referential in the literature review to elucidate this area through the experiences of lessons described in academic articles and recommendations of specialized entities.

2.1 Literature Review

A literature review study by [9] addresses the risk perspectives in implementing the smart cities concept. The authors maintain that if such risks are not understood and addressed, they can create problems in terms of privacy and security and, therefore, affect the functioning of smart cities. In this study, smart-cities risks are technology related, organizational, and external environmental.

Technical risks are related to technology and its implementation, such as risks associated with Internet of Thinks (IoT), BigData, and Artificial Intelligence (IA) as the most important.

The study shows that non-technical risks affect the implementation and operation of smart cities [9]. The authors maintain that socioeconomic hazards include the traditional mindset of stakeholders and decision-makers. Therefore, implementing the smart cities concept means managing multidisciplinary projects that require a considerable budget, trained staff, and technological exposure from citizens, decision-makers, and professionals.

Another concern is managing legal issues related to data privacy and protection risks. According to [10], potential risk sources include typical online activities such as web browsing, e-mail communication, messaging programs, social networks, and wireless networks. Therefore, identifying vulnerabilities and threats and their potential consequences as part of risk analysis is essential to implementing appropriate security measures.

The security concerns from public administration IT employees and citizens represent another risk factor that characterizes the state of cybersecurity in the public sector [8, 11].

According to [10], another risk factor refers to the lack of risk awareness of the higher authority. Although security awareness is to increase at all administrative levels, studies show that a lack of understanding regarding cybersecurity-related issues among staff at the executive level plays a crucial role in cybersecurity in the public sector because it is who approves and provides resources.

2.2 Strategic Goals for Alignment with European Cybersecurity Commission

European Union has a strategy to promote the security of networks and information systems in its member states. The main objective is to increase the ability of governments, citizens, and institutions to deal with cyber-attacks. The strategy includes the definition of procedures to be adopted by governments in preventing and responding to cyber-attacks and international cooperation and collaboration among states to combat cybercrime [3].

Another step was the adoption of the General Data Protection Regulation (GDPR), passed on April 15, 2016, and came into effect on May 25, 2018 [12]. GDPR is a European law regulating privacy and personal data protection, applicable to all individuals in the European Union and the European Economic Area (EEA).

Complementarily, the Network and Information Security Agency (ENISA) [13] establish to assist member states in dealing with cyber-attacks. ENISA's role is to set a primary network and data security level in the European Union, alert citizens to risks and promote an Internet security culture for EU citizens, consumers, businesses, and public authorities.

In Portugal, the Parliament approved the Cyberspace security regime (Law No. 46/2018, of August 13 [14]), transposing Directive (EU) 2016/1148 of the Europe-an Parliament and of the Council, of July 6, 2016, on measures to ensure a high standard level of network and information security across the European Union [15].

The Law [14] approved by the Parliament defines a "National Cyberspace Security Strategy, the national cyberspace security structure, including creating the Higher Council for Cyberspace Security. In addition, it identify the national point of contact for international cooperation (National Cybersecurity Center). She also defined the incident reporting obligations to the National Cybersecurity Center. The National Cybersecurity Center is the National Cybersecurity Authority, and the national cybersecurity incident response team is subordinate to it [15].

Law 46/2018 [14] establishes rules for the Public Administration, namely, as an operator of essential service, critical infrastructure operators, operators of essential services, digital service providers, and any entity that uses networks and information government systems.

2.3 Standard ISO/IEC 27002:2022

ISO/IEC 27002:2022 [16] is a document that provides a reference set of generic information security controls on cybersecurity and privacy protection. This document is for organizations of all types and sizes, including the public and private sectors. It refers to the ISO/IEC 27001-based information security management system (ISMS).

2.4 National Cybersecurity Strategies (NCSS)

ENISA through the National Cyber Security Centre (NCSC) published a 2021 guide [17] to help local authorities understand the security considerations needed to design, build, and manage intelligent cities. In addition, the manual recommends a set of cybersecurity principles that can help secure an infrastructure to be more resilient to cyber-attacks. According to the NCSC, the guide is particularly relevant for information security administrators, CISOs, cybersecurity architects and engineers, and the personnel who will manage the day-to-day operations of the cyber network infrastructure.

3 The Survey

The team designed a survey to be applied to organizations to check their procedures regarding the processes related to cybersecurity. This questionnaire is the checklist object of this study. This survey intends to collect data regarding the local reality of cybersecurity. The questionnaire contains ten dimensions and 120 questions:

- Dimension I - Organizational Security Structure: This topic consists of three questions (1 to 3) and intends to assess whether the organization has an internal area with competencies in cybersecurity and how that is structured.

- Dimension II – Asset Management: This topic consists of ten questions (4 to 13) and intends to know if IT area have knowledge of their IT assets covers.
- Dimension III – Cyber Risk Management: This topic consists of 19 questions (14 to 32) and intends to investigate if risk analysis is on the technology employed in the organization and whether there is a plan for managing the risk.
- Dimension IV – Architecture and Configuration: This topic consists of 20 questions (33 to 52). It aims to verify cyber security in its systems from the beginning of its projects and if these can be updated to adapt to emerging threats and risks.
- Dimension V – Vulnerability Management: This topic consists of five questions (53 to 57). The goal is to find out if security updates are installed as soon as they are available and if there is vulnerability management and penetration testing.
- Dimension VI – Identity Management and Access Control: This topic consists of 14 questions (58 to 71). The goal is to know whether access is protected. Whether methods are employed to keep user accounts Up To Date and if it is possible to prove the identity of users on devices or systems.
- Dimension VII – Corporate Data Security: This topic contains 15 questions (72 to 86). The objective is to investigate whether data is protected against unauthorized access, modification, or deletion. If exist protections and safeguards from third parties, as in the cases of cloud computing and backup copies of critical data.
- Dimension VIII – Logging and Monitoring: This topic consists of 12 questions (87 to 98). The objective is to find out if there are records of system use and if security monitoring or protection is carried out in case of known attacks or unusual system behavior. It also intends to assess whether incidents are detected.
- Dimension IX – Incident Management: This topic consists of 14 questions (99 to 112). The objective is to know if there is incident management to help prevent further damage, reducing the financial and operational impact.
- Dimension X – Supply Chain: This topic is composed of eight questions (113 to 120). The goal is to know if policies are in place to improve security across the entire supply chain, including commodity suppliers, cloud providers, and vendors with which the organization has contracts.

All questions are "yes" or "no". Answer "yes" indicates compliance and "no" represents a non-conformity. After a general assessment of the instrument, the project team concluded that the tool allows for revealing the current practices concerning cybersecurity in the Municipality. The full inquiry can be accessed at "https://smartcitiesandlaw.pt/questionarios/" (part 4).

4 Application and Results

Five Municipalities responded the survey from April to August 2022. For each Municipality, we adopted the letters A, B, C, D, and E to preserve their confidentiality. The questions in "compliance" received a valid score of one, and the questions in "non-compliance" obtained a score of zero. Referring to the answers given by the Municipality, it was possible to conduct the following analyzes referring to each of the survey dimensions:

a) Lack of cybersecurity practices represented by non-compliance.
b) A Percentual Conformity Score (pcs) in inquiry dimension by each Municipality calculated with the formula: pcs = x/y*100, when x = numbers of conformities and y = number of questions.

Dimension I - Organizational Security Structure: a) Four of the five Municipalities do not have an area dedicated to cybersecurity. However, three of them do conduct related activities. Furthermore, one Municipality has designated a Permanent Contact Point with the National Cybersecurity Center. b) Fig. 1 shows the pcs. Three do not meet the indicators appointed in the survey, and two met 33% of the recommendations.

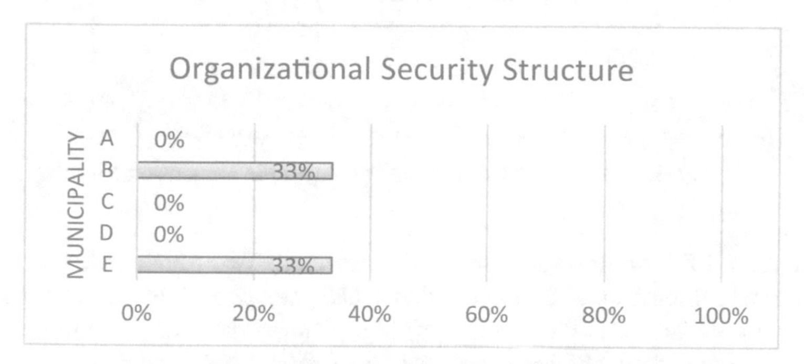

Fig. 1. Percentual Conformity Score (pcs) in Dimension 1

Dimension II – Asset Management: a) Four Municipalities reported having an updated inventory of the essential assets for their services. However, these do not report to the National Cybersecurity Center. Four informed us that their access accounts are duly cataloged. In these, the respondents declared that the IT area has tools to identify unknown assets. b) Fig. 2 shows the pcs. One did not reply to this topic, and the others obtained compliance of 50%, 90%, 10%, and 20%.

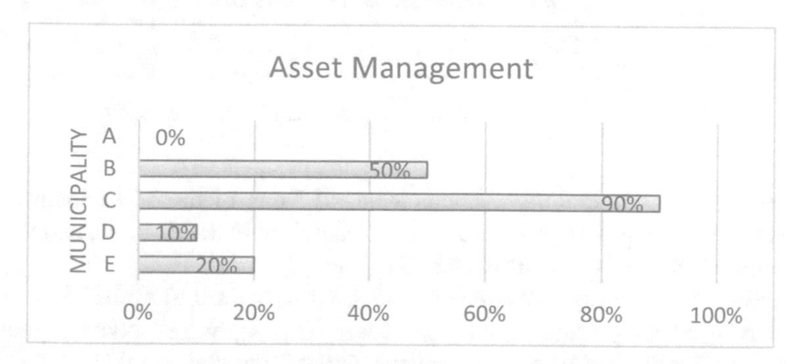

Fig. 2. - Percentual Conformity Score (pcs) in Dimension II

Dimension III – Cyber Risk Management: a) None of them performs risk analysis regarding the Computing assets. Although one of the Municipality claimed to have the means to track the "digital fingerprints" of its users. Two reported conducting awareness training in cybersecurity. b) Fig. 3 shows the pcs. One did not reply to this topic, and the others obtained conformities of 42%, 89%, 11%, and 16%.

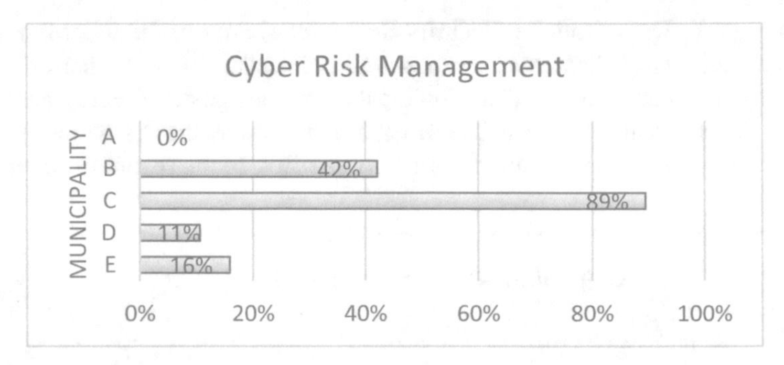

Fig. 3. - Percentual Conformity Score (pcs) in Dimension III.

Dimension IV – Architecture and Configuration: a) Two Municipalities claimed to have a standard threat model for their systems. Only one considers contracting a shared responsibility model for cloud services. b) Fig. 4 shows the pcs. Two did not reply to this topic, and the others obtained conformities of 75%, 100%, and 80%.

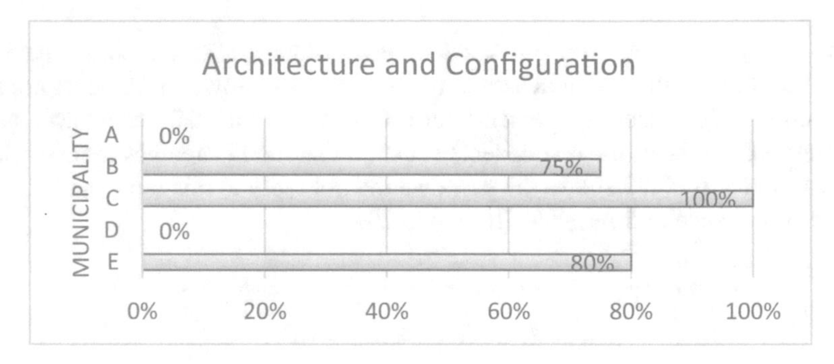

Fig. 4. - Percentual Conformity Score (pcs) in Dimension IV.

Dimension V – Vulnerability Management: a) None of the Municipalities adopt penetration tests. b) Fig. 5 shows the pcs. Two did not reply to this topic, and the others obtained conformities of 40%, 80%, and 20%.

Dimension VI – Identity Management and Access Control: a) Three Municipalities reported having identity management and access control policies. None of them uses multi-factor authentication to protect systems access. One claimed that establish non-disclosure data agreements with third-party entities. b) Fig. 6 shows the pcs. Two did not reply to this topic, and the others obtained conformities of 71%, 86%, and 71%.

Dimension VII – Corporate Data Security: a) None of the Municipalities reported performing periodic tests on the intentional data destruction equipment to gauge and

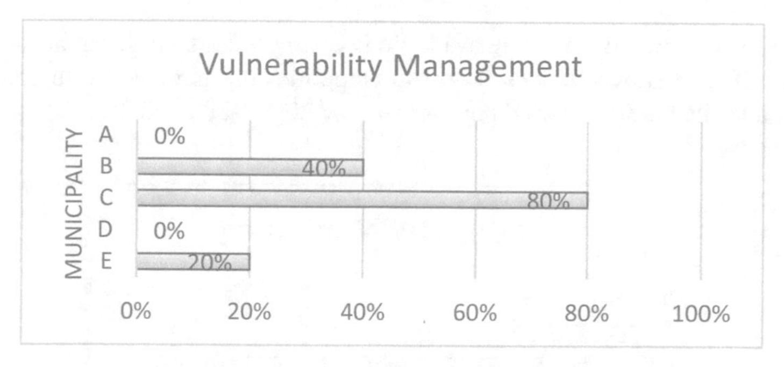

Fig.5. - Percentual Conformity Score (pcs) in Dimension V

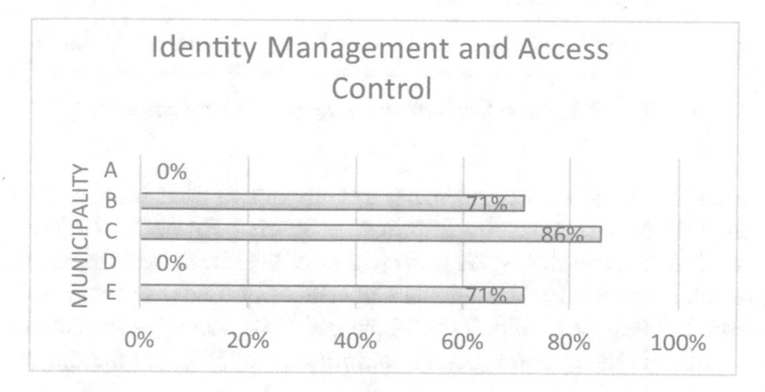

Fig. 6. - Percentual Conformity Score (pcs) in Dimension VI

evaluate its proper functioning. b) Fig. 7 shows that two Municipalities did not answer any question on this topic, and the others reported conformities of 53%, 93%, and 47%.

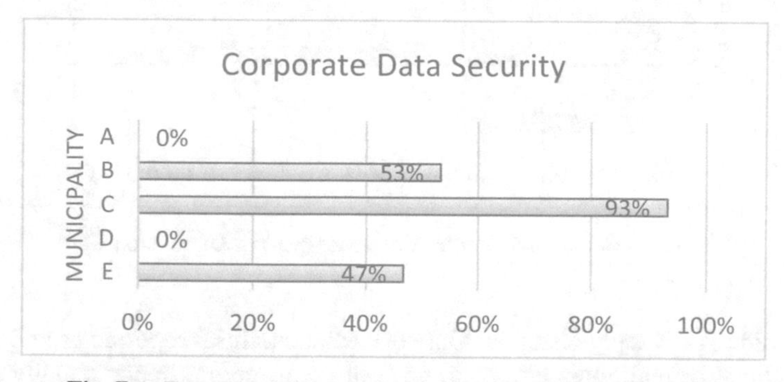

Fig. 7. - Percentual Conformity Score (pcs) in Dimension VII.

Dimension VIII – Logging and Monitoring: a) Two Municipalities reported performing monitoring of the systems' records. However, none of them could specify whether these monitoring records are made available in a simplified form for analysis and consultation. b) Fig. 8 shows the pcs. Two Municipalities did not answer any question on this topic, and the others reported conformities of 25%, 92%, and 42%.

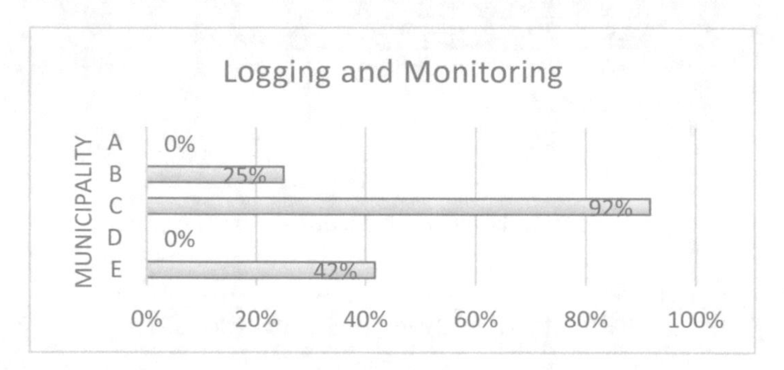

Fig. 8. - Percentual Conformity Score (pcs) in Dimension VIII

Dimension IX – Incident Management: a) Only one of the Municipalities reported being enrolled in the Cybersecurity Information Sharing Partnership (CiSP) and conducting threat information sharing. However, none of the Municipalities reported having internal procedures to notify the National Cybersecurity Center of incidents with relevant or substantial impact. b) Fig. 9 shows the pcs. Two Municipalities did not answer any question on this topic: two reported conformities of 7%, and the others 57% and 14%.

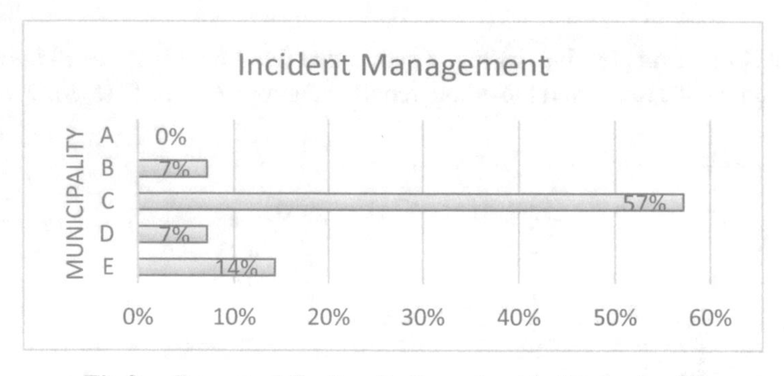

Fig.9. - Percentual Conformity Score (pcs) in Dimension IX

Dimension X – Supply Chain: a) Only one Municipalities reported provide to third-party suppliers' orientations. However, none of them reported contractually requiring the adoption of responsibility sharing. b) Fig. 10 shows the pcs. Three Municipalities did not answer any question on this topic, and the others two reported conformities of 13% and 75%.

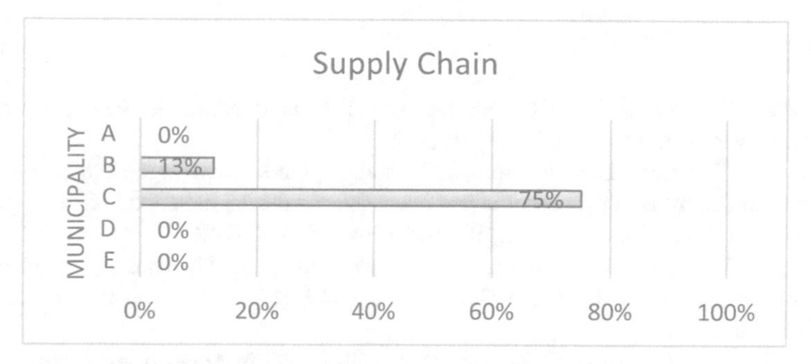

Fig.10. - Percentual Conformity Score (pcs) in Dimension X

5 Discussions and Conclusions

The analyses performed in this study allowed the team to develop an instrument to apply in target organizations to verify their maturity concerning local cybersecurity. In the analyses conducted by the project team, the concern with cybersecurity in Municipalities objects of this research still needs to be a consolidated practice. In addition, the team must also consider that the autarchies analyzed are in differentiated stages of compliance. Table 1 shows the general percentual conformity score (pcs) per Municipality in the third column. Although note some security practices by some Municipalities, the Percentual Conformity Score (pcs) was null in "A," Municipality "B" obtained 47%, "C" 76%, "D" 9%, and "E" 34%.

Table 1. Percentual Conformity Score (pcs) by dimensions and general average Municipality

	I	II	III	IV	V	VI	VII	VIII	IX	X	PCS
A	0%	0%	0%	0%	0%	0%	0%	0%	0%	0%	**0%**
B	33%	50%	42%	75%	40%	71%	53%	25%	70%	13%	**47%**
C	0%	90%	89%	100%	80%	86%	93%	92%	57%	74%	**76%**
D	0%	10%	11%	0%	0%	0%	0%	0%	70%	0%	**9%**
E	33%	20%	16%	80%	20%	71%	47%	42%	14%	0%	**34%**

6 Limitations of This Study

The security aspects related to equipment and physical device protection, notably when it comes to IoT use, were not objects of this study, for being the responsibility of manufacturers and hardware suppliers.

Neither were the aspects related to Cloud computing and Artificial Intelligence part of this study, for being these issues at a deeper technical level.

References

1. European Commission, DG CONNECT. eGovernment Benchmark 2021: Entering a New Digital Government Era: Insight Report (2021)
2. Lallie, H.S., Shepherd, L.A., Nurse, J.R.C., et al.: Cyber security in the age of COVID-19: a timeline and analysis of cyber-crime and cyber-attacks during the pandemic. Comput. Secur. **105**, 102248 (2021). https://doi.org/10.1016/j.cose.2021.102248
3. Carvalho, J.V., Carvalho, S., Rocha, Á.: European strategy and legislation for cybersecurity: implications for Portugal. Cluster Comput. **23**, 1845–1854 (2020). https://doi.org/10.1007/s10586-020-03052-y
4. Baig, Z.A., Szewczyk, P., Valli, C., et al.: Future challenges for smart cities: cyber-security and digital forensics. Digit. Investig. **22**, 3–13 (2017). https://doi.org/10.1016/j.diin.2017.06.015
5. Fonseca IC. Local E-Governance and Law: Thinking About The Portuguese Charter for Smart Cities: Challenges. IUS Publicom Netw. Rev. 1–24 (2021)
6. Walsham, G.: Doing interpretive research. Eur. J. Inf. Syst. **15**, 320–330 (2006). https://doi.org/10.1057/palgrave.ejis.3000589
7. Webster, J., Watson, R.: Analyzing the Past to Prepare for the Future: Writing a Literature Review. In: Manag. Inf. Syst. Res. Center, Univ. Minnesota (2002). https://www.google.com.br/#q=writing+a+literature+review+webster+and+watson+2002. Accessed 23 Jan 2015
8. Creswell, J.W.: Qualitative Inquiry and Research Design: Choosing Among Five Approaches. Nebraska, United States of America (2007)
9. Al, S.R., Pokharel, S.: Smart city dimensions and associated risks: review of literature. Sustain. Cities Soc. **77**, 103542 (2022). https://doi.org/10.1016/j.scs.2021.103542
10. Wirtz, B.W., Weyerer, J.C.: Cyberterrorism and cyber attacks in the public sector: how public administration copes with digital threats **40**, 1085–1100 (2016). https://doi.org/10.1080/01900692.2016.1242614
11. Pandey, P.: Making smart cities cybersecure. Deloitte Insights (2019)
12. European Parliament and Council of the European Union. General Data Protection Regulation - GDPR. Off J Eur Union 156 (2016)
13. ENISA EUAFC ENISA. https://www.enisa.europa.eu/
14. Assembleia da República. Lei n.º 46/2018 Regime jurídico da segurança do ciberespaço. Diário da República nº 155/2018 (2018). Série I 2018–08–13 4031–4037
15. Osório De Barros, G.: A Cibersegurança em Portugal (2018)
16. International Organization for Standardization - ISO ISO/IEC 27002:2022(en), Information security, cybersecurity and privacy protection — Information security controls. https://www.iso.org/obp/ui/#iso:std:iso-iec:27002:ed-3:v2:en. Accessed 14 May 2022
17. NCSC NCSC. Connected Places: Cyber Security Principles (2021)

Authentication, Authorization, Administration and Audit Impact on Digital Security

Fauricio Alban Conejo Navarro[1], Melvin García[2], and Janio Jadán-Guerrero[3](✉)

[1] Universidad Latina de Costa Rica, San José, Costa Rica
fauricio.rogers@ulatina.net

[2] Maestría en Gestión y Dirección de Proyectos, Universidad Galileo, Ciudad de Guatemala, Guatemala
melvingar@galileo.edu

[3] Centro de Investigación en Mectrónica y Sistemas Interactivos – MIST, Universidad Indoamérica, Quito, Ecuador
janiojadan@uti.edu.ec

Abstract. We can say technological security is in constant change. Security is constantly changing by the industry, because it is necessary to improve the integrity of technological security. These changes in technology come with new methods, processes, and technologies which lead to the need to be ahead in the new trends. The AAAA (Authentication, Authorization, Administration, Audit) is a topic of great relevance in organizations, it is a matter for everyone. This paper attempts to evaluate the variation presented in the Access Management after implementing digital security plans. The evaluation is visualized as an accompaniment for the improvement of the effectiveness in future decisions-making processes. The result of this analysis aims to find new improvement opportunities, due to the positive impact to growth in this field. The intention is to close the gap between the financial and technological security issues. Nowadays these issues are evaluated individually, and we pretend to unify these subjects.

Keywords: AAAA · Security · Authentication · Authorization · Audit · Planning · Procedures · Policies · Digital Security

1 Introduction

Informatics solutions and applications (software programs) have brought a major revolution about how permission control works. This occurs in practice with authentication, authorization, administration, and auditing (AAAA) processes, where the business demands security measures at the same pace of technological advances in the field of industry. This paper proposes a new methodology based on practice that is transferable to both physical and logical processes in existing software, but able to adjust to new digital trends. This methodology is based on the result to reach a good practice in technologies claiming that the developed software reaches the maximum benefit, less cost, wide portability, dependency with internal and external business and so on. This proposal is called AAAA which is considered as a technological or physical permission control resource in the management of services under this modality.

A. Rocha et al. (Eds.): WorldCIST 2023, LNNS 799, pp. 187–193, 2024.
https://doi.org/10.1007/978-3-031-45642-8_18

2 State of Art

According to security reports in organizations by the communication agency; Prensa y Comunicación, it is estimated that by 2020 there will be 50 to 100 million IoT (Internet of Things) devices [1]. The increase of devices in IoT rises the complexity in this spectrum and extend computing and communication capabilities, however there are many challenges in security issues [2]. There is no framework which works in the security aspects and, thus, no audit process in IoT so far. Entities who manage IoT create scheduled processes and tasks without a clear and focused strategy for control and monitoring. This lack organization ends up in lost money and internal security vulnerabilities.

Digital security has a constantly changing environment, as we have been seeing in recent years. Issues such as how and where information is stored, especially with the introduction of targeted services are main trending topics. Cloud computing provides notable advantages to the users' entities, but it also has significant risks in the internal control system and security. There is a clear need to change how the culture in IoT works and how it evaluates and monitors the information received and delivered to external entities and good practices in this environment [3].

Through the diagram in Fig. 1, it is shown how the different technological artefacts are interconnected. This causes security problems that could be considered the main causes, and threatens the security to the entities which, instead, have to opt for a digital security plan applying the four AAAA concepts. [4].

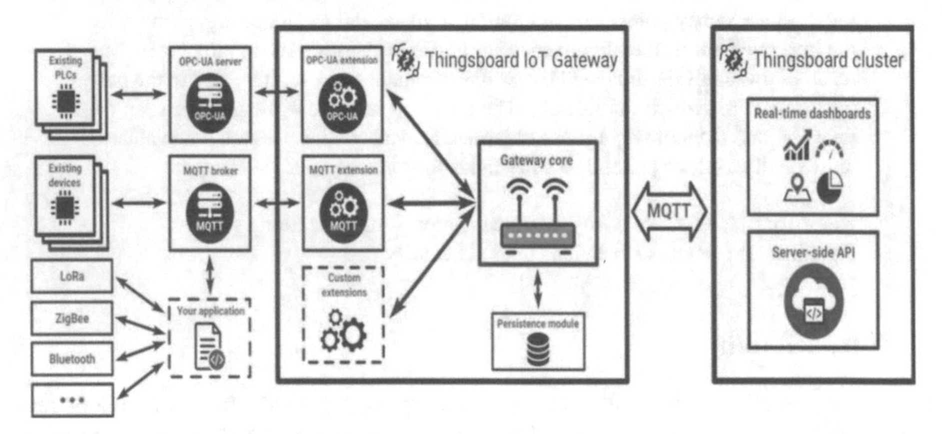

Fig. 1. IoT interconnection diagram, [8]

Digital security has become a hieroglyphic argument in organizations of the utmost importance. It is important to mention that public or private entities today require security processes and protocols in the form of tools to combat criminals in preventing harmful acts. According to the publication of El Nuevo Diario "To counteract these threats, the organization usually uses widely developed mechanisms and technologies that detect, correct and improve the information security of organizations." [4].

Additionally, for example in the Department of Defence of the United States (DoD) contractors were required to assess their compliance and complete any non-compliance by December 31, 2017. [5]. However, the intent of the US federal government is that all FARS contracts have to include a requirement to be compliant with NIST SP 800–171 for the years to come. Some regulations in these issues have already been put in place, with more coming. In December 2016, NIST publicly released the [6], Revision 1, Protection of Controlled Unclassified Information in Non-Federal Information Systems and Organizations.

3 Method

The scope of the research was skewed towards those organizations that implement digital security plans and those that do not. In order to forge a schematization of derived results and a proposal for a digital security plan for organizations that lack it.

Understand the effects of processes and protocols such as AAAA with digital security, analysing and relating through theoretical research, with this it is intended to generate an awakening of knowledge about IoT and digital security for organizations that lack it.

Understand the judgment and/or inexperience that exists in the entities in accordance with the AAAA applied to digital transformation strategies in order to measure results.

Correspond the good practices of implementing digital security plans with the traditional management model of the companies, this in order to guide it to a positive security transformation with AAAA as a good practice.

The limitations and restrictions in the investigation are. a) A proposal will be delivered for the development of a solution for the technology sector in digital format. b) Training on digital security and/or maturity level measurement is not provided (Fig. 2).

3.1 Planning, Policies and Procedures.

We ask ourselves the following questions as an initial reference point to start questioning the existence of digital security processes and protocols to follow:

1. Are there information security policies approved at a high management level?
2. Does the entity have a strategic plan for information systems?
3. Does the entity have a security investment budget in technological matters?
4. Briefly describe the main technological projects undertaken during the audited year and list the main projects planned for the next.
5. Is there a program so that users are aware of the security policies regarding information, procedures and use?
6. Is there a multi-year IT security awareness training plan?

Framework documentation:

- Copy of the entity's Security Policy.
- Copy of the Strategic Plan for Information Systems.
- Copy of the information security training and/or awareness plan.

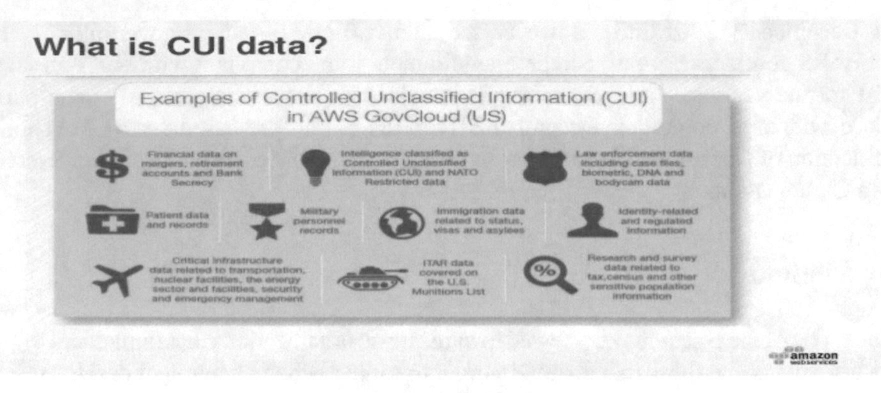

Fig. 2. What is CUI [8]

3.2 Defence Information Systems Agency

The Defence Information Systems Agency (DISA) is a combat support agency of the United States Department of Defence (DoD). To implement its standard, DISA developed the Cloud Computing Security Requirements Guide (SRG). The SRG establishes basic security requirements for cloud service providers (CSPs) that host DoD information, systems, and applications and for DoD use of cloud services [9].

3.3 Diagnosis

These and many other fascinating findings come from Wipro's 2018 State of Cybersecurity Report. The study is based on four main sources of data, including primary research from Wipro customers, primary research from the Cyber Défense Centre (CDC), secondary research sources, and content from Wipro partners. 42% of respondents are from North America, 10% from Europe, 18% from the Middle East, 21% from Asia and 8% from Australia. For additional details on the methodology, please see pages 3–5 of the report [10].

4 Proposal

The proposal to address the objectives of this paper is based on educating organizations about the benefits of implementing digital security plans in conjunction with AAAA and its direct impact in order to contribute positively to the decision-making process and identify points of improvement in the management model.

For the improvement proposal, the importance of the following strategic assignments is considered relevant:

Development Life Cycle [11]

- Requirements analysis: how user requests regarding security are received, who evaluates them, how they are approved, etc.
- Functional analysis.
- Development:

- Version management tool used
- Is there a development environment separate from the production environment?
- Test:
- Is there a test environment?
- Types of evidence to consider or Unitary, or Integration, or Certification.

Antivirus [11]

- Describe the antivirus solutions implemented in the entity.
- Describe the main management mechanisms:
- Update mechanisms.
- Update frequency: From the Firms Database, From client stations.
- Main functionalities installed:
- Personal Firewall option.
- On-demand scanning.
- Ways to prevent computers without antivirus software installed.

Physical Security [11]

- Are there access controls to the entity's dependencies (for those that are not freely accessible to the public: the people and the dates and times in which they access are noted)? Briefly describe existing controls.

List the Data Processing Centres where the entity's servers are located. For each of them, describe: [11]

- Inventory of Servers
- Non-exhaustive list of servers located in each rack.

Access controls [11]

- Location.
- Personnel authorized to access it.
- Access control method.
- Access procedure for external personnel (maintenance, cleaning, etc.).
- Enumeration of internal and external groups authorized to have physical access to the computing process environment.

Security Track [12]

- Network security
- Computer security
- Database security
- Web applications security
- Security policy, model and architecture
- Security social networks
- Security in parallel and distributed systems
- Security in mobile and wireless communications
- Security in grid/cloud/pervasive computing
- Authentication, authorization and accounting
- Miscellaneous security issues

Privacy Track [12]

- Privacy in Web-based applications and services
- Privacy in database systems
- Privacy in parallel and distributed systems
- Privacy in grid/cloud/pervasive computing
- Privacy in mobile and wireless communications
- Privacy in e-commerce and e-government
- Privacy in network deployment and management
- Privacy and trust
- Privacy and security
- Privacy and anonymity
- Miscellaneous privacy issues

The principle of least privilege maintains that an individual, process, or other type of entity should be given the minimum privileges and resources for the minimum period of time required to complete a task. This approach reduces the opportunity for unauthorized access to sensitive information. Separation of duties requires that the completion of a specified sensitive activity or access to sensitive objects is dependent on the satisfaction of a plurality of conditions. For example, an authorization would require signatures of more than one individual, or the arming of a weapons system would re-quire two individuals with different keys. Thus, separation of duties forces collusion among entities in order to compromise the system.

5 Conclusions

To sum up, the increase in interaction between people, objects and people with easily integrated objects, anywhere and at any time, is increasingly empowering the development of current societies with a more visible need for security.

In this sense, AAAA is already a reality, present at a global level, in all areas of development. With new technologies, there is a need to increase security processes and tools.

A paradigm shift is necessary in terms of the design of objects and services –based on the development of the AAA model– which should focus fundamentally on the approaches that define awareness of the context, as well as the application of intelligence, thereby allowing the design of resources according to particular situations, a situation that sets the tone for carrying out work and research that allows benefits to the development of AAAAs.

Cultural change represents a major obstacle to adapting these goals in the audit. Defining the goal, we want to achieve in this new phase is key when executing different actions.

6 Future Work

Existing the process of elaboration of the pleasant investigation, exclusively terms of use of those raised are requested. First, active use of the security practices of the AAAA applying the processes of the entity to follow for an assertive communication on the good practices of both the organization - collaborator - consumer. Second, Follow-up of the entity's communications about new changes to be up to date. Third, It is recommended to be following the update processes of the platforms of use to avoid possible security gaps. Fourth, with the Audit process, it is recommended to make periodic reviews of the users and their levels to grant the permissions according to the user's role.

References

1. Barbero, C.Z.: A Flexible Context Aware Reasoning Approach for IoT Applications (2011). Obtenido de ieeexplore.ieee.org: https://ieeexplore.ieee.org/document/6068446
2. Fourth International Conference on Intelligent Networking and Collaborative Systems (2019). www.ieeeconfpublishing.org: http://www.ieeeconfpublishing.org/cpir/AuthorKit.asp?Community=CPS&Facility=CPS_Sept&ERoom=iNCoS+2012
3. Yadav, S., Kalaskar, K.D., Dhumane, P.: Cyber security in IoT-based cloud computing: a comprehensive survey. Electronics **11**(1), 16 (2022). https://doi.org/10.3390/electronics11010016
4. Columbus, L.: (14 October 2018). www.forbes.com: https://www.forbes.com/sites/louiscolumbus/2018/10/14/the-current-state-of-cybersecurity-shows-now-is-the-time-for-zero-trust/#33c8042a5f15
5. comein.uoc.edu. comein.uoc.edu Clifford Stoll. Retrieved from comein.uoc.edu (2019). https://comein.uoc.edu/divulgacio/comein/es/numero36/articles/Article-Josep-Cobarsi.html
6. complianceforge. Complianceforge (2018). www.complianceforge.com/: https://www.complianceforge.com/reasons-to-buy/nist-800-171-compliance/
7. Federal Information Security Modernization Act of 2014 (P.L. 113–283). (December de 2014). Retrieved from gpo.gov: http://www.gpo.gov/fdsys/pkg/PLAW-113publ283/pdf/PLAW-113publ283.pdf
8. H2S Media. H2S Media (2019). www.how2shout.com: https://www.how2shout.com/tools/best-opensource-iot-platforms-develop-iot-projects.html
9. IEEE 11th International Conference on Trust, Security and Privacy in Computing and Communications. Retrieved from sn.committees.comsoc.org (2012). https://sn.committees.comsoc.org/journal-conference-publications/the-11th-ieee-international-conference-on-trust-security-and-privacy-in-computing-and-communications-trustcom-2012/
10. National Institute of Standards and Technology Framework for Improving Critical Infrastructure Cybersecurity (as amended). (2018). Retrieved from Cybersecurity Framework: https://www.nist.gov/cyberframework
11. NIST 800–171 Compliance Criteria. (2018). Retrieved from complianceforge: https://www.complianceforge.com/product/nist-800-171-compliance-criteria-ncc/
12. Tecnologia Empresarial. (29 de Octubre de 2013). El Nuevo Diario Tecnologia Empresarial. www.elnuevodiario.com.ni: https://www.elnuevodiario.com.ni/economia/300466-factor-humano-amenaza-interna-seguridad-informacio/
13. Krutz, R.L., et al.: Cloud Security: A Comprehensive Guide to Secure Cloud Computing, John Wiley & Sons, Incorporated, 2010. ProQuest Ebook Central, http://ebookcentral.proquest.com/lib/unintcostarica-ebooks/detail.action?docID=589027

Generation Z's [Incoherent] Approach Toward Instagram's Privacy Policy

Matilde Rocha[1,2], Natacha Grave[3], Juliana Reis[3], Maria Silva[3], Cicero Eduardo Walter[4], and Manuel Au-Yong-Oliveira[5](✉)

[1] Department of Physics, University of Aveiro, Aveiro, Portugal
matilde.rocha@ua.pt
[2] Instituto de Telecomunicações, University of Aveiro, Aveiro, Portugal
[3] Department of Languages and Cultures, University of Aveiro, Aveiro, Portugal
{natachagrave,julianareis,mariamsilva}@ua.pt
[4] Federal Institute of Education, Science and Technology of Piauí, Teresina, Brazil and DEGEIT, University of Aveiro, Aveiro, Portugal
eduardowalter@ifpi.edu.br
[5] DEGEIT, INESC TEC, GOVCOPP, University of Aveiro, Aveiro, Portugal
mao@ua.pt

Abstract. Social Networking Sites have not only become a dominant means of communication and information worldwide, but also a strategic source for cybercriminals to exploit users' personal information. This occurs when, for marketing purposes, online activities such as browsing and shopping, generate data for the algorithm to analyze and possibly interfere with users' on a daily basis, raising concerns about the users' privacy and data vulnerability. Therefore, this study carries out an analysis of Generation Z's awareness of the privacy policy (PP) terms and the vulnerability of their personal data, with particular emphasis on Instagram's PP, since this platform, due to its popularity, is now one of the most renowned networking tools, especially for younger generations. Thus, herein was constructed an online survey, which contained both demographic and social media questions, in order to collect data to analyze users' perceptions on this topic, having a sample comprised by 222 answers. The proposed study revealed that although 89.6% of the participants did not read Instagram's PP, only 21.2% of the respondents are not concerned about how their data is used by Instagram. Finally, this study reflects on the users' possible inertia towards reading the PP agreement, concluding that even if most of the respondents show concerns about the use of their personal data, they are not engaged in obtaining more information on how they can protect their data by reading the PP terms, thus revealing an incoherent posture as well as a dependency on Instagram. The Chi-square test of association was also applied to the data and various statistically significant associations were found, suggesting some incoherent behaviors, at the 5% level.

Keywords: Instagram · privacy policy · terms and conditions · social media · users' perception

A. Rocha et al. (Eds.): WorldCIST 2023, LNNS 799, pp. 194–204, 2024.
https://doi.org/10.1007/978-3-031-45642-8_19

1 Introduction

In recent years, the numerous technology improvements contributed to increase exponentially social media network popularity, it being estimated that almost 6 billion people around the world will use social media by 2027 [1–3]. Nowadays, Instagram is one of the leading social media platforms based on sharing photos and videos, with more than 500 million daily active users distributed worldwide [3], amongst whom 8.9% are aged between 13 to 17 years old and 30.2% aged from 18 to 24 years old [4], thus representing the Generation Z, a term often used to define the younger generation of people born between 1996 and 2012, according to the Pew Research Gate [5].

The year 2015 started with a significant update of Instagram's terms and data policies by employing new advertisement rules, in which users had to accept these changes in order to remain using the platform [6]. Thus, user-related advertising was implemented where the users' information is collected from various websites automatically, with the help of cookies. Therefore, this method exploits user data with the goal of personalizing the ads shown by considering elements of users' data such as language and location, browser history, through cookies saved in browsers, and social media activities, providing relevant data to advertisers [6].

The implementation of this new approach brought some benefits such as a vast profile that can help the user to become visible in the search of people with profile similarities (high school, university, workplace, living place, among others) [7]. Nevertheless, knowing that Instagram is used to socialize and find a friend and that most of its users do not reflect on who might be copying and using their information leads to frequent problems of fraud and cybercrimes on users' information [8]. Consequently, personal data protection, the users' manipulation through communication and spreading of fake news, and cyberbullying are some drawbacks raised by user-related advertising.

Aiming to decrease this negative impact, the privacy policy (PP) has emerged as a fundamental element of this process, since it consists of a principle of actions adopted by an organization to protect its users' information, acting as a guideline for the users who wish to share their information [7]. Although the PP has great potential to be a shield against privacy infringements, previous studies revealed several defects associated to its functionality [9]. For instance, users usually avoid reading PP texts if they are not presented by default or due to its inaccessibility and location on the website. As another example to diminish personal data intrusion, on June 2020, Instagram permanently disabled access to its application programming interface (API), moving forward to increase user privacy [10].

Thus, considering the low percentage of users reading the PP texts, herein we proposed to assess the users' perception of their own privacy on Instagram by relating it to their possible inertia towards reading the PP agreement, considering the age group that uses mostly the Instagram platform. Therefore, focusing mainly on the Portuguese Generation Z, our main goal is to study the relation between Instagram users reading the PP, and their level of concern about how their data is being used.

2 Literature Review

Social Networking Sites (SNS) have become a dominant mean of communication, information, interaction, and entertainment worldwide [11]. The opportunity of getting interconnected on global levels, by sharing experiences, knowledge, and news and getting updated in real time, makes the SNSs, such as Facebook, YouTube, WhatsApp, and Instagram, popular and addictive platforms for users [11].

Founded in 2010 by the software engineer Michel Krieger, Instagram gathered enough users to become one of the most renowned social media platforms [12]. In 2012, the company was acquired by Meta, a multinational technology conglomerate, and currently shares its systems and infrastructure with other Meta companies, such as Facebook, WhatsApp, and Oculus [13], reaching two billion active users in 2021, of whom 70% are under 35 years old [12].

Nevertheless, as its popularity continues to increase, a concern also arises for the users' awareness of the vulnerability of their personal data: if SNSs are one of the main tools for spreading information nowadays, they are also automatically representing a strategic source for cybercriminals to exploit users' personal information, thus compromising the security and privacy of the platform and of its users [8]. As many might be unaware of this situation [8], each SNS has the legal obligation of communicating the ways in which they collect, use, and manage users' data, as well as their intentions towards confidentiality, being expressed through their PP.

Considering Instagram's PP [14], it can be anticipated that they have access to information such as IP addresses, captured content, users' contacts, tags, cookie data, interactions, text message history, geolocation, transactional data from Facebook products and services, facial recognition data, different devices used to access the platform and places or things that the users are interested in [15]. The PP asserts that the cache of the collected data is used not only to communicate with its users and to investigate suspicious activity or harmful conduct, but also for strategic market research purposes, which includes the personalization of the ads that show up on the customers' feed, as aforementioned [16]. In addition, they imply that third-party companies (such as partners, publishers, advertisers, and app developers), can access any information that might have been shared with them and collect information on the users' purchases, devices, the ads they see, even if they are not logged in [17].

To perform this, and despite not being visibly evident to the users, the algorithm plays a crucial role in shaping their online experiences [18]. Online activities such as browsing, shopping, and game playing, generate data for the algorithm to analyse and possibly interfere with users' on a daily basis [18], by displaying products or information from third-party companies, that might be appealing to the users based on their personal tastes, in order to tempt the user. According to Instagram's PP, instead of paying to use this platform, by using the Service covered by these Terms, the unaware user, might not acknowledge that Instagram can show ads that businesses and organizations pay them to promote on and off the Meta Company Products, such as information about personal activity and interests, to show the users ads that are more relevant to them as previously mentioned [19].

Consequently, the PP is a lawful regulation meant to maintain users' trust in the SNSs [8], to let them know the limits of their protection and to display how their information is being used by the platform itself or by third-party organizations. According to Instagram's Data Policy, under the General Data Protection Regulation, the user can exercise their right to "access, rectify, port and erase" their data, as well as to "object to our processing of your data for direct marketing" or "where we are performing a task in the public interest or pursuing our legitimate interests or those of a third party" [20].

Although PPs are considered legally binding, they can be breached and doing so has its consequences just like what happened with Facebook, on more than one occasion [18]. The most recent, in April of 2021, when due to a vulnerability that had already been patched in August 2019, led to the personal information of 553 million users to be posted in a low-level hacking forum [21]. This information included "phone numbers, Facebook IDs, full names, locations, birthdates, bios, and, in some cases, email addresses" [21], enough information needed to scam users, and possibly risk real-life damage in several aspects of users' daily lives.

The first incident was when Cambridge Analytica scraped the data of over 80 million users to target voters with political ads in the 2016 election, once again providing support to the prior mentioned that these virtual events do hold the power to negatively impact users 'real, daily lives. The lack of care on the behalf of Facebook in regard to storing and mishandling users' data, should be handled accordingly and the Meta conglomerate might want to reevaluate the transparency of their social media services and their PP since, as this is a basic ethical issue for a large number of humans in general; when trust gets broken it may be hard to go back to a previously stable point of the past [21]. If however it is taken into account that the most recent incident has somehow injured the privacy rights of 553 million users, a number 6.9125 times higher than in the first breach in 2016, a positive evolution concerning the protection of their users' data seems to be a farfetched perspective, and although these breaches are in relation to Facebook, Instagram belongs to the same conglomerate, therefore it may cause a reason for concern amongst users, or not, due to the homogenous, interconnected way in which these services operate.

There are a few studies focused on this problematic, although they concentrate on different perspectives and aspects, such as the discussion of how the PPs are absorbed and understood by SNS's users, and what they would change to increase their interest in this type of reading [8]. Therefore, the focus of this study was not on the user and its perspective towards its content, but the way the PPs are written [8]. One of the main conclusions was that the usage of specific terms and legal jargon, non-accessible to most, can originate a dry PP of difficult understanding, which demotivates the user to read it. As another example, in [9] were proposed visualization techniques as a new PPs 'representation instead of traditional textual representation, also examining empirically their effects on users' information privacy awareness level. Thus, they concluded that visualized PPs lead to higher privacy awareness levels, contributing to a more vigilant posture in protecting their own privacy.

After the previously mentioned incidents, several questions concerning the users' personal data privacy and how it can be used for profit have been raised [8]. Therefore, the proposed study aims to mainly evaluate the Portuguese Generation Z's knowledge of the Instagram PP terms while relating it with their level of concern about how their data is being used by Instagram.

3 Methodology

As discussed in the sections above, there are previous studies on Instagram's PP that focus on the influence of the users' behavior as well as on Instagram's legal right to pursue its strategy. As a new approach, this work aims to mainly assess the relationship between Portuguese Generation Z reading the Instagram PP and their level of concern about how their data is being used. In order to accomplish this, a survey was created – using Google forms – as a research methodology to collect data and gain in-depth information about the respondents' perspective on this subject [22, 23].

The survey was developed based on its efficiency to collect mass volumes of data in a short amount of time, implicating low costs and resources, considering a method used in studies of similar nature. It aimed to discover Generation Z Instagram users' perspectives and behavior towards their PP by answering a series of multiple-choice questions, such as Yes/No/Maybe, to indicate their view and knowledge on the topic.

It consisted of two sections, with a total of thirteen quantitative questions and one qualitative question to assess the respondents 'knowledge about how their data is stored by this platform. The first section focused on collecting demographic information including the respondents' age and nationality. Then, the second was focused on gathering information about what social media platform the respondents use the most, considering also if the respondent read Instagram's PP, its perception on what information is stored and how it is used by this platform. As a complement, the respondents were questioned about if they would pay to remove ads and if they consider ethically correct the use of their data to personalize ads. Thus, the developed form was mainly shared in social media, such as Facebook and Instagram, being available from October 11th until October 24th, 2022 (in both English and Portuguese) in order to get significant answers from a fair-sized sample.

4 Data Analysis and Discussion – Descriptive Statistics

Aiming to assess the relationship between Instagram users reading Instagram's PP and their concern about how their data is used, the main goal of the proposed work was to collect survey answers, focusing mainly on the Portuguese Generation Z. That is, in this work 222 answers were collected from people aged between 11 and 25 years, as this is the generation that mostly uses social media platforms [4]. The data collected is mainly quantitative, apart from one survey question, which was qualitative. It should be noted that the sample is composed by 93.2% people with Portuguese nationality, as shown in Fig. 1 (Generation Z).

Also, according to the sample, the most used social media platform with 166 answers (74.8%) is Instagram as shown in Fig. 2.

Of the people who answered the survey, 89.6% did not read Instagram's PP and, therefore, agreed to it despite not knowing what they are agreeing to, since it is still mandatory to accept the terms to have an account. Although 110 users (49.5%) claim to

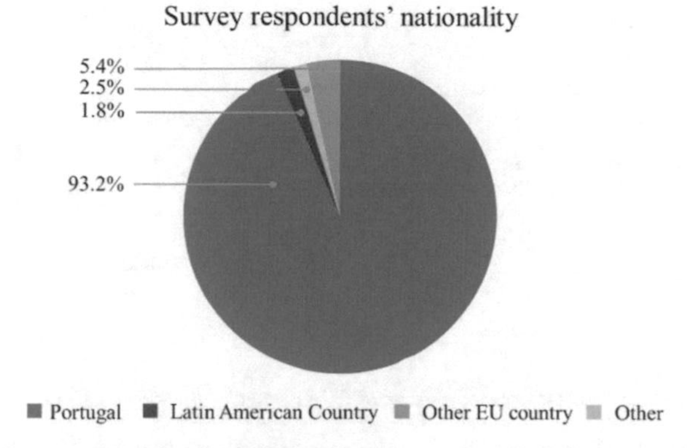

Fig. 1. Survey respondents' nationality.

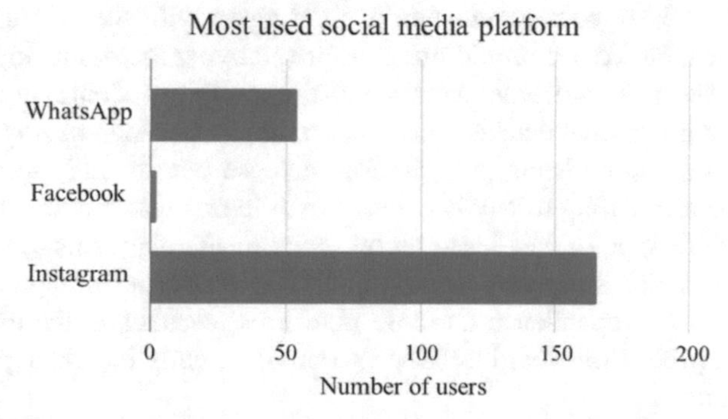

Fig. 2. The most used social media platforms between the survey respondents.

be concerned about the use of their personal data by social media platforms they did not read Instagram's PP.

Considering the low reading percentage, it was also found that 42.8% know what type of data is stored by Instagram, meaning that most of the information about PPs comes from external sources, mostly from the Media, as is visible in Fig. 3.

The results obtained demonstrated that only 21.2% of the respondents are not concerned about how their data is being used by Instagram, and that 15.2% of the other users, have claimed that they would delete Instagram if they knew that their data is being used for revenue. Albeit they continue using the app. 16.0% of the users that would delete Instagram, claimed to know how their data is used for profit, revealing a contrasting belief internalized by the users.

Furthermore, the results revealed that 163 (73.4%) of respondents, claimed to know how pop-up ads appear on their feed. Additionally, 85 (52.1%) of the 163, state that they know how the users' personal data is used by Instagram for profit. This attitude leads us to conclude that a significant percentage of users does not know how pop-up ads truly appear on their profile and do not realize that their personal data is being used to make these ads more customized.

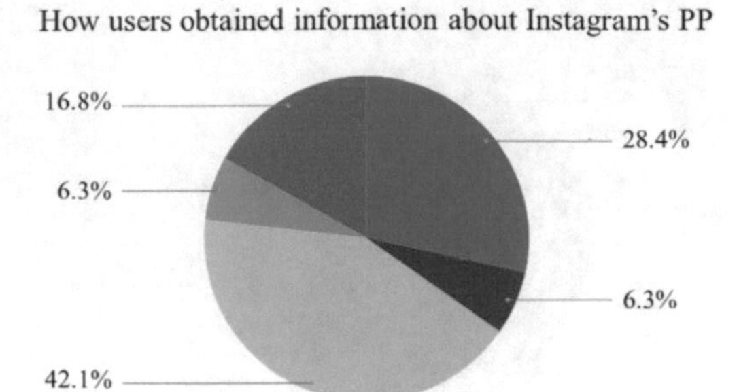

Fig. 3. How users obtained information about Instagram's PPs.

Additionally, 54.5% of respondents does not agree with the use of personal data to personalize ads, but 85.7% would not pay for removing them, making it possible to conclude that the users would rather remain with the possibility of enjoying an account for free, that uses their personal data as a revenue stream, rather than having their personal data protected. As a contrasting note, 95.3% of those who would pay to remove the personalized ads are willing to pay between 1 to 10 euros monthly. Even though users have access to what type of data is stored by social media platforms for free and could protect this information simply by signing out of the platform (through the deletion of their account), users' dependence on these platforms reveals that the more concerned users about data protection would rather pay a small monthly fee, than give their social media accounts up.

Although Instagram is lawfully protected, since the user must agree to Instagram's PP to have access to the app, the use of personal data has raised moral questions. Of the respondents, 57.7% of the users think that the use of personal data in ads is not an ethically correct practice though most remain on and continue to use the platform, as previously stated.

Overall, most users claim to be concerned about what type of personal data is stored and how it is used by social media platforms, specifically if this involves using it in a revenue stream, however they show a certain laziness and indifference in acting on these concerns by simply obtaining more knowledge on data protection, which is provided by the platform itself. Furthermore, when faced with this issue and with the ethical use of their personal data, as emphasized, most users would not delete the app. This hypocritical and contradictory posture by the user leads us to conclude that there is a high dependence on these platforms and great inertia towards acting on their concerns.

5 Data Analysis and Discussion – The Chi-Square Test of Association

Regarding the act of reading the terms of use of social networks as a function of the knowledge that the data are stored, the Chi-square test allows us to conclude, at the 5% significance level, that there is an association between reading the terms of use of the social networks used and the knowledge about how personal data are stored, since the value of proof obtained (p-value = 0.024) was lower than the level of significance considered.

However, when we evaluate the possible association between reading the terms of use of social networks with concern about data use, the Chi-square test reveals to us, at a 5% significance level, that there is no association between reading the terms of use of the social networks used and concern about the use of data by social networks, since the obtained proof value (p-value = 0.533) was greater than the assumed significance level of 5%.

Additionally, regarding the concern about the use of personal data, we performed the Chi-square test to verify the possible association between the said concern and users' knowledge about how personal information is used to generate revenue by social networks.

The Chi-square test allows us to conclude, at the 5% significance level, that there is no association between concern about data use and knowledge about how personal information is used to generate revenue by social networks since the obtained proof value (p-value = 0.479) was higher than the significance level considered.

When we evaluated the possible association between knowledge about how personal information is used to generate revenue by social networks with the act of deleting social networks when being aware that personal data is used for revenue generation, the Chi-square test allowed us to conclude, at the 5% significance level, that there is an association between knowledge about how personal information is used to generate revenue by social networks with the act of deleting social networks when being aware that personal data is used for revenue generation, since the obtained proof value (p-value < 0. 001) was less than the 5% significance level.

Concerning ads, we performed the Chi-square test to check the possible association between knowledge about how Pop-ups appear with agreement about the use of personal data for creating personalized ads.

The Chi-square test allows us to conclude, at the 5% significance level, that there is an association between knowledge about how Pop-ups appear with agreement about the use of personal data to create personalized ads since the obtained proof value (p-value = 0.002) was lower than the 5% significance level considered.

Still. Further, in relation to the agreement about the use of personal data to create personalized ads, we performed the Chi-square test to verify the possible association between this agreement and the perception that the use of data to create personalized ads is something ethical.

The Chi-square test allows us to conclude, at the 5% significance level, that there is an association between the agreement on the use of personal data for the creation of personalized ads and the perception that the use of personal data for the creation of personalized ads is ethically acceptable, since the obtained proof value (p-value < 0.001) was lower than the 5% significance level considered.

6 Conclusion

Aiming to present a newer and wider perspective focused on Instagram users, this research analyzed the knowledge on data protection of the respondents, as Instagram users, while relating it with their level of concern about how their data is being used for profit by this platform.

Thus, it was possible to conclude that Generation Z is indeed concerned about how their personal data is stored and used by Instagram, but not worried enough to act on it or to go as far as deleting their accounts, their lack of proactivity being consequently highlighted. This information has brought to light that although users nurture a concern for their confidentiality, this concern is somewhat hypocritical, but without noticing the extent of this hypocrisy themselves. As it was demonstrated, since most users of the sample do not take the time to read the Instagram's PPs, this behavior actually suggests a generalized inertia towards this issue, demonstrating an incoherent viewpoint regarding their level of worry and the respondent's actions regarding what they do consider a problem. Through the analysis of the respondents, it was possible to confirm that there is indeed a discrepancy between users' privacy concerns and their behavior, concluding that there is a privacy paradox present in the analyzed sample.

Furthermore, the fact that Generation Z Instagram users do not read its PP gravely affects their cybersecurity and makes them unable to understand the dangers they are facing when inserting personal data on SNSs and how their online activity is tracked and used by the platform, such as how this data is used to personalize ads and even by other users – for example, fishing scams.

Our results suggest that reading terms of use is associated with users' knowledge of how data is stored.

However, the independence between reading the terms of use of social networks relative to the use of that same data suggests a behavioral inconsistency.

Another behavioral inconsistency evidenced by our results is in the independence between concern about data use and the use of the same data for revenue generation. In other words, while respondents have shown some concern about data use, it is not sufficient to be related to concern about how this data is used for revenue generation by social media.

As future work, it would be interesting to replicate the present study in other SNSs such as on Tiktok and Twitch, to understand the users' approach towards their cybersecurity, since their main audience is also the Generation Z.

Acknowledgements. The authors would like to thank everyone who took their time to respond to the survey questionnaire. Author Matilde Rocha acknowledges the research grant BI/ Nº2021/00209 of the project Safe-Home (CENTRO-01–0247-FEDER-072082). Funding: This work was financially supported by the research unit on Governance, Competitiveness and Public Policies (UIDB/04058/2020) + (UIDP/04058/2020), funded by national funds through FCT - Fundação para a Ciência e a Tecnologia.

References

1. Kaplan, A., Haenlein, M.: Users of the world, unite! The challenges and opportunities of social media. Bus. Horiz. **53**(1), 59–68 (2010). https://doi.org/10.1016/J.BUSHOR.2009.09.003
2. Statista: Number of worldwide social network users from 2018 to 2027. https://www.statista.com/statistics/278414/number-of-worldwide-social-network-users/. Accessed 21 Oct 2022
3. Rejeb, A., et al.: The big picture on Instagram research: Insights from a bibliometric analysis. Telematics Inform. **73**, 101876 (2022). https://doi.org/10.1016/J.TELE.2022.101876
4. Statista: Distribution of Instagram users worldwide as of April 2022, by age group (2022). https://www.statista.com/statistics/325587/instagram-global-age-group/. Accessed 16 Oct 2022
5. Dimock, M.: Defining generations: Where Millennials end and Generation Z begins. Pew Research Center (2022). https://www.pewresearch.org/fact-tank/2019/01/17/where-millennials-end-and-generation-z-begins/. Accessed 30 Dec 2022
6. Bühler, J., et al.: Big data, big opportunities: revenue sources of social media services besides advertising. Lecture Notes in Computer Science (including subseries Lecture Notes in Artificial Intelligence and Lecture Notes in Bioinformatics) **9373**, 183–199 (2015). https://doi.org/10.1007/978-3-319-25013-7_15/FIGURES/5
7. Pelau, C., et al.: Big-Data and Consumer Profiles – The hidden traps of data collection on social media networks. Proc. Int. Conf. Bus. Excellence **13**(1), 1070–1078 (2019). https://doi.org/10.2478/PICBE-2019-0093
8. Talib, S., et al.: Perception analysis of social networks' privacy policy: instagram as a case study. In: The 5th International Conference on Information and Communication Technology for the Muslim World (ICT4M) 2014 (2014). https://doi.org/10.1109/ICT4M.2014.7020612
9. Soumelidou, A., Tsohou, A.: Effects of privacy policy visualization on users' information privacy awareness level: the case of Instagram. Inf. Technol. People **33**(2), 502–534 (2020). https://doi.org/10.1108/ITP-08-2017-0241/FULL/PDF
10. McCrow-Young, A.: Approaching Instagram data: reflections on accessing, archiving and anonymising visual social media **7**(1), 21–34 (2020). https://doi.org/10.1080/22041451.2020.1847820
11. Lewin, K., et al.: Social comparison and problematic social media use. Abbrev. Pers. Individ. Differ. **199**, 111865, (2022). https://doi.org/10.1016/j.paid.2022.111865
12. Iqbal, M.: Instagram Revenue and Usage Statistics. Business of Apps. https://www.businessofapps.com/data/instagram-statistics/. Accessed 16 Oct 2022
13. Reiff, N.: Top 5 Companies Owned by Facebook (Meta). Investopedia. https://www.investopedia.com/articles/personal-finance/051815/top-11-companies-owned-facebook.asp. Accessed 16 Oct 2022
14. Instagram: Data Policy. https://help.instagram.com/155833707900388. Accessed 16 Oct 2022
15. Data Policy: I. What kinds of information do we collect? (2022, January 4). Instagram. https://help.instagram.com/155833707900388. Accessed 16 Oct 2022
16. Data Policy: II. How do we use this information? (2022, January 4). Instagram. https://help.instagram.com/155833707900388. Accessed 16 Oct 2022
17. Data Policy: III. How is this information shared? (2022, January 4). Instagram. https://help.instagram.com/155833707900388. Accessed 16 Oct 2022
18. Cotter, K.: Playing the visibility game: how digital influencers and algorithms negotiate influence on Instagram. New Media Soc. **21**(4), 895–913 (2019). https://doi.org/10.1177/1461444818815684
19. Terms of Use, Instagram Help Center. https://help.instagram.com/581066165581870. Accessed 23 Oct 2022

20. Data Policy: VI. How can you exercise your rights provided under the GDPR? (2022, January 4). Instagram. https://help.instagram.com/155833707900388. Accessed 16 Oct 2022
21. Holmes, A.: "533 million Facebook users' phone numbers and personal data have been leaked online" (2021, April 3). Instagram. https://www.businessinsider.com/stolen-data-of-533-million-facebook-users-leaked-online-2021-4. Accessed 01 Nov 2022
22. Oakshott, L.: Essential quantitative methods - for business, management, and finance, 6th edn. Macmillan International Higher Education, London (2016)
23. Saunders, M.N.K., Cooper, S.A.: Understanding Business Statistics – An Active-Learning Approach. The Guernsey Press, Guernsey (1993)

Human-Computer Interaction

User eXperience (UX) Evaluation in Virtual Reality (VR)

Matías García[1], Sandra Cano[1], and Fernando Moreira[2,3](✉)

[1] School of Computer Engineering, Pontificia Universidad Católica de Valparaíso, Valparaíso, Chile
sandra.cano@pucv.cl
[2] REMIT, IJP, Universidade Portucalense, Porto, Portugal
fmoreira@upt.pt
[3] IEETA, Universidade de Aveiro, Aveiro, Portugal

Abstract. The potential of Virtual Reality (VR) devices lies in their ability to put the user in an interactive 3D environment as seen in simulations, prototyping and gaming. In order to create a more valuable experience in these environments, it is important to not only evaluate users' behavior, but also understand their preferences and emotions in this context. However the challenge of applying User eXperience (UX) evaluation in VR comes from the fact that behavioral methods may require the use of additional hardware to capture objective data unlike attitudinal methods which depend on what the user expresses during and after the experiment. This review lists the UX evaluation methods, comparing their applicability in VR devices, considering aspects such as complexity, data analysis and additional requirements. We also discuss the feasibility and benefit of combining these two types of evaluation methods in the same case of study.

Keywords: UX evaluation · Virtual Reality · Human-Computer Interaction

1 Introduction

Virtual Reality devices offer 3D worlds where users can interact with not only a controller but also their own body movement as inputs with virtual objects around them, reaching new levels of interaction compared to what can be achieved with other media. State-of-the-art applications of VR (online conferences [1], examination of forensic ballistics [2], surgery simulators [3]) prove that the utility and the capabilities of this technology are beyond the game industry.

Modern VR headsets such as Oculus Quest and HTC Vive integrate computing capabilities and advanced graphics, including mainstream game engines like Unity 3D and Unreal Engine, allowing small teams to develop their own VR applications without the need to additionally pay for Software Development Kits (SDK). This provides a fertile land where enthusiasts can experiment with their own ideas and explore the potential of this technology.

A. Rocha et al. (Eds.): WorldCIST 2023, LNNS 799, pp. 207–215, 2024.
https://doi.org/10.1007/978-3-031-45642-8_20

The problem lies in the shortcomings of exclusively using the design guidelines of Human-computer Interaction (HCI) in VR development due to it failing to cope their needs appropriately, for example HCI focuses on monoscopic display from an exocentric point of view while Human Virtual Environment Interaction (HVEI) [4] is mainly egocentric, 3D oriented and additionally addresses perception, navigation, object manipulation and user engagement as fundamental aspects in their designs that need special attention for users to successfully immerse themselves and effectively carry out tasks in Virtual Environments (VE). These HVEI aspects lack an empirical method to measure them but we can measure the effects on the user and these results may help to diagnose and track down flaws in the VE.

UX is treated as an extension of usability based on the CUE (Components User Experience) model [5]. Usability is regarded as part of the instrumental qualities related to the components and features that provide the system ease of use, control over the system behavior (e.g. letting expert users take shortcuts or allowing new users to step back in a task). Sutcliffe & Kaur [6] mention that standard usability evaluation methods can't be applied to VR, we extend this question including in our scope UX qualities. Non-instrumental qualities refer to visual aesthetics or haptic quality (the degree to which objects are perceived as graspable, closely related to object manipulation) that evoke appeal and attractiveness. Instrumental and non-instrumental qualities influence a third component, the user's emotional reaction.

Instrumental and non-instrumental qualities are crucial in establishing place *illusion* (the feeling of actually being in the VR world) and *plausibility* (performing actions effectively and reliably). The concepts of *place illusion* and *plausibility* describe the concept of **presence** [7]. The quality of presence directly affects human performance in VR systems [4]. Thus, UX evaluation may highlight elements of VEs that need to be fixed or redesigned to enhance human performance, notably for VR training simulations for workers in critical tasks with stressful factors involved.

UX evaluation methods are classified in two groups, attitudinal and behavioral methods [8]. This work aims to identify and compare the methods applicable to VR, assessing their implementation complexity, cost of equipment and their compatibility with other methods. It is discussed which methods complement each other, considering the benefits of combining quantitative and qualitative data [9]. Finally we present a chart with these methods that serves as a quick guide for future research interested in evaluating UX in VR.

2 Background

As exposed earlier, evaluating HVEI aspects such as presence, is no trivial task. Slater & Steed [10] discuss the definition of presence as more than just the subjective feeling of being in a VE, it rather serves as a dynamic selector that makes the person pay attention and become absorbed between the virtual environment and the real world. States of presence and no-presence can be triggered by simply external force (e.g. the weight of the headset) or internal VE cause such as visual stimuli lagging behind when changing view too quickly. The common method of evaluation relies on post-experiment questionnaires which can be affected by user expectations and subsequently alter the degree of presence, instead they employed a stochastic approach based on transitions between states of presence modeled as a Markov chain. The novel evaluation technique found a positive correlation between body movement and presence, implying that movement can be exploited in VE to achieve better user engagement.

Wienrich et al. [11] establishes a correlation between UX and VR aspects, coming up with a detailed chart, acknowledging that some UX aspects are objectively more strongly binded with VR aspects than others. Therefore, evaluating the UX of a VR system may help testers diagnose the flaws in their designs. Thankfully Bowman et al. [12] proposed a classification and comparison of usability methods for VR, but this survey doesn't review UX evaluation methods such as Eye Tracking or A/B testing. We take their work as the starting point where we add new methods focusing on their compatibility in a single case of study, particularly methods from both attitudinal and behavioral approaches, to complement their weaknesses and create a holistic view of the UX. There's also a need to update some information considering the cost of equipment and complexity of implementation has changed from 20 years ago.

3 Methodology

This research starts by reviewing the literature covering current UX evaluation methods and selecting those applicable to VR systems. Thus the framework in [13] is ideal, structuring the review in 5 stages: 1) identify the research questions, 2) identify relevant studies, 3) study selection, 4) chart the data and 5) report the results (Table 1).

3.1 Identify the Research Questions

Table 1. Research questions

ID	Question
RQ1	What UX evaluation methods are applicable to VR systems?
RQ2	What pros and cons each method has?
RQ3	How compatible are these methods in a single experiment?

3.2 Identify Relevant Studies

The keywords used were: "UX evaluation", "Usability evaluation" and "Virtual Reality" in the scientific database of Scopus and Web of Science. The articles recovered were filtered using the following criteria (Table 2):

Table 2. Inclusion criteria

Article	Criteria
Subject	The title and abstract must be related with applying UX or usability evaluation methods in VR systems
Publisher	It must be published in a scientific journal or be in a conference
Language	The article must be written in English
Publication year	It must be published from 2015 to 2022

3.3 Study Selection

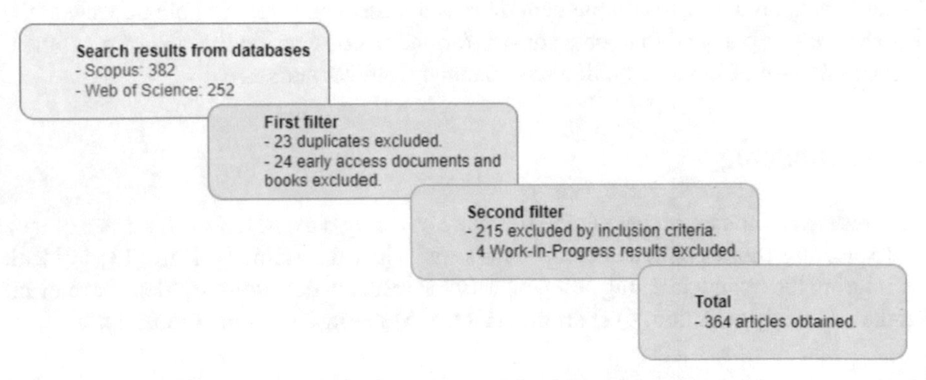

Fig. 1. Diagram illustrating the selection process. The 364 articles recovered were examined, finally quoting 7 of them as examples in the list below. It should be noted how questionnaires and interviews were the most popular methods in these articles (Source: authors' own).

4 Results

Answering **RQ1** and **RQ2**, this list provides the methods recovered from the literature review, referencing the original study with VR systems and other external sources for further reading on how to carry out testing. We also propose additional evaluation methods that haven't been used in VR studies so far (Fig. 1).

4.1 Attitudinal Methods

To put it simply, attitudinal methods rely on what users say and their stated beliefs, so these methods don't require buying specialized equipment to carry them out. However, users might not openly express all their thoughts about the system, either intentionally or unintentionally. Attitudinal methods that involve recording audiovisual media also demand more time (depending on the sample size) to analyze its results.

Questionnaires are very popular in UX evaluation due to its ease of implementation and the plethora of formats available. They are usually handed immediately after the interaction. An advantage of questionnaires is their capability of statistical analysis to model the sample feedback. To name a few disadvantages this method has limitations in evaluating presence as exposed by Slater [14] and sometimes the user doesn't comprehend the labels of the questions, resulting in misinterpretation. Rhiu et al. [15] used a variety of questionnaires in their experiment.

Interviews collect user feedback from an interviewer asking questions to the user, following a structured or semi-structured format, helping to understand the users' thoughts in depth. While interviews aren't expensive, they are time consuming hence why the recommended sample size isn't more than 15 participants. An example by Darin et al. [16] when evaluating Multimodal Interactive Virtual Environments.

Thinking Aloud protocol [17] consists in the user being instructed to verbalize their actions, choices and thoughts during the interaction with the system, which evaluators can take notes of. This method is very useful to learn about users' thoughts, feelings and their misconceptions of the UI that usually turns into concrete redesign recommendations. Disadvantages include making the users' express their thoughts may break their presence in the VE and the need to manually review audio media for assessing. Pettersson et al [18]. employs this method to evaluate in-vehicle experience in a VE.

Co-discovery [18] incorporates two participants interacting with the same system, exploring and freely discussing its features and what they think about it. This method provides more authentic experiential feedback than post-experiment interviews, but it's also harder to direct and control the discussion. Another consideration is that this method requires a proper setup for two people to be in the same VE (two headsets or one headset and one screen spectating). When the target end-users are children, the co-discovery evaluation is more comfortable and natural for young participants as they tend to be more willing to express their thoughts and feelings between peers.

4.2 Behavioral Methods

These methods collect data from the observation of what users do and their physical response when interacting with the system. Behavioral methods provide quantitative data from the interaction which is less influenced by subjectivity. However, the data is considerably more complex to analyze and often requires extra equipment.

A/B testing [19] consists in studying the user behavior when interacting with two variations of the same system with slight differences in their design. The user is then given a task and their performance is recovered. An advantage is that this method doesn't require specialized equipment, but it does need the development of two prototypes of the same VE.

Task tree models the task in VR into subtasks in a hierarchical structure that allows categorizing the actions of the user into effective and inefficient actions. Harms [20] proposed an automated evaluation that searches for patterns in these actions that represent signs of underlying usability issues. While this method doesn't require extra equipment, it does require programming this non-trivial automation algorithm.

Eyetracking is the technique of recording the user's eye movements during the interaction. Gaze behavior and eyetracking metrics are correlated with UX indicators [21]. The main challenge of Eyetracking is that not all headsets are compatible. Bala et al. [22] integrates Eyetracking analytics in an interactive dashboard for researchers to study VR storytelling.

Electroencephalography (or EEG) involves reading the user's brain activity during the experiment with specialized equipment placed around the head. Currently there are SDKs provided by the equipment's manufacturer that help visualizing brain activity as comprehensible data. This evaluation method may be costly to implement but it's unique as it allows researchers to study how the brain of the user behaves as exposed in Carofiglio et al. [23].

Electrodermal activity (or EDA) requires the user to wear equipment usually around the wrist. This device reads the skin inductance, also known as galvanic skin response (GSR), which serves as an indicator of psychological or physiological arousal. In other words, it detects the moment the user has a strong reaction (positive like excitement or negative like fear) from the electric response on their skin. This method is convenient when the objective of the study focuses on measuring significant physiological variations such as Niu et al. [24].

4.3 Comparison of Methods

Each UX evaluation method captures data from the interaction in a different way thus testers need to understand the limitations each method has and how to use them appropriately to extract as much information from the experience. Some methods may cost more than what the scope of their study justifies while others are dismissed for being harder to implement resulting in overuse of the reliable questionnaire. The chart below displays evaluation methods compared under two parameters: **cost** and **complexity**. This comparison rather than being strict and absolute, serves as a quick guide to know how each method stands next to the other under these two parameters (Fig. 2).

From the list of methods included in this review, only interviews and questionnaires are executed post experiment. These attitudinal methods are also the cheapest so there's little reason not to integrate them in short studies, however solely relying on them may also constrain the effectiveness of the UX assessment on the quality of the feedback given by the user. Combining interviews and questionnaires represent the cheapest and easiest way to evaluate UX but it should be noted that having an interview right after answering a questionnaire extends the post-experiment phase for the participant. Considering the most basic compatibility comes from during and post experiment evaluation, the only attitudinal methods that aren't compatible are Think Aloud with Co-discovery because users will likely focus on discussing together over explaining every single action and thought they individually have.

To summarize the benefits of incorporating attitudinal and behavioral methods: one retrieves explicitly what the user thinks/needs and the other backs it up with quantitative physiological data, therefore mitigating the risk of mistakenly using users' opinions as proof of hidden problems in design that, depending on the VR system, take significant resources and time to modify. Answering **RQ3**, some examples of compatibility between behavioral methods are A/B testing and Task trees, retrieving performance information and comparing effectiveness of two prototypes, useful when the product is still being refined and small differences influence the quality of the final product. Also Thinking Aloud with EDA and/or EEG helps to correlate physiological responses with specific

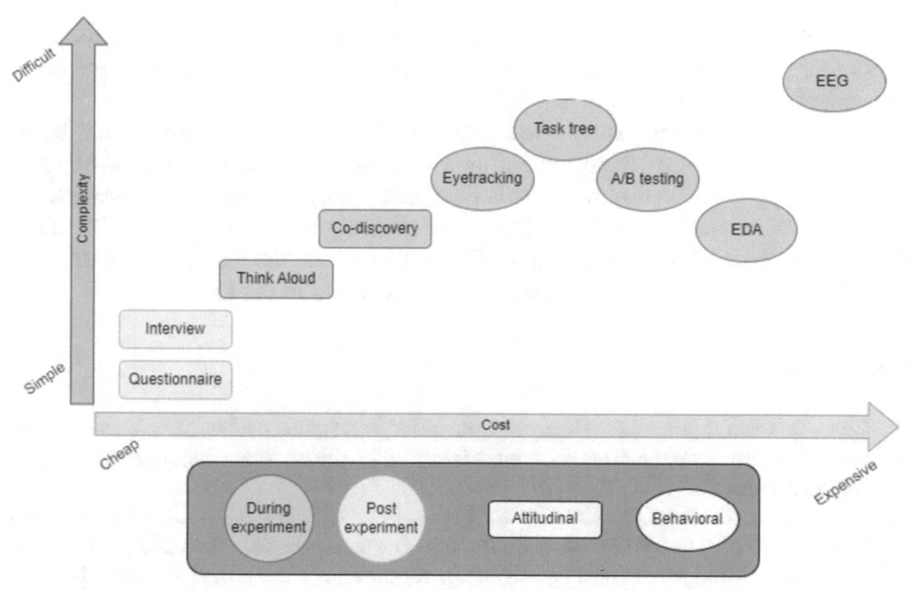

Fig. 2. This chart displays UX evaluation methods comparing their **complexity** vs the **cost**. Note that **complexity** factors in the specific skills needed for implementation and how trivial it is to translate data into results. **Cost** involves not only the price of extra equipment, but also the time required to develop, code and review results (Source: authors' own).

elements in the VE that cause arousal or stress in the participant. Finally Eyetracking with Co-discovery provides validation to the feedback given by young participants by showing how they visually analyzed the VE, what caught their attention first and their approach to solve problems.

5 Conclusions

In this paper we discuss the limitations when solely relying on the original design guidelines of HCI for the development and testing of VR systems. HCI doesn't detail HVEI aspects such as presence, object manipulation and perception that play a fundamental role in the construction of UX as its components are correlated to HVEI aspects [11]. Therefore being able to effectively assess UX allows testers to diagnose and address design flaws in VR systems.

The purpose of this research is to elaborate and offer a quick guide to learn about the applicable UX evaluation methods in VE; describing them with their pros and cons along with the equipment needed. The methods are classified into attitudinal and behavioral depending on how they collect data, also discussing the benefits of combining the two approaches (further reading [9]). We also include a chart with the layout of evaluation methods ordered using two parameters: cost and complexity. This chart highlights the reason why the most popular methods are questionnaires and interviews, however these attitudinal methods depend on the quality of the feedback given by the participants who might lie, misunderstand the questions or fail to express their experience to the fullest. Behavioral methods provide information based on performance and physiological responses, providing quantitative and objective data that helps to back up the observations recovered from attitudinal research.

We propose combinations between these methods in a single experiment, but the effectiveness of some of them remain untested, such as Task trees with A/B testing or Eyetracking with Co-discovery. This represents an opportunity for future research, notably in automated behavioral research. The search query can be improved to include more evaluation methods applicable to VR, for that reason the method list in this paper isn't absolute and new evaluation methods may be added in the future.

References

1. Roos, G., Oláh, J., Ingle, R., Kobayashi, R., Feldt, M.: Online conferences – towards a new (virtual) reality. Comput. Theor. Chem. **1189**, 112975 (2020). https://doi.org/10.1016/j.comptc.2020.112975
2. Guarnera, L., Giudice, O., Livatino, S., Paratore, A.B., Salici, A., Battiato, S.: Assessing forensic ballistics three-dimensionally through graphical reconstruction and immersive VR observation. Multimedia Tools Appl. (2022). https://doi.org/10.1007/s11042-022-14037-x
3. Mori, T., Ikeda, K., Takeshita, N., Teramura, K., Ito, M.: Validation of a novel virtual reality simulation system with the focus on training for surgical dissection during laparoscopic sigmoid colectomy. BMC Surg. **22**(1), 12 (2022). https://doi.org/10.1186/s12893-021-01441-7
4. Stanney, K.M., Mourant, R.R. Kennedy, R.S.: Human factors issues in virtual environments: a review of the literature. Teleoperators Virtual Environ. **7**(4), 327–351 (1998). https://doi.org/10.1162/105474698565767
5. Mahlke, S., Thüring, M.: Studying antecedents of emotional experiences in interactive contexts. In: Proceedings of the SIGCHI Conference on Human Factors in Computing Systems (2007).https://doi.org/10.1145/1240624.1240762
6. Sutcliffe, A.G., Kaur, K.D.: Evaluating the usability of virtual reality user interfaces. Behav. Inf. Technol. **19**(6), 415–426 (2000). https://doi.org/10.1080/014492900750052679
7. Slater, M.: Place illusion and plausibility can lead to realistic behaviour in immersive virtual environments. Philos. Trans. R. Soc. B: Biol. Sci. **364**(1535), 3549–3557 (2009). https://doi.org/10.1098/rstb.2009.0138
8. When to Use Which User-Experience Research Methods. (n. d.). Nielsen Norman Group. Retrieved January 3, 2023, from https://www.nngroup.com/articles/which-ux-research-methods/
9. Xu, J., Zhang, Z.: Research on user experience based on competition websites. J. Phys. Conf. Ser. **1875**(1), 012014 (2021). https://doi.org/10.1088/1742-6596/1875/1/012014
10. Slater, M., Steed, A.: A virtual presence counter. Teleoperators Virtual Environ. **9**(5), 413–434 (2000). https://doi.org/10.1162/105474600566925
11. Wienrich, C., Döllinger, N., Kock, S., Schindler, K., Traupe, O.: Assessing user experience in virtual reality – a comparison of different measurements. In: Marcus, A., Wang, W. (eds.) DUXU 2018. LNCS, vol. 10918, pp. 573–589. Springer, Cham (2018). https://doi.org/10.1007/978-3-319-91797-9_41
12. Bowman, D.A., Gabbard, J.L. Hix, D.: A survey of usability evaluation in virtual environments: classification and comparison of methods. Teleoperators Virtual Environ. **11**(4), 404–424 (2002). https://doi.org/10.1162/105474602760204309
13. Arksey, H., O'Malley, L.: Scoping studies: towards a methodological framework. Int. J. Soc. Res. Methodol. **8**(1), 19–32 (2005). https://doi.org/10.1080/1364557032000119616

14. Slater, M.: How colorful was your day? Why questionnaires cannot assess presence in virtual environments. Teleoperators Virtual Environ. **13**(4), 484–493 (2004). https://doi.org/10.1162/1054746041944849
15. Rhiu, I., Kim, Y.M., Kim, W., Yun, M.H.: The evaluation of user experience of a human walking and a driving simulation in the virtual reality. Int. J. Ind. Ergon. **79**, 103002 (2020). https://doi.org/10.1016/j.ergon.2020.103002
16. Darin, T., Andrade, R., Sánchez, J.: Usability evaluation of multimodal interactive virtual environments for learners who are blind: an empirical investigation. Int. J. Hum. Comput. Stud. **158**, 102732 (2022). https://doi.org/10.1016/j.ijhcs.2021.102732
17. Ericsson, K.A., Simon, H.A.: Protocol analysis: Verbal reports as data. The MIT Press, Cambridge, Massachusetts (1993). https://doi.org/10.7551/mitpress/5657.001.0001
18. Pettersson, I., Karlsson, M., Ghiurau, F.T.: Virtually the same experience? In: Proceedings of the 2019 on Designing Interactive Systems Conference (2019). https://doi.org/10.1145/3322276.3322288
19. Co-discovery Learning. (n.d.). Retrieved December 31, 2022, from http:/hci.ilikecake.ie/eval_codiscoverylearning.htm
20. Putting A/B Testing in Its Place. (n.d.). Nielsen Norman Group. Retrieved January 3, 2023, from https://www.nngroup.com/articles/putting-ab-testing-in-its-place/
21. Harms, P.: Automated usability evaluation of virtual reality applications. ACM Trans. Comput.-Hum. Interact. **26**(3), 1–36 (2019). https://doi.org/10.1145/3301423
22. Joseph, A.W., Murugesh, R.: Potential eye tracking metrics and indicators to measure cognitive load in human-computer interaction research. J. Sci. Res. **64**(01), 168–175 (2020). https://doi.org/10.37398/jsr.2020.640137
23. Bala, P., Nisi, V., Nunes, N.: Evaluating user experience in 360° storytelling through analytics. In: Nunes, N., Oakley, I., Nisi, V. (eds.) ICIDS 2017. LNCS, vol. 10690, pp. 270–273. Springer, Cham (2017). https://doi.org/10.1007/978-3-319-71027-3_23
24. Carofiglio, V., Ricci, G., Abbattista, F.: User brain-driven evaluation of an educational 3D virtual environment. In: 2015 10th Iberian Conference on Information Systems and Technologies (CISTI) (2015). https://doi.org/10.1109/cisti.2015.7170553
25. Niu, Y., Wang, D., Wang, Z., Sun, F., Yue, K., Zheng, N.: User experience evaluation in virtual reality based on subjective feelings and physiological signals. J. Imaging Sci. Technol. **63**(6), 60413–60421 (2019). https://doi.org/10.2352/j.imagingsci.technol.2019.63.6.060413

Natural Hand Gesture Interaction for 3D Transformations in Immersive VR

Naveed Ahmed(✉) and Mohammed Lataifeh

Department of Computer Science, University of Sharjah, Sharjah, United Arab Emirates
{nahmed,mlataifeh}@sharjah.ac.ae

Abstract. For 3D transformations in immersive VR, we introduce a brand-new hand gesture-based user interface. Our approach relies on natural gestures that only involve the thumb and index finger moving in conjunction with the rotation of the hand. These movements enable the user to scale a 3D model as well as rotate it around the three axes. We use Meta Quest to implement these movements and share the implementation details. To show how effective our user interface is in terms of usability and user experience, we conducted a preliminary user study. A gesture-based user interface can be used as a workable mechanism for interactive 3D transformation in immersive VR, according to the user study.

Keywords: Hand gestures · NUI · 3D transformation · Immersive VR

1 Introduction

There are now many methods to connect with software applications on a variety of devices because of the quick advancements in technology. Motion controls were introduced to the gaming community through gadgets like Kinect, PS Move, and Wii Remote. Among these, Kinect [1] enables the capture of motion without requiring any actual physical contact with the input hardware. Kinect [1] is used to implement a gesture-based UI for a variety of desktop applications since it can record the motion of the user in real-time. Although it can pick up on high level arm and body motion, it is unable to record fine hand movements as it does not track the motion of the fingers individually.

Users are more accustomed to using hand gestures, particularly multi touch interfaces using fingers, to control the programs with the introduction of touch-based mobile devices. An interface that permits comparable desktop environment interactions would be more effective and user-friendly. Virtual and augmented reality applications can also easily use any hand gesture control. For a user, utilizing a gesture-controlled system to communicate with any application feels more natural since it adheres to their intrinsic knowledge of using a touch-based interface on mobile devices.

The initial step in using gestures for user interaction is gesture detection. Various techniques employing various technologies have been proposed to detect gestures. Hidden Markov Models [2] have been used in a number of techniques to identify gestures. AdaBoost, multi-layer perception, principal component analysis, and Histogram of Oriented Gradients (HOG) characteristics are used to propose further approaches [2].

A. Rocha et al. (Eds.): WorldCIST 2023, LNNS 799, pp. 216–222, 2024.
https://doi.org/10.1007/978-3-031-45642-8_21

Cheng et al. presented a survey on 3D hand gesture recognition that discusses all the above approaches.

After the gesture recognition, a number of tests are carried out to assess the interfaces' usability and user experience. A usability study on the use of hand and arm gestures for typical desktop tasks was given by Farhadi-Niaki et al. [3]. Touch and air gestures for aiding developer meetings were demonstrated by Bragdon et al. [4]. Cabral et al. [5] presented a usability assessment of gestures in the virtual reality environment. Villaroman et al. proposed a Kinect-based gesture-based user interface design [6]. Their research demonstrates that, despite Kinect's restrictions on fine-grained gesture input, it is effective for common activities. Gesture-based controls for typical daily chores were given by Bhuiyan et al. [7], who also investigated how well they worked in various real-world situations. The limits of a gesture-based interaction for modifying CT scan data are demonstrated by Ebert et al. [8]. In order to demonstrate how well-suited gesture-based interfaces are for new technologies, Liao et al. [9] proposed a gesture based command system for interactive paper. A selection of vision-based hand gesture applications was given by Wachs et al. [10].

Leap Motion [11] provided a new method of interacting with desktop PCs utilizing hand- and finger-based movements for hand-based gestures. A study on the possibility of employing Leap Motion for recognition of sign language was presented by Potter et al. [12]. Leap Motion was used by Guerrero-Rincon et al. [13] to operate a robot utilizing a gesture control system. A general framework for detecting movements with both Kinect and Leap Motion was provided by Marin et al. [14]. Leap Motion was also incorporated by Khroub et al. [15] for a 3D user interface in immersive VR.

Recently, Meta Quest provided a natural way for detecting hand gestures in an immersive VR environment [16]. This work by Ahmed et al. [16] shows that it is possible to detect a grab or squeeze gesture in VR using Meta Quest hand tracking. Though, they did not implement any complex gestures for the 3D model transformations. Here, using Meta Quest, we demonstrate a brand-new user interface for interactive 3D transformations in immersive VR. One of the most common user interactions in computer graphics applications [17, 18] is the manipulation of a 3D model. A variety of programs, including 3D studio, maya, CAD, blender, etc., use 3D model manipulation. For user interaction, a mouse-based interface is typically employed. We employ four gestures, rotating and scaling the 3D model just using the index finger and thumb. The 3D model is rotated three times along the x, y, and z axes before being proportionally scaled with a fourth gesture. Following a preliminary user survey that assesses the usability and user experience of the gesture-based user interface, we go into detail on how the gestures are implemented.

The paper is structured as follows: The design and implementation of the gesture based used interface is presented in Sect. 2. Sect. 3 details the user study, results, followed by the discussion. The paper concludes in Sect. 4.

2 User Interface Design and Implementation

Meta Quest has been used to implement our gesture-based user interface. We used the Meta Quest as a tether-free VR HMD and collected training sets for gesture recognition using its integrated hand tracking (Fig. 1a). 24 bones are tracked dynamically by Meta

Quest (Fig. 1b). We utilized the technique developed by Ahmed et al. [16], who used tracked bones from Leap Motion to distinguish between static and dynamic motions, to the data supplied by Meta Quest. The hand tracking data is standardized and unaffected by the hand's size. As a result, the 98% accuracy of the gesture recognition algorithm ensures its dependability. We had no problems with input failure during the testing sessions, and test users had no trouble evaluating the system.

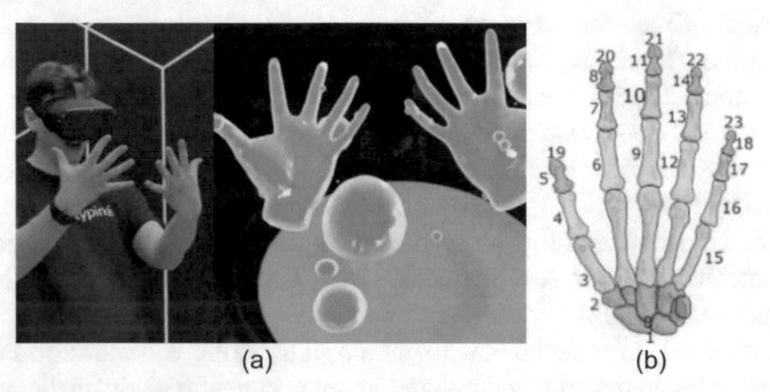

Fig. 1. (a) Real-time hand tracking as visualized using Meta Quest. (b) 24 bones tracked by the Meta Quest and used for gesture recognition training and recognition.

Using Meta Quest, we put four gestures for 3D transformations into practice. Rotation and scaling are the most common 3D data transformations. Due to the fact that we just use the uniform scaling, the rotation along the x, y, and z axes only requires one additional gesture. The following sections define the gestures.

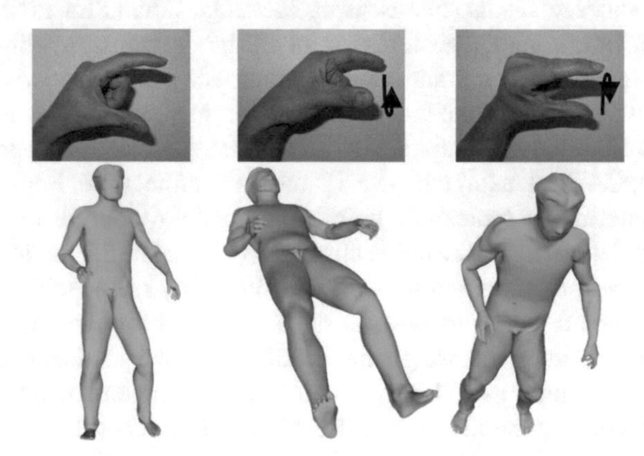

Fig. 2. Gesture for the rotation around the x-axis. First the neutral pose is identified, and the gesture is shown that performs rotation around the x-axis with its application on a 3D model.

2.1 Rotation

The three rotations in a 3D coordinate system are termed as pitch, yaw, and roll. These labels are given to rotations around the x, y, and z axes, respectively, and they make up the

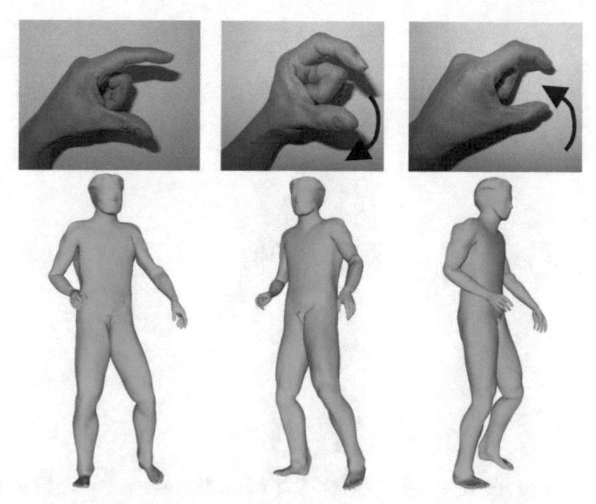

Fig. 3. Gesture for the rotation around the y-axis. First the neutral pose is identified, and the gesture is shown that performs rotation around the x-axis with its application on a 3D model.

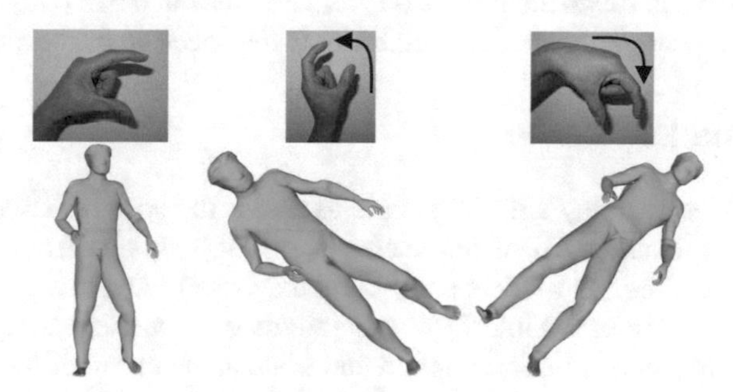

Fig. 4. Gesture for the rotation around the z-axis. First the neutral pose is identified, and the gesture is shown that performs rotation around the x-axis with its application on a 3D model.

complete 3D rotation. A neutral gesture is identified in the first stage. Our implementation system actively seeks for one of the three rotations or the scaling when it recognizes the Neutral Pose. Because it is determined by the actual 3D rotation of the hand in the real world as depicted in Figs. 2, 3, and 4, our rotation gesture is incredibly intuitive. After identifying the rotation gesture, we use the dynamic finger locations to pinpoint the axis of rotation. The appropriate 3D rotation is applied to a 3D model once the rotation's axis has been established.

2.2 Scaling

As our approach simply uses proportional scaling, detecting whether the model is being scaled up or down only requires one gesture. Following the neutral posture, the scaling gesture can be recognized by the shifting distance between the index finger and thumb. Figure 5 shows an illustration of the scaling gesture. Rotation and scale of a 3D model

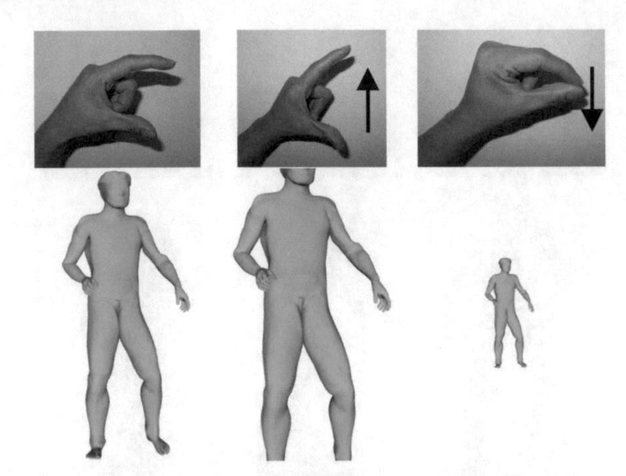

Fig. 5. Gesture for the uniform scaling. First the neutral pose is identified, and the gesture is shown that performs a zoom-in and zoom-out on a 3D model.

are controlled by the gestures mentioned above. The user study that confirms the efficacy of these gestures' user-experience and usability is described in the section that follows.

3 Results and Discussion

We carried out a user survey with 20 people to assess the gesture-based user interface (10 male, and 10 female). All of the users, who ranged in age from 18 to 30, were accustomed to handling 3D models using a mouse-based interface. None of the users had ever used a gesture-based interface to manipulate 3D models. They are instructed to record the 3D model's initial orientation and scale at the outset of the user testing. In the user study, they carried out the following procedures:

X-axis left rotation.
Y-axis down rotation.
Zoom-in scale.
Z-axis right rotation.
Zoom-out scale.
X-axis right rotation.
Free rotation and scale to match the original size and orientation of the 3D model.

Following the user study, the users were surveyed to gauge their satisfaction with the system's usability and overall user experience. Since this is a pilot study, we did not conduct a thorough survey; however, an extensive user study is now underway and will be included in future work.

Users completed the activities in less than a minute on average, and they were then taken directly to the questionnaire thereafter. 90% of users found the interface to be highly intuitive, and they were able to effectively accomplish all tasks, according to our usability analysis. According to the user experience analysis, 20% of users had trouble using the neutral gesture and 3D rotations, however everyone could effectively utilize the scale gesture.

As can be observed from the user survey, users strongly prefer the gesture-based interface both for its user experience, and usability. The results clearly show that the gesture-based interface is just as efficient at manipulating the 3D model as a mouse based interface. Users were able to simply rotate and resize the 3D model using gestures, and they were also able to finish the activity quickly and effectively. These first findings support the feasibility of using hand gestures to manipulate 3D scenes.

4 Conclusions

We demonstrated a brand-new hand gesture-based Meta Quest interface for 3D manipulations. We went into great length about how to use the four gestures. Our approach for identifying gestures first detects the Neutral Pose before actively looking for either the rotational or scaling gesture. Our user study demonstrates that gesture-based interface has a greater level of memorability and is simple to learn, use, and retain. The majority of users suggested the interface over a mouse-based interface as a consequence of the extremely positive user experience that it produced. The user research demonstrates the viability of our technology for 3D transformations and the wide range of applications it may be used in. In the future, we will extend the user study and thoroughly assess usability and user experience.

References

1. MICROSOFT, Kinect for microsoft windows. http://www.kinectforwindows.org/ (2010). Accessed Jan 2017
2. Cheng, H., Yang, L., Liu, Z.: Survey on 3D hand gesture recognition. IEEE Trans. Circuits Syst. Video Technol. **26**(9), 1659–1673 (2016)
3. Farhadi-Niaki, F., Etemad, S.A., Arya, A.: Design and usability analysis of gesture-based control for common desktop tasks. In: Proceedings of the 15th International Conference on Human-Computer Interaction: Interaction Modalities and Techniques - Volume Part IV, ser. HCI'13, pp. 215–224 (2013).https://doi.org/10.1007/978-3-642-39330-3_23
4. Bragdon, R., DeLine, K., Hinckley, Morris, M.R.: Code space: Touch + air gesture hybrid interactions for supporting developer meetings. In: Proceedings of the ACM International Conference on Interactive Tabletops and Surfaces, ser. ITS '11, pp. 212–221 (2011)
5. Cabral, M.C., Morimoto, C.H., Zuffo, M.K.: On the usability of gesture interfaces in virtual reality environments. In: Proceedings of the 2005 Latin American Conference on Human-computer Interaction, ser. CLIHC '05, 2005, pp. 100–108 (2005)
6. Villaroman, N., Rowe, D., Swan, B.: Teaching natural user interaction using openni and the microsoft kinect sensor. In: Proceedings of the 2011 Conference on Information Technology Education, ser. SIGITE '11, 2011, pp. 227–232 (2011)
7. Bhuiyan, M., Picking, R.: A gesture controlled user interface for inclusive design and evaluative study of its usability. JSEA **4**(9), 513–521 (2011)
8. Ebert, L.C., Hatch, G., Ampanozi, G., Thali, M.J., Ross, S.: You can't touch this: touch-free navigation through radiological images. Surg. Innov. **19**(3), 301–307 (2012)
9. Liao, C., Guimbretie're, F., Hinckley, K.: Papiercraft: a command system for interactive paper. In: Proceedings of the 18th Annual ACM Symposium on User Interface Software and Technology, ser. UIST '05. New York, NY, USA: ACM, pp. 241–244 (2005)

10. Wachs, J.P., Kölsch, M., Stern, H., Edan, Y.: Vision-based hand-gesture applications. Commun. ACM **54**(2), 60–71 (2011). https://doi.org/10.1145/1897816.1897838
11. LeapMotion, Leap motion controller. https://www.leapmotion.com/product/desktop (2012). Accessed Jan 2017
12. Potter, L.E., Araullo, J., Carter, L.: The leap motion controller: a view on sign language. In: Proceedings of the 25th Australian Computer-Human Interaction Conference: Augmentation, Application, Innovation, Collaboration, ser. OzCHI '13, 2013, pp. 175–178 (2013)
13. Guerrero-Rincon, C., Uribe-Quevedo, A., Leon-Rodriguez, H., Park, J.O.: Hand-based tracking animatronics interaction. In: Robotics (ISR), 2013 44th International Symposium on, Oct 2013, pp. 1–3 (2013)
14. Marin, G., Dominio, F., Zanuttigh, P.: Hand gesture recognition with leap motion and kinect devices. In: 2014 IEEE International Conference on Image Processing (ICIP), Oct 2014, pp. 1565–1569 (2014)
15. Kharoub, H., Lataifeh, M., Ahmed, N.: 3D user interface design and usability for immersive vr. Appl. Sci. **9**(22) (2019). https://www.mdpi.com/2076-3417/9/22/4861
16. Ahmed, N., Lataifeh, M., Junejo, I.: Visual pseudo haptics for a dynamic squeeze/grab gesture in immersive virtual reality. In: 2021 IEEE 2nd International Conference on Human Machine Systems (ICHMS), pp. 1–4 (2021)
17. Ahmed, N., Theobalt, C., Magnor, M., Seidel, H.-P.: Spatio-temporal registration techniques for relightable 3D video. In: 2007 IEEE International Conference on Image Processing, vol. 2, 2007, pp. II-501–II-504 (2007)
18. Naveed Ahmed, N.J.: A system for 3D video acquisition and spatio-temporally coherent 3D animation reconstruction using multiple RGB-D cameras. Int. J. Signal Process., Image Process. Pattern Recogn. **6**, 113–128 (2013). https://www.earticle.net/Article/A208880

VTubers, Their Global Expansion and Impact on Modern Society

An Exploratory and Comparative Study Between Portugal and the USA

Carina Miranda[1], Mariana Costa[1], Mariana Pereira[1], Selma Almeida[1], Frederico Branco[2,3], and Manuel Au-Yong-Oliveira[4](✉)

[1] Department of Languages and Cultures, University of Aveiro, Aveiro, Portugal
{carinamiranda,mariana.cos,marianahpereira,selma.almeida}@ua.pt
[2] INESC TEC, Porto, Portugal
fbranco@utad.pt
[3] Universidade de Trás-Os-Montes E Alto Douro, Vila Real, Portugal
[4] INESC TEC, GOVCOPP, DEGEIT, University of Aveiro, Aveiro, Portugal
mao@ua.pt

Abstract. Many companies have been trying to take advantage of the entertainment boom on the internet and are searching for the best ways to make money from content creators. Among these are Virtual YouTubers, who have been growing in English-speaking spaces. This paper compares their popularity and influence in Portugal and the USA, while reflecting on their impact on viewers. To accomplish this, we created a survey, which obtained 102 answers, 46 being from Portuguese people and 44 being from U.S. citizens. We noted that only 37% of Portuguese people had heard of VTubers before and only 13% watch them on a weekly basis. On the other hand, every respondent from the USA was familiar with VTubers, with 86.4% watching them regularly. When asked about how VTubers affected the respondents' personal lives, some stated that it was merely a form of entertainment, however, many U.S. citizens reported that they watch VTubers to fulfill their need for social interaction or cope with loneliness. Thus, it is evident that these creators are much more popular in the USA. Lastly, we conclude that VTubers can have a positive impact on people's lives, but they also serve as an unhealthy escape from reality at times.

Keywords: Virtual YouTubers · YouTube · Twitch · Online streaming · Content creation · United States · Portugal · Avatar · Mental health · Anime

1 Introduction

For over a decade, online video and streaming platforms have been a growing source of entertainment for people all over the globe. Through platforms such as YouTube and Twitch, just about anyone with a phone or computer in their hands can make a career out of content creation. Naturally, where there is opportunity for one independent creator to rise to fame, there is opportunity for business.

A. Rocha et al. (Eds.): WorldCIST 2023, LNNS 799, pp. 223–231, 2024.
https://doi.org/10.1007/978-3-031-45642-8_22

Since the early days of YouTube, many talent agencies have surfaced and attempted to boost multiple creators to the top. However, dealing with real people has proven to be risky, expensive, and unreliable. Whether it is getting into controversies or experiencing burnout, there are many reasons creators have fallen from grace and permanently cut off a stream of revenue for their agency. This created a market for a different type of content creator – the Virtual Influencer.

Virtual Influencers are great for branding. Prior to the mainstream use of video-sharing platforms, there had already been some attempts at fabricating virtual celebrities, be it bands, singers, or actors. Some, such as Gorillaz and Hatsune Miku, even found major success. It was only natural for there to be attempts at conquering video and streaming platforms next. This came with its fair share of obstacles, as there are a lot of limitations regarding how these types of celebrities can interact with fans. Fully fictional virtual influencers with entire teams coordinating their every move could never achieve the level of fan interaction and perceived authenticity that a conventional YouTube and Twitch creator could, which takes away a large part of the appeal of following an online content creator. That created the need for a middle ground, a bridge between real and virtual, and that could be one of the reasons why VTubers have managed to gain popularity.

A VTuber or Virtual Youtuber is defined as a type of online content creator who uses a virtually generated avatar while streaming to their audience [1]. These virtual avatars mimic the movement of the person behind them and usually have unique characteristics and backstories attributed to them to set them apart from others. It is important to note that there is a single person controlling the avatar and their identity is completely hidden to the public. When they are streaming, they put on a persona separate from their real self and see it as a sort of performance. Much like regular YouTubers, VTubers can be managed and promoted by talent agencies or entertainment companies, or they can freelance.

Originating in Japan, VTubers have been gaining traction around the world. So much so, that Japanese VTuber companies, such as Hololive Production, have been branching out and establishing English-speaking VTuber groups. As such, the purpose of this paper is to evaluate VTubers' success and popularity in an English-speaking country, namely the United States of America, and compare it to Portugal. The goal is to see how differently the Portuguese general public has been impacted by this niche trend, or if there has been any impact at all. To get the answers that we need, we created a survey, available in both Portuguese and English.

2 Literature Review

Japanese manga and anime have been gaining more and more traction over the last few decades [2]. Something that, not long ago, many people were made fun of for enjoying, now turned out to be a worldwide success, forming a unique subculture around it.

Naturally, as a genre starts to gain popularity and worldwide attention, new ways of entertaining the viewers and new businesses start appearing to fulfill the needs of that community. VTubers, being animated virtual avatars usually inspired by anime characters that deliver performances in live streams or recorded videos, are a very good example of this [2].

However, because this is a relatively new development, there are not many academic studies done about this growth. And a large part of these few studies do not approach VTubers specifically or are not thorough enough.

2.1 Background

VTubers launched their success in Japan in 2016 and since then have been gaining popularity worldwide [2]. The first time the term Vtuber was used was in 2016 by the first Vtuber to obtain major popularity, Kizuna AI. Since then, the VTubing community has been expanding worldwide, and in 2020 the viewership of this type of content nearly doubled its number [3]. With this kind of growth, the opportunities for new creators were limitless and the number of individual and company-managed VTubers rose exponentially.

As expected, when VTubers began to receive so much attention, many Japanese and Chinese companies started to invest large amounts of resources in this type of creator, and this led to the formation of agencies that began to manage these avatars. [2] Currently, company-managed VTubers are dominating the market and attracting more viewers than individual ones, the two most known and successful for-profit VTuber agencies being Hololive [4] and Nijisanji [5]. This first agency manages, at this date, the most popular VTuber right now – Gawr Gura -, 9 out of the top 10 most viewed VTubers in the world and 31 out of the top 50. Nijisanji is the second agency with more VTubers in the top 50 of the most viewed VTubers in the world, only having 5 VTubers in this ranking, and not having any in the top 10. There is also only one individually managed Vtuber in the top 20 of this ranking at present [6].

The reasons for company-managed VTubers being so much more popular than individual ones are, most likely, because they are designed and programmed by professional teams and have more expensive equipment, which gives them more realistic characteristics. They also have a much stronger promotion of their avatars compared to independent VTubers. However, while independent VTubers have full ownership over their projects, company-managed ones are entirely owned and controlled by the company and the employees of these agencies have usually almost no control over the changes that are made to these characters [2].

2.2 Agencies Expansion and Strategy

In modern society, where there is demand, we also need to have supply and, as one grows, the other is bound to grow with it. VTubers are no exception to the rule, because when they started to gain more and more popularity, eventually their agencies began to broaden their horizons and initiated a strategic worldwide expansion.

Hololive, the biggest VTuber agency at present, is a very good example of a company that understands the importance of supply and demand and that is perpetually innovating and expanding its brand. This company, which started by debuting one VTuber in 2017, has now branched out into three more markets with subsidiaries in Indonesia, China, and the USA. Additionally, what was until just a few years ago an exclusive female Vtuber agency, has since 2019 its first male Vtuber group, that was, at first, operating under a separate agency. However, in December 2019, it was announced that all of the separate agencies of Hololive would be merged, to form "Hololive production", being that the male group continues to operate under the name "Holostars". [7] Like this company, many others followed the expansion journey to achieve worldwide recognition and grow as a company and as a brand.

The world of VTubers is well known for its anonymity because, although we can watch and follow an avatar, we do not know who is the person behind it. These avatars are often voiced by an actor referred to, in Japanese, as Nakanohito [8]. This anonymity is said to be established to protect the creators' privacy, which to some extent is the case because they can live separate lives from their avatar persona and there is a more open space to be creative without the risk of being bullied [3]. However, this anonymity has a lot of benefits for the agencies that owned them and their stakeholders to avoid any type of problems or scandals related to the creators' personal lives. To avoid this kind of problem, the viewers are warned that any type of discussion about Nakanohito's personal life is strictly prohibited. Even though most viewers know that this anonymity is mostly used to support corporate interests, most of the community members follow and support this course of action [2]. Most VTuber followers are attracted to this kind of creator exactly because of their anonymity and have such a displacement to the creator itself that they mostly do not mind when the voice actor gets replaced by the company [8]. The most important thing to viewers is not the voice actor itself, but the quality of the content, and many viewers prefer when it is used by more than one actor because the content tends to be more diverse and for that more engaging. Another reason for them to prefer the anonymity of the creators is because this way they can maintain a perfect image of the VTubers in their mind. They also have different expectations than when watching more traditional content creators, being that they can tolerate certain language and behavior that they otherwise would not. This is because they feel like the avatars are living in the online world and that normal rules do not apply to them [2]. However, even though the online followers of these VTubers agree on the importance and appeal of maintaining their personal information secret and mostly do not mind when one voice actor is replaced, they still concern themselves when it seems that a company is not treating its creators correctly. For example, when a voice actor gets replaced without an apparent reason, the online community following their avatar wants the company in question to explain the reason for the switch of creators [2].

2.3 Companies Using VTubers for Marketing

With the exponential growth of VTubers in the last few years, many companies began to use them as a marketing strategy to attract and capitalize on a new wave of customers [3]. This strategy was used fairly recently by known brands like Netflix, Taco Bell, and Kellogg's.

In the case of Netflix, they created a VTuber and made it their anime ambassador on their YouTube account dedicated to anime content. This new Vtuber was named N-ko and it takes the form of a girl with sheep-like features. Her existence was announced by the company on April 27 of 2021, and it was also announced that she would host her weekly show "The N-Ko Show" which would premiere on April 30 of the same year [9].

As for Taco Bell, the company chose a more subtle way to use a VTuber to benefit them, using the most popular VTuber in one of their advertisements. In June of 2021, Taco Bell launched a fully animated advertisement that starred, in the background, the avatar of Gawr Gura, the most famous and followed Vtuber right now [10].

Kellogg's company transformed their mascot, Tony the Tiger, into an interactive VTuber, turning into the first brand to do so. Kellogg's is constantly following the latest trends, seeing that this mascot also appeared on a late-night show and recently joined TikTok. As for streaming, his first appearance as a VTuber was on Twitch on August 29, 2022 [11–13]. The director of Brand Marketing at Kellogg's said that this was a new way of engaging with the new generations in a fun way and that the brand is also looking for new ways to connect with the fans, either by releasing new flavors or appearing in new places [11].

3 Methodology

As mentioned above, the main purpose of this study is to try and determine whether Virtual Youtubers (VTubers) have an impact in Portuguese society and to compare the data with that of the United States of America. To do that, we created a survey, through Google forms, where we aimed to collect both quantitative and qualitative data regarding people's preferences and habits regarding the subject. The survey was left open for answers for 9 days and was shared through WhatsApp, Instagram, and Reddit. We were able to collect 102 answers, but, since the purpose of this essay is to compare Portugal to the USA, only 90 were considered.

The survey collected data from two types of samples: Convenience and Purposive [14]. The convenience sample came mostly from the Portuguese public since the survey was shared with our social circles, while the sample we got from U.S Citizens was more of a purposive one given that the answers we got from them mostly came from Reddit forums that are more connected with Japanese culture and VTubers. We acknowledge that our samples are not ideal, but we made sure to check if there were any Portuguese forums that were similar, but they were residual.

4 Findings/Discussion

One of the reasons that led us to do this study was our curiosity on how well the Portuguese know VTubers in comparison to U.S. citizens. Based on conversations with Portuguese friends, family and colleagues about the subject, almost no one knew what VTubers are about, which led us to believe that, in general, the Portuguese are not well acquainted with this form of entertainment. A belief that was proven to be true by the survey results.

With the questionnaire, we were able to gather 102 answers from people with 12 different nationalities, but, as our purpose is to compare Portugal to the USA, only the answers that came from the citizens of those countries will be taken into consideration.

In the end, we disregarded 12 answers, that came either from non-Portuguese and non-U.S. citizens or from people who chose not to specify their nationality, being left with 90 valid answers (Table 1).

Table 1. Chi-square test (SPSS output)

Chi-square Tests					
	Value	df	Asymptotic Significance (2-sided)	Exact Sig. (2-sided)	Exact Sig. (1-sided)
Pearson Chi-Square	40,927[a]	1	<,001	<,001	<,001
Continuity Correction[b]	38,091	1	<,001		
Likelihood Ratio	52,533	1	<,001	<,001	<,001
Fisher's Exact Test				<,001	<,001
N of Valid Cases	90				

a. 0 cells (0,0%) have expected count less than 5. The minimum expected count is 14,18.
b. Computed only for a 2x2 table.

Furthermore, it was found that there is a statistically significant association between the respondents' nationality and their familiarity with Vtubers. Please see Table 1, constructed using SPSS software (H0: independent, H1: not independent, confidence range = 95%, confidence level = 0.05). As Asymptotic Significance for the Continuity Correction (computed only for a 2x2 table) is at < 0.001 and the confidence level is set at 95%, and 0,001 is less than 0,05, H0 is therefore rejected. For this reason, we can conclude that the variables involved are not independent.

4.1 Portuguese Citizens

When asked if they had ever heard about VTubers only 37% of the 46 Portuguese valid answers responded positively, of which 64% were from the ages between 18 to 23. They had heard about them through social media and 53% identified themselves as female. Additionally, 88% stated they watch more YouTubers than VTubers.

Of the 37% that are familiar with VTubers, 65% stated that they do not watch VTubers, since 41% prefer to watch more conventional content creators and 47% just do not take interest in their content. On the other hand, 29% said that they watched less than 5 h per week.

In short, only 13% of 46 Portuguese participants actually consume VTubers' content on a weekly basis, having all declared to have never spent money on a VTuber. 99% watch less than 5 h per week and 1% watches 5 to 10 h per week.

4.2 U.S. Citizens

As per the 55 answers we got in English, 80% came from U.S Citizens, and we also got one person from each of the following nationalities: Portuguese, Mexican, British, Canadian, Danish, German, Cambodian, South Korean, Uruguayan and Russian. One person preferred not to specify their nationality. Of the 44 valid answers, 75% are between the ages of 18 and 29, which is congruent with the data recovered regarding both YouTube and Twitch's demographics [15, 16]. When asked about their gender, 75% identified themselves as male, 15.9% as female, and 9% as "Other".

Going into the questions specifically about VTubers, all of the people that answered were familiar with the term, and 72.7% came across VTubers through social media. Regarding the 7th question, "What do you watch more?", 54.5% answered VTubers and 48.6% answered YouTubers. In the following question, 90.7% said that they use YouTube as the platform to watch their content, leaving Twitch with 9.3%. We were also able to determine that 59.1% of the people had spent money on a VTuber before, but, out of that number, 73.1% said that they rarely do it. The next question asked the number of hours people spend watching VTubers weekly, and 13.6% said that they do not watch them because they prefer more conventional content creators, have no interest, do not have time or do not enjoy their content. Of the people who watch their content 31.8% answered that they watch them less than 5 h (per week), while 29.5% said they watch their content for 5 to 10 h, and 20.5% said 10 to 20 h. Now regarding what people find more interesting about VTubers 36.4% answered that it was the content, 34.1% the interactions, and 13.6% the avatar.

At the end of the survey, we added an open-ended question: "How do you think VTubers affect society and your personal life?", to which we got interesting answers both from the Portuguese and U.S.A. samples, some of which we share below.

(P27:PT) *"I think VTubers, especially streamers, create an interesting relationship with the audience. (…) And just having a streaming session going on in the background can provide a more relaxed atmosphere while you do your daily tasks, as people often do with podcasts. Besides, with the avatars, they are able to create an even closer relationship between the public and fictional characters, than otherwise impossible, with a 2D character or a real person. Unlike cosplay, which are characters with already built personalities, VTubers create a completely new character and often based on personality traits from the people behind the camera and I find this very interesting."*

(P34:PT) *"Many of them can be a positive influence to children."*

(P43:PT) *"I believe that people can create content without having to be judged by society for their appearance."*

(P49:USA) *"They have given me a community to belong to, even if online. I feel like I have a sense of place among the people I have met online as a part of the VTuber community."*

(P72:USA) *"People who would otherwise be self-conscious and shy are able to be closer to themselves behind a masked persona. Fans meanwhile can personify whatever they want with said content creator since we don't see a flesh and blood person but a caricature."*

5 Conclusion and Future Research

The VTuber market is set to continue to grow at a considerable rate for at least the next 6 years [17]. Considering a lot of the personal accounts we received from U.S. citizens who like VTubers they said they felt lonely and wanted to belong to a community, hence we can attribute a significant portion of this growth to the increase in social isolation in the U.S.A [18]. Comparatively, the Portuguese population does not struggle with loneliness as much [19], especially the younger generations, which could explain why this niche hobby has not taken off just yet – aside from other relevant factors such as language barriers. This can perhaps be explained by the individualistic nature of the United States, which was built around the idea of focusing on the nuclear family. With a score of 91 on the individualism scale on Hofstede Insight's country comparison tool (based on Hofstede's 5 cultural dimensions), people in the U.S.A tend to value independence, preferring not to rely on others for support. This often leads to social isolation and makes it more difficult to form deep bonds, especially among men [20, 23] – a relevant detail, considering VTubers' viewership is predominantly male [21]. In Portugal, on the other hand, there is a much stronger sense of community, due to the country's collectivist culture. The Portuguese are generally more committed to maintaining relationships with those outside of their direct family, which makes it less likely for them to have to rely on other ways of fulfilling their social needs, such as forming [perhaps] unhealthy attachments to online personalities.

Still, the phenomenon of VTubers is something worth looking into in the future, since its popularity is bound to grow outside of Japan and the USA, as multiple other niche hobbies have. It is also noteworthy that these creators have had both positive and negative impacts on the people who enjoy their content. Their content can help fans combat loneliness and participate in a community of like-minded individuals, just as easily as it can end up being an unsuitable replacement for social interaction. In an era of the internet where it is becoming increasingly common for people to establish "parasocial relationships" [22] with online content creators, virtual influencers such as VTubers can exacerbate this issue and take it to new extremes. As such, it is important to find ways to mitigate this problem and promote healthier viewing habits among fans of these creators.

There are a few ways this could be done. One possible solution would be to hold the companies that manage the biggest VTubers accountable by urging them to have their content creators establish healthy boundaries with viewers, which could prevent a good number of people from experiencing an illusion of friendship when consuming their content. Another good measure, which has already been implemented by some social media platforms, would be to encourage fans to take regular breaks from engaging with their content. However, it is worth noting that, although these methods are not unheard of, they have still not shown great results.

Seeing as the main issue the form of entertainment presented, the fact that people can become overly attached to the idea of maintaining a false friendship with their favorite VTubers, we perceive that, in the future, there should be a focus on finding more effective ways to break this illusion and to encourage the members of these online communities to interact more amongst themselves, so that they can establish true reciprocal relationships with likeminded people.

Funding. This work was financially supported by the research unit on Governance, Competitiveness and Public Policies (UIDB/04058/2020) + (UIDP/04058/2020), funded by national funds through FCT - Fundação para a Ciência e a Tecnologia.

References

1. How to geek. https://www.howtogeek.com/720841/what-is-a-vtuber/ Accessed 15 Oct 2022
2. Lu, Z., Shen, C., Li, J., Shen, H., Wigdor, D.: More Kawaii than a real-person live streamer: understanding how the otaku community engages with and perceives virtual youtubers. In: Proceedings of the 2021 CHI Conference on Human Factors in Computing Systems, pp. 1–14 (2021)
3. Choudhry, A., Han, J., Xu, X., Huang, Y.: "I Felt a Little Crazy Following a 'Doll'": investigating real influence of virtual influencers on their followers. In: Proceedings of the ACM on Human-Computer Interaction, pp. 1–28 (2022)
4. Hololive. https://www.hololive.tv/. Accessed 18 Oct 2022
5. Nijisanji. https://nijisanji.ichikara.co.jp/. Accessed 18 Oct 2022
6. Virtual Youtuber user local ranking. https://virtual-youtuber.userlocal.jp/document/ranking. Accessed 20 Oct 2022
7. Fandom. https://virtualyoutuber.fandom.com/wiki/Hololive. Accessed 22 Oct 2022
8. Turner, A.: Streaming as a Virtual Being: The Complex Relationship Between VTubers and Identity. Malmö University, pp. 16–29 (2022)
9. Netflix. https://about.netflix.com/en/news/n-ko-mei-kurono. Accessed 23 Oct 2022
10. CBR. https://www.cbr.com/taco-bell-anime-fry-force-olympics-real/. Accessed 23 Oct 2022
11. Kellogg's Company. https://newsroom.kelloggcompany.com/2022-08-15-Tony-the-Tiger-R-Tackles-Twitch-and-Becomes-the-First-Ever-Brand-Mascot-Working-with-Twitch-to-Transform-into-an-Interactive-VTuber. Accessed 23 Oct 2022
12. PR News Wire. https://www.prnewswire.com/news-releases/tony-the-tiger-tackles-twitch-and-becomes-the-first-ever-brand-mascot-working-with-twitch-to-transform-into-an-interactive-vtuber-301605720.html. Accessed 23 Oct 2022
13. Twitch. https://www.twitch.tv/videos/1605481209. Accessed 23 Oct 2022
14. Oakshott, L.: Essential quantitative methods - for business, management, and finance, 6th edn. Macmillan International Higher Education, London (2016)
15. Oberlo. https://www.oberlo.com/statistics/youtube-age-demographics. Accessed 28 Oct 2022
16. Business of Apps. https://www.businessofapps.com/data/twitch-statistics/. Accessed 28 Oct 2022
17. Open PR. https://www.openpr.com/news/2672625/vtuber-virtual-youtuber-market-size-2022-statistics-share. Accessed 28 Oct 2022
18. Berkeley Political Review. https://bpr.berkeley.edu/2021/03/25/america-the-lonely-social-isolation-public-health-and-right-wing-populism/. Accessed 28 Oct 2022
19. JPN. https://www.jpn.up.pt/2022/02/15/num-mundo-cada-vez-mais-so-portugal-e-apenas-um-pouco-solitario/. Accessed 28 Oct 2022
20. Hofstede Insights. https://www.hofstede-insights.com/country-comparison/portugal,the-usa/. Accessed 30 Nov 2022
21. Hololive. https://www.similarweb.com/pt/website/hololive.tv/#demographics. Accessed 30 Nov 2022
22. Kowert, R., Daniel, E., Jr.: The one-and-a-half sided parasocial relationship: the curious case of live streaming. Comput. Hum. Behav. Rep. **4**, 1–7 (2022)
23. Hofstede, G.: Culture's Consequences – Comparing Values, Behaviors, Institutions, and Organizations Across Nations, 2nd edn. Sage, California (2001)

Generation Y Health Care Professionals and Their Acceptance of Chatbots

Anja Zwicky, Valerio Stallone(✉), and Jens Haarmann

Zurich University of Applied Sciences, Winterthur, Switzerland
valerio.stallone@zhaw.ch

Abstract. Medical chatbots are already used regularly by patients and health care professionals (HCPs). By the end of 2020, the majority of practicing HCPs were born between 1980 to 1994. In other words, most HCPs currently belong to Generation Y, who are generally assumed to be more open towards new digital technologies. The aim of this study was to evaluate HCP acceptance and influencing factors on the latter of a chatbot-based information platform as a proxy for information that HCPs usually provide to patients. This research of 99 HCP showed that system relevance and the innovative elements of the chatbot itself were essential factors in perceiving the chatbot as useful. The innovative elements also had a positive effect on perceived ease of use. Our study provides insights into user acceptance of chatbots supporting the work of HCPs and is a starting point for further discussions and generational change in the labor market for HCPs.

Keywords: health care · health care professionals · telemedicine · chatbots · conversational agents · technology acceptance model

1 Introduction

In marketing, chatbots are implemented in campaigns, used in lieu of sales advisors or service staff, and to distribute editorial content [1]. Similarly, medical chatbots used by both patients and HCPs [2] have many benefits. In health care, a chatbot can act as a digital personal assistant to doctors and nurses, as well as helping patients and their families in many ways. It can help patients adhere to treatment and manage their medication [3]; it can help educate patients [4], provide support in emergencies [5], and offer solutions to more straightforward medical problems [6]. They can connect patients directly to a clinic for diagnosis and treatment and guide them through the treatment process [7]. Patients no longer have to wait for consultations in doctors' waiting rooms or hospitals [8]; chatbots reduce waiting time and improve the patient experience. And as the number of patients in waiting rooms decreases, patients with more complex medical issues may have to wait less long to be seen and treated [9].

Within the next decade, chatbots will be the first point of contact for patients to answer simple medical questions [7, 10]. Although medical chatbots are not entirely new, they have an innovative character because they serve as a previously unknown

A. Rocha et al. (Eds.): WorldCIST 2023, LNNS 799, pp. 232–241, 2024.
https://doi.org/10.1007/978-3-031-45642-8_23

means to an end [11]. However, so far, they have been unable to provide patients with the same level of communication as a doctor [10].

By the end of 2020, the majority of practicing HCPs were "millennials" (i.e., born between 1980 – 1994) [12, 13]. In other words, most HCPs currently belong to Generation Y (Gen Y), who are generally assumed to be more open towards new digital technologies because they have grown up with them [14, 15]. In the light of the current lack of data concerning the acceptance of chatbots by Generation Y HCPs, several questions arose about their attitude towards medical chatbots. This lead to our very research questions:

- Are Gen Y HCP willing to use chatbots in patient treatment?
- Which factors do directly and indirectly influence Gen Y HCP in their behavioral intention to use chatbots in patient treatment?

This study explores young health practitioners' perspectives on chatbot use in healthcare and develops a model that explains the factors that influence the adoption of chatbot among HCPs. With this novelty it will provide researchers/telehealth software developers with insight into this topic.

2 Methodical Approach

To fill the knowledge gap with our survey (second study), we opted to create our model based on constructs inherent to the field of digital health by conducting interviews with Gen Y HCP (first study). We explain our two-step approach in the next two sub-sections.

2.1 Study 1: Interviews

Data Collection. We developed a standardized questionnaire with open questions to enable the interviewees to express themselves in their own words. The categories of the questionnaire were inspired by the technology acceptance model (TAM). The technology acceptance model (TAM) is based on the theory of reasoned action model [16]. It explains why users do or do not use new information systems and is suitable for predicting technology acceptance and shows how external variables influence perceived usefulness (PU) and perceived ease of use (PEOU) [16]. Perceived usefulness defines a potential user's subjective assessment that, if used, the technology will enhance his or her work performance [16]. Perceived ease of use means that using the technology is seen as easy as possible. The model assumes that the two dimensions PU and PEOU influence attitude toward use (AT), which in turn affects behavioral intention to use (BI). Furthermore, the two determinants have a direct influence on BI [16].

Data Analysis. The interviews were conducted in German via Zoom video calls between 4 and 15 September 2020. They lasted between 30 min and one hour. The inclusion criteria for the sampling were HPCs of Generation Y practicing in Switzerland. We were then able to interview seven HCPs born between 1984 and 1992: three pharmacists, two senior physicians, one general practitioner, and one assistant physician. Using a semi-structured interview guide, the subjects were asked about the challenges for HCPs, their experience with chatbots, benefits, and possible applications, and the factors influencing their acceptance of chatbots. In order to analyze the conversational data systematically, we derived categories employing a deductive procedure [17]. We determined the influencing factors to be considered in answering the research questions, which, therefore, needed to be added to the basic conceptual model.

Intermediate Results. We relied on the findings of our qualitative research to find new indicators to add to the TAM for chatbot applications in health care. According to the findings from the qualitative research, we considered the following factors.

- *Experience (EX)*: None of the interviewees had personal experience with chatbots.
- *Risk (RI)*: Risks inherent to the adoption of chatbots in health care included the perceived danger of violating private and medically sensitive patient data (n = 5), the fear that chatbots could be misunderstood and incorrect self-diagnosis (n = 6).
- *Expected quality of outcome (EQ)*: In terms of quality of outcome, the subjects stated that the chatbot's statements would have to be unambiguous (n = 3), accurate (n = 4), and evidence-based (n = 2). For HCPs, chatbots must work effectively and be reliably accurate (n = 5).
- *Innovation (IV)*: Chatbots were perceived as innovative because they can reduce human error in tracking and documenting patient-related data (n = 4) and could automatically record logging (n = 6).
- *Functionality (FU)*: HCPs found chatbots useful if they were simple and intuitive to use (n = 5), clearly structured (n = 1), and comprehensible (n = 1). The HCPs confirmed that, in principle, both AI-based and simple rule-based chatbots can perform tasks but that their system relevance depended on the complexity of the questions asked (n = 4).
- *System relevance (SR)*: HCPs mentioned the low level of digitization in Swiss health care as the main reason (n = 3). The absence of a central database for patient information for the benefit of all HCPs is another one (n = 4).

2.2 Study 2: Survey

Chatbot Use Scenarios. A joint chatbot concept was created before moving on to chatbot development. Based on the qualitative survey, a use case for the chatbot was determined using Hundermark's chatbot canvas to define its goals, target group, personality, and functions [18]. A dialogue script was designed and integrated into the chatbot. We created a rule-based chatbot using the no-code tool "aiaibot". Aesculap (the chatbot's name) was integrated into a landing page, which was available on mobile as well as on desktop devices. In order to offer diverse interactions, we derived three different chatbot-based communication scenarios. In the first scenario, the chatbot answers questions from the interlocutor about the prevention of influenza. The second chatbot scenario provides information on the health checks offered by a pharmacy or hospital. The third chatbot scenario advises HCPs' patients on the vaccinations provided by a pharmacy or hospital, suggests dates for consultations, vaccinations, and health checks, and helps fill prescriptions and locate pharmacies and hospitals.

Hypotheses and Conceptual Model. In order to measure the acceptance of chatbots in patients treatment by HCP, the authors developed 14 hypotheses depicted in the conceptual model in Fig. 1.

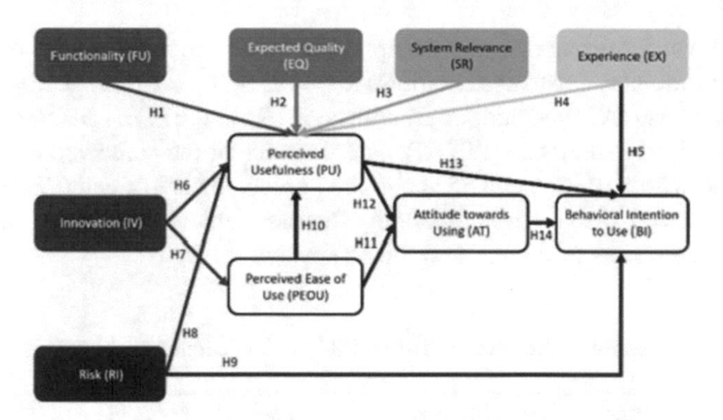

Fig. 1. Our conceptual model.

Operationalization. Indicators were formulated for all constructs used. In particular, the four variables of quality of outcome (EQ), experience (EX), system relevance (SR), and risk (RI) were adapted to the object of our study. Since we assumed that most test subjects had little or no experience with chatbots, we thought it would be difficult to assess the technology without actual use. Therefore, the constructs of expected quality of outcome and experience were related to the usability test. For the construct of system relevance (SR), the indicators were defined in terms of the type of system of rule-based or artificially intelligent chatbots. The indicator of risk (RI) was formulated based on the findings from the qualitative search regarding the reservations of the HCPs. The indicators for the external variables of innovation (IV), functionality (FI), and the latent variables of perceived usefulness (PU), perceived ease of use (PEOU), attitude to use (AT), and behavioral intention to use (BI) were taken from the original formulation of the corresponding theory [16].

Data Collection. The online survey was conducted in German using the survey platform Qualtrics. The data was collected between 9 October and 14 November 2020. The online survey took respondents between 9 to 15 min to complete if the scenarios control and acceptance questions were answered at a normal pace without interruption. In our study, participants were asked to take part in a research study involving a web-based self-report survey on HCPs' intentions of using chatbots in health care. Participants were asked to interact with a chatbot to ensure all participants had a common understanding of the technology. Next, participants were asked to complete a related survey. Before the chatbot was evaluated, the test subjects had to test it in the three different scenarios. This helped to make the expected quality of outcome measurable and, according to the concept of the control questions, to check whether the prototype had really been tested.

3 Results

A total of 212 HCP started the online survey, whereas 131 of them fully and validly completed it. Out of the 131 respondents, 32 had been born before 1980 and therefore could not be counted in, leaving us with 99 participants represented our final sample.

The survey showed a strong effect of the applied acceptance model on the mean value on HCPs' behavioral intention to use chatbots ($R^2 = 0.514$, $f^2 = 1.028$). The variance of the attitude towards use (AT) was supported with 68% ($R^2 = 0.677$) by perceived usefulness (PU) and perceived ease of use (PEOU). The variance of the perceived usefulness (PU) was explained with 55% ($R^2 = 0.555$) and the perceived ease of use (PEOU) with 19% ($R^2 = 0.555$, $f^2 = 0.189$) by the influencing factors. Table 1 shows the resulting values of the F-test, the results of the R^2, and the effect size f^2.

Table 1. Results, F-Test and R^2 of dependent variables.

Construct	Sig	F-Test	R^2	f^2
PU	0.000	18.455	0.555	1.116 = strong
PEOU	0.000	22.579	0.189	0.482 = strong
AT	0.000	103.502	0.677	1.447 = strong
BI	0000	43.593	0.635	1.318 = strong
Avg			0.514	1.028 = strong

EX had a poor Cronbach's alpha of 0.591 and was below the minimum value of 0.7 [11]. We, therefore, deleted indicators that had a value below 0.5 [11]. We retained both items because they (i) consisted of only two items, (ii) did not fall below 0.5, and (iii) deleting one item did not lead to better results. FU could not be tested with Cronbach's alpha as it only consisted of one indicator. SR showed a negative value. The most common reasons for this were that it had been forgotten to invert items or that items had been included that measured the opposite. On closer inspection, the indicators we used did not actually measure the same construct. SR_1 measured relevance in the context of the task of rule-based chatbots, and SR_2 measured relevance in the context of the task of artificially intelligent chatbots. This accounted for the negative correlation between the two items. We decided to keep SR_1 and omit SR_2. We justified our decision by determining that the usability test referred to a rule-based chatbot. The check of the indicator reliability using factor analysis showed that the Kaiser-Meyer-Olkin value of 0.875 can be classified as good and is above the prescribed acceptance value of 0.6 [11]. The analyzed p-value < 0.000 could be considered significant, and the null hypothesis was rejected. It showed that the variables correlate adequately and that the variable population is well suited for factor analysis [11]. The reliability analyses concluded that the selected constructs and the acceptance model were reliable and accurate. Within the model estimation, the established hypotheses were tested. The theoretically derived relationships between the latent variables were estimated based on the empirical data. The indicator SR_2 was eliminated because it had a negative value and, when examined closely, did not measure the same construct as SR_1. Consequently, 31 of 32 indicators remained to measure five constructs and test 14 hypotheses.

As Table 2 depicts, the path values show significant influences for our new construct's influence on perceived usefulness, whereas SR obtained a beta coefficient of 0.193 and IV a beta coefficient of 0.453.

Table 2. Results and significances.

Hypotheses		t-Test	Beta coeff	P values
H1	FU → PU	−1.366	−0.122	0.175
H2	EQ → PU	0.272	0.031	0.787
H3	SR → PU	2.064	0.193	0.042
H4	EX → PU	0.901	0.076	0.370
H5	EX → BI	0.941	0.057	0.349
H6	IV → PU	4.287	0.453	0.000
H7	IV → PEOU	4.752	0.409	0.000
H8	RI → PU	−1.877	−0.174	0.064
H9	RI → BI	0.724	−0.025	0.724
H10	PEOU → PU	3.458	0.382	0.001
H11	PEOU → AT	9.310	0.296	0.000
H12	PU → AT	3.494	0.603	0.001
H13	PU → BI	0.439	−0.038	0.661
H14	AT → BI	7.228	0.720	0.000

As shown in Fig. 2, the survey showed a strong effect of the acceptance model applied on the intention of HCPs to use chatbots. The variance in the behavioral intention to use chatbots is explained by 63% of the attitude towards use.

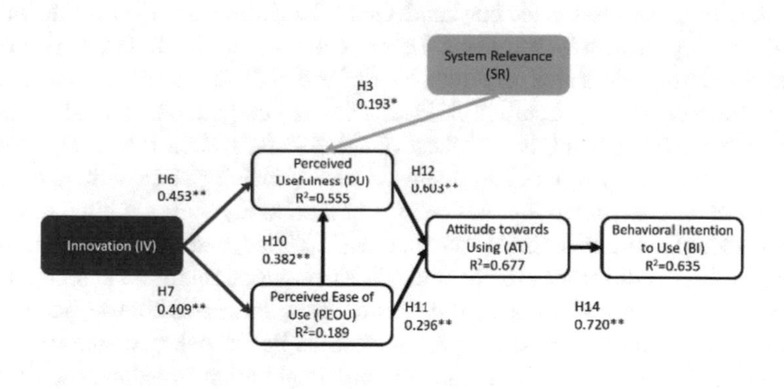

Fig. 2. Significant path coefficients.

At a significance level of 95%, a positively significant correlation between system relevance (t = 2.064, p = 0.042), innovation (t = 4.287, p = 0.000), perceived ease of use (t = 3.458, p = 0.001), and perceived usefulness was detected. Likewise, the perceived usefulness was positively and significantly influenced by innovation (t = 4.752, p = 0.000).

4 Discussion

In our study, attitude towards using chatbots appeared to be the most significant factor affecting HCPs' acceptance of chatbots. The path coefficient from this indicator was the highest in the model examined. It served as a proxy by transferring the importance of perceived usefulness and perceived ease of use towards the behavioral intention to use. Perceived usefulness, however, did not show a direct impact on the behavioral intention to use.

This finding have several implications. First, although being pragmatic in their technology-accepting decisions when assessing whether or not to adopt new technology, HCPs did not focus on the usefulness of chatbots alone. It was not enough for an HCP to accept a technology considered useful to his or her practice. Second, perceived usefulness was still a critical determinant of attitude, exhibiting tremendous influences on individual attitude formation. Consistent with other findings, perceived usefulness had a greater impact than perceived ease of use [19].

Perceived ease of use appeared to have a significant effect on both perceived usefulness and attitude. This makes sense: HCPs have a relatively high general competence and comprehend the use of new technology quickly, familiarizing themselves with its operations without going through the intense training that might be necessary among other user populations.

Of the indicators added to the TAM in our conceptual model, only system relevance and innovation were found to have a significant impact on other constructs. This finding had different implications. On the one hand, Gen Y HCPs appear to be open to innovative and novel development, which reflects their openness in general. The strong effect of both, perceived usefulness and perceived ease of use, indicates that HCPs with an innovative mindset have strong cognitive abilities, enabling them to understand the chatbot's functions. On the other hand, the weakest of the effects, system relevance, reflects the need for more openness towards intelligent systems, which is in line with Powell (2019).

Palanica et al. confronted the participants of their study with a written definition of chatbots without having them interact with one [15]. This, and the fact that they did not focus on age, is a big difference to our study. The behavioral intention in their research is comparable to our own. The same applies to attitude towards chatbot use, perceived ease of use, and perceived usefulness. The big difference lies in risk perception: According to our results, Gen Y HCPs' reservations that patients might misdiagnose themselves are lower. The same applies to the subjects' perception of the risk of misunderstanding the chatbot output. Concerning issues raised concerning patient privacy [15], our Gen Y HCPs did not consider them significant. This difference might be related to our subjects' age group and years of experience. On the other hand, no significant differences across the factors of age or years of practice have been found [15].

Our model showed that the more positive the attitude towards chatbot use, the higher the behavioral intention to use a chatbot. Gen Y HCPs see the usefulness of chatbots, especially in providing supplementary patient information, preparing and supporting differential diagnoses, and saving time by allowing the outsourcing of repetitive and administrative processes. Chatbot developers are, therefore, advised to be customer-centric in their efforts (in this case, 'practitioner-centric'). Such customer-centricity translates into a thorough understanding of HCPs' daily interactions with patients, which should lead to better health-care-specific features for the new medical chatbots.

To foster individual acceptance of a newly adopted or implemented technology, the management team in a professional organization needs to devise strategies for cultivating a positive attitude toward using the technology. Medical applications will likely benefit from spillover effects as significant investments in chatbot technology are undertaken in various industries. In this context, favorable perception of the technology's usefulness is crucial, whereas ease of use might not be of equal importance. Once an organization decides to adopt chatbots, its management must make every effort to emphasize, demonstrate, and communicate their usefulness for individual HCPs in helping with routine tasks and services. Initial information sessions and training programs should first focus on how the technology can improve the efficiency or effectiveness of individual HCPs' patient care and services before familiarizing HCPs with detailed procedures for operating chatbots. Courses on technology-aided HCP-patient relationships might become an integral part of every medical student's curriculum.

Our findings show that for a medical institution to promote its innovativeness by demanding cutting-edge patient-related services, it must consider the technological affinity of its HCPs to chatbot technology. Technology has penetrated everyone's lives – HCPs' and patients' - and current knowledge about it is an indicator for preparedness and attitude towards future developments in the medical field.

5 Limitations

This study has several limitations. As its focus is on a particular technology involving a specific group of professionals, caution is advised when generalizing the findings and discussion to other technologies or professional groups. Responses on behavioral intention were collected using a self-reporting method, a common data collection technique with inherent limitation.

Since no patients were interviewed in the context of this study, we had to make assumptions on a typical persona representing the targeted user group. For the prototype we developed to be used, this persona would have to be defined in more detail by systematically interviewing patients. For further research, we recommend exploring medical chatbots from the Swiss patient perspective. An international view will also be useful. A similar study involving other countries could focus on country-specific differences in Gen Y HCPs' attitudes towards new, patient-oriented communication technology and, specifically, towards chatbots.

In our study, we focused on the acceptability of applying chatbots from the perspective of HCPs, but we did not validate the outcome the HCPs would generate in real-life situations with patients. We also were not able to compare outcome differences between HCPs applying these very chatbots and HCPs not doing so. We asked the HCPs to interact with chatbots for hypothetical cases and not actual patient requests. In further studies, it would be of crucial importance to test the application of chatbots in a field experiment.

6 Conclusion

This study shows that Swiss HCPs belonging to Generation Y perceive chatbots as useful, especially by answering standardized questions from patients and assisting in HCPs' communication with patients. We took a mixed-methods approach by exploring additional indicators for behavioral intention as a measure of technology acceptance. The findings of our study suggest several potential implications for HCPs and their organizations to differentiate themselves and be ready for the technological future. A benefit of our study is that it evaluates individual technology acceptance in a real-world professional setting, including principal target users who assessed chatbots in light of their routine clinical tasks and services. In addition, this study represents a much-needed and timely effort to extend the HCP focus to chatbot research, a field demonstrating increasing IT investment and penetration [15].

References

1. Denk, M., Khabyuk, O.: Wie relevant sind Chatbots als Kommunikations- und Marketinginstrument für Hochschulen? Konzeption und Akzeptanz eines Chatbot-Prototyps für den Master-Studiengang „Kommunikations-, Multimedia- und Marktmanagement" der Hochschule Düsseldorf (2019). https://econpapers.repec.org/paper/ddfwpaper/52.htm. Accessed 01 Mar 2021
2. Senaar, K.: Chatbots for Healthcare – Comparing 5 Current Applications, emerj.com (2019). https://emerj.com/ai-application-comparisons/chatbots-for-healthcare-comparison/. Accessed 08 Mar 2021
3. Battineni, G., Chintalapudi, N., Amenta, F.: AI chatbot design during an epidemic like the novel coronavirus. Healthcare **8**(2), 154 (2020). https://doi.org/10.3390/healthcare8020154
4. Hauser-Ulrich, S., Künzli, H., Meier-Peterhans, D., Kowatsch, T.: A smartphone-based health care chatbot to promote self-management of chronic pain (SELMA): Pilot randomized controlled trial, JMIR mHealth and uHealth, **8**(4) (2020). https://doi.org/10.2196/15806
5. Te Pas, M.E., Rutten, W.G.M.M., Bouwman, R.A., Buise, M.P.: User experience of a chatbot questionnaire versus a regular computer questionnaire: Prospective comparative study. JMIR Med. Inform. **8**(12) (2020). https://doi.org/10.2196/21982
6. Zand, A., et al.: An exploration into the use of a chatbot for patients with inflammatory bowel diseases: retrospective cohort study. J. Med. Internet Res. **22**(5), e15589 (2020). https://doi.org/10.2196/15589
7. Mesko, B.: The Top 12 Healthcare Chatbots. The Medical Futurist (2020). https://medicalfuturist.com/top-12-health-chatbots/. Accessed 08 Mar 2021
8. Ouerhani, N., Maalel, A., Ghézela, H.B.: SPeCECA: a smart pervasive chatbot for emergency case assistance based on cloud computing. Cluster Comput. **23**(4), 2471–2482 (2020). https://doi.org/10.1007/s10586-019-03020-1
9. Bao, Q., Ni, L., Liu, J.: HHH: an online medical chatbot system based on knowledge graph and hierarchical bi-directional attention. In: ACM International Conference Proceeding Series, vol. 10 (2020). https://doi.org/10.1145/3373017.3373049
10. Bhirud, N., Tatale, S., Tataale, S., Randive, S., Nahar, S.: A literature review on chatbots in healthcare domain. Int. J. Sci. Technol. Res. **8**(7) (2019)
11. Diers, T.: Akzeptanz von chatbots im consumer-marketing. wiesbaden: Springer Fachmedien Wiesbaden (2020). https://doi.org/10.1007/978-3-658-29317-8
12. McKinsey: Digital health ecosystems: a payer perspective (2019)

13. VEEVA/Across Health, Intelligent HCP Engagement in Europe (2018). https://www.across.health/publication/engaging-intelligently-with-european-hcps. Accessed 01 Mar 2021
14. Kochhan, C., Elsässer, A., Hachenberg, M.: Marketing- und Kommunikationstrends. Springer Fachmedien Wiesbaden (2020).https://doi.org/10.1007/978-3-658-30848-3
15. Palanica, A., Flaschner, P., Thommandram, A., Li, M., Fossat, Y.: Physicians' perceptions of chatbots in health care: Cross-sectional web-based survey. J. Med. Internet Res. **21**(4), e12887 (2019). https://doi.org/10.2196/12887
16. Davis, F.D.: Perceived usefulness, perceived ease of use, and user acceptance of information technology. MIS Q.: Manage. Inf. Syst. **13**(3), 319–339 (1989). https://doi.org/10.2307/249008
17. Mayring, P., Fenzl, T.: Qualitative Inhaltsanalyse. In: Baur, N., Blasius, J. (eds.) Handbuch Methoden der empirischen Sozialforschung, pp. 633–648. Springer Fachmedien Wiesbaden, Wiesbaden (2019). https://doi.org/10.1007/978-3-658-21308-4_42
18. Hundertmark, S.: Digitale Freunde (2020). Accessed 11 Oct 2022. https://www.wiley-vch.de/de/fachgebiete/finanzen-wirtschaft-recht/wirtschaft-und-management-13ba/marketing-u-vertrieb-13ba8/strategisches-marketing-13ba83/digitale-freunde-978-3-527-51036-8
19. Chau, P.Y.K., Hu, P.J.H.: Investigating healthcare professionals' decisions to accept telemedicine technology: an empirical test of competing theories. Inf. Manage. **39**(4), 297–311 (2002). https://doi.org/10.1016/S0378-7206(01)00098-2
20. Powell, J.: Trust me, i'm a chatbot: how artificial intelligence in health care fails the turing test. J. Med. Internet Res. **21**(10), e16222 (2019). https://doi.org/10.2196/16222

TV Notifications to Promote Trustworthy News and Local School Newspapers

Simão Bentes[1], Luísa Júlio[1(✉)], Ana Velhinho[1], João Encarnação[1], Martinho Mota[1], Diogo Miguel Carvalho[1], Rita Santos[2], and Telmo Silva[1]

[1] Digimedia, Department of Communication and Arts, University of Aveiro, 3810-193 Aveiro, Portugal
{bentes,luisaamj,ana.velhinho,jrangel29,m.vaz.mota, diogocarvalho28,tsilva}@ua.pt

[2] Digimedia, Águeda School of Technology and Management, University of Aveiro, 3754-909 Aveiro, Portugal
rita.santos@ua.pt

Abstract. The growth of social media and user-targeted information reinforces the need to check sources to prevent the dissemination of unreliable content. In this context, an innovative solution is presented based on the established proliferation of push notifications and the potential identified in the TV ecosystem, since it connects multiple devices and platforms. The proposed approach aims to send news notifications with a high level of credibility through television as well as the latest news from local school newspapers. This strategy expects to engage audiences of different generations to promote family discussion and critical thinking about current topics. Furthermore, the dissemination of local events and releases of school newspapers intends to create a sense of community beyond the school while promoting social and local cohesion. This solution combines two on-going R&D projects that bring together the academia and well-established partners of the Portuguese industry, namely a news media company and an Internet Protocol Television (IPTV) provider. After the good results from a preliminary field trial to assess the receptiveness to news notifications on the TV, a second evaluation will be carried out with school students (who will create the news) and their families (who will receive the notifications at home). Finally, this solution aims to contribute with more knowledge for preventing fake news dissemination, and also to demonstrate the potential of a personalized notifications mechanism for delivering trustworthy information on the TV ecosystem.

Keywords: Digital Platforms · Fake News · News Creation Tool · Push Notifications · TV Ecosystem

1 Introduction

The spread of information through different media (i.e., newspapers, television, radio, and websites) facilitates people's access to information about society and worldwide issues. Since there isn't a standard system for accessing information, it is essential

A. Rocha et al. (Eds.): WorldCIST 2023, LNNS 799, pp. 242–248, 2024.
https://doi.org/10.1007/978-3-031-45642-8_24

to analyse the credibility of the displayed information on every device. According to Obercom's report [1], in Portugal, social media sites are the main channel where young people access information making it easier to spread fake news among this audience.

Despite fake news already existing in traditional media, nowadays people are more limited in distinguishing credible and unreliable content [2] because of the proliferation of online sources and the usage of digital platforms [3]. Furthermore, it is necessary to stimulate critical thinking in youngsters to help them avoid believing every news content they read. For that purpose, some initiatives have already been implemented, such as the introduction of media literacy in schools [4] and the presentation of the news media codes of conduct [3].

Furthermore, push notifications as a message mechanism that catches users' attention multiple times a day on their devices (mainly mobile devices), are becoming a prevalent way of accessing news information among younger audiences [5]. Beyond priority information and reminders which can also be delivered to connected devices by the Internet of Things (IoT), these kinds of messages are usually associated with a well-defined call to action related to content (e.g., upcoming releases, new emails, etc.) and promotions that bring benefits to users. Either way, notifications often lead users to apps for generating online engagement but may also alert them to outdoor events and appointments. In this sense, many notifications are location-based and time-sensitive, which are key factors to their personalized efficacy.

Academic research regarding news notifications has also been focusing on the importance of temporality, by identifying the moments when people value news and information messages the most. When considering specifically news consumption by young people, [5] highlights four key time slots: a) dedicated moments (evenings and weekends); b) moments of update (morning); c) time fillers (commuting or queuing); d) intercept moments with notifications while performing other tasks.

In terms of devices, notifications are mostly used in personal and mobile devices, including wearables (e.g., smartwatches and smart glasses). Television is a less explored mechanism to deliver personal information because TV continues to be used collectively which may raise some privacy issues [6]. Nevertheless, the potential of notifications in the TV ecosystem is device connectivity, namely the use of second screens, which allow exploring companion apps to provide additional information for several purposes, like content discovery and access to marketing actions which may benefit users. As for OTT apps, they typically use the users' profiles to inform them about new releases based on personal viewing habits and to foster binge-watching (marathon viewing).

Based on the established proliferation of news notifications and the potential identified in the TV ecosystem to connect multiple devices and OTT platforms, an innovative solution with a social focus is presented in this article. The proposed approach consists of sending useful and high-credibility information through television, including school newspapers news produced by young people. With this solution, it is expected that the TV notifications raise the interest of audiences of different generations and promote family discussion as well as critical thinking about current topics. Furthermore, the dissemination of local events and giving visibility to school newspapers intends to create a sense of community beyond the school while promoting social and local cohesion.

The proposed solution results from the combination of two ongoing R&D projects, which bring together the industry and academia: (i) the Trustworthy news and Related content for a Unified writing Environment (TRUE) project, which consists of a news creation tool to help Portuguese students write school news and share then in a school newspaper; and (ii) the Personalized notifications for the TV ecosystem (OverTV) project regarding notifications directed to the TV ecosystem. Both projects will be detailed in the following sections.

Lastly, the article is structured into three sections: (i) introduction; (ii) contextualization of the proposed solution and details the system architecture and (iii) final remarks.

2 TV Notifications Strategies to Promote Local Community and Social Engagement with School Newspapers

For companies and service providers, push notifications have become a powerful strategy to attract audiences' attention throughout the day. As for the users, checking notifications is an embedded behaviour across devices to get updates and quick access to personalized information, namely, news feeds [5, 7]. Therefore, the scope of this article is based on the pertinence of sending notifications of reliable information based on a credibility news evaluation system. This can be achieved by combining the strengths of two R&D projects that result from partnerships with well-established and credible news media and Internet Protocol Television (IPTV) providers from the Portuguese industry. The goals and features of the TRUE project and the OverTV project are contextualized separately in the following subsections.

2.1 A News Creation Tool for School Newspapers

As mentioned previously, fake news can negatively influence the daily lives of young people, mainly because they consume information from sources with a low level of credibility. To fill this gap, the TRUE project [8], in partnership with the Portuguese newspaper Público (https://www.publico.pt/) and the provider of digital media technology MOG Technologies (https://www.mog-technologies.com/), is currently developing a news creation tool to help Portuguese students write school news. While the student is writing this tool shows a list of related news from various sources ranked by its credibility. The integration of this tool may decrease misinformation in younger audiences and encourage the students to critically analyse media content.

In this sense, the news creation tool enables students to create their news with audiovisual content (audio, video, and images); visualize the possible spelling errors and their respective corrections; change repeated words through the presentation of a list of related words; search news by keywords ranked by their credibility rating; and obtain news related to the written text from a credible newspaper source.

The related news articles are presented with a set of progress bars that allow students to identify the credibility level of each news before visualizing it. In this sense, young people can be more selective about the news sources they use for their newspapers.

Lastly, it is expected that this project brings more knowledge to other R&D projects, mainly about fake news, which is an aggravating problem nowadays.

2.2 A Personalized Notification System for the TV Ecosystem

The OverTV project, developed in partnership with the IPTV provider Altice Labs (https://www.alticelabs.com/), consists of a system for sending personalized notifications to the TV ecosystem, that allows to generate, schedule, and monitor several types of messages delivered through the television set-top box (STB) of an IPTV operator and connected mobile devices [9]. The web application used for management developed within this project differentiates itself by allowing both the company managers and end-users to create custom messages to be sent to the TV. These can be based on geo-referenced data and their preferred content regarding information, entertainment, and services, as well as calendar and health reminders. The system includes: (i) morning and evening daily routines consisting of sequenced notifications that aim to prepare users for the current and following day and (ii) thematic notifications based on the type of actionable event with scheduled timings according to the user's preferences.

The same event may generate several notifications depending on the settings pre-defined in the management platform for each category (which are customizable). The parameters of the notifications' settings are related to STB triggers and timings.

2.3 Goals of the Proposed Solution

Digital news formats, specially provided by social media and news aggregators, are dominant in younger audiences, as older audiences tend to be more focused on traditional media, like the printed press and television. Thus, news notifications from credible sources displayed on the TV and second screens of connected devices can bring a surplus value to all members of the households.

On the one hand, the TRUE project offers a cohesive news community and robust instruments for the selection and credibility assessment of information sources, as well as a news writing tool, developed within the project, that helps young students to produce trustworthy news content, aiming to promote critical thinking and awareness about fake news. On the other hand, the OverTV project provides a TV-first approach to push notifications, by using a notification management platform, custom-built for this project, that allows scheduled personalized information to be displayed on-screen in the shared television device as well as on personal connected devices, such as smartphones and tablets. The solution proposed in this article was based on integrating news notifications in the OverTV project information flow, to foster family social engagement around news, including the ones published by the local school newspapers of their children and grandchildren.

2.4 System Architecture

The proposed solution includes two types of approaches for displaying news notifications on the TV screen: (i) a pop-up message integrated into the morning and evening routines of the OverTV project's system with headlines of general news and local news, with the identification of its source in the notifications' QR code; (ii) a content card notification displayed whenever a news post is released on the schools' newspapers, delivered according to a personalized preferable schedule. This richer format of notification adopts the look and feel of social media and OTT platforms' thumbnail cards to appeal to younger audiences, aiming to foster pride in their family members and generate engagement, to reinforce a sense of community.

The solution was designed based on a flow of information that automates the selection and dispatch of credible news to people's households. Thus, to better understand the exchanges between both projects, Fig. 1 presents the system architecture.

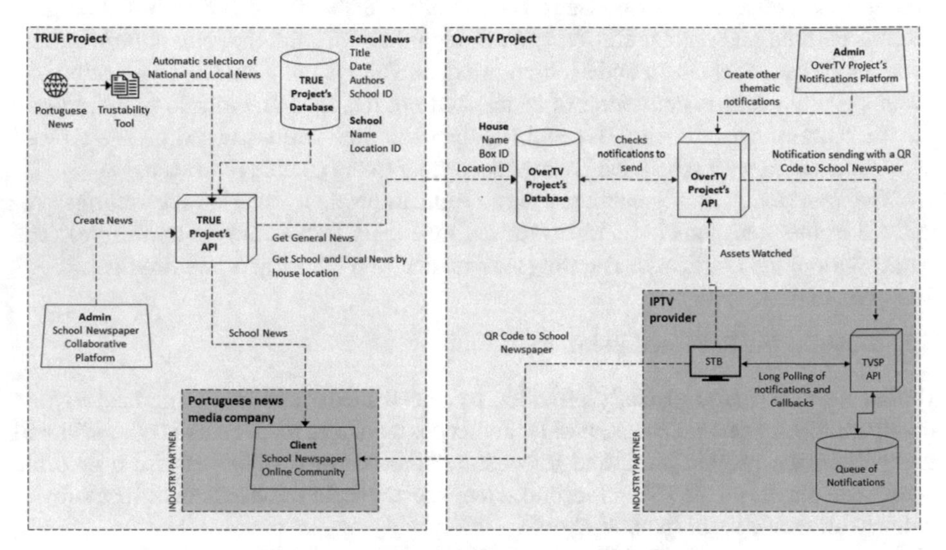

Fig. 1. Proposed merged solution: System architecture

The first step of this solution is to obtain the content used to create a notification which is the TRUE project's main role. For the pop-up messages, the TRUE project integrates an API that detects the key points of the news and returns similar results from the main Portuguese news outlets. Then, the TRUE project's API returns the articles' information and their credibility level. Next, according to the viewers' house location, the two most credible and relevant news are sent to the OverTV project's database.

For the content card notification, the TRUE project will provide information about the release of one or more news posts of previously selected school newspapers, to the OverTV project' database.

In the second step, the OverTV project is responsible for creating and sending notifications to the user's TV, based on pre-determined parameters. In this way, the OverTV project's API checks all the saved information to determine if any new notifications need to be sent and generates them according to the available information.

Afterwards, if the user's TV STB is on, the notifications are sent to the TV's service provider API. Finally, this API sends the notifications to the designated household TV STB.

Despite aiming to address the entire household to generate debate, the two types of news notifications abide by different time-sending criteria: (i) the general and local news notifications are integrated into the morning and evening; and (ii) the school newspapers' notifications are generated if new posts are published.

2.5 Results Achieved

A field trial was already conducted to assess if personalized notifications on the TV were well received by potential users. These trials gathered 25 participants, who received generic and personalized notifications at home throughout a period of 44 days. The field trial aimed to evaluate the receptiveness and usefulness of notifications on the TV, namely of general and local trustworthy news. Users expressed that scheduled notifications on the TV were "very useful" (32%) or "useful" (52%) to their routines. Especially the notifications received later in the day after 7 pm (52%) or when they turn on the TV early in the morning (36%). In the case of general news notifications, the participants expressed a preference for news integrated into the morning routine (40%), although the platform analytics showed a highest delivery rate at night (70%) when compared to the morning delivery rate (54,3%).

Based on these positive results, and to promote community cohesion and intergenerational dynamics, the integrated solution to send notifications of news from local school newspapers to the TV has considerable potential. Hence, the main goals of this solution are: to deliver trustworthy information; disseminate and generate engagement in local school newspapers; and transform moments of the shared experience of watching TV into opportunities to foster family debates about current news topics. To achieve these goals, a set of strategies were designed to take advantage of the strongest aspects of both platforms, namely the selection and credibility assessment of general news of the TRUE project and the full customization of the look and scheduling of notifications of the OverTV project. These strategies will culminate in another stage of field trials, which will include the on-going iterations of both platforms (the news creation tool and the notifications manager), the improved schedules of the notifications, and aiming to validate the proposal of creation and delivery of school newspapers articles on the TV, fully readable using mobile devices, by working with local schools in Aveiro, Portugal. It is intended to replicate this approach in other cities in a next stage of evaluation.

3 Final Remarks

In a context of ubiquitous and unregulated information, younger generations are a vulnerable audience, so it is crucial to develop services and systems that allow them to recognize credible content to prevent misinformation and the proliferation of fake news. Furthermore, complementary to global approaches that reach broader audiences, it is also important to address segmented needs and local contexts by delivering personalized information.

This article addresses a solution of push notifications on the TV, including news and local school newspapers releases. The TV is a medium preferred by older adults, while school newspapers are generally only distributed on the school's inner channels. By gathering this information and sending it to the TV ecosystem, it would be possible for multiple generations to have easy access to it and thus promoting an intergenerational discussion about current topics.

Despite the privacy issues related to the TV being a shared device, the proposed solution may counter the growing seclusion of individuals magnetized by their personal devices, by trying to revitalize the family moments to instigate debate about what is happening in the world and in their communities. Beyond that, it can promote young students to share their school news with friends and family members.

These smaller-scale solutions can facilitate fostering social cohesion and consolidate the sense of community, which includes several groups: the family, the schools, and proximity events and venues. After the field trials already conducted which included high credibility news on TV notifications, in future work this merged workflow of delivering notifications over the TV of credible news and the latest released news from local schools' newspapers will be tested in pilot schools and field trials at home with potential users.

Acknowledgements. This work was developed and funded under the scope of the project TRUE (POCI-01–0247-FEDER-046923) financed by COMPETE 2020, POCI 2020 and Portugal 2020.

References

1. Cardo, G., Paisana, M., Pinto-Martinho, A.: Digital News Report Portugal 2022 (2022)
2. Olan, F., Jayawickrama, U., Arakpogun, E.O., Suklan, J., Liu, S.: Fake news on Social Media: the Impact on Society. Inf. Syst. Front. **1**, 1–16 (2022). https://doi.org/10.1007/s10796-022-10242-z
3. Martens, B., Aguiar, L., Gomez-Herrera, E., Mueller-Langer, F.: The digital transformation of news media and the rise of disinformation and fake news (2018)
4. Farmer, L.S.J.: News literacy and fake news curriculum: school librarians' perceptions of pedagogical practices. J. Media Lit. Educ. **11**, 1–11 (2019)
5. Nielsen, R.K., Newman, N., Fletcher, R., Kalogeropoulos, A.: Reuters institute digital news report 2019. Rep. Reuters Inst. Study J. (2019)
6. Silva, L.A., Leithardt, V.R., Rolim, C.O., González, G.V, Geyer, C.F.R., Silva, J.S.: PRISER: managing notification in multiples devices with data privacy support (2019). https://doi.org/10.3390/s19143098
7. Wheatley, D., Ferrer-Conill, R.: The temporal nature of mobile push notification alerts: a study of European news outlets' dissemination patterns. Digit. J. **9**, 694–714 (2021). https://doi.org/10.1080/21670811.2020.1799425
8. Carvalho, D.M., Júlio, L., Silva, T.: Students as creators of online newspapers: a tool's user-interface proposal. Lect. Notes Educ. Technol. In Press (2023)
9. Encarnação, J., Velhinho, A., Bentes, S., Silva, T., Santos, R.: A management system to personalize notifications in the TV ecosystem. Procedia Comput. Sci. In Press (2023)

Students' Habits, Strategies, and Decisive Factors Around Course Selection

Gonzalo Gabriel Méndez(✉) and Patricio Mendoza

Escuela Superior Politécnica del Litoral, Guayaquil, Ecuador
{gmendez,pdmendoz}@espol.edu.ec

Abstract. Before the beginning of a new academic term, university students carry out a series of activities to decide which courses they will be attending. The specific combination of chosen courses and the academic load these courses impose have a direct impact on several aspects of a student's life, including their academic performance. In this paper, we report the results of a survey that inquired about the activities students engage in when making course enrollment decisions in planning an upcoming academic term. Our survey asked about how students use some of the resources provided by their institution to inform their enrollment decisions, particularly when it comes to selecting specific instructors. We discuss the implications of our observations both for the student population and other university stakeholders. Our findings can inform the development of better tools to assist students in term planning.

1 Introduction

When planning a new academic term, university students must take a number of steps to complete their enrollment. The process often begins right after the academic offer is published. Students must review the courses available considering their academic history (e.g., to verify the fulfillment of prerequisites) and interests (e.g., to decide how to spend their optional credits). Logistic considerations must also be made. Although the timetables of lectures are often predefined, the schedules of other activities associated with a course might be less certain. The details of tutorial sessions, for example, may be unknown before a new term begins, as the university may organize these activities based on a myriad of criteria (number of students enrolled, tutors available, lab space, and other organizational aspects). When planning a new term, students must also consider their extracurricular activities both within the academic context (e.g., teaching or research assistantships) as well as family and work duties.

After considering these and many other factors, the enrollment process continues with the definition of an initial set of courses of interest. More often than not, however, this set is not definitive. Unanticipated events—whose control is beyond the students' reach—often reshape their initial decisions. For example, as other students complete their enrollment, a desired course taught by a desired instructor may become unavailable. Events of this kind imply revising the set of courses initially considered. Ultimately, unanticipated changes of the initial conditions may translate into having to go back to

A. Rocha et al. (Eds.): WorldCIST 2023, LNNS 799, pp. 249–258, 2024.
https://doi.org/10.1007/978-3-031-45642-8_25

re-explore the academic offer and starting the planning and course selection processes all over again.

In this paper, we explore how students of a Latin American engineering- oriented university make course selection decisions in scenarios like the one described above. More specifically, we investigate the students' habits and strategies around enrollment and the factors students consider decisive when choosing their courses' instructors. This includes how much students value the course recommendations of academic advisors as well as how they use university-provided data on student-generated ratings of instructors to inform their enrollment decisions. We discuss the results of our survey to provide recommendations that may inform the design and development of tools to assist university students when exploring their enrollment options for upcoming academic terms.

2 Related Work

Student enrollment may refer to actions and decisions students make at different levels. At a large scale, enrollment may mean the act of arranging to attend an educational institution. This type of enrollment has been extensively investigated. Previous research has focused, for example, on understanding the factors that play a role in enrolling in an educational institution. Examples include motivation and engagement [2], sex and racial differences [5, 13, 20], existing educational debts [7], sociological and economic factors [8], advising [11], academic performance [12], and family influences [16].

At a smaller scale, student enrollment comprises the steps students take when deciding which courses to sign up for in an upcoming academic term. This may include reviewing the academic catalog, considering academic goals, considering schedule constraints, consulting with an academic advisor, and reviewing the course descriptions and syllabi. This is the type of enrollment we focus on in this paper.

Previous work has explored specific aspects that influence students' course enrollment decisions in varied contexts and settings (e.g., [9, 19]). Jacqmin studied the demand factors that drive enrollment in massive open online courses [10]. Along the same lines, Apel et al. report a project to transform the enrollment experience at the University of Wisconsin-Madison [3]. The authors followed a design thinking strategy to produce an intuitive enrollment experience in line with the students' and institutional goals. Wagner et al. explored enrollment in the context of a German University [22] by incorporating the perspectives of students in the design of a novel course enrollment system based on student performance data. In the same direction, Vialardi et al. presented a recommender system to support decision making during course enrollment [21]. Similar databased efforts have considered students' skills, knowledge, and interests to help them with their course enrollment decisions (e.g., [1, 6, 15]).

We share with the body of work referred above the overall goal of understanding students' course enrollment habits, approaches, and rationale. More closely related to our work, Rivas et al. used a questionnaire to identify the enrollment habits of Computer Engineering students in their first and second semesters [14]. The authors found that, when students choose courses, their tutors' recommendations are not as relevant as the information they encounter in institutional web pages. A subsequent analysis showed that the most decisive factors for students were study time availability and their experience in previous terms. Our research builds upon and expands these findings by also investigating how students use university-provided data-based tools.

Understanding the different aspects of the process of academic enrollment constitutes an important step toward improving the students' enrollment experience and toward supporting other institutional stakeholders more effectively. Ultimately, this can lead to a more positive and productive academic experience in which students achieve their academic goals.

3 Motivating Context and Research Questions

Our research site is a Latin American engineering-oriented university with over 10,000 students and 32 undergraduate programs. The enrollment period of a new academic term opens about two weeks before the lectures begin. The registration dates for each student are decided based on their academic record: students with higher GPAs[1] have earlier priority every term, whereas students with lower GPAs are assigned later dates. Students may register on or after their registration date, and may choose up to 15 credits (typically, between 4 and 6 courses) from several categories (compulsory, optional, and elective courses). This number of credits is enforced by the country's public education policies on credits that are *free* (i.e., paid by the government) for courses that are taken for the first time. Additional credits, as well as those of previously failed courses, must be paid by the students.

Prior to the beginning of the enrollment period, students have a meeting with a designated academic advisor. The meeting, that has a recommended duration of 15 min, is intended to discuss the student's enrollment options and is compulsory for first-year students and for those with low GPAs or who must retake previously failed courses. The outcome of this advising session is a set of recommended courses that academic advisors suggest to students based on their academic record, interests, and personal circumstances.

Students carry out their enrollment using a web tool that enables the exploration of the available courses, instructors, and timetables. The tool also allows students to enter the waiting list of courses with no more available spaces. Students can also use this tool to change their enrollment (e.g., by leaving a chosen course) up to two weeks into the new academic term.

The university also provides access to historical student-generated ratings of the university's instructors. This information is available through another web tool that contains the students' assessment of each instructor in all the courses they have ever taught. This information is collected at the end of each academic term through compulsory surveys that students must complete.

In this context, we seek to investigate **what approaches do students take when enrolling in a new academic term?** To answer this research question, we designed and conducted a survey. The section that follows provides details on our participants,

[1] A student's GPA is calculated by adding up all the numbered grades they have received and dividing them by the number of courses the student has taken. Grades are given on a numeric scale from 0 to 10 and a course is passed with a minimun grade of 6.

procedure, and a description of the data we collected. This is followed by details on our analyses and our findings.

4 The Survey

In this section, we describe the survey we conducted to uncover the students' enrollment habits and the factors they consider when deciding their courses.

4.1 Participants & Procedure

We conducted a survey with students enrolled in a computer science (CS) undergraduate program. Out of the 105 students who received the invitation, 91 volunteered to answer the survey (74 male, 17 female; 19–32 years old—median 22). These participants were at different stages of their degree: second or third year (n = 39), fourth (n = 27), and higher years (n = 25). All participants had gone through the enrollment process at least once and all were familiar with the support provided by the university around the academic enrollment process (e.g., advising sessions, access to student-generated ratings of instructors).

4.2 Data Collection & Analysis

We distributed the survey as an electronic form. Access to the form was granted through institutional credentials but all answers were collected anonymously. The survey included an initial questionnaire to collect the participants' consent and demographic information. This was followed by several multiple-choice questions organized in four categories: a) the term planning sessions students have with their academic advisor, b) the importance students give to the opinions about courses that advisors and other students have, c) the activities students perform to decide their courses, and d) the characteristics of courses and instructors that students deem important when deciding their enrollment.

4.3 Findings

The survey responses reveal that most students make their enrollment decisions on their own, before talking to their academic advisors (Fig. 1(a)). However, 33 students (36%) reported making enrollment decisions also during or after these meetings. The decision process of most students takes between 15 and 45 + minutes (Fig. 1(b)). This duration is in stark contrast to the time they usually talk with their advisors (Fig. 1(c)), which is limited to 15 min due to logistic constraints and to the advising policies defined by the university.

The time students spend deciding their courses is mostly devoted to three activities (Fig. 2): inquiring other fellows about the reputation of courses and instructors, and reviewing the student-generated ratings of the instructors under consideration. Asking instructors about the reputation of the course is the less frequent activity. Note that the reported frequency of these activities correlates with the importance ratings that the students' assigned.

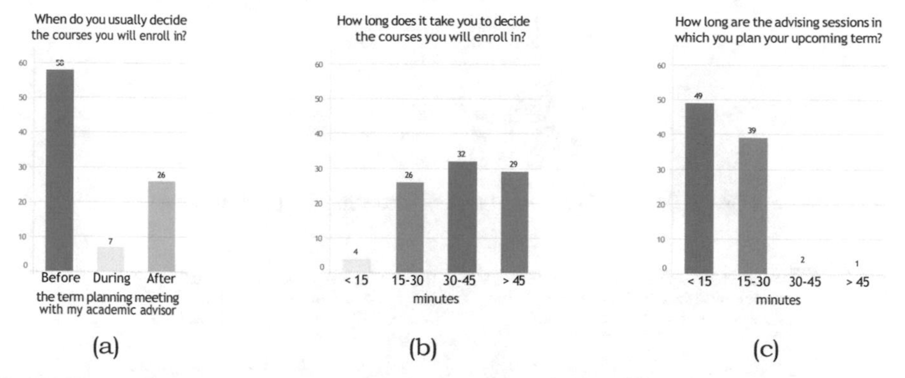

Fig. 1. When students decide the courses they will take (a) and how long they spend in this process with (b) and without their academic advisor (c).

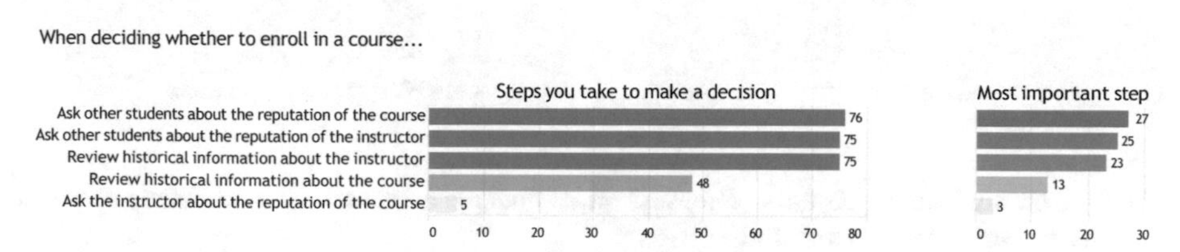

Fig. 2. Steps students take to make their enrollment decisions.

On the motivations to select a course, students seem to prioritize those that open paths for future enrollments. That is, courses that are prerequisite of others (Fig. 3). This criterion prevails over others such as having enjoyed or successfully passed similar courses. Interestingly, the least important factor to select a course was the fact that it is deemed as easy by other students.

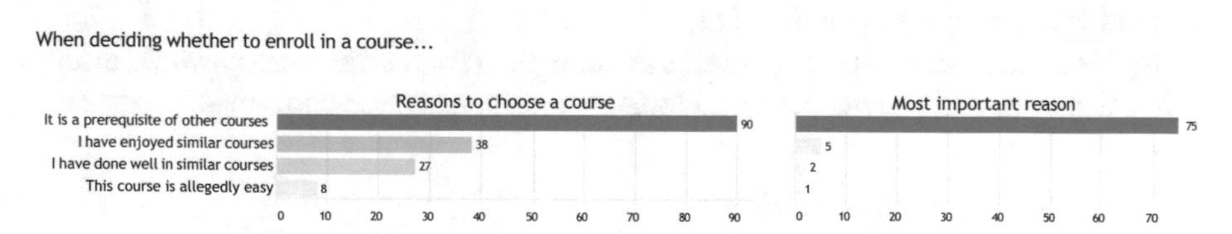

Fig. 3. Reasons students deem important when choosing a course.

The survey also showed that most students prefer to enroll in courses taught by instructors that explain the subject well, over instructors that are generous markers (Fig. 4).

Finally, when deciding on their courses, students value the advise of other fellow students as much as that of their academic advisors (see Figs. 5(a) and 5(b)).

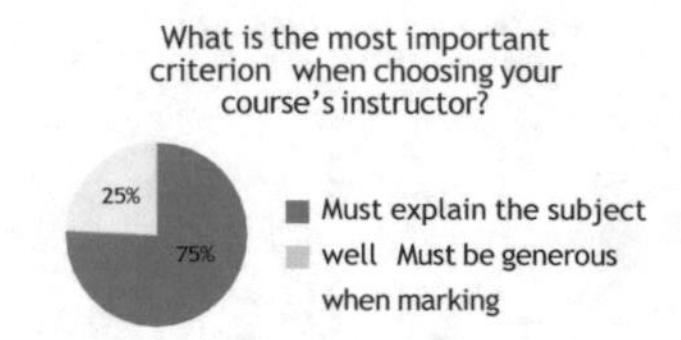

Fig. 4. Students' preferences when choosing course instructors.

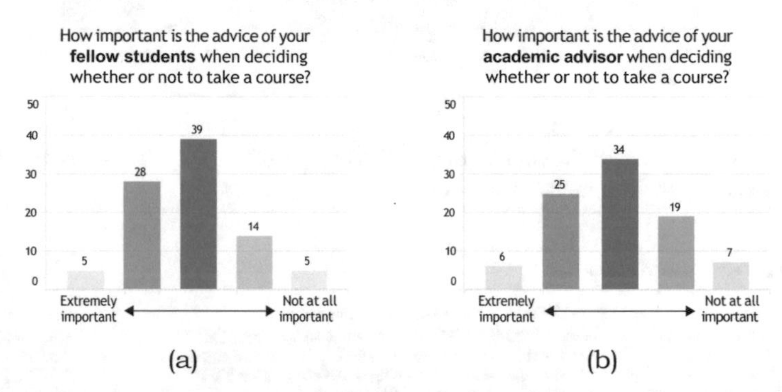

Fig. 5. How students value the opinions of other students and their academic advisors when considering which courses to enroll in.

5 Discussion

Through our survey, we were able to characterize the approaches that students take when enrolling in a new academic term. We identified some of their planning habits, the opinions they value, and the factors that play a role in their course enrollment decisions. Our observations align with previous research on what students want from learning analytics tools [17]; in particular, regarding the features students expect from tools oriented to support their planning [23].

In this section we interpret our results as we explain the relevance of our work, discuss its implications, and provide a critical look at current academic enrollment processes.

5.1 The Value of Timely Advising

Among other types of support, the academic advising system of our research site seeks to provide students with enrollment guidance. However, the results of our survey suggest that, quite often, students make up their minds about the courses they will take even before talking to their academic advisors. This observation has important implications. First, it means that students often make enrollment decisions informed almost exclusively by the *vox populi* (i.e., by what other students say), which may result in a wide ecosystem of unofficial, overwhelming (and sometimes contradicting) information. Second, this observation also implies that students do not always benefit from an expert perspective that goes beyond the academic aspects of an academic program. Because of their vast experience, advisors are able to assess the potential consequences of choosing a particular set of courses for a student with a specific academic history. However, for this advise to be effective, it has to be delivered in a timely manner.

These implications suggest that, as students wait for their enrollment advising meetings, they would benefit from some sort of preliminary guidance. This could be provided, for example, via data-based tools that make available objective information on their courses (e.g., historical distribution of grades, passing and failing rates, typical level of dropouts). Such tools may help reduce the effect of the *vox populi* and could support students in making more informed decisions. These tools, however, should be designed carefully, as they could also motivate students to avoid courses with certain reputation [4] (e.g., those with a low historical passing grade) or make shortsighted choices regarding their careers [18].

5.2 Students' Preferences Around Courses and Instructors

Our data shows that students rely heavily on information from their peers and student-generated ratings to inform their course selection decisions. This suggests that the reputation of a course and instructor is a key factor in students' decision-making process. Additionally, the survey results indicate that students prioritize courses that open paths for future enrollments over other factors like enjoying or successfully completing similar courses. That is, students are willing to move faster across their academic program and to reduce their time to graduation. This highlights the importance of course sequencing to ensure that students are taking the right courses at the right time.

It is also important to note that some student preferences may give rise to conflicting interests when it comes to course selection. A desired course may be taught by a poorly rated instructor, or it may not be the prerequisite of other courses. This highlights, again, the role of the academic advisors in helping students navigate these scenarios in light of reliable data and advice. In terms of instructors, the data indicate that students prefer those who explain a subject well over those who are generous markers. This suggests that the quality of the teaching and learning experience is more important to students than grades alone. The take away message of these observations is that proper human guidance is required during the enrollment process and such guidance may not be easily replaceable by automated tools.

5.3 Better Support for Enrollment

Academic enrollment is comprised by a series of steps that take place over a period of time. It is not an action that takes place *just* on the student's registration date. Rather, it involves the execution of activities before, during, and after that moment. Our observations suggest that higher-educational institutions should provide proper support for each of these stages. Doing so is not only in the best interest of the students, but it also has benefits for the mission of universities. A student who enrolls in the right courses at the right time is most likely to succeed than a student who feels forced to take undesired courses. Students may feel frustrated when their enrollment does not go as planned or deviates too much from their original plan. Thus, improving the students' academic enrollment can also translate in improving their general well being at the university.

There will always be a gap between what students want and what an educational institution can—and should—provide. This is especially true for databased technological solutions, as their development imposes restrictions on data privacy and ethics. It is also important to note that no tool can support *all* the activities students perform during their enrollment process. There are informal conversations with fellow students that will unavoidably take place outside of *any* official tool. Interaction and UX designers working on this space should consider this when designing novel tools for academic enrollment.

6 Limitations and Future Work

The generalizability of our findings is limited by the sample size of our survey. With a total of 91 respondents, our sample is relatively small and not fully representative of all potential audiences. Additionally, the survey relied on selfreport data, which may be subject to personal biases. Because our survey only contained close-ended questions with limited sets of answers, the students' responses and, in turn, the conclusions that can be drawn from them, are also limited. In particular, our survey is insufficient to understand what are the activities that the students *want* to perform—beyond those they already perform with the institutional support at their disposal.

Although our characterization of the students' habits during the academic enrollment process advance our understanding in the area, there is much left to do. Our survey only included students from a single university. Additional studies are needed to investigate the enrollment habits of other populations of students. Future work could focus on characterizing the enrollment process of educational institutions with different contexts (e.g., with different academic advising systems and enrollment policies). Finally, it could be valuable to conduct qualitative research to gather more in-depth insights into students' decision- making processes and the factors that influence their enrollment choices.

7 Conclusion

This paper provides insight into students' habits, strategies, and decisive factors when it comes to course selection. Our survey data shows that most students make their enrollment decisions independently, relying on information from peers and student-generated ratings. Our findings also uncover the importance that students give to course sequencing and to the quality of the teaching and learning experience. All this highlights the value of academic advising to provide guidance during the course selection process. Overall, these findings contribute to our understanding of the factors that influence students' course selection decisions and the role of academic advisors in the process. These results can inform the development of academic practices and policies to better support student enrollment.

References

1. Ajanovski, V.V.: Context aware recommendations in the course enrolment process based on curriculum guidelines. International Association for Development of the Information Society (2013)

2. An, B.P.: The role of academic motivation and engagement on the relationship between dual enrollment and academic performance. J. High. Educ.uc. **86**(1), 98–126 (2015)
3. Apel, A., Hull, P., Owczarek, S., Singer, W.: Transforming the enrollment experience using design thinking. Coll. Univ. **93**(1), 45–50 (2018). https://www.proquest.com/scholarly-journals/transforming-enrollment-experience-using-design/docview/2032395521/se-2
4. Bar, T., Kadiyali, V., Zussman, A.: Quest for knowledge and pursuit of grades: information, course selection, and grade inflation. Behav. Experimental Econ. (2008). https://doi.org/10.2139/ssrn.1019580
5. Borrego, M., Knight, D.B., Gibbs, K., Jr., Crede, E.: Pursuing graduate study: factors underlying undergraduate engineering students' decisions. J. Eng. Educ.ng. Educ. **107**(1), 140–163 (2018)
6. Byd˘zovská, H.: Course enrollment recommender system. Int. Educ. Data Mining Soc. (2016)
7. Chen, R., Bahr, P.R.: How does undergraduate debt affect graduate school application and enrollment? Res. High. Educ. **62**(4), 528–555 (2021)
8. Douglas, S.: Understanding the decision to enroll in graduate business programs: influence of sociological and economic factors and gender. Ph.D. thesis, Colorado State University (2017)
9. Haris, N.A., Abdullah, M., Hasim, N., Rahman, F.A.: A study on students enrollment prediction using data mining. In: Proceedings of the 10th International Conference on Ubiquitous Information Management and Communication, IMCOM 2016. Association for Computing Machinery, New York (2016). https://doi.org/10.1145/2857546.2857592
10. Jacqmin, J.: What drives enrollment in massive open online courses? EMOOCs **2021**, 1 (2021)
11. Kot, F.C.: The impact of centralized advising on first-year academic performance and second-year enrollment behavior. Res. High. Educ. **55**(6), 527–563 (2014)
12. López Guarín, C.E., Guzmán, E.L., González, F.A.: A model to predict low academic performance at a specific enrollment using data mining. IEEE Revista Iberoamericana de Tecnologias del Aprendizaje **10**(3), 119–125 (2015). https://doi.org/10.1109/RITA.2015.2452632
13. Perna, L.W.: Understanding the decision to enroll in graduate school: sex and racial/ethnic group differences. J. High. Educ. **75**(5), 487–527 (2004)
14. Rivas, N., Minguillón, J., Chacón, J.: Enrolling habits in higher education. what sources of information do students have and what are missing? In: INTED2021 Proceedings, pp. 4980–4988. 15th International Technology, Education and Development Conference, IATED, 8–9 March, 2021. https://doi.org/10.21125/inted.2021.1025
15. Rivera, J.E.H.: A hybrid recommender system to enrollment for elective subjects in engineering students using classification algorithms. International Journal of Advanced Computer Science and Applications **11**(7) (2020). https://doi.org/10.14569/IJACSA.2020.0110752
16. Rosas, M., Hamrick, F.A.: Postsecondary enrollment and academic decision making: family influences on women college students of mexican descent. Equity Excellence Educ. **35**(1), 59–69 (2002)
17. Schumacher, C., Ifenthaler, D.: Features students really expect from learning analytics. Computers in Human Behavior **78**, 397–407 (2018). https://doi.org/10.1016/j.chb.2017.06.030. https://www.sciencedirect.com/science/article/pii/S0747563217303990
18. Smith, P.: On the unintended consequences of publishing performance data in the public sector. Int. J. Public Adm. **18**(2–3), 277–310 (1995). https://doi.org/10.1080/01900699508525011
19. Stiber, G.: Characterizing the decision process leading to enrollment in master's programs: further application of the enrollment process model. J. Mark. High. Educ.rk. High. Educ. **11**(2), 91–107 (2001)
20. Strayhorn, T.L.: When race and gender collide: Social and cultural capital's influence on the academic achievement of African American and Latino males. Rev. High. Educ. **33**(3), 307–332 (2010)

21. Vialardi, C., Chue, J., Peche, J.P., Alvarado, G., Vinatea, B., Estrella, J., Ortigosa, Á.: A data mining approach to guide students through the enrollment process based on academic performance. User modeling and user-adapted interaction **21**(1), 217– 248 (2011)
22. Wagner, K., Hilliger, I., Merceron, A., Sauer, P.: Eliciting students' needs and concerns about a novel course enrollment support system (2021)
23. Whitelock-Wainwright, A., Gašević, D., Tejeiro, R., Tsai, Y.S., Bennett, K.: The student expectations of learning analytics questionnaire. J. Comput. Assisted Learn. **35**(5), 633–666 (2019). https://doi.org/10.1111/jcal.12366. https://onlinelibrary.wiley.com/doi/abs/10.1111/jcal.12366

Contrasting Analysis Between Motion Capture Technologies

Fabian Arun Panaite, Emanuel Muntean, Monica Leba(✉), and Marius Leonard Olar

University of Petrosani, Petrosani, Romania
monicaleba@upet.ro

Abstract. This research has explored capturing the motion of the human arm, and after a series of approaches, the two most vigorous methods have been proven to be: motion capturing method using the video camera Microsoft Kinect, that was initially developed for use as game controller and motion capturing method that does not involve video, but a group of rotary linear potentiometers, known for their major performance in the domain of robotics. In the video-based motion capture, the interpretation of the arm's joints and elements, with their movements on the three spatial axis is done with the help of the MATLAB environment, while the potentiometer-based motion capture, the angles are measured by the potentiometers themselves, the data being sent from the potentiometers to a computer's serial monitor through an Arduino development board. In all motion capture cases, the arm from which the motion is captured, has a well-established set of movements and trajectories. For the potentiometer-based motion capture, the arm is wearing an exoskeleton type of device, designed specifically for the task of motion capture, which includes the rotary potentiometers, for each joint and its possible motions, so that the shoulder has three potentiometers, the elbow has one, and the wrist has another one that measures the rotation of the forearm on the three spatial axes.

Keywords: Orientation · Kinect · Arm motion · Exoskeleton · Simulation

1 Introduction

In this section, the paper will largely expose the simple principles of motion capture and offer a brief explanation for the following sections. The main reason for this study is to find and employ the correct motion capture system, by analyzing its features, functionalities and degrees of freedom. And to this purpose, the destination application type and the degrees of freedom have been studied, and neatly sorted in the table, in Sect. 3, for the reader's convenience.

The simplest way to explain the notion of motion capture system, is to say that they employ the use of sensors to estimate the human body's motion, in joint angles, bone segments and express their kinematic measurements in a digital form.

In the second section the current state of research at a global scale will be presented in a general form, that shows the basic principles and results on the subject. In this section

A. Rocha et al. (Eds.): WorldCIST 2023, LNNS 799, pp. 259–268, 2024.
https://doi.org/10.1007/978-3-031-45642-8_26

there is also a short literature review of research that has involved or was based on video cameras, Microsoft Kinect, Laser, EMG (electromyogram sensors), IMU (Inertial Measurement Units), and other potentiometers, including a review table.

In the third section of this paper, the experiments and methods that have been used to capture the motion data, and the approaches of this paper are presented, along with a brief schematic of the motion capture system, a classification and comparison of the main features of the two proposed methods, that includes a versus between the two methods, with schematics and a quick analysis of the two methods.

In the fourth section of this paper, the details of the experiments, the data, both explained, and illustrated, providing the approaches for motion capture, with the principle schematics and figures that show the motion capture designs, and diagrams of the motion capture data, the MATLAB simulation results for each approach, for both motion captures, the one that involves the Microsoft Kinect, and the one based on rotary linear potentiometers.

In the fifth and last section, there are the compared results, the case selection rules for every method, which allows and favors the selection of the motion capture system for each application it could be used for. Furthermore, for every type of application, such as the control of a robotic arm, the control of an actuated exoskeleton, like the ones used to amplify the strength of a movement, the motion control of a 3D virtual object, character or avatar.

2 Literature Review

In this section the current state of research at a global scale will be presented in a general form, that shows the basic principles and results on the subject of motion capture.

Motion capture systems usually depend on either stationary sensors, such as cameras placed around in a room, or wearable sensors, such as inertial sensors (IMUs) worn on the body. Optical motion capture technology is the latest industry standard. Precise calibration can favorize camera systems to follow the position of markers with millimetric precision [1]. In related research the kinematic model and marker placement (on the areas on the body, segments and joints) relied upon on ISB guidelines. Moreover, groups of 3 markers were well placed on the arm and the forearm in a bilateral manner. Motion capture was done with 7 iR Vicon MX3 cameras, sampled at 200 Hz [2].

Statistical analysis [3] can be performed using MATLAB. Microsoft Kinect V1 was launched in 2010 [4], to be used as a control device [5] for the Xbox video game console. The Microsoft Kinect V2 is the second generation [6] of a cost effective, depth-estimating device, considered to be a successful marker-less [7] human physical activity tracking device. The Microsoft Kinect can also be used in software that provides datasets and databases of adult or even child motion [8] that can be used in character animation, gait analysis [9], marker [10] and marker less motion tracking [11].

For the accuracy [12] of the Microsoft Kinect, the 3D acquisition of the human body [13] has very small mean errors, with a precision up to the resolution of 1280×720 at 12 fps [14] for the RGB camera, 640×480 at 30 fps for the depth camera, with Infrared Depth technology and a Field of View of 57° H, 43° V, and a specified measuring distance between 40 cm and 4 m and a USB 2 and USB 3 connectivity.

The Microsoft Kinect is also a cost-effective [15], clinician friendly assessment tool. The latest scientific and technological trends, allow the Wi-fi motion tracking [16] which, for now has its limitations to a set of predefined movements.

Other technologies have employed LIDAR, which as a difference to camera motion capture, extracts the 3D pose for subjects [17] with gait motion from the 3D dense body joints. The comparison affirms that the proposed computational modelling provides higher quality human identification, when compared to OpenPose video software.

Data collected using LIDAR returns low reflection from skin because of low skin reflectivity at the operating laser wavelength.

Approaches based on EMG are also widely used for limbs' motion capture [18]. However, a resolver for robust shoulder capturing has not been employed to this date. Human motion posture estimation and muscle analysis capture the organic motion of bone joints by multiple sorts of sensors or many sorts of cameras to thoroughly measure the range of muscle activity in the body [19]. This is why EMG is used, mostly, as an instrument to figure out muscular activity, recently developed as a way to determine force in EMG signals and relying on normalization from activation during high voluntary contraction to record high muscle force [20].

By this time, it can be said that EMG sensors, the Microsoft Kinect, and IMU sensors can be considered to be effective instruments for human assisting robotic systems. There is an expanding number of research studies and experiments that employ the use of EMG controllers for applications in robotics [21]. Many human motions can be tracked with EMG signals with procedures such as fuzzy control, SVMs or CNNs.

Using bionic hands has no limit regarding medical applications, but has been vastly employed also for industrial applications. Man-made bionic hands are able to perform multiple types of tasks in hazardous environments or restricted areas, at the same time, maintaining the pilot's dexterity, grip, precision and quick timing. In these circumstances, gestures can be recognized with image detection [22] and this could suffice in providing the corresponding hand motion.

More user-friendly ways of control and communication must be projected for people with motion impairments or neuromuscular conditions, such as cerebral palsy, stroke or other injuries. The regression offers multiple degrees of freedom to be controlled at once, with proportional controls [23]. But its control is of low precision for more than two degrees of freedom. Electromyography based bionic and robotic arm experiments for control in past research speculated that the control precision of a bionic or robotic arm could be enhanced by proprioceptive modulation. Experimenting with more circumstances would aid in obtaining better playback of proprioceptive modulation and would help the interpretation of the way proprioceptive feedback of the bionic and robotic arm may aid body's scheme extension to the bionic and robotic arm [24].

Solving the kinematics equations forms the foundation of all research regarding humanoid manipulation and are crucial to the dynamic equations necessary for gait, balance and the manual use of any tools. When given the spatial configuration of a robot, inverse kinematics provide a set of joint angle values that reach a designated position for the robot's absolute motion. Inverse Kinematics can be solved in both computational and analytical ways. The difference between analytical solutions and computational methods is that the analytical solutions are unable to be used in tool employment scenarios or modifications in robot configuration, and the computational solver must be designed and coded before the numerical values are introduced [25].

3 Experiment Development

The first experiment, the one using the Arduino and linear-rotary potentiometers was developed for capturing the motion angles of the human arm (see Fig. 1). Its objective was to find the most cost-effective motion capture system and was made possible by using a 3D printed device, shaped as an exoskeleton, having potentiometers in each joint. These capture the motion for the shoulder and for the elbow (see Fig. 2).

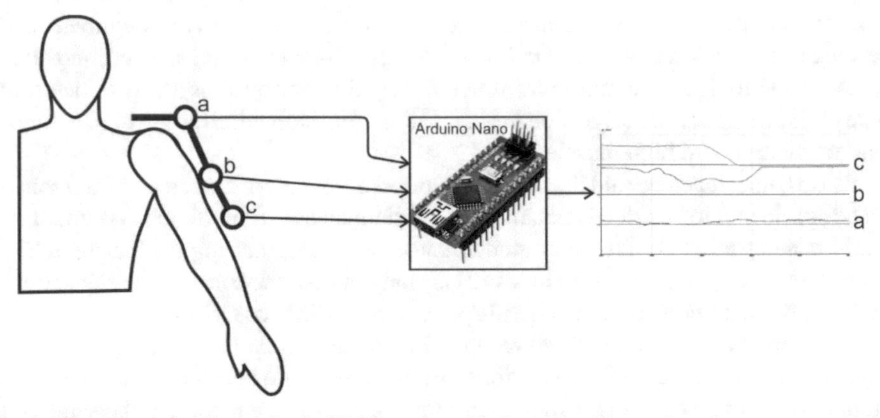

Fig. 1. The principle schematics in 3D blueprint with the rotary linear potentiometers, connected to an Arduino NANO development board.

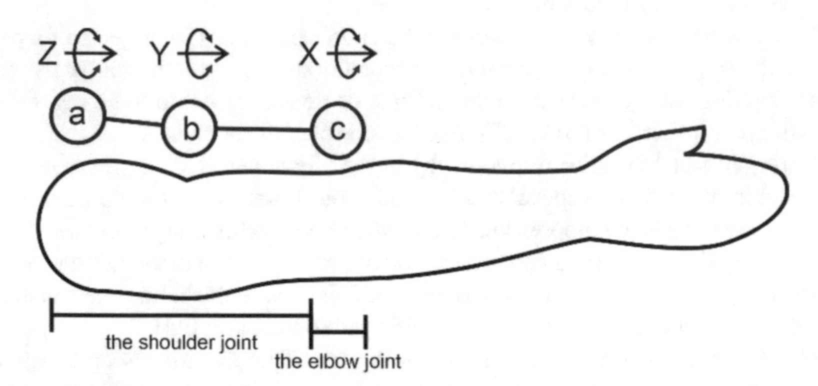

Fig. 2. Positioning of the 3 potentiometers, two for the shoulder joint and one for the elbow

The obtained results, three angles captured, first two for shoulder movement and third for elbow movement, are presented in Fig. 3, as follows:

The second experiment for capturing the motion angles of a human arm was developed using the video camera technique that involves using a Microsoft Kinect V1,

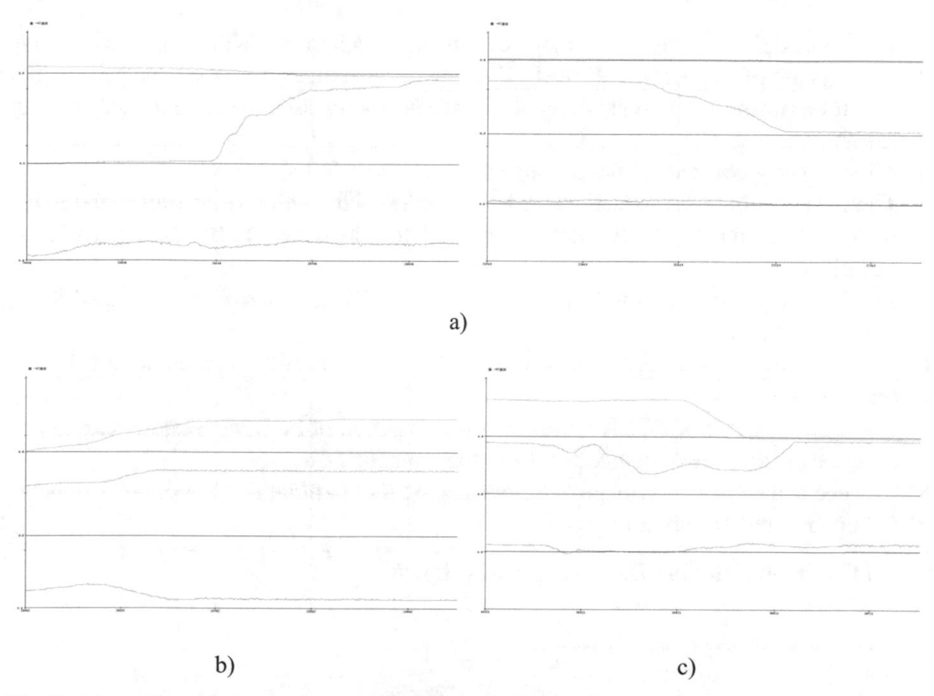

Fig. 3. The values of the angles acquired from the potentiometer, during the running of the experiment with potentiometers. a) The angular values for the arm motion from the shoulder in the abduction/adduction (up/down direction) in the Frontal plane and Sagittal axis. b) The angular values for the arm's motion in the flexion/extension (forward/backward direction) in the Transverse plane and Longitudinal axis.; c) The angular values for the arm's elbow flexion, movements on the Sagittal plane and Frontal axis.

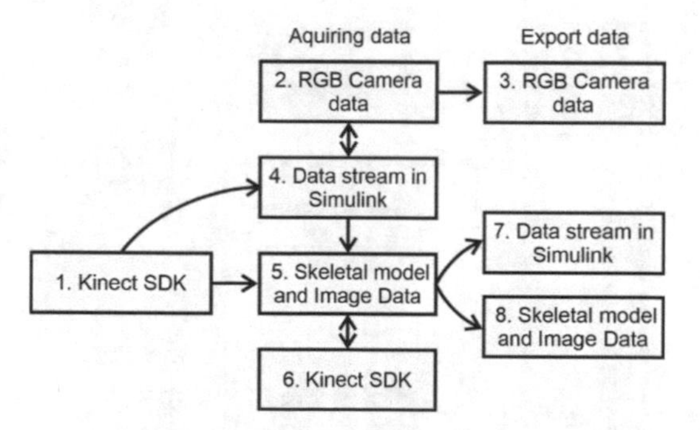

Fig. 4. Logical scheme of operation of the method of capturing the movement of an element of the human skeleton, in our case the upper limb, in the shoulder and elbow joints.

together with MATLAB/Simulink and its main objective was to develop the motion capture system that provided the highest set of degrees of freedom for the person whose motion is being captured.

In Fig. 4 the motion capture using the Kinect is being explained.

1. The Kinect SDK receives the image data from the Microsoft Kinect in a raw format.
2. In this phase of acquiring data, the RGB Camera data comes from the Kinect controller as if it was normal video camera data with the normally colored elements within every frame.
3. The normally colored frames are also sent and exported to the PC.
4. The data stream is composed of the RGB camera data which is later analyzed in the Kinect SDK, to generate the skeletal model and image data, to be sent to MATLAB Simulink.
5. The Skeletal model and Image Data from both the Kinect controller and Kinect SDK, are sent to MATLAB Simulink
6. The data stream from the Kinect SDK and, Kinect controller camera sensor have a bidirectional flow.
7. The data stream in MATLAB Simulink is composed of both the Skeletal model and image data, and the data received from the Kinect SDK.
8. The skeletal model and Image data, can also be sent to other devices or environments for control and communication.

In Fig. 5 the Simulink Block diagram is shown.

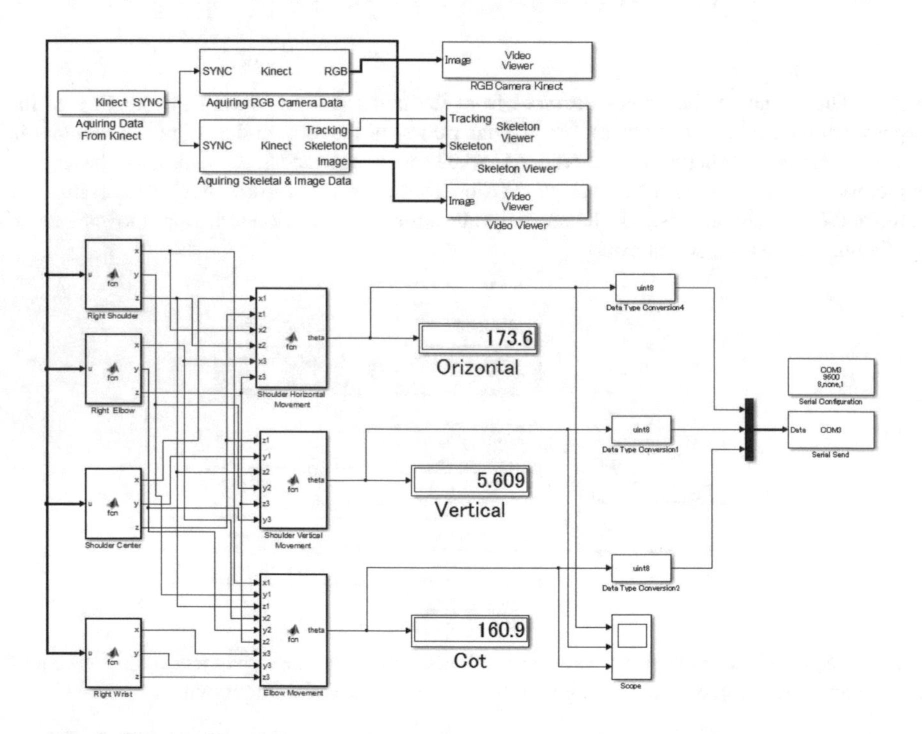

Fig. 5. The MATLAB Simulink block diagram for motion capture using the Kinect V1

The Acquired data can be considered to be a collection of the RGB Camera data, the Data stream in Simulink, the Skeletal model and Image Data, and the Kinect SDK data.

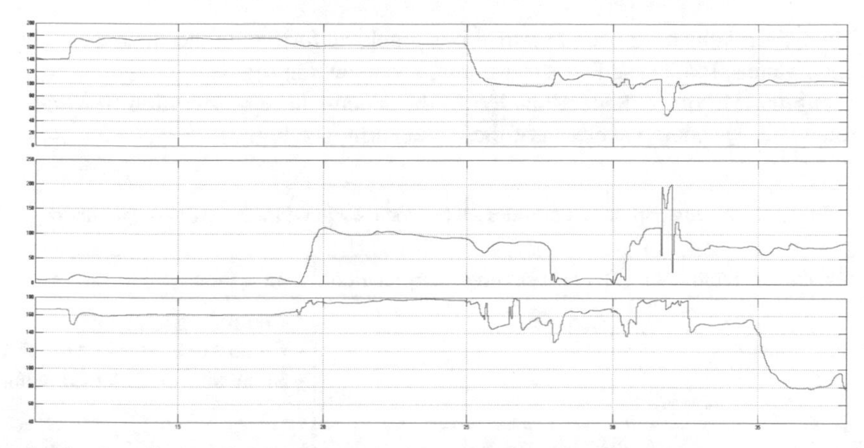

Fig. 6. The data diagram for the three rotary linear potentiometers during the human arm motion

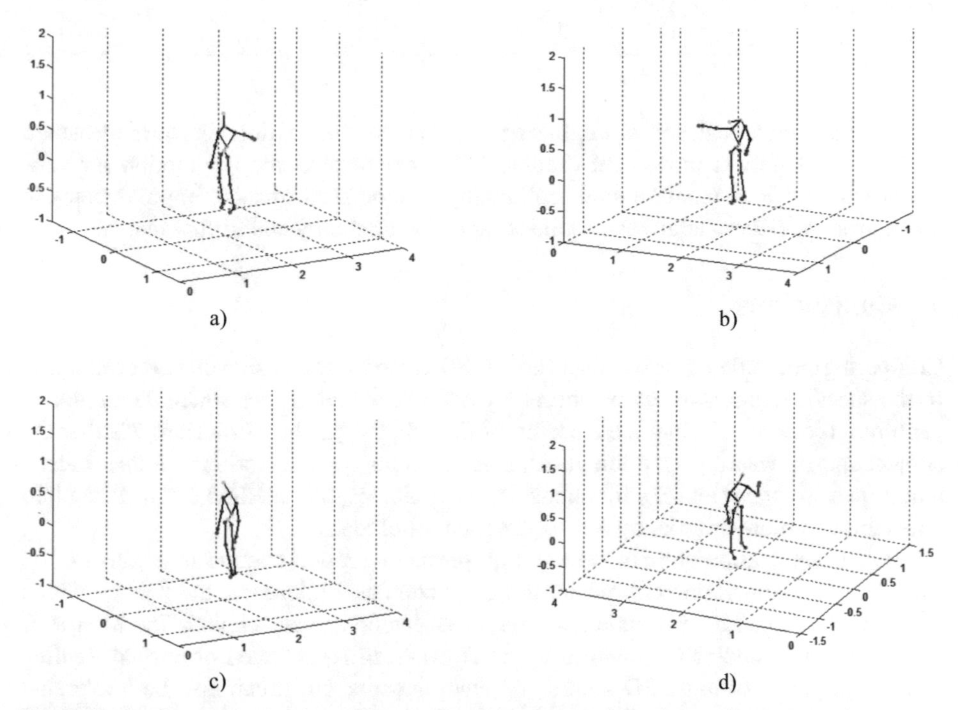

Fig. 7. In the first representation of the position of the arm, it is found horizontally and vertically (a) and b)), and in c. in the relaxed position (a). In d), the elbow is found in the flexed position: a) Arm in horizontal position; b) Arm in vertical position, c) Arm in relaxation position, d) Elbow in flexion position.

The Exported data can be considered a composition of the RGB Camera data, the Data stream in Simulink and the Skeletal model and image Data.

Thus, the results have been obtained in a graphical form with the same three angles, and their values, like in the first experiment, and shown in Fig. 6.

Using the Microsoft Kinect has led to the schematic representation of a subject's motion as a 3D rendered skeleton of the human arm (see Fig. 7).

Table 1. Motion capture techniques, data processing and their destined application

Motion capture technique	Data processing/interpretation	Application
Video: • Video camera • Kinect • Laser	Open CV	• Controlling a robotic arm • Control an avatar in AR or VR • Simulations (Matlab-Simulink)
Non-video: • IMU sensor • Potentiometer • LIDAR	Matlab/Simulink	

Most of the clinical, sports, engineering, and entertainment domains have measured kinematics of human movement (Table 1) in order to replicate the motion for various reasons such as rehabilitation, optimizing athletic performance, improvements in mechatronics, electric actuators, robotics, or cinematography and virtual reality.

4 Conclusions

Comparing the results of the two experiments has led to the conclusion that the destination of the specific equipment can be chosen by the type of activity for which the motion is captured. It can be said that the freedom of the subject may be a criterion, whether the subject opts for wearing a motion tracking device or not, yet the motion capture is indeed a non-invasive operation. So, for choosing the motion capture method can be taken into consideration more criteria, by a case selection set of ideas.

By the application domain, where high precision motion capture is not necessary, for example in entertainment, where the error rate has a tolerance, the Kinect, which was also developed for entertainment purposes, can be an easy choice. The Microsoft Kinect can be a choice for controlling a 3D avatar in Augmented or Virtual Reality, or a video game, or other 3D simulation environments, but it can also be a tolerable choice for controlling a robotic arm, the control of an actuated exoskeleton for medical rehabilitation purposes.

The rotary linear potentiometer exoskeleton based motion capture is of a much higher precision and takes a lot less resources, has a higher direct data transmission speed, in exchange for the disadvantage that it is a wearable device, yet its use is a lot more favorable choice for controlling a robotic arm, or an actuated exoskeleton, and even more, for a medical rehabilitation purpose, but does not exclude its use in entertainment at all, like the control of a Virtual Reality, or Augmented Reality Avatar, or video game element or accessory, or for animating a character.

Future research may involve a more extended exoskeleton, to capture the motion of multiple elements of the human body, or of the entire body, using the rotary linear potentiometers.

References

1. Slade, P., Habib, A., Hicks, J.L., Delp, S.L.: An open-source and wearable system for measuring 3D human motion in real-time. IEEE Trans. Biomed. Eng. **69**(2), 678–688 (2021)
2. Schlagenhauf, F., Sreeram, S., Singhose, W.: Comparison of kinect and vicon motion capture of upper-body joint angle tracking. In: 2018 IEEE 14th International Conference on Control and Automation (ICCA), pp. 674–679, June 2018
3. Assi, A., Bakouny, Z., Karam, M., Massaad, A., Skalli, W., Ghanem, I.: Three-dimensional kinematics of upper limb anatomical movements in asymptomatic adults: dominant vs. non-dominant. Human Movement Sci. **50**, 10–18 (2016)
4. Albert, J.A., Owolabi, V., Gebel, A., Brahms, C.M., Granacher, U., Arnrich, B.: Evaluation of the pose tracking performance of the azure kinect and kinect v2 for gait analysis in comparison with a gold standard: a pilot study. Sensors **20**(18), 5104 (2020)
5. Bilesan, A., Komizunai, S., Tsujita, T., Konno, A.: Improved 3D human motion capture using Kinect skeleton and depth sensor. J. Robot. Mechatronics **33**(6), 1408–1422 (2021)
6. Steinebach, T., Grosse, E.H., Glock, C.H., Wakula, J., Lunin, A.: Accuracy evaluation of two markerless motion capture systems for measurement of upper extremities: Kinect V2 and Captiv. Hum. Factors Ergonomics Manuf. Serv. Ind. **30**(4), 291–302 (2020)
7. Naeemabadi, M., Dinesen, B., Andersen, O.K., Hansen, J.: Influence of a marker-based motion capture system on the performance of Microsoft Kinect v2 skeleton algorithm. IEEE Sens. J. **19**(1), 171–179 (2018)
8. Aloba, A., et al.: The UF kinect database of child and adult motion. In: Eurographics (Short Papers), pp. 13–16, April 2018
9. Roy, G., Bhuiya, A., Mukherjee, A., Bhaumik, S.: Kinect camera-based gait data recording and analysis for assistive robotics-an alternative to goniometer-based measurement technique. Procedia Comput. Sci. **133**, 763–771 (2018)
10. Bilesan, A., et al.: Marker-based motion tracking using Microsoft Kinect. IFAC-PapersOnLine **51**(22), 399–404 (2018)
11. Bilesan, A., Behzadipour, S., Tsujita, T., Komizunai, S., Konno, A.: Markerless human motion tracking using microsoft kinect SDK and inverse kinematics. In: 2019 12th Asian Control Conference (ASCC), pp. 504–509, June 2019
12. Yu, K., Barmaki, R., Unberath, M., Mears, A., Brey, J., Chung, T.H., Navab, N.: On the accuracy of low-cost motion capture systems for range of motion measurements. In: Medical Imaging 2018: Imaging Informatics for Healthcare, Research, and Applications, vol. 10579, pp. 90–95, May 2018
13. Pasinetti, S., Hassan, M.M., Eberhardt, J., Lancini, M., Docchio, F., Sansoni, G.: Performance analysis of the PMD camboard picoflexx time-of-flight camera for markerless motion capture applications. IEEE Trans. Instrum. Meas.Instrum. Meas. **68**(11), 4456–4471 (2019)
14. Guzsvinecz, T., Szucs, V., Sik-Lanyi, C.: Suitability of the kinect sensor and leap motion controller—a literature review. Sensors **19**(5), 1072 (2019)
15. Tanaka, R., Takimoto, H., Yamasaki, T., Higashi, A.: Validity of time series kinematical data as measured by a markerless motion capture system on a flatland for gait assessment. J. Biomech.Biomech. **71**, 281–285 (2018)
16. Ren, Y., Wang, Z., Tan, S., Chen, Y., Yang, J.: Winect: 3d human pose tracking for free-form activity using commodity wifi. Proceedings of the ACM on Interactive, Mobile, Wearable and Ubiquitous Technologies **5**(4), 1–29 (2021)

17. Glandon, A., Vidyaratne, L., Sadeghzadehyazdi, N., Dhar, N.K., Familoni, J.O., Acton, S.T., Iftekharuddin, K.M.: 3D skeleton estimation and human identity recognition using lidar full motion video. In: 2019 International Joint Conference on Neural Networks (IJCNN), pp. 1–8, July 2019
18. Samper-Escudero, J.L., Contreras-González, A.F., Ferre, M., Sánchez-Urán, M.A., Pont-Esteban, D.: Efficient multiaxial shoulder-motion tracking based on flexible resistive sensors applied to exosuits. Soft Rob. **7**(3), 370–385 (2020)
19. Chen, Z., et al.: Analyzing human muscle state with flexible sensors. J. Sensors (2022)
20. Gowtham, S., Krishna, K. A., Srinivas, T., Raj, R. P., Joshuva, A.: EMG-based control of a 5 DOF robotic manipulator. In 2020 International Conference on Wireless Communications Signal Processing and Networking (WiSPNET), pp. 52–57, August 2020
21. Liao, L. Z., Tseng, Y. L., Chiang, H. H., Wang, W. Y.: EMG-based control scheme with SVM classifier for assistive robot arm. In 2018 International Automatic Control Conference (CACS), pp. 1–5. (2018, November)
22. Said, S., Boulkaibet, I., Sheikh, M., Karar, A.S., Alkork, S., Naït-Ali, A.: Machine-learning-based muscle control of a 3d-printed bionic arm. Sensors **20**(11), 3144 (2020)
23. Nougarou, F., Campeau-Lecours, A., Massicotte, D., Boukadoum, M., Gosselin, C., Gosselin, B.: Pattern recognition based on HD-sEMG spatial features extraction for an efficient proportional control of a robotic arm. Biomed. Signal Process. Control **53**, 101550 (2019)
24. Rangwani, R., Park, H.: Vibration induced proprioceptive modulation in surface-EMG based control of a robotic arm. In: 2019 9th International IEEE/EMBS Conference on Neural Engineering (NER), pp. 1105–1108, March 2019
25. Beeson, P., Ames, B.: TRAC-IK: an open-source library for improved solving of generic inverse kinematics. In: 2015 IEEE-RAS 15th International Conference on Humanoid Robots (Humanoids), pp. 928–935, November 2015

Health Informatics

ASCAPE - An Intelligent Approach to Support Cancer Patients

Mihailo Ilić, Mirjana Ivanović(✉), Dušan Jakovetić, Vladimir Kurbalija, Marko Otlokan, Miloš Savić, and Nataša Vujnović-Sedlar

University of Novi Sad, Faculty of Sciences, Novi Sad, Serbia
{milic,mira,dusan.jakovetic,kurba,otlokanmarko,svc, natasa.vujnovic}@dmi.uns.ac.rs
https://www.dmi.uns.ac.rs/

Abstract. Nowadays the number of people living with cancer is constantly increasing. Numerous multidisciplinary research teams are working on development of powerful intelligent systems that will support medical decisions and help patients with critical diseases, including cancer, to keep and even increase their quality of life (QoL). ASCAPE (Artificial intelligence Supporting CAncer Patients across Europe) is an H2020 project which main objective is to use powerful techniques in Big Data, Artificial Intelligence and Machine Learning in processing cancer (breast and prostate) patients' data in order to support their health status. A key result of the project is the implementation of an Artificial Intelligence/Machine Learning (AI/ML) infrastructure. It will allow the deployment and execution of AI/ML algorithms locally in a hospital on patients' private data, producing new knowledge. Newly generated knowledge will be sent back to the infrastructure and will be available to other users of the system keeping private patients' data locally in hospitals. In this paper we will briefly present the structure of an open AI/ML infrastructure and how federated learning is employed in it.

Keywords: Federated Learning · Machine Learning · Personalized Services in Medicine · Patients' Quality of Life

1 Introduction

In modern society with constantly growing but also turbulent economies, different serious epidemic/pandemic situations, and increasing number of local wars people are under specific stressful living conditions. Such dynamic life is one of key factors in triggering critical health problems like cancer, cardiovascular, neurological, and others. On the other hand, the latest cancer statistics[1] show that the absolute number of people living with cancer is constantly increasing. Cancer is a specific disease and there are a lot of factors that affect its treatment and quality of life of patients.

[1] https://doi.org/10.1016/S1470-2045(16)00091-7.

A. Rocha et al. (Eds.): WorldCIST 2023, LNNS 799, pp. 271–277, 2024.
https://doi.org/10.1007/978-3-031-45642-8_27

Rapid advances and developments in the ICT area especially AI, ML, use of wearable devices, efficient techniques for processing Big Data and others are very promising to cancer patients and support the development of reliable and sophisticated solutions/services for before, during and after the treatment.

To employ powerful AI/ML methods it is necessary to have big medical datasets and to find ways to collect the wide spectra of patients' complex data. Apart from standard clinical data different health parameters should also be collected from wearable devices, sensors, everyday activities, nutritional data, environmental data and so on. Very important is also patients' feedback obtained through different validated questionnaires that depict the particular QoL parameters. All kinds of patients' data are necessary for training and obtaining reliable predictive AI/ML models [8]. Achieved results after successful processing of patients' data should be presented to doctors in an understandable way [5] which help them propose adequate treatment and give better recommendations that could improve patients' health status. The EU-funded research and innovation H2020 project ASCAPE[2] is focused on the use of contemporary techniques in Big Data, AI and ML as a support to cancer (breast and prostate) patients in keeping and even improving their health status and quality of life parameters [1,7]. The ASCAPE Consortium[3] is jointly working on the realization of a powerful AI/ML open infrastructure which will offer great opportunities for clinicians to make adequate personalized treatments and suggest interventions for patients [3].

ASCAPE infrastructure allows deployment and execution of AI/ML algorithms locally in a hospital on patients' private data. Newly produced knowledge is sent back to the infrastructure and is made available to other users and hospitals while preserving the privacy of patients' data. Open AI/ML infrastructure is highly based on advantages that federated learning offers. University of Novi Sad (UNSPMF), as a member of the consortium, has a crucial role within the ASCAPE project in implementing and realizing ML and federated learning aspects of ASCAPE architecture.

The main intention of the paper is to give a brief overview of the ASCAPE architecture but also to present key aspects of federated learning as an original scientific contribution of the UNSPMF team.

The rest of the paper is organized as follows. The second section gives a brief overview of the ASCAPE infrastructure which was developed as a joint effort of the entire ASCAPE consortium. The third section is devoted to federated learning aspects incorporated in the ASCAPE architecture as a result of UNSPMF activities. Concluding remarks are given in the last section.

2 ASCAPE Architecture

The ASCAPE architecture incorporates both edge and cloud computing in order to make innovative AI/ML models available to doctors and patients. All of this

[2] https://ascape-project.eu/.

[3] https://ascape-project.eu/the_project/consortium.

is achieved while also preserving the privacy of patients' sensitive data. The first prototype of ASCAPE architecture was developed and is currently in phase of intensive testing in three hospitals, members of the ASCAPE consortium [1–3, 7]. The system is based on microservices as this paradigm is a natural fit for the processes within the ASCAPE framework [7]. All software components of the ecosystem form 3 distinct modules: the ASCAPE edge and cloud, external data aggregators and pilot site specific integration components [3, 7] (see Fig. 1).

ASCAPE Edge and Cloud – The edge components are located at the hospital and their main task is to synchronise the data collected from the Healthcare Information Systems (HIS) and external data aggregators. These components also enable the training of local and global models, while preserving data privacy. Finally, they offer prediction making functionality.

External Data Aggregators – The main task of these components is to collect data which is not collected directly by the HIS, which includes weather data, data collected by wearable devices like Fitbit data etc.

Pilot Site Specific Integration Components – The main task of these components is collection and transfer of data to the ASCAPE edge components for use in training and inference actions.

The modules support workflows such as: data ingestion, predictive model training, personalized AI predictions and simulations, and user interaction.

Data Ingestion – It is the process of transferring data from the Healthcare Information Systems (HIS) to the ASCAPE edge component installed at the healthcare institution. Conversion to a suitable format for use in the ASCAPE environment must be executed first. Moreover, data enrichment may be carried out by including data not collected by the HIS, for example weather data.

Predictive Model Training – The aggregated data may be used to train numerous machine learning models. Three distinct cases are supported: local model training, global federated model training and training of homomorphically encrypted global models.

Local model training covers a wide variety of algorithms like support vector machines, random forest classifiers/regressors etc. All models are validated against local validation data. In the case of local model training, only the model which performed the best on validation data is persisted and used later on for inference requests. Each of these approaches sees preservation of privacy as a key objective.

Personalized AI Predictions and Simulations – Personalized QoL predictions can be made for individual patients upon requests made from doctors. Inference requests are made to all available models: local, global federated and homomorphically encrypted models. The effects of various interventions can also be simulated by using the available models.

User Interaction – Doctors are granted access to the system through the ASCAPE dashboard. Here they are able to view and update their patients' info, run predictions based on this data and suggest further treatment.

Apart from the main goal of providing support to doctors and patients using advanced ML methods, data security and preservation of privacy are crucial. Edge components (nodes) are deployed at the medical institutions/hospitals where integration with the existing healthcare information systems is achieved through ASCAPE integration components. Through this integration patient data is aggregated and enriched with external ASCAPE compatible data. This is done locally at the healthcare institution in order to preserve data security. Following this, local ML models can be trained which then provide predictions for each patient upon requests made by physicians. Patient data along with QoL predictions and simulations are made available to doctors through the ASCAPE dashboard in a user-friendly manner based on explainable AI techniques. Edge computing allows healthcare providers to protect the privacy of their patients' data, while also assisting doctors with proposing treatments.

The ASCAPE cloud also enables all members of the ASCAPE ecosystem to exchange knowledge. Institutions can collaborate and train so called "global models" by using two privacy preserving ML methods.

Federated Deep Learning – By participating in FL, institutions train deep models on local data, after which they share the learned weights with the ASCAPE cloud. The FL Manager aggregates all of the received weights, after which it sends them back to the edge nodes for further training. Intermediate global models are persisted at the cloud and are made available to all members of the ecosystem.

Homomorphic Deep Learning – Patient data is homomorphically encrypted and uploaded to the cloud. Global models trained on this kind of data produce homomorphically encrypted outputs, which can be decrypted by using the secret key known only to the edge node.

3 Federated Learning in ASCAPE

The ASCAPE framework provides healthcare institutions the possibility of knowledge sharing through the creation and training of global FL models. Confidential patient data must not leave its respective healthcare institution, which is why federated learning is introduced. Federated learning mechanisms which are incorporated in ASCAPE architecture represent original scientific contributions of UNSPMF team [8]. Adequate components and parts of the architecture responsible for the realization of FL were developed and tested in the first ASCAPE prototype [2]. Promising results of effectiveness of ML predictive models in increasing patients' QoL are achieved using retrospective patient data collected from hospitals. For details please look at [8].

Each member of the ASCAPE ecosystem may participate in training of global federated models, and this process is orchestrated by the ASCAPE feder-

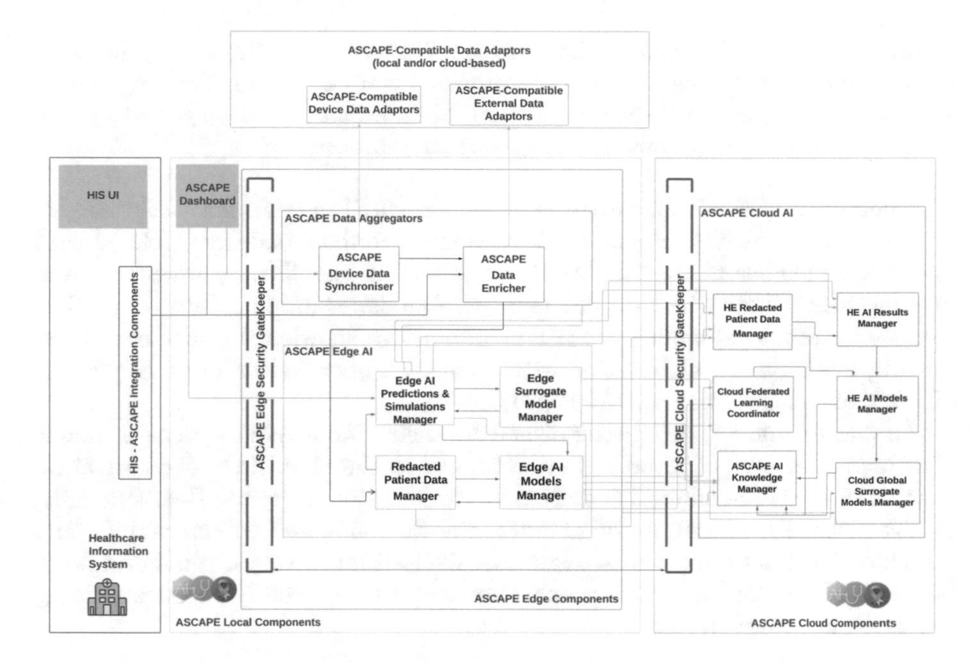

Fig. 1. The ASCAPE architecture.

ated learning coordinator. The only prerequisite is the compatibility of datasets between all institutions willing to train a ML model for a particular use case.

In ASCAPE FL each edge node makes local updates to the global FL model using locally available data. These updates are then sent to the FL coordinator where they are aggregated. Following this, the global FL model is persisted in the cloud and returned to each edge node for further training (see Fig. 2) [4]. Multiple modes of FL exist, some of which are supported in the ASCAPE framework. Each edge node E_i participating in FL has its own corresponding local training set D_i, where $i \epsilon [1..k]$.

Incremental Federated Learning – The FL model is improved incrementally from the first to the last client. The model M is created by E_1 and trained on D_1 after which it is saved in the ASCAPE cloud. Following this, edge node E_2 retrieves model M and updates it using D_2. The entire process is finished once E_k submits its updates to the global FL model.

Concurrent Federated Learning – All edge nodes E_1 ... E_k train model M simultaneously. Training of M is split into rounds, where in most cases a round is equivalent to an epoch. At the end of each round edge nodes submit their updates to the federated learning coordinator in the form of updated hidden layer weights, where they are aggregated using the federated averaging [6] method. This updated model M is sent back to the edge nodes for a subsequent round of training. The process is repeated for an arbitrary number of rounds, until a viable model is produced.

Semi-concurrent Federated Learning – The ASCAPE framework also supports seamless switching between the previous two FL modes. Training can be started by a single healthcare institution, after which it can be joined by any number of other institutions during the training process.

Regardless of the FL strategy used, all global FL models and their weights are stored in the ASCAPE cloud. These models are then made available for each and every member of the ASCAPE ecosystem to use. This ensures that even medical institutions which do not have an abundance of local data which they can use to train local models are still able to utilize knowledge gained elsewhere. Fueling this type of collaboration with a focus on preservation of privacy is one of the main goals of the ASCAPE project.

Global FL models are distinguished by their name which acts as a unique identifier. The first edge node that connects to the cloud in order to train a specific model is the one which defines its hidden layer structure. Before defining a new global FL model, an edge node may be configured to run architecture search on local data in order to determine the best hidden layer configuration to be used. The execution of this preparation step can be greatly beneficial during federated learning later on.

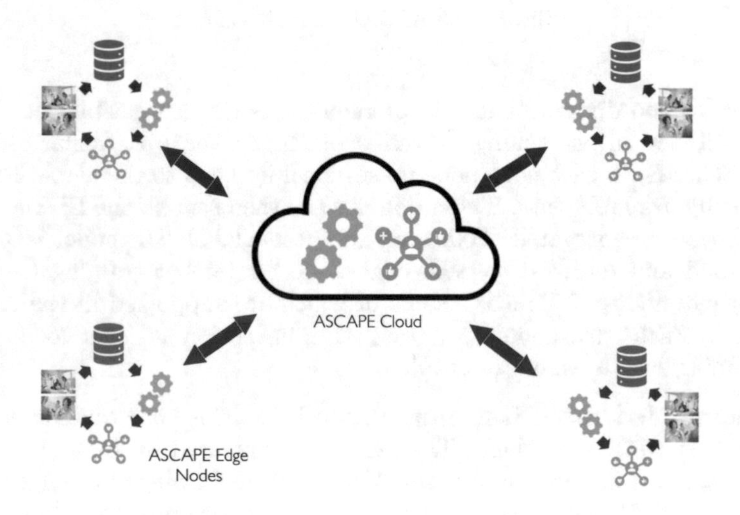

Fig. 2. Federated Learning within the ASCAPE framework.

4 Conclusion

Treatment of cancer patients usually negatively influences their everyday activities. To make their life easier and bearable it is important to develop adequate services to support their health-related quality of life parameters. Patients' data collected from different sources coupled with employing powerful AI/ML and Data Science methods to process such data are essential in producing accurate and reliable prediction models. Contemporary technologies and their combination offer great opportunities for clinicians to make adequate personalized treatments and suggest interventions that can increase patients' QoL.

The ASCAPE project offers an implementation of an open AI infrastructure which will support cancer patients and improve their QoL through predictive modeling and providing better medical interventions [1–3,7,8]. Currently the first prototype, which was successfully tested with rich patient retrospective data [8] is deployed in three hospitals. First results achieved in a hospital with more than 120 newly recruited patients are very promising.

Acknowledgment. This research was supported by the EU H2020 ASCAPE project under grant agreement No 875351.

References

1. ASCAPE Deliverable - D1.1 Positioning ASCAPE's open Al infrasetructure in the after cancer-care Iron Triangle of Health (2022). https://ascapeproject.eu/node/57
2. ASCAPE Deliverable - D2.4 ML-DL Training and Evaluation Report (2022). https://ascape-project.eu/node/118
3. ASCAPE Deliverable - D4.1 Personalized interventions and user-centric visualizations (2022). https://ascape-project.eu/node/120
4. ASCAPE framework and technical innovations (2022). https://ascape-project.eu/marketing-material/ascape-framework-and-technical-innovations
5. Holzinger, A., Langs, G., Denk, H., Zatloukal, K., Müller, H.: Causability and explainability of artificial intelligence in medicine. Wiley Interdisciplinary Rev. Data Mining Knowl. Discovery **9**(4), e1312 (2019)
6. Kairouz, P., et al.: Advances and open problems in federated learning. Found. Trends Mach. Learn.**14**(1-2), 1–210 (2021)
7. Lampropoulos, K., et al.: ASCAPE: an open AI ecosystem to support the quality of life of cancer patients. In: 2021 IEEE 9th International Conference on Healthcare Informatics (ICHI), pp. 301–310. IEEE (2021)
8. Savić, M., et al.: The application of machine learning techniques in prediction of quality of life features for cancer patients. Comput. Sci. Inf. Syst. **20**(1), 381–404 (2023). https://doi.org/10.2298/CSIS220227061S

Factors Influencing the Effectiveness of Medical Products Stock Management in Polish Hospitals - Research Results

Anna Gawronska[1](✉), Adam Kolinski[1], and Michal Adamczak[2]

[1] Lukasiewicz Research Network – Poznan Institute of Technology, Poznan, Poland
{anna.gawronska,adam.kolinski}@pit.lukasiewicz.gov.pl
[2] Poznan School of Logistics, Poznan, Poland
michal.adamczak@wsl.com.pl

Abstract. The subject of health care is fundamental to all of us. However, logistic tasks performed in health care are becoming more and more important. One of the logistics issues is the medicinal product stocks management, the process of their preparation and delivering to the patient, and IT tools supporting these activities. The authors of this article conducted research among hospitals located in the territory of the Republic of Poland. 45 hospitals participated in the study, which allows the results to be representative. The CAWI method was used in the study. The Kruskal-Wallis method was used in the analysis of the results. The research objective of the article is to identify the impact of the three distinguished factors (stock replenishment system, organization of medicinal product dose preparation and IT systems used) on the percentage of medicinal product transferred for disposal due to expiry of the date and the time between the moment the medicinal product is delivered to the moment it is given to the patient. Six hypotheses were put forward in the study. The conducted research allows for setting conclusions about the factors influencing the percentage of medicinal product transferred for disposal due to expiry of the date and the time between the moment the medicinal product is delivered to the moment it is given to the patient. Significant factors include the adopted stock replenishment system and the organization of preparation of medicinal product administered to the patient. The provider of the IT system supporting these processes is an irrelevant factor.

Keywords: Medicinal product stock management · Survey research · Reduction of disposed medicinal product

1 Introduction

Hospitals play an important role in the health care system by providing medical services to citizens. Pursuant to the Polish provisions of the Act on Medical Activity, a hospital is "an enterprise of a medical entity in which this entity carries out medical activities such as hospital services". The concept of hospital services, on the other hand, is understood as "comprehensive 24-h health services consisting in diagnosis, treatment, care and

A. Rocha et al. (Eds.): WorldCIST 2023, LNNS 799, pp. 278–287, 2024.
https://doi.org/10.1007/978-3-031-45642-8_28

rehabilitation, which cannot be provided under other stationary and round-the-clock health services or outpatient health services" (Act of April 15 2011). According to the data of the Central Statistical Office, there were 898 stationary general hospitals in Poland at the end of 2020. In 2021, the National Health Found spent nearly PLN 50 billion on medical services and medicinal product that can be tailored to a specific patient or service. (www.prawo.pl).

One of the therapies used in the hospital treatment process is pharmacotherapy, i.e., treatment with medicinal products. A medicinal product - according to Polish law - is "a substance or a mixture of substances presented as having the properties of preventing or treating diseases occurring in humans or animals or administered to make a diagnosis or to restore, improve or modify the physiological functions of the body through pharmacological and immunological effects. or metabolic" (Act of September 6 2001). According to estimates, the average hospital in Poland manages approx. 130 thousand. Medicinal products (Karkowski 2015, p. 173), and medicinal products constitute the second - after wages and salaries - cost-intensive element of hospital functioning (Religioni 2016). Taking also the aspect related to the need to ensure an adequate level of patient service, and mainly its safety, especially in the field of pharmacotherapy, the management of logistic processes related to the flow of medicinal products in a hospital is an extremely important area of hospital operation.

The pharmacy is the heart of the hospital from the point of view of managing the circulation of medicinal products from suppliers to individual departments or clinics. The pharmacy's tasks are diverse and result from the provisions of the Pharmaceutical Law. Many of the tasks carried out by a hospital pharmacy concern medicinal product management, and thus have a direct and indirect impact on the shaping of the supply chain and in-hospital distribution of medicinal products.

The hospital pharmacy in a given unit fulfills the demand for medicinal products only according to the hospital list of medicinal products. Exceptionally, he brings the medicinal product to the rescue on the order of the head of the ward or the senior doctor of the hospital. In order to ensure patient safety in terms of medicinal product availability, he must properly manage medicinal product stocks, and the choice of a management model influences the shaping of the external distribution chain. Basic replenishment models include: an information level model and a continuous order cycle model. It may be helpful to analyze the various variants of these models, both in the system of external procurement and at the level of departments / clinics. The selection of the appropriate model for a given hospital is influenced, among others, by medicinal product delivery conditions. Therefore, it is important to consider the hospital's delivery needs when negotiating with suppliers. Classic models work well for hospitals whose assortment is numerous and diverse. A hospital pharmacy can also introduce a JiT model (just-in-time). This model consists in placing orders only for a specific assortment, in the amount reported by the department / clinic.

The research objective of the article is to identify the impact of the three distinguished factors (restocking system, organization of medicinal product dose preparation and IT systems used) on the percentage of medicinal products transferred for disposal due to expiry of the date and the time between the moment the medicinal product is delivered to the moment it is given to the patient. Due to the research goal, six hypotheses were formulated:

H1: The inventory replenishment system affects the percentage of medicinal products transferred for disposal due to expiry of the date.

H2: The inventory replenishment system affects the time between the moment the medicinal product is delivered to the moment it is given to the patient.

H3: The hospital department (clinic or pharmacy) where medicinal product doses are prepared for patients has an impact on the percentage of medicinal products transferred for disposal due to expiry of the date.

H4: The hospital department (clinic or pharmacy) where medicinal product doses are prepared for patients has an impact on the time between the moment the medicinal product is delivered to the moment it is given to the patient.

H5: The use of the same IT system in a hospital pharmacy and clinics has an impact on the percentage of medicinal products transferred for disposal due to expiry of the date.

H6: The use of the same IT system in a hospital pharmacy and clinics has an impact on the time between the moment the medicinal product is delivered to the moment it is given to the patient.

2 Literature Review

It is difficult to find in the literature the results of research conducted on the im-pact assessment of the replenishment system for the supply chain efficiency of medical products to hospitals, in such a comprehensive scope. Fragmented anal-yses can be found in the literature on replenishment to monitor medical products considering storage space requirements (storage locations) in hospitals (William et al. 2020), management of medical supplies based on patient's health status (Saha and Ray 2019), application of medicine management optimization concpecs to improve supply and dispatch of medicines to hospitals (Jordon et al. 2019; Green and Hughes 2011; el Farouk and Jawab 2020; Atika and Irwan 2020), the reorganisation of logistics processes for medicines replenishment within a dedi-cated hospital (Farouk et al. 2018; Ghatari et al. Ghatari et al. 2013), the application of IT technologies in medicines supply chain management (Nabelsi and Gagnon 2017; Gawronska and Nowak 2017), or within the broader framework of hospital logistics management (Pan and Pokharel 2007). Such a fragmented range of analyses only confirms the complexity of the problem in terms of both academic research and observations of business practice.

Recent years have seen an increase in research interest in the issue of logistics and supply chain management in hospitals. This may be due to the search for ongoing solutions to improve logistics processes in the medical supply chain, due to the COVID-19 pandemia. Therefore, the next step of the literature analysis in this area was to verify the trend of the number of scientific publications in Web of Science and SCOPUS databases in the field of hospital logistics and the phar-maceutical supply chain to hospitals.

The tables shown are based on the ScienceDirect.com compilation, which in-cludes articles from both databases. Only articles classified as review and re-search articles

Table 1. Analysis of publications on hospital logistics.

	Scope	2018	2019	2020	2021	2022	TOTAL
Article type	Review articles	205	252	280	351	358	1446
	Research articles	974	1034	1388	1854	1922	7172
Subject areas	Medicine and Dentistry	682	729	922	1168	1174	4675
	Social Sciences	148	156	226	301	291	1122
	Decision Sciences	99	139	143	191	249	821
	Engineering	100	133	127	214	190	764
	Environmental Science	95	87	153	211	193	739
	Nursing and Health Professions	104	95	140	178	166	683
	Computer Science	97	101	111	178	172	659
	Business, Management and Accounting	61	67	119	155	207	609
	Biochemistry, Genetics and Molecular Biology	69	75	99	153	140	536
	Immunology and Microbiology	86	74	93	127	124	504

were included in the literature analysis. Table 1 presents an analysis of the number of scientific publications on hospital logistics.

Analysing the general interest in hospital logistics research, it is important to point out a continuous increasing trend in each year. The scope of the research conducted is mainly in the medical sciences, but also in the social sciences. Sub-sequently, the decision-making process and the engineering of logistics processes in hospitals are important research problems.

Table 2 presents an analysis of the number of scientific publications in the pharmaceutical supply chain.

Analysis of the scientific literature on the pharmaceutical supply chain confirms a trend of growing research interest in optimising logistics processes in hospitals. The analysis of the pharmaceutical supply chain shows a concentration of research findings on: (1) Biochemistry, Genetics and Molecular Biology; (2) Environmental Science; (3) Pharmacology, Toxicology and Pharmaceutical Science.

It should be noted that there are significantly more articles presenting research results for the pharmaceutical supply chain than there are research results for hospital logistics, which may indicate a greater research focus on the pharmaceutical supply chain as a more dedicated segment of business needs.

Table 2. Analysis of publications on the pharmaceutical supply chain

	Scope	2018	2019	2020	2021	2022	TOTAL
Article type	Review articles	694	787	1080	1738	1881	6180
	Research articles	2888	3101	3819	4680	4969	19457
Subject areas	Biochemistry, Genetics and Molecular Biology	929	919	1123	1293	1256	5520
	Environmental Science	564	658	808	1300	1424	4754
	Pharmacology, Toxicology and Pharmaceutical Science	703	649	855	1059	1049	4315
	Chemistry	702	702	836	921	1045	4206
	Chemical Engineering	495	579	694	1011	1154	3933
	Agricultural and Biological Sciences	484	509	611	821	861	3286
	Medicine and Dentistry	336	354	464	640	611	2405
	Materials Science	306	369	447	567	631	2320
	Energy	276	294	382	474	523	1949
	Engineering	205	204	263	333	361	1366

3 A Survey Study

3.1 A Survey Methodology

The study was conducted using the CAWI (Computer-Assisted Web Interview) method. The form was made available electronically to hospital employees holding managerial positions, in particular positions responsible for medicinal product stocks in the hospital. Questionnaires were sent to 250 randomly selected hospitals from the list provided by the Ministry of Health in Poland. The return rate was 18%. The study was conducted on a sample of 45 hospitals located in the territory of the Republic of Poland. Taking into account the number of hospitals in Poland (898), the examined sample gives representativeness of the research results, assuming a 95% confidence level and 15% maximum error. The research was conducted in February and March 2022.

The questionnaire consisted of two parts: the record and the research part. In the research part, the interval and ratio scale were used.

3.2 A Survey Results

The first stage of the analysis of the results was to check the normality of the distribution of the results (answers to questions related to the percentage of medicinal products transferred for disposal due to expiry of the date and the time between the moment the medicinal product is delivered to the moment it is given to the patient.) In both cases the Anderson-Darling test indicated no normal distribution.

This result allowed for a decision to use in the analysis the results of the Kruskal-Wallis test, which is the equivalent of one-way analysis of variance for parametric tests. Hypothesis 0 in the Kruskal-Wallis test assumes equality of medians in the compared populations. Thus, in the event of rejection of the null hypothesis, one can talk about the difference in median, i.e. differences between the tested samples.

H6: The use of the same IT system in a hospital pharmacy and clinics has an impact on the time between the moment the medicinal product is delivered to the moment it is given to the patient.

The results of the Kruskal-Wallis tests (p-value adjusted for ties) are presented in Table 3.

Table 3. Kruskal-Wallis test results

Factors	Percentage of medicinal products transferred for disposal due to expiry of the date	The time between the moment the medicinal product is delivered to the moment it is given to the patient
Medicinal products stocks replenishment system	0,015	0,022
Organization of preparing medicinal product doses for patients (in clinics or in pharmacy)	0,053	0,006
Using the same IT system in the hospital pharmacy and clinics	0,118	0,196

The interpretation of the results and their influence on the verification of the research hypotheses are presented in Table 4.

Summarizing the results of the conducted research, it should be stated that the use of the same IT system in a hospital pharmacy and in clinics does not affect both the percentage of medicinal products transferred for disposal due to expiry of the date and the time between the moment the medicinal product is delivered to the moment it is given to the patient. The replenishment system used has the greatest impact on the phenomena studied. The use of the re-order point (ROP) system causes that the percentage of medicinal products transferred for disposal due to expiry of the date is lower and the time between the moment the medicinal product is delivered to the moment it is given to the patient is shorter. The organization of medicinal product preparation affects the time between the moment the medicinal product is delivered to the moment it is given to the patient (when medicinal products are prepared in clinics, it is shorter than when they are prepared in a hospital pharmacy). The organization of medicinal product preparation, however, has no influence on the percentage of medicinal products transferred for disposal.

Table 4. Verification of research hypotheses

Factors	Percentage of medicinal products transferred for disposal due to expiry of the date		The time between the moment the medicinal product is delivered to the moment it is given to the patient	
Medicinal products stocks replenishment system	H1	confirmed	H2	confirmed
Organization of preparing medicinal product doses for patients (in clinics or in pharmacy)	H3	rejected	H4	confirmed
Using the same IT system in the hospital pharmacy and clinics	H5	rejected	H6	rejected

4 Discussion

The obtained results of the study confirm the previously indicated trend concerning the growing interest of hospitals in the improvement of logistic processes. Multidimensional changes in the area of broadly understood health care have made hospital managers take decisions to revise the current ways of conducting their business. They not only want to be responsible for administering the hospital, but also the actual management of it, often with the use of methods and strategies proven in other industries (Bloom et al. 2015; Ghanem et al. 2015). More and more hospitals adopt management methods such as lean management (Hussain & Malik 2016), just in time (Karkowski et al. 2017), or reorder point (Parildar and Akyürek 2021) in the field of stock management in hospitals.

In order to implement the above mentioned management methods, hospitals use various IT systems to help them in the field of stock management. The lack of quick access to reliable real-time data disrupts the flow of information on the level of inventory of medicinal products, which poses a potential threat to patient safety. On the one hand, this may result in a shortage of certain products or a problem related to over-inventory and / or overdue inventory. The reason for this is, on the one hand, the inability to update data in the IT system in real time and, on the other hand, the collection of insufficient data on medicinal products as well as on patients. Moreover, the lack of comprehensive data and the possibility of their sharing between the participants of in-hospital logistics processes leads to information gaps, duplicate activities, as well as a longer response time to the patient's needs. More and more often these systems are based on interoperability standards that make it possible to integrate IT solutions provided by different solution providers. That is may be the reason why the provider of the system is irrelevant from the point of view of the subject of the research, e.g., stock management.

It should also be remembered that in order to implement various stock management methods, it is necessary to take a number of actions that will result in the availability of a specific medicinal product and with an appropriate expiry date. Automatic data collection techniques (e.g. barcode scanning) can significantly contribute to the improvement of stock management in hospitals. Research carried out over 20 years ago showed that when you manually enter data using the keyboard, you get 1 error in 100 characters, and when scanning barcodes, you get 1 error in 10 million characters (Puckett 1995). Therefore, the implementation of even the most advanced IT system will not contribute to the improvement of logistics processes in the hospital, if the data to this system are mainly entered manually. Moreover, there are research results according to which the implementation of IT solutions in health care does not always result in an improvement in the level of patient service (Chaudhry et al. 2006).

Among the data carriers used for ADC purposes, barcodes are the foundation of solutions, the most widespread in corporate logistics (Goundrey-Smith 2013) and are also more and more often implemented by hospitals, supporting the medicinal product distribution process from the moment of placing an order, through acceptance of deliveries, internal distribution, ending with the registration of the administration of the medicinal product to the patient. It is worth remembering that in order for the read data to be understandable for each entity in the supply chain, bar codes must be standardized both in terms of symbology and the type of data presented (ICMRA, 2021). Internal solutions popular for years, also known as private ones, give way to standard solutions using GS1 standards. Automatic data collection techniques and GS1 standards significantly contribute to the reduction of errors resulting from incorrect administration of a medicinal product, and thus facilities are not forced to incur additional costs that would be associated with prolonged hospitalization of patients (McKinsey & Company, 2012).

5 Conclusions

The conducted research allows us to draw conclusions on the factors that affect the efficiency of managing medical products. The choosen replenishment system (ROP or ROC) has a significant impact on the effectiveness of trade in medical products. Further improvement of the efficiency of trade in medical products in hospitals is possible through more precise tracking of the circulation of these products in the hospital. In this aspect, more advanced use of IT tools and ADC techniques may be helpful.

The conducted research, although on a representative sample, was limited spatially only to the territory of the Republic of Poland. Only the influence of three selected factors on the effectiveness of donating medical products in hospitals was examined.

In the future, the authors plan to extend the research both in the spatial scope to hospitals located in countries other than Poland and in the subject scope. Extending the scope of the subject is a more complex task. It is planned to expand both the definition of efficiency in the management of medical products (so far it was defined only by two factors: the percentage of medical products transferred for disposal and the time between the admission of the medicinal product to the hospital and its administration to the patient) and the range of factors influencing it. The extended factor catalog should also include factors related to the degree of use of selected functionalities of IT systems and the use of ADC techniques.

References

Act of April 15, 2011: Ustawa z dnia 15 kwietnia 2011 r. o działalności leczniczej. https://isap.sejm.gov.pl/isap.nsf/ (28.10.2022)

Act of September 6, 2001: Ustawa z dnia 6 września 2001 r. Prawo farmaceutyczne. https://isap.sejm.gov.pl/isap.nsf/ (28.10.2022)

Atika, F., Irwan, D.: Pharmacy Inventory System Design Using Agile Methods. Al'adzkiya International of Computer Science and Information Technology (AIoCSIT) Journal, 1(2) (2020)

Bloom, N., Propper, C., Seiler, S.: The impact of competition on management quality: evidence from public hospitals, Centre for Economic Performance Discussion Paper, 983 (2015)

Chaudhry, B., et al.: Systematic review: impact of health information technology on quality, efficiency, and costs of medical care. Ann. Internal Med. **144**(10), 742–752 (2006)

el Farouk, I., Jawab, F., Arif, J.: Reconfiguration of replenishment process of resuscitation unit of an public hospital in Morocco. In: Proceedings of the International Conference on Industrial Engineering and Operations Management Bandung, Indonesia, pp. 2660–2669 (2018)

el Farouk, I.I., Jawab, F.: Improving sustainability in public hospital through Medicines Supply chain management. In: 2020 IEEE 13th International Colloquium of Logistics and Supply Chain Management (LOGISTIQUA), pp. 1–5. IEEE (2020)

Gawronska, A., Nowak, F.: Modelling medicinal products inventory management process in hospitals using a methodology based on the BPMN standard. LogForum **13**(4), 455–464 (2017)

Ghanem, M., Schnoor, J., Heyde, C.E.: Management strategies in hospitals: scenario planning, GMS Interdisciplinary plastic and reconstructive surgery DGPW, no. 10. https://www.researchgate.net/publication/283294473_Management_strategies_in_hospitals_scenario_planning [access: 25.09.2015] (2015)

Ghatari, A.R., Mehralian, G., Zarenezhad, F., Rasekh, H.R.: Developing a model for agile supply: an empirical study from Iranian pharmaceutical supply chain. Iranian J. Pharmaceutical Res. IJPR **12**(Suppl), 193 (2013)

Goundrey-Smith, S.: Barcodes and Logistics. In Information Technology in Pharmacy, pp. 175–192, Springer, London (2013)

Green, C., Hughes, D.: Medicines supply and automation. Hospital Pharmacy, 41 (2011)

Jordon, K., Dossou, P.E., Junior, J.C.: Using lean manufacturing and machine learning for improving medicines procurement and dispatching in a hospital. Procedia Manufacturing **38**, 1034–1041 (2019)

Karkowski, T.A.: Świadczenia szpitalne w powiązaniu z procesami zaopatrzenia medycznego i niemedycznego [Hospital services in connection with the processes of medical and non-medical supplies]. Wolters Kluwer (2015)

Karkowski, T.A., Karkowska, D., Skoczylas, P.: Just-in-Time method in the management of hospital medication stock. Przedsiębiorczość i Zarządzanie [Entrepreneurship and Management], 18(10, cz. 3) (2017)

Nabelsi, V., Gagnon, S.: Information technology strategy for a patient-oriented, lean, and agile integration of hospital pharmacy and medical equipment supply chains. Int. J. Prod. Res. **55**(14), 3929–3945 (2017)

Na budżecie NFZ w 2022 roku zyskają uzdrowiska. Kolejki do specjalistów będą nadal [Health resorts will benefit from the NFZ budget in 2022. The queues for specialists will continue] (2021). www.prawo.pl (28.10.2022)

Pan, Z.X.T., Pokharel, S.: Logistics in hospitals: a case study of some Singapore hospitals. Leadership in Health Services (2007)

Parildar, O., Akyürek, Ç.: Determination of The Safety Stock Level and Reorder Point in Hospitals with Probabilistic Inventory Model: An Example of a Public Hospital. SOSYOEKONOMI, 29(47) (2021)

Puckett, F.: Medication-management component of a point-of-care information system. Am. J. Health Syst. Pharm. **52**(12), 1305–1309 (1995)
Religioni, U.: Optymalizacja kosztów leków – wskazówki dla szpitali [Optimizing Drug Costs - Tips for Hospitals] (2016). https://www.medexpress.pl/optymalizacja-kosztow-lekow-wskazowki-dla-szpitali/63156. (28.10.2022)
Saha, E., Ray, P.K.: Patient condition-based medicine inventory management in healthcare systems. IISE Trans. Healthcare Syst. Eng. **9**(3), 299–312 (2019)
Strength in unity, McKinsey&Company (2012). www.gs1.org (27.10.2022)
The promise of global standards in healthcare Recommendations on common technical denominators for traceability systems for medicines to allow for interoperability (2021). https://www.icmra.info (28.10.2022)
William, W., Ai, T.J., Lee, W.: A joint replenishment inventory model to control multi-item medicines with consideration of space requirements in the hospital. Int. J. Ind. Eng. Eng. Manage. **2**(2), 39–48 (2020)

Generation of Synthetic X-Rays Images of Rib Fractures Using a 2D Enhanced Alpha-GAN for Data Augmentation

Mariana Lindo[1,2](✉), André Ferreira[1,2], Jan Egger[2], and Victor Alves[1]

[1] ALGORITMI Research Centre/LASI, University of Minho, Braga, Portugal
{a88360,id10656}@alunos.uminho.pt, valves@di.uminho.pt

[2] AI-guided Therapies, Institute for Artificial Intelligence in Medicine (IKIM), University Medicine Essen (AöR), Essen, Germany
jan.egger@uk-essen.de

Abstract. X-rays are the most commonly performed medical imaging tests to detect fractures. However, some fractures are difficult to detect and may go unnoticed by physicians. In addition, there are very few public X-rays datasets of rib fractures. Although, the creation of such datasets is very time-consuming because of the bureaucratic and ethical issues involved, it is very useful, because these images can be used for teaching and data enhancement without privacy issues. Recently, generative models have been used to synthesize images with high quality and realism. In this work, a generative model was developed to generate synthetic X-ray images with hard-to-detect rib fractures. These images were evaluated using quantitative metrics and a Turing test. It was found that the images generated were not realistic enough due to the large heterogeneity of the dataset used for training, which made it impossible for the model to correctly evaluate the most important features.

Keywords: alpha-GAN · data augmentation · X-ray · fractures · rib

1 Introduction

Medical imaging is a process that aims to identify the internal or external structure of the human body. It consists of observing medical diagnoses, analyzing diseases, and developing data sets of normal and abnormal images [1]. One of the most commonly used modalities, especially for visualization of the chest and trauma, is X-rays. Although, X-rays are the most commonly performed imaging examination for the detection of fractures, each X-rays examination exposes the patient to radiation that is harmful to health. In addition, some fractures, namely on ribs, are difficult to detect on X-rays and may go unnoticed by physicians. This can lead to later complications, either for the patient, who may suffer infections and other problems related to the fracture or for the doctor and the hospital, who may be sued. Nowadays, machine learning is increasingly used in the field of X-rays to train models for the diagnosis of lung diseases. These models, which

A. Rocha et al. (Eds.): WorldCIST 2023, LNNS 799, pp. 288–297, 2024.
https://doi.org/10.1007/978-3-031-45642-8_29

are validated and accepted by regulatory agencies such as the American FDA [2] and the European Commission [3], typically require a large amount of data to train properly and produce good results. However, there are several problems related to the acquisition of real data, namely the fact that this acquisition is difficult due to all the ethical and bureaucratic issues related to data protection and also expensive due to the costs associated with labeling. Although, conventional data augmentation techniques already exist to overcome these difficulties, these techniques can produce unrealistic images, for example, by exaggerating elastic deformations that can alter the actual morphological shape [4]. Currently, generative models have been used to generate medical images in synthetic data. In the case of X-rays modality, there are very few publicly available X-ray datasets with hard-to-detect rib fractures. The main purpose of this work is to develop a generative model to generate synthetic X-rays with hard-to-detect rib fractures. These images can later be used for teaching and research without privacy issues since they are not associated with any patient.

2 Related Work

Conventional data augmentation techniques are often applied to augment the minority class data of unbalanced datasets. However, these techniques may lead to further overfitting [5]. In 2021, Venu et al. [6] proposed a DCGAN-based model for generating chest X-rays and augmenting the original data. The synthetic images were evaluated using the Frechet Inception Distance (FID) score. This metric summarizes how similar the generated and real images are, with a perfect value *0.0* indicating the two sets of images are identical [7]. The developed model achieved a value of 1.289, which means that the generated images are very realistic and very close to the original images. In addition, Venu et al. [6] trained a neural network classifier using DCGAN-enhanced datasets, which showed higher accuracy than when trained with datasets created using conventional methods.

Prezja et al. [8] proposed a WGAN-GP to synthesize realistic images of knee joint X-rays with varying degrees of osteoarthritis severity, to address privacy and legal issues that complicate medical data sharing and collection. The generated images were validated by surveying medical experts with real and fake images, and the results showed that the synthetic images were realistic enough to fool the experts. The fake images were also used for data augmentation in a classification task for osteoarthritis severity, and the results revealed that the synthetic images improved classification accuracy in the presence of scarce real data and in Transfer Learning (TL). Furthermore, replacing the real training data with simply generated data in the same classification task resulted in a loss of only 3.79% of the baseline accuracy in classifying real osteoarthritis X-rays.

In the field of MR imaging, Ferreira et al. [9] developed an alpha GAN-based model to generate realistic MRI brain scans in order to synthesize more data that can later be used to train appropriate Deep Learning (DL) models. To evaluate the quality of the synthetic scans, a Turing test was performed, which showed that these scans were able to outsmart almost all the experts. The synthetic scans were also used to determine their impact on the performance of an existing DL model. This model was developed for segmenting the rat brain into white matter, cerebrospinal fluid (CSF), and gray matter. The Dice score was used to compare the models, and the results showed improvements in all segmentations, especially CSF segmentation. The use of synthetic data generated by the new data expansion model improved the segmentation model more than the use of traditional data expansion. An overview of GAN-based approaches for generating realistic 3D data, can be found here [10].

3 Generative Model Architecture

Generative Adversarial Network (GAN) was introduced by Goodfellow et al. [11] to create fake images that look exactly like real images. GANs are suitable for training on large data sets and can produce visually compelling example images. However, this flexibility introduces instabilities in optimization that lead to the collapse mode problem, where the generated images do not fully reflect the distribution of the available data. To improve the GAN architecture, Rosca et al. [12] developed the alpha-GAN. It consists of a GAN and a Variational AutoEncoder (VAE). VAE are auto-encoders with regularized training that avoid overfitting and guarantee that the latent space has good properties suitable for generative processes.

The alpha-GAN architecture, which was taken as a starting point for the development of the model used in this work to generate synthetic X-ray images, is constituted by four networks: Code Discriminator, Encoder, Generator, and Discriminator. The code discriminator is a latent classifier that distinguishes between the vectors generated by the encoder (it must consider them as false) and the vectors obtained randomly from a standard Gaussian distribution (it must consider them as true). That is, at each training cycle, the code discriminator is given two vectors, of which it does not know which is true and which is false, and must assign a value from *0* to *1* (a probability) to each of the vectors, where *0* is completely false and *1* is completely true. The encoder receives as input a real image and generates a vector containing the main features of this image. The generator receives the encoder vector as input and creates a reconstruction of the real image from it. The generator also receives the Gaussian distribution vector and generates a random image from it. Finally, the discriminator receives the reconstructed image, the real image, and the generated image, and the objective is that it can classify which is the true image (with a value close to or equal to *1*) and which is the false image (with a value close to or equal to *0*). It is also worth noting that the random vector size is very important, as this value must be sufficiently large to represent the data set, but not too large to avoid overfitting.

4 Material and Methods

In order to generate images with difficult-to-detect rib fractures obtained from frontal X-rays, it is necessary to train the model with images that have these features. However, the dataset to be used for this purpose contained only 260 images, which is insufficient for training the model. To address the lack of training data, a TL scheme was applied, where first a model was trained with normal X-rays of the ribs, from a collection of 12,512 images. Those images included normal X-rays or X-rays with lung pathologies, namely pneumonia, and it included PNG images from several public datasets: the *MontgomerySet* [13], the *Chest_xray*, the *Chest_xray2* [14], and the *ChinaSet* [15]. After training the model with these data, the training process of the TL schema was complemented with the 260 X-rays of rib fractures from the *PadChest* dataset [16].

This implementation was developed in Python using the DL framework PyTorch 1.12.1 [17]. The overall process flow of training and evaluation is illustrated in Fig. 1. The Image Resources block (Fig. 1A) describes the acquisition of the X-ray images and the different datasets used. The preprocessing block (Fig. 1B) represents the preprocessing of the dataset. In this step, after obtaining the images from the data collection, those that contained only X-ray images of the frontal surface without or with slight inclination were selected for training the model. Ultimately, only 11,504 of the 12,512 images analyzed were selected. Two experiments were then conducted, with each model used in the respective experiment formally defined as M_d, where d is the size ($width \times height$) of the images used. Since the images had different dimensions, a first experiment with the model (M_{384}) was done, where those images were resized to 384×384. This size was chosen because, among the 11,504 images, the smallest had a size of 384×384. Then, a standardization of the scans was performed, with *STD* equal to *0.1* and mean equal to *0*, followed by a normalization between *0* and *1*. Then, each of the arrays representing an image was stored in a single array of all images, which in turn was stored in a compressed zip file. This allowed that each time this array needed to be accessed, it only needed to be loaded into the memory instead of having to run the entire process again, thus saving time and CPU. Finally, after loading the zip file into memory, this was used by a dataloader to train the model.

A second experiment with the model (M_{1024}) was done, where the developed model presented an architecture very similar to the M_{384} architecture, but with more inner layers and with different inputs, which consisted of X-ray images with dimensions 1024×1024. For this resizing process, only images whose width and height were equal or greater than 1024 were considered. The reason for this is that upsampling images with smaller dimensions would favor the appearance of artifacts in the images, which would affect the model training. Therefore, only 5297 images were selected at the end of the resizing process. For the dataset containing X-rays with hard-to-detect fractures, the same preprocessing was performed as described above, except that it was not necessary to select the X-rays of the entire frontal surface, since all of them already had these features. The DL application block (Fig. 1C) describes the training and qualitative evaluation of the models and the selection of the best model. After the final model was selected, it was evaluated qualitatively and quantitatively, as presented in Fig. 1D. Finally, the Development Environment block (Fig. 1E) represents the development environment with all libraries, frameworks, and other dependencies for training the models.

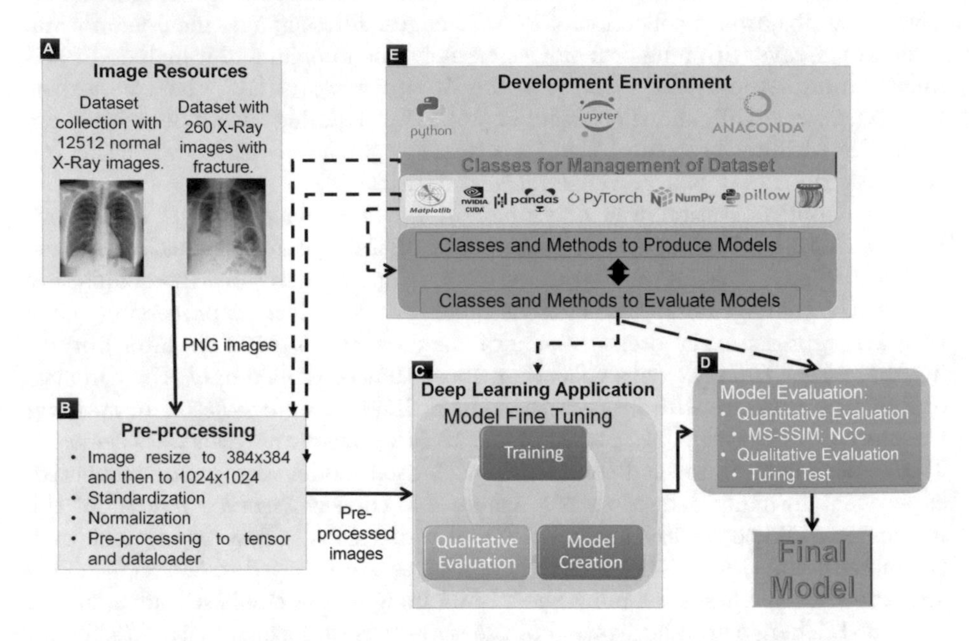

Fig. 1. Overall training and evaluation process workflow.

The architecture of the model M_{1024} is illustrated in Fig. 2 and is based on the model of Ferreira et al. [9], in which instance normalization (IN) was added in the generator and encoder after each convolution, except in the first and last layers. The LeakyReLU activation function was used to accelerate the training and improve the results [18]. It was not used IN the code discriminator and discriminator to avoid artifacts and decrease computational costs. The loss functions employed in this training process were (1) and (2) for generator/encoder, (3) for code discriminator, and (4) for discriminator with $\lambda_1 = 100$ and $\lambda_2 = 1/100$. In the loss function (2), Ferreira et al. [9] used the Mean Squared Error (*MSE*) loss function instead of the *L1* loss function of the alpha-GAN and also added the gradient difference loss (*GDL*) [19] loss Function. However, this *GDL* loss only applies when the input has 3D dimensions and was therefore replaced by the perceptual loss (PER) described in recent work on super-resolution [20].

$$L_{GD} = -E_{z_e}[D(G(_{z_e}))] - E_{z_r}[D(G(_{z_r}))] \quad (1)$$

$$L_G = L_{GD} - E_{z_e}[C(_{z_e})] + \lambda_1 \parallel x_{real} - G(_{z_e}) \parallel_{\text{MSE}} + \lambda_2 \parallel x_{real} - G(_{z_e}) \parallel_{\text{PER}} \quad (2)$$

$$L_C = E_{z_e}[C(_{z_e})] - E_{z_r}[C(_{z_r})] + \lambda_1 L_{GP-C} \quad (3)$$

$$L_D = -L_{GD} - 2E_{x_{real}}[D(x_{real})] + \lambda_1 L_{GP-D} \quad (4)$$

where L_{GD} symbolizes the feedback from the discriminator, D represents the discriminator, G refers to the generator and C symbolizes the code discriminator. Z_e represents the encoder's latent vector, Z_r denotes the random input vector and x_{real} represents a real scan. L_{GP-D} symbolizes the gradient penalty of the discriminator, L_{GP-C} denotes the gradient penalty of the code discriminator, *MSE* represents the *MSE* loss function, *PER* denotes the perceptual loss function, and E symbolizes the overall distribution. The *AdamW* optimizer [21], which has a more stable weight drop than *Adam*, was used with betas of *0.9*, *0.999*, a learning rate of *0.0002*, a weight decay of *0.01*, and eps of *10-8*. The model M_{1024} was trained for one million iterations with a stack size of *1*, and the input random vector size was *1024*. This value was chosen after a latent vector with a size of *5120* was used in a first experiment and it turned out that the model was not capable of learning because it had too much capacity. Therefore, the size of the latent vector was reduced to *1024*. In addition, to increasing the vector size, other changes were made from the baseline model of Ferreira et al. [9], most notably replacing 3D convolutions with 2D convolutions and increasing the number of learning parameters, because the images being generated had much higher resolutions than those generated by the baseline model.

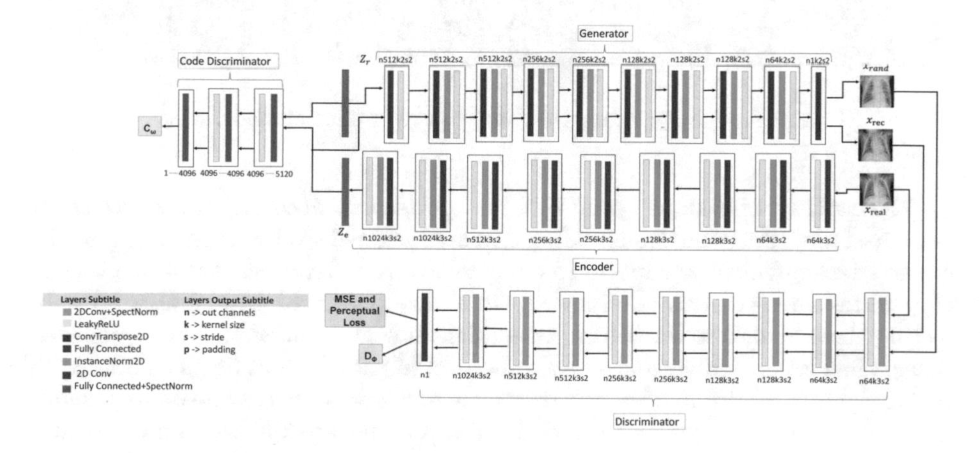

Fig. 2. Architecture of the model M_{1024}.

5 Results

Two categories of evaluation metrics were selected: quantitative (Multi-Scale Structural Similarity - MS-SSIM and Normalized Cross Correlation - NCC) and qualitative (visual Turing test conducted by experts). Starting with M_{384}, several synthetic images were created after training this model. In order to evaluate them qualitatively, the Turing test was performed by two radiologists on eight images illustrated in Fig. 3, of which four were real and four were synthetic. After evaluating the X-rays, one of the experts stated that the resolution of the images was not sufficient to conclude the presence or absence of fractures. In addition, the other radiologist confirmed that the images produced were not very realistic, as there were broken lines between the bones and ribs, and white areas were present in places where the image should have been black.

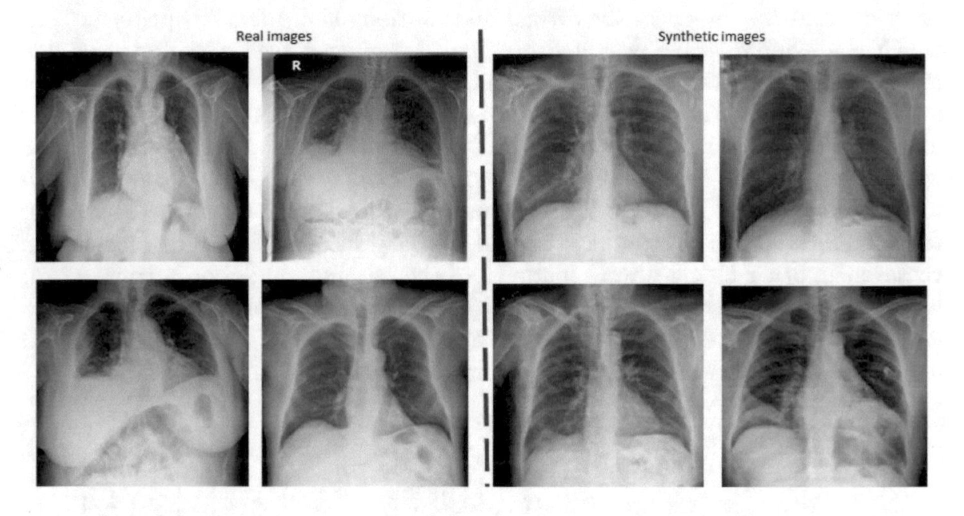

Fig. 3. Real and synthesized X-ray images from the model M_{384}.

Taking the evaluations, given by the experts, into account, the model M_{384} was discarded and the model M_{1024} was developed. With these images, the experts had sufficient resolution to draw conclusions. After the training process, they obtained new synthetic images, which are presented in Fig. 4. After subjecting these images to the Turing test, the radiologists found that the generated images had lower quality and lower resolution in the ribs and lungs compared to real images. In addition, although they were much smoother, some discontinuous lines were still noted between the ribs. However, the specialists emphasized that despite these limitations, the images showed relatively hard-to-detect fractures, proving that the model could distinguish well enough between images with and without fractures during training.

The quantitative evaluation of the generated images was performed only for the images generated with M_{1024}, as these were considered the best by the experts. The *STD* and mean were calculated for each quantitative metric. For the NCC, *26,000* (number of images in the dataset * 100 epochs) comparisons were made and a value of 0.6752 ± 0.006 was obtained, where the first value is the mean and the second is the *STD*. The closer the NCC value is to *1*, the more realistic the images produced. For the MS-SSIM, only *50* comparisons were made for both the generated and real data, because the computational power required was much higher than for the other metrics. A stack size of *2* was used to calculate the MS-SSIM value. For the real images, the mean and *STD* values were 0.9822 ± 0.0003 and for the generated images 0.7258 ± 0.0054, respectively. The MS-SSIM value should be identical to the MS-SSIM value of the real dataset to confirm that the distribution is similar.

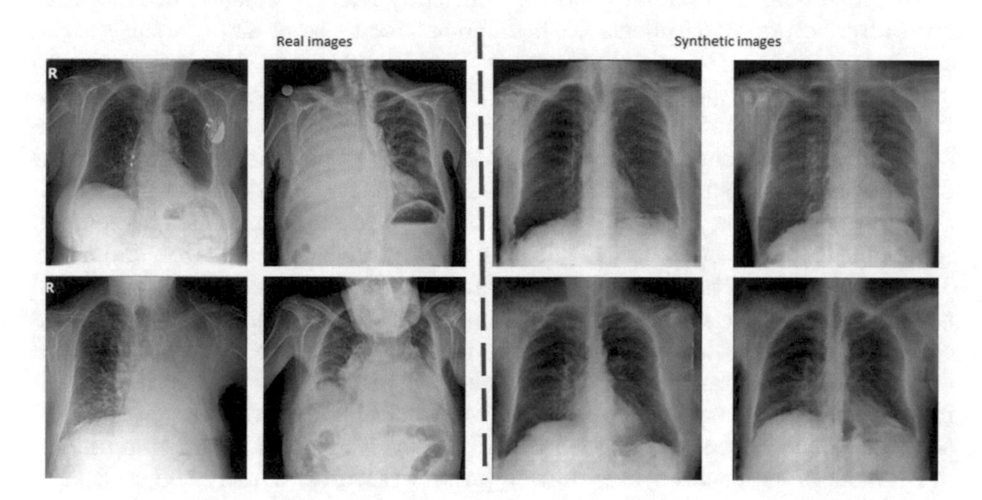

Fig. 4. Real and synthesized X-ray images from the model M_{1024}.

In the loss function plots of the M_{1024} (Fig. 5), it can be seen that the training of the discriminator was stable, and after 80,000 iterations, the training of the generator also stabilized. This stabilization was possible due to the use of the alpha-GAN architecture, which avoids mode collapse caused by the divergence between the loss function diagrams of the discriminator and the generator, one tending to -∞ and the other to +∞.

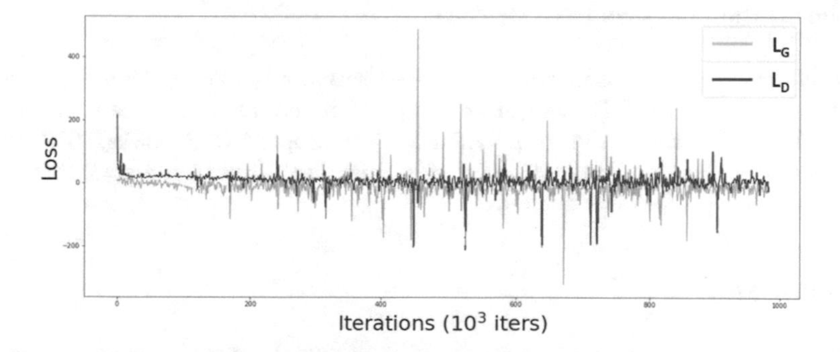

Fig. 5. Discriminator (red) and Generator (orange) loss function plots of the M_{1024} training process. (Color figure online)

6 Discussion and Conclusion

Analysis of the results provided by the experts suggests that the models did not learn properly and could not produce sufficiently realistic images. The fact that even after several experiments with different latent vector sizes, architectures, and preprocessing, the images shown in Figs. 3 and 4 were the best is due to problems with the data set. One of the main problems was the fact that the images were all from different sources, with different types of acquisition, preprocessing and dimensions. There were also images of patients with and without pathologies, and some images had different artifacts, such as the letter R in the upper left corner, medical instruments, the presence of too much fat mass in some images, the presence of a single lung or lungs with different sizes, and more. The fact that the dataset was so diverse and contained so many artifacts, as well as the small number of images, meant that the developed models could not learn correctly, because if they learned the features of one image correctly, they did not learn the features of other images correctly. It was also observed that the models focused on less important features in their learning process, such as the fact that some of the generated images had pathologies. However, the models did not focus as much on the more important features, namely the realism of the lungs and ribs, which should be sharper, more uniform, and smoother. The quantitative results show that the real and generated MS-SSIM values are not very close, just as the NCC value is not very high. In this way, both metrics agree with the results of the Turing test, i.e., the synthesized X-ray images are not realistic enough.

One of the goals of this study was to use generative models to produce synthetic X-rays with difficult-to-detect fractures to avoid the cost of expert labeling and difficult public access to X-rays with these features. Although, it was shown that a model with an alpha GAN-based architecture can produce realistic X-rays and thus can be used as a data augmentation technique, the main objective was not achieved, because experts are still needed to validate/guarantee that the synthetic X-rays have fractures.

Acknowledgments. We acknowledge the Plattform für KI-Translation Essen (KITE) project from the REACT-EU initiative (https://kite.ikim.nrw/). This work was also supported by FCT within the R&D Units Project Scope: UIDB/00319/2020. Finally, we thank the Gulbenkian Foundation for New Scientific Talent in Artificial Intelligence, under which this project was started.

References

1. Chowdhary, C.L., Acharjya, D.P.: Segmentation and feature extraction in medical imaging: a systematic review. Procedia Comput. Sci. **167** (2020)
2. Artificial Intelligence and Machine Learning in Software as a Medical Device FDA. https://www.fda.gov/medical-devices/software-medical-device. Accessed 4 May 2021

3. Artificial Intelligence Act Think Tank European Parliament. https://www.europarl.europa.eu/thinktank/en/document/EPRS_BRI(2021)698792. Accessed 4 May 2021
4. Nalepa, J., Marcinkiewicz, M., Kawulok, M.: Data augmentation for brain-tumor segmentation: a review. Front. Comput. Neurosci. **13** (2019)
5. Shorten, C., Khoshgoftaar, T.M.: A survey on image data augmentation for deep learning. J. Big Data **6**, 1–48 (2019)
6. Venu, S.K., Ravula, S.: Evaluation of deep convolutional generative adversarial networks for data augmentation of chest X-ray images. Future Internet **13** (2021)
7. How to Implement the Frechet Inception Distance (FID) for Evaluating GANs. https://machinelearningmastery.com/how-to-implement-the-frechet-inception-distance-fid-from-scratch/. Accessed 6 May 2021
8. Prezja, F., Paloneva, J., Pölönen, I., Äyrämö, S.: DeepFake knee osteoarthritis X-rays from generative adversarial neural networks deceive medical experts and offer augmentation potential to automatic classification. Sci **12** (2022)
9. Ferreira, A., Magalhães, R., Meriaux, S., Alves, V.: Generation of synthetic rat brain MRI scans with a 3D enhanced alpha-GAN (2021)
10. Ferreira, A., Li, J., Pomykala, K.L., Kleesiek, J., Alves, V., Egger, J.: GAN-based generation of realistic 3D data: a systematic review and taxonomy. arXiv (2022)
11. Goodfellow, I.J., et al.: Generative adversarial nets, vol. 3 (2014)
12. Rosca, M., Lakshminarayanan, B., Warde-Farley, D., Mohamed, S.: Variational approaches for auto-encoding generative adversarial networks (2017)
13. Candemir, S., et al.: Lung segmentation in chest X-rays using anatomical atlases with nonrigid registration. IEEE Trans. Med. Imaging **33**, 577–590 (2014)
14. Kermany, D.S., Zhang, K., Goldbaum, M.: Labeled optical coherence tomography (OCT) and chest X-ray images for classification (2018)
15. Jaeger, S., Candemir, S., Antani, S., Wáng, Y.-X.J., Lu, P.-X., Thoma, G.: Two public chest X-ray datasets for computer-aided screening of pulmonary diseases. Quant. Imaging Med. Surg. **4**, 475 (2014)
16. Bustos, A., Pertusa, A., Salinas, J.-M., de la Iglesia-Vayá, M.: PadChest: a large chest X-ray image dataset with multi-label annotated reports. Med. Image Anal. **66** (2019)
17. Paszke, A., et al.: PyTorch: an imperative style, high-performance deep learning library. In: Advances in Neural Information Processing Systems 32, pp. 8024–8035. Curran Associates, Inc., Red Hook (2019)
18. Agrawal, N., Katna, R.: Assessment of cutting forces in machining with novel neem oil-based cutting fluid. In: Mishra, S., Sood, Y.R., Tomar, A. (eds.) Applications of Computing, Automation and Wireless Systems in Electrical Engineering. LNEE, vol. 553, pp. 859–863. Springer, Singapore (2019). https://doi.org/10.1007/978-981-13-6772-4_74
19. Mathieu, M., Couprie, C., LeCun, Y.: Deep multi-scale video prediction beyond mean square error. In: 4th International Conference on Learning Representations, ICLR 2016 - Conference Track Proceedings (2015)
20. Sims, S.D.: Frequency domain-based perceptual loss for super resolution. In: IEEE International Workshop on Machine Learning for Signal Processing, MLSP 2020, September 2020 (2020)
21. Loshchilov, I., Hutter, F.: Decoupled weight decay regularization. In: 7th International Conference on Learning Representations, ICLR (2019)

Children's Hearing Loss Assessment Application Development and Execution Processes Using Systems Engineering Approach

Syed Nasirin[1](✉), Abdul Kadir[2], Esmadi A. A. Seman[1], Hadzariah Ismail[1], Abdullah M. Tahir[1], Suddin Lada[1], and Soffri Yussof[1]

[1] Universiti Malaysia Sabah, 87000 Labuan, Malaysia
dssrg@ums.edu.my
[2] Universitas Sari Mulia, Banjarmasin City, Indonesia

Abstract. This study discusses the challenges faced by children with impaired hearing, particularly in regions like East Malaysia (e.g., Sabah), where limited access to full hearing tests at primary referral centers leads to delays in receiving necessary care due to a shortage of audiologists. To address this issue, audiologists suggest implementing an integrated online hearing loss assessment system within their existing appointment system. The proposed online intervention aims to reduce delays and improve access to care for children with hearing issues. The methodology includes requirements engineering, functional analysis, functional allocations, design synthesis, system testing and verification, system execution, system maintenance, and system training. Case descriptions further detail the collaborative process involved in developing and executing the system, highlighting iterative refinement as a critical element.

Keywords: Audiology · Child Hearing · Children Hearing Loss · Hearing Loss Assessment · Systems Development · Systems Execution · Systems Engineering

1 Introduction

Children with impaired hearing may encounter challenges across different domains of life, including communication with family members, academic performance, and enjoyment of auditory activities. Despite the existence of hearing screenings, their impact on public health remains relatively constrained [2, 10]. Additionally, there is a lack of established management strategies for children with hearing loss, resulting in prolonged wait times and unequal access to care.

To address this issue, the audiologists at the referral center propose implementing an online hearing loss assessment system and integrating it into the existing appointment system by adding a hearing loss checklist. The literature on children's hearing loss, particularly in developing countries, is also reviewed. The methodologies used to develop the proposed online intervention are discussed, along with case studies and relevant discussions. The study concludes that implementing an online hearing loss assessment

A. Rocha et al. (Eds.): WorldCIST 2023, LNNS 799, pp. 298–307, 2024.
https://doi.org/10.1007/978-3-031-45642-8_30

system can help alleviate the long wait times and improve access to care for children with hearing concerns. Future research can explore strategies to improve the screening and management of children with hearing loss in developing countries.

2 Review of the Literature

In many instances, children with hearing loss are behind their coequals in speech and growth, often capping their participation in educational movements [2]. Hearing loss involves one in six people in the UK and is a noteworthy disease burden [10]. In addition to communication troubles, there is also an association with unhappiness. With targeted audiological examination, a clinical inspection can determine the characteristics and reasons leading to hearing loss [18].

With an appropriate intervention coupled with contemporary hearing technologies, children with hearing loss can still develop their speech capability at rates proportionate to their hearing peers. Hence, an early intervention application that furnishes thorough information to families and authorizes talented professionals should be devised to attain satisfactory outcomes for these children.

Children's hearing loss screening management applications must be fully developed for Malaysia to avoid considerable losses of children's hearing. For example, an application with a hearing loss assessment capability may detect hearing-loss degrees in various frequency bands for both ears [5]. Nonetheless, the conventional face-to-face hearing test method demands an audiologist to perform the test process, which hinders the patient's cost and travelling time.

The hearing loss assessment application always measures how well a child can hear. Although hearing loss can occur at any age, hearing troubles in children can have profound effects. That is because average hearing is paramount for language growth in babies. Hence, even a less severe hearing loss can make it difficult for a child to comprehend spoken language.

Hearing loss screenings may have limited documented impact as Malaysia (i.e., the State of Sabah, in particular) is still in its infant stage in combating the sickness. Even a single case should be considered one public health achievement. On the other hand, the hearing loss screening assessment would boost the public's confidence in the community. The online children hearing loss assessment systems (CHLAS) have eased these potential dangers and enabled the multi-agency partnership to combat the illness.

2.1 Consultation Team

Systems consultants provide technology advice and maintenance to support proprietors and users in planning, configuring, and operating their systems. They analyze needs, recommend solutions, and offer practical fixes for issues. Consultants also act as catalysts in information system projects, steering groups, and approving project proposals. Their key responsibility is to provide strategic advice based on their understanding of the client's requirements and industry developments.

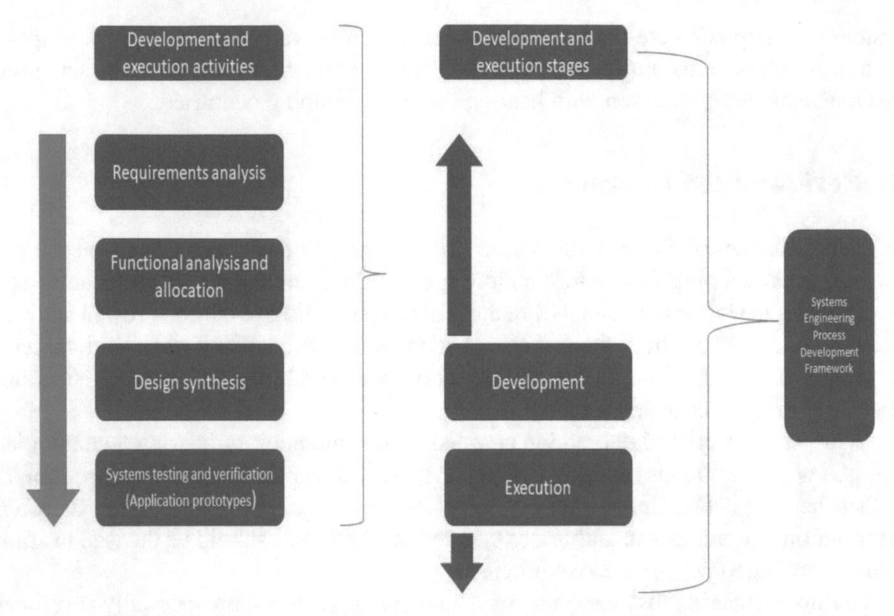

Fig. 1. The theoretical framework formed based on the familiar themes found in the extent of local healthcare.

2.2 Requirements Engineering

Requirement engineering defines and maintains the client's needs [17]. In addition, it is a process of specifying the assistance supplied by the system. Deciding on the requirement of the owners and users is one of the problematic yet crucial parts of developing and executing an IS. Identifying the correct requirements can demonstrate to be evasive. In addition, forming a set of requirements based on departmental purposes takes much work. In addition, a few studies infer that systems execution victory is attributed to user requirements identification [4, 12, 14]. The requirements engineering process consists of the following main activities; 1) Requirements elicitation, 2) Requirements specification, and 3) Requirements confirmation.

2.3 Functional Allocations, Design Synthesis, Systems Testing and Integration

Functional analysis is the next step in the systems engineering process. The functional analysis diverges a system into smaller parts, known as functional elements, illustrating what one would like each design and development element to do [6]. System testing also referred to as system-level examinations, is the process in which a development team assesses how the various segments of an application interact in the integrated system or application. It is the quantifying method of verification required to verify the system design. On the other hand, system integration ensures the integration of the newly developed application's functional and physical interfaces. Furthermore, it ensures that system design reflects the needs of all system elements (i.e., hardware, software, facilities, people, and data). During this activity, the systems engineering guideline focuses on integrating the functional specialties to ensure that every vital omission has been made that shall render the system less practical than it might otherwise be [11].

In short, much has been written about systems design, development and execution in various established children hearing loss applications. This is because the solutions to these questions permit one to bypass the problems upon executing the systems in question. Nonetheless, given the diverse nature of children hearing loss application in Sabah, results obtained from previous studies are likely to provide only a partial picture of the key issues surrounding the true system's development and execution. Furthermore, many characteristics contribute to these children hearing loss applications. Nevertheless, unfortunately, the results of these prior studies are still inconclusive. A wider view of what comprises the system development and execution approach is thus needed (see Fig. 1).

3 Theoretical Underpinning

The gist of the theoretical framework for encapsulating the children hearing loss assessment system is guided by the systems engineering process (SEP) development framework [7] instead of using the user-centered design approach. This is a technique endorsed by many studies [7, 8, 13], which brings user considerations into application development initiative. The SEP development framework was preferred as it can furnish a structure that sets clear limits for matters to venture out of the vast array of topics that impinge upon the development and execution processes. The framework characterizes two key segments in two development perimeters: 1) external and 2) internal. The external atmosphere evolves around the external surroundings that may mold the proposed system (i.e., the need to apprehend technological constraints). The internal environment consists of development and execution issues that may shape the development and execution of the system (i.e., the necessity to integrate with other systems). The framework provides a clear structure for solving the design and development problems of the CHLAS.

Moreover, one of the common assumptions made by the supporters of a user-centered approach is that system developers have unlimited access to users. This is achieved only in those organisations producing large bespoke systems and consequently have access to those end-users. The SEP development framework solves this acute proprietor and user access problem by inviting them over to be the main part of the development and execution team. Nevertheless, hospital administrators and their specialists, such as those audiologists, may like to explore new ways to make judgments or may want a better approach to handling their system development and execution processes.

On the other hand, while scholars have addressed system development and execution issues, there have been different interpretations of the terms. From a technological diffusion perspective, execution is an effort to diffuse appropriate Information Technology (IT) within the user community [9]. They offer a stage model of IT development and execution movements based on Lewin's change model [3, 15], which incorporates six principal stages (initiation, adoption, adaptation, endorsement, routinization and infusion).

Traditionally, the process of system development and execution has been perceived as starting with system and design efforts, and concluding once the system is functional.

However, it's important to acknowledge that this process also involves assessing actions post-execution. Agarwal & Lucas [1] assert that the execution of computer-based information systems is an ongoing endeavor, encompassing the entire system development lifecycle. In this investigation, "development and execution" denote a continuous process spanning from the pre-execution stage (e.g., planning) to the post-execution stage (e.g., maintenance).

4 Research Methodology

Stage 1 (algorithm development): The assessment model for children's hearing loss shall be developed using the principal components analysis (PCA) method. PCA is a statistical procedure that permits one to summarize the information in extensive data tables utilizing a smaller set of "summary indices" that can be easily interpreted. PCA is very useful for speeding up computation by reducing the dimensionality of the data. In addition, when one has high dimensionality with the highly correlated variable one, the PCA can enhance the accuracy of a classification model.

Stage 2 (requirements engineering): The outcome from Stage 1 shall be embedded into the proposed system design and development. At this stage, key activities such as requirements engineering, functional analysis and allocation and design synthesis will occur. The first step is to examine the inputs required for the system's requirements. Next, needs analysis is used to develop functional requirements; it was rephrased into requirements representing what the system must do. Subsequently, the system executors will ensure that the requirements are coherent and concise (Stage 3). Finally, requirements analysis must refine and define the proposed system's requirements and design constraints.

On the other hand, functions are analyzed by decomposing higher-level functions identified into lower-level functions under the functional analysis and allocation substage. Moreover, the lower functions have defined the application (CHLAS) according to what the application's performance requires. In addition, the functional analysis and allocation performance results in a better insight into the requirements and should encourage reconsidering the needs analysis. This iterative process of revisiting requirements analysis resulting from functional analysis and allocation is called the requirements loop. Finally, the last part of this stage is the application's design synthesis. It defines the application in terms of the physical elements (i.e., producing the functional architecture of the system).

Stage 3 (test and verification): The outcome from Stage 2 will go through a test and verification stage. System testing is the detailed quantifying verification method, and it is required to verify the system design and development. System testing is how objective judgments are made, while verification aims to inspect, examine, and estimate data obtained and other means to make systematic decisions.

Stage 4 (dissemination): The dissemination phase will commence once the research project has entered the verification phase. New scientific knowledge within the children hearing loss domain will be disseminated initially to the main beneficiaries of the research work through a series of workshops and executive development programs (EDPs). Other main dissemination channels are regional conferences and academic and practitioner journals related to hearing loss assessments. In addition, web documents will also be placed on UMS's Faculty of Computing & Informatics web server, describing the research work and its main results that interested public members can apprehend. It is not anticipated that the results of this work will immediately be commercially exploitable. Still, the research team will consult with the university's commercialization office to safeguard intellectual property rights.

Before the full hearing test, the children must be first assessed at the local health clinics (i.e., "Klinik Kesihatan"). However, there is only a manual assessment system available at these clinics. Accordingly, the medical officers aid the children with hearing problems by going through the hearing assessment set by the hospital. Unfortunately, this manual approach has thus far caused a tremendous number of appointment delays as those audiologists are absent on-site. Hence, the audiologist at the hospital believes that the long delays can be minimized by executing an online hearing assessment system.

The audiologists are trained to test children's hearing status. They can do many tests to determine if a person has a hearing loss, how much of a hearing loss there is, and what type it is. For instance, hearing screening. It is a test to tell if the children might have hearing loss. Accordingly, this research seeks to carefully formulate hearing assessment algorithms embedded in a web-based system to assess the children's hearing status.

5 Case Descriptions and Analysis

The Audiology Unit, part of the Department of Otorhinolaryngology at HQE, is the primary center for audiology services across multiple hospitals in Sabah and Labuan. Their services include hearing screening and rehabilitation for those with auditory function problems and being a referral center for Cochlear Implants. However, there has been a long waiting time of about ten months for infant and child patients seeking appointments. Many of these patients are late speech cases, and delays in addressing hearing issues can negatively affect learning, reading, and cognitive function. To address this issue, the Child Hearing Loss Appointment System (CHLAS) was developed to simplify and speed up the appointment process for deaf and hard-of-hearing children in the Kota Kinabalu area. This online application helps health clinic staff prepare appointment data and obtain previous condition history information, ultimately saving time for all involved.

5.1 Consultation Team Collaborations

Developing a web-based Children's Hearing Loss Assessment System (CHLAS) involved five major teams. The first team consisted of systems consultants and developers from a local university with knowledge of CHLAS, who was brought in to aid with the basic set-up of the system. Next, the system's owner, led by a chief audiologist and audiologist, promoted the system's benefits to the state health officials and the hospital's senior managers. Finally, the state health officials, led by the public health director, were happy with the system's proposed benefits and approved its development and execution to reduce the number of children with hearing loss in the state.

End-users, mainly medical officers, and senior nurses from local clinics within the City of Kota Kinabalu, played a critical role in the system's requirements, design, synthesis, and verification stages. They were also instrumental in supporting the development and execution of the CHLAS. In addition, officials from the hospital's IT Department monitored the application's development to ensure it was in tandem with other applications and requested integration through a single sign-on mode. They also provided hardware support, such as application and web servers, for the CHLAS to be properly executed. The development and execution effort of the CHLAS prototype utilized the Joint Application Design (JAD) approach. An application prototype was created to evaluate the technical aspects of CHLAS and accommodate the proprietors' and users' requirements.

5.2 Requirements Engineering

The introduction of the application prototype has changed the focus of owner and user perspectives as it supplies a complete evaluation during the design, development and execution processes. Nonetheless, it can be challenging for system consultants and developers if they must throw away several months of hard work. Thus, it is significant to realise that the prototype was not an end application.

A CHLAS development and execution framework was established through a series of requirements engineering. Discussions were held with the system proprietors, end-users, hospital administrators, hospital ICT heads, and state ICT heads and observations were made on the children's hearing loss assessment system. The audiology head ascertained continuous formal and informal support was held with users in their attempt to improve the system.

A common stumbling block took time for the consultants and developers to identify the system owners, end-users, hospital administrators, hospital ICT heads, and state ICT heads' needs. Thriving CHLAS Execution happens when owners and users perceive that their systems are functional for their hospital because they are easy to embody. In addition, a CHLAS is likely to be successful when owners and users perceive that the system provides extra benefits than their previous approach to hearing loss assessments (i.e., reducing the length of time for the children with hearing loss concerns to see the audiologists).

5.3 Functional Allocations, Design Synthesis and System Testing

Functional allocation issues are technical issues of the CHLAS development process. The functions were focused on the design of the user applications. It was designed to include pull-down menus and graphical user interfaces (GUIs) that will meet the needs of all potential users. Users reported problems and dissatisfaction and requests for improved features to system consultants and developers responsible for managing the progress effort. System testing means testing the system. All the segments are integrated to verify whether the system works as expected. System testing was initiated once the application prototype was launched to the users. The testing was in the form of a front-end interface examination done by the system's users and owners. At the same time, the hospital's IT Department was inspecting the back-end system configuration.

5.4 System Integration

System integration (SI) is an initiative that joins different subsystems or features as one large system and element crucial for the hospital as it hosts many applications (Olivieri, Bocchi, Rizzo, Dobre, & Xhafa, 2016). It ensures that each merged subsystem functions as needed. SI is also used to add value to a system through new functionalities provided by connecting functions of different systems. In this instance, the system integration effort was aimed at streamlining CHLAS and the hospital's other applications. As a result, the hospital ICT team were initially concerned about the integration possibility of the proposed application. Yet, the concerns were eliminated as their understanding towards CHLAS has grown.

6 Discussions

The Systems Engineering Process (SEP) was developed in the 1970s to provide a structure for the development and execution of extensive government systems. The iterative process utilizes requirements and design loops for system developers to revisit partially completed segments. However, the success of developing and executing a functional CHLAS (Comprehensive Hospital Laboratory Automation System) depends on conveying engineering requirements that accurately reflect the needs of all stakeholders. Therefore, stakeholder feedback is critical to the process, and a division cannot exist between owner needs, user needs, and how system consultants and developers interpret those requirements.

Traditional systems development and execution involve lengthy requirements and functional analyses, but this approach is unsuitable for a dynamic system like CHLAS. Instead, the successful development and execution of CHLAS require a diverse approach, where system consultants and developers start with a small but discernible segment that can be incorporated into an application prototype. This prototype may be flawed, but it garners the curiosity of owners and users.

The development and execution of the CHLAS requires a unique approach compared to other information systems projects in hospital settings. The consultants and developers of the system acknowledge that owners have limited time available to support the project due to their busy schedules. Our examination of both the field and existing literature reveals that the expansion of the system in question is restricted by six fundamental concepts (see Fig. 2).

7 Conclusions

This article presents a case study of a Malaysian audiology unit's experience developing and executing the Children's Hearing Loss Assessment System (CHLAS). The study utilized two phases of execution activities to capture the development and execution initiatives and answer the research questions on how the audiologists and their team members executed the system. The CHLAS can offer confidence to the public, especially during a pandemic, by providing systematic hearing loss screening under strict infection

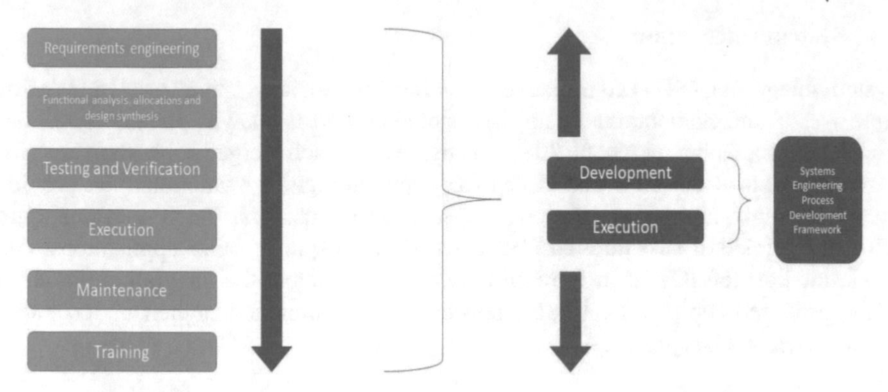

Fig. 2. The emerging development and execution framework of CHLAS

control measures. The study's results can benefit academic researchers and have long-term implications for public health communities. It can serve as a basis for further investigation in developing and executing CHLAS in other countries. Additionally, the study's results are of direct interest to international organizations such as the World Health Organization (WHO), as Malaysia is obliged to control children's hearing loss. It is important to note that the case studies involved Malaysian health organizations, and replications in other nations are appealing.

References

1. Agarwal, R., Lucas, H.C., Jr.: The information systems identity crisis: focusing on high-visibility and high-impact research. MIS Q. 381–398 (2005)
2. Bobsin, L.L., Houston, K.T.: Communication assessment and intervention: implications for pediatric hearing loss. Otolaryngol. Clin. North Am. **48**(6), 1081–1095 (2015)
3. Burnes, B.: Kurt Lewin and the planned approach to change: a re-appraisal. J. Manage. Stud. **41**(6), 977–1002 (2004)
4. Calkins, J., et al.: Sharing of geographic information: research issues and a call for participation. I-9 Specialist Meeting Report. National Center for Geographic Information an Analysis, Santa Barbara, CA (1991)
5. Chen, F., Wang, S., Li, J., Tan, H., Jia, W., Wang, Z.: Smartphone-based hearing self-assessment system using hearing aids with fast audiometry method. IEEE Trans. Biomed. Circuits Syst. **13**(1), 170–179 (2018)
6. Chen, Y.M., Fan, K.S., Chen, L.C.: Requirements and functional analysis of a multi-hazard disaster-risk analysis system. Hum. Ecol. Risk Assess. Int. J. **16**(2), 413–428 (2010)
7. Doolin, B., McLeod, L.: Sociomateriality and boundary objects in information systems development. Eur. J. Inf. Syst. **21**(5), 570–586 (2012)
8. Haberfellner, R., Nagel, P., Becker, M., Büchel, A., von Massow, H.: Systems Engineering, p. 5. Springer, Cham (2019)
9. Kim, S.B., Kim, D.: ICT implementation and its effect on public organisations: the case of digital customs and risk management in Korea. Sustainability **12**(8), 3421 (2020)
10. Lee, J.W., Bance, M.L.: Hearing loss. Pract. Neurol. **19**(1), 28–35 (2019)
11. Nielsen, C.B., Larsen, P.G., Fitzgerald, J., Woodcock, J., Peleska, J.: Systems of systems engineering: basic concepts, model-based techniques, and research directions. ACM Comput. Surv. (CSUR) **48**(2), 1–41 (2015)

12. Campbel, H.: Theoretical perspectives on the diffusion of GIS technologies. In: GIS Diffusion, pp. 23–45. CRC Press (2020)
13. Lee, J., Runge, J.: Adoption of information technology in small business: testing drivers of adoption for entrepreneurs. J. Comput. Inf. Syst. **42**(1), 44–57 (2001)
14. Pinto, J.K., Onsrud, H.J.: In search of the dependent variable: toward synthesis in GIS implementation research. In: Geographic Information Research, pp. 129–145. CRC Press (2020)
15. Shulga, L.V.: Change management communication: the role of meaningfulness, leadership brand authenticity, and gender. Cornell Hosp. Q. **62**(4), 498–515 (2021)
16. Sarker, S.: Toward a methodology for managing information systems implementation: a social constructivist perspective. Inf. Sci. **3**, 195 (2000)
17. Stroh, F., Winter, R., Wortmann, F.: Method support of information requirements analysis for analytical information systems. Bus. Inf. Syst. Eng. **3**(1), 33–43 (2011)
18. Xia, S., et al.: Altered brain functional activity in infants with congenital bilateral severe sensorineural hearing loss: a resting-state functional MRI study under sedation. Neural Plast. (2017)

Smartphones and Wristbands Detect Stress as Good as Intrusive Physiological Devices

Gema Bello-Orgaz[1] and Héctor D. Menéndez[2](✉)

[1] Departamento de Sistemas Informáticos, Universidad Politécnica de Madrid, ETSI de Sistemas Informáticos, Madrid, Spain
gema.borgaz@upm.es

[2] King's College London, Department of Informatics, London, UK
hector.menendez@kcl.ac.uk

Abstract. Detecting stress timely is relevant to help people's mental and general health. Multiple researchers explored different aspects of this problem by analysing physiological responses of the sympathetic nervous system (changes in heart rate, breathing rate variability, skin temperature or conductance) or by studying the person's activity levels throughout the day by taking data from their smartphone use. Currently, smart devices can provide relevant information associated with stress depending on their usage. Combined with machine learning systems, it is possible to obtain high accuracy levels in detecting stress, some reaching up to 90% under ideal conditions. This study aims to detect stress at different intrusion levels (low for smartphones and medium or high for devices recording physiological data) to focus on reducing intrusiveness and keeping a reasonable accuracy level. Our stress detection models obtain up to 75% in the wild for low intrusion and 97% for medium or high intrusion levels by using public datasets. Moreover, high intrusion devices do not improve the quality significantly with respect of the medium intrusion ones (between 1 and 3 points). These promising results show that, by using only smartphones and wristbands, we can obtain confident detection, similar to invasive methods, by providing a non-intrusive procedure to help millions of people to detect and deal with stress.

Keywords: Stress Detection · Mental Health · Smartphones · Wearable Device · Machine Learning · Classifiers

1 Introduction

The World Health Organisation (WHO) declared mental health issues as one of the main causes of disability in Europe [1]. In fact, the current public health crisis caused by COVID-19 is leading to anxiety and depression in the short term, aggravating existing mental health issues, and generating new clinical cases in the medium and long term [2]. This effect is remarkable among teenagers, where

A. Rocha et al. (Eds.): WorldCIST 2023, LNNS 799, pp. 308–319, 2024.
https://doi.org/10.1007/978-3-031-45642-8_31

suicide became one of the main causes of death in several European countries [1].

According to multiple studies, some habits can lead to mental illness. For instance, stress leads to depression under specific circumstances [3]. Considering the consequences of depression, it is important to study how different social factors increase unnecessary stressful situations. Under these circumstances, the accumulation of stress can hurt multiple people in our society [4]. Multiple studies focused on detecting stress via new technologies such as smart devices. Data collection from smartphones usage and wearable devices (such as smartwatches or smart wristbands) can be a great opportunity for early detection of mental health-related problems, especially stress [5], because people tend to carry them around as an integral part of their lives in modern society, and they do not find them uncomfortable or invasive like other more accurate but intrusive devices that can measure health status, such as head-bands [6].

These smart devices collect sensor data from people that can go through different levels of intrusion. For instance, when the data is collected from a mobile phone, it takes information from the different applications that are opened, the GPS information, the WiFi, and Bluetooth, among others. However, it is also possible to recover information from the pulse, heartbeat or breathing rate. In this case, the device would be more intrusive, but the provided information would be more complete with respect to the general condition of the person or patient which is under analysis.

In this paper, we study how intrusion levels affect stress detection. We focus on studying different levels of intrusion: smartphones, wristbands and wearable devices. For reproducibility purposes, we acquire the data for each level from public sources (Sect. 2.2). For each of these levels, we construct prediction models based on classifiers (Sect. 3.2) that we evaluate according to our research objectives (Sect. 4). Our results show that using a low level of intrusion, i.e. smartphones, we obtain accuracy levels up to 85% accuracy in cross-validation and 75.1% in test, which is significantly good considering that in one of the case studies analysed with a high level of intrusion (where the stress is induced), the accuracy achieved is similar (75.6%). However, it is remarkable that the quality increment from a low to a medium level of intrusion is very significant, from 75.7% to 97.7%, while from medium to high the increment does not reach more than 1 point (Sect. 5). Therefore, according to the results, the level of intrusion that is most accurate for detecting stress is the medium (wristbands).

2 Mental State Detection

Modern technologies allow us to use sensory information to identify mental states. This led to multiple works in detecting anxiety and depression by using data from multiple devices that can read this sensory information (Sect. 2.1). Hickey et al. [5] collect some of the tools and approaches used in the field, although in this work we had studied different databases focused on detecting stress at different intrusion levels (Sect. 2.2).

2.1 Detecting Mental Health Issues from Sensors

Detecting mental health issues have been successfully performed with the use of electroencephalogram (EEG) devices [7]. These devices obtain high accuracy [5], but their level of intrusiveness is significant. However, and also considering that some of these devices are not widely available, it is important to identify other ways to detect stress that are not intrusive. For that reason, this study considers mobile phones and wearable devices (such as smartwatches or smart wristbands) as the data sources.

The way people interact with their mobile phones leaves a signature about their mental state [8]. Using this signal as a side channel to understand people's psychological state reduces the level of intrusiveness, although it can violate people's privacy if the data is not properly anonymised according to the legal standards provided by the GDPR. For that reason, during our research, we aim to use publicly available datasets that have been already prepared according to the legal and ethical standards.

Providing a ground truth for mental health conditions –as depression, stress, or anxiety– is challenging and requires experts that performed these tasks via surveys [9]. These surveys are subjective, considering that they strongly depend on people's self-perception. Also, the surveys aim to measure different levels of depression that might not be strongly correlated with the data extracted. For that reason, this work considers only the detection, in a binary way (Sect. 4).

During the detection process, some works apply machine learning to detect patterns within the sensory information. An example is the work of Alić et al. using neural networks for detecting stress [10] trained with data from Physionet database that achieves results with accuracy of 99%. In the work of Can et al. [11] that combine multiple classifiers to detect stress based on wearable devices (concretely smart watches), they obtain 90.40% accuracy by using Empatica E4 devices with high data quality, whereas the accuracy with Samsung S devices was 84.67%. Other options focus on mobile phones, such as the work of Ciman and Wac [12], who analyse how the users interact with their mobile phones and perform the prediction with different machine learning classifiers. Their global model has a F-measure of 70–80% showing how this method enables an accurate stress assessment without being too intrusive. Similarly, Vildjiounaite et al. [13] perform the detection from smartphone data using hidden Markov models which its accuracy was evaluated using two real-life datasets and achieving 59 and 70% accuracy, respectively. Unfortunately, several of these studies did not publish their datasets. Those who have a dataset published that we had identified during this work are described in Sect. 2.2.

Evaluating the different levels of intrusiveness that we need to identify mental health issues based on the sensor data requires checking different datasets that provide information from smartphones, smart watches, EEG devices, or other sensors. We have investigated several datasets previously used in the literature for this purpose.

2.2 Public Databases Based on Smart Devices

Several databases can be found in the literature related to multiple sensory information and prediction targets. During this research, we have identified databases that use smartphones or wearable devices data to predict the users' state. Considering that several of the identified datasets focus on depression or other mental states, and that we focused this study on stress, many of them have been discarded. For that reason, we have selected the datasets of StudentLife as the less intrusive one, due to its features relate only to mobile devices, and PASS and WESAD as the most intrusive, because their features are extracted from wearable physiological devices:

- **StudentLife**: This dataset provides a complete analysis of different levels of stress based on smartphone activities of 48 students across a 10 week term at a college. The researchers used the Ecological Momentary Assessment (EMA) for measuring the level of stress [14].
- **PASS**: This database provides different stress scenarios for users, where they play video games. The data is collected through different wearable devices. The stress level is measured with a questionnaire answered at the end of each experiment [6].
- **WESAD** (Wearable Stress and Affect Detection): This dataset collects physiological data from wrist and chest devices of 15 subjects of performing 3 different types of tasks (from non-stressful tasks such as reading magazines and watching video clips to stressful tasks such as public speaking and mental arithmetic tasks). The mental states are measured with different self-assessment questionaries, concretely: Positive and Negative Affect Schedule (PANAS), State-Trait Anxiety Inventory (STAI), Self-Assessment Manikin (SAM) and Short Stress State Questionnaire (SSSQ) [15].

3 Detecting Mental Health Issues

Detecting mental health issues is a sensitive problem where the data needs to be selected appropriately to respect the user's privacy. For that reason, during this set of experiments, we guarantee that all the data are anonymous (Sect. 2.2). This section explains how the data has been preprocessed (Sect. 3.1), and the classification algorithms used during the analysis process (Sect. 3.2).

3.1 Data Aggregation and Preprocessing

Considering the selected datasets (StudentLife, PASS and WESAD), we have constructed aggregated features in order to manage the information provided:

- From **StudentLife**, we performed a feature selection process allowing us to aggregate multiple features from different perspectives: app usage, sensing (activity, noise, GPS, bluetooth, WiFi, light, phone locked, and charging), call logs, and sms logs. These aggregations perform averages of these features during a specific time frame (we considered 1, 2, 5, 10 and 24 h).

- In **PASS** case, different devices extract signals from the subjects: an Empatica E4 wristband and a BioHarness3 chest band. The Empatica E4 device provides the following sensor data: blood volume pulse (BVP), electrodermal activity (EDA), body temperature, and three-axis acceleration. And the BioHarness3 device provides data about electrocardiogram (ECG) and breathing rate (BH).
- **WESAD** dataset was recorded with two devices as well: a chest band (RespiBAN) and a wristband (Empatica E4). The data obtained using the Empatica E4 are the same, however in the case of the RespiBAN the sensor data provides the following: ECG, EDA, electromyogram (EMG), BH, body temperature, and three-axis acceleration. The aggregations of all data from both datasets include the mean, standard deviation, maximum and minimum.

In the case of PASS and WESAD there are no time frames since they correspond to the duration of the test cases carried out. Once the data is properly aggregated we have performed the following preprocessing before the training process:

1. We have removed every constant column, considering that they do not contribute to the learning process.
2. We have transformed every categorical feature into a binary representation for StudentLife. This mainly affects the applications used by the users.
3. We have normalized the data using a min-max normalization process [16].
4. We have balanced the classes by using resampling, considering that we have more instances of one class on top of the other. The resampling methodology that we have used is the Synthetic Minority Over-sampling Technique (SMOTE) [17].

After preprocessing, the number of features and size for each dataset was:

- StudentLife (at different times): 1071 for 24 h, 989 for 10 h, 958 for 5 h, 900 for 2 h, and 862 for 1 h. The original number of features was 1421. The number of rows is 2289.
- PASS (for each device): 26 for the Empatica E4 device and 8 for Bioharness3 device. The size of the final dataset, after the cleaning process, is 182 rows.
- WESAD (for each device): The same number for the Empatica E4 device (26) and 32 for RespiBAN device. The final number of rows is 1178.

3.2 Classifiers

Once the data is ready and considering that its nature is very diverse, we have tested multiple classifiers in order to evaluate which approach is more suitable to identify patterns. Our experiments used the following classifiers, whose description can be found in [16]: Logistic Regression, Linear Discriminant Analysis, Naïve Bayes, K-Nearest Neighbours, Decision Trees, Random Forest, Support Vector Machine, Neural Networks, Ada Boost, Gradient Boosting, and Extreme

Gradient Boosting. We have selected these classifiers because they cover a significant spectrum of the state of the art on classification methods, and they create different statistical models that can be adapted to understand people's behaviour.

4 Experimental Setup

For evaluating our study, we measure the abilities of different classifiers to identify stress on devices' data at different intrusion levels. Considering that the literature only allow us to identify three suitable datasets –StudentLife, PASS and WESAD– we have considered the former as the low level, as it uses information from mobile devices, while the second and the third one have a higher level of intrusion, as they uses information of the physiological responses. Our experiments focus on answering the following research questions:

RQ: Which Level of Intrusion is More Accurate to Detect Stress? We select 11 machine learning classifiers to evaluate their abilities to extract patterns from the three datasets. And we have defined 3 levels of intrusion depending on how intrusive are the devices used to obtain the data (see Fig. 1). These levels are low for mobile phones, medium for wristbands and high for chest bands.

Initially, we aim to answer **RQ1: What is the accuracy for each level of intrusion?** and then, **RQ2: Which dataset contribute the most to detect stress?** Depending on the dataset type used to train the classifiers, we can identify at which level of intrusion it is easier to obtain a good prediction, and which classifiers deal better with the specific datasets.

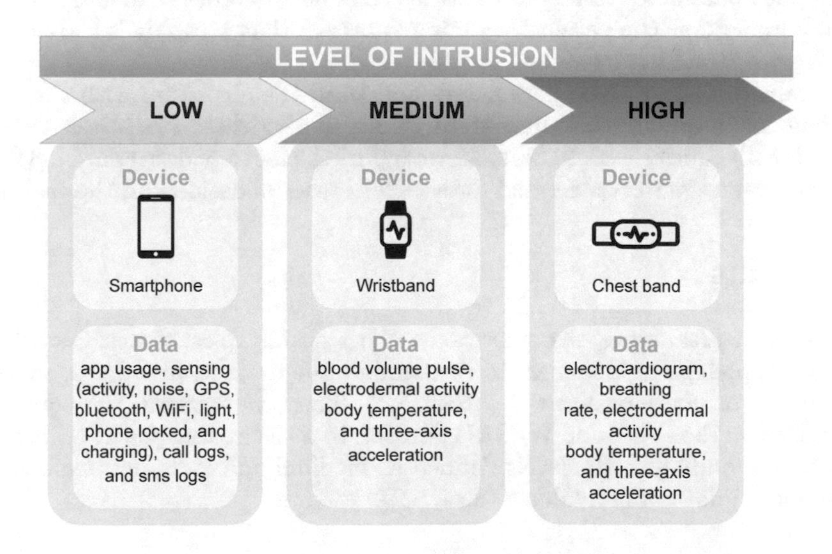

Fig. 1. Levels of intrusions depending on the devices used.

4.1 The Experimental Environment

All the experiments have been carried out in a cloud machine using an NVIDIA V100 card with a cloud instance of Ubuntu 22.04, 32 cores (Xeon Processor), 256 GB RAM and 400 GB of SSD hard drive. Our data has been stored in a Mongo dataset where the queries performed the aggregation. For reproducibility purposes, you can find the datasets, the queries and the classifiers in the following Github repository[1]. This information should be sufficient to perform the whole set of experiments, nonetheless, if you have any queries, please, contact the authors.

4.2 The Classifiers

The classifiers and the parameter tuning methods that we have used in the implementation correspond to the *scikit-learn* library in Python[2]. Our prediction method simplifies the detection process to a binary problem (stress, non-stress), where we consider that even a low level of stress is an indicator. The training process is divided into 3 steps: we divided the dataset into 80% for training and 20% for testing, selecting the data samples uniformly at random. Then, we use a 10-fold cross-validation on the training data to select the best classifiers. We select the 3 top classifiers in terms of accuracy:

$$Accuracy = \frac{TP + TN}{TP + TN + FP + FN} \quad (1)$$

At the same time, we measure the training performance of these classifiers. We applied parameter tuning to the three classifiers in order to identify the best set of parameters. The parameter tuning approach that we apply is Halving grid search with 10-fold cross-validation [18].

Finally, we apply these parameters and train the algorithm with the whole training set, applying this process 10 times, due to these statistical methods depend on the initial seed. In Sect. 3.2 we report the mean and standard deviation accuracy values per classifier and dataset for cross-validation and the test data.

5 Results

This section establishes the answers to the research questions in Sect. 4. We start by applying our classifiers to the StudentLife dataset, in order to measure the quality of non-intrusive data (Sect. 5.1). Then, we perform the equivalent evaluation on the PASS and WESAD dataset, to measure the quality of intrusive data corresponding to the levels defined as medium and high and dealing with physiological data of individuals (Sect. 5.2).

[1] https://github.com/hdg7/mentalhealth.
[2] https://scikit-learn.org/stable/.

5.1 Evaluating Non-intrusive Smartphone Data (StudentLife)

For the evaluation of the StudentLife dataset, we start by applying 10-fold cross-validation on the 11 classifiers described in Sect. 3.2. We also collect information about the individual features for the time-frames 1, 2, 5, 10 and 24 h before the stress measure is retrieved. Table 1 (top) shows the results for the cross-validation phase. The table highlights in bold the top 3 classifiers for the different datasets generated depending on the collection time. We can see that there is a consistent set of results for all the different time-frames, where the classifiers are ranked in a similar fashion.

Table 1. Mean accuracy values for the StudentLife dataset (low level of intrusion) and the 11 classifiers with respect to 10-fold cross-validation (top) and the three best classifiers results with respect to the test data (bottom). The numbers in brackets are the standard deviations and the best accuracy results obtained by each dataset are marked in bold.

Classifier	1 h	2 h	5 h	10 h	24 h
Decision Trees	75.9 (2.0)	75.0 (3.1)	75.5 (2.1)	74.5 (2.9)	76.1 (1.5)
K-Nearest Neighbours	72.4 (2.5)	72.2 (2.4)	74.4 (3.3)	73.1 (2.4)	72.9 (2.5)
Linear Discriminant Analysis	73.8 (2.4)	72.9 (1.7)	75.1 (2.9)	74.1 (2.3)	76.2 (2.6)
Naïve Bayes	63.4 (1.4)	64.7 (2.6)	64.6 (1.5)	65.6 (1.4)	66.1 (2.4)
Support Vector Machine	57.4 (4.0)	53.8 (1.8)	55.9 (4.2)	55.8 (5.6)	54.7 (3.1)
Logistic Regression	71.3 (2.5)	72.3 (3.5)	74.3 (2.1)	73.1 (2.2)	75.3 (3.3)
Random Forest	**84.9** (2.1)	**83.7** (1.8)	**85.5** (2.2)	**85.6** (2.3)	**84.8** (3.0)
Boost	78.5 (2.5)	79.4 (2.3)	80.2 (1.5)	80.2 (2.5)	79.9 (3.0)
Gradient Boosting	72.0 (2.8)	73.3 (1.9)	73.9 (2.5)	73.1 (2.7)	74.5 (1.5)
Extreme Gradient Boosting	**82.5** (1.6)	**82.5** (2.6)	**83.4** (1.5)	**83.8** (2.7)	**83.9** (2.7)
Neural Networks	**83.9** (2.9)	**83.4** (2.2)	**84.6** (1.6)	**85.1** (2.1)	**83.4** (2.5)
Random Forest	71.4 (0.6)	**73.4** (0.8)	**70.8** (0.7)	**73.1** (0.7)	**75.1** (0.5)
Extreme Gradient Boosting	69.2 (0.0)	73.1 (0.0)	69.2 (0.0)	72.1 (0.0)	73.6 (0.0)
Neural Networks	**71.7** (0.8)	70.2 (1.2)	67.4 (0.8)	72.1 (0.7)	73.1 (1.1)

The worst classifiers generally are Support Vector Machine (whose accuracy results range from 53.8% to 57.4%) and Naïve Bayes (ranging from 63.4% to 66.1%). Several classifiers obtain accuracy results ranging between 70% and 77% (Decision Trees, K-Nearest Neighbours, Linear Discriminant Analysis, Logistic Regression, and Ada Boost), while Gradient Boosting obtain results ranging from 78.5% to 80.2%. The top 3 classifiers in all cases are Random Forest [83.7%–85.6%], Extreme Gradient Boosting [82.5%–83.9%], and Neural Networks [83.4%–85.1%]. These classifiers are selected for the parameter tuning process using halving grid search with 10-fold cross-validation.

After the parameter tuning process, we used the test data for comparing (in terms of accuracy) the tuned classifiers with those created by using the default parameters (Table 1 bottom). The results show that the classifiers provide similar results, significantly worse than the cross-validation ones (as expected). In several cases, the tuning process does not contribute to the accuracy (in some cases it is even lower, or there is no a significant difference, such as for Neural Network or Random Forest). Therefore, based on these results obtained in the parameter tuning process, we use the test data obtained by the classifiers with the default parameters (their values can be found in the documentation for thc *scikit-learn* library in Python) to compare in terms of accuracy the results obtained. And, as can be seen from the results obtained in the test phase, any of these three classifiers looks promising for identifying stress with low intrusion levels, at any time-frame.

5.2 Evaluating Intrusive Physiological Data (PASS and WESAD)

For the evaluation of the PASS and WESAD datasets, we started by applying 10-fold cross-validation on the 11 classifiers described in Sect. 3.2 for each of the 2 feature vectors created in the pre-processing phase respective to each dataset (the Empatica E4 and BioHarness3 for PASS dataset and the Empatica E4 and RespiBAN for WESAD). Table 2 shows the results where the best values of accuracy for the different classifiers and datasets used are highlighted in bold.

Table 2. Mean accuracy values for the PASS and WESAD datasets (medium and high level of intrusion) and the 11 classifiers in 10-fold cross-validation (top) and the three best classifiers in test (bottom). The numbers in brackets are the standard deviations and the best accuracy results obtained by each dataset are marked in bold.

Classifier	PASS		WESAD	
	EmpaticaE4	BioHarness3	EmpaticaE4	RespiBAN
Decision Trees	74.2 (5.2)	73.5 (10.9)	96.5 (1.4)	93.8 (2.5)
K-Nearest Neighbours	73.4 (6.9)	70.1 (10.8)	97.7 (0.9)	97.5 (1.2)
Linear Discriminant Analysis	66.0 (4.3)	63.2 (11.9)	91.3 (2.0)	88.0 (1.5)
Naïve Bayes	63.3 (7.2)	56.3 (10.7)	80.7 (3.8)	72.5 (2.7)
Support Vector Machine	63.4 (3.2)	50.5 (6.0)	89.1 (3.5)	84.6 (2.5)
Logistic Regression	65.1 (5.3)	68.7 (7.4)	92.5 (2.5)	85.1 (2.8)
Random Forest	**80.2** (4.7)	**77.1** (11.8)	**98.6** (1.0)	**98.7** (1.1)
Gradient Boosting	**78.3** (3.2)	**76.6** (10.3)	**98.0** (1.2)	**97.4** (1.4)
Ada Boost	73.2 (5.8)	70.6 (8.4)	96.0 (1.8)	96.9 (1.4)
Extreme Gradient Boosting	**82.3** (3.7)	**77.1** (8.5)	**98.3** (1.0)	**97.5** (1.3)
Neural Networks	75.6 (3.4)	68.7 (12.7)	97.2 (1.5)	92.1 (2.1)
Random Forest	71.7 (2.2)	69.2 (1.3)	**97.3** (0.4)	96.9 (0.3)
Gradient Boosting	72.4 (0.3)	**70.3** (0.0)	97.2 (0.2)	96.8 (0.0)
Extreme Gradient Boosting	**73.5 (0.0)**	64.9 (0.0)	97.0 (0.0)	**97.7** (0.0)

Analysing the results obtained in relation to the different classifiers used, we can see that there is a consistent set of results in the validation phase with respect to the StudentLife dataset. The worst classifiers generally are again Support Vector Machine (whose accuracy results range from 50.5% to 63.4% for PASS and from 84.6% to 89.1% for WESAD) and Naïve Bayes (ranging from 56.3% to 63.3% for PASS and from 72.5% to 80.7% for WESAD). Several classifiers obtain accuracy results ranging between 60% and 77% for PASS and between 80% and 97% for WESAD (Decision Trees, K-Nearest Neighbours, Linear Discriminant Analysis, Logistic Regression, Ada Boost and Neural Networks). The top 3 classifiers in almost all cases are Random Forest ([77.1%–80.2%] for PASS and [98.6%–98.7%] for WESAD), Extreme Gradient Boosting ([77.1%–82.3%] for PASS and [97.5%–98.3%] for WESAD), and Gradient Boosting ([76.6%–78.3%] for PASS and [97.4%–98.0%] for WESAD). These three classifiers are selected for the parameter tuning process using halving grid search with 10-fold cross-validation. Comparing these results with StudentLife, the main difference is that the top 3 classifiers include Gradient Boosting instead of Neural Networks.

As in the case of StudentLife, the tuning process does not contribute to the accuracy. So, we use the test data obtained by the untuned classifiers (using the default parameters of the scikit-learn library implementation) to compare in terms of accuracy the results obtained, first by level of intrusion, then by dataset. In terms of intrusion level, both datasets provide data from medium level devices (Empatica E4 wristband) and high level devices (BioHarness3 chest bands for PASS and RespiBAN for WESAD). In both datasets we can see that increasing the intrusion level does not contribute to a significant increase in accuracy ([73.5–70.5%] for PASS and [97.3%–97.7%] for WESAD), on the contrary even in the case of PASS better results are obtained with the medium level (73.5%). Therefore, the data obtained with devices corresponding to a medium level of intrusion look promising for identifying stress with a good accuracy. Furthermore, in the case of WESAD, a very high accuracy of 97% is achieved. On the other hand, there is a significant difference in the maximum accuracy achieved with respect to the dataset used, being 73.5% for PASS and 97.7% for WESAD. This is because the study scenarios proposed in each case are very different (people performing different types of daily tasks, such as reading or public speaking in the case of WESAD, and playing video games in PASS).

5.3 Discussion

Our results allow us to identified the following research answers concerning the three level of intrusions:

RQ1. According to the results for the StudentLife dataset, considering different time-frames (1, 2, 5, 10 and 24 h) where the level of intrusion is low, the accuracy is good (up to 85.6% in cross-validation and 75.1% in test) for detecting stress. According to the results for the PASS and WESAD datasets, considering data extracted by a wristband where the level of intrusion is medium, the accuracy is remarkable (up to 98.8% in cross-validation and 97.7% in test) for detecting stress. The results obtained with the chest-bands (corresponding to a high intrusion level) do not improve the quality significantly (in the case of WESAD they do not increase it more than 1%, and for PASS they are even worsen). However, it is worth noting that the quality increment from a low to a medium level of intrusion is very significant, from 75.7% to 97.7%. Therefore, according to the analysis of the results, the level of intrusion that is most accurate for detecting stress is the medium. At this level, the system requires physiological data although it does not need an intrusive device that affects people's daily lives.

RQ2. WESAD is the dataset with the best accuracy results. Its predictions are remarkable (97.7% of accuracy) considering an medium level of intrusion (using only information from the wristband). In the case of StudentLife, good predictions are obtained for a low level of intrusion, using only data from mobile-user interactions (73.4% of accuracy). However, the predictions of PASS obtained for both medium and high intrusion level are the worst (73.5% and 70.3% respectively). Furthermore, this analysis gives us valuable insight about the type of data which is most relevant to stress/non-stress classification. In real-world scenarios, where the data is extracted from people's daily lives (StudentLife and WESAD), the results are significantly more accurate than in simulated environments (PASS). In addition, the best two classifiers that outperform all others for all datasets are: Random Forest and Extreme Gradient Boosting.

6 Conclusions and Future Work

Depending on the scenario where stress is identified, there are significant differences on stress detection between low and high intrusion, or both methods are equivalent in terms of detection and accuracy (RQ1 of Sect. 5). This allows replacing intrusive methods with smartphone data, in some case, although in other case, we showed that a medium intrusion device (i.e. a wristband) provides the same results as the highest intrusive device. According to our results, a high level of intrusion is not needed.

For future work, we aim to create a system for extracting sensory and app features from smartphones which allows configuring the users' privacy, too. We also aim that our system can provide a timely response to the users. For improving the system's performance and decision-making, we aim to apply time-series analysis to personalise the predictions.

Acknowledgments. This research was funded by the UK Research and Innovation Trustworthy Autonomous Systems Node in Verifiability (EP/V026801/2).

References

1. World Health Organization: The European mental health action plan 2013–2020 (2015). https://www.euro.who.int/__data/assets/pdf_file/0020/280604/WHO-Europe-Mental-Health-Acion-Plan-2013-2020.pdf

2. Brooks, S.K., et al.: The psychological impact of quarantine and how to reduce it: rapid review of the evidence. Lancet **395**(10227), 912–920 (2020)
3. Van Praag, H.: Can stress cause depression? Prog. Neuropsychopharmacol. Biol. Psychiatry **28**(5), 891–907 (2004)
4. World Health Organization: World mental health report: transforming mental health for all (2022). https://www.who.int/publications/i/item/9789240049338
5. Hickey, B.A., et al.: Smart devices and wearable technologies to detect and monitor mental health conditions and stress: A systematic review. Sensors **21**(10), 3461 (2021)
6. Parent, M., et al.: Pass: a multimodal database of physical activity and stress for mobile passive body/brain-computer interface research. Front. Neurosci. **14**, 542934 (2020)
7. van Kraaij, A.W.J., Schiavone, G., Lutin, E., Claes, S., Van Hoof, C.: Relationship between chronic stress and heart rate over time modulated by gender in a cohort of office workers: cross-sectional study using wearable technologies. J. Med. Internet Res. **22**(9), e18253 (2020)
8. Kolenik, T.: Methods in digital mental health: smartphone-based assessment and intervention for stress, anxiety, and depression. In: Integrating Artificial Intelligence and IoT for Advanced Health Informatics. IT, pp. 105–128. Springer, Cham (2022). https://doi.org/10.1007/978-3-030-91181-2_7
9. Narziev, N., Goh, H., Toshnazarov, K., Lee, S.A., Chung, K.M., Noh, Y.: STDD: short-term depression detection with passive sensing. Sensors **20**(5), 1396 (2020)
10. Alić, B., Sejdinović, D., Gurbeta, L., Badnjevic, A.: Classification of stress recognition using artificial neural network. In: 2016 5th Mediterranean Conference on Embedded Computing (MECO), pp. 297–300. IEEE (2016)
11. Can, Y.S., Chalabianloo, N., Ekiz, D., Ersoy, C.: Continuous stress detection using wearable sensors in real life: algorithmic programming contest case study. Sensors **19**(8), 1849 (2019)
12. Ciman, M., Wac, K.: Individuals' stress assessment using human-smartphone interaction analysis. IEEE Trans. Affect. Comput. **9**(1), 51–65 (2016)
13. Vildjiounaite, E., et al.: Unobtrusive stress detection on the basis of smartphone usage data. Pers. Ubiquit. Comput. **22**(4), 671–688 (2018)
14. Wang, R., et al.: StudentLife: assessing mental health, academic performance and behavioral trends of college students using smartphones. In: Proceedings of the 2014 ACM International Joint Conference on Pervasive and Ubiquitous Computing, pp. 3–14 (2014)
15. Schmidt, P., Reiss, A., Duerichen, R., Marberger, C., Van Laerhoven, K.: Introducing wesad, a multimodal dataset for wearable stress and affect detection. In: Proceedings of the 20th ACM International Conference on Multimodal Interaction, pp. 400–408 (2018)
16. Larose, D.T., Larose, C.D.: Discovering Knowledge in Data: An Introduction to Data Mining, vol. 4. John Wiley & Sons, Hoboken (2014)
17. Chawla, N.V., Bowyer, K.W., Hall, L.O., Kegelmeyer, W.P.: Smote: synthetic minority over-sampling technique. J. Artif. Intell. Res. **16**, 321–357 (2002)
18. Brazdil, P., Rijn, J.N.v., Soares, C., Vanschoren, J.: Metalearning for hyperparameter optimization. In: Brazdil, P., Rijn, J.N.v., Soares, C., Vanschoren, J. (eds.) Metalearning, pp. 103–122. Springer, Cham (2022). https://doi.org/10.1007/978-3-030-67024-5_6

Open-Source Web System for Veterinary Management: A Case Study of OSCRUM

Nancy Rodríguez[1], Lucrecia Llerena[1](✉), Pamela Guevara[1], Rosa Baren[1], and John W. Castro[2]

[1] Facultad de Ciencias de la Ingeniería, Universidad Técnica Estatal de Quevedo, Quevedo, Ecuador
{nrodriguez,lllerena,pguevarat,rosa.baren2015}@uteq.edu.ec
[2] Departamento de Ingeniería Informática y Ciencias de la Computación, Universidad de Atacama, Copiapó, Chile
john.castro@uda.cl

Abstract. Nowadays, even though technology has advanced by leaps and bounds, companies still carry out part of their management without using a computer system. Most veterinaries keep track of their pets' medical records manually. Therefore, an open-source web system called SysVet has been proposed to support veterinaries' management in administering pet health records. The Systematic Mapping Studying (SMS) methodology was used to search for relevant information, and the OSCRUM methodology was used to carry out the development process of the open-source web system, which has allowed the organization and timely fulfill the system requirements. It can be concluded that the use of the OSCRUM methodology has allowed for obtaining favorable results in terms of fulfilling the tasks for the implementation of the SysVet web system.

Keywords: System Web · Open-Source · Veterinary

1 Introduction

For more than 30 years, we have been accustomed to the fact that whoever sells software imposes the conditions under which it can be used. Because of these challenging conditions, the open source software (OSS) movement has arisen. OSS is a software development community where anyone can view or modify the source code of the systems they develop [1, 2, 3]. Many OSS projects are hosted in repositories such as GitHub, where they can be accessed or participated in community projects. Without these OSS communities, users and developers would not be able to offer extra knowledge to those software systems that are under development free of charge. Linux tools, Ansible, and Kubernetes are popular OSS projects [4] where many users and developers have been integrated into the development.

Currently, the problem in veterinary medicine is that most clinics do not have technological tools for centralized and digitized information management. Tasks are performed manually, which makes the organization of stored records difficult and, in many cases,

A. Rocha et al. (Eds.): WorldCIST 2023, LNNS 799, pp. 320–329, 2024.
https://doi.org/10.1007/978-3-031-45642-8_32

leads to the loss or duplication of information [5]. In addition, the results were almost nil among the OSS searches for veterinarians, which may be because veterinary-medical processes are usually extensive. Thus, the objective of the present work is to create a web-based OSS system for the management of veterinary processes. Implementing a software system that manages these processes ensures an improvement in everything related to the administration of veterinary clinics.

After analyzing the studies, the limited use of OSS development methodologies was identified, meaning OSS development methodologies have not been implemented to create web software systems. Therefore, a case study is proposed to evaluate the OSCRUM methodology used for OSS development. Finally, this methodology will be used to develop the SysVet web system.

SysVet is a web application conceived as an OSS that facilitates interaction with the source code in this community. In this way, the system can evolve according to the technology, innovations, knowledge, and needs of the people and/or companies that need this web system. This SysVet web system will be used in veterinary medicine to manage veterinary medical consultations. In addition, SysVet will manage the information and the clinical history of the health of the animals or events that have occurred to each animal that goes to the veterinarian. Also, the SysVet web software system can receive improvements or updates from external people who wish to add some functionality to the system by modifying the source code available in the GitHub repository.

This research has as a scientific novelty the purpose of contributing to future software projects that use the OSCRUM methodology in creating OSS web systems. OSCRUM helps to achieve the best planning in the OSS development process, and this helps us to propose an appropriate design for our project and motivate the OSS community, becoming an acceptable way of working for the developers of such tools, becoming outstanding in the software development industry. Among the contributions presented in this work are the following: (i) For the OSS community, software development problems are identified, and solutions are proposed through the design of a web system for the veterinary health area by applying the OSCRUM development methodology; (ii) for the scientific community, web artifacts are proposed to facilitate the planning of activities to be completed within a specific time frame and to obtain feedback from users testing the system.

This paper is organized as follows. Section 2 describes the related work. Section 3 describes the development method followed for implementing the open-source web system. Section 4 describes the case study. Section 5 describes the results of the case study. Section 6 discusses the results and the limitations encountered. Finally, Sect. 7 describes the conclusions and future research.

2 Related Work

Some research works have reported on open-source developments in the area of animal health, and these studies will allow us to have a more precise notion of what is intended to be done, how it should be done, and what we should take into account when working on the development of the open-source web system.

The work of Zaninelli [6] details a system that manages the clinical history of patients (animals) that are attended. In addition to containing a file area where you can view any test or prescription given to the animal, this system used a type of Emotional Intelligence (EI) model that consists of a series of competencies that facilitate people to manage different operations according to their knowledge. This model promotes the integration of information systems that support modern healthcare institutions and identifies a subset of functional components of healthcare. In addition, the multilayer methodology is used, which is a way of working in sections, i.e., we worked on each of the parts of the system to later integrate them and make them a single one.

Hutchison's study [7] refers to a surveillance system to alert to animal health problems. This system uses seven mobile data collection tools to improve speed and minimize errors during data submission, allowing veterinarians to know the diseases affecting the population of animals. The study uses a methodology called Lean, which enables efficient and productive systems development.

Kariuki [8] shows a novel Kenyan animal biosurveillance system (KABS) that includes a mobile application for data collection, a web-based application for account management and editing of data collection forms with a sequential methodology, where the most complicated part was the search for information to analyze how to apply biosurveillance. The Scrum development methodology was used to realize this system.

Ariff [9] proposed a technological approach to livestock care, an electronic system that helped farmers and veterinarians share information to prevent early on dangerous diseases from infecting their livestock. This project used radio frequency identification (RFID) technology for livestock health management to prepare the animal profile. In addition, we worked with Kanba methodology, achieving work flexibility and prioritizing activities to be developed.

It is concluded that none of the works were found to apply a development methodology for creating an open-source web system in the area of veterinary medicine, so it is considered that this work will be helpful for future research.

3 Description of the Proposed OSS Development Methodology: OSCRUM

OSCRUM, a modified Scrum development methodology, is adapted to develop OSS. This methodology motivates developers to rely on five values for the success of their projects: (i) individual commitment, (ii) open-mindedness, (iii) being highly competent, (iv) focus on the goal, and (v) collaboration [10]. The OSCRUM methodology differs from the SCRUM development methodology in certain concepts such as (i) time box, (ii) stakeholders, and (iii) some scrum rules, among others. At the same time, similarities have also been found between the two, such as: (i) the peculiarities of the team of developers (i.e., highly self-managed, cross-functional), (ii) the acceptance of feedback in a short loop, the usual release of the working version, integration and collaboration, accountability to the customer, among others [10].

The development of the open-source web system was carried out under the framework of the OSCRUM methodology through the activities outlined according to Rahman et al. [10]. The OSCRUM methodology is set out in a defined process of 11 activities, summarized in Table 1.

Table 1. OSCRUM methodology activities.

#	CRITERIA	DEFINITION
1	See the problem and look for volunteers	A group of people, whether or not they are the main-maintainer or core-contributors, give an idea and description of it, which will be published on the web to gather contributors
2	Establishing Communication	Communicates with key contributors to generate the list of software product requirements
3	Plan meetings for the initial launch	Several sprints have been created: • backlog of product features • interface design • complete system
4	Launch plan and product status	Upon completion of the requirements sprint backlog, the activities are listed in the release plan, with their respective status
5	Resetting of characteristics	During this sprint, if some new requirements are approved by the main-maintainer and core-contributors, the requirements backlog will be updated
6	Test the source code	This is done through a repository and the community tests it
7	Report errors	The community gets and gives the errors found
8	Contributions of external members to the project	The contributions that are given are integrated, previously approved by the main maintainer of the software
9	Repairs	Repairs are made to the errors found
10	Approval	After submitting and resolving the requirements where errors were found, they are rechecked, and the revision is accepted
11	Iteration	The process is repeated and continued according to product requirements

4 Case Study

The results of applying the OSCRUM development methodology to create the SysVet web system are detailed below.

Problem discovery and search for volunteers. The OSCRUM development methodology indicates that the problem should be discovered, and volunteers sought to develop a solution [10]. For this reason, a web artifact was created, specifically a blog, to collect information from the OSS community (regular OSS users registered in the community).

The link to the blog is the following: https://veterinaryclin.blogspot.com/. In addition, this blog was used as a communication channel between the team members and the community using this system. The initial interface of the SysVet web system blog is presented in Fig. 1.

Fig. 1. SysVet blog, for requirements gathering through open-source contributors.

Communication. In this activity, a meeting was held with participants, such as experts in the veterinary area and interested users. This was done to establish communication and plan the initial launch of the SysVet web system. For it was used tools such as email and Google Meet to discuss the healthcare area's needs regarding the systematization of the processes handled by the veterinarians. In addition, the backlog of product features, described in Table 2, has been generated.

Initial launch planning meeting. In the second meeting, the requirements and ideas given by each person involved in the system's development were presented—the link to access the SysVet system hosting: https://sysvet-web.herokuapp.com/.

Release procedure and status. Once the requirements sprint backlog is completed, the activities are listed in the release plan. In the blog: https://veterinaryclin.blogspot.com/, the link to the system has been inserted where contributors can use the software and make suggestions or observations for improvements to the Sysvet web system.

Feature update. The OSCRUM methodology suggests that if the main maintainer or core contributors approve new features or solutions, the feature backlog of the open-source project will be updated [10]. In this case, in the various meetings that have been held, all the project features have been taken into account, so it has not been modified.

Test Source Code. SysVet, is an OSS project that is publicly and freely available to all users. The source code of the SysVet web system is hosted in a very popular repository in the developer environment called GitHub. This repository allows users to test the program and its functionalities and leave comments to improve the system or fix bugs. Below is the link to the open-source web system: https://github.com/JoseSolorzanoC/sysvet-api.

Table 2. SysVet feature backlog, development prioritization and estimated size.

Product feature backlog item (SysVet)	Prioritization	Estimated Size
Login	High	Small
User regpistration (Animal or pet guardian)	High	Small
Animal or pet registration	High	Medium
Registration of veterinarians	High	Medium
Medical consultation record	High	Medium
Medical history generation	Media	Great
Schedule medications	Media	Medium
Veterinary appointment scheduling	High	Great
Schedule of veterinary care	Media	Great
Delete appointment	Download	Small
Send medical history by mail to users (animal or pet owners)	High	Great

Error Reporting. The OSS community has evaluated the open-source web system. It has been established that there are several errors in the SysVet web system, so a template was designed to register these errors and sent to the evaluators by e-mail. The following link presents the bug reports made to improve the SysVet system: https://goo.su/gvejJ5y.

Community collaboration. The OSCRUM development methodology indicates that new contributors can be integrated as long as the main maintainer approves them to add new features and fix bugs to the project's development [10]. In this case, expert contributing developers were used, the same ones who downloaded the system's source code from GitHub and proceeded to evaluate it. Once this evaluation was made, the experts could argue or describe the possible problems the Sw had. These problems were reported in the blog created, among other things, to report the findings during the development and tests of SysVet.

Repair. When bugs or missing functionalities were detected, each was corrected and added to the SysVet project, and the main maintainer approved these.

Approval. With the bug fixes, several proposals for solutions to problems raised in the source code have been checked. Once the proposed solutions have been registered, the main maintainer proceeds to review and approve them for the improvement of the system.

Iteration. The process is repeated iteratively as many times as necessary to obtain all the functionalities of the system [10], development methodology, the process is repeated iteratively as many times as necessary to obtain all the functionalities of the system, with the purpose that all the functionalities can be presented from the first delivery of the open-source web system. This part could be seen reflected when the pet registration and scheduling module was finished, which corresponds to the patient's part, to resume with the other module, which was the veterinarians.

5 Case Study Results

This section reports the results of implementing the OSCRUM methodology to develop the open-source web system SysVet and its source code. The results of this case study are mainly based on the main functionalities and requirements established in the backlog features. These features are (i) registration of guardians and pets, (ii) registration of veterinarians, (iii) appointment scheduling, and (iv) clinical history of pets. With those mentioned above, a veterinary care system has been implemented to provide efficient, timely, and quality service to its clients.

The SysVet web system is easy to use for users. Therefore, several types of users have been defined: (i) Tutor user, the pet owner, who can only register, log in, and schedule an appointment. (ii) Veterinarian user, which is the specialist. This Veterinary User has the same functionalities as the guardian user and can request the pet's or animal's medical history, generate a medical prescription, and generate an appointment schedule. It should be noted that the Veterinarian user can register new patient-type users, who will be the pet or animal being treated, (iii) Administrator User, who has access to all the system's functionalities, including the ability to create new Veterinarian-type users. Obtain reports on medical-veterinary consultations and medical history reports, among other functionalities.

Regarding the OSCRUM development methodology, the first activity is the discovery of the problem and the search for volunteers. Together with the second part, communication, these activities were carried out through video calls via Google Meet. In addition, an agreement was reached to define how the community would be informed about the development process. A web artifact (blog) was created, resulting in the backlog of product features (SysVet), such as: (i) scheduling pet appointments, (ii) registering medical prescriptions, (iii) generating the pet's clinical history, among others. Subsequently, a meeting was held to plan the initial launch of the SysVet open-source web system, where the priorities of each of the features mentioned in the list of functionalities of the software product were specified. A follow-up of the development of the features or functionalities that SysVet should have was carried out. This follow-up made it possible to know the details of the development status of each of the features of the SysVet web system, which facilitated the observation of which features were already developed and which were still to be developed, having as an advantage mainly an orderly software development. Regarding updating features in the requirements backlog, no new functionalities have been added since no new needs have arisen. However, if necessary, according to the use required by the community, the backlog of features can be updated before the main maintainer approves it and analyzes whether the feature, functionality or set of features is scalable with the system.

To test the source code, the project has been published on GitHub. It is essential to mention that, with the development of this work, it has been detected that one of the disadvantages of the free test servers was that they have many limitations of use, so it has been a real challenge to publish the code on the test server. Also, some templates were made to register error reporting. Once the community tested the SysVet web system deployed on the test server, these templates were provided for them to report their errors. Thus, one of the difficulties that have arisen and that the OSS community has emphasized is the validation of forms, but not of the functionality as such. Another

OSCRUM activity involves adopting a new participant to the project, where a contributor contributes individually with improvements to the open-source project's source code [1]. However, this activity has not been possible to report since no collaborators have been shown who want to contribute to the project with solutions or improvements in the source code.

Error reports from the community have allowed us to know the SysVet system's shortcomings and identify the problems that could be solved to improve it. Therefore, once these errors were identified, it was possible to repair these software problems and provide corrective maintenance to the functionalities by writing and modifying the source code. The errors reported by the users were several, such as (i) unsecured login, (ii) error when scheduling a veterinary appointment, and (iii) error when viewing the pet's clinical history, among others. Thus, with the information reported by users, a list of errors has been created, and corrective maintenance has been proposed to the source code. However, all these changes are not immediately incorporated into the source code. These changes or proposed solutions first go through a process where the main maintainer analyzes, verifies, and approves the software features before allowing the changes in the source code. The last step of the OSCRUM methodology refers to iteration [10], where this process is repeated as many times as necessary to obtain new features for the software project and provide maintenance to the functionalities.

The implementation of the SysVet web system was successfully completed thanks to the OSCRUM methodology. This development methodology has allowed this system to manage the information and medical consultations of the animal and/or pet. The interfaces created for the SysVet system can be reviewed at the following hosting link: https://sys-vet-web.herokuapp.com. Figure 2 shows the main interface of the SysVet web application, as well as the form where people must log in to access the system.

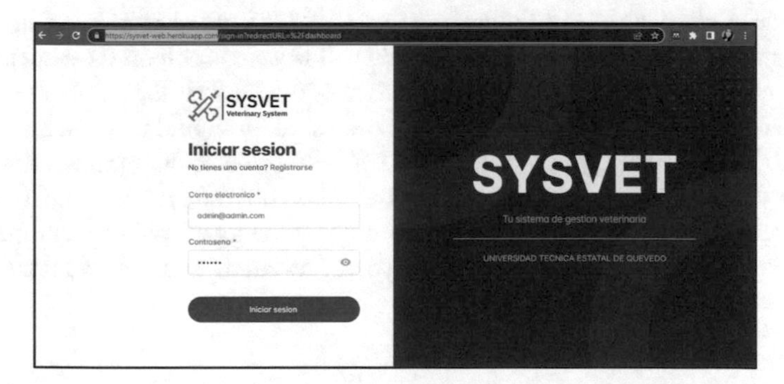

Fig. 2. SysVet web system startup interface and authentication.

Figure 3 shows the interface where the guardian or owner of the animal and/or pet can enter the information about the animal and/or pet that the veterinarian will attend. The SysVet source code is hosted in the GitHub repository, and this code can be viewed and/or downloaded at the link: https://github.com/JoseSolorzanoC/sysvet-api.

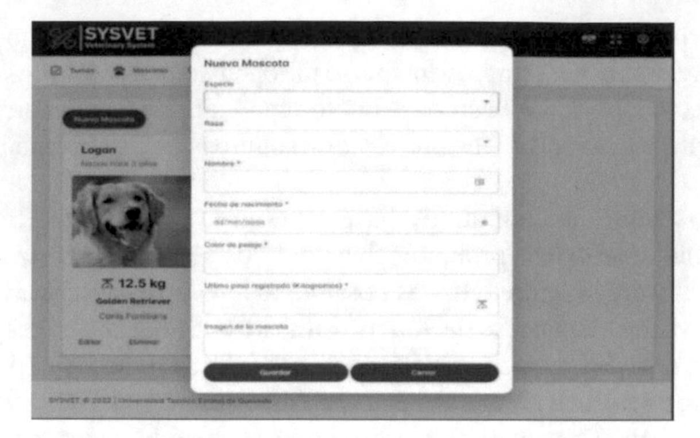

Fig. 3. Entering pet data in SysVet.

6 Discussion and Limitations

As a result of applying the OSCRUM methodology, an open-source web system called SysVet has been created to register the data of the pet and/or animal to be treated by the veterinarian. After this, the veterinarian can make the prescription of medicines for the animal and/or pet that is being treated. Another objective of implementing the system was managing the veterinarian's data to keep track of the pets treated during the day. It is essential to mention that the use of a development methodology appropriate to the creation of OSS has dramatically facilitated the planning and implementation of this SysVet software project, also allowing the generation of a quality open-source web system. Therefore, OSCRUM has proved to be a practical methodology for developing OSS projects.

The case study has several limitations that could affect the validity of its contributions. Regarding construct validity and internal validity, no problems have been detected. Regarding external validity, the critical limitation of our study is the number of case studies (only one OSS project). Therefore, more case studies should be conducted to apply the OSCRUM methodology in developing other OSS projects. Regarding reliability, the case study should include all types of OSS users, i.e., not only the participation of OSS users. Therefore, before applying the methodology, we suggest looking for other options to get users who want to participate in this type of research, such as social networks.

7 Conclusion and Future Work

This research aimed to develop an open-source web-based system (SysVet) using the OSCRUM methodology. SysVet allows: (i) veterinarians to record each pet consultation and (ii) pet caregivers to obtain online information about their pet's condition and what remedies have been prescribed. In addition, working with the OSCRUM methodology allowed us to comply with the facets of the SysVet web system, where we used free tools such as NestJS and Angular, fundamental tools in the project, making the web system completely OSS. In this research, we have established the general requirements that a system should meet to improve the quality of the code, and we have designed a series of prototypes to refine these requirements iteratively. The result is the SysVet web system centered on the OSS development environment to work collaboratively with new developers.

As future work, it is proposed to perform the usability evaluation of the SysVet web system with users and, from the results obtained, make improvements that can be implemented to meet users' needs.

References

1. DeKoenigsberg, G.: How successful open source projects work, and how and why to introduce students to the open source world. In: 21st Conference on Software Engineering Education and Training, pp. 274–276, IEEE, Charleston, SC, USA (2008)
2. Llerena, R., Rodríguez, N., Llerena, L., Castro, J.W., Acuña, S.T.: Adoption of the HTA technique in the open source software development process. In: Stephanidis, C., Marcus, A., Rosenzweig, E., Rau, P.-L., Moallem, A., Rauterberg, M. (eds.) HCII 2020. LNCS, vol. 12423, pp. 184–198. Springer, Cham (2020). https://doi.org/10.1007/978-3-030-60114-0_13
3. Reyes Ch., R.P., Fonseca C., E.R., Castro, J.W., Vaca, H.P., Calderón, M.P.: An empirical evaluation of open source in telecommunications software development: the good, the bad, and the ugly. In: Rocha, Á., Guarda, T. (eds.) ICITS 2018. AISC, vol. 721, pp. 508–517. Springer, Cham (2018). https://doi.org/10.1007/978-3-319-73450-7_48
4. Miller, K.W., Voas, J., Costello, T.: Guest editors' introduction: free and open source software. IT Professional **12**(6), 14–16, IEEE (2010)
5. Felmer, R., Chávez, R., Catrileo, A., Rojas, C.: Current and emergent technologies for animal identification and their use in animal traceability. Arch. Medicina Vet. **38**(3), 197–206 (2006)
6. Zaninelli, M., et al.: The O3-Vet project: a veterinary electronic patient record based on the web technology and the ADT-IHE actor for veterinary hospitals. Comput. Methods Programs Biomed. **87**(1), 68–77 (2007)
7. Hutchison, J., et al.: New approaches to aquatic and terrestrial animal surveillance: the potential for people and technology to transform epidemiology. Prev. Vet. Med. **167**, 169–173 (2019)
8. Bell, S.M., et al.: An integrated chemical environment to support 21st-century toxicology. Environ. Health Perspect. **125**(5), 1–4 (2017)
9. Ariff, M.H., Ismarani, I., Shamsuddin, N.: RFID based systematic livestock health management system. In: 2014 IEEE Conference on Systems, Process & Control (ICSPC'14), pp. 111–116, IEEE, Kuala Lumpur, Malaysia (2014)
10. Rahman, S.S.M.M., et al.: OSCRUM: A modified scrum for open source software development. Int. J. Simul. Syst. Sci. Technol. **19**(3), 20.1-20.7 (2018)

Smart Stress Relief – An EPS@ISEP 2022 Project

Gema Romera Cifuentes[1], Jacobine Camps[1], Júlia Lopes do Nascimento[1], Julian Alexander Bode[1], Abel J. Duarte[1,2], Benedita Malheiro[1,3], Cristina Ribeiro[1,4], Jorge Justo[1], Manuel F. Silva[1,3(✉)], Paulo Ferreira[1], and Pedro Guedes[1,3]

[1] ISEP/PPorto - School of Engineering, Polytechnic of Porto, Porto, Portugal
mss@isep.ipp.pt
[2] REQUIMTE, ISEP, Porto, Portugal
[3] INESC TEC - Institute for Systems and Computer Engineering, Technology and Science, Porto, Portugal
[4] INEB - Institute of Biomedical Engineering, Porto, Portugal
https://www.eps2022-wiki5.dee.isep.ipp.pt/

Abstract. Mild is a smart stress relief solution created by DSTRS, an European Project Semester student team enrolled at the Instituto Superior de Engenharia do Porto in the spring of 2022. This paper details the research performed, concerning ethics, marketing, sustainability and state-of-the-art, the ideas, concept and design pursued, and the prototype assembled and tested by DSTRS. The designed kit comprises a bracelet, pair of earphones with case, and a mobile app. The bracelet reads the user heart beat and temperature to automatically detect early stress signs. The case and mobile app command the earphones to play sounds based on the user readings or on user demand. Moreover, the case includes a tactile distractor, a scent diffuser and vibrates. This innovative multi-sensory output, combining auditory, olfactory, tactile and vestibular stimulus, intends to sooth the user.

Keywords: Engineering Education · European Project Semester · Mild · Smart Companion · Design · Stress Relief

1 Introduction

The European Project Semester (EPS) is a capstone design semester offered by several European higher education institutions[1], including the Instituto Superior de Engenharia do Porto (ISEP). In the spring of 2022, four EPS@ISEP students worked together as team DSTRS to tackle the problem of anxiety through the creation of a smart companion. The students, from diverse countries and majors, contributed with different interests, expertise and values to the project.

[1] http://europeanprojectsemester.eu/.

A. Rocha et al. (Eds.): WorldCIST 2023, LNNS 799, pp. 330–339, 2024.
https://doi.org/10.1007/978-3-031-45642-8_33

Anxiety is a severe mental health problem, which was aggravated by the COVID-19 pandemic, affecting many people daily. DSTRS chose to try to make their daily life more liveable. The solution must comply with the applicable European Union directives, use open source software and technologies, adopt processes, techniques, materials and components based on sustainability and ethical criteria, and meet the identified requirements.

Mild aims to help the user to overcome anxiety in public in an easy and discreet way. The objective is to have a companion that provides relief quickly at the first signs of anxiety. The design should be fashionable and based on daily usage items that camouflage its purpose. The production of Mild should use integrated electronics, be kept simple and adopt best practices. The product should be high standard and enjoy a good market reputation.

This document comprises five additional sections covering the preliminary research, proposed solution, prototype development, product evaluation and conclusion.

2 Preliminary Studies

Anxiety is one of the most well-known and common mental health diseases. As a natural body response to stress, it can happen to everyone at anytime. Severe anxiety can last longer than six months and may become a chronic disorder, interfering seriously with the patient life quality.

2.1 Related Work

Existing commercial products take the form of mobile apps or smart devices. Examples of mobiles apps are Pacifica and WEconnect. Pacifica is a control anxiety mobile app that uses cognitive behavioural therapy, relaxation and wellness principles (tips and goal setting) to break the cycle of anxiety [1]. WEconnect is a mobile app with widgets and web components that provides support for those trying to combat drug or alcohol addiction [10]. Spire and the Fisher Wallace stimulator are illustrative smart devices. The Spire smart device detects moods, breathing patterns, and other psychological cues that indicate how the user is feeling. The device sends a notification to the mobile phone with suggestions on how to relax or wellness tips to improve mood [7]. The Fisher Wallace stimulator is a headband-shaped device that stimulates the brain to release serotonin and dopamine, thereby reducing stress and increasing feelings of happiness [9].

The scientific community has also focused on anxiety relief devices. Lith *et al.* explore art making as a companion to the mental health recovery process [4]. The DSTRS team, inspired on the reported and recommended methods for smart devices chose to design a solution with a smart device. Harwood *et al.* analyse the effects of smart-devices on mental health, such as excessive Internet-browsing, gaming, texting, emailing, social networking, and phone calling [2]. This study led the team to decide against the development of just another app since employing smartphones to track and treat mental health issues can be counterproductive. Serin *et al.* evaluate the therapeutic effect of bilateral alternating tactile stimulation technology (BLAST) on the stress response [5]. This work evolved into the commercial Touchpoints solution, comprising two skin touching devices for bilateral stimulation, physiological and environmental sensors, a control system and a mobile app [6]. Touchpoints rates the levels of emotional stress and bodily distress on a scale of 0 to 10 and applies BLAST for 30 s. Results show a statistically significant reduction in the levels of both emotional stress and bodily distress. Inspired by Touchpoints, the team included, among other components, a sensing bracelet, a vibrating case and a mobile app. Khan *et al.* propose a smart companion agent for mental well-being supported by deep learning and natural language processing [3]. They developed a chatbot that holds a quality conversation with a mental illness patient.

2.2 Ethics

Engineering ethics is the branch of applied ethics that brings together the set of moral principles established for the practice of engineering. It examines and establishes the obligations of engineers to society, clients and the profession.

Sales and marketing ethics brings clear benefits like prestige, reputation, add value to the product, establish long-term relationships with stakeholders and customers, improve staff quality, team cohesion, commitment and, ultimately competitiveness. Mild's advertising should follow the eight principles of ethical marketing proposed by the Institute of Advertising Ethics: appropriate design of labels, respect for laws and standards, sell efficient and beneficial products, be as clear and accurate as possible in order to had a good communication with customers and with all members of the project, ensure transparency and promote responsibility, fairness and honesty.

Environmental ethics provides a set of arguments related to the conservation of the planet and directed at people's practices. These are fundamental considerations for environmental decision-making, setting priorities in research and studies, publishing results of environmental impacts, and setting policy. The team was committed to design a product with a large product-life, environmentally friendly materials, increased energy efficiency, reduced waste, reusable packaging and reusing working components.

Finally, in terms of liability, Mild must comply with the applicable European Union Directives.

2.3 Marketing

The marketing studies started with a deep market research, concluding that the high anxiety numbers got worse with the COVID-19 pandemic. To identify the micro and macro environment that affects Mild as a product, DSTRS performed a Political, Economic, Social, Technological, Environmental, and Legal analysis. The micro environment encompasses important strategic partners, like suppliers, customers, and influencers. An example of an important strategic partner are psychologists since they can recommend Mild to their patients. The competitor analysis showed that Mild, although possibly more expensive than other anxiety tools, presents extra functions and advantages.

Mild, as a product, has several strengths: provides immediate relief since it automatically monitors the anxiety level; combines different devices (earphones and bracelet tracker and case); is a daily use product, unlike an anxiety tool; and addresses a taboo topic. The weaknesses of Mild are the: difficulty to reach the target audience since it is hard to talk about; need to use the tracker bracelet; use of several gadgets; and price. The opportunities are the growing prevalence of anxiety and depression worldwide; and the use of earphones, a very common daily item, to play appeasing sounds. The main Mild threats are the price of the technical solution and the fact that anxiety is a still a taboo topic.

The team distributed a survey and talked to a psychologist to establish that the target audience are people between 18–30 years, who struggle with anxiety and long for stress relief. This lead to defining the persona and market position (functionality and price). Finally, the team specified the marketing budget, strategy, goals and objectives, as well as the strategy control method.

2.4 Sustainability

According to the United Nations World Commission on Environment and Development, sustainable development is development that meets the needs of the present without compromising the ability of future generations to meet their own needs [8]. Sustainability is often divided into three pillars: environmental, social, and economic (informally known as planet, people, and profits).

The environmental pillar refers to the impact of the company on the environment and the measures taken to mitigate the risks. The company must take actions to minimise the negative impacts of its activity, namely by using raw materials and renewable energy sources efficiently, recycling and minimising waste. The social pillar focuses on improving the quality of life of people directly or indirectly connected to the company, inside or outside the company. This means that the company has to build a responsible and sustainable work environment, respecting human rights and privacy issues. The attention of economic pillar is on a profitability that contemplates both environmental and social sustainability.

With Mild, the team is committed to improve the users' quality of life, use resources parsimoniously and establish lasting partnerships with stakeholders, building a brand strongly associated with social responsibility and sustainability.

3 Proposed Solution

Given the conducted preliminary studies, this section introduces the developed concept and the proposed product design.

3.1 Concept

Based on the prior state-of-the-art, ethics, marketing and sustainability analyses, DSTRS decided to create Mild (Fig. 1): a unique smart companion in the form of earphones, an earphone case, a stress tracking bracelet and a monitoring app intended to help young adults to tackle anxiety. The tracker monitors user physiological data allowing the early identification of stress signs. Mild monitors the user stress level, plays music (earphones) accordingly or on demand, offers breathing exercises (earphone case vibration), scent diffusion (earphone case) and tactile distraction (earphone case). Moreover, the user can control the sound and vibration through the app.

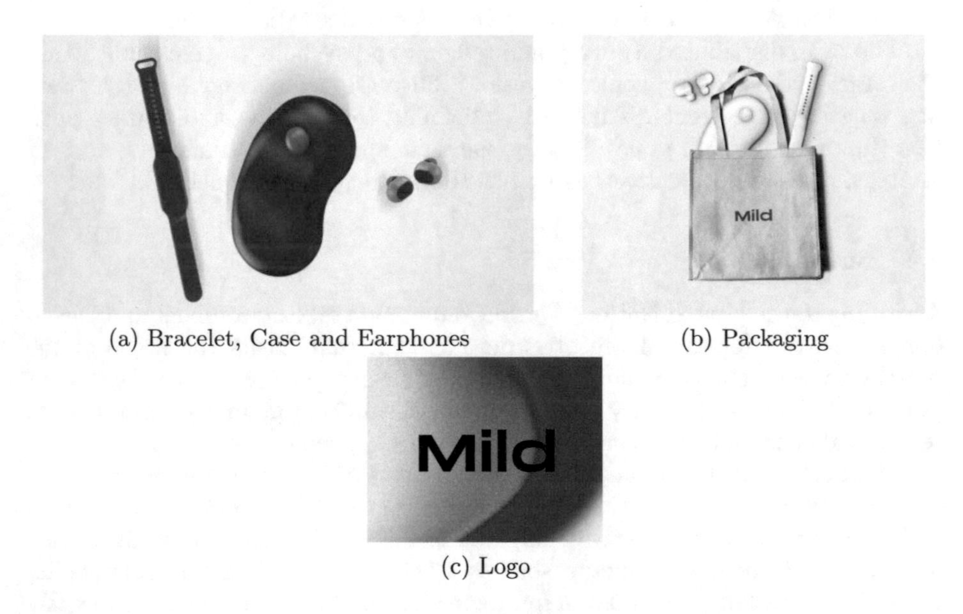

(a) Bracelet, Case and Earphones (b) Packaging

(c) Logo

Fig. 1. Concept Mockups

3.2 Design

The goal was to create a clean, timeless design for the brand (Fig. 1c) and product. In terms of requirements, Mild must be soft, light, ergonomic – the bracelet, earphones and case must adapt to the wrist, ears, hand and thumb – and the colour scheme induce a calming effect to ensure comfort, usability, and a soothing look and feel. The colour scheme relies on blue, orange and pink. According to colour research, blue is a calming colour, creating a peaceful, tranquil, secure, and orderly feeling. Orange suggests happiness and spirituality as well as captures the attention. Pink resembles kindness, calmness and creativity. The symbolism of this colour pallet is deeply aligned with Mild's purpose.

Structure. The choice of materials plays an essential role in the design and production of a wearable. Wearables are expected to be fashion items and endure constant stress through the contact with clothes, skin, and aggressive substances like sweat or soap. Moreover, Mild has to be soft, light, ergonomic and produce a calming effect. The structural material must meet all of the above requirements and allow a simple, affordable, environmentally friendly production technique. For the structural design of the bracelet and the earphone case, design requirements must be applied. An ergonomic shape and enough space to fit all the electrical components are necessary, besides room for all Mild functions and the earphones are to be fitted inside of the earphone case.

The product is expected to withstand load forces that resemble falls from standard heights or forces that could be easily put on by the user's hand. This was achieved through structural numerical simulations using the finite element method of the earphone case. A simple load example of 50 N static force was simulated, using two different y-directions, to check for deformation and stress distribution on the upper and lower part of the case.

Control. The bracelet, earphones and the earphone case are controlled by the mobile application via Bluetooth. The soft texture bracelet contains a battery, sensors and a microcontroller. When the microcontroller detects increasing stress levels, it reports the finding to the mobile app. This allows the user to choose how Mild can help: case vibration and/or earphone playing. The soft texture case comprises a battery, a vibration motor and a microcontroller. The case vibration, controlled by the app, helps the user to breathe at a pace that lowers his/her stress level. In addition, the case provides analogue distraction and scent tools for further relaxation. Finally, the mobile app controls the earphones.

Mobile Application. The mobile app implements the following use cases (Fig. 2): registration, login, logout, as well as music, sounds, and vibration control. The app wireframe allows the user to register, for maintaining a database of users and their profiles, a login function, for accessing the app, a logout function, and the main home screen with three buttons: (*i*) the music button to create personalised playlists, select, start and stop playing a playlist; (*ii*) the sounds button to select, start and stop playing relaxing sounds from a library of different natural sounds, such as ocean sounds, birds, and forest sounds; (*iii*) the paced breathing vibration button to switch on and off the case vibration. This option is effective in situations without earphones. The application design reuses the concepts and graphic elements from the Mild product. Furthermore, to enhance user experience, the components and the interface of the mobile app were created in the most intuitive way possible, facilitating use in stressful situations.

4 Prototype Development and Assembly

Based on the design proposed in the previous section, here is described the prototype development, considering the product assembly, the preliminary tests conducted and the results achieved, and, finally, are discussed the main project achievements.

The team chose to 3D print the bracelet and case prototypes with dimensions adjusted to accommodate the off-the-shelf electronic components, minimising the

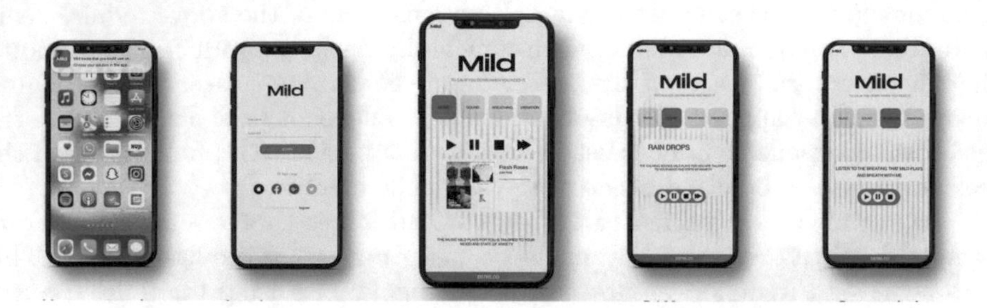

Fig. 2. Mobile App

Fig. 3. Assembled Prototype

quantity of material and, thus, the printing costs and time. The prototype was 3D printed with black Polylactic Acid (PLA) for the case and black Thermoplastic Polyurethane (TPU) for the bracelet. The key aspects to choose both materials were their mechanical properties, namely the elasticity and durability of TPU, and the variety of colours available. Finally, the case and bracelet were sanded and, then, the electronic components assembled and placed inside. Figure 3 displays the resulting prototype. In case of production, the bracelet dimensions will reduce significantly by using dedicated integrated electronic boards.

Structure. Prototype requirements differ from those of the final product since the goal was to quickly build a simpler proof of concept version. In the case of the real product, the 3D printing with thermoplastics can be substituted by casting or injection molding with Polycarbonate (PC). PC allows mass production while retaining good mechanical properties, like tensile strength, elasticity module or hardness. Another major requirement is the housing of the electronics inside the structure. This requirement lead, *e.g.*, to a bulky bracelet prototype. Nonetheless, since mass production typically relies on dedicated electronic printed circuit boards, the dimensions of the internal electronics reduce considerably in the production stage. Moreover, the material for the bracelet needs to be light, comfortable and adaptable to the wrist, *e.g.*, silicone.

Control. The app controls the bracelet, earphone case and earphones. It offers all user options on a single screen, including the configuration of the Bluetooth connections with the bracelet and case. Upon entrance, the app presents the user with five buttons: three to reduce the level of stress (music, vibration and breathing, and sound) and two to manage the Bluetooth connections. When the user presses the music button or the vibration and breathing button, the app sends a command to the earphones or the case. After the pressing of the sound button, the shows four additional buttons, corresponding to different sounds that the user can choose (cat purr, ocean, rain, jungle sounds). The smartphone and earphones are paired by default.

The earphone case consists of a charging module (USB-C port), a rechargeable battery, a vibration motor, a transistor, and an ESP32 microcontroller (including a Bluetooth module). All these elements are connected and programmed using the Arduino Integrated Development Environment (IDE). The code produces a vibration designed to guide the breathing and lower the stress levels of the user. This vibration is activated through the app. The case connects to the app via Bluetooth, with the case as the server and the app as the client.

Finally, the bracelet encompasses a heart rate sensor, a rechargeable battery and a microcontroller, also with a Bluetooth connection to the app and programmed using the Arduino IDE. The code of the bracelet analyses the user stress level by determining the heart rate variability, called root mean square of successive differences between normal heartbeats (RMSSD). The individual baseline RMSSD is determined during the first weeks of use. When the measured RMSSD undercuts the individual baseline RMSSD, the bracelet sends a message to the app to warn and suggest the use of the app calming options. This time the app acts as the Bluetooth server and the bracelet as the client.

Mobile App. The app was developed with the MIT App Inventor mobile application development environment[2]. The music and the vibration buttons check whether the required Bluetooth connection is established and then send to the corresponding command – to play the desired music (already accessible to the program) and to vibrate, respectively. When the user selects the sound button to play calming sounds, the code checks whether there is Bluetooth connection between the app and the earphones and shows the buttons corresponding different calming sounds. Finally, when the bracelet identifies increased levels of user anxiousness, the code notifies the user through a smartphone notification.

5 Product Evaluation

After completing the prototype several tests were conducted to evaluate its functioning according to the requirements.

5.1 Tests and Results

The testing of the Mild prototype was mostly functional. The structural functions were tested after the assembly of the 3D printed prototype. This included

[2] https://appinventor.mit.edu/.

testing the hinge of the earphone case and with various kinds of Bluetooth in-ear earphones. The testing of the different devices control systems, such as the earphone case, the bracelet and the app were done virtually and physically. For the earphone case, a circuit with the vibration motor, a transistor, and the LOLIN 32 Lite (ESP 32) microcontroller was set up. The Bluetooth connection to the developed App was tested with success. The vibration motor was activated to produce a vibration pattern with 6 s of raising intensity and 6 s of fading intensity. The test of the bracelet, which entails the calculation of the RMSSD value from the readings of the heart rate sensor, failed due to the malfunctioning of the heart rate sensor (always outputted 0 beat/min). This problem needs to be addressed in future work. The testing of the app, involving the Bluetooth connection and the four calming functions, was a success.

5.2 Discussion

Mild was created after extensive research and idea forming. In the team's opinion, the prototype doesn't completely reach their expectations, because, for example, the components do not completely fit in the earphone case. The look of the prototype is also a little different than their expectations, but, globally, they are happy with how it turned out.

There are some differences between how Mild was envisioned as a product, and how it was build as a prototype. The ceramic ball is intended to act as a smell diffuser and as a moving distraction tool. However, the 3D printed case does not support ball motion nor smell diffusion. The bracelet in the prototype is considerable thick and big. The real bracelet is to be made of highly wearable and comfortable material, containing miniaturised custom-designed electronic components instead of the off-the-shelf components stacked inside the bracelet prototype, resulting in a small, fashionable bracelet. The Plenthysmography iHaospace MAX30102 sensor used to measure the heart rate variability and blood pressure in the prototype can be substituted by a cortisol (stress hormone) electrochemical sensor in the product. Mild, as a product, obviously contains its own earphones, while the prototype reuses existing earphones.

6 Conclusion

The project outcomes comprise the design of a stress relief solution driven by the market, ethics and sustainability, followed by the building and test of a proof of concept prototype.

6.1 Achievements

The team was able to design a smart companion to automatically detect and provide quick stress relief, as well as create a clean, fashionable design based on daily use devices and accessories. These two goals are the basis for a future high quality product, produced with best components, materials and practices, contributing to a solid and enduring reputation.

6.2 Future Development

Mild, as any other product, requires continuous refinement. To further meet user needs and preferences, and improve usability, the kit should be redesigned to become slimmer, lighter and increasingly customisable (personalised colours, scents, shapes and functionalities).

The creation of the DSTRS company would enable the production of Mild and the expansion of the brand in the market by designing new mental health aid products, driven by ethics, sustainability and the user community.

Acknowledgements. This work was partially financed by National Funds through the Portuguese funding agency, FCT – Fundação para a Ciência e a Tecnologia, within project UIDB/50014/2020.

References

1. Buhigas, I.: Pacífica, la aplicación para acabar con el estrés desde el móvil (2015). https://www.eleconomista.es/apps/noticias/6840968/07/15/Pacifica-la-app-para-decir-adios-al-estres.html. Accessed Oct 2022
2. Harwood, J., Dooley, J.J., Scott, A.J., Joiner, R.: Constantly connected - the effects of smart-devices on mental health. Comput. Hum. Behav. **34**(6), 267–272 (2014). https://doi.org/10.1016/j.chb.2014.02.006
3. Khan, R., Sohel, A.A., Shreyashee, F., Azad Hossain, S., Fiaz, M.: Smart companion agent for mental well-being through Deep Learning and NLP. Bachelor thesis, Brac University (2021). http://dspace.bracu.ac.bd/xmlui/handle/10361/14973
4. Lith, T.V., Fenner, P., Schofield, M.: The lived experience of art making as a companion to the mental health recovery process. Disabil. Rehabil. **33**(8), 652–660 (2011). https://doi.org/10.3109/09638288.2010.505998
5. Serin, A., Hageman, N.S., Kade, E.: The therapeutic effect of bilateral alternating stimulation tactile form technology on the stress response. J. Biotech. Biomed. Sci. **1**(2), 42–47 (2018). https://doi.org/10.14302/issn.2576-6694.jbbs-18-1887
6. The TouchPoint Solution: Wearable stress relief device | TouchPoints (2022). https://thetouchpointsolution.com/. Accessed Oct 2022
7. Ulanoff, L.: Spire smart pebble watches your breathing and mental health (2016). https://mashable.com/article/spire-review. Accessed Oct 2022
8. United Nations: Report of the world commission on environment and development - our common future (1987). https://sustainabledevelopment.un.org/content/documents/5987our-common-future.pdf. Accessed Oct 2022
9. Wade, T.: What about the Fisher-Wallace stimulator for anxiety and depression? (2017). https://www.tomwademd.net/what-about-the-fisher-wallace-stimulator-for-anxiety-and-depression/. Accessed Oct 2022
10. WEconnect Health Management: WEconnect (2022). https://www.weconnectrecovery.com/. Accessed Oct 2022

LIDia: A Serious Game for Type 1 Diabetes Education

Esperança Amengual-Alcover[1](✉), Miquel Mascaró-Oliver[1], and Maria Caimari-Jaume[2]

[1] Universitat de les Illes Balears, Ctra. de Valldemossa, Km. 7.5, 07122 Palma, Spain
{eamengual,miquel.mascaro}@uib.es

[2] Hospital Universitari Son Espases, Ctra. de Valldemossa, 79, 07120 Palma, Spain
maria.caimari@ssib.es

Abstract. In this paper an ongoing work whose goal is to develop an educational serious game for type 1 diabetes (T1D) is introduced. A serious game is a video game that has been designed with a main purpose that goes beyond entertainment. The ultimate goal of a serious game is to help users to reach their objective in a playful way through the game. Initially, the target audience is children and/or adolescents with T1D. The first version of the game represents a hypoglycemia situation. The objective is to offer players a tool so that they can learn in an entertaining way to manage the different scenarios that can take place in a situation of hypoglycemia. The goodness of this first version will be validated with patients. It is intended to evolve the video game in future versions that will expand the functionality of the game with new aspects related to diabetes education. The ultimate goal is to cover as many scenarios as possible to offer patients an educational tool that helps them to better control diabetes and improve their quality of life.

Keywords: type 1 diabetes mellitus · serious games · diabetes self-management

1 Introduction

Diabetes, or diabetes mellitus, is a group of metabolic disorders characterized by the loss of the ability of the pancreas to produce insulin, the hormone that regulates blood glucose. Type 1 diabetes (T1D), once known as juvenile diabetes or insulin-dependent diabetes, is a chronic condition in which the pancreas produces little or no insulin by itself [1]. As a consequence, people with T1D need to use insulin shots to control blood glucose or blood sugar.

The prevalence and incidence of T1D are increasing in the world [2]. According to the World Health Organization [1], in 2017 there were 9 million people with T1D; the majority of them live in high-income countries. Neither its cause nor the means to prevent it are known. Treatment and care aim to control blood glucose levels. This control must be carried out by the patient who has to learn many concepts related to diabetes. For new patients, the amount of information to assimilate is enormous. In addition, T1D debuts in childhood or adolescence. Therefore, many patients are young people.

A. Rocha et al. (Eds.): WorldCIST 2023, LNNS 799, pp. 340–346, 2024.
https://doi.org/10.1007/978-3-031-45642-8_34

Video games are increasingly popular among the young population, but also among adults. Since they began to become popular in the early 70 s, video games have evolved to become products available to everyone. Serious games are computer games designed for a primary purpose other than pure entertainment. The ultimate purpose of a serious game is to allow users to reach a specific goal in an entertaining and engaging manner through the experience of playing the game. In the healthcare field, serious games are largely focused on treatment, recovery, and rehabilitation.

Based on the experience acquired in the development of serious games for healthcare [3], our current goal is the development of an educational serious game whose purpose is diabetes education. Initially, the target audience is children and adolescents with T1D.

In this work a preliminary version of LIDia (Learn and Improve Diabetes) is presented. This first version of this serious game represents a hypoglycemia situation. Hypoglycemia, or low blood sugar, is a fall in blood sugar to levels below normal, typically below 3.9 mmol/L. The objective is to offer players a tool so that they can learn in an entertaining way how to manage the different scenarios that can take place in a situation of hypoglycemia.

It is important to highlight that this initiative is supported by a medical team from Hospital Universitari Son Espases, a public hospital in the Balearic Islands [4]. They are responsible for validating both the definition of the requirements and the design, as well as the experience of use with patients. For the first version of LIDia, it is expected that, after a period of use, the number of hypoglycemic events experienced by patients decreases.

2 Background

T1D requires continuous medical care and educational support so that the patient can self-manage, as far as possible, the control of diabetes, prevent decompensations (hypoglycemia and hyperglycemia) and reduce the risk of chronic complications. The control of T1D is complex, going beyond glycemic control, and requires interdisciplinary care where therapeutic education is a fundamental pillar.

The World Health Organization [5] defines "Therapeutic Education in diabetes and other chronic diseases" as an ongoing process and an integral part of person-centered care. It includes raising awareness, information, education for the learning of self-management and psychosocial support in the different situations related to the disease and its treatment. The goal is to help patients and their families to develop the ability to self-manage treatment to prevent complications maintaining or improving quality of life. The American Diabetes Association (ADA) [6] defines Diabetes Self-Management Education (DSME) as a continuous process that facilitates the knowledge, skills and attitudes for diabetes self-management.

2.1 Serious Games in Education

Although different definitions for "serious games" exist, they all agree with Abt's original definition of the term [7]: games that have and "explicit and carefully thought-out educational purpose and are not intended to be played primarily for amusement". The author also recognized that this "does not mean that serious games are not, or should not be, entertaining." It has been shown that the cognitive activities required by serious games attract the attention of users [8] and that they provide a relevant number of education benefits [9].

There is a large number of serious games for education in the literature. In [10] a systematic review assessing the effectiveness of serious games in improving knowledge and/or self-management behaviors in young people with chronic conditions is conducted. If we look at serious games for diabetes education in particular, we can find a review of video games for diabetes education in which 9 works with a total of 11 different video games where studied [11]. In [12] several trends in the field of serious games for T1D are analyzed. The conclusion is that the development of serious games for children with diabetes is insufficient.

3 LIDia: Learn and Improve Diabetes

The purpose of LIDia is twofold. On the one hand, the serious game is intended to contribute to diabetes education in a playful way. On the other hand, the final goal is to contribute to the improvement of glycemic control and quality of life of people with T1D thanks to the knowledge about diabetes management acquired with the game.

It is well known that, from the onset of T1D, education on how to manage diabetes starts. Thereafter, patients will have to participate actively to acquire the necessary knowledge for a good management of the disorder. Classical education offered in public hospitals is sometimes insufficient for various reasons: lack of availability of experts in the sector or lack of predisposition of patients to take part in the education in an active way.

3.1 LIDia v1

The first version of LIDia will be a playable prototype that implements a first stress situation. The purpose of this prototype is to have a tool that allows to start an effective collaboration both with health professionals and with possible future users. This collaboration is absolutely essential to design a serious game that fulfills its objective in a safe and entertaining way.

LIDia v1 is a non-cooperative individual game [13]. Conceptually it is an adventure-type game where the game character must face an adverse situation. LIDia v1 stages a hypoglycemia scenario. The decision to implement this scenario in this first version of the video game was clear: hypoglycemia is a key component of diabetes care that interferes with activities of day-to-day living and poses a constant danger to patients.

To deal with the situation, the game character will be able to choose between three options that the game randomly offers. Depending on the user's choice, the game moves to a more favorable or worse state.

The game shows the character's blood glucose curve, simulating continuous glucose monitoring, similar how life or energy bars are displayed in video games. The player's decisions affect the glucose curve. This cause a change from one state to another. Figure 1

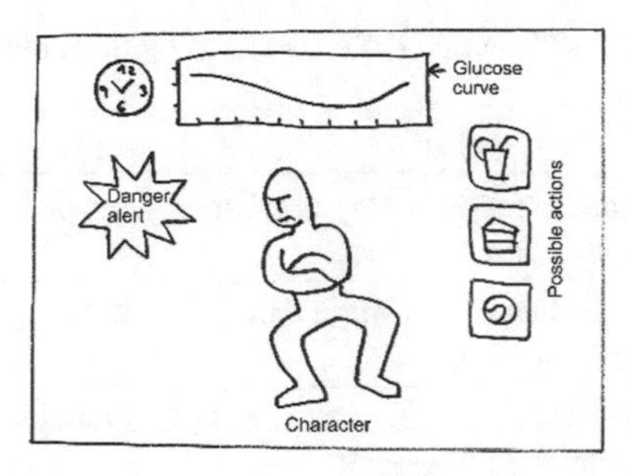

Fig. 1. Prototype of the distribution of the elements of the game

shows the definition of the graphical elements of the game and their distribution on the screen.

The scenario shown in Fig. 1 represents a certain situation of any given day for a person with T1D. A simple animation is used so that the character shows the current situation, also reflected in the glucose curve. The three possible actions appear randomly and the player can choose one.

Hypoglycemia States. Following the ISPAD guidelines for clinical practice consensus for the assessment and management of hypoglycemia in children and adolescents with diabetes [15], for the hypoglycemia scenario initially three different states have been identified: *clinical hypoglycemia alert*, *serious hypoglycemia*, and *severe hypoglycemia*. Table 1 summarizes the different levels of hypoglycemia considered as possible states in de video game.

Table 1. Levels of hypoglycemia

Level of hypoglycemia	Glucose value	Description
Clinical hypoglycemia alert	< =3.9 mmol/L	Threshold value for identifying and treating hypoglycemia
Serious hypoglycemia	<3.0 mmol/L	Clinically important hypoglycemia. Neurogenic symptoms and cognitive dysfunction occur below this level
Severe hypoglycemia	Not defined	Severe cognitive impairments (including coma and convulsions) requiring external assistance

As mentioned before, the game stages the three levels of hypoglycemia as different states. In addition to these three states, other three states are included for the hypoglycemia scenario: *Initial*, *Successful*, and *Final* (see Fig. 2).

Game dynamics has been defined in collaboration with the medical team. At each state the game character can choose between three actions that the game will show randomly. Table 2 shows the six actions for the hypoglycemia scenario.

Table 2. Possible actions to be selected by the game character

Action	Game response
Play	The character is playing. Time passes on the clock. Glucose value goes down
Have a fruit juice	The character drinks a fruit juice. 15 min pass on the clock. Glucose value goes up
Eat cookies	The character eats some cookies. 5 min pass on the clock. Glucose value doesn't change
Administer glucagon	A person administers glucagon to the character. 5 min pass on the clock
Use wildcard	A secondary character performs the correct action depending on the stage
Listen to the bad guy	A secondary character appears. He starts playing or doesn't know what to do

Figure 2 shows the game possible states and transitions designed for the game.

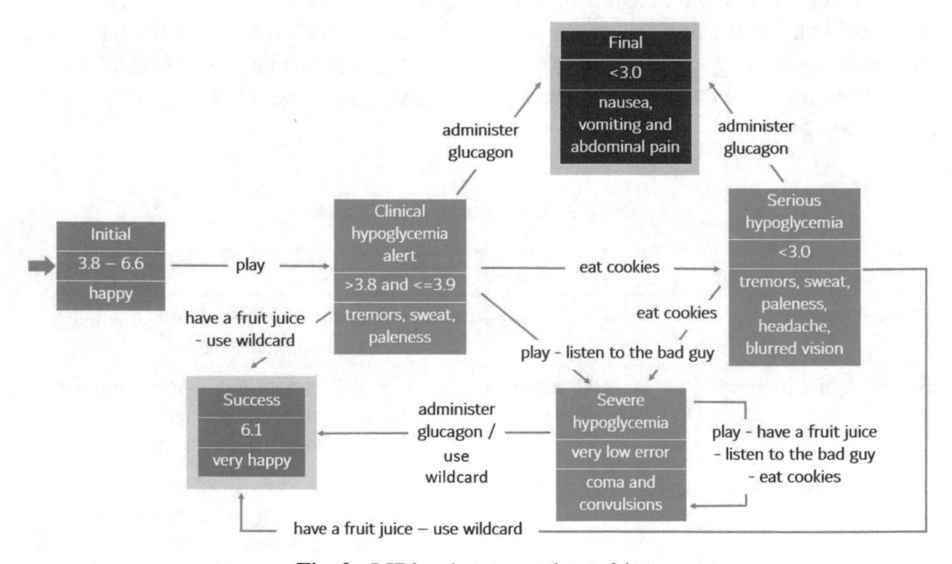

Fig. 2. LIDia v1 states and transitions

In Fig. 2 states are represented as rectangles with three sections: the name of the state, the glucose value (mmol/L) to be shown on the game, and the symptoms the character should display. Transitions are shown with arrows and depend on the action selected by the character at each moment.

Game Design. Figure 3 shows the main graphical interface of LIDia v1 as planned in the prototype (see Fig. 1). Figure 4 represents two storyboards to implement the action named "Play" (see Table 2).

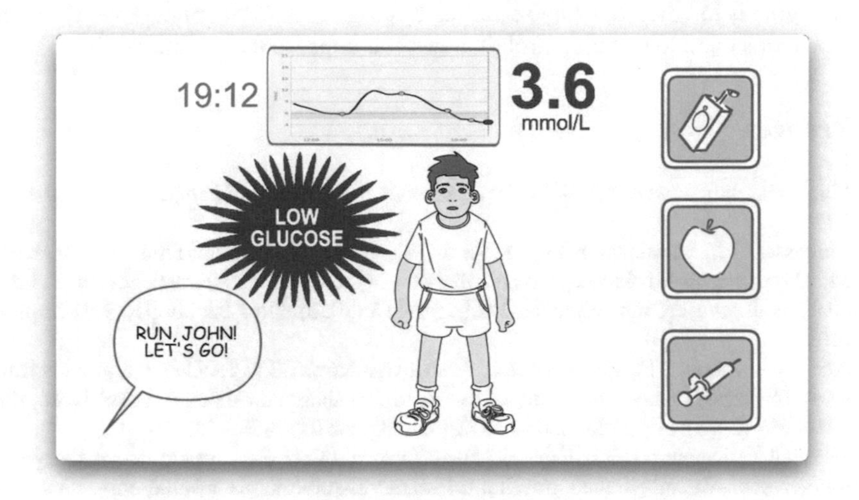

Fig. 3. LIDia v1 interface

Fig. 4. LIDia v1 storyboard examples

4 Development Status and Future Work

LIDia v1 is currently under development. Unity3d is used as a game engine for the development of the first prototype. This tool allows to build mobile games for both iOs and Android. Based on the game objectives, it has been decided to create a full 2D game with flat graphic elements, usually called sprites, and with a camera without perspective. The tool has a 2D editor that facilitates the integration of code and graphics.

A designer and a programmer are currently integrating the sprites that make up the main interface of the game. This initiative is waiting for financial support to advance more efficiently in the development of the serious game.

Once we have an executable version of the game, it will be validated, through a clinical trial, in patients from various public hospitals in the Balearic Islands.

As future work we plan to develop new versions of LIDia that will expand the functionality of the first version with new aspects related to diabetes education. The ultimate goal is to cover as many scenarios as possible to offer patients an educational tool that helps them to better control diabetes and improve their quality of life.

References

1. 'Diabetes'. https://www.who.int/es/news-room/fact-sheets/detail/diabetes. Accessed 10 Oct 2022
2. Mobasseri, M., Shirmohammadi, M., Amiri, T., Vahed, N., Hosseini Fard, H., Ghojazadeh, M.: Prevalence and incidence of type 1 diabetes in the world: a systematic review and meta-analysis. Health Promot. Perspect. **10**(2), 98–115 (2020). https://doi.org/10.34172/hpp.2020.18
3. Amengual Alcover, E., Jaume-i-Capó, A., Moyà-Alcover, B.: PROGame: a process framework for serious game development for motor rehabilitation therapy. PLoS ONE **13**(5), e0197383 (2018). https://doi.org/10.1371/journal.pone.0197383
4. Hospital Universitario Son Espases. https://www.ibsalut.es/es/servicio-de-salud/organizacion/gerencias-ibsalut/gerencia-hospital-universitario-son-espases/hospital-universitari-son-espases. Accessed 18 Oct 2022
5. Therapeutic patient education : continuing education programmes for health care providers in the field of prevention of chronic diseases : report of a WHO working group, p. 90
6. Haas, L., et al.: National standards for diabetes self-management education and support. Diab. Care **35**(11), 2393–2401 (2012). https://doi.org/10.2337/dc12-1707
7. Abt, C.C.: Serious games. University Press of America (1987)
8. The Invisible Computer. MIT Press. https://mitpress.mit.edu/9780262640411/the-invisible-computer/. Accessed 18 Oct 2022
9. Tsekleves, E., Cosmas, J., Aggoun, A.: Benefits, barriers and guideline recommendations for the implementation of serious games in education for stakeholders and policymakers. Br. J. Educ. Technol. **47** (2014). https://doi.org/10.1111/bjet.12223
10. Charlier, N., Zupancic, N., Fieuws, S., Denhaerynck, K., Zaman, B., Moons, P.: Serious games for improving knowledge and self-management in young people with chronic conditions: a systematic review and meta-analysis. J. Am. Med. Inf. Assoc.: JAMIA **23**(1), 230–239 (2016). https://doi.org/10.1093/jamia/ocv100
11. DeShazo, J., Harris, L., Pratt, W.: Effective intervention or child's play? A review of video games for diabetes education. Diab. Technol. Ther. **12**(10), 815–822 (2010). https://doi.org/10.1089/dia.2010.0030
12. Makhlysheva, A., Arsand, E., Hartvigsen, G.: Review of serious games for people with diabetes, pp. 412–447 (2015). https://doi.org/10.4018/978-1-4666-9522-1.ch019
13. The Art of Game Design, 2nd Edition. https://www.oreilly.com/library/view/the-art-of/9781466598645/. Accessed 18 Oct 2022
14. ISPAD clinical practice consensus guidelines 2014 compendium: type 2 diabetes in the child and adolescent. Pediatr. Diab. **16**(5), 392 (2015). https://doi.org/10.1111/pedi.12239
15. ISPAD Clinical Practice Consensus Guidelines 2018 - International Society for Pediatric and Adolescent Diabetes. https://www.ispad.org/page/ISPADGuidelines2018. Accessed 18 Oct 2022

Automatic Hemorrhage Detection in Magnetic Resonance Imaging in Cerebral Amyloid Angiopathy

Tiago Jesus[1(✉)], Cláudia Palma[1], Tiago Gil Oliveira[2], and Victor Alves[1]

[1] ALGORITMI Research Centre/LASI, University of Minho, Braga, Portugal
tiago.jesus@algoritmi.uminho.pt, a85401@alunos.uminho.pt, valves@di.uminho.pt

[2] ICVS – Life and Health Sciences Research Institute, University of Minho, Braga, Portugal
tiago@med.uminho.pt

Abstract. Cerebral hemorrhages, or intracranial hemorrhages, can be caused by the rupture of blood vessels inside the skull. One of the most common causes of brain hemorrhage is cerebral amyloid angiopathy, which can also be associated with Alzheimer's disease. However, the Magnetic Resonance Imaging (MRI) scans in which the hemorrhages can be found must be manually examined and segmented which is very challenging due to the difficult differentiation between hemorrhages and iron deposits or calcifications which makes this a tiresome process susceptible to human error. To improve and automate the detection of brain hemorrhages, deep artificial neural networks were used as they have shown good results in similar applications. The dataset used in this work contains MRI data from 65 patients, using only T2* scans. We propose a two-stage approach in which manual annotation of T2* examinations were used to train a network to identify exams that contain a specific type of hemorrhage, either micro- or macro-hemorrhages. The developed approach achieved an accuracy of 0.94 and 0.9 in detecting slices containing macro and micro hemorrhage, respectively, by using a DenseNet architecture. In the second stage, the identified scans are segmented using an AttentionUnet network which achieved a Dice Score of 0.77 for macro hemorrhages and 0.66 for micro was achieved. This solution provides good results, which proved sufficient for individual detection of hemorrhages however, it still has room for further improvement.

Keywords: Macrohemorrhage · Microhemorrhage · Deep Learning · MRI · Object detection · Segmentation

1 Introduction

Brain hemorrhages, or intracranial hemorrhages, can be caused by blood vessel rupture within the skull. There are various types of hemorrhages such as epidural, subdural, subarachnoid, and intraparenchymal which vary in size, shape, and location of their occurrence in the brain [1, 2].

A. Rocha et al. (Eds.): WorldCIST 2023, LNNS 799, pp. 347–356, 2024.
https://doi.org/10.1007/978-3-031-45642-8_35

Cerebral amyloid angiopathy (CAA) is a cerebrovascular disease characterized by the accumulation of amyloid beta peptides within the walls of blood vessels in the brain [3]. This disorder is reportedly the second most frequent cause of spontaneous intracranial hemorrhage accounting for 5% to 20% of nontraumatic cerebral hemorrhages in elderly patients. These hemorrhages usually occur in the lobar, cortical, or cortical-subcortical locations frequently extending to the subarachnoid space [4, 5].

It is reported that more than 80% of patients with Alzheimer's disease (AD), a neurodegenerative disease that slowly progresses and causes regional atrophy, exhibit CAA, although more widespread and advanced than in non-demented patients [6, 7, 8]. Therefore, it is known that CAA is a key pathological feature in brains with AD.

The detection and assessment of intracranial hemorrhages is very complex and a continuous challenge for physicians. However, this process is decisive for the diagnosis of CAA which is heavily based on the presence of brain hemorrhages as following the Boston criteria that take into consideration clinical symptoms, radiographic features, and pathologic criteria [9].

Magnetic resonance imaging (MRI) scans, particularly, the T2* sequence, can be very advantageous in the visualization and therefore in the detection of these hemorrhages since it detects hemosiderin debris, which causes local magnetic field disturbances [10]. However, the MRI slice images must be segmented manually which is very challenging due to the difficult differentiation of hemorrhages, particularly, microhemorrhages or microbleeds, from regions such as iron deposits, calcifications, and veins making this a tiresome process susceptible to human error and inaccurate results. Therefore, there is a necessity for automated segmentation methods to improve the process of hemorrhage detection.

It is important to note that the relevance and impact of brain hemorrhages in AD patients are still unknown, so a successful automated segmentation method could potentially provide valuable knowledge concerning that domain.

Even so, both the automated segmentation of brain structures and the detection of abnormalities in the brain are persistent challenges. During the last decade, there have been various studies based on deep learning which attempted to find a solution to this problem. Kuijf et al. (2011) utilized radial symmetry transform (RST) [11]. Ghafaryasl et al. (2012) proposed a method based on thresholding and sequentially applied two different classifiers to remove the false positives based on geometrical information and local image descriptors [12]. Bian et al. (2013) developed a semi-automated method with a two-dimensional fast RST (2D-FRST) [13]. Kuijf et al. (2013) utilized the 2D version of RST [14]. Fazlollahi et al. (2014) used a novel cascade of random forest (RF) classifiers trained on radon transform [15]. Chen et al. (2015) trained a 3D convolutional neural network (CNN) and used support vector machine (SVM) [16]. Dou et al. (2015) proceeded with the detection via an unsupervised stacked convolutional Independent Subspace Analysis (ISA) and an SVM [17]. Van den Heuvel et al. (2015) developed a random forest model with shape features [18]. Roy et al. (2015) utilized a 3D-RST [19]. Dou et al. (2016) proposed a two-stage CNN [20]. Zhang et al. (2016, 2017) proposed deep neural networks (DNN) that included autoencoders (SAE) [21, 22]. Wang et al. (2017) trained a five-layer CNN and used a stochastic gradient descent with momentum (SGDM) [23]. Lu et al. (2017) trained an eight-layer CNN with SGDM [24]. Zhang et al.

(2017) used early stopping and leaky rectified linear unit (LReLU) [25]. Pszczolkowski et al. (2018) utilized shape and intensity analysis and thresholding of hyperintensities [26]. Chen et al. (2019) developed a 3D residual network (ResNet) [27]. Liu et al. (2019) proposed a 3D-FRST [28]. Hong et al. (2019) adapted the 2D-ResNet-50 [29]. Al-masni et al. (2020) proposed an approach integrating a regional-based You Only Look Once (YOLO) and a 3D CNN [30]. Myung et al. (2021) proposed an approach integrating YOLO with both single and double labels, as well as CSF filtering [31]. Li et al. (2021) applied feature enhancement in Single-Shot Detector (SSD) algorithms [32]. Rashid et al. (2021) developed a U-Net model [33]. Lee et al. (2022) implemented a method for detection of Cerebral Microbleeds (CMB) in MRI scans using a Single-Stage Triplanar Ensemble Detection Network (TPE-Det) [34]. Suwalska et al. (2022) propose CMB-HUNT, a deep hybrid neural network (HUNT) for automated detection of CMB using susceptibility-weighted imaging (SWI) [35].

Here, we present a two-stage deep learning model for automatic detection of hemorrhages in T2* MRI scans of the brain.

2 Materials and Methods

For brain hemorrhage detection, a two-stage deep learning based solution was developed (Fig. 1). It integrates a DenseNet to identify potential slices containing hemorrhages (Fig. 1 – First Stage) and an AttentionUnet for segmentation of the positively identified slices (Fig. 1 – Second Stage). This process was applied for both micro and macro hemorrhages due to their clinical importance in CAA patients.

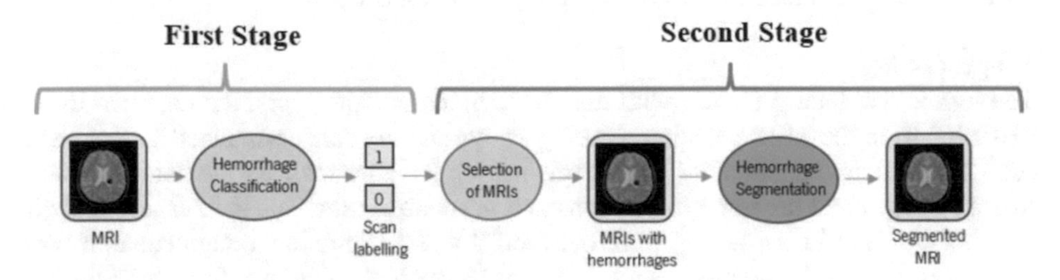

Fig. 1. Overview diagram of the proposed two-stage deep learning approach.

First Stage: DenseNet for Candidate Slice Identification

The DenseNet network architecture from the Monai framework was used to classify the slices (Fig. 1 – First Stage) as it allows maximum information flow between layers and, as it requires fewer parameters, it does not need to learn redundant feature maps. In this phase, each slice went through the DenseNet architecture trained to classify it as 1 or 0, depending on whether that slice is identified as containing at least one hemorrhage of the type for which it was trained (1) or no hemorrhage (0). If 1 is selected, i.e., the corresponding slice is identified as containing one or more hemorrhages, this slice continues with the second stage, segmentation, where the second neural network segments the slice to find out the hemorrhages' location.

Second Stage: AttentionUnet for Hemorrhage Segmentation
As stated before, the selection process for this stage began as soon as the slices were classified as 0 or 1 in the previous step. This means that of all the slices that went through the classification process, only the slices that were identified as 1 (contained hemorrhages) were selected for the segmentation process (Fig. 1 – Second Stage). For this purpose, an AttentionUnet from the Monai framework was used, whose attention mechanism allowed selectively focusing on relevant details regarding hemorrhages and ignoring other redundant details. After using this model, a mask is obtained, which highlights the hemorrhages' location on each slice segmented.

Dataset
To train the proposed deep learning model for hemorrhage detection a dataset of patients with non-traumatic ICH was acquired. This data was obtained in accordance with the Health Ethics Committee by 1.5-T MRI between 2014 and 2017 at Hospital de Braga. The dataset consisted in 971 MRI exams, which included T1-, T2-, T2*-weighted images, and T2-FLAIR, concerning 81 subjects with CAA imaging criteria, however, upon quality-control image selection, only 65 cases were included in this work. The reason for removing the remaining cases was that only T2*-weighted MRI exams were considered, which amounted to only 77 cases, and therefore, the others were excluded. Among the 77 cases, twelve were excluded because it was decided to only utilize one T2* exam per patient and for this reason, three exams were removed, the other nine removed exams contained artifacts that would influence the detection of the hemorrhages.

Preprocessing
To prepare the dataset for use with the above model architectures, the data had to be extracted from the MRI scans into NumPy arrays and the data normalized so that each scan had the same conditions. This normalization includes resizing the data to ensure that all images had the same dimensions and normalizing the values to ensure that all values contained in the arrays were between 0 and 1. Some data augmentation was performed during the preprocessing step as well, such as random x-flips that allowed the images to be flipped while maintaining their shape, random rotations of the images, and random affine. All of these data normalization measures were necessary to ensure that the data input to the model could produce useful results and to ensure that there was sufficient diversity of images to make the model as successful as possible. The data was also divided, for training and testing the model in an 80/20 ratio, i.e., 80% for training and 20% for validation.

Ground Truth Labeling
The labeling of the brain hemorrhages was performed by an expert neuroradiologist on all the slices of the T2* scans. This process was accomplished using the software application ITK-SNAP[1] [36] which allowed to manually segment the hemorrhages with

[1] www.itksnap.org.

the paintbrush mode tool. With this tool, it was possible to paint with an active label which was very useful since four labels were utilized: micro (microhemorrhage) in red contouring, macro (macrohemorrhage) in green contouring, linear (linear hemorrhage) in blue contouring and IV (intraventricular hemorrhage) in yellow contouring.

3 Experimental Results

Classification Process

Various metrics were used to evaluate the effectiveness of the classification process. The different values used were precision, recall, and F1-score. The results obtained are presented in Table 1 and Table 2. It is also important to note that there was some imbalance in the dataset (Table 3), as there were more slices in the dataset that contained hemorrhage, which could affect the results, and for this reason, the macro average and weighted average were also included. As seen below in the Segmentation Process subchapter, it is harder to detect microhemorrhages than it is to detect macrohemorrhages, which was proven by the metrics on the tables. The microhemorrhages can be confused with the blood vessels, or even with iron deposits or calcifications, which leads to a higher error possibility whereas macrohemorrhages, due to their larger size, are usually easier to identify in the exam.

Table 1. Metrics used for the evaluation of the classification process in macrohemorrhages.

	Precision	Recall	F1-score	Support
0	0.94	0.81	0.87	21
1	0.94	0.98	0.96	66
Accuracy			0.94	87
Macro avg	0.94	0.90	0.92	87
Weighted avg	0.94	0.94	0.94	87

Table 2. Metrics used for the evaluation of the classification process in microhemorrhages.

	Precision	Recall	F1-score	Support
0	0.94	0.81	0.87	37
1	0.88	0.96	0.92	54
Accuracy			0.90	91
Macro avg	0.91	0.89	0.89	91
Weighted avg	0.90	0.90	0.90	91

Segmentation Process

To evaluate the segmentation process Dice Score metric was used. This metric measures the intersection of two areas, in this case, the segmentation area predicted by the model and the ground truth segmented area. It has a maximum value of 1 when both areas are completely overlapped, and a minimum of 0 when the areas are fully separate. After training the model, the average Dice Score values were calculated, yielding a value of 0.769 for macrohemorrhages and 0.664 for microhemorrhages. Visual inspection was also performed to check the performance of the model, for both types of hemorrhages, as shown in Fig. 2 and Fig. 3, representing examples of segmentations of macro and micro hemorrhages, respectively.

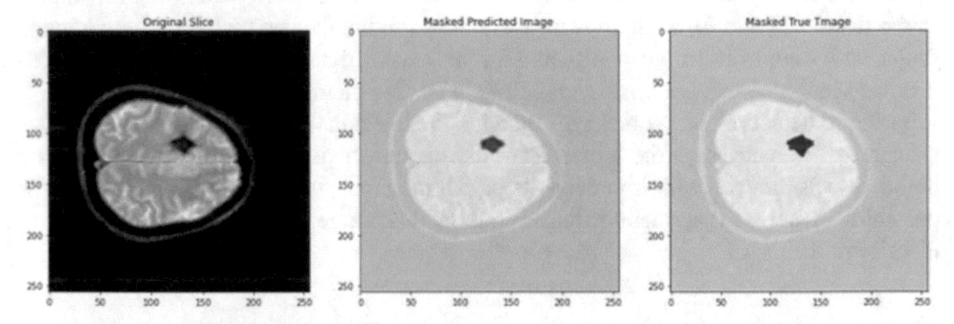

Fig. 2. Example segmentation results for macrohemorrhages.

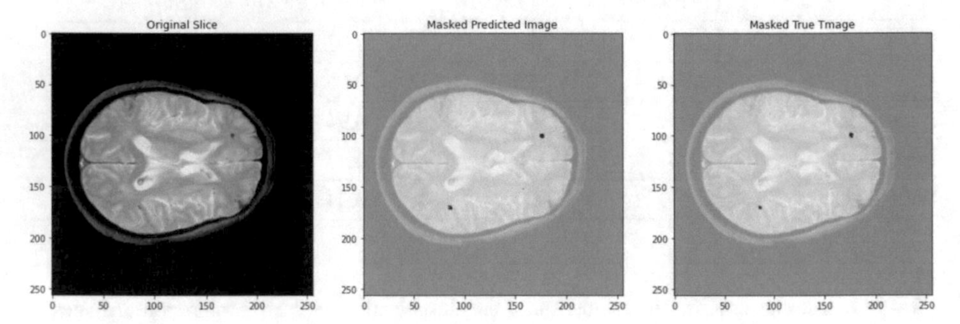

Fig. 3. Example segmentation results for microhemorrhages.

Table 3. Distribution of slices with and without hemorrhage in the models.

	Training (Macro)	Validation (Macro)	Training (Micro)	Validation (Micro)
No hemorrhage	84	21	158	37
Hemorrhage	260	66	205	54

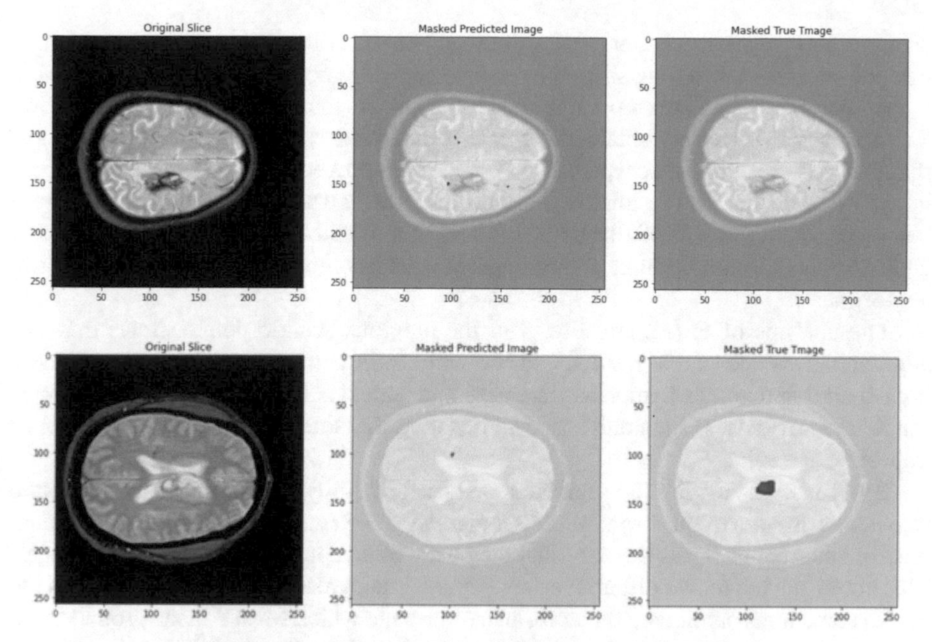

Fig. 4. Example of macrohemorrhage bad segmentation.

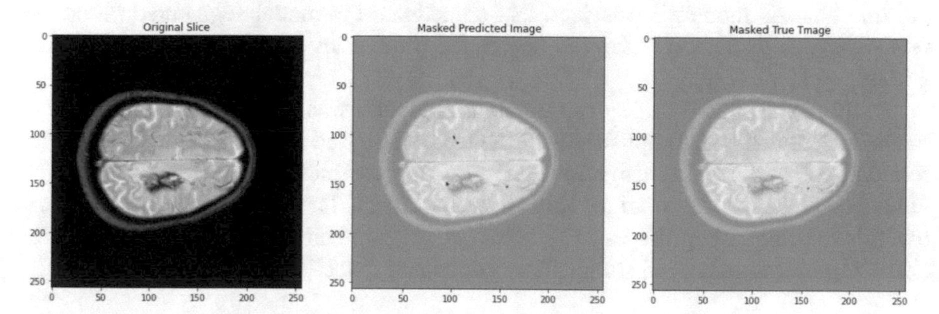

Fig. 5. Example of microhemorrhage bad segmentation.

4 Discussion and Conclusion

Based on the results presented in Sect. 3, it indicates that the classification step of this approach gave good results and immediately excluded most of the slices that did not have hemorrhages, which would otherwise consume an unnecessary amount of computational power in the second stage. As a potential immediate application, it could be used to screen for hemorrhages even in the absence of a neuroradiologist observation and it could identify which scans need an immediate specialist assessment. In this first stage, the accuracy in detecting macro and micro hemorrhages was 0.94 and 0.90, respectively, which is close to perfection. Although not every slice without hemorrhages was eliminated in this stage, almost all slices with hemorrhages made it to the second stage. This is important because it is better to pass slices without bleeding to the second stage than to lose the slices with bleeding since the second stage does not segment in the first option.

As for the second and most important stage, segmentation, the Dice Score shows that a value of 0.769 was obtained for macrohemorrhages and 0.664 for microhemorrhages. These results are acceptable but could be improved by increasing the diversity in the dataset with more patients. Furthermore, the introduction of patients without hemorrhages would potentially be helpful, as this would allow the model to observe what features a brain without hemorrhages has and compare them to those that have hemorrhages.

The analysis of Fig. 2 and Fig. 3 in the previous section demonstrates that this approach gives acceptable results, and is able to train with a small dataset to segment both macro and micro hemorrhages easily and quickly. However, not every slice is segmented correctly (Fig. 4 and Fig. 5), indicating that there is still work to be done to improve the results.

By visually analyzing the segmentations performed by the models, it was possible to detect some recurrent errors in the segmentations. Regarding the model to segment macrohemorrhages it was detected that they were often confused with IV hemorrhages. This confusion led to two different errors in the segmentation results. The first is that the model sometimes predicted, as macro, an hemorrhage which was IV in the ground truth. The other was that the model sometimes predicted nothing in a region that could be IV but was macrohemorrhage in the ground truth (as was captured in Fig. 4). Regarding the microbleeds model, it was seen that sometimes the model segmented blood vessels which, as discussed before, is one of the difficulties for this type of small-sized hemorrhage.

In this work, a two-stage deep learning approach was presented in which hemorrhage detection was accomplished using a DenseNet-based neural network in each slice, followed by candidate segmentation using an AttentionUnet network. This solution provides good results with room for further improvement. The main objective of this approach is to detect hemorrhages to be subsequently confirmed by experts. The accuracy of the segmentation proved sufficient for individual detection of microhemorrhages.

Acknowledgments. Tiago Jesus was supported by a scholarship from Fundação para a Ciência e Tecnologia (FCT), Portugal (Scholarship number 2021.05068.BD). This work was also supported by FCT within the R&D Units Project Scope: UIDB/00319/2020.

References

1. Al-Ayyoub, M., Alawad, D., Al-Darabsah, K., Aljarrah, I.: Automatic detection and classification of brain hemorrhages. WSEAS Trans. Comput. **12**, 395–405 (2013)
2. Tenny, S., Thorell, W.: Intracranial Hemorrhage. Simwars Simulation Case Book: Emergency Medicine, pp. 159–163 (2021)
3. Yamada, M., Naiki, H.: Cerebral Amyloid Angiopathy, 1st ed. Elsevier Inc. (2012)
4. Fischbein, N.J., Wijman, C.A.C.: Nontraumatic intracranial hemorrhage. Neuroimaging Clin. N. Am. **20**, 469–492 (2010). https://doi.org/10.1016/j.nic.2010.07.003

5. Knudsen, K.A., Rosand, J., Karluk, D., Greenberg, S.M.: Clinical diagnosis of cerebral amyloid angiopathy: validation of the Boston criteria. Neurology **56**, 537–539 (2001). https://doi.org/10.1212/WNL.56.4.537
6. Alzheimer's Association: 2019 Alzheimer's disease facts and figures. Alzheimer's Dementia **15**, 321–387 (2019).https://doi.org/10.1016/j.jalz.2019.01.010
7. Greenberg SM Cerebral amyloid angiopathy and Alzheimer disease — one peptide, two pathways. Nat. Rev. Neurol. https://doi.org/10.1038/s41582-019-0281-2
8. Thal, D.R., Griffin, W.S.T., de Vos, R.A.I., Ghebremedhin, E.: Cerebral amyloid angiopathy and its relationship to Alzheimer's disease. Acta Neuropathol. **115**, 599–609 (2008). https://doi.org/10.1007/s00401-008-0366-2
9. Weber, S.A., Patel, R.K., Lutsep, H.L.: Expert review of neurotherapeutics cerebral amyloid angiopathy : diagnosis and potential therapies cerebral amyloid angiopathy : diagnosis and potential therapies. Expert Rev. Neurother. **1** (2018). https://doi.org/10.1080/14737175.2018.1480938
10. Mark, E., Tkach, A., Parrish, B.: Reduction of T2* Dephasing in Gradient Field-Echo Imaging. Radiology **170**, 457–462 (1989)
11. Kuijf, H.J., de Bresser, J., Geerlings, M.I., et al.: Efficient detection of cerebral microbleeds on 7.0 T MR images using the radial symmetry transform. Neuroimage **59**, 2266–2273 (2012). https://doi.org/10.1016/j.neuroimage.2011.09.061
12. Ghafaryasl, B., van der Lijn, F., Poels, M., et al.: A computer aided detection system for cerebral microbleeds in brain MRI. In: Proceedings - International Symposium Biomedical Imaging, vol. 138–141 (2012). https://doi.org/10.1109/ISBI.2012.6235503
13. Bian, W., Hess, C.P., Chang, S.M., et al.: Computer-aided detection of radiation-induced cerebral microbleeds on susceptibility-weighted MR images. Neuroimage Clin. **2**, 282–290 (2013). https://doi.org/10.1016/j.nicl.2013.01.012
14. Kuijf, H.J., Brundel, M., de Bresser J., et al.: Semi-automated detection of cerebral microbleeds on 3.0 T MR images. PLoS One **8,** e66610 (2013)
15. Fazlollahi, A., Meriaudeau, F., Villemagne, V., et al.: Efficient machine learning framework for computer-aided detection of cerebral microbleeds using the radon transform, pp. 113–116 (2014). https://doi.org/10.1109/ISBI.2014.6867822
16. Chen, H., Yu, L., Dou, Q., et al.: Automatic detection of cerebral microbleeds via deep learning based 3D feature representation. In: Proceedings - International Symposium on Biomedical Imaging 2015, pp. 764–767 (2015). https://doi.org/10.1109/ISBI.2015.7163984
17. Dou, Q., Chen, H., Yu, L., et al.: Automatic cerebral microbleeds detection from MR images via independent subspace analysis based hierarchical features. In: Proceedings of the Annual International Conference of the IEEE Engineering in Medicine and Biology Society, EMBS 2015, pp. 7933–7936 (2015). https://doi.org/10.1109/EMBC.2015.7320232
18. van den Heuvel, T.L.A., Ghafoorian, M., van der Eerden, A.W., et al.: Computer aided detection of brain micro-bleeds in traumatic brain injury. In: Medical Imaging 2015: Computer-Aided Diagnosis, vol. 9414, pp. 608–614 (2015). https://doi.org/10.1117/12.2075353
19. Roy, S., Jog, A., Magrath, E., et al.: Cerebral microbleed segmentation from susceptibility weighted images. In: Medical Imaging 2015: Image Processing , vol. 9413, pp. 364–370 (2015). https://doi.org/10.1117/12.2082237
20. Dou, Q., Chen, H., Yu, L., et al.: Automatic detection of cerebral microbleeds from MR images via 3D convolutional neural networks. IEEE Trans. Med. Imaging **35**, 1182–1195 (2016). https://doi.org/10.1109/TMI.2016.2528129
21. Zhang, Y.D., Hou, X.X., Lv, Y.D., et al.: Sparse autoencoder based deep neural network for voxelwise detection of cerebral microbleed. In: Proceedings of the International Conference on Parallel and Distributed Systems – ICPADS, pp. 1229–12322016). https://doi.org/10.1109/ICPADS.2016.0166

22. Zhang, Y.-D., Zhang, Y., Hou, X.-X., Chen, H., Wang, S.-H.: Seven-layer deep neural network based on sparse autoencoder for voxelwise detection of cerebral microbleed. Multimed. Tools Appl. **77**(9), 10521–10538 (2017). https://doi.org/10.1007/s11042-017-4554-8
23. Wang, S., Jiang, Y., Hou, X., et al.: Cerebral micro-bleed detection based on the convolution neural network with rank based average pooling. IEEE Access **5**, 16576–16583 (2017). https://doi.org/10.1109/ACCESS.2017.2736558
24. Lu, S., Lu, Z., Hou, X., et al.: Detection of cerebral microbleeding based on deep convolutional neural network. In: 2016 13th International Computer Conference on Wavelet Active Media Technology and Information Processing, ICCWAMTIP 2017, pp. 93–96 (2017). https://doi.org/10.1109/ICCWAMTIP.2017.8301456
25. Zhang, Y.D., Hou, X.X., Chen, Y., et al.: Voxelwise detection of cerebral microbleed in CADASIL patients by leaky rectified linear unit and early stopping. Multimed. Tools Appl. **77**, 21825–21845 (2017). https://doi.org/10.1007/s11042-017-4383-9
26. Pszczolkowski, S., Law, Z.K., Gallagher, R.G., et al.: Automated segmentation of haematoma and perihaematomal oedema in MRI of acute spontaneous intracerebral haemorrhage. Comput. Biol. Med. **106**, 126–139 (2019). https://doi.org/10.1016/j.compbiomed.2019.01.022
27. Chen, Y., Villanueva-Meyer, J.E., Morrison, M.A., Lupo, J.M.: Toward automatic detection of radiation-induced cerebral microbleeds using a 3D deep residual network. J. Digit. Imaging **32**, 766–772 (2019). https://doi.org/10.1007/s10278-018-0146-z
28. Liu, S., Utriainen, D., Chai, C., et al.: Cerebral microbleed detection using susceptibility weighted Imaging and deep learning. Neuroimage **198**, 271–282 (2019). https://doi.org/10.1016/j.neuroimage.2019.05.046
29. Hong, J., Cheng, H., Zhang, Y.D., Liu, J.: Detecting cerebral microbleeds with transfer learning. Mach. Vis. Appl. **30**, 1123–1133 (2019). https://doi.org/10.1007/s00138-019-01029-5
30. Al-masni, M.A., Kim, W.R., Kim, E.Y., et al.: Automated detection of cerebral microbleeds in MR images: a two-stage deep learning approach. Neuroimage Clin. **28**, 102464 (2020). https://doi.org/10.1016/j.nicl.2020.102464
31. Myung, M.J., Lee, K.M., Kim, H.G., et al.: Novel approaches to detection of cerebral microbleeds: single deep learning model to achieve a balanced performance. J. Stroke Cerebrovasc. Dis. **30**, 105886 (2021). https://doi.org/10.1016/j.jstrokecerebrovasdis.2021.105886
32. Li, T., Zou, Y., Bai, P., et al.: Detecting cerebral microbleeds via deep learning with features enhancement by reusing ground truth. Comput. Methods Prog. Biomed. **204**, 106051 (2021). https://doi.org/10.1016/j.cmpb.2021.106051
33. Rashid, T., Abdulkadir, A., Nasrallah, I.M., et al.: DEEPMIR: a deep neural network for differential detection of cerebral microbleeds and iron deposits in MRI. Sci. Rep. **11**, 1–14 (2021). https://doi.org/10.1038/s41598-021-93427-x
34. Lee, H., Kim, J.H., Lee, S., et al.: Detection of cerebral microbleeds in MR images using a single-stage triplanar ensemble detection network (TPE-Det). J. Magn. Reson. Imaging (2022). https://doi.org/10.1002/JMRI.28487
35. Suwalska, A., Wang, Y., Yuan, Z., et al.: CMB-HUNT: automatic detection of cerebral microbleeds using a deep neural network. Comput. Biol. Med. **151**, 106233 (2022). https://doi.org/10.1016/J.COMPBIOMED.2022.106233
36. Yushkevich, P.A., Piven, J., Hazlett, H.C., et al.: User-guided 3D active contour segmentation of anatomical structures: significantly improved efficiency and reliability. Neuroimage **31**, 1116–1128 (2006). https://doi.org/10.1016/J.NEUROIMAGE.2006.01.015

Subthalamic Nucleus and Substantia Nigra Automatic Segmentation Using Convolutional Segmentation Transformers (Conv-SeTr)

Juan Nebel[1], Franklin E. Magallanes Pinargote[1], Colon Enrique Peláez[1(✉)], Francis R. Loayza Paredes[2], and Rafael Rodriguez-Rojas[3]

[1] Electrical and Computer Science Engineering Department Escuela Superior Politécnica Del Litoral - ESPOL University, Guayaquil, Guayas 090150, Ecuador
epelaez@espol.edu.ec

[2] Mechanical and Production Science Engineering Department Escuela Superior Politécnica del Litoral - ESPOL University, Guayaquil, Guayas 090150, Ecuador

[3] HM CINAC (Centro Integral De Neurociencias Abarca Campal), Hospital Universitario HM Puerta Del Sur, Madrid, Spain

Abstract. The Subthalamic Nucleus and Substantia Nigra have an important role in the treatment of Parkinson's Disease (PD); however, they are difficult to identify on magnetic resonance imaging (MRI) and are of paramount importance in PD, which requires their precise localization. We present a pipeline methodology that allows autonomous segmentation of both structures, based on MRI T2-weighted images and Deep Learning techniques. Three segmentation architectures were compared: CLCI-Net, 3D U-Net and Conv-SeTr. All models were trained in two instances: the first with 60 T2-weighted standard protocol images from 1.5T MRI. Transfer learning was applied for the second instance in which the models were trained with 20 T2-weighted adjusted protocol images from 3T MRI. In all cases, the ground truth was obtained through manual segmentation by experts. All models produced an image mask with the segmented labelled structures co-registered to native space. The proposed transformer-based model segmented the volumes of interest with a DICE coefficient of 0.81 and an AVD of 0.06, which outperformed the other architectures. The Conv-SeTr presented promising results for segmenting Subthalamic Nucleus and Substantia Nigra, key structures in Parkinson's disease research and treatment.

Keywords: Subthalamic Nucleus · Substantia Nigra · Segmentation · Deep Learning

1 Introduction

Parkinson's disease (PD) is a progressive neurological disorder which causes tremors, rigidity, balance, sleeping and memory problems, among others. The

A. Rocha et al. (Eds.): WorldCIST 2023, LNNS 799, pp. 357–367, 2024.
https://doi.org/10.1007/978-3-031-45642-8_36

Substantia Nigra (SN) is a nucleus of the basal ganglia that gets its name from the pigmented dopaminergic neurons or cells that comprise it. The pathological process of PD progresses as these cells which the SN is made of, are lost due to the disease [16]. At large, PD is a complex health issue that requires more research and communication channels to minimize the effects of the disease [10].

The Subthalamic Nucleus (STN) is a small structure located at the junction between the midbrain and the diencephalon [6], and it is part of the subthalamus, which regulates movement. The STN contains a significant number of glutaminergic neurons, which play an excitatory neurotransmitter role for the central nervous system [2]. The STN is connected to the SN via limbic connections. Both the STN and the SN are part of a bigger chain for motor regulation in which each of it's links excite or inhibit signals that are expressed as movement [6].

Current treatment options for PD aim to mitigate the crippling symptoms associated with the disease because there is not a definitive cure yet [1,7]. Such options range from ablations [19], Deep Brain Stimulation (DBS) [22] to recent High-intensity Focused Ultrasound (HIFU) [24].

These interventions require precise localization of those components as well as technological innovation and better image integration [3]; however, the problem exacerbates when identifying the STN and the SN using medical imaging, due to several factors: Low contrast between targets and the surrounding structures; high amounts of iron present in the region, which affects the quality of the resulting image; and, a high signal to noise ratio present mainly in resonators below 1.5 T, a common MRI for medical purposes [12].

In this work a pipeline methodology for identifying these brain structures is proposed, using Deep Learning (DL) techniques for automatic segmentation.

2 Related Work

The automatic segmentation process for the STN used the optimized sub-cortical non-linear registration between the atlas-based segmentation and the patient native space [8], and it was carried out by Advanced Normalization Tools (ANTs) [5]. However, up to today, it has not been found in the literature methods based on DL segmentation techniques, such as those performed for other brain structures. Chronic stroke lesion segmentation has been achieved by using a Cross-Level Fusion (CLF) and Context Inference Network (CLCI-Net) [11,18], which outperformed state-of-the-art methods. Other brain structures, such as brain tumours have also been segmented using 3D U-Net convolutional architectures, using federated learning procedures to achieve the training process [20]. Other U-Net convolutional-based models have also been successfully used, such as in [23], where a fully CNN architecture was adapted and tailored for segmenting White Matter lesions related to cognitive impairment. That model used T1-weighted and FLAIR images. Among other similar approaches, as those proposed in [4,13,15,17].

As an alternative to the typical convolutional architectures, Zheng S. et al. [21] proposed a model for performing semantic segmentation, as a sequence-to-sequence prediction task based on transformers, which is known as SETR (SEgmentation TRansformer), where the input images were encoded as a sequence of patches, then a simple decoder paired the completed model yielding a high-performing segmentation technique. Later, this method was also proposed for localizing and segmenting prostate cancer [25].

3 The Dataset and Methods

The dataset included 60 3D T2-weighted MRIs with 2-mm separation between slices and 1-mm interpolation and 20 3D T2-weighted MRIs with native 1-mm between slices. All images were in standard space. The ground truth was performed by experts radiologist, segmenting and labelling manually the bilateral STN and SN with four labels. The proposed pipeline methodology is illustrated in Fig. 1.

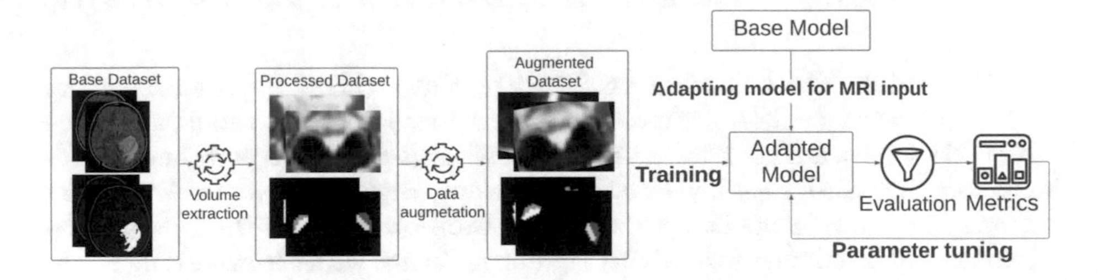

Fig. 1. Data flow and life cycle

3.1 Dataset Pre-processing

To reduce the size of the volume of interest and focus the search on the brain's target region, the coordinates enclosing the slices of interest were recorded, and then a plotted distribution of values was constructed to isolate the volume containing the bilateral SN and STN.

3.2 Data Augmentation

With the reduced volume of interest, a horizontal-vertical flipping, bias correction, intensity normalization and image rotation, with 4 possible degrees (0, 90, 180 and 270) were applied. The Cartesian product between these transformations produced 64 possible combinations, i.e., if the dataset's cardinality was 60 examples, it grew up to 3.840 examples after these data augmentation operations were applied.

3.3 The Base Models and Their Performance

In this work, 3 base models were chosen to be tested and compared using a similar set of hyper-parameters; then, the best model was selected for further tuning to improve its performance when executing a segmentation task on images from 3T MRI. The 3 models were based on the following architectures:

3D U-Net Model: An encoder-decoder architecture, where the encoder section is responsible for receiving an image and returning a feature map, while the decoder section reconstructs the segmented image [14]. This architecture has been widely used for segmenting brain structures from 2D and 3D images [25].

CLCI-Net Model: This architecture is also based on an encoder-decoder paradigm. It features 3D convolutional layers and 1D filters [11]. However, its main difference lies in which type of layer the skip connections are carried out. These connections are not performed by a concatenation operation, but by an LSTM (Long Short Term Memory)-convolutional layer, to capture the patterns within the hidden states; that is, the latent feature maps in skip connections [11].

SETR Model: The behaviour exhibited by this architecture is analogous to CNN layer-based models, except that the encoder is based on multi-attention layers [21]. In this architecture, a series of patches go through a positional encoding before the transformer layers, and then to an upsampling layer as the encoder output. The input image is encoded by the transformer layer, then decoded by convolutional layers that reconstruct the output image with the same dimensions and channels as the input [21].

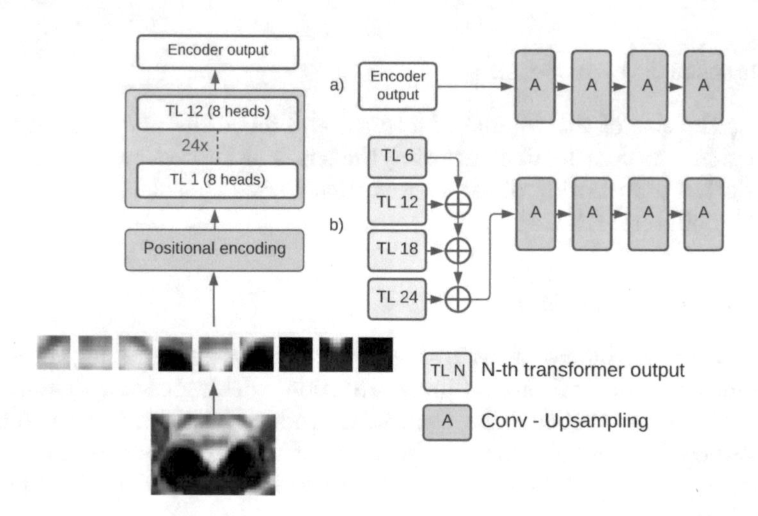

Fig. 2. SETR architecture diagram. (a) PUP (b) MLA

The encoder section is a transformer with positional encoding and codification of patches. That is, a transformation block with normalization, multi-attention and normalization layers, then an MLP layer. The decoder section has two options: Progressive Upsampling (PUP) and Multilevel Feature Aggregation (MLA) [21]. The PUP consists of taking the last encoder output as the decoder input and applying four upsampling convolution blocks to reconstruct the image with the segmented objects of interest [21]. The MLA option of the SETR architecture utilizes skip connections to keep relevant patterns in the hidden state during the learning process, as shown in Fig. 2.

Models Refinements. As shown in Table 1, some hyper-parameters and the models' architectures were changed to fit our segmentation task. In particular in the output layer.

Table 1. Changes made to the models

	Before	After
U-Net	Returns a matrix of shape $80 \times 80 \times 3$, where 3 is the number of channels	Returns a matrix of shape $80 \times 80 \times 56$, where 56 is the number of channels, that represent the amount of MRI slices
CLCI-Net	Returns a matrix of shape $80 \times 80 \times 3$, where 3 is the number of channels	Returns a matrix of shape $80 \times 80 \times 56$, where 56 is the number of channels, that represent the amount of MRI slices
SETR	Pytorch implementation	Tensorflow implementation

Evaluation Metrics. For evaluating all models we used the metrics as specified in [9]: Precision, to measure how correctly the voxels were classified as part of the volume of interest; recall, to evaluate the model's ability to find all voxels belonging to the same volume; F1 Score, to determine the harmonic mean between precision and recall and evaluate the model's trade-off between these 2 metrics without getting blind sighted by focusing on just one; a DICE Coefficient, to measure the overlapping between the ground truth segmentation and the model's output; the Hausdorff Distance, to evaluate the distance between two subsets belonging to the same metric space; and, the Average Volume Distance (AVD) to show the percentage difference in volume, between the ground truth and the model's segmentation, relative to the ground truth segmentation.

3.4 Comparison of the Base Models' Performance

As Fig. 3 shows, the SETR base model performed better in segmenting the volumes of interest, as compared with the other reference models. The DICE coefficient as depicted in Fig. 3b, as well as the F1-Score in Fig. 3e are noticeably better.

CLCI-Net and U-Net based models' performance was inferior to the observed performance of the SETR base model; in particular, the metrics showed that the U-Net model was not able to learn the composing 3D structures.

The loss function for both the SETR and the CLCI-Net converged, however, the SETR model was faster, as can be seen in Fig. 3a. Precision also performed better for both models, while the CLCI-Net showed a decrease in the recall, as it can be drawn from Figs. 3c and 3d respectively.

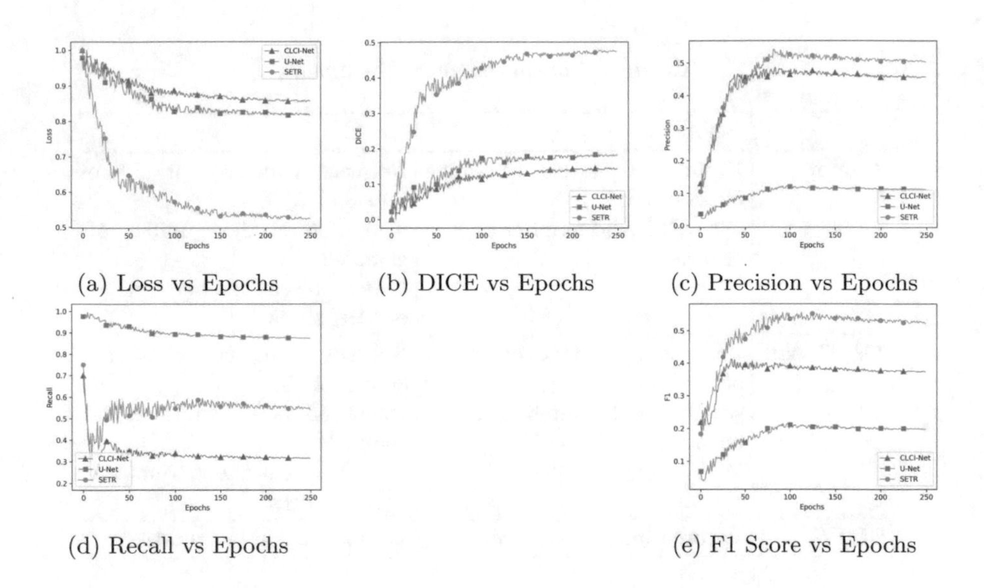

(a) Loss vs Epochs (b) DICE vs Epochs (c) Precision vs Epochs

(d) Recall vs Epochs (e) F1 Score vs Epochs

Fig. 3. Training dataset base models' metrics

During training, cross-validation was performed, then testing, showing similar performance behaviour in all metrics for the SETR base architecture. However, for the other architectures, the performance was lower; for the CLCI-Net case, although the tendency followed the training behaviour, there were cycles in the testing phase that overshot the target.

Once these based models were analyzed, the SETR architecture was chosen to be further tuned for segmenting the STN and SN structures, as described in the next section.

4 The Proposed Model Architecture

We propose a Convolutional Segmentation Transformer (Conv-SeTr), which builds upon the proven ability to learn effective image representations using Convolutional Neural Networks, combined with the ability of Transformers to effectively capture attention dependencies in its inputs. The methodology processes the input images in two stages: First, four convolutional layers create a feature map that summarizes the presence of the target structures. Then, the transformer captures the hierarchical global attributes from the features.

The proposed architectural changes on the transformer-based model are shown in Fig. 4; the relevant redesign features and hyper-parameters are explained in the Table 2.

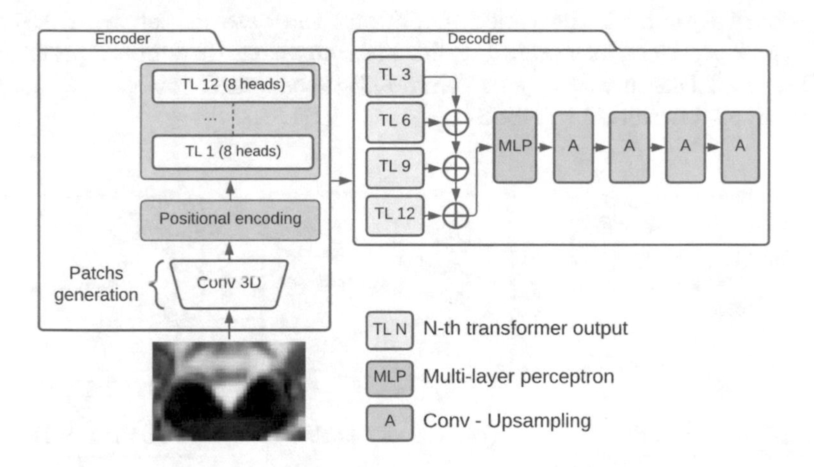

Fig. 4. Proposed Conv-SeTr architecture

Table 2. Changes made to the SETR model

Base SETR	Proposed SETR
Patches were obtained with direct slices in the 3D MRI volume	The method for obtaining the patches was redefined from slicing the input to using a 3D convolutional layer first, this allowed the model to collect the patterns' features early on
The output of the transformer layers was concatenated and passed directly to the 3D Convolution layers	The output from the transformer was passed through a fully connected layer. This layer processes the hidden state of the intermediate layers of the transformer. As a result, a new hidden state with $10 \times 10 \times 7 \times 25$ tensors is reconstructed

This Conv-SeTr model was trained using cross-validation and then testing. As Fig. 5 shows, the set of metrics for both the base and variant models were contrasted, showing better results for the Conv-SeTr than those obtained with the base model. The DICE coefficient improved up to 0.7, as shown in Fig. 5b. In addition, convergence to these values was faster than in the base model.

With this improved architecture, the model was also tested using 20 3D T2-weighted MRIs with native 1-mm separation between slices, the results are shown in Fig. 5, yielding a DICE coefficient of 0.8, as seen in Fig. 5b, with an F1-Score and a convergence rate similar to the scores obtained during training and testing, as Fig. 5e shows. Fast convergence of the loss function in all three models was also observed before the 50 epochs mark; however, the proposed variant model managed to outperform the base models early on, in terms of precision as seen in Figs. 5a and 5c. For the 20 3D T2-weighted MRIs test, the recall was also outperformed for both the tuned model and the base model; with 0.87, 0.70 and 0.74 respectively, as seen in Fig. 5d. For comparing the models' performance the Hausdorff Distance, Average Volume Distance and Cross-entropy were also measured, as described in Table 3.

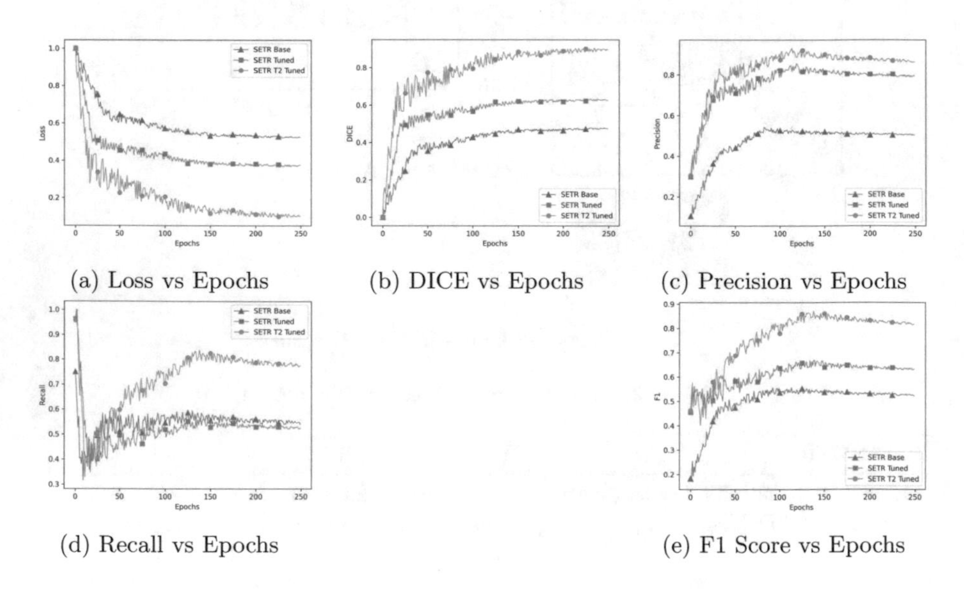

(a) Loss vs Epochs (b) DICE vs Epochs (c) Precision vs Epochs

(d) Recall vs Epochs (e) F1 Score vs Epochs

Fig. 5. Conv-SeTr model's metrics compared to 3D-Unet and CLCI-Net

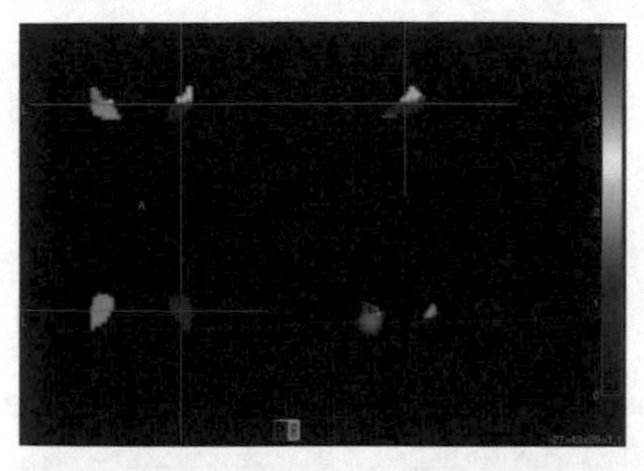

(a) SETR base model segmentation as seen in MRICron 1.

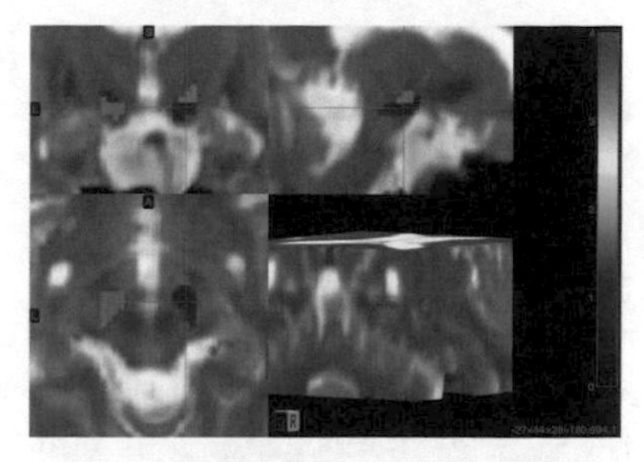

(b) SETR base model segmentation as seen in MRICron 2.

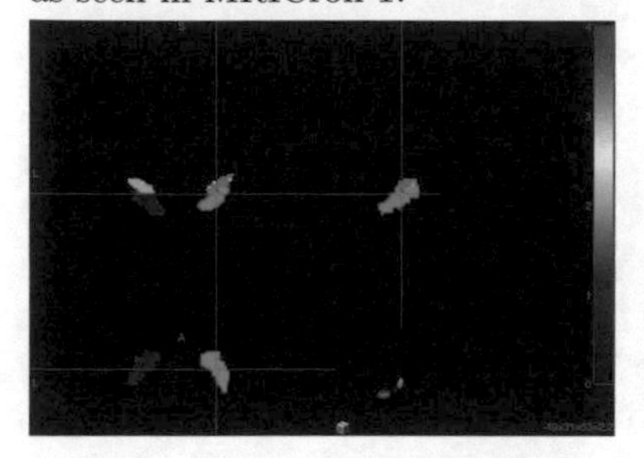

(c) SETR variant model segmentation as seen in MRICron 1.

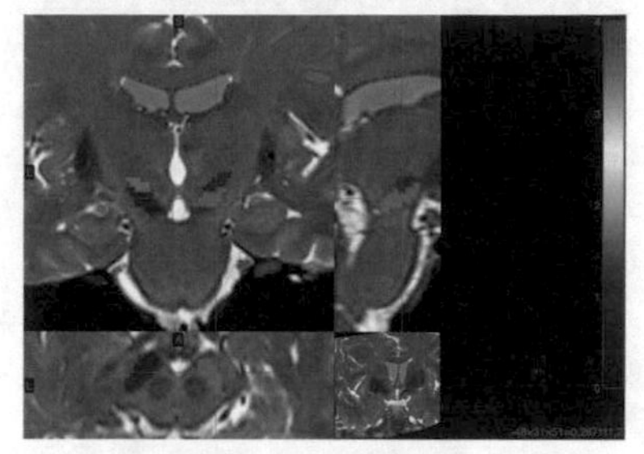

(d) SETR variant model segmentation as seen in MRICron 2.

Fig. 6. 3D segmentation results on the different models tested

Table 3. Results for the testing portion of the dataset.

Metric	CLCI-Net 3D	U-Net 3D	SETR 3D base	Conv-SeTr
DICE	0.12	0.17	0.59	**0.85**
Recall	0.32	0.89	0.51	**0.81**
Precision	0.43	0.10	0.67	**0.84**
F1 Score	0.35	0.19	0.62	**0.85**
Loss	0.88	0.83	0.41	**0.15**
AVD	0.74	429.99	0.16	**0.06**
HD	45.36	51.95	3.46	**2.40**
Cross Entropy	0.03	16.10	0.01	**0.001**

The DICE coefficient shows that the overlapping of the predicted space also improved significantly with the Conv-SeTr architecture; from 0.59 to 0.85, after tuning and 0.81 on the test dataset. Precision and recall were also outperformed by the Conv-SeTr proposed variant, with 0.83 and 0.87 respectively. As for the F1-Score, the proposed variant had a 0.85 score, better than the others.

Figures 6a, 6b, 6c and 6d show the segmentations created by the SETR base and variant models, which clearly delineate the components of interest, and group them correctly as defined in the ground truth; where the left STN is in red, right STN in yellow, left SN in green and right SN in blue, along the coronal plane.

5 Conclusions

For the 3D image segmentation task, the Conv-SeTr architecture showed the best performance, and it is suited for STN and SN segmentation tasks using 3D MRIs. Metrics such as DICE, precision, recall and loss all outperformed as compared with the U-Net and CLCI-Net models.

Changes in the SETR architecture significantly influenced the performance of the baseline architecture, from an F1 Score of 0.63 to 0.70, and to 0.85 when the model was evaluated using better-quality images. Changes included using a 3D convolutional layer and feeding the output of the transformer to a fully connected layer.

The Conv-SeTr architecture turned out to be a better choice as compared to the CLCI-Net model, mainly because with the latter we faced unacceptably slow training time due to its LSTM section, which did not allow for its parallelization, as with the Conv-SeTr model.

Also, since we used 3D images the 3D U-Net architecture demanded considerably more RAM at the GPU, mainly due to their 3D convolutional layers, which was notably higher than the number of convolutions in the proposed version of Conv-SeTr model.

References

1. Abteen Mostofi, F.M., et al.: Pain in Parkinson's disease and the role of the subthalamic nucleus. Brain **144**, 1342–1350 (2021). https://doi.org/10.1093/brain/awab001
2. Alexander, G.E., DeLong, M.R., Strick, P.L.: Parallel organization of functionally segregated circuits linking basal ganglia and cortex. Annu. Rev. Neurosci. **9**, 357–381 (1986). https://doi.org/10.1146/annurev.ne.09.030186.002041
3. Andres, M., Lozano, N.L., et al.: Deep brain stimulation: current challenges and future directions (2019). https://doi.org/10.1038/s41582-018-0128-2
4. Anindya A, N.I., et al.: UNet-vgg16 with transfer learning for MRI-based brain tumor segmentation. Telkomnika (Telecommunication Computing Electronics and Control) **18**, 1310–1318 (2020). https://doi.org/10.12928/TELKOMNIKA.v18i3.14753
5. Avants, B.B., Tustison, N., Song, G., et al.: Advanced normalization tools (ANTs). Insight J. **2**(365), 1–35 (2009)
6. Basinger, H., Joseph, J.: Neuroanatomy, Subthalamic Nucleus (2022)
7. Braak, H., Del Tredici-Braak, K., Gasser, T.: Special issue Parkinson's disease. Cell Tissue Res. **373**(1), 1–7 (2018). https://doi.org/10.1007/s00441-018-2863-5
8. Ewert, S., Horn, A., Finkel, F., Li, N., Kühn, A.A., Herrington, T.M.: Optimization and comparative evaluation of nonlinear deformation algorithms for atlas-based segmentation of DBS target nuclei. Neuroimage **184**, 586–598 (2019)

9. Goutte, C., Gaussier, E.: A probabilistic interpretation of precision, recall and F-score, with implication for evaluation. In: Losada, D.E., Fernández-Luna, J.M. (eds.) ECIR 2005. LNCS, vol. 3408, pp. 345–359. Springer, Heidelberg (2005). https://doi.org/10.1007/978-3-540-31865-1_25
10. Grimes, D., et al.: Canadian guideline for Parkinson disease. CMAJ **191** (2019). https://doi.org/10.1503/cmaj.181504
11. Yang, H., et al.: CLCI-Net: cross-level fusion and context inference networks for lesion segmentation of chronic stroke. In: Shen, D., et al. (eds.) MICCAI 2019. LNCS, vol. 11766, pp. 266–274. Springer, Cham (2019). https://doi.org/10.1007/978-3-030-32248-9_30
12. Hutchinson, M., Raff, U.: Structural changes of the substantia nigra in Parkinson's disease as revealed by MR imaging. Am. J. Neuroradio. **21**, 697–701 (2000)
13. Hwang, H., Rehman, H.Z.U., Lee, S.: 3D U-Net for skull stripping in brain MRI **9** (2019). https://doi.org/10.3390/app9030569
14. Iglovikov, V., Shvets, A.: TernausNet: U-Net with VGG11 encoder pre-trained on imagenet for image segmentation (2018)
15. Jethi, A.K., Murugesan, B., et al.: Dual-encoder-Unet for fast MRI reconstruction. Institute of Electrical and Electronics Engineers Inc. (2020). https://doi.org/10.1109/ISBIWorkshops50223.2020.9153453
16. Prakash, K.G., Bannur, B.M., et al.: Neuroanatomical changes in Parkinson's disease in relation to cognition: an update. J. Adv. Pharm. Technol. Res. **7** (2016). https://doi.org/10.4103/2231-4040.191416
17. Li, H., et al.: Fully convolutional network ensembles for white matter hyperintensities segmentation in MR images. Neuroimage **183**, 650–665 (2018)
18. Mena, R., Macas, A., Pelaez, E., Loayza, F., Franco-Maldonado, H.: A pipeline for segmenting and classifying brain lesions caused by stroke: A machine learning approach. In: Rocha, A., Adeli, H., Dzemyda, G., Moreira, F. (eds.) World Conference on Information Systems and Technologies, vol. 470, pp. 415–424. Springer, Cham (2022). https://doi.org/10.1007/978-3-031-04829-6_37
19. Obeso, J., Rodriguez, M., Gorospe, A., Guridi, J., Alvarez, L., Macias, R.: Surgical treatment of Parkinson's disease. Bailliere's Clin. Neurol. **6**(1), 125–145 (1997)
20. Pati, S., et al.: Federated learning enables big data for rare cancer boundary detection. arXiv preprint arXiv:2204.10836 (2022)
21. Sixiao Zheng, J.L., et al.: Rethinking semantic segmentation from a sequence-to-sequence perspective with transformers (2021). https://doi.org/10.48550/arXiv.2012.15840
22. Uma, V., Mahajan, V.K.R., et al.: Bilateral deep brain stimulation is the procedure to beat for advanced Parkinson disease: a meta-analytic, cost-effective threshold analysis for focused ultrasound. Neurosurgery **88**, 487–496 (2021). https://doi.org/10.1093/neuros/nyaa485
23. Viteri, J., Pelaéz, E., Loaiza, F., Layedra, F.: U-Net CNN model for segmentation of white matter hyperintensities (2020)
24. Zahra Izadifar, Z.I., et al: An introduction to high intensity focused ultrasound: systematic review on principles, devices, and clinical applications. J. Clin. Med. **9** (2020). https://doi.org/10.3390/jcm9020460
25. Zhu, L., Han, C., et al.: U-Net deep learning network for automatic segmentation and localization of prostate cancer on MRI apparent diffusion coefficient map. Chin. J. Radiol. **54**, 974–979 (2020). https://doi.org/10.3760/cma.j.cn112149-20191004-00745

Intelligent and Decision Support Systems

Application of Binary Petri Nets to Knowledge Representation and Inference

Zbigniew Suraj(✉)

Institute of Computer Science, College of Natural Sciences, University of Rzeszów, Rzeszów, Poland
zsuraj@ur.edu.pl

Abstract. In this paper, we present a binary Petri net model designed for knowledge representation and inference in rule-based systems. We assume that knowledge and inference in such systems is expressed in the formalism of classic logic. The usefulness of binary Petri nets is analyzed as an example of a simple rule-based train traffic control system. The exemplary model was additionally subjected to experimental verification and assessed using artificially generated test data for this model.

Keywords: Binary Petri nets · Knowledge representation · Inference · Rule-based systems · Control

1 Introduction

Petri nets are a well-known model of concurrent systems [1]. They have also become an important and popular computational paradigm for intelligent systems because they provide a pictorial language for visualizing, communicating, and interpreting engineering problems [2,3], and are relatively easy to describe in the precise language of mathematics. The concept of a Petri net has its origin in [4]. A lot of extensions of Petri nets have been proposed improving such aspects as hierarchical nets, high-level nets or temporal nets [3].

Over the past decades, Petri nets have been gaining a growing interest among people in Artificial Intelligence [5] due to its adequacy to represent the reasoning process as a dynamic discrete event system [6]. A number of further modifications were made towards the so-called fuzzy Petri nets (FPNs) using the fuzzy set methodology [7], its generalizations or extensions. The first work in this direction is [8]. It is also worth mentioning and briefly discussing several other works in this field. For example, Chen *et al.* [9] used the fuzzy Petri net model for knowledge representation. Li & Lara-Rosano [10] developed adaptive fuzzy Petri nets for dynamic knowledge representation and inference. Pedrycz & Gomide [11] introduced a generalized fuzzy Petri net model. Skowron & Suraj [12] developed a parallel algorithm for real-time decision-making based on rough set theory and classic Petri nets [1]. Peters *et al.* [13] combined the theory of FPNs, rough sets, and colored Petri nets to develop a rough fuzzy Petri net model. Suraj & Hassanien [14] combined the theory of FPN and interval fuzzy

A. Rocha et al. (Eds.): WorldCIST 2023, LNNS 799, pp. 371–381, 2024.
https://doi.org/10.1007/978-3-031-45642-8_37

sets [15]. Bandyopadhyay *et al.* [16] proposed to link Petri nets and soft sets [17]. Liu *et al.* [18] proposed to integrate intuitionistic fuzzy sets [19] with FPN. Suraj [20] presented an approach to building a fuzzy Petri net modeling a real-time decision-making system based on knowledge extracted from empirical data. Liu *et al.* [21] reviewed the literature for the use of fuzzy Petri nets for knowledge representation and reasoning, and Zhou & Zai [22] for the application of these net models in practice.

The aim of this paper is to present one of the simplest classes of Petri nets called binary Petri nets (BP-nets) and to show how this model can be used more economically than those known from the literature for the representation of rule-based knowledge, as well as in inference in rule-based systems. The net model, unlike the fuzzy Petri nets, is based on classic logic. A simple mathematical apparatus is used to describe the structure of the net model, while the behavior of BP-nets can only be described by two logical operators, *And* and *Or*. The area of application of these nets is also important. BP-nets can be successfully used not only for knowledge representation and inference in rule-based systems, but also in control modeling. To the best of our knowledge, there is no study in the literature devoted to such the net model.

The organization of this paper is as follows: Sect. 2 introduces the BP-net formalism. In sect. 3 the transformation of production rules in the BP-net is described. Section 4 presents an algorithm that constructs a BP-net model based on production rules. An example illustrating the proposed approach is given in sect. 5. In Sect. 6 the results of the experiments carried out on the net model from Sect. 5 are discussed. Section 7 contains final comments and remarks.

2 Binary Petri Nets

This section discusses the basics of binary Petri nets with particular emphasis on their structure, dynamics and the interpretation of their elements in the area of knowledge representation and inference. The more important terms defined in this section are illustrated by an example. Additional information on Petri nets can be found in [1].

A *binary Petri net* (BP-net for short) is a tuple $N_B = (P, T, I, O, M_0, S, \alpha, \beta, \gamma, Op, \delta)$, where: $P = \{p_1, p_2, \ldots, p_n\}$ is a finite set of *places* ($n > 0$); $T = \{t_1, t_2, \ldots, t_m\}$ is a finite set of *transitions* ($m > 0$); $I\colon P \times T \to \{0, 1\}$ is the *input function*; $O\colon T \times P \to \{0, 1\}$ is the *output function*; $M_0\colon P \to \{0, 1\}$ is the *initial marking*; $S = \{s_1, s_2, \ldots, s_n\}$ is a finite set of *statements* (P, T, S are pairwise disjoint); $\alpha\colon P \to S$ is the *statement binding function*; $\beta\colon T \to \{0, 1\}$ is the *truth degree function* $\gamma\colon T \to \{0, 1\}$ is the *threshold function*; $Op = \{And, Or\}$ (*And*, *Or* - classic logic operators) is the set of *operators*; and $\delta\colon T \to Op \times Op \times Op$ is the *operator binding function.*

Graphically, places are represented by circles and transitions by rectangles. The function I describes the oriented arcs connecting places with transitions, and the function O - the oriented arcs connecting transitions with places. We assume that arcs with weights 0 will be omitted in the drawings, and arcs with weights 1 are drawn, but their weights are not shown at the appropriate arcs.

If $I((p,t)) = 1$ then a place p is called an *input place* of a transition t, and if $O((t,p')) = 1$, then a place p' is called an *output place* of t. The initial marking M_0 is an initial distribution of numbers in the places. For $p \in P$, $M_0(p)$ can be interpreted as a truth value of the statement s bound with a given place p by means of the function α, i.e., $\alpha(p) = s$. Pictorially, the token is represented by a grey "dot" with the number 1 in the circle corresponding to appropriate place. We assume that if $M_0(p) = 0$ then the token does not exist in the place p. The number $\beta(t)$ is placed in a net picture under the transition t. This number is interpreted as the truth degree of an implication corresponding to a given transition t. The meaning of the function γ is explained below. For this class of net, the set O_p contains only two logical operators, *And* and *Or*. However, the function δ allows to attach to each transition three operators from the set O_p, where the first element of a given triple can be *And* or *Or*, the second element should be *And*, and the third - *Or*. The type of the first operator depends on the type of production rule represented by the transition. This aspect of representation of production rules by transitions will be discussed in detail in the next section.

By ${}^\bullet t$ we denote the set of all input places of t, and $t^\bullet$ - the set of all output places of t, i.e., ${}^\bullet t = \{p : I(p,t) = 1\}$, and $t^\bullet = \{p' : O(t,p') = 1\}$.

Let N_B be a BP-net. A *marking* of N_B is a function $M \colon P \to \{0,1\}$.

The binary Petri net dynamics defines how new markings are computed from the current marking when transitions are fired.

Let N_B be a BP-net, $t \in T$, ${}^\bullet t = \{p_{i1}, p_{i2}, \ldots, p_{ik}\}$ be a set of input places for a transition t, and M - a marking of N_B.

A transition t is *enabled* for marking M if the following condition is true:

$$And(M(p_{i1}), M(p_{i2}), \ldots, M(p_{ik})) \geq \gamma(t) > 0.$$

Only enabled transitions can be fired. In the paper we define two operating modes of the binary Petri nets.

Let $N_B = (P, T, I, O, M_0, S, \alpha, \beta, \gamma, Op, \delta)$ be a BP-net, $t \in T$, ${}^\bullet t = \{p_{i1}, p_{i2}, \ldots, p_{ik}\}$ be a set of input places for a transition t, $\beta(t)$ be a value of the truth degree function β corresponding to t, $\gamma(t)$ be a value of threshold function γ corresponding to t, and M be a marking of N.

Mode 1. If M is a marking of N_B enabling transition t and M' is the marking derived from M by the firing transition t, then for each $p \in P$:

$$M'(p) = \begin{cases} Or(And(And(M(p_{i1}), M(p_{i2}), \ldots, M(p_{ik})), \beta(t)), M(p)) \text{ if } p \in t^\bullet, \\ M(p) \text{ otherwise.} \end{cases}$$

Mode 2. If M is a marking of N_B enabling transition t and M' is the marking derived from M by the firing transition t, then for each $p \in P$:

$$M'(p) = \begin{cases} 0 \text{ if } p \in^\bullet t, \\ Or(And(And(M(p_{i1}), M(p_{i2}), \ldots, M(p_{ik})), \beta(t)), M(p)) \text{ if } p \in t^\bullet, \\ M(p) \text{ otherwise.} \end{cases}$$

The difference between the definitions of the two net modes is as follows: in Mode 2, tokens are removed from all input places of the fired transition t, while in Mode 1 all tokens are merely copied (not removed) from input places of t.

Example: Consider a BP-net such that: the sets $P = \{p_1, p_2, p_3\}$ and $T = \{t_1\}$, the functions $I((p_1, t_1)) = I((p_2, t_1)) = 1$, $I((p_3, t_1)) = 0$, $O((t_1, p_3)) = 1$, $O((t_1, p_1)) = O((t_1, p_2)) = 0$, the initial marking $M_0 = (1, 1, 0)$, the set $S = \{s_1, s_2, s_3\}$, the functions α: $\alpha(p_1) = s_1$, $\alpha(p_2) = s_2$, $\alpha(p_3) = s_3$, β: $\beta(t_1) = 1$, γ: $\gamma(t_1) = 1$, the set $O_p = \{And, Or\}$ and the function δ: $\delta(t_1) = (And, And, Or)$.

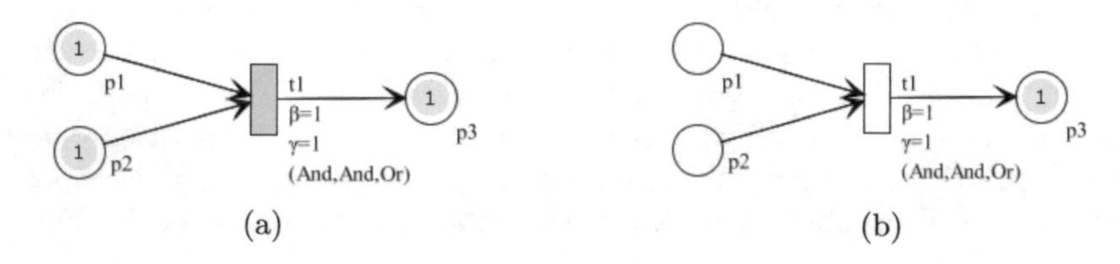

Fig. 1. BP-net after firing t_1 by the initial marking M_0: (a) Mode 1, (b) Mode 2

Transition t_1 is enabled by the initial marking M_0. Firing transition t_1 by M_0 according to Mode 1 transforms M_0 to the marking $M' = (1, 1, 1)$ (Fig. 1(a)). However, firing t_1 by M_0 according to Mode 2 results in the marking $M'' = (0, 0, 1)$ (Fig. 1(b)).

Remark: Here and in all net drawings in this paper, instead of $\beta(t) = b, \gamma(t) = c$, where t is the transition and b,c is 0 or 1 for the abbreviation we write $\beta = b, \gamma = c$. Moreover, the figure omits the statements s_1, s_2, s_3 corresponding to the places p_1, p_2, p_3, respectively.

3 Net Representation of Production Rules

Modeling a rule-based system with a Petri net is done by transforming its production rules (rules for short) into a BP-net model depending on the form of transformed rules. We consider two types of rules with parameters.

Type 1. r_1 : IF s THEN s' $[a; \beta(r_1) = b; \gamma(r_1) = c]$.

This rule can be modeled as shown in Fig. 2(a). The values of the parameters characterizing the rule r_1 are interpreted as follows: a is the logical value of the statement s, b is the value of the certainty factor β of the rule r_1, and c is the value of the threshold γ of the rule r_1. These parameters take the values 0 or 1. According to Fig. 2(b) the token value d in the output place p' of a transition t corresponding to the rule r_1 is calculated as follows: $d = And(a, b)$. The rule r_1 can fire if the condition $a \geq c$ is true.

Type 2. r_2 : IF s_1 And/Or s_2 ... And/Or s_n THEN s' $[a_1, a_2, \ldots, a_n; \beta(r_2) = b; \gamma(r_2) = c]$.

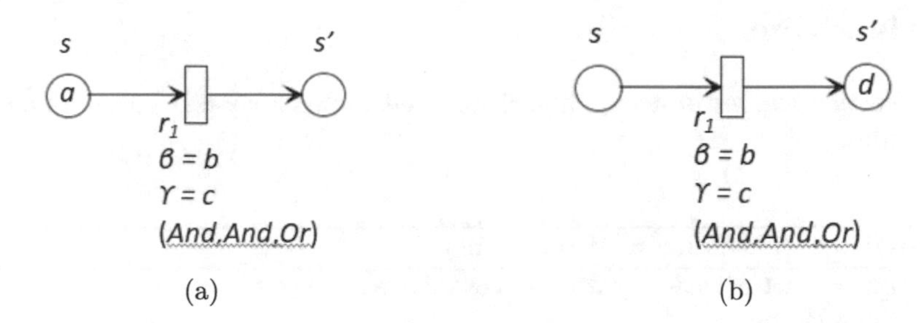

Fig. 2. BP-net representation of the rule type 1: (a) before firing the rule r_1, (b) after firing r_1 (Mode 2)

This rule type can be modeled by a BP-net as shown in Fig. 3(a). The value of the token d in the output place p' corresponding to the rule r_2 (Fig. 3(b)) is computed as follows: $d = And(And/Or(a_1, a_2, \ldots, a_n), b)$ if $And/Or(a_1, a_2, \ldots, a_n) \geq c$ is true.

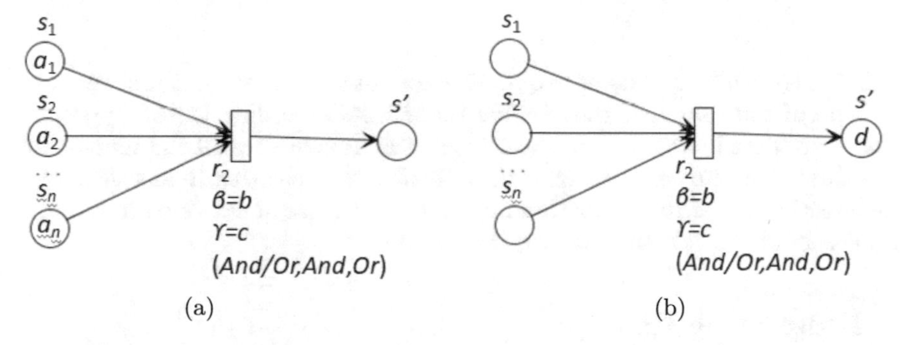

Fig. 3. BP-net representation of the rule type 2: (a) before firing the rule r_2, (b) after firing r_2 (Mode 2)

Remarks: (1) The notation *And*/*Or* used in the rule of type 2 means that *And* or *Or* operator must be explicitly selected in the rule. Then, depending on your choice of operator in the rule, you need to reformulate all formulas in which this notation occurs. (2) If the markings of the output places are nonzero, then the operator *Or* attached to net transitions should also be taken into account. Therefore, in each of the above formulas, the final value d' of the token assigned to the output place p' of the transition t representing the modeled rule r should be calculated as follows: $d' = Or(d, M(p'))$, where d' is the modified token value, d is the existing token value computed as described above for each rule type, and $M(p')$ is the current marking of the output place p'. (3) The proposed net representation of rule type 2 with operator *Or* in the rule premise allows to reduce the size of the net and the time of calculating the degree of truthfulness of its conclusion (cf. [9]).

4 Algorithm

In this section, we introduce an algorithm that builds a BP-net based on a given set of rules.

Algorithm 1: Building a BP-net using a set of rules.

Input : A set of rules R with a list of their parameters
Output: BP-net N_B
$F := \emptyset$;
for each rule $r \in R$
if r *is type 1* **then**
  build the N_r subnet according to the scheme shown in Fig. 2(a);
if r *is type 2* **then**
  build the N_r subnet according to the scheme shown in Fig. 3(a);
$F := F \cup \{N_r\}$;
Build the resulting net N_B by merging all subnets from F in identical places;
return N_B;

This algorithm consists of two main steps. In the first step, it first recognizes the type of the rule, and then builds the subnet according to the appropriate scheme presented in Sect. 3. In the second step, it combines all the subnets built in the first step into one net, connecting them with each other in identical places. The effectiveness of this algorithm depends on the size of the rule set R and the complexity of the rule structure in this set [9].

5 Illustrating Example

To illustrate our methodology, let us consider a simple problem of controlling train traffic [23].

Problem: A train B waits at a certain station for a train A to arrive in order to allow some passengers to change train A to train B. Now, a conflict arises when the train A is late. In this situation, the following alternatives can be taken into consideration: (1) Train B waits for train A to arrive. In this case, train B will depart with delay. (2) Train B departs in time. In this case, passengers disembarking train A have to wait for a later train. (3) Train B departs in time, and an additional train is employed for the train $A's$ passengers.

In order to describe the traffic conflict, we propose to consider the following three rules with a list of their parameters:

- r_1: IF s_2 *Or* s_3 THEN s_6 $[0, 1; \beta(r_1) = 1; \gamma(r_1) = 1]$
- r_2: IF s_1 *And* s_4 *And* s_6 THEN s_7 $[1, 1, 0; \beta(r_2) = 1; \gamma(r_2) = 1]$
- r_3: IF s_4 *And* s_5 THEN s_8 $[1, 1; \beta(r_3) = b[; \gamma(r_3) = 1]$

where: s_1 = 'Train B is the last train in this direction today', s_2 = 'The delay of train A is huge', s_3 = 'There is an urgent need for the track of train B', s_4 = 'Many passengers would like to change for train B', s_5 = 'The delay of train A is short', s_6 = '(Let) train B depart according to schedule', s_7 = 'Employ an additional train C (in the same direction as train B)', and s_8 = 'Let train B wait for train A'.

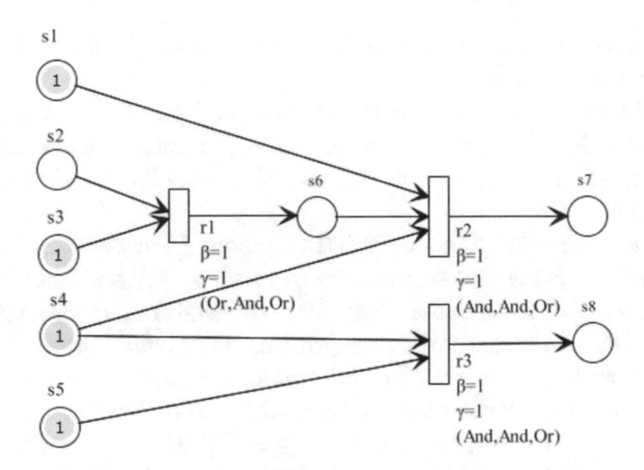

Fig. 4. An example of BP-net model of train traffic control with the initial marking

Figure 4 shows a BP-net model built using Algorithm 1 and the above three rules. In the following considerations, places p_1 to p_5 of this net will be considered as starting places and places p_7 and p_8 as the goal places. Note that the statements s_1, s_3, s_4, s_5 in this figure are true because the net places to which these statements are attached have a token value of 1. Assessing the statements s_1 through s_5 shows, that the rules r_1 and r_3 in the initial marking shown in this figure can be fired. By firing, for example, the sequence $r_1r_2r_3$ according to the firing rules for BP-net in Mode 1, we get the final marking where both statements s_7 and s_8 are true. This means that in this situation it is not possible to make an unique decision with regard to the alternatives considered, since both alternative situations are possible to the same extent. Therefore, it is necessary to look for other possible executions of this net model or to obtain additional information enriching the knowledge about its initial marking. If we choose, for example, the sequence r_3r_1 of rules and we fire it in Mode 2, we get the final marking corresponding to the statements s_7, s_8 equal to 0, 1, respectively. This means that the final marking now allows you to make an unambiguous decision that corresponds to the statement s_8 because this statement is true and the statement s_7 is false. For both operating modes, due to lack of space, BP-nets with the final markings p_7 and p_8 corresponding to statements s_7 and s_8, respectively, are omitted.

This example clearly shows that different BP-net operation modes can lead to completely different decision outcomes. By choosing the appropriate mode of operation in the net, we can also obtain satisfactory support in making decisions. The rest in this case depends a lot on the experience of the modeler.

6 Experiments and Results

This section presents the results of the experiments performed with the net model described in Sect. 5.

The aim of the experiments was to determine all possible states (markings) in which the net model may occur, with particular emphasis on the states of places responsible for making decisions in this model on the basis of input information supplied to its starting places.

Given that the model has five starting places (denoted from p_1 to p_5 and representing respectively the statements from s_1 to s_5) and that each marking can be 0 or 1, we had to consider 32 ($= 2^5$) combinations of binary five-element sequences. We call this method of generating data *combinatorial*. Besides this method of generating data, we also use *random method*. Thanks to it, we generated three data files with sizes of 200, 1200, and 7200 binary five-element sequences, which we used in this research (see Fig. 6).

Using our own software [24], we calculated all final markings of goal places (i.e., p_7, p_8 to which statements s_7 and s_8 are attached) of the net model together with the information provided in the goal place, whether the marking calculated for it allows for making an unambiguous decision or not. In our experiments we also took into account the fact that the net model can operate in two different modes of operation, i.e., Mode 1 or Mode 2, described in Sect. 2. The simulations of the net model were carried out with the use of net analysis methods derived from the Petri net theory [1]. They mainly concern the construction and analysis of the reachability graphs of simulated nets. A characteristic fragment of a text file containing several lines with input data obtained by the random method, names of data analysis methods, their parameters and calculation results for the net model from Fig. 4 are shown in Fig. 5.

By analyzing the results of the experiments (see Fig. 6), it can be seen that in both operating modes of the net model (regardless of the amount of input data) the total number of decisions is the same for each data file and amounts to 11 (77, 446, 2531), respectively. In Mode 1, the net model proposed 8 unambiguous and 3 ambiguous decisions for combinatorial input data, 56, 21 for random data of size 200, 319, 127 for random data of size 1200, and 1818, 713 for random data of size 7200. However, in Mode 2, all decisions made are de facto unambiguous, although 3 were selected in a non-deterministic manner for combinatorial input data, 21 for random data of size 200, 127 for random data of size 1200, and 713 for data of size 7200 (see Fig. 5). Note that any decision made in Mode 2 in a non-deterministic manner directly corresponds to ambiguous Mode 1 decisions for each of the input sizes. This situation is perfectly clear since in Mode 1 there are no conflict situations when rules are fired. Each true premise contained in

```
#Uploaded file: TTCtrl.txt
#Uploaded net: TTCtrl.pnml
#Reachability graph settings:
#Operating mode: Leaving the marking (Mode 1)
#Search duplicate mode: Search on the path
#Step mode: Step
#Step strategy: Minimum
#Building strategy: Single use of the transition
#Goal places: P7,P8
#Conflict resolution method: Max
p1,p2,p3,p4,p5
0,1,0,1,1,Decision P8 = 1 %1%
0,0,1,1,1,Decision P8 = 1 %2%
0,1,1,1,0,No decision
0,0,0,0,0,No decision
1,0,0,0,0,No decision
0,1,1,1,1,Decision P8 = 1 %6%
0,1,0,0,0,No decision
1,1,1,1,0,Decision P7 = 1 %8%
1,1,1,1,1,Non-deterministic decision P7 = P8 = 1
0,0,0,1,0,No decision
1,1,1,1,1,Non-deterministic decision P7 = P8 = 1
0,0,0,1,0,No decision
0,0,1,0,0,No decision
1,0,0,1,1,Decision P8 = 1 %14%
1,1,0,0,1,No decision
```

(a)

```
#Uploaded file: TTCtrl.txt
#Uploaded net: TTCtrl.pnml
#Reachability graph settings:
#Operating mode: Removing the marking (Mode M2)
#Search duplicate mode: Search on the path
#Step mode: Step
#Step strategy: Minimum
#Building strategy: Single use of the transition
#Goal places: P7, P8
#Conflict resolution method: Max
p1,p2,p3,p4,p5
0,1,0,1,1,Decision P8 = 1 %1%
0,0,1,1,1,Decision P8 = 1 %2%
0,1,1,1,0,No decision
0,0,0,0,0,No decision
1,0,0,0,0,No decision
0,1,1,1,1,Decision P8 = 1 %6%
0,1,0,0,0,No decision
1,1,1,1,0,Decision P7 = 1 %8%
1,1,1,1,1,Decision P7 = 1 %9%
0,0,0,1,0,No decision
1,1,1,1,1,Decision P7 = 1 %11%
0,0,0,1,0,No decision
0,0,1,0,0,No decision
1,0,0,1,1,Decision P8 = 1 %14%
1,1,0,0,1,No decision
0,0,1,1,0,No decision
```

(b)

Fig. 5. A fragment of the final information in goal places p_7 and p_8: (a) Mode 1, (b) Mode 2

Mode/Criteria	Method of generating data	Number of inputs	Number of no decisions	Number of decisions	Number of unambiguous decisions	Number of ambiguous decisions
Mode 1	combinatorial	32	21	11	8	3
Mode 1	random	200	123	77	56	21
Mode 1	random	1200	754	446	319	127
Mode 1	random	7200	4669	2531	1818	713
Mode 2	combinatorial	32	21	11	11(3)	0
Mode 2	random	200	123	77	77(21)	0
Mode 2	random	1200	754	446	446(127)	0
Mode 2	random	7200	4669	2531	2532(713)	0

Fig. 6. Comparison of the results of experiments for two modes of operation of the net model from Fig. 4

at least two rules can be used only once during the simulation of the net model, because the place corresponding to this premise loses its token value after firing such a rule.

It should also be noted that in both modes of operation of the net model, the total number of inputs for which a decision could not be proposed is the same and is 21 for combinatorial inputs, 123 for random inputs of 200, 754 for random inputs of 1200, and 4669 for random inputs of 7200. This situation results from the lack of sufficient input information (true premises) to fire at least one rule, which would generate a token value for some goal place in the net model.

Summarizing the considerations concerning the conducted experiments, it can be stated that the net model used here will always generate the same number of unambiguous decisions regardless of the net operation mode, as long as the rule-based system modeled with rules does not include conflict rules in its description.

7 Summary

The BP-net model combining the graphical power of Petri nets and the theory of classic logic to model rule-based expert knowledge in a rule-based system have been described in the paper. In the experimental part of the work, attention was paid to the similarities and differences in the behavior of the net model depending on the choice of its operation mode. At the same time, it was explained what causes different observable behaviors of the analyzed nets.

The proposed model is simple and relatively easy to use. The simplicity and efficiency of this model is due to the fact that it is described by a not very advanced mathematical apparatus. In addition, such a description has a simple and intuitive graphical representation, which greatly facilitates the perception of the model structure, as well as understanding of its behavior. Only two simple logical operators and three parameters characterizing the production rules are used to describe the behavior of the created model. The consequence of the simplicity of the description of such models is also the fact that these models are very efficient computationally, i.e., they are characterized by low computational complexity. The approach proposed in the paper may - as it seems - find application not only in modeling and analysis of rule-based systems, but also in modeling, inter alia, controllers [25].

The classic logic is insufficient to apply when descriptions of objects or situations of interest are incomplete, unclear, or inaccurate. Thus, the research work described in this paper opens the field for further expansion and deeper exploration. In further research, we intend to generalize the net model proposed in this paper by using fuzzy logic [7,25] or its generalizations and modifications, and propose some methods to optimize the model in terms of the effectiveness and quality of the proposed decisions.

Acknowledgments. The author is grateful to the anonymous referees for their helpful comments.

References

1. Peterson, J.L.: Petri Net Theory and the Modeling of Systems. Prentice-Hall Inc, Englewood Cliffs, N.J. (1981)
2. David, R., Alla, H.: Petri Nets and Grafcet: Tools for Modelling Discrete Event Systems. Prentice-Hall, Hoboken (1992)
3. Jensen, K., Rozenberg, G.: High-level Petri Nets. In: Pagnoni, A., Rozenberg, G. (eds.) Applications and Theory of Petri Nets, Informatik-Fachberichte, vol. 66, pp. 166–180. Springer, Cham (1991). https://doi.org/10.1007/978-3-642-69028-0_12

4. Petri, C.A.: Kommunikation mit Automaten. Schriften des IIM Nr. 2, Institut für Instrumentelle Mathematik, Bonn (1962)
5. Munakata, T.: Fundamentals of the New Artificial Intelligence. Springer, Cham (1998)
6. Cardoso, J., Camargo, H.: Fuzziness in Petri Nets. Springer, Cham (1999)
7. Zadeh, L.A.: Fuzzy sets. Inf. Control **8**, 338–353 (1965)
8. Looney, C.G.: Fuzzy petri nets for rule-based decision-making. IEEE Trans. Syst., Man, Cybern. **18**(1), 178–183 (1988)
9. Chen, S.M., Ke, J.S., Chang, J.F.: Knowledge representation using fuzzy Petri nets. IEEE Trans. Knowl. Data Eng. **2**(3), 311–319 (1990)
10. Li, X., Lara-Rosano, F.: Adaptive fuzzy Petri nets for dynamic knowledge representation and inference. Expert Syst. Appl. **19**, 235–241 (2000)
11. Pedrycz, W., Gomide, F.: A generalized fuzzy Petri net model. IEEE Trans. Fuzzy Syst. **2–4**, 295–301 (1994)
12. Skowron, A., Suraj, Z.: A parallel algorithm for real-time decision making: a rough set approach. J. Intell. Inf. Syst. **7**, 5–28 (1996)
13. Peters, J., et al.: Approximate real-time decision making: concepts and rough fuzzy Petri net models. Int. J. Intell. Syst. **14**, 805–839 (1999)
14. Suraj, Z., Hassanien, A.E.: Fuzzy Petri nets and interval analysis working together. In: Bello, R., Falcon, R., Verdegay, J.L. (eds.) Uncertainty Management with Fuzzy and Rough Sets. SFSC, vol. 377, pp. 395–413. Springer, Cham (2019). https://doi.org/10.1007/978-3-030-10463-4_20
15. Liang, Q., Mendel, J.M.: Interval type-2 fuzzy logic systems: theory and design. IEEE Trans. Fuzzy Syst. **8**, 535–550 (2000)
16. Bandyopadhyay, S., Suraj, Z., Nayak, P.K.: Soft petri net. In: Mihálydeák, T., et al. (eds.) IJCRS 2019. LNCS (LNAI), vol. 11499, pp. 253–264. Springer, Cham (2019). https://doi.org/10.1007/978-3-030-22815-6_20
17. Molodtsov, D.: Soft set theory - first results. Comput. Math. Appl. **37**, 19–31 (1999)
18. Liu, H.-C., et al.: Fuzzy Petri nets using intuitionistic fuzzy sets and ordered weighted averaging operators. IEEE Trans. Cybern. **46**(8), 1839–1850 (2016)
19. Atanassov, K.: Intuitionistic fuzzy sets. Fuzzy Sets Syst. **20**, 87–96 (1986)
20. Suraj, Z.: A hybrid approach to approximate real-time decision making. In: 2021 IEEE International Conference on Fuzzy Systems (FUZZ-IEEE 2021), Luxembourg, 11-14 July 2021, pp. 71-78. IEEE (2021)
21. Liu, H.-C., et al.: Fuzzy Petri nets for knowledge representation and reasoning: a literature review. Eng. Appl. Artif. Intell. **60**, 45–56 (2017)
22. Zhou, K.-O., Zain, A.M.: Fuzzy Petri nets and industrial applications: a review. Artif. Intell. Rev. **45**, 405–446 (2016)
23. Fay, A., Schnieder, E.: Fuzzy Petri nets for knowledge modelling in expert systems, vol. 6, pp. 300–318
24. Suraj, Z., Grochowalski, P.: PNeS in modelling, control and analysis of concurrent systems. In: Ramanna, S., Cornelis, C., Ciucci, D. (eds.) IJCRS 2021. LNCS (LNAI), vol. 12872, pp. 279–293. Springer, Cham (2021). https://doi.org/10.1007/978-3-030-87334-9_24
25. Pedrycz, W.: Fuzzy Control and Fuzzy Systems. Wiley, Hoboken (1993)

Credit Risk Scoring: A Stacking Generalization Approach

Bernardo Raimundo[1] and Jorge M. Bravo[2(✉)]

[1] NOVA IMS - Universidade Nova de Lisboa, Lisbon, Portugal
braimundo@novaims.unl.pt
[2] NOVA IMS - Universidade Nova de Lisboa & MagIC & Université Paris-Dauphine PSL & ISCTE-IUL Business Research Unit (BRU-IUL) & CEFAGE-UE, Lisbon, Portugal
jbravo@novaims.unl.pt

Abstract. Forecasting the creditworthiness of customers in new and existing loan contracts is a central issue of lenders' activity. Credit scoring involves the use of analytical methods to transform historical loan application and loan performance data into credit scores that signal creditworthiness, inform, and determine credit decisions, determine credit limits, and loan rates, and assist in fraud detection, delinquency intervention, or loss mitigation. The standard approach to credit scoring is to pursue a "winner-take-all" perspective by which, for each dataset, a single believed to be the "best" statistical learning or machine learning classifier is selected from a set of candidate approaches using some method or criteria often neglecting model uncertainty. This paper empirically investigates the predictive accuracy of single-based classifiers against the stacking generalization approach in credit risk modelling using real-world peer-to-peer lending data. The findings show that stacking ensembles consistently outperform most traditional individual credit scoring models in predicting the default probability. Moreover, the findings show that adopting a feature selection process and hyperparameter tuning contributes to improving the performance of individual credit risk models and the super-learner scoring algorithm, helping models to be simpler, more comprehensive, and with lower classification error rates. Improving credit scoring models to better identify loan delinquency can substantially contribute to reducing loan impairments and losses leading to an improvement in the financial performance of credit institutions.

Keywords: Credit scoring · Ensemble learning · Probability of default · Stacking generalization · Risk management

1 Introduction

The critical role of the credit market in causing the most recent global financial crisis led to a strengthening in banking regulation and increased academic research and policy interest in this area. The banking and accounting regulatory changes brought by the revised Basel Committee on Banking Supervision Accords and the International Financial Reporting Standards (IFRS) #9 and Financial Accounting Standards Board (FASB)

A. Rocha et al. (Eds.): WorldCIST 2023, LNNS 799, pp. 382–396, 2024.
https://doi.org/10.1007/978-3-031-45642-8_38

Current Expected Credit Loss (CECL) standards introduced stronger risk management obligations for banks, with capital requirements closely linked with estimated credit portfolio losses over the entire life of the exposures, conditional on macroeconomic factors, on a point-in-time basis [1, 2]. As a result, banks now devote substantial resources to developing internal credit risk models to better support decisions when granting loans, quantify expected credit losses, and assign mandatory economic capital [3, 4]. Credit scoring involves the use of analytical methods to transform traditional historical and non-conventional loan application and loan performance data into credit scores that signal creditworthiness, inform, and determine credit decisions, determine credit limits, and loan rates, assist in fraud detection, delinquency intervention (e.g., loan restructuring) or loss mitigation. Originally credit scoring used a pure judgmental approach to accept or reject an application, known as the 5Cs of credit, composed of the following metrics: (i) the character of the applicant, (ii) the capital, (iii) the collateral, (iv) the capacity and (v) the condition [5]. However, due to its subjectivity, only using professional expert judgment, would result in inconsistent measures [6].

Since this initial approach, several techniques have been developed to further help decision-makers and financial analysts to better support their decisions by considering traditional statistical learning methods and machine learning and deep learning modelling techniques. Simple linear classification models remain a popular choice among credit institutions, largely due to their adequate accuracy, straightforward implementation, and interpretability of results [7]. The most popular single-period classification techniques are regression models which include, but are not limited to, Linear Discriminant Analysis (LDA) [8] and Logistic Regression Analysis (LR) [9–11]. Empirical results show that the classification accuracy of traditional scoring models is a function of (i) explanatory risk variables, (ii) the training dataset, and (iii) the cut-off point [12]. In the meantime, advancements made in credit risk modelling allowed for the development of newer and more contemporary techniques to assess credit risk amid advances in computer technology. Since credit risk analysis is similar to pattern-recognition problems, machine learning algorithms can be used to classify the creditworthiness of counterparties [13]. Some noteworthy algorithms include Decision Trees (DT) [14], Support Vector Machines (SVM) [15], K-Nearest Neighbours (KNN) [16], and Artificial Neural Networks (ANN) [17, 43].

Despite the usage of single classifiers, it is hard to overstate the importance of model uncertainty for economic modelling. The empirical work in economics and social modelling is subject to a large amount of uncertainty about the model specification [18]. The standard approach to risk modelling is to pursue a "winner-take-all" perspective by which, for each dataset, a single believed to be the "best" model is selected from a set of candidate approaches using some method or criteria often neglecting model uncertainty for statistical inference purposes. The use of different lookback periods, diverse selection procedures, alternative accuracy metrics, misspecification problems, and the presence of structural breaks in the data-generating process can lead to different model choices and predictive accuracy [19–24]. When presented with multiple candidate models or algorithms picking one model that can give optimal performance for future data can result in unstable or useless information. Recently, there is a growing interest in homogenous and heterogeneous model combinations in credit risk modelling due to its ability to produce a more robust classifier since it includes the predictions from all the considered models while circumventing any potential preference [25, 26]. The widely used ensemble methods are bagging, boosting, and stacking. Bagging and boosting, who consider homogenous weak learners, combine the results of multiple models to get a generalized result. Bagging (short for bootstrap aggregation) is a parallel procedure; where randomly sampled datasets are produced from the original data taking the average of these predictions to make a final prediction. Boosting is a sequential procedure; it follows a sequential ensemble technique where each subsequent model corrects the error of the earlier model. By identifying these misclassifications in previous iterations, more weight is given to them thus in the next iteration the learner will focus more on these misclassifications.

Traditional model classifications have some limitations. First, the model weights are often calculated using a single validation set. Second, the approach is sensitive to the choice of the model set and the prior distribution of each model. Third, current approaches often use the same weights for all forecasting horizons. Stacked regression ensembles offer an alternative approach to tackling some of the shortcomings of traditional model combination methods. Stacking considers heterogeneous weak learners combining models of diverse types. The architecture of a stacking ensemble involves two or more base models, often referred to as level-0 models with different weight combinations, and a meta-model combining individual models predictions.

This paper empirically investigates the predictive accuracy of single-based classifiers against the stacking generalization approach in credit risk modelling. Four individual classifiers were considered in the analysis as level-0 models and inputs for the meta-model. The model set includes classifiers using machine learning methods (KNN, SVM, and DT) and traditional statistical learning methods (LR). The datasets considered in this paper are provided by Lending Club, a peer-to-peer lending company, and include all accepted and rejected loans from 2007 to 2018. The findings suggest that despite the overall good performance of individual classifiers, the application of ensemble learning techniques can contribute to improving the predictive accuracy of traditional classifiers in credit risk scoring. However, when applying an ensemble model combination, the researcher should pay close attention to these considerations: (i) a reduction of model interpretability when using an ensemble-based approach due to the increase in model complexity, which may make it harder to draw any crucial business insights and (ii) Computation time, an ensemble is a very computationally expensive approach which might not be suitable for real-life applications.

The remainder of the paper is organized as follows. Section 2 summarises the materials and methods used in this study, including the empirical strategy adopted. Section 3 reports the empirical findings of the study. Finally, Sect. 4 critically discusses the results and concludes, pointing directions for future research directions.

2 Materials and Methods

2.1 Stacked Regression Ensemble for Credit Scoring

Ensembles are sets of learning machines that combine data from different modelling approaches to obtain more dependable and more accurate predictions in supervised and unsupervised learning problems [27]. An ensemble contains several learners, called based learners, generated by a base learning algorithm. Most ensemble methods produce homogeneous base learners, i.e., learners with the same characteristics, leading to homogeneous ensembles, but some methods utilize multiple learning algorithms to reproduce heterogeneous learners, i.e., learners with distinctive characteristics leading to heterogeneous ensembles techniques [28]. The generalization ability of an ensemble is usually much stronger compared to that of a single classifier [29, 44].

Stacking combines point forecasts from multiple heterogeneous base learners using weights that optimize a cross-validation criterion [30]. The base learners can, for example, be alternative statistical learning, machine learning, and deep learning credit scoring methods. The customary stacked regression ensemble approach comprises two modules, base learners (level-0) and a meta-model (level-1), and proceeds in two steps in generating the final credit scoring predictions. The first step (level-0), base learning, consists of multiple base credit risk models which separately generate cross-validated predictions from the training data. The predictions from various models and the observed response variable compose the metadata. In the second step (level-1), a meta-learner is trained on the metadata (low-level output) to estimate the optimal weights for combining multiple base learners while minimizing some optimization (e.g., cross-validation) criterion. The trained meta-classifier is then used to make an overall final prediction by considering each prediction made by the base-level classifier.

For blending the individual forecasts, several choices are possible such as simple averaging, weighted averaging, Bayesian model ensembles, model confidence set, linear or logistic regression, adaptative blending, and neural network (see, e.g., [23, 45–49]). For instance, stacked regression ensemble credit scoring is a linear combination of individual base learners' forecasts in which the model combination weights are coefficients of a linear regression model with observed default rates as the dependent variable and the point forecasts from the individual learners as the independent variables.

2.2 Individual Credit Scoring Models and Stacking Framework

An ensemble is composed of several base learners. This paper considers four widely used base learners, i.e., LR, SVM, DT, and KNN to investigate the predictive accuracy of stacking in credit risk modelling. The LR [30] is regarded as an industry standard because of its simplicity and balanced error distribution. LR models are fitted via maximum likelihood estimation which involves maximizing a likelihood function to find the probability distribution and parameters that best explain the data. In other words, during the fitting process, this method will estimate the coefficients in such a way it will maximize the probability of labelling an applicant as default as well as maximize the probability of labelling an applicant as non-default.

The SVM [31] algorithm is a state-of-the-art technique with a good generalization performance and computational efficiency. By applying the concept of margin, given a labelled training set, the SVM constructs and uses a discriminant hyperplane or a set of hyperplanes to identify classes. Intuitively, a good separation is achieved by the hyperplane with maximum distance to the nearest training data samples (i.e., maximum margin) therefore an optimal hyperplane maximizes the margin. Both the SVM and LR algorithms were trained using Stochastic Gradient Descent (SGD), an iterative procedure for optimizing an objective function with suitable smoothing properties.

The DT algorithm is a graphical representation with the primary goal of creating a model that predicts the value of a target variable based on several input variables. The algorithm is based on a measure of data impurity that determines the split of each node, using different metrics. These metrics are applied to each node, and the resulting values are combined (i.e., averaged) to provide a measure of the quality of the split [32]. Then, the impurity of each test is compared, and the split with the lowest impurity is chosen. This process is then continued for each node of the tree to find the "purest" node in terms of the target variable.

The KNN is a distance-based algorithm, taking a majority vote between the "K" closest observations. The main idea of the KNN algorithm is that whenever there is a new point to predict, its nearest neighbours are chosen from the training data [33]. It is easy to understand and implement, with no need to estimate parameters or training. During this phase, the use of machine learning pipelines facilitated the implementation of different transformations such as dealing with missing values, and outliers, encoding categorical variables, and proper scaling. Furthermore, cross-validation and hyperparameter tunning were performed to ensure robustness in the classifiers. Finally, principal component analysis was performed to reduce the number of dimensions in the dataset and discard irrelevant information. For the setup of the stacking ensemble, we considered the LR model as the meta classifier while the remainder classifiers were considered the base learners. Figure 1 schematizes the implementation of the stacking ensemble framework when forecasting the default probability.

Four different model combinations have been investigated in this paper. Table 1 provides a synopsis of the ensemble combinations considered. The final estimator is trained using cross-validated predictions of the base estimators. The meta-learner model used in this classification exercise is set to logistic regression.

2.3 Sample Credit Dataset and Variables

The data considered in this paper is provided by Lending Club, a peer-to-peer lending company that matches lenders and borrowers through an online platform without the need for any financial intermediation. The datasets comprise all loan applications issued between 2007 and 2018, including both accepted loans and reject loans, totalling 2.26 million records and 151 features. The dataset is unbalanced, so we follow for instance [35], and use a combination of both undersampling and oversampling combined with stratified K-fold cross-validation to achieve a more balanced distribution.

One of the biggest challenges in building credit scoring models is to select the most relevant features to be used in model calibration. The feature selection process is important to mitigate the overfitting in the curse of dimensionality of classification algorithms

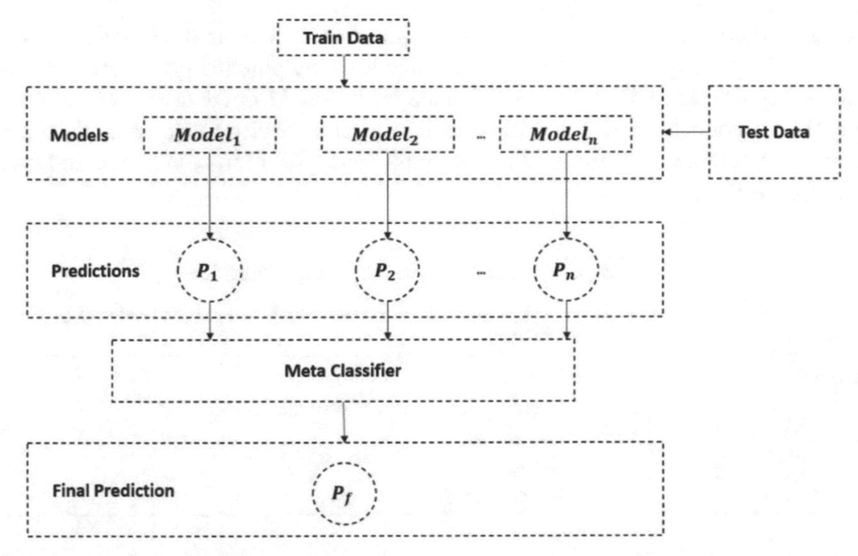

Fig. 1. Stacking ensemble framework when forecasting the default probability Source: Author's Preparation based on [34].

Table 1. Model Combinations

Combination	Base Learners	Meta Model
SC1	KNN and SVM	LR
SC2	DT and SVM	LR
SC3	DT and KNN	LR
SC4	DT, KNN, and SVM	LR

Source: Author's preparation

and eliminate irrelevant and/or highly correlated variables. This may ultimately lead to a misleading interpretation of the credit scoring model and poor predictive performance. The feature selection process used in this paper includes Principal Component Analysis (PCA) a technique used to eliminate irrelevant, incomplete, and/or unreliable information while also increasing algorithm performance [36]. Out of the 151 features available in the data, the feature selection process identified twenty-three for modelling purposes. These features focus on: (i) Personal details (e.g., annual income, address, and homeownership); (ii) Credit history (e.g., the balance of accounts, revolving and current past due accounts); and (iii) Loan characteristics (e.g., purpose, grade, term). A summary of these characteristics is reported in Table 2.

To calibrate the individual credit scoring models, all structures are trained on the training set comprising 80% of the data, with the remaining 20% used for testing the model's predictive performance. Following the split, additional pre-processing steps were implemented to treat missing values, encode categorical features, and standardize the data for modelling. To achieve this, a pipeline was built using the libraries provided by the python package `sklearn`. The machine learning pipeline procedure automates the workflow it takes to produce a machine learning model from data extraction and pre-processing to model training and deployment. In addition to cross-validation, the tunning of hyperparameters was performed using a parameter grid to maximize the underlying score of the estimator.

Table 2. Chosen features for modelling credit risk

Characteristics	Selected features
Borrower	Annual income Homeownership (rent, own, mortgage, other) Verification status Address State Last payment amount
Credit	Debt-to-Income-Ratio Delinquency status in the last 2 years Number of inquiries in the last 6 months The number of open credit lines in the borrower's credit file The number of current open credit lines Number of derogatory public records Total credit revolving balance Revolving line utilization rate Number of mortgage accounts FICO Score The number of public record bankruptcies
Loan	Loan amount Loan term Loan interest rate LC assigned loan grade Purpose

Source: Author's preparation

The performance obtained in each combination of hyperparameters was measured using the F1-score. Table 3 summarizes the hyperparameters chosen for model optimization and the tuned parameters used for predictions.

2.4 Predictive Performance Evaluation Criteria

The evaluation criteria for our experiments are adopted from established standard measures in the field of credit scoring [37]. These measures include confusion matrices, the ROC (Receiver Operating Characteristic) curve, and AUC (Area under the ROC Curve) score. A confusion matrix is a table layout that allows the visualization of the performance of an algorithm [38] where the rows represent instances of the actual class, and the columns represent instances of the predicted class. From the confusion matrix, four

Table 3. Hyperparameter optimization

Models	Hyperparameters	Values	Chosen Values
Logistic Regression	Penalty	L1, L2	L2
	Model Alpha	10**−3, 10**−2, 10**−1	10**−2
Support Vector Machine	Penalty	L1, L2	L1
	Model Alpha	10**−3, 10**−2, 10**−1	10**−3
K-Nearest Neighbours	Neighbourhood Size	3, 5, 7, 9, 11	9
	Metric	Minkowski	Minkowski
Decision Tree	Criterion	Gini, Entropy	Entropy
	Minimum Sample Splits	20, 30, 50, 100	100
	Maximum Depth	3, 5, 10, 15, 20	15

Source: Author's preparation

different metrics can be derived: (i) accuracy, (ii) precision, (iii) recall, and (iv) F1-score. These metrics are computed as follows: (Table 4).

Table 4. Confusion matrix for credit scoring

		Predicted Values	
		Negative = 0	Positive = 1
Actual Values	Negative = 0	True Negative (TN)	False Positive (FP)
	Positive = 1	False Negative (FN)	True Positive (TP)

Source: Author's preparation

$$Accuracy = \frac{TP + TN}{TP + TN + FP + FN} \tag{1}$$

$$Precision = \frac{TP}{TP + FP} \tag{2}$$

$$Recall = \frac{TP}{TP + FN} \tag{3}$$

$$F1 - Score = 2 * \frac{(Precision * Recall)}{Precision + Recall} \tag{4}$$

A ROC curve is a graphical plot for visualizing, organizing, and selecting classifiers based on their performance. It shows the relationship between recall and false positive rate (also known as the probability of false alarm) at different threshold points. The AUC score represents the area enclosed by the ROC curve. It is a measure of the ability of a classifier to distinguish between both classes with a value range between 0 and 1. The model fitting, forecasting, simulation procedures, and additional computations have been implemented using a python software routine.

3 Results

The classification algorithms were trained with the selected features. Tables 5 and 6 summarize the predictive accuracy metrics obtained for the individual credit-scoring learners before and after hyperparameter optimization with no sampling techniques applied, respectively.

Table 5. First Experimental Results (Before hyperparameter tunning)

	Accuracy	Precision	Recall	F1-score	AUC Score
LR	0.7438	0.4221	0.7676	0.5446	0.8325
SVM	0.7374	0.4156	0.7766	0.5415	0.7528
KNN	0.8315	0.6012	0.4639	0.5237	0.7845
DT	0.9135	0.7876	0.7757	0.7816	0.5768

Source: Author's preparation

Table 6. First Experimental Results (After hyperparameter tunning)

	Accuracy	Precision	Recall	F1-score	AUC Score
LR	0.7489	0.4276	0.7607	0.5474	0.8329
SVM	0.7454	0.4238	0.7667	0.5459	0.7564
KNN	0.8396	0.6372	0.4566	0.5320	0.8453
DT	0.9258	0.7492	0.9433	0.8351	0.9764

Source: Author's preparation

The results from the first experiment suggest that the usage of hyperparameter tuning can increase the overall performance of a classifier. The DT classifier had the best overall performance among the different classifiers. It had the highest F1-score of 0.8351 which is the proportion of charged-off applicants correctly classified and the overall capability of the classifier to predict default. Furthermore, the AUC score of the DT algorithm, which represents the classifier's ability to distinguish classes, outperformed the remaining learners in terms of predictive performance. The LR and the SVM, trained using the SGD algorithm, exhibit similar results with the LR coming slightly ahead with an F1-score of 0.5474 compared to 0.5459, respectively. Regarding the KNN, despite its relatively good performance in the training data, it was evident that the algorithm was overfitting. As training examples increase, the model is not able to perform well on unseen data. It is capturing specific blunders from the data that prevent a good generalization. After this initial analysis, the sampling technique was applied to the dataset. Table 7 summarizes the individual learners' predictive accuracy obtained in the second experiment:

Table 7. Second Experimental Results (Sampling applied)

	Accuracy	Precision	Recall	F1-score	AUC Score
LR	0.8387	0.5636	0.8514	0.6783	0.9185
SVM	0.8379	0.5618	0.8584	0.6782	0.8450
KNN	0.8607	0.6003	0.8829	0.7167	0.9265
DT	0.9262	0.7514	0.9422	0.8361	0.9732

Source: Author's preparation

The results show an increase in predictive accuracy across all four models. The DT algorithm, despite the loss in its recall score, continues to outperform the other credit scoring models after applying sampling techniques. Interestingly, the DT algorithm benefited the least from the sampling approach compared to the remaining classifiers. During the implementation of the second experiment, the problem of overfitting that was present in the KNN algorithm vanished, highlighting the importance of considering sampling methods in learning models. Having a set of heterogeneous base-level learners is essential for a successful stacking solution. After performing both hyperparameter optimization and sampling, the best base-level learners and meta-learners were taken to perform the ensemble approach experiments. Table 8 summarizes the model combinations predictive accuracy metrics.

Table 8. Results for each combination

	Accuracy	Precision	Recall	F1 Score	AUC Score
SC1	0.8623	0.6072	0.8790	0.7182	0.9335
SC2	0.9265	0.7525	0.9415	0.8365	0.9726
SC3	0.9284	0.7587	0.9404	0.8399	0.9739
SC4	0.9277	0.7564	0.9409	0.8386	0.9731

Source: Author's preparation

The results in Table 8 suggest the stacked regression ensembles consistently outperform most individual credit scoring models in predicting the default probability in the sample loan data used in this study. The best ensemble is SC3 – stacking with DT and KNN as the base level models followed by SC4, the combination that includes all models. These results suggest that the usage of the SVM algorithm as a base learner in the stacking ensemble may be confusing the meta learner, lowering the importance of base learners' predictions, and decreasing the overall performance of the stacking ensemble. Furthermore, SC1 – stacking with SVM and KNN performed much worse than the single best classifier, the DT algorithm further reinforcing its robustness and predictive capabilities.

This improved predictive accuracy suggests that by assigning model weights that maximize out-of-sample prediction accuracy, the stacking ensemble tends to optimally combine the features of alternative credit scoring models to form a meta-learner credit scoring model, which captures the features of the loan data more adequately than any individual credit risk classifier. The stacking ensemble approach assigns model weights to individual learners that express the out-of-sample performance of the credit scoring classifiers and incorporate future loan data uncertainty into the weights by minimizing the cross-validation criterion. Although the stacking ensemble approaches improvements over the individual base learners, the increased improvement may not always be worth the increased complexity. A stacking ensemble is a good approach for pushing the performance limits, but the improvements of the approach and the costs need to be independently evaluated by the researcher.

4 Conclusions and Discussion

Credit scoring methods serve as a risk management tool that allows a financial institution to check applicant reliability to pay off the debt in time. With the developments made in machine learning applications, credit scoring replaces the reliance on "gut feeling" with statistical analysis reducing the potential risks associated with personal judgment. In this study, a comparative assessment between an ensemble approach and individual classifiers is carried out. Several lessons have been learned which are summarized as follows. The first is that one of the biggest challenges of building robust credit scoring models using real-world data is to select the most relevant credit risk features to be used in the modelling. Data must be properly pre-processed to mitigate overfitting in the curse of dimensionality and eliminate irrelevant and/or highly correlated variables. The results obtained in this study suggest that the feature selection process contributed to improving the performance of individual credit risk models and the super-learner scoring algorithm, helping models to be simpler, more comprehensive, and with lower classification error rates. Secondly, the results suggest that the usage of hyperparameter tuning can increase the overall performance of a credit risk classifier. This result is consistent with previous literature (see, e.g., [39]) and means hyperparameters can impact the algorithm's behaviour by affecting its properties, structure, or complexity, and that the customary use of the default values in the statistical packages adopted to implement super learning raises concerns over the performance of meta-learners. In adopting machine learning methods to credit scoring, hyperparameters should thus be tuned using a grid search procedure whereby the algorithm's cross-validated performance is continually measured over a grid of possible hyperparameter values to optimally identify the best combination. Third, identifying a set of heterogeneous base-level learners is essential for a successful stacking solution. Fourth, the findings suggest the stacked regression ensembles consistently outperform most traditional individual credit scoring models in predicting the default probability. This improved predictive accuracy suggests that the stacking ensemble tends to optimally combine the features of heterogeneous credit scoring models to form a meta-learner credit scoring model which captures the features of the loan application and loan performance data more adequately than any individual classifier.

From a managerial point of view, our findings reinforce the main conclusion that improving credit scoring models to better identify loan delinquency can substantially contribute to reducing loan impairments and losses leading to a significant improvement in the financial performance of credit institutions. Although managers may still be reluctant to implement advanced machine learning and deep learning models to score credit applications due to their higher complexity and the need for qualified experts, the fact that more sophisticated models adopt a data-driven approach to credit decisions, reducing subjective human intervention, judgment and errors may be sufficient to convince them that investing in more robust credit scoring models will ultimately pay-off.

Future research should focus on experimenting with other state-of-the-art machine learning and deep learning model combinations and different sampling techniques such as Synthetic Minority Oversampling Technique (SMOTE) [40]. Unlike random oversampling which only duplicates random instances of the minority class, SMOTE creates synthetic data points that are slightly different from the original data points. These new instances are generated based on the distance of each data point and the minority class's nearest neighbours. This procedure is executed until balancing is achieved. By using SMOTE the possibility of overfitting is minimized and no information is lost during sampling. However, by introducing these new synthetic points, the possibility of adding noise to the data increases. Additionally, given the challenges that come with unbalanced data, it would be interesting to evaluate a cost-sensitive learning approach since it allows for different misclassification costs for both Type I and Type II errors. Rather than artificially balancing the data distribution via sampling techniques, cost-sensitive learning solves the imbalanced problem by utilizing cost matrices that outline the costs associated with the misclassifications of the various classes [41]. Previous research showed that cost-sensitive learning generates enhanced performance in applications where the dataset has a skewed class distribution [42].

Acknowledgements. This work has been supported by Fundação para a Ciência e a Tecnologia, grants UIDB/04152/2020 - Centro de Investigação em Gestão de Informação (MagIC) and UIDB/00315/2020 (BRU-ISCTE-IUL).

References

1. Ashofteh, A., Bravo, J.M.: A conservative approach for online credit scoring. Expert Syst. Appl. **176**, 114835 (2021)
2. Ashofteh A., Bravo J.M.: A non-parametric-based computationally efficient approach for credit scoring. In: CAPSI 2019 - 19th Conference of the Portuguese Association for Information Systems, Lisbon, Code 160805 (2019)
3. Chamboko, R., Bravo, J.M.: On the modelling of prognosis from delinquency to normal performance on retail consumer loans. Risk Manage. **18**, 264–287 (2016)
4. Chamboko, R., Bravo, J.M.: A multi-state approach to modelling intermediate events and multiple mortgage loan outcomes. Risks **8**(2), 64 (2020). https://doi.org/10.3390/risks8020064

5. Thomas, L.C.: A survey of credit and behavioural scoring: forecasting financial risk of lending to consumers. Int. J. Forecast. **16**(2), 149–172 (2000)
6. Saunders, A., Allen, L.: Credit Risk Measurement-New Approaches to Value at Risk and Other Paradigms. Wiley, New York (2002)
7. Lessmann, S., Baesens, B., Seow, H., Thomas, L.C.: Benchmarking state-of-the-art classification algorithms for credit scoring: an update of research. Eur. J. Oper. Res. **247**(1), 124–136 (2015)
8. Altman, E.I.: Financial ratios, discriminant analysis and the prediction of corporate bankruptcy. J. Financ. **23**(4), 589–609 (1968)
9. Chamboko, R., Bravo, J.M.: Modelling and forecasting recurrent recovery events on consumer loans. Int. J. Appl. Decis. Sci. **12**(3), 271–287 (2019)
10. Chamboko, R., Bravo, J.M.: Frailty correlated default on retail consumer loans in developing markets. Int. J. Appl. Decis. Sci. **12**(3), 257–270 (2019)
11. Altman, E.I., Haldeman, R.G., Narayanan, P.: ZETATM analysis a new model to identify bankruptcy risk of corporations. J. Bank. Finan. **1**(1), 29–54 (1977)
12. Ala'raj, M., Abbod, M.F.: Classifiers consensus system approach for credit scoring. Knowl.-Based Syst. **104**, 89–105 (2016)
13. Barboza, F., Kimura, H., Altman, E.: Machine learning models and bankruptcy prediction. Expert Syst. Appl. **83**, 405–417 (2017)
14. Zhang, D., Zhou, X., Leung, S., Zheng, J.: Vertical bagging decision trees model for credit scoring. Expert Syst. Appl. **37**, 7838–7843 (2010)
15. Huang, C.L., Chen, M.C., Wang, C.J.: Credit scoring with a data mining approach based on support vector machines. Expert Syst. Appl. **33**(4), 847–856 (2007)
16. Mukid, M., Widiharih, T., Rusgiyono, A., Prahutama, A.: Credit scoring analysis using weighted k-nearest neighbour. J. Phys. Conf. Ser. **1025**, 012114 (2018)
17. West, D.: Neural network credit scoring models. Comput. Oper. Res. **27**(11–12), 1131–1152 (2000)
18. Steel, M.F.J.: Model averaging and its use in economics. J. Econ. Lit. **58**, 644–719 (2020)
19. Ashofteh, A., Bravo, J.M., Ayuso, M.: A new ensemble learning strategy for panel time-series forecasting with applications to tracking respiratory disease excess mortality during the COVID-19 pandemic. Appl. Soft Comput. **128**, 109422 (2022)
20. Bravo, J.M., Ayuso, M., Holzmann, R., Palmer, E.: Addressing the life expectancy gap in pension policy. Insur. Math. Econ. **99**, 200–221 (2021)
21. Bravo, J.M.: Pricing participating longevity-linked life annuities: a Bayesian model ensemble approach. Eur. Actuar. J. **12**, 125–159 (2021)
22. Ayuso, M., Bravo, J.M., Holzmann, R., Palmer, E.: Automatic indexation of the pension age to life expectancy: when policy design matters. Risks **9**(5), 96 (2021). https://doi.org/10.3390/risks9050096
23. Bravo, J.M., Ayuso, M.: Mortality and life expectancy forecasts using Bayesian model combinations: an application to the Portuguese population. RISTI - Revista Ibérica de Sistemas e Tecnologias de Informação, E40, 128–144 (2020). https://doi.org/10.17013/risti.40.128–145
24. Bravo, J.M., Ayuso, M.: Linking pensions to life expectancy: tackling conceptual uncertainty through Bayesian model averaging. Mathematics, **9**(24), 3307 (2021). 1–27
25. Feng, X., Xiao, Z., Zhong, B., Qiu, J., Dong, Y.: Dynamic ensemble classification for credit scoring using soft probability. Appl. Soft Comput. J. **65**, 139–151 (2018)
26. Xia, Y., Liu, C., Da, B., Xie, F.: A novel heterogeneous ensemble credit scoring model based on bstacking approach. Expert Syst. Appl. **93**, 182–199 (2018)
27. Re, M., Valentini, G.: Ensemble methods: A review. Advances in Machine Learning and Data Mining for Astronomy, pp. 563–594. Chapman & Hall (2012). https://doi.org/10.1201/B11822-34

28. Zhou, Z.: Ensemble Methods: Foundations and Algorithms, pp. 15-16. Chapman and Hall (2012).https://doi.org/10.1201/b12207
29. Dietterich, T.G.: Ensemble methods in machine learning. In: Multiple Classifier Systems. MCS 2000, LNCS, pp. 1–15 (2000). https://doi.org/10.1007/3-540-45014-9_1
30. Wolpert, D.: Stacked generalization. Neural Netw. **5**, 241–259 (1992)
31. Cox, D.R.: The regression analysis of binary sequences. J. R. Stat. Soc. Ser. B (Methodol.) **20**(2), 215–232 (1958)
32. Cortes, C., Vapnik, V.: Support vector network. Mach. Learn. **20**, 273–297 (1995)
33. Jijo, B.T., Abdulazeez, A.M.: Classification based on decision tree algorithm for machine learning. J. Appl. Sci. Technol. Trends **2**(01), 20–28 (2021)
34. Zhang, Y., Wang, J.: K-nearest neighbors and a kernel density estimator for GEFCom2014 probabilistic wind power forecasting. Int. J. Forecast. **32**(3), 1074–1080 (2016)
35. Jiang, W., Chen, Z., Xiang, Y., Shao, D., Ma, L., Zhang, J.: SSEM: a novel self-adaptive stacking ensemble model for classification. IEEE Access **7**, 120337–120349 (2019)
36. Marqués, A.I., García, V., Sánchez, J.S.: On the suitability of resampling techniques for the class imbalance problem in credit scoring. J. Oper. Res. Soc. **64**(7), 1060–1070 (2013)
37. Mishra, S., Sarkar, U., Taraphder, S., Datta, S., Swain, D., Saikhom, R., et al.: Multivariate statistical data analysis- principal component analysis (PCA). Int. J. Livestock Res. **7**(5), 60–78 (2017)
38. Abdou, H., Pointon, J.: Credit scoring, statistical techniques and evaluation criteria: a review of the literature. Int. Syst. Acc. Finan. Manag. **18**, 59–88 (2011)
39. Powers, D.M.W.: Evaluation: From precision, recall and f-measure to ROC., informedness, markedness & correlation. J. Mach. Learn. Technol. **2**, 37–63 (2011)
40. Luo, G.: A review of automatic selection methods for machine learning algorithms and hyper-parameter values. Netw. Model. Anal. Health Inform. Bioinform. **5**(1), 1–16 (2016). https://doi.org/10.1007/s13721-016-0125-6
41. Chawla, N., Bowyer, K., Hall, L., Kegelmeyer, W.: SMOTE: synthetic minority over-sampling technique. J. Artif. Intell. Res. (JAIR) **16**, 321–357 (2002)
42. Mienye, D., Sun, Y.: Performance analysis of cost-sensitive learning methods with application to imbalanced medical data. Inform. Med. Unlocked **25**, 1–10 (2021)
43. Yu, H., Sun, C., Yang, X., Zheng, S., Wang, Q., Xi, X.: LW-ELM: a fast and flexible cost-sensitive learning framework for classifying imbalanced data. IEEE Access **6**, 28488–28500 (2018)
44. Ampountolas, A., Nyarko Nde, T., Date, P., Constantinescu, C.: A machine learning approach for micro-credit scoring. Risks **9**(3), 50 (2021)
45. Bravo, J.M., Ayuso, M.: Forecasting the retirement age: a Bayesian model ensemble approach. In: Rocha, Á., Adeli, H., Dzemyda, G., Moreira, F., Ramalho Correia, A.M. (eds.) WorldCIST 2021. AISC, vol. 1365, pp. 123–135. Springer, Cham (2021). https://doi.org/10.1007/978-3-030-72657-7_12
46. Ashofteh, A. Bravo, J.M.: Life table forecasting in COVID-19 times: an ensemble learning approach. In: Rocha, A., Gonçalves, R., Penalvo, F.G., Martins, J. (eds.), Proceedings of CISTI 2021 - Iberian Conference on Information Systems and Technologies. IEEE Computer Society Press (2021). https://doi.org/10.23919/CISTI52073.2021.9476583
47. Bravo, J.M., El Mekkaoui, N.: Short-term CPI Inflation forecasting: probing with model combinations. In: Rocha, A. et al. (eds.) Information Systems and Technologies. WorldCIST 2022. Lecture Notes in Networks and Systems, vol. 468, pp. 564–578. Springer, Cham (2022). https://doi.org/10.1007/978-3-031-04826-5_56
48. Ashofteh, A., Bravo, J.M. Ayuso, M.: A novel layered learning approach for forecasting respiratory disease excess mortality during the COVID-19 pandemic. In: CAPSI 2021 Proceedings, Volume 2021 – October 2021, Code 183080 (2021)

49. Bravo, J.M.: Longevity-linked life annuities: a Bayesian model ensemble pricing approach. In: CAPSI 2020 Proceedings. 29. https://aisel.aisnet.org/capsi2020/29 (Atas da 20ª Conferência da Associação Portuguesa de Sistemas de Informação 2020) (2020)
50. Bouttier, F., Marchal, H.: Probabilistic thunderstorm forecasting by blending multiple ensembles. Tellus A **72**(1), 1–19 (2020)

Camera- and Sensor Data-Based Process Observation in Liquid Steel Production

Maria Thumfart[1], Roman Rössler[2], Christine Gruber[1], and Johann Wachlmayr[1](✉)

[1] K1-MET GmbH, 4020 Linz, Austria
johannes.wachlmayr@k1-met.com
[2] Voestalpine GmbH, 4020 Linz, Austria

Abstract. The Ruhrstahl-Heraeus (RH) degassing plant is of vital importance for ultra-clean steel production, due to its effective degassing and homogenizing, but its vacuum-based nature and the high temperatures of liquid steel renders the process rather inaccessible to direct observations. We developed a camera-based image analysis solution, which in combination with enhanced sensor data evaluation (e.g., chamber pressure, offgas composition or O_2-injection) determines the status of the process and detects irregularities and malfunctions of the plant. The proposed research helps to deeper understand the RH process in steelmaking and provides the basis for a future on-line decision support system for operation.

Keywords: image analysis · data analysis · secondary metallurgy

1 Introduction

The steel industry faces major challenges in the upcoming years including increasing production costs and CO_2 reduction goals. The EU's climate targets and the associated reduction in CO_2 emission render a switch from the highly CO_2-intensive blast furnace (BF) to the electric arc furnace (EAF) production route inevitable.

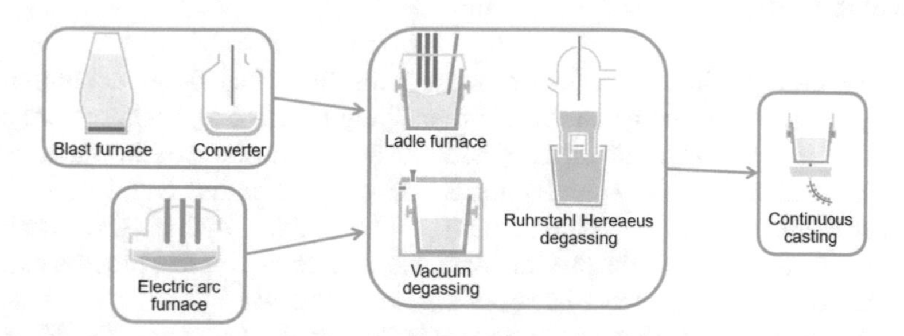

Fig. 1. Major steps in the steelmaking process.

An essential process step in the production of ultraclean steel (see Fig. 1) for both the BF and the EAF route is the degassing treatment [1–4]. Contaminating elements

A. Rocha et al. (Eds.): WorldCIST 2023, LNNS 799, pp. 397–403, 2024.
https://doi.org/10.1007/978-3-031-45642-8_39

like carbon, hydrogen, oxygen or nitrogen are extracted from the crude steel bath by application of vacuum with either a vacuum degasser or a RH plant [5, 6]. In case of crude steel from an EAF the expected content of dissolved gases in the melt is higher and the expected variety of alloys is larger implicating a larger need for treatments in secondary metallurgy. Two main categories of treatments are performed with RH plants: decarbonization and degassing. While the latter is tendentially calmer, decarbonization treatments show more reactivity, related to a higher oxygen content of the steel bath, which results in higher reaction rates with carbon or alloy materials [7]. Due to the extreme process conditions – high temperatures of the liquid steel, low pressure in the vacuum chamber (< 5 mbar) and thick refractory walls, the RH process is rather inaccessible to direct measurement. However, there are process variables which are recorded as time series, including offgas composition, chamber pressure and purging gas injection rate to drive circulation and stirring. Furthermore, a camera, which is mounted in a water cooled, nitrogen flushed casing, is installed at the top of the vacuum chamber.

Of vital importance for the lifetime of the vacuum chamber and the quality of the liquid steel is the detection of process abnormalities and malfunctions. Supply of wrong alloying material or ambient air leakage in the lower part of the vacuum chamber can lead to a significant change in liquid steel composition or even to massive, exothermal reactions which can result in a deflagration like behavior of the melt. This behavior can damage the vacuum chamber itself and equipment further downstream.

The research presented in this paper suggests an algorithm to detect abnormalities in the process based on the linear data gathered at the plant and the image data from the camera installed at the top of the vacuum chamber. The linear data can be exploited to determine the current process step, while the image data shows the reactivity and surface structure of the melt in the chamber. This surface structure can range from single drops in fog to bubbly and clear view. The goal of the data processing presented in this paper is to detect these differences so they can be related to the process step and generate an alert signal for the operator.

2 Linear Data

Linear data refers to time dependent process parameters, including vacuum chamber pressure, mass flow rate, composition of offgas, purging gas rate, backpressure of each sparging nozzle separately, data on alloy addition and the position of the vacuum chamber. Process data evaluation (using Python 3.9.13 and its libraries NumPy and Pandas) shows that there are three key parameters (listed in Table 1) which give particularly important information about the process state. Figure 2 shows typical process parameter plots in case of a degassing and a decarbonization treatment.

The chemical composition of the offgas, including H_2, CO, CO_2, O_2, N_2 and Ar, is measured with a GAM 300 mass spectrometer. The resulting mass fractions of the constituents can be converted into units of kg/s by using the total offgas flow (in Nm^3/h).

The purging gas rate indicates the injection of purging gas (argon) into the up snorkel to induce circulation and mixing of the liquid steel.

Table 1. Relevant process parameters.

Process parameters	Unit	Description
Vacuum pressure	[mbar]	Pressure in the vacuum chamber p_{vac}
Offgas composition	[% kg]	Chemical composition of offgas from the chamber in the form of mass percentage
Purging gas rate	[Nm^3/h]	Quantity of purging gas (Ar or N_2) used to drive the circulation of liquid steel into the vacuum chamber

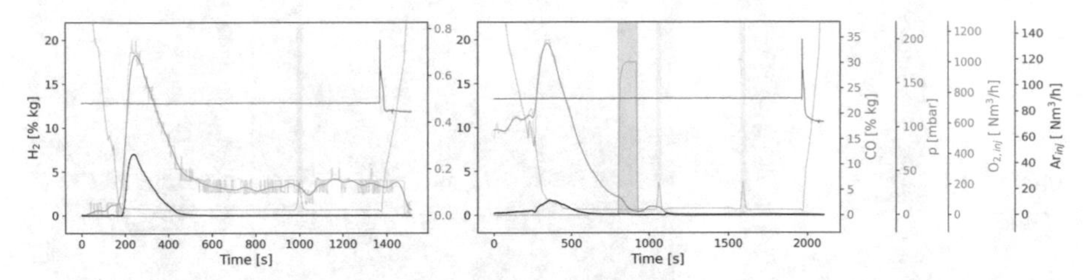

Fig. 2. Sensor data evaluation of degassing (left) and decarbonization (right) treatments. Yellow and green areas represent alloy additions and oxygen injections, respectively.

RH treatments start with the application of vacuum. Already at a chamber pressure of 300 mbar, the type of treatment (degassing or decarbonization) can be identified according to the content of $CO + CO_2$. Until the pressure reaches its minimum at $p_{vac} < 5$ mbar, the view to the melt might still be blurred or covered by smoke and residual slag covers the steel melt. This slag layer is then slowly emulsified in the steel and transported into the ladle. Furthermore, the chamber pressure is used to identify alloying events, as it slightly increases during alloy additions. The evolution of hydrogen in the offgas can be linked to the stirring efficiency at the beginning of the treatment.

The linear process data evaluation detects RH treatments, degassing and decarbonization events as well as alloy additions and O_2 injections and is used to identify those time regions, when good image quality is expected.

3 Integral Data

Integral data refers to image material from process surveillance using a camera, which is mounted in the vacuum chamber approximately 9 m above the melt surface. Figure 3a) shows a schematical representation of the lower part of an RH plant. After immersion of the vacuum chamber into the steel filled ladle, vacuum is applied and argon injection induces circulation of the liquid steel up through one snorkel into the vacuum chamber, down the other snorkel and back into the ladle [8]. The offgas is removed at the top of the vacuum chamber. The inner diameter of a new vacuum chamber is approx. 1.5 m. Figure 3b) to d) show three example images acquired with the original camera. Figure 3b) shows clear view conditions, Fig. 3c) the state of alloying and Fig. 3d) shows foggy conditions which occur at the beginning of the treatment and in case of reactions as well as after O_2 injections. Situations like in Fig. 3c) and d) can be excluded from evaluation based on the linear data available. The bright orange area in Fig. 3b) shows melt surface, all the other parts show skull. In the region of the melt barely any features can be seen. On close view these images show strong compression patterns. Accordingly, an enhanced camera and lens have been temporarily installed at the plant to increase the image quality. Figure 3e) and h) show two examples of the improved image quality.

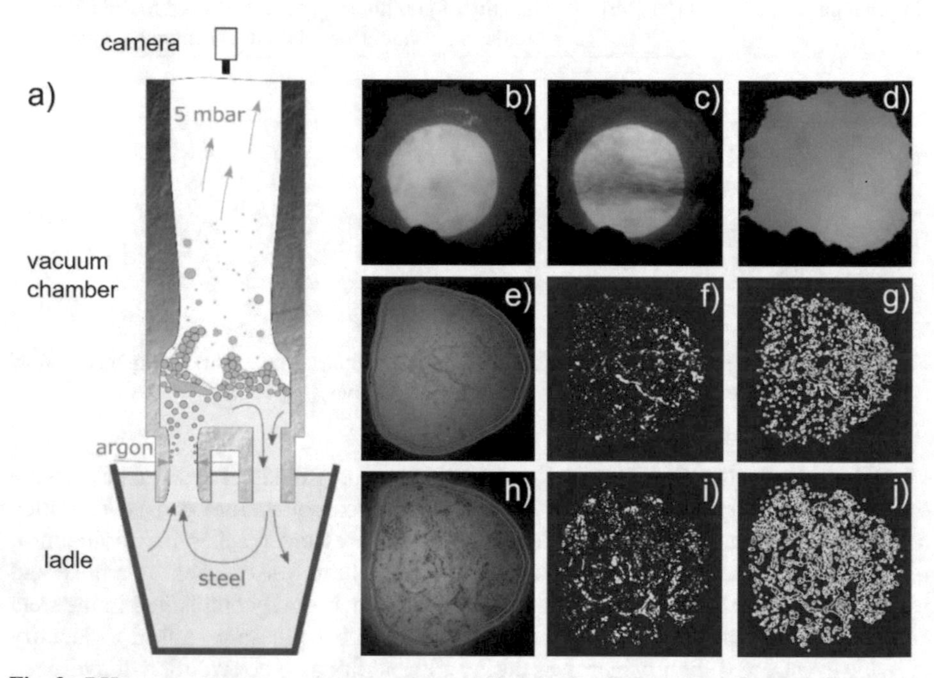

Fig. 3. RH treatment stages and process states a) Sketch of an RH vacuum chamber. The steel is lifted from the ladle into the vacuum chamber by applying an absolute pressure of approximately 5 mbar. Argon is injected in the left snorkel to lift the liquid steel into the chamber where the refining process takes place. The steel exits through the right snorkel. The deviation of the up snorkel from the cylindrical shape is caused by erosion. The hourglass shape of the vacuum chamber stems from the formation of skull (solidified splashes of melt). The color of the wall material qualitatively indicates its temperature at the inside of the vacuum chamber; b) normal process conditions with clear view. c) alloy material addition; d) smoke at the end of a treatment. e) improved image of melt surface during a reactive phase of decarbonization with skull boundary (blue line) and region of interest (orange line). The structures seen are drops surrounded by light emitting fog. f) Detected features of the droplet-dominated regime and g) the surroundings of the features. h) Improved image with marked ROI (orange) of foam-like surface of degassing treatment in nonreactive state with extracted features i) and corresponding surroundings j).

While Fig. 3h) represents a calm degassing state, Fig. 3e) refers to a reactive state during a decarbonization treatment where oxygen is still present in the chamber. In calm degassing states the melt surface shows a foamlike consistency. In contrast, in reactive states there are more drops visible which are brighter compared to their surroundings. In case of a lower vessel leakage ambient air can enter the vacuum chamber, leading to reactive conditions as well. So, if the melt surface appears reactive in a process step which should only be a degassing step, it is a clear sign that there is an unexpected issue, most likely a leak. The goal of the image processing steps presented in this paper is to detect these differences. The development of the image processing has been performed in Matlab 2021b using the signal processing toolbox and the image processing toolbox [9, 10].

3.1 Region of Interest

The first step in image evaluation is the definition of the region of interest (ROI). In case of the RH process, the region of interest is the melt surface. As the skull in the vacuum chamber continuously changes its appearance, the ROI cannot be set as a constant, but needs to be adapted for each treatment. The region of the image showing melt is usually brighter and more dynamic than the part showing skull. This is represented in the time average of the grey level and the averaged absolute of the time derivative of the grey values. Based on these two values the blue lines in Fig. 3e) and h) have been detected. To account for minor errors and possible skull buildups during one treatment the ROI is eroded, indicated by the orange lines. For the calculation of texture parameters, which require a rectangular ROI, the largest possible square is inscribed into the ROI [11].

3.2 Texture Parameters

There is an abundance of texture parameters available in literature. For this case three approaches have been employed:

- Feature detection based: Darker objects in the ROI are detected. Their brightness compared to their surroundings and their total area relative to the ROI area are evaluated.
- Fast Fourier transform (fft): The magnitudes in different frequency bands are compared.
- Gray-level co-occurrence matrix: Image Energy and contrast are calculated based on images normalized by their median value and by 255.

To narrow the number of texture parameters, the density-based spatial clustering algorithm dbscan (radius: 0.06, minimal cluster size: 35 points) has been used on different parameter combinations. Several promising combinations of the smoothed values of the texture parameters have been tested. The best clustering result on the given data could be achieved using the six parameters shown in Fig. 4. These parameters are based on the feature-based approach and the gray-level co-occurrence matrix. The fft based values resulted in no further improvement of the clustering results.

The feature detection-based algorithm targets the gray level differences between the bubbles or drops and their immediate surroundings. For this approach the ROI indicated in Fig. 3e) or h) is used. Features are detected using the standard settings of adaptive threshold [12]. Their immediate surroundings are achieved by dilatating the features with a disk of diameter 5 pixels. The average gray value of all feature pixels is divided by the average gray value of all immediate surrounding pixels resulting in a relative gray

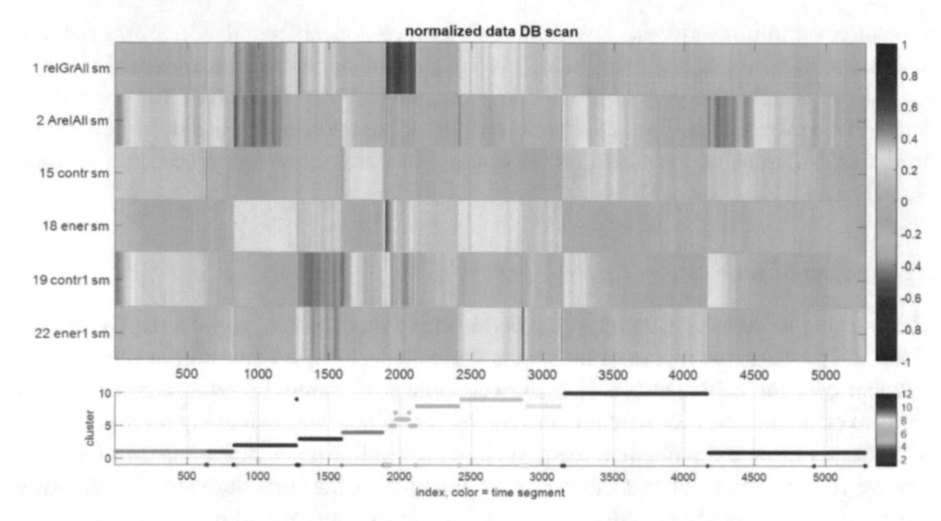

Fig. 4. Most relevant texture parameters and clustering result.

value (relGrAll, Fig. 4). The difference in the pixels covered by features is expressed by the total area of all features divided by the ROI area (ArelAll, Fig. 4).

The second approach proven to be fruitful in this study is based on the gray-level co-occurrence matrix. Two versions of image normalization are used as preprocessing. One case is normalized using the median gray level in each ROI resulting in the third and fourth parameter in Fig. 4. As a second approach the gray values (ranging from 0 to 255) are divided by 255 resulting in parameter five and six. From each gray-level co-occurrence matrix the contrast (contr) and the energy (ener) are calculated.

All parameters described above were smoothened with respect to time using a Hamming window of a length of 20 datapoints and the rows were normalized to range from 0 to 1. The lower part of Fig. 4 shows the cluster each image was assigned to by dbscan. The colors of the dots indicate the video segment the frames are from. Almost all video segments can be assigned to their own cluster and the ones being identified as the same category are similar. So even smaller differences can be detected based on the six proposed parameters. Accordingly, an early detection of any deviation from the normal state can be done based on the six proposed parameters and the linear data.

The long-term goal is the development of an online decision support system based on these findings. Currently the offline characterization of long-term plant data to determine parameters like the necessary lens cleanness and general viewing conditions is ongoing and showing good results.

4 Conclusion and Outlook

The data presented in this paper show the high potential of camera-based surveillance in combination with linear data evaluation in the RH process. A robust algorithm for the detection of the region of interest is proposed and it is shown that texture parameters can be used to distinguish different process steps. The comparison of these texture parameters with the process data can support the early detection of irregularities in the process and thus help to reduce the probability of malfunctions, damage to the plant and down time. Based on the results of this paper, an online decision support system is being developed.

Acknowledgments.  The INEVITABLE project, carried out in a collaboration with voestalpine Stahl GmbH, has received funding from the European Union's Horizon 2020 research and innovation program under grant agreement No 869815. This paper reflects only the authors' views, and the European Commission is not responsible for any use that may be made of the information it contains. The authors moreover gratefully acknowledge the funding support of K1-MET GmbH, whose research program is supported by COMET (Competence Center for Excellent Technologies), the Austrian program for competence centers. COMET is funded by the Austrian ministries BMK and BMDW, the provinces of Upper Austria, Tyrol, and Styria, and the Styrian Business Promotion Agency (SFG).

References

1. Chen, G., He, S.: Mixing behavior in the RH degasser with bottom gas injection. Vacuum **130**, 48–55 (2016)
2. Wei, J.H., Yu, N.W.: Mathematical modelling of decarburisation and degassing during vacuum circulation refining process of molten steel: application of thc modcl and results. Steel Res. **73**(4), 143–148 (2002)
3. Pirker, S., Puttinger, S., Rössler, R., Lichtenegger, T.: Steel alloy homogenization during rheinsahl–heraeus vacuum treatment: conventional computational fluid dynamics, recurrence computational fluid dynamics, and plant observations. Steel Res. Int. **91**(12), 2000214 (2020)
4. Wang, M., Bao, Y.P., Yang, Q., Zhao, L.H., Lin, L.: Coordinated control of carbon and oxygen for ultra-low-carbon interstitial-free steel in a smelting process. Int. J. Miner. Metall. Mater. **22**(12), 1252–1259 (2015)
5. Kai, F.E.N.G., He, D.F., Xu, A.J., Wang, H.B.: End temperature prediction of molten steel in RH based on case-based reasoning with optimized case base. J. Iron. Steel Res. Int. **22**, 68–74 (2015)
6. Zhang, J., Liu, L., Zhao, X., Lei, S., Dong, Q.: Mathematical model for decarburization process in RH refining process. ISIJ Int. **54**(7), 1560–1569 (2014)
7. Gruber, C., Bückner, B., Schatzl, M., Thumfart, M., Eßbichl, R., Rössler, R.: Big data handling in process surveillance and quality control of secondary metallurgical processes. Steel Res. Int. **93**(12), 2200060 (2022)
8. Ahrenhold, F., Pluschkell, W.: Mixing phenomena inside the ladle during RH decarburization of steel melts. Steel research **70**(8–9), 314–318 (1999)
9. MathWorks Signal Processing Toolbox: User's Guide (R2021b) (2021). Accesed 6 Oct 2022 https://nl.mathworks.com/help/pdf_doc/signal/signal.pdf
10. MathWorks Image Processing Toolbox: User's Guide (R2021b). (2021). Accessed 6 Oct 6 2022 https://nl.mathworks.com/help/pdf_doc/images/images_ug.pdf
11. Jaroslaw Tuszynski Inscribed_Rectangle (2023). Accessed 17 Jan 2023 (https://www.mathworks.com/matlabcentral/fileexchange/28155-inscribed_rectangle), MATLAB Central File Exchange
12. Bradley, D., Roth, G.: Adaptive thresholding using the integral image. J. Graph. Tools **12**(2), 13–21 (2007)

Risks in the Process of Transparency in the Transmission of Live Sessions by Internet: A Case Study in the Brazilian Chamber of Deputies

Newton Franklin Almeida[1(✉)], José Fábio de Oliveira[1(✉)], Paulo Evelton Lemos de Sousa[1(✉)], and Lucas da Silva Almeida[2(✉)]

[1] Universidade de Brasília-CIC, Brasília, DF 70910-900, Brazil
newton.franklin@gmail.com, josefabioo@gmail.com, pevelton@gmail.com

[2] Northeastern University-NSI, Boston, MA 02115, USA
almeida.lu@northeastern.edu

Abstract. Organizing debate and voting sittings is a critical process for the Brazilian Chamber of Deputies. There are several transparency processes linked to legislative activities and projects, based on long-term strategic guidelines and that facilitate the monitoring of society through news on the web, radio and television. This study applies the ISO ABNT 31000 risk management approach to the live transmission process via the Internet of the Chamber's plenary sessions. The context is provided by the application of Business Process Modeling and Value Chain, and the analysis is performed with the support of ISO 31010, Pareto diagram and FMEA. The study concludes that the application of risk management principles and techniques contributes to more effective governance and supports the institution to strengthen the fulfillment of its mission.

Keywords: Risk Analysis · Transparency · Legislative Process · ISO 31000 · Live Streaming

1 Introduction

The Brazilian Chamber of Deputies has three main elements in its Mission, established in the Federal Constitution of 1988 [1]: (i) Represent the People; (ii) to elaborate the laws and (iii) to supervise the other two Powers of the Republic.

The Chamber adopts in its Strategic Planning [2] the Balanced Scored Card from Kaplan & Norton [3] and, in the scope of the "Internal Processes" perspective, it is aligned with the strategic objective "To increase the visibility of the legislative activity of administrative acts".

The use of information and communication technologies and instruments that enable the visibility and transparency of legislative activity and administrative acts is a result

A. Rocha et al. (Eds.): WorldCIST 2023, LNNS 799, pp. 404–416, 2024.
https://doi.org/10.1007/978-3-031-45642-8_40

of Guideline number 2 "Improving the transparency of the activities and information of the Chamber of Deputies and Public Policies" [4].

The question that guides this study is: What is the impact on the transparency of the legislative process due to the possible risks of live streaming of Chamber sittings? The answer can provide elements to the Legislative Branch to enable improvements in the transmission process, as well as provide the identification of events related to these risks. In the computational field, the study can consolidate the effective use of business process modeling and the use of risk analysis and identification tools.

The purpose of this article is to map the current process (As-is), analyze, identify and address the risks inherent to the process and, finally, suggest improvement measures.

There is also, in this work, a survey of the supporting literature, followed by a detailing of the methodology and the description of the structured steps to carry out the research, in addition to the presentation of the results collected, comments and analyzes to support the conclusions and suggestions for further studies.

2 Theoretical Reference

2.1 Political Transparency

Studies in the European Union and the United States have shown that Political Transparency is a strengthening element in cooperation between border countries [5] and reduces political risks by stimulating foreign direct investments (FDI-Foreign Direct Investments) [6].

Transparency is seen as a positive concept, generally desirable and associated with democratic and ethical values and good organizational performance. On the other hand, however, there are authors who claim that it is not a pure good [7] because there are some negative effects associated with its application. There are records of damages to certain governmental actions, as well as damages to the reputations of individuals and/or organizations. There are also reports that transparency can cause a certain "Bureaucratic Tension" due to an oscillation between transparency and guilt prevention in public administration [8].

This case study adopts a positive view on the concept of transparency, based on the delimitation of this construct established by Strategic Directive nº 2 - Transparency of the Chamber of Deputies, which is established in 3 lines of action:

(i) Facilitate society's access to information in clear language and in an inclusive manner.
(ii) Promote active transparency and social control.
(iii) Expand the dissemination of parliamentary activity and institutional actions on different platforms. [4].

2.2 Legislative Process

According to the Federal Constitution [1] the legislative process is formed by a set of acts carried out by the organs of the Legislative Power from previously established rules, in order to elaborate legal norms. Such rules are defined in the Constitution: item "On the

Legislative Power". In Brazil, this power is exercised by the National Congress, made up of the Chamber of Deputies and the Federal Senate. Both define their own rite for processing the legislative process.

2.3 ISO 31000 - Risk Management

Risk management is increasingly recognized as a means of optimizing the probability of success in complex projects, however the lack of publicity of best practices may explain the low adoption of risk management in projects [9]. In support of risk management, there is the ISO 31000 standard [10] which provides common guidelines for managing any type of risk faced by organizations, in addition to enabling the reflection of needs and realities. This standard is a globally accepted standard for risk management, as it supports a straight and simple way of thinking about the topic, in addition to indicating the process of resolving inconsistencies and ambiguities that exist between different approaches [11].

The risk management standard defined by the ISO 31000 can, in the same way, be applied to critical infrastructure resiliency management frameworks. It has different aspects and categories and can be adopted at different levels at the same time [12]. The applicable objectives of the standard can be financial, quality, information security and at different levels such as services, products, projects and processes [13].

The Corporate Risk Management Policy in the Chamber of Deputies [14] is mainly aimed at supporting its decision-making processes and is governed by the principles of efficiency, parsimony and uncertainty reduction.

2.4 Live Broadcast

Live streaming or live transmission is the transmission of audio and video, today one of the most used actions and that most consume Internet bandwidth [16], in real time of an event from the Web [17]. Powered by rich media generation tools and associated with the rapid development of cyberspace, as well as collaborative productions and streaming services, live streaming allows anyone to launch a stream or watch someone else in real time, which has garnered great popularity [18]. However, the capillarity achieved by live broadcasts is based on two contexts of trust: on the platform and on security, which are important for customers to adopt and reuse a particular Streaming service [19] and [20].

2.5 Incident Management

Incident Management is the process by which IT support organizations are able to restore the normal operation of a service after an interruption has occurred [21] and [22]. Service outages can have a considerable impact on the business operations of organizations and require the implementation of specific processes to manage these incidents and to enable the restoration of these services to operational normality [23].

When citing incident management, the literature, as well as best practices, present ITIL as a tool, as cited by Iden [24]. ITIL consists of collections of approaches best suited to implementing, for example, service support processes such as and Problem Management [25].

3 Methodology

3.1 Research Method

The research developed is exploratory based on combined procedures of document survey, correlation to technical standards, database search and bibliographic support. The central theme is based on the discipline of Risk Management applied to a case study of the public transmission process, via live streaming, of the debates and voting sessions held in the Plenary of the Chamber of Deputies. The following steps are covered in structuring the research.

3.2 Research Structuring

The composition of the research is divided into four stages: documentary survey, mapping, data collection and application of the FMEA.

The documentary survey consists of two phases. The first, a bibliographic survey, encompasses the identification of theoretical subsidies relevant to the concepts developed in the Literature Review item, in the context of the institution under study. Then, there is the collection and identification of documentation related to strategic planning, guidelines, policies and other normative acts connected to risk management processes, as well as to the transmission process of plenary sessions.

Step 2 consists of three actions. The first activity lies in mapping the SIPOC matrix (Supplier, Input, Process, Output, Customer) [26], which will allow a detailing of the scenario of the process under study. The second is the creation of the As-is model, which allows the visualization of the flow of information inherent to the chain of activities of the process and its actors. With the Sipoc matrix created and the As-is model conceived, the third action models the Value Chain, according to Porter [27], which allows an association of the process to the added value to customers.

The next stage consolidates the data and information cycle, which allows the qualification of incidents linked to the live transmission process. For this, a selection and recording of these process events is made in the database of the Chamber´s Incident Management tool - OTRS [28]. From the selection, there is the categorization of incidents, the creation of a Pareto Diagram, which allows a visualization of the categories of events that demand greater technical effort for business continuity.

The last step consists of applying the FMEA [29] within the cycle proposed by the ISO 31000 standard [10]. The initial stage consists of identifying risks with the support of Brainstorming and Delphi techniques, according to the literature [29]. The identification of risks aims to produce a list that includes different sources, as well as the factors, their causes, possible effects and the affected areas in the organization [30]. These techniques are applied through meetings and interviews with the technical support team for the process, composed of four analysts and four technicians. With the participation of all the specialists involved in the immersion of risk identification.

Subsequently, the risks are qualified by assigning weights to the indices of severity, occurrence and detection of risks [31]. With the qualification performed, the RPN – Risk Priority Number – is identified, which is the calculated risk associated with the failure mode used in the categorization of risks (low, medium, high, very high) [31]. Finally, the treatment of risk, backed by the execution of the previous steps, involves the selection of one or more options for modifying the risks. Once implemented, treatments provide or modify controls. Responses are options for certain actions with a view to increasing opportunities and reducing threats [32].

4 Results and Analysis

4.1 Document Survey

The basis of the collection was the search for documents in the Institutional Portal of the Chamber and in the Edoc process and document processing system, where data and organizational information, process management and strategic planning were collected. The following documents were prospected: Strategic Planning 2012–2023 [2]; Process that requests improvements in capturing and encoding the video signal. (No. 502,883/2018) [33]; Procedure for defining critical business processes and value chain (nº 117.351/2014 [34]); Corporate Risk Management Policy – Board Act 233/2018 [14]; Corporate Risk Management Methodology of the Chamber of Deputies [15].

4.2 Mapping

The information gathered allowed the identification of the value chain, Porter [27]. Its composition is formed by six finalistic processes, which are supported by six managerial and support processes [34].

From the creation of the Chain of Values, it was possible to model the As-is process of live transmission. It was carried out in a process of research and interviews, which had the cooperation of specialists, responsible for the live transmission service of the plenary sessions of the Chamber of Deputies. The model created used the technique BPM - Business Process Modeling [35], according to the literature followed by the market. For Aalst [36] BPM is the discipline that combines approaches to design, execution, control, measurement and optimization of business processes. The result can be seen in Fig. 1.

4.3 Data Collection

With the As-is process mapped, the next step was responsible for the search and collection of incidents linked to the transmission of plenary sessions. This information was selected from the databases of the incident management tool - OTRS [28], which are consolidated in a management information environment via the Pentaho data warehouse [38] and recorded from August/2016 to November/ 2020 (Total – not a sample). The records were not ordered or categorized, but only registered by the professionals of the Call Center, without any cataloging criteria.

To group the data, categories were created, where the incidents were gathered, namely: transmission error; preview error, help; help support; improvement and problems.

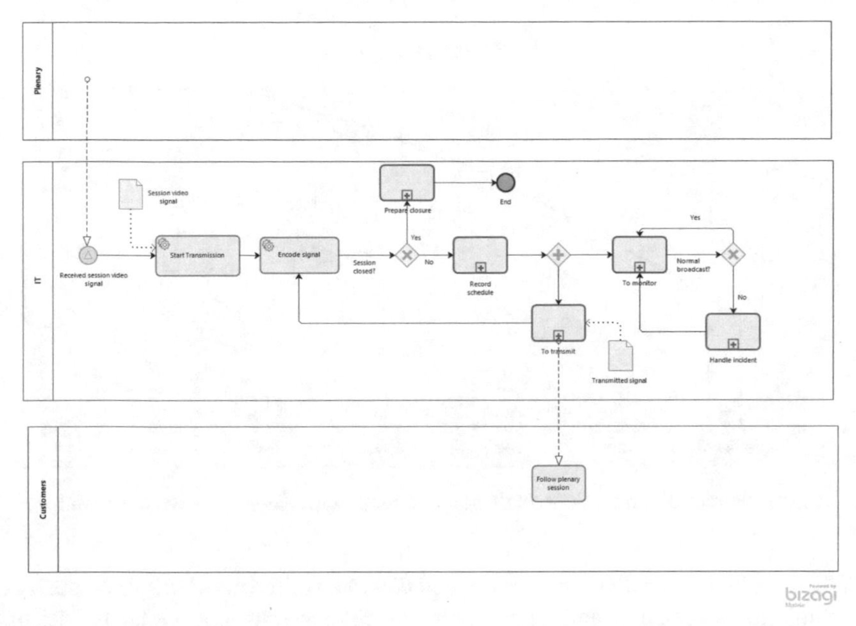

Fig. 1. Process transmission of live streaming - As-is. Source: Own authorship, with the help of software Bizagi [37]

This categorization was based on the literature [39] and according to the best practices contained in ITIL [25], described in Table 1:

Table 1. Categories of selected incidents

Categories – All years	Quantity	%	Q. Accumulated	% Accumulated
Transmission errors	74	35.07	74	35.07
Simple help	45	21.33	119	56.40
Support help	37	17.54	156	73.93
Visualization error	30	14.22	186	88.15
Problem	21	9.95	207	98.10
Upgrade	4	1.90	211	100.00
Total	211	100.00		

Source: Own authorship

Once the categories were created, it was possible to order them from the highest to the lowest incidence, perform the calculation of the calculated values and, finally, determine the percentage accumulated in each category. With this, a diagram was created that represents the Pareto Principle (80/20 rule) [40], according to Fig. 2.

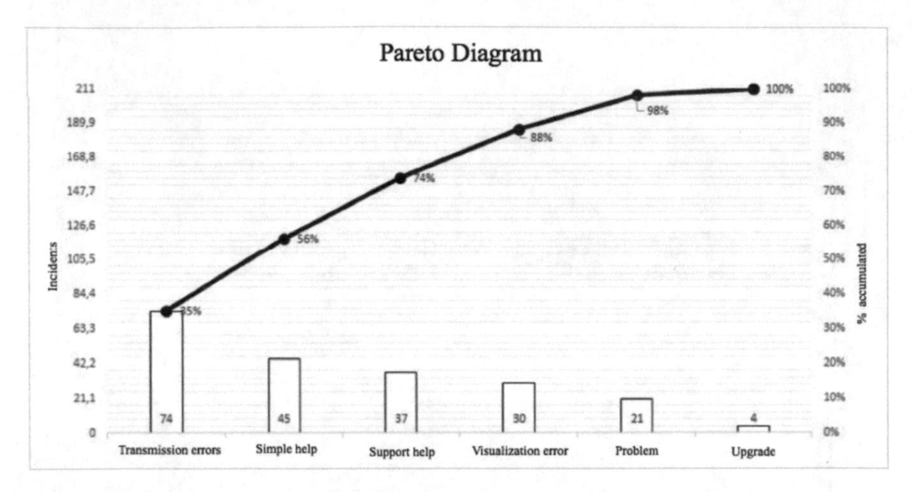

Fig. 2. Pareto diagram. Source: Own authorship, with the help of Excel software.

When analyzing the diagram, it is noticed that most of the problems faced are related to transmission errors, help and support help, which represent approximately 74% of the total incidents.

4.4 Application of FMEA

After steps one, two and three, the identification of risks was elaborated as described above, in addition to being supported by the contributions extracted from the literature of [41, 42]. This information is input for the application of Failure Mode and Effect Analysis, the FMEA, which aims to anticipate, identify and avoid failures in the operation of a new system while the project is still on the drawing board [31], since it is a technique developed to analysis of potential failures that may occasionally generate risks. In addition, the tool proposes improvement actions in the construction of projects, products or processes [43].

FMEA is widely used in risk assessment to ensure that the results of a project provide consistency to the intended level of performance [44] and also to generate information regarding the constraints of processes and their environments. This method is one of the tools listed in the ABNT ISO 31010 standard [29]. In this work, the FMEA techinique was used, which takes into account the failures in the planning and execution of the examined process, whose bases are provided by the non-conformities of the product [45].

The method was applied along the lines listed by Mock [46], which describes the divergent development of risk analysis conducted by operators of sites and IT infrastructures. The result of applying the FMEA recorded the identification, qualification, categorization, treatment and response to the 16 risks listed by the experts. Table 2 shows the risks with the most significant RPN values that presented a "very high" degree of criticality (above 200).

Table 2. Partial FMEA table of risks with higher RPN

Risk ID	R01	R16
Risk Event	Non-isolation of the Video Coding Layer (VCL) on the network (NAL)	Error recording signal
Effect	Loss of compressed data during transport	Loss of recording of the plenary session
Causes	Lack of isolation from the network layer VCL (NAL)	Unmapped event; equipment failure or even interruption of service
Current controls	Unidentified	Unidentified
S	6	7
O	4	4
D	9	8
RPN	**216**	**224**

Source: Own authorship

4.5 Identified Improvements

After the document analysis, with the elaboration of the SIPOC matrix, value chain, As-is model, identification and categorization of risks, Pareto diagram and, finally, the application of the FMEA, relevant overload points were identified, such as "errors of transmission" and "help", which accumulate in the area responsible for the process approximately 56% of the treatment of incidents, as shown in Fig. 2. Therefore, it is appropriate to propose an evolution in the current process of live transmission, in order to allow the customer himself makes the indication of problems. Thus, the Service Desk team can categorize risks correctly, the person responsible for handling the risk can be communicated in an automated way, in addition to enabling integration with incident monitoring tools. The To-be process proposed in Fig. 3 improves the previous model by automating the gaps identified from the incidents.

Other needs identified during the analysis refer to the creation of indicators and service level agreements. According to Kaplan [3], the indicators are relevant for signaling the state of normality or abnormality of the services in operation. Service level agreements define minimum admissible service quality limits and entail contractual penalties if they are not complied with, see Kaplan [3].

In turn, service level agreements are important, according to ITIL [25], in order to guarantee objective conditions so that the IT areas can apply the due treatment to failure occurrences, so that the customer considers the operation and quality of the service satisfactory. Table 3 summarizes the indicators and SLAs aligned with the To-be process in order to improve business continuity, since these monitoring and control elements were not identified in the survey carried out.

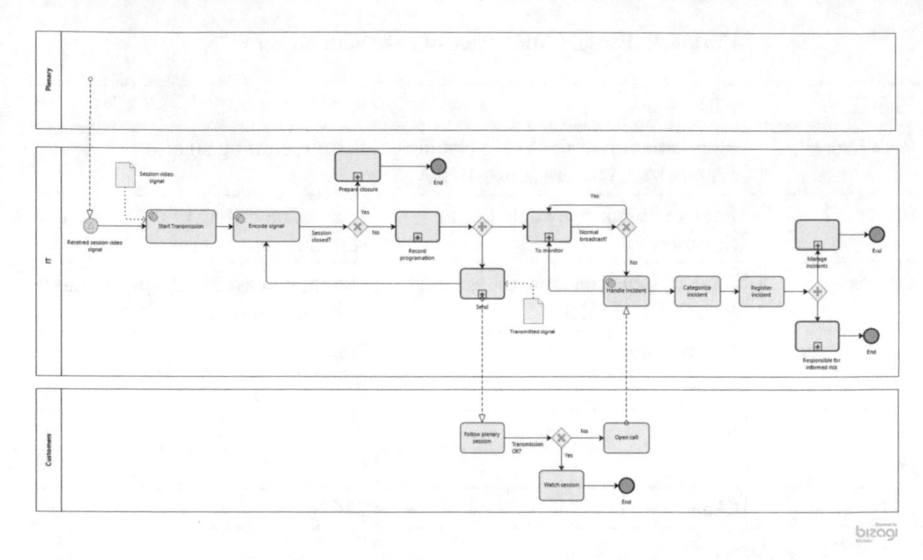

Fig. 3. Live streaming process – To-be. Source: From the authors, with the help of the Bizagi software [37]

Table 3. Suggested Indicators

Indicator	Description	SLA	Calculation criteria
Availability	Percentage of time services remain in normal production conditions	Initial monthly target of 98% on a 7x24 basis	# of minutes the service was available in a month, according to the formula below: TD % = {[(DR + IJ) / DP] × 100}
Latency	Time it takes a packet to travel across the network connection (from source to destination)	Latency must be less than or equal to 50 ms At each 5-min interval	The average latency time is measured between sending and receiving a packet with 500 bytes
Execution delay	Delay in the delivery of the solution of the tickets beyond 1 business day	For each partial or total delivery scheduled	TAE = PEE - PPE (planned - measured)
Quality	Amount of more than 5 errors or situations out of specification	Monthly	No. of errors and nonconformities found

Source: Own authorship

5 Conclusion and Future Work

The use of combined process analysis and risk management techniques enabled the identification of problems and possible improvements in the live transmission process of the plenary sessions. The categorization of risks made possible to characterize the main events related to failures in the process. In this way, the creation of identifiers and service levels are ways of monitoring that provide an increase in quality in the delivery of the final product. Automation also allows an increase in the performance of handling these incidents, since the person responsible for the risk will be communicated immediately and automatically in an integrated way with the process monitoring mechanisms [47]. Also, the improvements presented can offer an improvement in the service customers experience (Quality of Experience - QoE), which can also apply training and qualifications to users in the use of the live viewing tool of the plenary sessions, in order to reduce the number of of incidents related to "help" requests.

It is understood that the main objective of the study was achieved and the research problem, within the scope of the analyzed process, was solved. For future studies, it may be advisable to use other risk elicitation techniques and categorization tools to ratify the approach to using BPM and refine the risk management process [48].

And finally, for the purpose of temporal reference, it is pointed out that research was conducted just at the beginning of the Covid 19 pandemic, complying with the adoption of remote system voting for Deputies [49]. With this perspective, the live broadcasts process have gained a much more significant character and signaling that this theme deserve reviews to evaluate its subsequent developments.

References

1. BRASIL: Federal Constitution (1988). http://www.planalto.gov.br/ccivil_03/constituicao/constituicaocompilado.htm. Accessed 25 Oct 2020
2. BRASIL, Câmara dos Deputados: Strategic Planning 2012–2023. https://www2.camara.leg.br/a-camara/estruturaadm/gestao-na-camara-dos-deputados/gestao-estrategica-na-camara-dos-deputados/arquivos/planejamento-estrategico-2012-2023#:~:text=A%20gest%C3%A3o%20estrat%C3%A9gica%20na%20C%C3%A2mara,dos%20princ%C3%ADpios%20constitucionais%20de%20legalidade. Accessed 11 Oct 2020
3. Kaplan, R.S., Norton, D.P.: The Balanced Scorecard Measures That Drive Performance. Harvard Business Press, Boston, MA (1992)
4. BRASIL, Câmara dos Deputados: Strategic Guidelines 2012–2023. https://www2.camara.leg.br/a-camara/estruturaadm/gestao-na-camara-dos-deputados/gestao-estrategica-na-camara-dos-deputados/diretrizes-estrategicas-2012-2023. Accessed 11 Oct 2020
5. Castanho R.A., Vulevic A., Naranjo Gómez JM, et al: (2019). Political commitment and transparency as a critical factor to achieve territorial cohesion and sustainable growth. European cross-border projects and strategies. Reg. Sci. Policy Pract. **11**(2), 423-435https://doi.org/10.1111/rsp3.12201
6. Barry, C.M., DiGiuseppe, M.: Transparency, risk, and FDI. Polit. Res. Q. **72**(1), 132–146(2019).https://doi.org/10.1177/1065912918781037
7. Bannister, F., Connolly, R.: The trouble with transparency: a critical review of openness in e-government. Policy Internet 3(1), 8. (2011). https://doi.org/10.2202/1944-2866.1076

8. COSO, Committee of Sponsoring Organizations of the Treadway Commission: Enterprise Risk Management Applying enterprise risk management to environmental, social and governance-related risks (2018)
9. Olechowski, A., et al.: The professionalization of risk management: what role can the ISO 31000 risk management principles play? (2016). https://doi.org/10.1016/j.ijproman.2016.08.002
10. ABNT, Norma ISO 31.000: Gestão de Riscos – Princípios e Diretrizes, Rio de Janeiro, Brasil. (2009)
11. Broodleaf, G.P.: ISO 31000:2009—setting a new standard for risk management. Risk Anal. **30**(6), 2010 (2010). https://doi.org/10.1111/j.1539-6924.2010.01442
12. Rød, B., Lange, D., Theocharidou, M., Pursiainen, C.: From risk management to resilience management in critical infrastructure. J. Manage. Eng. **36**(4), 04020039 (2020)
13. Barafort, B., Mesquida, A.L., Mas, A.: ISO 31000-based integrated risk management process assessment model for IT organizations. J. Softw. Evol. Process **31**(1), e1984 (2019). https://doi.org/10.1002/smr.1984(2019)
14. BRASIL, Câmara dos Deputados, Ato da Mesa 233/2018 – Política de Gestão Corporativa de Riscos (2018). https://www2.camara.leg.br/legin/int/atomes/2018/atodamesa-233-24-maio-2018-786753-publicacaooriginal-155674-cd-mesa.html. Accessed 25 Oct 2020
15. BRASIL, Câmara dos Deputados: Corporate Management Risk Methodology v1.1 (2017). https://www2.camara.leg.br/a-camara/estruturaadm/gestao-na-camara-dos-deputados/governanca/documentos/metodologia-corporativa-de-gestao-de-riscos-1.1 Accessed 09 Nov 2020
16. Bilal, K., Erbad, A.: Impact of multiple video representations in live streaming: a cost, bandwidth, and QoE analysis. In: Proceedings - 2017 IEEE International Conference on Cloud Engineering, IC2E 2017 7923791, pp. 88–94 (2017). https://doi.org/10.1109/IC2E.2017.20
17. Chen, C., Lin, Y.: What drives live-stream usage intention? The perspectives of flow, entertainment, social interaction, and endorsement. Telematics Inform. 35(6), 1794 (2018). https://doi.org/10.1016/j.tele.2017.12.003
18. Zhao, Q., Chen, C.-D., Cheng, H.-W., Wang, J.-L.: Determinants of live streamers' continuance broadcasting intentions on Twitch: a self-determination theory perspective. Telematics Inform. **35**(2), 406–420 (2018). https://doi.org/10.1016/j.tele.2017.12.018
19. Li, D., et al.: Modelling the roles of cewebrity trust and platform trust in consumers' propensity of live-streaming: an extended TAM method. Comput. Mater. Continua **55**(1), 137–150 (2018)
20. Guarnieri, T., Drago, I., Vieira, A.B., Cunha, I., Almeida, J.: Characterizing QoE in large-scale live streaming. In: 2017 IEEE Global Communications Conference, GLOBECOM 2017-Proceedings 2018-January, pp. 1–7 (2017). https://doi.org/10.1109/GLOCOM.2017.8254062
21. Bartolini, C., Stefanelli, C., Tortonesi, M.: SYMIAN: analysis and performance improvement of the IT incident management process. IEEE Trans. Netw. Service Manage. 7(3), 5560569 (2010) https://doi.org/10.1109/TNSM.2010.1009.I9P0321. pp. 132–144
22. Bartolini, C., Stefanelli, C., Tortonesi, M.: Business-impact analysis and simulation of critical incidents in IT service management. In: 2009 IFIP/IEEE International Symposium on Integrated Network Management, IM 2009 5188781, pp. 9–16 (2009). https://doi.org/10.1109/INM.2009.5188781
23. Pollard, C., Cater-Steel, A.: Justifications, strategies, and critical success factors in successful ITIL implementations in US and Australian companies: an exploratory study. Inf. Syst. Manag. **26**(2), 164–175 (2009). https://doi.org/10.1080/10580530902797540
24. Iden, J., Eikebrokk, T.R.: Implementing IT service management: a systematic literature review. Int. J. Inf. Manage. **33**(3), 512–523 (2013). https://doi.org/10.1016/j.ijinfomgt.2013.01.004
25. Lahtela, A., Jäntti, M., Kaukola, J.: Implementing an ITIL-based IT service management measurement system (2010)

26. Fleaca, B., Corocaescu, M.: Process map to create added value to customer based on quality deployment function. In: 17th International Multidisciplinary Scientific Geo Conference Surveying Geology and Mining Ecology Management, SGEM vol. 17. no. 53, pp. 675–682 (2017). https://doi.org/10.5593/SGEM2017/53/S21.081
27. Porter, M: Competitive advantage. review by William B. Gartner. The Academy of Management Review, vol. 10, no. 4, pp. 873–875 (1985)
28. OTRS. Open-Source Ticket Request System. Accessed 03 Dec 2020. https://otrs.com/pt/home/
29. ABNT. Normas ISO 31.010. Gestão de Riscos – Técnicas para o processo de avaliação de Riscos, 2012. Rio de Janeiro, Brasil
30. de Oliveira, U.R., Marins, F.A.S., Rocha, H.M., Salomon, V.A.P.: The ISO 31000 standard in supply chain risk management. J. Clean. Prod. **151**, 616–633 (2017)
31. Sevcik, F., Ketron Inc.: Current and future concepts in FMEA Inc., USA; January 1; 8p; Annual symposium on reliability and maintenance, 27–29 January,1981, Philadelphia, PA; See also A81–24251 09–38 (1981)
32. Barafort, B., Mesquida, A.L., Mas, A.: Integrating risk management in IT settings from ISO standards and management systems perspectives. Comput. Stand. Interfaces **54**(3), 176–185 (2017)
33. Brasil, Câmara dos Deputados: Administrative Process nº 502.883/2018. (2018),
34. Brasil, Câmara dos Deputados: Administrative Process nº 117.351/2014 (2014)
35. BPM CBOK Version 4.0: Guide to the Business Process Management Common Body Of Knowledge, ABPMP International ISBN: 978–1704809342 (2019)
36. Aalst, W.V.D.: Process Mining-Data Science in Action. https://doi.org/10.1007/978-3-662-49851-4
37. Bizagi, Processes Modeling (2016). https://www.bizagi.com/. Accessed 04 Dec 2020
38. Pentaho Data Warehouse Software. https://sourceforge.net/projects/pentaho/. Accessed 03 Dec 2020
39. 4th International Conference on Digital Society ICDS Includes CYBERLAWS 2010: The 1st International Conference on Technical and Legal Aspects of the e-Society 5432788, pp. 249–254 (2010). https://doi.org/10.1109/ICDS.2010.48
40. Chen, Y., Li, K.W., Levy, J., Hipel, K.W., Kilgour, D.M.: A rough set approach to multiple criteria ABC analysis. Trans. Rough Sets **VIII**, 35-5 (2008)
41. Nawaz, O. Minhas, T., Fiedler, M.: Optimal MTU for realtime video broadcast with packet loss - a QoE perspective. In: The 9th International Conference for Internet Technology and Secured Transactions (ICITST) (2014)
42. Nawaz, O. Minhas, T., Fiedler, M.: QoE based comparison of H.264/AVC andWebM/VP8 in an error-prone wireless. In: IFIP/IEEE IM 2017 Workshop: 1st International Workshop on Quality of Experience Management (2017)
43. Koncz, A.: Failure mode and effect analysis types in the automotive industry. In: Proceedings of the Mini Conference on Vehicle System Dynamics, Identification and Anomalies 2019-November, pp. 321–328 (2019)
44. Rivera Torres, P.J., Serrano Mercado, E.I., Anido Rifón, L.: Probabilistic Boolean network modeling and model checking as an approach for DFMEA for manufacturing systems. J. Intell. Manuf. **29**(6), 1393–1413 (2018). https://doi.org/10.1007/s10845-015-1183-9
45. Shirani, M., Demichela, M.: Integration of FMEA and human factor in the food chain risk assessment World academy of science, engineering and technology, international journal of social, behavioral, educational, economic, business and industrial. Engineering **9**(12), 4247–4250 (2015)
46. Mock, R., Straumann, H., Fischer, A.: A second chance for risk assessment in IT system analysis?, Safety, reliability and risk analysis: beyond the horizon. In: Proceedings of the European Safety and Reliability Conference, ESREL 2013, pp. 2237-2244 (2014)

47. Aven, T.: Perspectives on the nexus between good risk communication and high scientific risk analysis quality. Reliab. Eng. Syst. Safety **178**, 290–296 (2018)
48. Olechowskia, A., Oehmenb, J., Seeringa, W., Ben-Daya, M.: The professionalization of risk management: what role can the ISO 31000 risk management principles play? Int. J. Project Manage. **34**(8), 1568–1578 (2016)
49. Williamson, A.: How are parliaments responding to the Coronavirus pandemic? (2020). https://www.hansardsociety.org.uk/blog/how-are-parliaments-responding-to-the-coronavirus-pandemic?. Accessed 27 Dec 2022

Demand Driven Material Requirements Planning: Using the Buffer Status to Schedule Replenishment Orders

Nuno O. Fernandes[1,7](✉), Nelson Guedes[2], Matthias Thürer[3,4], Luis P. Ferreira[2,5], P. Avila[2,6], and Silvio Carmo-Silva[7]

[1] Instituto Politécnico de Castelo Branco, 6000-767 Castelo Branco, Portugal
nogf@ipcb.pt
[2] School of Engineering (ISEP), Polytechnic of Porto, 4200-465 Porto, Portugal
[3] University of Applied Sciences Upper Austria, 4400 Steyr, Austria
[4] Jinan University, 519070 Zhuhai, People's Republic of China
[5] INEGI - Instituto de Ciência e Inovação em Engenharia Mecânica e Engenharia Industrial, 4200-465 Porto, Portugal
[6] INESC TEC - Instituto de Engenharia de Sistemas e Computadores, Tecnologia e Ciência, 4200-465 Porto, Portugal
[7] ALGORITMI Research Centre, University of Minho, 4710-057 Braga, Portugal

Abstract. Demand Driven Material Requirements Planning argues that production replenishment orders should be scheduled on the shop floor according to the buffers' on-hand inventory. However, the actual performance impact of this remains largely unknown. Using discrete event simulation, this study compares scheduling based on the on-hand inventory, with scheduling based on the inventory net flow position. Results of our study show that scheduling based on the former performs best, particularly when multiple production orders are simultaneously generated and progress independently on the shop floor. Our finds give hints that are important to both, industrial practice and software development for production planning and control.

Keywords: Production Control · DDMRP · Priority Scheduling

1 Introduction

In the current context of volatility, uncertainty, complexity, and ambiguity (VUCA), inventory management is of utmost importance and the choice of an appropriate Production Planning and Control (PPC) system becomes even more crucial [1].

In the last decades, several PPC approaches have been used, the most widespread being Material Requirements Planning (MRP), Just-In-Time (JIT) and Theory of Constraints (TOC). However, each of these approaches alone were not developed for VUCA environments. According to Ptak and Smith [2], MRP implements a push paradigm that usually leads to long lead times, JIT may reduce the inventory level to the point of making supply chains less agile and TOC does not consider the *bill of material* explosion and, hence, does not deal well with complex product structures.

A. Rocha et al. (Eds.): WorldCIST 2023, LNNS 799, pp. 417–424, 2024.
https://doi.org/10.1007/978-3-031-45642-8_41

Considering these weaknesses, Ptak & Smith [2, 3] designed a new PPC approach known as Demand Driven Materials Requirements Planning (DDMRP). This approach seeks to reduce delivery times by aligning production with demand. DDMRP provides planning and execution performance improvements in VUCA environments, where customer tolerance times are shorter than delivery lead times - a situation that is becoming more and more common in current days.

Given its importance, a large literature on DDMRP recently emerged, e.g. [4–7] and [8]. Most of this literature is focused on comparing DDMRP with alternative PPC systems. Only a subset is focused on the DDMRP parametrization and, to the best of our knowledge, only a previous study by [13] focusses on the impact of the scheduling rule on the execution performance of DDMRP. However, this study does not consider the scheduling rule proposed in the DDMRP literature by Ptak & Smith [3]. Therefore, we analyze the impact of two buffer-oriented scheduling rules in a multi-stage assembly system controlled by DDMRP: one based on the on-hand inventory status as suggested by Ptak & Smith and the other based on the inventory net flow position.

The remainder of this study is structured as follows. Section 2 shortly reviews the execution priority in the context of DDMRP. Section 3 then presents the simulation model considered in this study, before the results are presented and discussed in Sect. 4. Finally, conclusions are summarized in Sect. 5, where managerial implications, limitations and future research directions are also put forward.

2 Background

It is common to manage the execution priority of production in complex systems through priority scheduling rules. Reviews on priority scheduling research, as those by [9, 10] and [11], found hundreds of priority rules, varying from very simple, such as the First-Come-First-Served or Earliest Due Date, to more sophisticated ones.

The DDMRP literature suggests using the buffer status for priority dispatching. It advocates that the lower the on-hand level of inventory, the higher should the priority be. Inventory buffers in DDMRP are divided into three zones (Fig. 1): green, yellow and red. Each zone has a specific management purpose. The green zone determines order generation, the yellow zone deals with the demand coverage and the red zone refers to the safety stock embedded in the buffer.

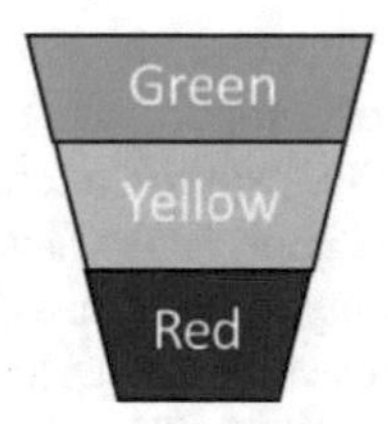

Fig. 1. DDMRP inventory buffer (adapted from [3]).

If DDMRP is used, planners can quickly judge the relative priority of production replenish orders, based on the amount of on-hand inventory divided by the top of the entire red zone, or alternatively simply based on the amount of on-hand (for cases with identical buffer zones across product types). However, while information concerning on-hand inventory is important, from the perspective of determining the relative priority across multiple product types, it ignores the relative priority across products of the same type. For this context, the net flow position given by *on hand inventory + open orders – customer orders*, may provide a better alternative. This leads to the following research question:

RQ: Should DDMRP orders be prioritized by the on-hand inventory status as suggested in the DDMRP literature, or by the inventory net flow position?

To the best of our knowledge this study analyses for the first time the impact of the scheduling rule on the performance of DDMRP.

3 Simulation Study

Discrete event simulation is used as research method to give an answer to our research question. A simulation model of a multi-stage assembly system was developed in Arena software.

3.1 Production System

The production system considered is based on [7] and [12], allowing for a comparation to be made with previous literature. However, part quantities per product have been adjusted to create a balanced shop floor situation. The product structure and part quantities are indicated in Fig. 2.

The production system consists of six assembly stations, where eight different product types are assembled. Each product requires six assembly operations to be completed, and each operation requires a different station. Station 1 refers to the gateway station where components F1 and F2 are made from raw parts, while station 6 performs the final assembly operation. Raw materials at the gateway station are assumed to be always available, and consequently do not constrain production. Operation processing times follow a 2-Erlang distribution with a mean of 1.0 time units at stations 1, 2, 3 and 4, and 0.5 time units at stations 5 and 6.

In our study, we consider an average demand arrival rate of 9 products per period of 10 time units. The demand rate follows a Poisson distribution, and all products have the same probability of being assigned to a customer order. The resulting utilization rate at all stations is 90%.

Finally, customer due dates are determined by adding a constant allowance to the order entry time. Two settings for this due date allowance are considered in our study: tight, i.e. 30 time units; and, loose, i.e. 40 time units. These values are based on the realized average throughput times in preliminary simulation experiments and allow to show the impact of the due date allowance on the performance of the production system.

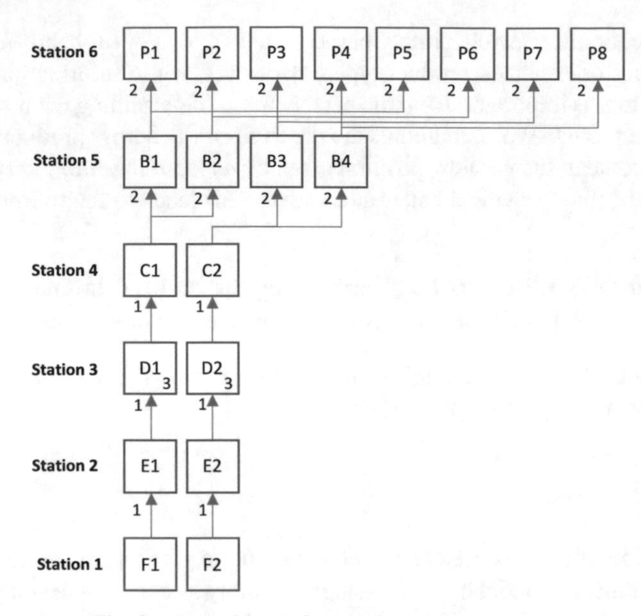

Fig. 2. Assembly stations and products structure.

3.2 DDMRP Control

As in Azzamouri et al. (2022), DDMRP has been implemented with two buffers, a raw parts buffer (RPB) and a finished goods buffer (FGB), as illustrated in Fig. 3. Demand (D) for a final product is satisfied from the FGB at the order due date and replenishment orders are scheduled on the gateway station as soon as they are generated.

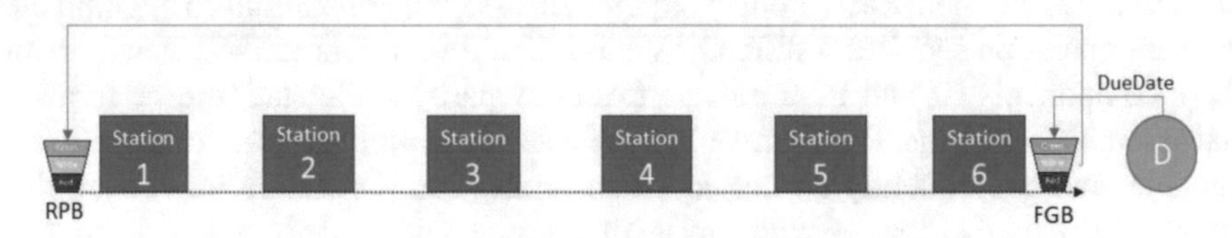

Fig. 3. DDMRP assembly system.

In DDMRP the buffer color needs to be established for each product and it is essential for the order generation process. If the net flow position of a product falls in the yellow or red zones of the buffer a production order is generated. The priority of this order is determined by the relation of the net flow position to the top inventory level of the green buffer, expressed in percentage. This percentage is an indication of the degree of inventory coverage. So, the lower this percentage for an order is, the higher its priority. The re-order quantity of the replenishment order is given by the difference between the top of green buffer value and the net flow position.

In our study, a production order is generated, for the replenishment of a specific product type, when the net inventory position of the FGB reaches the re-order point

(ROP), i.e. falls to the yellow or red zones. Given the characteristics of the production system used, we set identical inventory levels for the top of the green buffer across product types. The re-order quantity (ROQ) is considered an experimental factor and tested at values of 1 and 4 units. In this study, the re-order quantity is not expressed in terms of the lot size, but rather in terms of the number of production orders generated. Therefore, a re-order quantity of 4 means the simultaneously generation of four production orders of one product unit each.

3.3 Experimental Design and Performance Measures

We used a full factorial design [14] with four experimental factors that are tested at different levels as follows: (i) two levels for the scheduling rule (NFP – lowest Net Flow Position and OHI – lowest on-Hand Inventory), (ii) three levels for the ROP (3, 4 and 5 units), (iii) two levels for the ROQ (1 and 4 units), and (iv) two levels for the due date tightness (30 and 40 time units). This results in a total of 24 ($2 \times 3 \times 2 \times 2$) experimental scenarios.

Results were collected over 30,000 time units following a warm-up period of 20,000 time units. Three performance measures have been considered: (i) the service level (SL), i.e. the fraction of customer orders that is fulfilled until the due date; (ii) the customer waiting time (CW), i.e. the time a customer has to wait when the required end product is not in stock; and (iii) the finished goods inventory (FGI), i.e. the average number of end products in stock.

4 Results

Performance results together with the 95% confidence interval are summarized in Table 1 and Table 2 for NFP and OHI, respectively.

Results indicate that the due date tightness has a positive impact on the service level (as somewhat expected). But a closer look reveals that this can be explained by higher FGI values. Comparing loose and tight settings for similar FGI setting no impact on the SL can be observed. A looser due date simply creates higher FGI levels which then result in higher SL values. Similar observation can be drawn for the ROQ and the ROP. There appears to be frontier constituted by FGI and Service level, influencing the parameter settings mainly the position on this frontier. However, comparing results in Table 1 and Table 2 an important difference can be observed for FGI and, thus for the customer waiting time.

To better highlight this performance difference, Fig. 4 and Fig. 5 plot the service level against the finish good inventory for both rules when ROQ is 1 and 4, respectively. The following can be observed from the results:

- *Performance Impact*: priority scheduling based on the inventory *net flow position* performs identically to priority scheduling based on the *on-hand* inventory when ROQ is 1. However, when ROQ is 4, the *net flow position* clearly outperforms priority scheduling based on the *on-hand* inventory. That is, a high service level and a lower customer waiting time can be obtained with a lower finish good inventory.

Table 1. Experimental results for the Net Flow Position (95% confidence interval).

Controls			Performance Measures		
ROP	ROQ	DD	FGI (units)	SL (%)	CW (time)
3	1	tight	26.0 ± 0.55	88.8 ± 0.68	11.3 ± 0.82
3	1	loose	34.3 ± 0.61	94.7 ± 0.50	9.24 ± 0.78
4	1	tight	33.5 ± 0.61	93.3 ± 0.56	10.3 ± 0.77
4	1	loose	42.0 ± 0.65	97.0 ± 0.36	8.2 ± 0.70
5	1	tight	41.1 ± 0.65	96.1 ± 0.39	9.2 ± 0.70
5	1	loose	49.9 ± 0.68	98.3 ± 0.25	7.2 ± 0.64
3	4	tight	35.8 ± 0.54	96.4 ± 0.44	9.7 ± 0.69
3	4	loose	44.5 ± 0.55	98.4 ± 0.28	9.0 ± 0.68
4	4	tight	43.6 ± 0.56	98.2 ± 0.31	8.3 ± 0.69
4	4	loose	52.5 ± 0.56	99.2 ± 0.19	7.5 ± 0.76
5	4	tight	51.5 ± 0.58	99.1 ± 0.21	6.9 ± 0.76
5	4	loose	60.4 ± 0.57	99.6 ± 0.12	5.6 ± 0.87

Table 2. Experimental results for On-hand Inventory (95% confidence interval).

Controls			Performance Measures		
ROP	ROQ	DD	FGI (units)	SL (%)	CW (time)
3	1	tight	26.4 ± 0.50	90.1 ± 0.56	12.6 ± 0.51
3	1	loose	34.3 ± 0.61	94.1 ± 0.43	12.7 ± 0.70
4	1	tight	33.7 ± 0.65	94.3 ± 0.44	11.5 ± 0.57
4	1	loose	42.0 ± 0.70	96.6 ± 0.31	11.6 ± 0.67
5	1	tight	41.2 ± 0.60	96.8 ± 0.32	11.1 ± 0.72
5	1	loose	50.5 ± 0.59	98.1 ± 0.23	10.5 ± 0.62
3	4	tight	36.1 ± 0.57	93.1 ± 0.43	13.4 ± 0.56
3	4	loose	44.2 ± 0.72	96.3 ± 0.30	13.2 ± 0.62
4	4	tight	43.6 ± 0.63	96.0 ± 0.31	12.2 ± 0.64
4	4	loose	52.2 ± 0.59	97.9 ± 0.22	11.5 ± 0.53
5	4	tight	50.8 ± 0.64	97.8 ± 0.19	10.6 ± 0.56
5	4	loose	60.3 ± 0.61	98.9 ± 0.14	10.5 ± 0.70

- *Performance robustness*: performance differences concerning the service levels of the two rules, i.e. the *net flow* position and the *on-hand inventory,* become higher for a lower ROP and tighter due date. This can be observed in Fig. 5. In the situation that orders are generated with low buffer inventory levels and tight due dates, correctly prioritizing jobs on the shop floor seems to become particularly important.

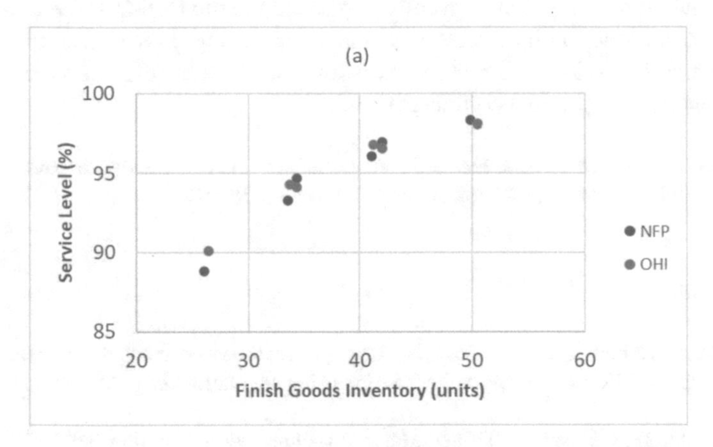

Fig. 4. FGI vs Service level for the two scheduling rules when ROQ is 1.

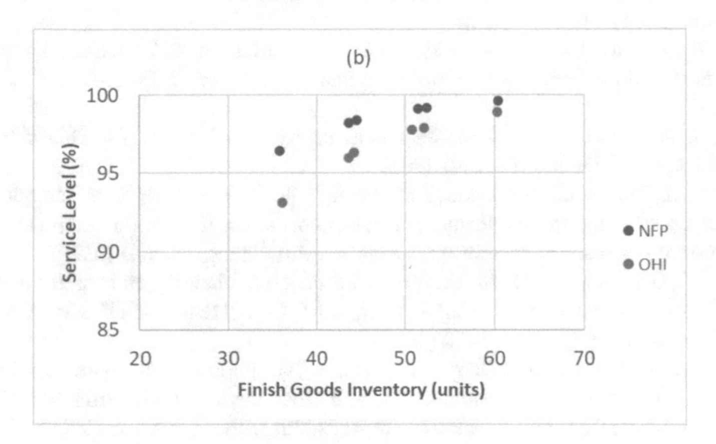

Fig. 5. FGI vs Service level for the two scheduling rules when ROQ is 4.

5 Conclusions

The objective of this study was to compare the performance impact of two different scheduling rules based on the inventory buffers status in a DDMRP controlled multi-stage assembly system. We asked: Should DDMRP orders be prioritized by the on-hand inventory status as suggested in the DDMRP literature, or by the inventory net flow position as proposed here? Simulation results indicate that scheduling based on the *net flow position* outperforms scheduling based on the *on-hand inventory* when multiple production orders are simultaneously generated and progress independently on the shop floor. Meanwhile, the relative performance differences across both rules seems to be impacted by operation parameters, such as the re-order point and level of due date tightness. These findings are particularly relevant and can be an advantageous alternative to the scheduling based on-hand inventory suggested in the DDMRP literature. A main limitation of our study is its limited environmental setting. It would be of interest for future research to explore the relative performance of these rules in other production scenarios and further product complexity.

Acknowledgements. This work has been supported by FCT – Fundação para a Ciência e Tecnologia within the R&D Unit Project Scope UIDB/00319/2020.

References

1. Abuhilal, L., Rabadi, G., Sousa-Poza, A.: Supply chain inventory control: a comparison among JIT, MRP, and MRP with information sharing using simulation. Eng. Manag. J. **18**(2), 51–57 (2015)
2. Ptak, C., Smith, C.: Demand Driven Material Requirements Planning (DDMRP). Industrial Press, Norwalk, CT (2016)
3. Ptak, C., Smith, C.: Orlicky's Material Requirements Planning 3/E. McGraw Hill Professional, New York (2011)
4. Miclo, R., Lauras, M., Fontanili, F., Lamothe, J., Melnyk, S.A.: Demand Driven MRP: assessment of a new approach to materials management. Int. J. Prod. Res. **57**(1), 166–181 (2019)
5. Lee, C.J. and Rim, S.C.: A mathematical safety stock model for DDMRP inventory replenishment. Mathem. Prob. Eng. (2019)
6. Dessevre, G., Martin, G., Pierre, B., Lamothe, J., Pellerin, R., Lauras, M.: Decoupled lead time in finite capacity flowshop: a feedback loop approach. In: The 8th International Conference on Industrial Engineering and Systems Management, Shangai, China (2019)
7. Thürer, M., Fernandes, N.O., Stevenson, M.: Production planning and control in multi-stage assembly systems: an assessment of Kanban, MRP, OPT (DBR) and DDMRP by simulation. Int. J. Prod. Res. **60**(3), 1036–1050 (2022)
8. Azzamouri, A., Baptiste, P., Pellerin, R., Dessevre, G.: Impact of the continuous and periodic assessment of a buffer replenishment on the DDMRP method. Int. J. Prod. Res. (2022)
9. Panwalkar, S.S., Iskander, W.: A survey of scheduling rules. Oper. Res. **25**(1), 45–61 (1977)
10. Blackstone, J.H., Philips, D.T., Hogg, G.L.: A state-of-the-art survey of dispatching rules for manufacturing job shop operations. Int. J. Prod. Res. **20**(1), 27–45 (1982)
11. Pickardt, C.W., Branke, J.: Setup-oriented dispatching rules – a survey. Int. J. Prod. Res. **50**(20), 5823–5842 (2012)
12. Jodlbauer, H., Huber, A.: Service-level performance of MRP, kanban, CONWIP and DBR due to parameter stability and environmental robustness. Int. J. Prod. Res. **46**(8), 2179–2195 (2008)
13. Fernandes, N.O., Guedes, N., Thürer, M., Ferreira, L.P., Ávila, P.: DDMRP: how to prioritize open orders? Fexible automation and intelligent In: Manufacturing International Conference, Faim (2023). Submitted
14. Montgomery, D.C.: Design and Analysis of Experiments, 10th edn. Wiley, Hoboken (2019)

Detecting and Predicting Smart Car Collisions in Hybrid Environments from Sensor Data

Hector D. Menendez(✉) and David Kelly

King's College London, London, UK
{hector.menendez,david.a.kelly}@kcl.ac.uk

Abstract. Self-driving or smart cars are becoming the next step forward for safety and autonomy of driving, aiming to provide the best travelling experience. Guaranteeing their trustworthiness requires understanding of different critical situations, such as collisions. Smart cars need to detect collisions to perform recovery actions and, even in some cases, call an ambulance. Ideally, smart cars need to predict collisions, not merely detect them after the event. This prediction is difficult, especially when human drivers are part of the environment. This work proposes a system for collision prediction and detection based on machine learning: it focuses only on the information that the car receives through its sensors. We evaluate our method using a multi-car hybrid system designed to generate realistic and complex scenarios for the smart cars, thereby providing sensory information that is close to a realistic road. Our methodology detects collisions with 99% accuracy and predicts 5 different levels of collision risk with up to 43% accuracy.

Keywords: Smart Cars · Collision Detection · Collision Prediction · Carla Simulator · Machine Learning

1 Introduction

Smart cars are currently under active development. Their introduction will create a disrupting social change in terms of autonomy and security. Their quality is crucial, as they will be responsible of human lives. Several works have studied different technical aspects of these systems, such as their autonomy [1], intelligence [2], security [3] and different aspects related to trust [4].

Analysing these studies, it becomes clear that we need more work related to multi-agent environments. These are situations where the car does not just face a single road under varying conditions, but also must respond to peers that will have similar behaviour with different destinations. In these scenarios, evaluating their reliability is crucial, as they will have to make complex decisions that we humans find overwhelming; where time, environmental and emotional constraints and stresses lead to error and potential collisions.

Collision detection in smart cars might save lives; predicting collisions in advance might even more. Some work is already focused on understanding how

A. Rocha et al. (Eds.): WorldCIST 2023, LNNS 799, pp. 425–435, 2024.
https://doi.org/10.1007/978-3-031-45642-8_42

sensor information can be used to detect the complex circumstances under which collisions occur [5]. The information that the sensors read, however, depends significantly on the environment. To create realistic scenarios, it is important to consider that even if a collision is normally between two entities (such as two cars), there are other elements that might impact on the sensor information.

This problem is rarely studied; its significance should lead to the creation of accurate methods not only for detection but also for prediction of collisions, seconds before they happen. This work evaluates collision detection under a realistic scenario where there are multiple cars driving in a simulated city. An intelligent agent controls the cars to guarantee that they respect traffic regulations. In this scenario, we extract local information from the cars' sensors independently. We investigate whether each car can understand if it has suffered a collision or if it can actually predict that one will happen soon.

We implement our methodology in the Carla simulator [6], which helps us to collect the data, and our analysis employs different types of classifiers in order to detect collisions from sensor information. Our methodology can detect collisions with 99% accuracy. Considering the significant ability to detect collisions, we also evaluate whether the car can predict collisions by using different time windows for events. In these scenarios, however, the cars can hardly distinguish whether they are at high, low or no risk of collisions. Their predictive abilities go up to 43% in the best case scenario, although we discover that providing more memory to the prediction system significantly increases detection abilities. The main contributions of this work are:

- We have created the first approach for detecting and predicting collisions in in a multi-car simulated environment (Sect. 3).
- We created a dataset of different scenarios and collisions, with 6,308,617 records for experimental reproduction and further research (Sect. 4).
- Combining car sensors with machine learning reaches accuracy levels of 99% at collision detection, with 100% precision. The most relevant sensors are the accelerometer and gyroscope. Regarding the ability to predict collision within a time frame, our accuracy is up to 43% at different risk levels (Sect. 5).

2 Related Work

Multiple studies work on detecting and predicting collisions for smart cars in multiple contexts. Some collision-detection systems based on images aim to detect animals, specifically animal-vehicle collision, by combining a cascade of classifiers [7]. However, images require extra processing, delaying a system's decisions. Our work focuses on using real-time sensors, and relying on just one classifier, thereby giving the smart car more reaction time. Another system which relies on image processing is proposed by Shaik *et al.* [8]. This system automatically calls the police when an accident is detected. The evaluation of the techniques was performed indoors, with three people waling, instead of cars running. Image processing may work better at the slower speeds of pedestrians, but they also require considerable efforts on image calibration and scaling for detecting objects and distances. One of our sensors (see Sect. 3.2), LiDAR, can do this automatically.

In terms of real-time sensors, Krishnan [5] applied a sensor system on top of Arduino cars, for detecting object proximity through light and ultrasonic sensors. Although the idea is interesting, it is not applied to smart cars or simulators directly, but only evaluated theoretically. On the other hand, Jeon *et al.* [9] use radars to prevent collisions between bicycles and automobiles by proximity. However, bicycles have a more limited capability for carrying complex sensors, due to weight and energy considerations [10]. Car-to-car collision offers a greater variety of different sensors, individually and in cooperation, which we investigate.

As an alternative to sensors mounted directly on the cars themselves, Tripathi and Singh [11] aim to predict potential inter-vehicle collisions by using smart devices attached to street lights. These light-mounted devices directly control the cars on the road. This system, simulated in Virtual Crash[1], obtains a strong accuracy at detecting how many vehicles might collide at the same time. It is not clear, however, how much the system can provide a time window to prevent the collision. By giving a collision prediction some time in advance, which is our main goal, we aim to reduce the number of collisions, not just quantify them.

Zhang *et al.* [12] and Chen *et al.* [13] work on the braking problem. The braking problem considers, in addition to collision avoidance, the comfort and safety of the human occupants in the smart vehicle. Both works propose a system for controlling collisions based on distances, by using either a braking prediction system or a hardware-in-the-loop device, respectively. These methods are only evaluated in two-car scenarios, where, in at least some cases, a human is controlling one of the cars. Chen *et al.* use CarSim[2] for the simulation scenarios. In other braking problem work, Ho *et al.* [14] study how humans react to braking in order to prevent collisions.

In work related to multiple vehicles, such as we use in our experimental design, we find Alonso *et al.* [15], who create a system for avoiding collision in bicycles by using trajectories. Jones [16] explores the use of cruise control, a collaborative method among multiple cars driving in similar directions. This extends the work of Li *et al.* [17]. Modern approaches often add collision avoidance to deep learning models, such as Chang *et al.* [18], who look to avoid collisions with car doors by finding the proper timing for opening doors, or by adding human driving policies to reinforcement learning models [19].

3 Methodology

Our main goal is to create a system that can detect and predict potential collisions, giving a warning of at least a few seconds to the smart car but using very limited memory. To achieve this, we use a set of sensors that generate real time information with the aim of fast decisions. The advantages of using limited memory models is evident: far less data need be stored by the smart car, and there is

[1] https://www.vcrashusa.com/.
[2] https://www.carsim.com/.

Fig. 1. Carla showing a nighttime environment, and a pre-collision environment

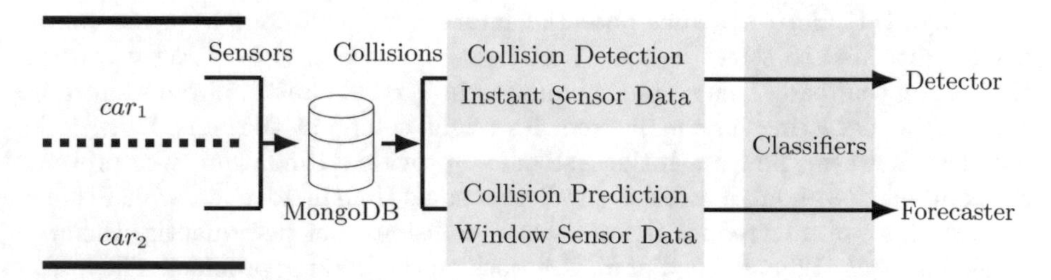

Fig. 2. Architecture of the whole data workflow. We experiment with a variety of different classifiers (Sect. 3.3).

also less data to process in general, over time windows of known size. Decision making speed is vital to practical smart cars and their successful deployment. We train a learning system to set responses depending on the sensors data.

3.1 Data Extraction Process

We extracted the data from a set of simulated environments populated with smart cars. The simulation system that we used is named Carla and it provides different location maps where the cars perform their activity (see Fig. 1). Carla provides a richly detailed simulated environment and uses a traffic management system to run the simulations. It makes the cars follow their routes, respecting traffic lights, lanes, signals, pedestrians and other cars.

In order to provide a smart/human hybrid environment, we select the sensors for each car (see Sect. 3.2) and add a human-driven car to the simulation. The human car provides a necessary chaotic element to Carla's simulations to generate collisions. We then analyse all sensor information to ascertain whether any of the collisions could have been predicted from the data. The information is extracted by simulation frame, except for those sensors that activate under specific circumstances (i.e. the collision sensor and obstacle sensor). Our system sends all the sensor data to a MongoDB database[3] in real-time for further

[3] https://www.mongodb.com/.

Fig. 3. LiDAR's point cloud as viewed from above and its semantic interpretation overlaid on the simulated environment.

analysis. For a deeper analysis post experiment, we also record every simulation. The full system workflow and architectural design is in Fig. 2.

3.2 Sensor Information and Collision Cases

We selected sensors that provide a fast reading to give sufficient time for a useful reaction. A sensor unable to provide data until a collision is completely unavoidable does not provide us with the necessary leeway to react. We rely on the collision detection sensor to provide labelled data for our learning process. Given these constraints, the sensors our system uses are:

- **LiDAR**: this creates a point cloud around the car (see Fig. 3). In addition, it provides the number of different wavelengths used to image objects and their angle to the sensor. It includes semantic information about the objects associated with the points.
- **Object Detection**: this detects objects close to the car and provides semantic information about the object and its distance.
- **Collision Detection**: it only activates when a collision occurs. It provides information about the agents involved and the crash impact. We use this type of sensor to identify collisions and create labelled data, but its information is excluded for training because this is what we want to predict.
- **IMU Measurement**: provides the car's acceleration and its gyroscope.
- **GNSS Sensor**: this sensor provides the car position within the Carla map.

Among the most important sensors that we do not include are the cameras. We exclude these due to the comparatively long time window associated with the processing of visual information. Cameras have already been investigated in other research (see Sect. 2). We also exclude radars, as LiDAR provides richer information (Fig. 3).

For the final classification, we create a labels based on the following cases: **Collision** (Level 0): the collision is underway and is no longer avoidable; **High collision risk** (Level 1): the collision will happen in the last step of the memory window; **Medium collision risk** (Level 2): the collision will happen in the two last steps of the memory window; **Low collision risk** (Level 3): the collision will happen in the three last steps of the memory window; and finally, **No collision risk** (Level 4): if there is a collision risk then it will not occur for at least four more steps of the memory window.

3.3 Classifiers

The data gathered from the simulations is very diverse. Given this diversity, we tested multiple classifiers to identify the most suitable for this problem.

Our experiments used the following classifiers, descriptions of which can be found in [20]: **Logistic Regression**: performs a learning process based on a linear prediction of the features; **Linear Discriminant Analysis**: separates features linearly by performing a previous selection on top of them; **Naïve Bayes**: considers the features as independent contributions, creating Gaussian models for each of them that are then aggregated to perform predictions; **K-Nearest Neighbours (KNN)**: assigns the class to each sample according to the K closest neighbours; **Decision Trees**: defines a tree of decisions by dividing the space linearly per feature; **Random Forest**: aggregates several trees into a voting system, where each tree is trained using different sets of features chosen uniformly at random; **Support Vector Machine (SVM)**: creates a hyperplane to separate the feature space into classes, using some feature vectors for defining the margin for this hyper-plane, and kernels when the separation is non-linear; **Neural Networks (NNET)**: commonly used for deep learning algorithms, it connects perceptrons or neurons into multiple layers, associating an activation function with each layer; **Ada Boost**: sequences multiple classifiers in a similar fashion as a neural network; **Gradient Boosting**: it performs a voting system on top of multiple weak classifiers, optimised by a gradient method; **Extreme Gradient Boosting (XGBoost)**: extension of gradient boosting that optimises the features chosen to train the different classifiers.

We selected these different classifiers as they cover a significant spectrum of the state of the art of classification methods.

4 Experimental Setup

Our system baseline goal is to detect collisions through sensor information, but if its detection abilities are good, then the overarching aim is to predict collisions, to either prevent them, or to minimise them. Therefore, the evaluation of the system focuses on the following research questions:

- **RQ1**: *Can our sensors detect a collision using no memory? If they can, what are the most relevant features?*
- **RQ2**: *Is it possible to predict that a collision is about to happen? How long does the system need to make a collision prediction?*

To answer **RQ1**, we isolate collision data from our scenarios (Sect. 3.1). We separate this data into different moments in which the car is close to a collision (from 1 to 4 s before a detection), as described in Sect. 3.2. We train the classifiers (Sect. 3.3) and measure the detection accuracy. We use 10-fold cross validation with 80% of the data extracted. Using the cross-validation results, we select the best classifier which we then optimised and evaluate using the test data, the remaining 20%. From this last classifier, we study which features are the most relevant.

To address **RQ2**, as to whether we can predict a collision from sensor data, we explore different time windows. Our time windows go from 1 to 5 s, and from these time windows, we extract the most recent relevant information from the sensors. Every window corresponds with different reaction reflexes of the car. We aim to predict the last 5 moments for each time window from the collision (moment 0) passing through three risk levels, up to no immediate risk (moment 4) (see Sect. 3.2).

For the experiments, we set 7 different scenarios of different smart cars driving in the Carla simulator. Each scenario has 50 cars and an extra car driven by a human. The Carla traffic manager controls all of the smart cars, by using omniscient information of the environment, traffic signals, lights and surrounding elements. The simulations last 5 min each and we extract information only from the cars' sensors instead of the environment (Sect. 3.2). We record 30 frames per second, taking information from every sensor. We identify the collisions in the system and create a label directly for imminent collisions, with the rest of the labels following the description given in Sect. 3.1.

5 Results

The system evaluation starts by measuring its ability to detect collisions (Sect. 5.1) and then for forecasting them (Sect. 5.2). During the whole run of experiments, our system collected 1,904,500 records from the LiDAR, 586,581 from the obstacle sensor, 1,904,968 from the GNSS, 1,905,299 from the IMU and 7,269 records about collisions, of which 1,320 are vehicle to vehicle records that correspond to 65 different collisions. The obstacle sensors readings at much lower as this tends to activate only when an obstacle is very near.

5.1 Detecting Collision

Detecting collisions requires first evaluating whether the sensors are rich enough to understand the collision itself. For that, we use the data acquired from the different scenarios to assess the classifiers (Sect. 3.3) and to identify the most accurate at detecting collision versus non-collision in collision-immanent situations. In this case, we chose the data related to the 1 s time window, it being the closest to the collision time.

Figure 4 (left) shows the cross-validation results for all the different classifiers. In this case, the Naïve Bayes, SVM, LDA and Linear Regression classifiers

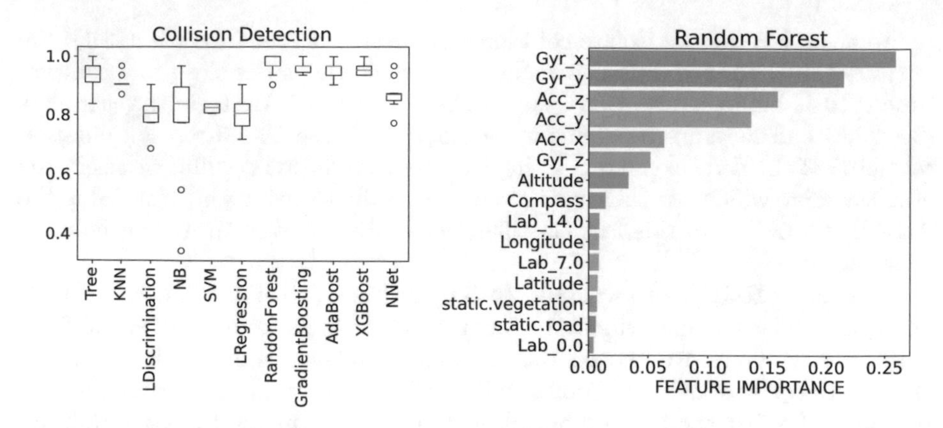

Fig. 4. Cross-validation results for the different classifiers trained with the collision data (left) and more relevant features (right) for the best classifier (Random Forest).

show the worst prediction results (around 80% accuracy), followed closely by NNET (87%) and KNN (90.7%). The boosters, and tree-based classifiers show significant improvements (from 93% to 96%), with the best being Random Forest (97%). Random Forest seems to do best partially because of its ability to choose features and their boosting design combining trees, which we see, in Fig. 4, are already very accurate (93%).

We perform a retraining through parameter tuning for the Random Forest classifier that reaches 98.7% accuracy in test, having a precision of 100% for non-collision detection and 99% for collision detection, and a recall of 90% for non-collision detection with a recall of 100% for collisions. These results show that the system accurately detects collisions, and it is able to retrieve all of them. Figure 4 (right) shows the most relevant features that influence the final decision for the retrained Random Forest classifier. The most relevant sensors are the accelerometer and gyroscope, which are the sensors that are most affected by impact.

RQ1: *Our methodology successfully predicts collisions with 99% accuracy and 100% precision by using a Random Forest classifier. The most relevant features for detecting collisions are related to the accelerometer and gyroscope of the car.*

5.2 Predicting Collisions

In order to predict collisions, we set different time windows in order to evaluate the prediction abilities given a system memory budget. The windows sizes go from 1 to 5 s.

Table 1 shows the 10-fold cross-validation results for the classifiers under the different memory levels. Generally, we can see that the data quality understanding grows when the memory capacity grows, passing from a maximum of 27.6% of accuracy when the system only remembers the last second to 31.7% when it remembers the last 5 s. When the memory capabilities of the system are low

Table 1. Accuracy results of the 11 classifiers with respect to 10-fold cross-validation. XGBoost, given a 5 s memory, provides the best collision prediction result.

Classifier	1 s	2 s	3 s	4 s	5 s
Tree	22.1 (5.1)	**24.4** (4.7)	26.2 (7.5)	**30.8** (9.1)	29.1 (6.0)
KNN	20.5 (7.2)	19.9 (7.4)	26.6 (5.3)	23.7 (5.8)	26.6 (5.8)
LDiscrimination	12.2 (3.5)	11.5 (3.8)	12.2 (7.6)	16.0 (2.1)	15.3 (6.3)
NB	23.4 (6.8)	**25.9** (5.3)	25.3 (4.7)	29.1 (5.3)	26.3 (6.5)
SVM	21.2 (3.9)	23.1 (2.7)	22.1 (2.3)	22.4 (4.4)	21.8 (2.8)
LRegression	9.9 (6.0)	12.2 (7.6)	14.1 (7.2)	17.0 (6.2)	16.0 (7.8)
RandomForest	20.8 (3.9)	22.1 (5.2)	**28.8** (7.8)	**31.4** (4.2)	**30.4** (6.9)
GradientBoosting	21.2 (7.0)	21.8 (2.7)	26.9 (4.4)	27.9 (3.5)	**30.8** (8.2)
AdaBoost	**27.6** (4.7)	**27.6** (4.5)	**27.2** (6.9)	28.5 (3.6)	28.2 (2.0)
XGBoost	**26.6** (4.1)	21.8 (4.7)	**27.2** (7.8)	**29.8** (5.9)	**31.7** (6.5)
NNet	**23.7** (6.3)	21.8 (8.5)	26.3 (4.4)	23.4 (3.7)	24.1 (7.7)

(only up to 2 s), the booster-based algorithms show better results in general (ranged between 26.6% and 27.6%), although, when the system memory is high (up to 5 s), Random Forest also provides good results (30.8%). The best prediction system is XGBoost with 5 s memory (31.7%), which also shows good results for all of the other cases, apart from 2 s.

We selected the XGBoost algorithm with a 5 s memory for the final training with all the training data and applying the test dataset. Its accuracy improved to 39.7% with default parameters. In order to improve it, we applied a Halving Search with 10 cross-fold validation with the parameters of gamma and regulation alpha, ranging these values between 0 and 0.5 and 0 to 0.05, respectively. The optimized classifier, which used a gamma of 0.0 and a regularized alpha of 0.05, obtained 43.6% accuracy for the 5 classes. The 15 most relevant features were objects detected, such as poles, roads, buildings, bridges and vegetation becomes relevant, in tandem with the information from the LiDAR, specially when it detects walls, guardrails, fences, buildings, roads, and static objects. The last relevant features, that detected the collision itself, were the gyroscope and the accelerometer.

RQ2: *Our methodology predicts with up to 43.6% accuracy the 5 levels of risks for collision when the distance among them is about 5 seconds. This would provide a long window in which a car can take action to avoid a potential collision.*

6 Discussion and Future Work

The human driver was involved in all but 6% of the collisions in our simulation. We need further analysis on the small number of collisions not involving the human in order to determine whether there is any meaningful difference between human-caused collisions and otherwise. It would also be interesting to investigate, in those collisions involving the human, how often the human is the active participant in the collision (i.e. the cause) rather than the victim. Our experiments were limited to one human-driven vehicle; it would be interesting to add further human agents to see how this affects both our collision detection results and prediction abilities.

While it is difficult to be precise, human short term memory is generally considered to be in the region of 15–30 s [21]. Our prediction system effectively models a short-term memory of 5 s, with a reasonable prediction ability. It would be interesting to extend the analysis time window, balancing processing time and car reaction abilities against the size of the processing time window.

Finally, we want to extend the monitoring system to different applications, concretely: 1) measuring cars ethical decisions, for example, in scenarios where some of the vehicles might be an ambulance in an emergency situation; 2) measuring car fairness by using introducing different kinds of pedestrians into the simulation, and 3) measuring the system robustness by setting different environmental conditions that might change the information of the sensors.

7 Conclusions

Independent collision detection for smart cars is simply performed by using sensor information, allowing cars to react to these situations by, for instance, calling an ambulance or performing an emergency stop. In terms of predicting collisions, smart cars need more effort, considering that real roads have multiple cars surrounding and interacting with each other. This confuses their risk and hazard assessment, making it unclear when, and how, the car need best react. Some potential improvements in this direction can go into creating a memory-based system, utilising approaches such as state machines that control car decisions.

Acknowledgments. This research was funded by the UK Research and Innovation Trustworthy Autonomous Systems Node in Verifiability (EP/V026801/2).

References

1. Hussain, R., Zeadally, S.: Autonomous cars: research results, issues, and future challenges. IEEE Commun. Surv. Tutorials **21**(2), 1275–1313 (2018)
2. Lipson, H., Kurman, M.: Driverless: Intelligent Cars and the Road Ahead. MIT Press, Cambridge (2016)
3. Kong, H.K., Hong, M.K., Kim, T.S.: Security risk assessment framework for smart car using the attack tree analysis. J. Ambient. Intell. Humaniz. Comput. **9**(3), 531–551 (2018)
4. Pacheco, J., Satam, S., Hariri, S., Grijalva, C., Berkenbrock, H.: IoT security development framework for building trustworthy smart car services. In: 2016 IEEE Conference on Intelligence and Security Informatics (ISI), pp. 237–242. IEEE (2016)

5. Krishnan, P.: Design of collision detection system for smart car using Li-Fi and ultrasonic sensor. IEEE Trans. Veh. Technol. **67**(12), 11420–11426 (2018)
6. Dosovitskiy, A., Ros, G., Codevilla, F., Lopez, A., Koltun, V.: CARLA: an open urban driving simulator. In: Proceedings of the 1st Annual Conference on Robot Learning, pp. 1–16 (2017)
7. Mammeri, A., Zhou, D., Boukerche, A., Almulla, M.: An efficient animal detection system for smart cars using cascaded classifiers. In: 2014 IEEE International Conference on Communications (ICC), pp. 1854–1859. IEEE (2014)
8. Shaik, A., et al.: Smart car: an IoT based accident detection system. In: 2018 IEEE Global Conference on Internet of Things (GCIoT), pp. 1–5. IEEE (2018)
9. Jeon, W., et al.: A smart bicycle that protects itself: active sensing and estimation for car-bicycle collision prevention. IEEE Control Syst. Mag. **41**(3), 28–57 (2021)
10. Jeon, W., Rajamani, R.: A novel collision avoidance system for bicycles. In: 2016 American Control Conference (ACC), pp. 3474–3479 (2016)
11. Tripathi, S., Singh, D.: Smart streetlight framework for collision prediction of vehicles. Expert Syst. Appl. **208**, 118030 (2022)
12. Zhang, R., Li, K., He, Z., Wang, H., You, F.: Advanced emergency braking control based on a nonlinear model predictive algorithm for intelligent vehicles. Appl. Sci. **7**(5), 504 (2017)
13. Chen, S.L., Cheng, C.Y., Hu, J.S., Jiang, J.F., Chang, T.K., Wei, H.Y.: Strategy and evaluation of vehicle collision avoidance control via hardware-in-the-loop platform. Appl. Sci. **6**(11), 327 (2016)
14. Ho, C., Reed, N., Spence, C.: Multisensory in-car warning signals for collision avoidance. Hum. Factors **49**(6), 1107–1114 (2007)
15. Alonso-Mora, J., Breitenmoser, A., Beardsley, P., Siegwart, R.: Reciprocal collision avoidance for multiple car-like robots. In: 2012 IEEE International Conference on Robotics and Automation, pp. 360–366. IEEE (2012)
16. Jones, W.D.: Keeping cars from crashing. IEEE Spectrum **38**(9), 40–45 (2001)
17. Li, L., Peng, X., Wang, F.Y., Cao, D., Li, L.: A situation-aware collision avoidance strategy for car-following. IEEE/CAA J. Automatica Sinica **5**(5), 1012–1016 (2018)
18. Chang, W.J., et al.: A deep learning-based intelligent anti-collision system for car door. In: 2020 IEEE 9th Global Conference on Consumer Electronics (GCCE), pp. 148–149. IEEE (2020)
19. Emuna, R., Borowsky, A., Biess, A.: Deep reinforcement learning for human-like driving policies in collision avoidance tasks of self-driving cars. arXiv preprint arXiv:2006.04218 (2020)
20. Larose, D.T., Larose, C.D.: Discovering Knowledge in Data: An Introduction to Data Mining, vol. 4. Wiley, Hoboken (2014)
21. Atkinson, R.C., Shiffrin, R.M.: The control of short-term memory. Sci. Am. **225**(2), 82–90 (1971)

Detection of Racism on Multilingual Social Media: An NLP Approach

Ikram El Miqdadi[1](✉), Jamal Kharroubi[1], and Nikola S. Nikolov[2]

[1] Laboratory of Intelligent Systems and Applications, University of Sidi Mohamed Ben Abdellah, Fez, Morocco
{ikram.elmekdadi,jamal.kharroubi}@usmba.ac.ma
[2] Department of CSIS, University of Limerick, Limerick, Ireland
Nikola.Nikolov@ul.ie

Abstract. This paper presents a comparison between various text vectorization and machine learning algorithms for solving the problem of detection of racism on multi-lingual social media. We train classification models on Facebook comments and tweets in three different languages: English, French and Arabic. Our findings suggest that for the English-language comments, the combination of KNN with TF-IDF works best with an accuracy of 78.34%, while for French, the use of the SVM classifier with BOW provides an accuracy of 82.56%. For Arabic we obtain an accuracy of 91.13% when KNN is coupled with BOW. Overall, our results suggest that the combination of SVM and TF-IDF is the best choice for detection of racism on social media that contains content in English, French and Arabic at the same time. As part of this work, we also present a new annotated dataset of social media comments in three languages.

Keywords: NLP · machine learning · detection of racism

1 Introduction

The detection of racism is a task that is not always straightforward due to the ambiguity in the definition of racism. Often, racism is not expressed only by specific words, but it is rather inferred from the context. It can also manifest differently across languages and cultures. Thus, the development of an automatic system for the detection of racism on social media is particularly difficult.

Despite all the attempts made by social media platforms such as Facebook and Twitter, these platforms continue to provide a fertile environment for the spread of racist content, according to the American Psychological Association [1].

In this work we aim at contributing to the detection of racism on social media using various combinations of text vectorization techniques and machine learning algorithms, in order to train an effective classifier.

We evaluate multiple models on three datasets of comments and tweets gathered from Facebook and Twitter. One of them consists of comments in French, the second in English and the third in Arabic, thus covering the languages most widely spread in the

A. Rocha et al. (Eds.): WorldCIST 2023, LNNS 799, pp. 436–445, 2024.
https://doi.org/10.1007/978-3-031-45642-8_43

North African region. In fact, there are multiple types of Arabic [2]: Quranic or Classical Arabic, Modern Standard Arabic as well as approximately 30 different dialects [3]. For the purpose of this study, we work with the dialects of Arabic spoken in Morocco and Algeria, which are almost identical [4].

In the next section we review related research. In Sect. 3, we introduce our dataset and describe the annotation process we undertook. Section 4 outlines the methodology used for training a classifier for the detection of racism on social media, and Sect. 5 presents our results. In Sect. 6, we draw conclusion from our work.

2 Related Research

To detect racism, it is necessary to grasp what it is, therefore defining racism was the initial challenge we faced when addressing this subject. The concept of racism has been mentioned by several philosophers and sociologists over the course of history. Multiple definitions have been attributed to it, influenced by the colonial history of various countries around the world. Yet, there is not an universally accepted definition of racism [5–7].

In the Oxford Dictionary of Sociology, Scott and Marshall define racism as "the deterministic belief-system which sustains the unequal treatment of a population group purely because of its possession of physical, social or psychological characteristics" [5]. Indeed, many people believe that racism can only be identified by skin color, yet there are numerous other indicators of racism. According to Grosfoguel, skin color has been the dominant marker of racism since colonial times, but it is not the only marker. He strongly believes that: "racism can be marked by color, ethnicity, language, culture and/or religion" [6]. In another work, Bonilla-Silva and Razavi [7], believe that racism today is no longer limited to physical or ethnic characteristics, but can even include social and cultural aspects.

In our work, we adopt this wider understanding of racism and consider a piece of text racist if it contains any discrimination or insult based on *physical*, *ethnic*, *linguistic*, *cultural*, or *religious* differences.

A considerable amount of study has already been conducted on the procedure and techniques for the automated detection of racism on social media. Early work [8] involves a dictionary-based approach to classify racist content in Dutch social media comments, by building three racism dictionaries of unigrams: original, expanded and cleaned, and using them as input to a Support Vector Machine (SVM) classifier. Their best performing model is based on the expanded and cleaned version of the dictionary with F-score values of 45% and 46%, respectively.

The phase of data encoding in digital format has always been essential in the process of training a classifier and may have a direct impact on the performance of the trained model. Therefore, Bag of Words (BOW), Part of Speech (POS) and bigram vectorization techniques have been tested [9] to identify the most relevant one. With an accuracy of 91.50% and an F-score of 91.37%, the combination of the BOW technique and SVM as a classifier has produced highly satisfactory results.

In another study Kwok and Wang focus specifically on the widespread racism against black people on social media especially on Twitter [10]. They report 76% accuracy rate in their experiments with Naïve Bayes and BOW vectorization. However, Kwok and Wang claim that the ambiguity in the meaning of words implies that using keywords or unigrams alone cannot provide an accurate estimation of racism against black people.

Even though Arabic is one of the most widely used languages in the world, with 420 million native speakers [11], the classification of racism in this language has not received as much study attention as for other languages, such as English. We were able to identify one study by Alotaibi and Abul Hasanat who employ deep learning techniques for the detection of racist Arabic text [12].

3 Datasets and Annotations

3.1 Dataset

For the work presented here we collected a dataset of 4,191 Facebook comments and tweets, 1,919 of which are in English, 1,254 are in French and 1,018 in Arabic. We did this with the help of the open-source libraries Facebook Scraper [13] and Tweepy [14] for data gathering from Facebook and Twitter, respectively.

We use three different dictionaries of known racist words, each one corresponding to a different language, in order to collect data related to racist content from Twitter. For the Arabic language, we have performed a search only in the Moroccan and Algerian regions, since we target the North African region and the dialects used in these two countries are almost identical [4]. We noticed that there is not much data shared on Twitter by Moroccan and Algerian users, only 7.78% of the Algerians who use social media prefer Twitter vs 53.03% who prefer Facebook. Similarly, in Morocco 11.15% of all social media users prefer Twitter vs 44.78% who prefer Facebook [15].

Thus, unlike some previous studies which use only Twitter data, we chose to collect data from both Twitter and Facebook. To do so, we targeted some Facebook pages that are known to contain racist comments and publications in Moroccan dialect. We picked data where each racist keyword can be expressed in both racist and non-racist contexts to allow our system to be contextually flexible and not dependent only on the keywords. For example, the racist keyword "wetback"[1] can occur in racist context as in the sentence: "*I knew these three wetback immgrants right from the amazon jungle, running around naked, banging pots and pans like these zipperhead koreans*", or in a non-racist one as in: "*The number of times I heard the words "border hopper" and "wetback" from the age of nine and no one did a damn thing – my GOD, it makes me tremble!*" (These sentences are taken from the dataset collected for this study).

3.2 Annotation

The annotation of the dataset was made by three persons who understand the three languages, English, French and the Algerian and Moroccan dialects of Arabic. Two of them performed an annotation of the entire dataset based on three classes: *racist*, *non-racist* and *unclear*. The label *unclear* refers to comments which contain sarcasm or are ambiguous in some other sense, these comments are then removed. The third

[1] Wetback: an ethnic slur used in the USA to refer to illegal Mexican immigrants.

annotator was assigned the data on which there was a disagreement between the other two annotators. The label distribution of the whole dataset is shown in Table 1.

Table 1. Distribution of labels after removing unclear comments.

Language	Percentage of racist data	Percentage of non-racist data
English	59.30	40.70
French	49.93	50.07
Arabic	31.40	68.60

Before starting the data preprocessing step, we employed oversampling with RandomOverSampler [16–18], which involves increasing the number of instances of a minor class, which in our case is the class *racist* for the French and Arabic parts of the dataset and the class *non-racist* for the English part of the dataset.

4 Method

In this section we present the process of training a model for the automatic detection of racism on the dataset of English, French and Arabic text documents, introduced in the previous section.

4.1 Preprocessing

As a first step we converted all textual data to lowercase. We noticed that French and English writing users regularly employ acronyms and abbreviations, which could be misinterpreted during training, therefore we replaced them by their complete form. We then employed the RegEx [19] library to clean the data by removing usernames, URLs, consecutive duplicated characters, punctuation, special characters, and duplicated whitespace. Then, we deleted the emojis and all stop words except for those conveying negation. For the Arabic data we introduced an extra step involving diacritics removal. Next, we utilized spaCy [20] to perform a lemmatization based on part-of-speech (POS) tags for the French and the English data, and for Arabic we used ISRIStemmer [21] in order to reduce each word to its meaningful root.

4.2 Vectorization

One of the most challenging tasks in text processing is finding the best possible way of vectorizing textual input, i.e., representing text in numerical form.

In our study, we evaluated three distinct vectorization techniques: the standard Term Frequency – Inversed Document Frequency (TF-IDF) and BOW methods, and also the word embedding technique Doc2Vec.

Considering a corpus of documents, TF-IDF [22, 23] is a statistical technique for assigning a score to each term, i.e., word or combination of words, in a document according to its frequency of occurrence in the document multiplied by its inversed document frequency. The inversed document frequency of a term, on the other hand, is inversely proportional to the number of documents in the corpus that contain the term. Thus, a term would have a high TF-IDF score in a document if it occurs frequently in it but does not occur in many other documents in the corpus.

BOW [24] is another vectorization technique renowned for its simplicity and ease of application, which is frequently employed in document classification and language modeling. According to BOW, the representation of a document is a vector the elements of which are the number of occurrences of each term in the document.

Doc2Vec was first introduced in 2014 [25] as an extension to the Word2Vec method [26]. It employs a shallow neural network with two hidden layers to generate a model capable of computing the vector representation of a document while capturing the semantic information of the terms so that words with the same meaning have close vector representations; these representations are known as *embeddings*.

4.3 Learning

State-of-the-art deep learning models, such as LSTM and transformers, reportedly achieve excellent results [27], however, they require large training databases [28]. Since our dataset is relatively small, we perform experiments using traditional machine learning algorithms, namely Support Vector Machine (SVM), K-Nearest Neighbors (KNN), Decision Tree (DT), Random Forest (RF), and Multilayer Perceptron (MLP).

We apply each of these algorithms on our dataset after it has been vectorized and its dimensionality reduced by Truncated Singular Value Decomposition (TSVD). We chose to apply TSVD since it reportedly performs well on sparse matrices [29], which is the case with our dataset after the vectorization step. We also employed grid search to fine tune the hyperparameters of each of the machine learning algorithms.

Cross-validation was chosen for this study for evaluating the learning models' reliability. In more detail, we applied the StratifiedKFold method because it provides the same distribution of target classes across all folds. We used 3 folds, i.e., 3-fold cross-validation.

5 Experimental Results

In our experiments, we start by passing the dataset through the pre-processing step, described in detail in the previous section. Then we separately applied three vectorization techniques TF-IDF, BOW, and Doc2Vec.

TF-IDF and BOW produced vectors with high dimensionality, which could have a significant impact on the learning time. To tackle this issue, we employed the dimensionality reduction technique TSVD, as mentioned in the preceding section, using 170 components. On the other hand, by employing the Doc2Vec technique, we were able to obtain 170-dimensional vectors that served as input to the learning model. The choice of vectors with a size of 170 was based on tests performed using vectors of different size.

Then we used the prepared and vectorized data to train a variety of machine learning models, including SVM, KNN, DT, RF and MLP models with grid search for hyperparameter tunning. We present the obtained results in Tables 1, 2 and 3, respectively, for the English, French and Arabic parts of our dataset.

Google Colab was utilized during the implementation procedure with Python as a programming language and the open-source library Scikit-learn.

Table 2. Cross-validation results of training a classification model for the detection of racism on the English language part of dataset.

Method	Model	Accuracy	Precision	Recall	F-score
TF-IDF	KNN	**78.34**	73.83	**81.06**	77.27
	SVM	75.66	73.72	76.91	75.28
	RF	70.21	75.57	67.35	71.22
	DT	70.39	**76.78**	68.47	72.38
	MLP	71.53	76.66	78.82	**77.72**
BOW	KNN	**75.22**	79.67	**89.49**	**84.29**
	SVM	72.45	77.54	81.85	79.63
	RF	72.67	78.59	64.80	71.03
	DT	72.01	**79.89**	76.75	78.28
	MLP	72.19	78.17	77.23	77.69
Doc2Vec	KNN	60.28	62.42	63.52	62.96
	SVM	57.95	56.64	**81.37**	66.78
	RF	**61.95**	**63.16**	72.29	**67.41**
	DT	58.34	58.67	67.66	62.84
	MLP	58.43	57.42	69.90	63.04

The combination of TF-IDF and KNN appears to be the winner for the English part of the dataset (see Table 2) with the highest accuracy of 78.34% and the second highest (but very close to the highest) F-score of 77.27%. The best parameters for KNN are Euclidean distance and number of neighbors equal to 30. We obtained very close results by coupling TF-IDF with SVM reaching an accuracy of 75.66%, and an F-score of 75.28%. BOW with KNN, while having lower accuracy, reports a much higher F-score of 84.29% due to its high recall value.

SVM clearly outperforms the other models for the French data (see Table 3) when paired with BOW, achieving an accuracy of 82.56% and an F-score of 82.62%. Its best parameters reported by grid search are rbf kernel, gamma equal to 0.1 and C equal to 1.

Furthermore, the combination of SVM and TF-IDF yields close results, with an accuracy of approximately 81.76% and an F-score equal to 81.01%.

With an accuracy of 91.13% and an F-score of 91.39%, BOW coupled with KNN model yields the best results for the Arabic part of dataset (see Table 4), compared to the other models.

Also, the combination of SVM with TF-IDF gives satisfactory results achieving an accuracy of 90.91% and an F-score of 86.54%.

Table 3. Cross-validation results of training a classification model for the detection of racism on the French language part of dataset.

Method	Model	Accuracy	Precision	Recall	F-score
TF-IDF	KNN	73.64	71.44	**81.06**	75.94
	SVM	**81.76**	**85.59**	76.91	**81.01**
	RF	76.03	81.86	67.35	73.89
	DT	74.67	79.04	68.47	73.37
	MLP	78.10	77.92	78.82	78.36
BOW	KNN	75.39	70.53	**89.49**	78.88
	SVM	**82.56**	83.41	81.85	**82.62**
	RF	76.19	**83.89**	64.80	73.11
	DT	75.39	74.78	76.75	75.75
	MLP	78.66	79.59	77.23	78.39
Doc2Vec	KNN	59.39	59.60	63.52	61.49
	SVM	61.07	58.21	**81.37**	67.86
	RF	**65.84**	**64.42**	72.29	**68.12**
	DT	55.81	54.70	67.66	60.49
	MLP	63.77	62.39	69.90	65.93

Table 4. Cross-validation results of training a classification model for the detection of racism on the Arabic language part of dataset.

Method	Model	Accuracy	Precision	Recall	F-score
TF-IDF	KNN	90.77	**99.61**	**81.06**	**89.38**
	SVM	**90.91**	98.95	76.91	86.54
	RF	86.83	87.87	67.35	76.25
	DT	84.33	81.46	68.47	74.40
	MLP	86.48	83.48	78.82	81.08
BOW	KNN	**91.13**	93.38	**89.49**	**91.39**
	SVM	89.62	**95.20**	81.85	88.02
	RF	86.69	86.47	64.80	74.08
	DT	81.04	75.63	76.75	76.18
	MLP	84.97	80.77	77.23	78.96
Doc2Vec	KNN	70.38	64.25	63.52	63.88
	SVM	58.72	58.03	**81.37**	67.74
	RF	**86.40**	**90.90**	72.29	**80.53**
	DT	71.88	68.69	67.66	68.17
	MLP	69.45	67.76	69.90	68.81

6 Discussion

Comparing the results of the various experiments performed on our dataset consisting of French-, English- and Arabic-language text documents collected from social-media, we observe that the TF-IDF and BOW vectorization methods produce comparable performance, whereas Doc2Vec method is the least effective. This can be explained by the fact that the use of a relatively small dataset prevents Doc2Vec from capturing the semantic relations between terms and achieve its potential.

On the other hand, as we have already mentioned, during the data collecting process, we selected data where certain words appear both in a racist context as well as in a non-racist context. None of the vectorization methods TF-IDF, BOW and Doc2Vec, can represent appropriately the different meanings of a word depending on the context, and this can have a negative impact on the performance of the model.

Our experimental results demonstrate that, for all languages, the combination of SVM and TF-IDF achieves results comparable to those of the best model for a particular model. Thus, our results show evidence that this combination, i.e., SVM coupled with TF-IDF can be the most appropriate model for detecting racism in a multilingual corpus consisting of documents in English, French and North African dialects of Arabic, i.e., covering the languages used by social media users in the Nort African region.

7 Conclusion and Future Work

In this work we leveraged Facebook and Twitter for collecting a dataset that can be utilized for training a classification model for the detection of racist texts in three languages: English, French and Arabic.

We employed several text vectorization and machine learning techniques to train multiple classification models, with the hyperparameters of the machine learning algorithms being fine-tuned by grid search.

Our results suggest that KNN trained on either TF-IDF or BOW data is the best approach for the English part of our dataset with TF-IDF leading to a better accuracy of 78.34%, and BOW to a better F-score of 84.29%. For the French part of dataset, SVM paired with BOW reports the best results with an accuracy of 82.56% and an F-score of 82.62%. Finally, KNN trained on BOW data proves to be the most successful approach for the Arabic part of dataset with an accuracy of 91.13% and an F-score of 91.39%. We also observe that SVM on TF-IDF data, while not being the best model for any of the three languages, consistently performed well, and it might be the best choice for detecting racism in a multilingual social media environment where the three languages, English, French and Arabic are used at the same time. Such is the case with social media in the North African region.

It is likely that even better results can be achieved by considering the concept of which words are being used. To this end, we consider employing the BERT algorithm in a future study. We also need to extend our dataset in order to make it possible to train effective deep learning models on it.

Acknowledgements. We are very grateful to Sarah Ismail, an Algerian PhD student at the University of Limerick in Ireland and to Saad Mboutayeb, a PhD student at the University of Sidi Mohamed Ben Abdellah in Morocco, for their collaboration on the dataset annotation process.

References

1. American Psychological Association (APA). https://www.apa.org. Accessed 16 Nov 2022
2. What Are The Different Forms Of Arabic? Your Guide To Learning Arabic. https://arabicgoals.com/types-of-arabic/. Accessed 09 Nov 2022
3. Arabic Dialects: Different Types of Arabic Language – Tarjama. https://www.tarjama.com/arabic-dialects-different-types-of-arabic-language/. Accessed 05 Nov 2022
4. Benkato, Adam: Maghrebi Arabic. Presented at the April 8 (2020)
5. Scott, J., Marshall, G. (eds.): A Dictionary of Sociology. Oxford University Press, Oxford (2005)
6. Grosfoguel, R.: What is racism? J. World-Syst. Res. **22**, 9–15 (2016)
7. Bonilla-Silva, E.: The linguistics of color blind racism: how to talk nasty about blacks without sounding "racist". Crit. Sociol. **28**, 41–64 (2002)
8. Tulkens, S., Hilte, L., Lodewyckx, E., Verhoeven, B., Daelemans, W.: A dictionary-based approach to racism detection in Dutch social media (2016)
9. Greevy, E., Smeaton, A.F.: Classifying racist texts using a support vector machine. In: Proceedings of the 27th Annual International Conference on Research and Development in Information Retrieval - SIGIR 2004, p. 468. ACM Press, Sheffield (2004)
10. Kwok, I., Wang, Y.: Locate the hate: detecting tweets against blacks. In: Proceedings of the AAAI Conference on Artificial Intelligence, vol. 27, pp. 1621–1622 (2013)
11. Arabic Speaking Population In the World : How many people speak Arabic?. https://www.protranslate.net/blog/en/arabic-speaking-population-in-the-world-2/. Accessed 09 Nov 2022
12. Alotaibi, A., Abul Hasanat, M.H.: Racism detection in twitter using deep learning and text mining techniques for the arabic language. In: 2020 First International Conference of Smart Systems and Emerging Technologies (SMARTTECH), pp. 161–164. IEEE, Riyadh (2020)
13. facebook-scraper PyPI. https://pypi.org/project/facebook-scraper/. Accessed 09 Nov 2022
14. Tweepy Documentation—tweepy 4.12.1 documentation. https://docs.tweepy.org/en/stable/. Accessed 09 Nov 2022
15. Social Media Stats Worldwide | Statcounter Global Stats. https://gs.statcounter.com/social-media-stats/. Accessed 09 Nov 2022
16. Mohammed, R., Rawashdeh, J., Abdullah, M.: Machine learning with oversampling and undersampling techniques: overview study and experimental results. In: 2020 11th International Conference on Information and Communication Systems (ICICS), pp. 243–248 (2020)
17. RandomOverSampler—Version 0.9.1. https://imbalanced-learn.org/stable/references/generated/imblearn.over_sampling.RandomOverSampler.html. Accessed 15 Nov 2022
18. Makienko, D., Seleznev, I., Safonov, I.: The effect of the imbalanced training dataset on the quality of classification of lithotypes via whole core photos. Data Sci. **5** (2020)
19. re—Regular expression operations—Python 3.11.0 documentation. https://docs.python.org/3/library/re.html. Accessed 19 Nov 2022
20. spaCy Industrial-strength Natural Language Processing in Python. https://spacy.io/. Accessed 15 Nov 2022
21. nltk.stem.ISRIStemmer. https://tedboy.github.io/nlps/generated/generated/nltk.stem.ISRIStemmer.html. Accessed 19 Nov 2022

22. Luhn, H.P.: TF. IBM J. Res. Dev. **1**, 309–317 (1957)
23. Sparck Jones, K.: A statistical interpretation of term specificity and its application in retrieval. J. Doc. **28**, 11–21 (1972)
24. Harris, Z.S.: Distributional structure. WORD **10**, 146–162 (1954)
25. Le, Q.V., Mikolov, T.: Distributed representations of sentences and documents (2014)
26. Mikolov, T., Chen, K., Corrado, G., Dean, J.: Efficient estimation of word representations in vector space (2013)
27. Vaswani, A., et al.: Attention is all you need (2017)
28. Chen, H., et al.: Assessing impacts of data volume and data set balance in using deep learning approach to human activity recognition. In: 2017 IEEE International Conference on Bioinformatics and Biomedicine (BIBM), pp. 1160–1165 (2017)
29. Akritidis, L., Bozanis, P.: How dimensionality reduction affects sentiment analysis nlp tasks: an experimental study. In: Maglogiannis, I., Iliadis, L., Macintyre, J., Cortez, P. (eds.) AIAI 2022. IFIP Advances in Information and Communication Technology, vol. 647, pp. 301–312. Springer, Cham (2022). https://doi.org/10.1007/978-3-031-08337-2_25

Integrated Information Theory with PyPhi: Testing and Improvement Strategies

Luz Enith Guerrero[1], Jeferson Arango-López[1], Luis Fernando Castillo[1], and Fernando Moreira[2,3](✉)

[1] Departamento de Sistemas e Informática, Facultad de Ingenierías, Universidad de Caldas, Calle 65 # 26-10, Edificio del Parque, Manizales, Caldas, Colombia
{luzenith_g,jeferson.arango,luis.castillo}@ucaldas.edu.co

[2] REMIT, IJP, Universidade Portucalense, Rua Dr. António Bernardino Almeida, 541-619, 4200-072 Porto, Portugal
fmoreira@upt.pt

[3] IEETA, Universidade de Aveiro, Aveiro, Portugal

Abstract. The study of consciousness has increased in relevance in recent years in the scientific community. In the same line, integrated information theory (IIT) is making inroads into the understanding of consciousness. However, the enormous computational costs required make it difficult to apply IIT to experimental data. In this work we intend to explore and design efficient computational algorithms applying techniques such as Divide and Conquer and parallel computation that can provide some kind of solution to the challenges proposed by IIT based on optimizations and approximations used by PyPhi toolbox to reduce the complexity of the calculations. This software package allows users to easily study cause-effect structures of discret dynamical systems of binary elements, for causal analysis, serves as an up-to-date reference implementation of the formalisms of integrated information theory, and has been used in our research on efficient algorithms.

Keywords: IIT · MIP · PyPhi · Improvement Strategy · integrated information · Phi · efficient algorithms

1 Introduction

Integrated information theory (IIT) attempts to model consciousness mathematically both in quantity and quality [1, 2], more specifically it provides a theoretical-mathematical framework that defines consciousness, quantifies it, and allows determining which systems are conscious and which are not. As stated by [3] for IIT the relationship between the conceptual structure specified by a complex of elements, such as a brain, and the environment to which it is adapted, is not a relationship of "information processing", but of "coincidence" between internal and external cause-effect structures. IIT considers that for consciousness to exist, information must be generated and integrated. In this sense, the theory establishes a way of measuring or quantifying consciousness according to the degree of integrated information in a system [3].

A. Rocha et al. (Eds.): WorldCIST 2023, LNNS 799, pp. 446–456, 2024.
https://doi.org/10.1007/978-3-031-45642-8_44

The theory relates the amount of integrated information to the degree of awareness and establishes a mathematical measure of information integration, which once applied to a system produces a number known as Phi (Φ) that denotes the degree of integrated information, or how much awareness that system has and proposes that, to quantify the integration of information in a system as a whole, different partitions of the system are analyzed until the partition known as the Minimum Information Partition (MIP) is found. Evaluating the irreducibility of a system consists of finding the MIP and this is a computationally intractable problem due to the combinatorial explosion generated by the number of partitions that result [3].

Only a few studies have been able to quantify the integrated information in real neural data because of the computational costs involved making it difficult to apply IIT to experimental data. Recently, [4] proposed a computational solution, applying Queyranne's algorithm, of order O(N3) taking advantage of the submodularity of mutual information (ΦMI). The difficulty lies in the fact that the first version of Phi is submodular but more recent versions are not [5], thus the problem remains. On the other hand, [6] proposed an approximate method based on spectral clustering with correlation of neural time series data, but this method uses a graph theoretic measure that differs from Φ in some cases and limits the application of the results achieved, generating uncertainty.

In this research, we explore the integrated information Φ, in a practical way, with the help of PyPhi, a toolbox written in the Python language that implements this conceptualization provided by IIT and models the cause-effect structure of discrete binary element dynamic systems [7]. This software allows studying causal analysis by understanding these cause-effect structures. Moreover, this toolbox provides an environment for the implementation of the different updates that IIT presents. Taking into account the above, we will explore PyPhi options that will serve as a basis for our proposal of efficient ways to find the MIP and consequently, Φ.

2 Background

2.1 Integrated Information Theory (IIT)

To quantify the integration of information in a system as a whole, different partitions of the system should be analyzed until the partition known as the Minimum Information Partition (MIP) is found. This means, that the integrated information should be measured across the partition of the system in which the information loss caused by system cutoff or reduction is minimized, and that partition is the MIP [8]. Integrated information is designed to quantify the degree of spatiotemporal interactions between subsystems. Many measures of integrated information have been proposed such as mutual information (ΦMI), stochastic interaction (ΦSI) and integrated geometric information (ΦG) and those measures are generally expressed as the Kullback-Leibler divergence between the current probability distribution p(X, X') and a disconnected probability distribution $q(X, X')$ (X and X' are the past and present states of the system respectively) where the interactions between the sub-systems are removed [9].

The Kullback-Leibler divergence measures the differences between probability distributions and that can be interpreted as the loss of information when $q(X, X')$ is used to approximate $p(X, X')$ [10]. In conclusion, the integrated information is interpreted as the loss of information caused by removing interactions. It can be said that there are many ways to re-move interactions between units and that leads to different distributions of disconnected probabilities q and also different measures of the integrated information (ΦMI, ΦSI, ΦG among others). In the first version of IIT (IIT1.0) the integrated information measure used was mutual information and in the most recent version, IIT3.0, effective information was used. IIT uses a perturbative approach that requires extensive knowledge about the physical mechanisms of a system to evaluate probability distributions.

2.2 Minimum Information Partition (MIP)

MIP, is the partition that divides a system into the less interdependent subsystems so that the loss of information caused by removing the interactions between the subsystems is minimized. The MIP, πMIP, is defined as a partition where integrated information is minimized, although there could be more than one partition that complies.

with the request.

$$\pi_{MIP} := \arg\min \Phi(\pi), \pi \in P$$

For a set of partitions, in general, is the universal set of partitions, which includes the different k-partitions. However, bi-partitions are the simplest way to partition a system since it is only determined by specifying a subset S (since by partitioning a set C into two, one obtains S and S′ = C\S), the integrated information can be considered as a function of a set S, Φ(S), so finding the MIP is the same as finding the subset SMIP, which achieves the minimum integrated information.

$$S_{MIP} := \arg\min \Phi(S), S \subset \Omega, S \neq \varnothing$$

In this sense as explained by [5] searching MIP is an optimization problem of a set function, in which the number of partitions for the system with N elements is 2N−1 − 1, which for large N makes the problem computationally intractable, because the number of partitions grows exponentially as system size increases making it difficult to be able to apply IIT to experimental data. That is, practical limitations, such as the large number of computational resources needed to estimate Φ for real systems, have made its application to the brain quite difficult and have also questioned the usefulness of IIT in general [11].

2.3 PyPhi

PyPhi is a Python software package that implements this framework for causal analysis and unfolds the full cause-effect structure of discrete dynamical systems of binary elements. The software allows users to easily study these structures, serves as an up-to-date reference implementation of the formalisms of integrated information theory, and has been applied in research on many areas. PyPhi can be installed on linux, macOS and windows (with some adjustments). This software was developed by the work of [7].

3 Interventions

3.1 Minimum Information Partition (MIP)

One of the most computationally expensive processes is the exhaustive search for the MIP, which must make all possible partitions in order to find the one that produces the least loss of information. A mechanism-purview pair will be tested by partitioning it, cutting the connections, this is done by injecting noise rather than eliminating the connections, to generate the partitions., and then a partitioned repertoire is obtained. The MIP lookup procedure is implemented by the methods Subsystem.causeMip() and Subsystem.efectMip(). Each returns a repertoireIrreducibilityAnalysis object containing the MIP, as well as the phi value, mechanism, purview, temporal direction (cause or effect), non-partitioned repertoire and partitioned repertoire.

3.2 Current PyPhi Solution

Following, we present the elements considered for the optimization of the library. It is pertinent to analyze the different classes of the toolbox (See Fig. 1).

Fig. 1. Files of PyPhi library.

Once the class structure has been reviewed, we work on the partition.py class and its method all_partitions(), because of the large number of operations it performs, since this function is invoked to generate tripartitions and bipartitions.

3.3 "all_partitions" Method

The method receives as parameters a mechanism and a purview. Also, the node labels (optional) and returns all possible partitions of the mechanism and the purview.

The steps of the main algorithm are presented below. It is in the partitions.py file.

1. A tuple of integers corresponding to the mechanism is received.

2. An integer tuple corresponding to the purview is received.
3. A call is made to the partitions function with the received mechanism, and this function returns a collection of partitions.
4. For each partition collection an empty list is added.
5. -n mechanism parts- variable is defined with the value of the length of the split mechanism according to the previous collection.
6. A variable named -max purview partition- is set with the smaller value between the length of the purview tuple and the variable -n_ mechanism_parts.
7. Then for each possible purview, in which iterates from size 1 to -n mechanism parts-plus-one.
8. For each iteration the variable n_empty is instantiated with the value resulting from subtracting n_mechanism _parts with n purview parts.
9. For each possible partition of the purview according to the function k_partitions that receives the tuple purview and -n_purview parts - the value of the parts into which it will be **partitioned.**
10. -purview_partition- **variable is** defined with the result of the tuple generated within each iteration, containing all partitions with the same parts (e.g. all tri-partitions).
11. Using pythons extend method, empty tuples are added so that the purview partition has the same size as the purview **mechanism.**
12. Unique permutations of the partitioned purview are made to avoid duplicate gaps.
13. An object of the class -Part- is generated; a call to the zip method is made.
14. Then, Partitions are performed on the mechanism unless the purview is completely separated from the **mechanism.**
15. Finally, a call is made to the python function -Yield- which returns a list in memory, which has the parts performed in the previous step.

4 Suggested Strategy

Dividing a large problem into a number of smaller scale problems that make it easier to solve is one of the key ideas of the divide and conquer strategy. A problem of scale n is decomposed into k smaller-scale subproblems that are independent of each other. Repeated application of the divide and conquer method can cause the problem to be downscaled to a trivially solvable problem.

The improvement proposal focused on the Parallel_Cut_Evaluation strategy. This option controls whether the system cuts are evaluated in parallel, which is faster but requires more memory. If the cuts are evaluated sequentially, only two instances of SystemIrreducibilityAnalysis need to be in memory at a time [7]. This option is chosen since according to [7] in the PyPhi configuration information they state that it is the most efficient for most networks, because the algorithm has exponential performance depending on the number of nodes being evaluated, so that most of the time is spent on the largest subsystem.

We propose an agile way to generate the partitions and for this we intend to apply the divide and conquer strategy together with the potential that can provide the techniques of parallelism or parallel processing, and it was decided to work on the function all_partitions, since as mentioned above is a widely used function and performs a large number of operations. The function is located in the partitions.py file. First, the processes that seek to return all possible partitions of the mechanism - purview - are segmented. For this, the function all_partitions were modified by decomposing it into four methods, including the original method. This was intended so that the main method would take care of defining the main partitions of the mechanism and the other methods would focus on partitioning the purview and permutation of the mechanism-purview into parts. Thus, with the tasks defined, for each possible partition the input was divided and a call was made for each part of the input making it run in parallel. Once all processes are finished, a tuple of two lists is returned, implementing a merge of both (see Fig. 2).

```
@partition_types.register("ALL")
def all_partitions(mechanism, purview, node_labels=None):

    for mechanism_partition in partitions(mechanism):
        mechanism_partition.append([])
        n_mechanism_parts = len(mechanism_partition)
        max_purview_partition = min((len(purview)/2), n_mechanism_parts)
        check,check2 = checkEmptyValuesPurview(max_purview_partition)
        if(check == 0 and check2 == 0):
            break

def checkEmptyValuesPurview(max_purview_partition, n_mechanism_parts, purview, mechanism_partition, mechanism, node_labels):

    for n_purview_parts in range(1, max_purview_partition + 1):
            n_empty = n_mechanism_parts - n_purview_parts
            if(n_empty == 0):
                return 0
            else:
                for purview_partition in k_partitions(purview, n_purview_parts):
                    purview_partition = [tuple(_list) for _list in purview_partition]

                    purview_partition.extend([()] * n_empty)

                    return setPermutations(purview_partition, mechanism_partition, mechanism, node_labels)

def setPermutations(purview_partition, mechanism_partition, mechanism, node_labels):
    checkList1 = []
    checkList2 = []
    for purview_permutation in set(permutations(purview_partition)):

                    parts = [
                        Part(tuple(m), tuple(p))
                        for m, p in zip(mechanism_partition, purview_permutation)
                    ]

                    if parts[0].mechanism == mechanism and parts[0].purview:
                        continue

                    yieldKpartitions(parts, node_labels)
                    checkList1.append(parts[0].mechanism)
                    checkList2.append(parts[0].purview)
                    return checkList1,checkList2

def yieldKpartitions(parts, node_labels):
    yield KPartition(*parts, node_labels=node_labels)
```

Fig. 2. Proposed code in the improvement of the process.

4.1 Test Plan

The initial step is to create a network object and here we can provide the transition probability matrix (TPM) in node-state form. We know that the TPM is the only argument required by PyPhi but we also provide the connectivity matrix (CM) because we already know that there are no automatic loops in the system and the Pyphi library is the one that will be in charge of using this information for the calculation.

For each of the networks to be tested in PyPhi, we will initially start with the declaration of the network. The application requires the TPM of the network, the initial state and optionally the connectivity matrix, with this data the respective tests will be performed. The following is a more detailed explanation of the fundamental aspects to be considered for the tests.

4.2 PyPhi Configuration

For initial tests, the calculation of the integrated information (Φ) for all the test graphs (systems) was performed using the functionalities implemented by the PyPhi library. For each system to be tested, it is required to know which are its nodes (node_labels), how they are connected between them (CM), and how they will change state over time (TPM). As for the CM, the library shows that it is not a required parameter, while the TPM is mandatory and also allows to enter it in one of 3 ways (complying with the Little-endian convention):

- State per state: It is a square matrix of size 2n (with n as the number of nodes),
- State per node: It is a 2-dimensional matrix, with 2n rows (with n as the number of nodes) and n columns.
- Multidimensional form state per node: This matrix has n + 1 dimensions, the first n dimensions represent the nodes in the current state and the last dimension represents the probability of each node to be in the ON state.

4.3 Distance Measures Supported by PyPhi

PyPhi supports different distance measures, all of them allowing to calculate how different are 2 probability distributions, each one with a characteristic algorithm; besides allowing to enter customized algorithms to calculate this parameter:

- EMD: Also known as Wasserstein metric
- L1: Also known as the cab driver metric.
- ENTROPY DIFFERENCE: Measures the difference in entropy between two probability distributions.
- PSQ2: Pseudo-quasimetric type distance measure.

4.4 Partitioning Schemes Supported by Pyphi

For the calculations of φ (Integrated information of a mechanism) Pyphi allows all possible partitioning schemes, i.e., a mechanism or repertoire can be divided in 2, in 3, in k, or in all possible parts, customizing the value of k as required with a valid integer for the number of nodes of the mechanism; moreover, only partitions that actually divide the mechanism are considered. For the tests performed, these 3 partitioning schemes will be taken:

- Bipartitions: Divide the repertoire into groups of 2 parts, e.g., to calculate φ of the AC mechanism on the ABC repertoire, the distance of all the following partitions would have to be evaluated:

$$\frac{AC}{ABC} \rightarrow \left\{ \frac{\varnothing}{A} \times \frac{AC}{BC\prime}, \frac{A}{\varnothing} \times \frac{C}{ABC'}, \frac{A}{BC} \times \frac{C}{A'}, \ldots \right\}$$

- Tripartitions: Divide the repertoire into groups of 3 parts, e.g., for calculating φ of the AC mechanism over the ABC repertoire, one would have to evaluate the distance of all the following partitions:

$$\frac{AC}{A} \rightarrow \left\{ \frac{\varnothing}{A} \times \frac{A}{B} \times \frac{C}{C'}, \frac{A}{AB} \times \frac{C}{\varnothing} \times \frac{\varnothing}{C'}, \frac{A}{\varnothing} \times \frac{C}{A'} \times \frac{\varnothing}{BC'}, \ldots \right\}$$

– All partitions: Divides the repertoire into all possible partitions.

To configure the software to perform the calculations with the required partition scheme the following instruction was used: pyphi.config.PARTITION_TYPE = type, being "type" the selected partition scheme.

5 Results

For the tests, changes were made for each of the schemes of each network, entering the CM, since with it a shorter execution time is obtained. Although in some occasions tests were made without the CM, the TPM was used in its node-state form, and some of the distance measures and partitioning method were revised. The following information is available in the test table (file Results1.xlx):

- **graph1** represents the system to be considered
- **emd** represents the distance measure
- **bi** represents the partitioning method
- parallel represents the name of the Python method used in case a specific method is used.
- **no_cm** is used to indicate that the connectivity matrix is not used.
- **cm** connectivity matrix is used, by default the connectivity matrix is always used.
- **MIP** shows the partition of the minimum information found.
- **Phi** has the value of the integrated information for the represented system.
- **Time** records the time taken to calculate Phi.

The Table 1 presents the information concerning the different optimization methods and/or PyPhi approximations, which were tested with the EMD distance measure and with the Bipartition partitioning scheme that showed the best results in the initial tests. The tests were developed for 10 different networks, but we abbreviate by presenting only the information for graph 3, the complete results are in the appendix A.

The Table 2 presents the results after applying our improvement strategy, which as explained above was based on PyPhi's Parallel_Cut_Evaluation method, which was modified and yielded better results than those presented by PyPhi. The detailed results for all systems (graphs) are presented in the Appendix B, and the results for graph 3 are presented in Table 2 and Table 3 presents a summary for all graphs:

After analyzing the results, we found that our proposal improved the Phi calculation times in all the graphs, which varied in including from 3 to 6 nodes, characterizing networks with different topologies. Before, in the original method, 4 nested for loops that should iterate each permutation of the partitions for each part of the purview. On each permutation a k-Partition is returned and execution is paused. With the application of our method, the solution was structured in point tasks to return the possibilities, the data input was divided and a call was made for each part of the input to be executed parallelly, and once the processes were finished, a tuple with two lists is returned and a mixture of both is made. Consequently, it is a faster generation of possibilities, so that later in a next stage, they can be analyzed in order to find the MIP.

Our proposal presents a possibility to solve this kind of combinatorial problems by using techniques for algorithm design such as divide and conquer coupled with parallel

Table 1. Test results PyPhi for graph 3.

System	Results without method	PyPhi method applied	Results with method applied
	MIP: Cut [A, B] ——/ /——➤ [C] Phi: Φ = 0.520834 Time: 1.102348 s	*Cut_one_approximation*	MIP:Cut [B, C] —/ /——➤ [A] Phi:Φ = 0.520834 Time:0.524636 s
		"No new concepts"	MIP:Cut [A] —/ /——➤ [B, C] Phi:Φ = 0.520834 Time:0.558397 s
		Memoization and caching	MIP:Cut [A, B] —/ /——➤ [C] Phi:Φ = 0.520834 Time:0.637821 s
		Connectivity optimizations	MIP:Cut [A, B] —/ /——➤ [C] Phi:Φ = 0.659722 Time:0.820733 s
		Parallel_Cut_Evaluation	MIP:Cut [A, B] —/ /——➤ [C] Phi:Φ = 0.520834 Time:0.565997 s

Table 2. Test results for graph 3

System	PyPhi method selected	PyPhi results	Results with our improved method
	Parallel_Cut_Evaluation	MIP: Cut [A, B] —/ /——➤ [C] Phi:Φ = 0.520834 Time:0.565997 s	MIP: Cut [A, B] —/ /—➤ [C] Phi:Φ = 0.520834 Time:0.432971 s

Table 3. Summary of results for all graphs (method Parallel_Cut_Evaluation).

System	PyPhi results		Results with Our improved method	
Graph 1	Phi= 0	Time: 7.839227 s	Phi = 0	Time: 7.516224 s
Graph 2	Phi= 1.916665	Time: 1.186331 s	Phi= 1.916665	Time: 1.066230 s
Graph 3	Phi= 0.520834	Time: 0.565997 s	Phi= 0.520834	Time: 0.432971 s
Graph 4	Phi=0.308035	Time:7.011258 s	Phi=0.308035	Time: 6.315788 s
Graph 5	Phi=0.048788	Time:74.070221 s	Phi=0.048788	Time:73.865130 s
Graph 6	Phi=0.250981	Time:1034.164633 s	Phi=0.250981	Time:1032.875204 s
Graph 7	Phi=4.805374	Time:2376.588076 s	Phi=4.805374	Time:2375.698527 s
Graph 8	Phi=0.009583	Time:398.610314 s	Phi=0.009583	Time:398.418324 s
Graph 9	Phi=0.531249	Time:480.839672 s	Phi=0.531249	Time:480.547253s
Graph 10	Phi=0.003057	Time:367.031683 s	Phi=0.003057	Time:366.435743 s

computing, which allow tackling larger and more complex problems with input data that exceed the memory capacity of a machine.

6 Conclusions and Future Work

The study of efficient algorithms is a permanent challenge, which as in the case of IIT would allow its application to experimental data and as [1] puts it "it could be said that theories without experiments are unconvincing, but experiments without theories are blind". In such a sense, a science of consciousness needs both experimentation and theory, by the very nature of the problems presented; only a theoretical framework can provide a grounded explanation of areas of the brain relevant to consciousness.

Since the problem of finding the MIP is a complex problem, in our proposal we focus on an agile method to find all the partitions. For this we have used a divide and conquer approach coupled with parallelism techniques to take advantage of the independence of the generated subproblems.

Finally, we should mention some of the limitations of the present work. Foremost among them is that our examples used small networks, so it is unclear to what extent the method suggested by our partial explorations would scale with larger networks. This is because we focused on reviewing different types of architectures to look at the efficiency and effectiveness of the method.

The purpose of the research is to continue with the study of efficient algorithms that indicate a solution path to find the MIP, according to IIT. In future work we will explore how techniques such as backtracking coupled with parallel computing, heuristic functions [12] and metaheuristics can help reduce the combinatorial explosion in the partitions of a system. This will ultimately result, for example, in simulations that can be run in more detail and a physical phenomenon can be modeled more closely to reality.

Acknowledgements. This work was supported by the FCT – Fundação para a Ciência e a Tecnologia, I.P. [Project UIDB/05105/2020].

Appendix

A - The file can be found at the following link http://shorturl.at/cfX46.
B - The file can be found at the following link http://shorturl.at/wyAP7.

References

1. Tononi, G.: Consciousness as integrated information: a provisional manifesto. Biol. Bull. **215**(3), 216–242 (2008)
2. Tononi, G.: Integrated information theory of consciousness: an updated account. Arch. Ital. Biol. **150**, 56–90 (2012)
3. Tononi, G.: Integrated information theory. Scholarpedia **10**(1), 4164 (2015)
4. Hidaka, S., Oizumi, M.: Fast and exact search for the partition with minimal information loss. PLoS One **13**(9), e0201126 (2018)

5. Kitazono, J., Kanai, R., Oizumi, M.: Efficient algorithms for searching the minimum information partition in integrated information theory. Entropy **20**, 173 (2018)
6. Toker, D., Sommer, F.T.: Information integration in large brain networks. PLoS Comput. Biol. **15**(2), e1006807 (2019)
7. Mayner, W.G.P., Marshall, W., Albantakis, L., Findlay, G., Marchman, R., Tononi, G.: PyPhi: a toolbox for integrated information theory. PLoS Comput. Biol. **14**(7), e1006343 (2018)
8. Oizumi, M., Albantakis, L., Tononi, G.: From the phenomenology to the mechanisms of consciousness: integrated information theory 3.0. PLOS Comput. Biol. **10**(5), e1003588 (2014)
9. Oizumi, M., Tsuchiya, N., Amari, S.: Unified framework for information integration based on information geometry. Proc. Natl. Acad. Sci. **113**(51), 14817–14822 (2016)
10. Kenneth, P., David, R.A.: Model Selection and Multimodel Inference: A Practical Information-Theoretic Approach, 2nd edn., p. 488. Springer, New York (2002). https://doi.org/10.1007/b97636
11. Kim, H., et al.: Estimating the integrated information measure Phi from high-density electroencephalography during states of consciousness in humans. Front. Hum. Neurosci. **12**, 42 (2018)
12. Sevenius Nilsen, A., Juel, B.E., Marshall, W.: Evaluating approximations and heuristic measures of integrated information. Entropy **21**(5), 525 (2019)

Recursive Least Squares Identification with Extreme Learning Machine (RLS-ELM)

Alanio F. Lima[1(✉)], Laurinda L. N. dos Reis[1], Darielson A. Souza[1], Josias G. Batista[1], Antonio B. S. Júnior[2], Francisco Heleno V. Silva[1], and Vinícius R. Cortêz[1]

[1] Department of Electrical Engineering, Federal University of Ceará, Fortaleza, Ceará, Brazil
allanio007@gmail.com

[2] Federal Institute of Education Science and Technology of Ceará, Maracanaú, Ceará, Brazil

Abstract. This work investigates the artificial neural networks models in systems identification. The search for process optimization applied to robotics in industry has constantly increased with the course of the computerization of the industry over the years. Identify an automation and control of a process in a manipulator are requirements to have a better quality for the industry final product with a continuous improvement using system identification. To obtain an optimized control in a system identification it is necessary to exist the premise (the output of the model of the system is closer to the real output). This work aims to demonstrate the identification by methods: Least Squares (LS), Recursive Least Square (RLS) and a hybrid model that takes a RLS with Extreme Machine Learning (ELM), applied to the robotic manipulator joint model. The Coefficient of Determination (R^2) results are used in the follow identification models: LS, RLS and RLS-ELM.

Keywords: Artificial Neural Networks · System Identification · Single Layer Feedforward Network · Least Square · Recursive Least Square

1 Introduction

Most industry robots perform tasks of handling loads, materials, assembly, welding, among other purposes. Given the advances in robots moving systems, the need for more robust systems and well identified by mathematical models emerged.

Notably, mathematical models have been proposed in the literature using RLS [1–4] and ELM for several purposes: time series prediction, [5], pattern recognition [6], and in industrial applications [7,8].

A. Rocha et al. (Eds.): WorldCIST 2023, LNNS 799, pp. 457–466, 2024.
https://doi.org/10.1007/978-3-031-45642-8_45

In the works in [8,10] the applicability of the OS-ELM-RLS algorithm called Extreme Online Sequential Machine Learning combined in Recursive Least Squares. In these two works there was a shorter execution time of the algorithm in relation to the sequential algorithms common in the literature. It is important to note that in these works, the common use of RLS and ELM is strongly aimed at obtaining promising results.

In [12] a hybrid optimization strategy of ELM with PSO (Particle Examination Optimization) is presented, both combines the advantages of the chaos optimization strategy, the adaptive update strategy and the mutation strategy. I highlight that the chaos optimization strategy optimizes the initial swarm distribution to improve the PSO optimization efficiency. Then, it demonstrated that the ELM has robustness when using the hybrid optimization strategy.

It is noteworthy that in [1] the applicability of RLS with Kalma Filter (KF) is mentioned in the identification of a system. I emphasize the contribution of a hybrid algorithm for the identification of industrial robotic manipulators based on (RLS), which has its matrix of regressors and vector of parameters optimized via (KF). The KF method was proposed to adjust the RLS parameters. The proposed hybrid method (KF to adjust the RLS parameters) obtained a better determination coefficient than the RLS an ERLS when valuated in training and test steps.

This work purpose a hybrid identification method that combine the RLS and ELM for Single Hidden Layer Feedforward Networks (SLFN). The comparison of the LS, RLS and RLS-ELM algorithms with their transfer equations stands out. The methods are evaluated by the R^2 metric.

2 The Problem Characteristics

Different types of manipulators are used by different applications directly related to their geometry, thus, in this work it's a cylindrical robotic arm. In this manipulator there are three degrees of freedom (3-DOF), where the first is the basis for rotary movements, the second is linear, which is the trunk that makes the vertical movements and the third degree to horizontal movements [1].

2.1 Experimental Data

An experiment was used to collect the input data (speed and current) for the system training. Each experiment contains 1000 s. As input for application of the methods, the signal *PRBS* was used.

The current signal, in Fig. 1, is an experimental data from the input of a manipulator joint in [1] and velocity data, Fig. 2, from the output of this manipulator joint.

2.2 Least Squares (LS)

The LS mathematics formalism can be read in [9]. The input $u(t)$ and output $y(t)$) are parameters used in the manipulator identification process.

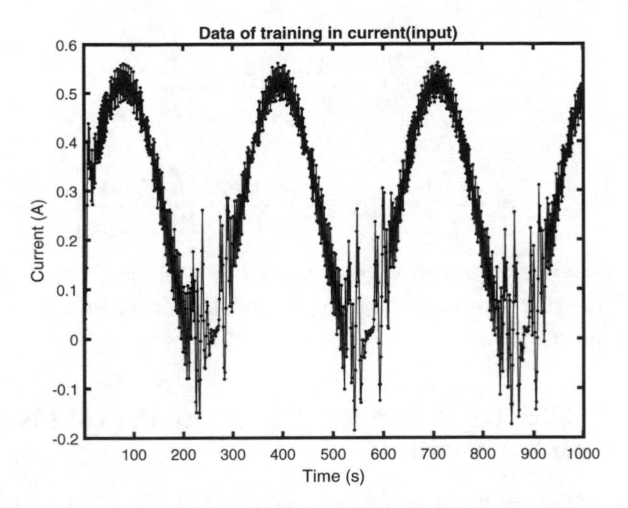

Fig. 1. System data input.

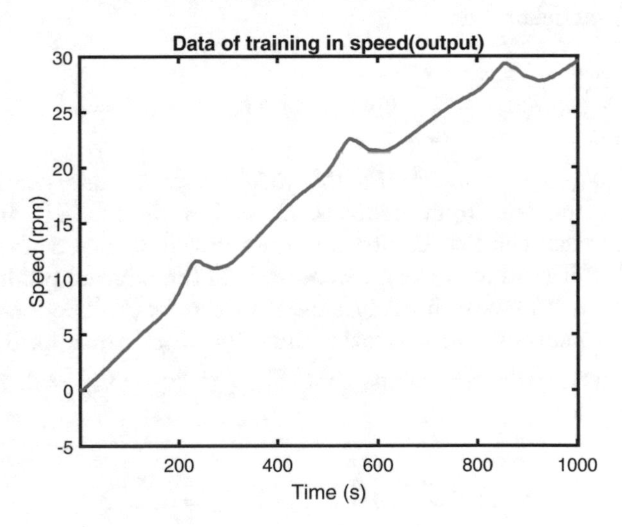

Fig. 2. System data output.

2.3 Recursive Least Squares (RLS)

The RLS is used in a real time system models calculations and it is in operation [9]. Supposing that the model is in a data set that's represented by a regress equation defined by,

$$\hat{y} = \varphi\hat{\theta} + \xi \tag{1}$$

where φ is a regress matrix and ξ is the residual. The estimated parameters are calculated by,

$$\theta_{[k]} = \theta_{[k-1]} + L_{[k]}\left[y_{[k]} - \varphi_{[k]}^{T}\theta_{[k-1]}\right] \tag{2}$$

where

$$L_{[k]} = \frac{P_{[k-1]}\varphi_{[k]}}{\lambda_{[k]} + \varphi_{[k]}^T P_{[k-1]}\varphi_{[k]}} \tag{3}$$

and

$$P_{[k]} = \frac{1}{\lambda_{[k]}} \left[P_{[k-1]} - \frac{P_{[k-1]}\varphi_{[k]}\varphi_{[k]}^T P_{[k-1]}}{\lambda_{[k]} + \varphi_{[k]}^T P_{[k-1]}\varphi_{[k]}} \right]. \tag{4}$$

where P is a matrix where the regress vector is initialized with values close to zero. In the loop this values are updated and the final better parameters are obtained through a fit.

2.4 Single Hidden Layer Propagation Networks (SLFNs) Whit Random Hiden Neurons

To N distinct arbitrary samples $(\mathbf{x}_i, \mathbf{t}_i)$, where $\mathbf{x}_i = [x_{i1}, x_{i2}, ..., x_{in}]^T \in \mathbf{R}^n$, e, $\mathbf{t}_i = [t_{i1}, t_{i2}, ..., t_{im}]^T \in \mathbf{R}^m$, standard SLFNs with $\tilde{N}$ neurons and activation function $\mathbf{g(x)}$ modelated as:

$$\sum_{i=1}^{\tilde{N}} \beta_i g_i(\mathbf{x}_j) = \sum_{i=1}^{\tilde{N}} \beta_i g(\mathbf{w}_i \cdot \mathbf{x}_j + b_i) = \mathbf{o}_j, \ \ j = 1, ..., N. \tag{5}$$

where $\mathbf{w}_i = [w_{i1}, w_{i2}, ..., w_{in}]^T$ it's the weight vector thats connects the i'th hidden neuron and the input neurons, $\beta_i = [\beta_{i1}, \beta_{i2}, ..., \beta_{in}]^T$ it's the wight vector thats conects the i'th hidden neuron and the output neurons, and $\mathbf{b}_i$ is the threshold of i'th hidden neuron. $\mathbf{w}_i.\mathbf{x}_i$ denote the internal product of $\mathbf{w}_i$ and $\mathbf{x}_i$. The output neurons are linearly chosen in this work. The standard SLFNs with $\tilde{N}$ hidden neurons and activation function $\mathbf{g(x)}$ can approximate the N samples with zero error, it's means that $\sum_{j=1}^{\tilde{N}} = \|\mathbf{o}_j - \mathbf{t}_j\| = \mathbf{0}$, that is, exist, β_i, $\mathbf{w}_i$ e b_i such as:

$$\sum_{i=1}^{\tilde{N}} \beta_i g(\mathbf{w}_i \cdot \mathbf{x}_j + \mathbf{b}_i) = \mathbf{t}_j, \ \ j = 1, ..., N. \tag{6}$$

This $\tilde{N}$ can be presented in compact mode: $\mathbf{H}\beta = \mathbf{T}$, where,

$$\mathbf{H}(\mathbf{w}_1, ..., \mathbf{w}_{\tilde{N}}, b_1, ..., b_N, \mathbf{x}_1, ..., \mathbf{x}_N) = \begin{bmatrix} g(\mathbf{w}_1.\mathbf{x}_1 + b_1) & ... & g(\mathbf{w}_{\tilde{N}}.\mathbf{x}_1 + b_{\tilde{N}}) \\ \vdots & ... & \vdots \\ g(\mathbf{w}_1.\mathbf{x}_N + b_1) & ... & g(\mathbf{w}_{\tilde{N}}.\ \mathbf{x}_N + b_{\tilde{N}}) \end{bmatrix} \tag{7}$$

$$\beta = \begin{bmatrix} \beta_1^T \\ \vdots \\ \beta_{\tilde{N}}^T \end{bmatrix} \ and \ \mathbf{T} = \begin{bmatrix} \mathbf{t}_1^T \\ \vdots \\ \mathbf{t}_{\tilde{N}}^T \end{bmatrix} \tilde{N}. \tag{8}$$

H it's the output matrix from the neural network hidden layer; the i'th column of **H** is the i'th output from the hidden neuron in relation to the inputs $\mathbf{x}_1, \mathbf{x}_2, \mathbf{x}_N$.

If the activation function 'G' is infinitely differentiable, we can prove that the necessary number of hidden nodes $\tilde{N} \leq N$.

strictly, we have,

Theorem 2.1: *Given a default SLFN with N hidden neurons and activation function* $\boldsymbol{g}$*:*$\boldsymbol{R} \rightarrow \boldsymbol{R}$*, that is infinitely differentiable in any range, to N arbitrary distinct samples* $(\mathbf{x}_i, \mathbf{t}_i)$*, where* $\mathbf{x}_i \in \mathbf{R}^n$*, and* $\mathbf{t}_i \in \mathbf{R}^m$*, to any* $\mathbf{w}_i$ *e* $\mathbf{b}_i$ *randomly chosen from any interval of* $\mathbf{R}^n$ *and* $\mathbf{R}$*, respectively, according with any continuous probability distribution, so with probability 1 (one), the output matrix of the hidden layer* $\mathbf{H}$ *from SFLN is a non reversible matrix and* $\|\boldsymbol{H}\beta - \boldsymbol{T}\| = \mathbf{0}$.

Theorem 2.2: *Give a even value* $\varepsilon > \mathbf{0}$*, and the activation function* $\boldsymbol{g}$*:*$\boldsymbol{R} \rightarrow \boldsymbol{R}$ *that is infinitely differentiable in any range, exist* $\tilde{\mathbf{N}} \leq \boldsymbol{N}$*, such that, for N arbitrary distinct samples* $(\boldsymbol{x}_i, \boldsymbol{t}_i)$*, where* $\mathbf{x}_i \in \mathbf{R}^n$*, and* $\mathbf{t}_i \in \mathbf{R}^m$*, to any* $\mathbf{w}_i$ *e* $\mathbf{b}_i$ *randomly chosen samples from any interval of* $\mathbf{R}^n$ *e* $\mathbf{R}$*, respectively, according with any continuous probability distribution, so with probability 1 (one),* $\|\boldsymbol{H}_{Nx\tilde{N}}\beta_{\tilde{N}x\,m} - \boldsymbol{T}_{Nx\,m}\| = \mathbf{0}$.

2.5 Methodology

The work proposal conjecture on the utilizing the RLS-ELM to system identification. In Fig. 3 is presented the diagram of the proposed interaction flux of the system identification.

The ELM is a effective solution with minimal weight norm, since the solution $\beta = \mathbf{H'T}$ has the minimal norm among all the solutions of the RLS of a general linear system $\mathbf{H}\beta = \mathbf{T}$. The main idea from the ELM is used to estimate recursively of the parameters $(\theta, \mathbf{x} = \phi)$ from the RLS methode in each interaction. It's important to point that as the parameters $(\theta, \mathbf{x} = \phi)$ are being update on the algorithm 1 loop, implies that we will have two delays. The algorithm 1 was initialized with the parameters from the matrix W_w with 4×4 dimensions and main diagonal equals to $W_v = 0.99$ and the others elements equal to zero.

I emphasize that in algorithm 1, where the hybrid method (RLS-ELM) is applied, in order to obtain a better optimization, the parameters P, φ, was defined with values next to zero.

Presents a transfer functions algorithms study: LS, RLS, e RLS-ELM. Where the RLS-ELM method is the main proposal in this work. We point that when utilizes the RLS method to adjust a huge variability of parameters aiming the optimization, the term's adjust don't obtain a efficient adjust coefficient, thus, to overcome this shortcoming we used the hybrid algorithm (ELM combined with the RLS).

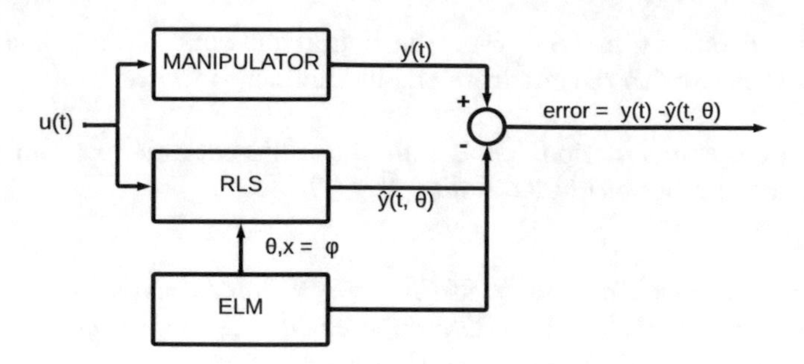

Fig. 3. The Hybrid system RLS-ELM.

Algorithm 1: Hybrid Method (RLS-ELM)
Initialize the values of: θ, φ, x *and* P;
for $i = delay$ *to* t, *where* $t = length(y)$ **do**
$\varphi = [y_{(t-1)}; y_{(t-2)}; u_{(t-1)}; u_{(t-2)}]$
$\hat{y}_{(t)} = x'\theta$
$P = PW_w$, *propagation step*
$M = \frac{Px}{W_v + x'Px}$, *propagates correlation*
$E = y_{(t)} - \hat{y}_{(t)}$
$\theta_{(t)} = \theta + ME$
$P = (I - Mx')P$, *where I is the identify matrix with 4x4 dimensions*
$a_1 = \theta_{(1)}$
$a_2 = \theta_{(2)}$
$b_0 = \theta_{(3)}$
$b_1 = \theta_{(4)}$
$x = \varphi$
end for

3 Experimental Results

In this section the identification approach results by the LS, RLS and RLS-ELM methods.

3.1 LS

In Fig. 4 a second order identification system through LS. The regression vector θ was defined in the train step and used to the validation on test step, described as follow: $\theta = [-1.8293\ -0.8293\ -0.0054\ 0.0200]$.

Equation 9 presents the transfer function in discrete time with 2 s of sampling time, generated by the identified model (LS), as follows:

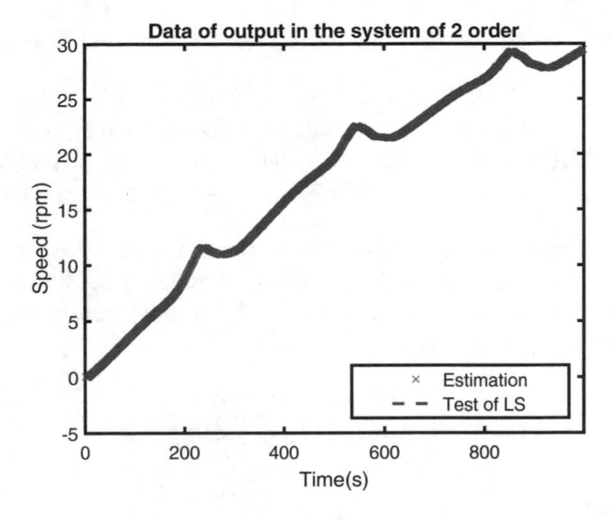

Fig. 4. LS System

$$\frac{-0.005421z + 0.02004}{z^2 - 1.8293z + 0.8293} \tag{9}$$

3.2 RLS

In Fig. 5 contains the identified system by the RLS that's represents the second order system:

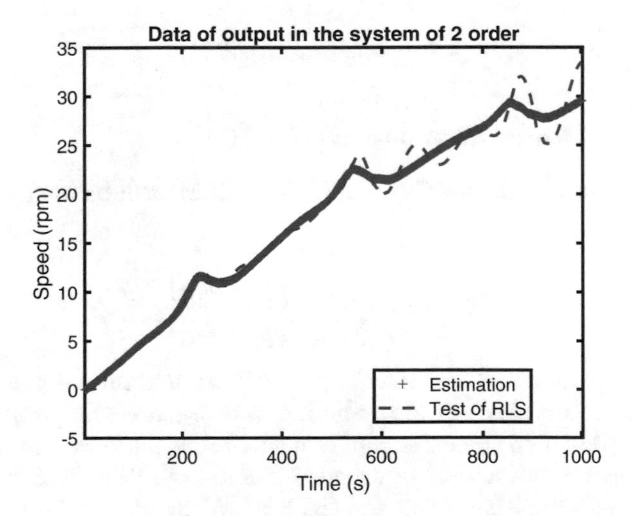

Fig. 5. RLS System

The strictly defined matrix with values from the recursive search:

$$P = \begin{bmatrix} 0.0389 & -0.0389 & 5.0844e-04 & 0.0029 \\ -0.0389 & 0.0389 & -4.2186e-04 & -0.0028 \\ 5.0844e-04 & -4.2186e-04 & 0.0441 & -0.0363 \\ -0.0029 & -0.0028 & -0.0363 & 0.0444 \end{bmatrix} \quad (10)$$

and θ the regression vector, generated in the train step and used in the validation test step, it can be described as follow: $\theta = [-0.5477 \;\; -0.4460 \;\; 0.2611 \;\; 0.2604]$.

Equation 11 presents a sampling time of 0.2 s, from the identified model (RLS) in the transfer function equation:

$$\frac{0.2611z + 0.2604}{z^2 - 0.5477z - 0.446} \quad (11)$$

3.3 RLS-ELM

In the figure Fig. 6 we have, the strictly defined matrix P with values from the recursive search:

$$P = \begin{bmatrix} 56.6490 & -57.5759 & -4.7214 & 4.3989 \\ -57.5759 & 58.5537 & 5.5174 & -2.6255 \\ -4.7214 & 5.5174 & 141.0717 & 1.1613 \\ 4.3989 & -2.6255 & 1.1613 & 140.3845 \end{bmatrix} \quad (12)$$

and like before, θ is the regression vector, that's comes from the training step and used in the validation step, where $\theta = [-0.5223 \;\; -0.4684 \;\; 0.7036 \;\; 0.8086]$.

The Eq. 13 is the identified transfer function equation:

$$\frac{0.7036z + 0.8086}{z^2 - 0.5223z - 0.4684} \quad (13)$$

3.4 The Algorithms Comparative Study

In this section the algorithms (LS, RLS e RLS-ELM) will be analyzed by the $\mathbb{R}^2$ metric:

$$\mathbb{R}^2 = 1 - \frac{\sum_{i=1}^{n} \left(y_{(i)} - \hat{y}_{(i)}\right)^2}{\sum_{i=1}^{n} \left(y_{(i)} - \bar{y}_{(i)}\right)^2} \quad (14)$$

The Eq. 14 parameters are described as follow: determination coefficient $\mathbb{R}^2$, $y_{(i)}$, the observation, $\hat{y}_{(i)}$ is the prediction and $\bar{y}_{(i)}$ the observations mean. In the Table 1 is presented the determination coefficient performance $\mathbb{R}^2$ in the test step and the computation cost of each algorithm (LS, RLS, RLS-ELM).

We have as a promising result the (RLS-ELM) hybrid method. So, using the $\mathbb{R}^2$ metric and evaluating by the test step thought the 3 methods: LS, RLS, RLS-ELM, their respective results 0.9984, 0.9835 e 0.9999.

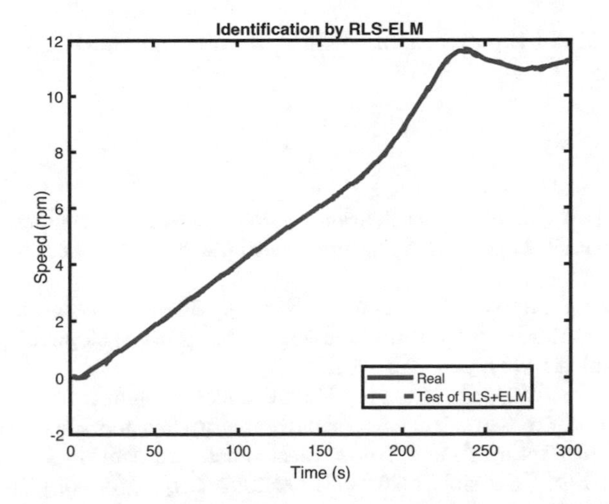

Fig. 6. RLS with ELM System.

Table 1. Output of $\mathbb{R}^2$ and computational cost application in the stages

Test stage		
$\mathbb{R}^2$		
LS	RLS	RLS-ELM
0,9984	0,9835	0,9999
Test stage		
Computacional cost [s]		
LS	RLS	RLS-ELM
0,8643	0,8791	0,8234

4 Conclusion

We highlight that the hybrid method (RLS-ELM) surpass the (LS, RLS) methods and that the determination coefficient R^2 in RLS-ELM had the better identification robustness because it is a only one layer Neural Network, as we note in the literature in others applications, it have better results when compared with others neural networks applied in identification system.

The proposed hybrid method to adjust the RLS parameters, in the test step, presented better performance by the determination coefficient equal to 0.9999 compeared to the others methods: LS and RLS equal to 0.9984, 0,9835, respectively. This Hybrid method (RLS-ELM) prooved that's efficient for high performance complex problems for identification system.

In future works it can be used to evaluate the results prediction error, for effects of comparisons it can used in others metrics: The Mean Absolute Error (MAE), Mean Absolute Percent Error (MAPE), Mean Square Error (MSE)

and the Root Mean Square Error (RMSE). And also the ELM combined with extended RLS.

References

1. De Souza, D.A., et al.: Identification by Recursive Least Squares With Kalman Filter (RLS-KF) Applied to a Robotic Manipulator. IEEE Access **9**, 63779–63789 (2021)
2. Yao, J., Wang, T., Wan, Z., Chen, S., Niu, Q., Zhang, L.: Identification of acceleration harmonics for a hydraulic shaking table by using Hopfield neural network. Scientia Iranica **25**(1), 299–310 (2018)
3. Wang, Q., Xu, W., Li, X., Liu, Y.: A novel resonant frequency estimation method based on recursive least squares algorithm for linear compressor. In: 2020 IEEE 9th International Power Electronics and Motion Control Conference (IPEMC2020-ECCE Asia), 2020, pp. 1992-1997. https://doi.org/10.1109/IPEMC-ECCEAsia48364.2020.9367828
4. Lee, D.: Motor resistance detection method using RLS technique. In: 8th International Electrical Engineering Congress (iEECON). IEEE 2020, pp. 1–3 (2020)
5. Wang, L., Zeng, Y., Chen, T.: Back propagation neural network with adaptive differential evolution algorithm for time series forecasting. Expert Syst. Appl. **42**(2), 855-863 (2015)
6. Siniscalchi, S.M., Vendsen, T., Lee, C.-H.: An artificial neural network approach to automatic speech processing. Neurocomputing **140**, 326–338 (2014)
7. Soares, S., Araújo, R., Sousa, P., Souza, F.: Design and application of Soft Sensor using Ensemble Methods. In: ETFA2011, 2011, pp. 1–8. https://doi.org/10.1109/ETFA.2011.6059061
8. Liang, N., Huang, G., Saratchandran, P., Sundararajan, N.: A fast and accurate online sequential learning algorithm for feedforward networks. IEEE Trans. Neural Networks **17**(6), 1411–1423 (2006). https://doi.org/10.1109/TNN.2006.880583
9. Coelho, A.A.R., dos Santos Coelho, L.: Identificação de sistemas dinâmicos lineares (2004)
10. Huang, G.B., Liang, N.Y., Rong, H.J., Saratchandran, P., Sundararajan, N.: Online sequential extreme learning machine. Comput. Intell. **2005**, 232–237 (2005)

A Data-Driven Cyber Resilience Assessment for Industrial Plants

Francesco Simone[1(✉)], Claudio Cilli[2], Giulio Di Gravio[1], and Riccardo Patriarca[1]

[1] Department of Mechanical and Aerospace Engineering, Sapienza University of Rome, Rome, Italy
{francesco.simone,giulio.digravio, riccardo.patriarca}@uniroma1.it

[2] Department of Computer Science, Sapienza University of Rome, Rome, Italy
claudio.cilli@uniroma1.it

Abstract. Cyber-Physical Systems (CPSs) are becoming more integrated into smart industrial assets. CPS however are increasingly exposed to external vulnerabilities, beyond technical failures, i.e., cyber attacks, due to their complex structure that combines cyber, cyber-physical, and physical components. Operationalizing the notion of cyber resilience represents a solution to deal with this family of problems. In this study, a methodology to assess the resilience of CPSs is presented and instantiated on dataset referred to SWaT (Secure Water Treatment) testbed data to show its effectiveness in a particular application. Two metrics are calculated to quantify the resilience of the system under analysis. The obtained results gives suggestions for the resilient design and management of CPSs.

Keywords: Resilience · Cyber attacks · Big data

1 Introduction

The shift towards Industry 4.0 has favored digitalization and connectivity of industrial processes, leading industries to embrace new paradigms and technologies to improve their operations and functionality [1]. Cyber-physical systems (CPSs) represent a keystone of the sustainable growth of economy, manufacturing and smart and connected communities, with a transversal application on almost all sectors [2].

Their usage, though, opens to a series of completely new disruption scenarios with potential disastrous impacts. In this context, a cyber security issue may not just refer to data or information leakage, but it may lead to a modification of the physical system causing real-world faults. Systems are becoming more software-centric, and more vulnerable to cyber attacks resulting in physical consequences, too [3]. The focus on the so-called cyber resilience is expected to deal with such perspective, where cyber resilience is defined as the ability of a system to absorb, adapt, and recover to adverse events caused by cyber-attacks [4]. Making a system cyber resilient will result in a better management and response in case a cyber attack occurs, not simply hardening the cybersecurity aspects. A common approach to study this type of problems rely on the usage

A. Rocha et al. (Eds.): WorldCIST 2023, LNNS 799, pp. 467–476, 2024.
https://doi.org/10.1007/978-3-031-45642-8_46

of simulation models to reproduce the system functioning in a fail-safe digital environment, e.g., [5–7]. The exchange of data between the physical system and the cyber system in any CPSs generates large amount of data that carry out useful information on both normal functioning and abnormal status, as affected by potential cyber attacks. Keeping track of changes that occur in system operations offers thus useful insights on the cyber resilience of the system itself. Accordingly, the aim of this paper is to present the results of a data-driven assessment of cyber resilience in an industrial plant. Starting from a 5-step methodology previously discussed in [8], this paper investigates simple metrics to assess system resilience. The methodology is presented in detail in Sect. 2. In Sect. 3 the case study to instantiate the methodology is presented, real data from a Seawater Desalination Plant testbed have been used for this sake. Section 3 also contains results and a discussion on them. In Sect. 4, conclusion and further developments and applications are discussed.

2 Methodology for Data-Driven Cyber Resilience Assessment

This paper relies on a step-by step methodology to assess cyber resilience based upon the availability of historical data about cyber attacks targeting the system under analysis (see Fig. 1). The five steps can be summarized as follows:

- Step 1. Identify problem. This phase is about identify the purpose of the analysis, the scope and the boundaries (i.e., what to be included, what to be considered out of scope). The problem statement should contain information about: the main objective of the assessment, information about the system – or the systems – of interest, an assumption about one or more disruption that occur in the system.
- Step 2. Describe system. Dealing with CPSs, the system description should span from the pure physical aspect of the processes involved, to the communication and controls elements (i.e., physical components, sensors, actuators, communication links, and control algorithms).
- Step 3. Retrieve data. Valuable data should be gathered from the system under analysis and the components identified in Step 2. A set of Extraction Transformation Loading (ETL) operations are usually needed to have data in a data mart that would be actually meaningful and representative of the problem at hand [9].
- Step 4. Define resilience metrics. The metrics being defined should account for absorption (attitude to tolerate a certain disturbance from the acceptable status), recovery (attitude to bounce back after a certain disturbance), adaptation (attitude to reach an acceptable status after a certain disturbance) [10].
- Step 5. Assess cyber resilience. This assessment should be grounded on the metrics already defined, which are dependent from the data available and accessible. The assessment can indeed be performed at different granularity levels, i.e., at overall level (system's outcome) or atomic level (specific components' outcome).

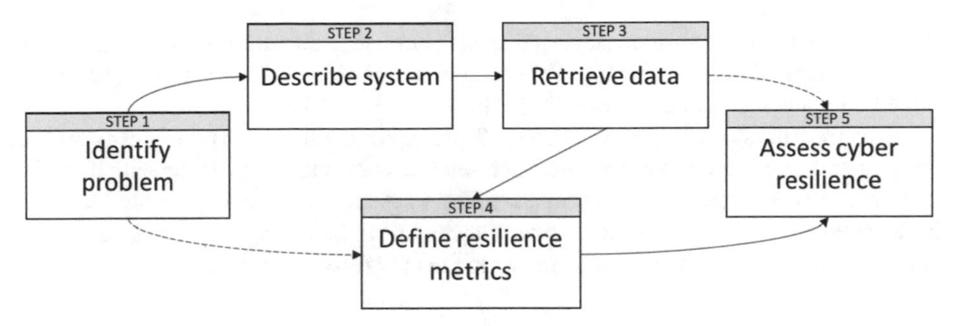

Fig. 1. Steps of the proposed methodology.

3 Results

The data-driven resilience assessment methodology presented in Sect. 2, has been instantiated through a case study referred to data from the iTrust Centre of Research in Cyber Security at the Singapore University of Technology. More specifically, Singapore University of Technology detains the Secure Water Treatment (SWaT) testbed [11], a scaled-down model of a real time water treatment plant designed for cyber security research. The purpose of SWaT testbed is to help the research community in investigating potential attack scenarios against industrial facilities – mainly, water treatment plants – and in developing recommendations for the protection of the critical infrastructures in general. This work focuses on the reverse osmosis process and on defining its resilience capacity against cyber attacks.

To proceed with the problem statement, it is at first necessary to set the boundaries of the analysis by describing the system and its component to be considered. The SWaT testbed represents a scaled-down version of a real water treatment plant [12], producing almost 20 [l/m] of filtered water through an ultrafiltration process made by the reverse osmosis membrane units. The reverse osmosis units make use of semi-permeable membranes that overcome the osmotic pressure allowing pure fluid to pass through while rejecting impurities. Thus, the fluid is passing from the more concentrated side to the less concentrated one. Two products are obtained: the permeate stream, consisting of water with low quantity of unwanted particles (mainly salts and suspended solids), and the brine, a reject stream of highly saline water. To force the passage through the reverse osmosis units (and to overcome the osmotic pressure of course), the fluid is pressurized by high pressure pumps up to $6 \div 8$ [MPa].

The process can be divided into six main phases:

- Intake of raw water
- Chemical disinfection
- Ultrafiltration
- Dechlorination
- Reverse osmosis
- Storage and backwash

Each stage is conducted by specific components which are controlled by couples of PLCs, to ensure the system functioning with a backup if the main controller fails. The storage of water at the various stages takes place in tanks, all equipped with sensors for the monitoring of the water level. An extended description of the six phases, and their linked components can be retrieved in [11].

Data from the system described in Sect. 3.2 have been collected in two main operation scenarios: system under normal condition, and system under attack. Specifically, the plant has been left operating in normal condition for seven days continuously, and then it has been targeted with different cyber attacks in the following four days. About the data used in the plant, the reader can refer to [11] for further information.

3.1 Base Scenario Analysis

The processes normal functioning condition is described by the set of data collected in the first seven days of operations. In this time frame, the system runs its operations always producing an acceptable output. Raw data have been stored in a data matrix *N*, containing normal condition data. Each component state value is reported for each time step of normal operations. The matrix contains the n_{ti} elements, and it has dimension *TN*x*I*, where: the $1 < t < TN$ represents the time step at which the normal operation data have been collected (*TN* is the total duration of normal condition operations that have been collected), C_1, C_2,..., C_i,..., C_I are the *I* components in the system, and the various n_{ti} are the values of data for each *i*-th component at each *t*-th time step.

$$N = n_{ti} = \begin{pmatrix} n_{11} & \cdots & n_{1I} \\ \vdots & \ddots & \vdots \\ n_{TN1} & \cdots & n_{TNI} \end{pmatrix} \tag{1}$$

Data in *N* are not static and constant, and some variations occurs in the system's stages even if the final output remains good. For this reason, the base scenario analysis has been made with a statistical approach by considering the frequency of each state (at component level) that have occurred in the normal operations time frame. The classification of an acceptable range of functioning for each component relies on the following values computed for each component C_i:

$$\mu_{Ci} = \frac{\sum_{t=1}^{TN} n_{ti}}{TN} \tag{2}$$

$$M_{Ci} = \max_{0 \le t \le TN} n_{ti} \tag{3}$$

$$m_{Ci} = \min_{0 \le t \le TN} n_{ti} \tag{4}$$

$$P_{Ci}^{x} = \vec{C}_i\left(p_{Ci}^{x}\right) : p_{Ci}^{x} = \frac{x \cdot (TN + 1)}{100} \tag{5}$$

where $\vec{C}_i$ is the ordered vector referred to C_i that contains the values n_{ti} sorted in an ascending way, and *x* is a number ranging from *1* to *100* representing the percentile to cover in P_{Ci}^{x}. In this case study two different values of P_{Ci}^{x} have been considered, specifically, for $x_1 = 25$ and $x_2 = 75$.

3.2 Attack Scenarios Analysis

Available data contains information about the system being stressed with 36 cyber attacks [11], to be differentiated depending on the assault point (i.e., the component being targeted). It is assumed that every attack attempt was successful and that only wireless links between sensors and corresponding PLCs could represent assault points. Accordingly, four types of cyber attacks can be highlighted:

- Single stage single point attacks. Attacks that focus on a specific single point of the system impacting only one stage of the process.
- Single stage multi points attacks. Attacks that focus on multiple points of the system impacting only one stage of the process.
- Multi stages single point attacks. Attacks that focus on a specific single point of the system impacting multiple stages of the process.
- Multi stages multi points attacks. Attacks that focus on multiple points of the system impacting multiple stages of the process.

The cyber attacks also vary by means of their duration. Also, the duration of system stabilization changes between different cyber attacks: some attacks had a stronger effect on system dynamics and took longer for the system to stabilize, while other attacks less time. Some attacks that will not have any visible deviation from system nominal functioning are present, too. An additional assumption is made at this point: the system always manage to regain an acceptable state (with respect to its normal condition functioning) between a cyber attack and the following. As for the normal conditions data, raw data have been stored in a disrupted condition – under attack – data matrix *D*. Each component's state value is reported for each time step of disrupted operations. The disrupted operations do not only involve the time steps in which a cyber attack is acting, but also the moments in which the cyber attack is not performing any operation but the system is still suffering a perturbation. Accordingly, all the data from the four days in which cyber attacks target the system has been stored in *D*, and the duration of each cyber attack goes from its starting time step, to the starting time step of the following attack. In *D*, each component state value is reported for each time step. The *D* matrix contains the d_{ti} elements, and it has dimension TNxI, where: the $1 < t < TN$ represents the time step at which the data have been collected (TN is the total duration of normal condition operations that have been collected), C_1, C_2,..., C_i,..., C_I are the I components in the system, and the various d_{ti} are the values of data for each i-th component at each t-th time step.

$$D = d_{ti} = \begin{pmatrix} d_{11} & \cdots & d_{1I} \\ \vdots & \ddots & \vdots \\ d_{TN1} & \cdots & d_{TNI} \end{pmatrix} \quad (6)$$

Additionally, the data related to each single cyber attack are retrieved from the SWaT database. Data are stored in the matrix containing attacks' specifications, namely *A*. This latter contains data in Table 1: for each j-th cyber attack, it is reported the start time ts_j, the end time te_j (moment in which the cyber attack is not active anymore, different from the moment in which the system regains an acceptable state), and a vector containing the component – or the components – under attack $\overline{C_{i_j}}$. Notice that each occurrence in the table is reported for $1 < j < J$ being J the total number of cyber attacks considered ($J = 36$ in this case), and so A is a matrix which has dimension Jx3.

Table 1. Table of attacks' specifications data, to be inserted in matrix A.

Attack	Start time	End time	Components
1	ts_1	te_1	$\overline{C_{i_1}}$
2	ts_2	te_2	$\overline{C_{i_2}}$
…	…	…	…
J	ts_J	te_J	$\overline{C_{i_J}}$

3.3 Assess Resilience Metrics

Three cyber resilience metrics are proposed to model absorption and recovery [13]. To delve into the metrics definitions, suppose to deal with a situation as in Fig. 2, where a time dependent system performance measure $P(t)$ is plotted over time t, following data collected at every Δt. The two horizontal dashed lines define the acceptable thresholds of performance (P_{max} and P_{min} in Fig. 2); every point laying outside those thresholds is representative of an anomalous performance (grey points in Fig. 2). The two vertical dashed lines define the period of interest for the analysis, which goes from t_0 to t_1 (cf. Fig. 2); every point that lays outside those boundaries is not taken into account in calculations.

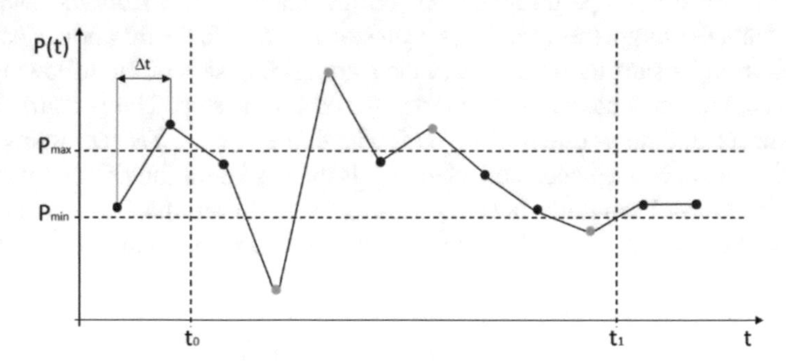

Fig. 2. Exemplary plot of a generic time performance $P(t)$.

On this basis, the absorption capacity must be related to the distance between the points outside the thresholds and the thresholds values themselves. Accordingly, this quantity $\delta(t)$ is:

$$\delta(t) = \begin{cases} |P(t) - P_{max}| & \forall t : P(t) > P_{max} \wedge t_0 \leq t \leq t_1 \\ |P(t) - P_{min}| & \forall t : P(t) < P_{min} \wedge t_0 \leq t \leq t_1 \\ 0 & \forall t : P_{min} \leq P(t) \leq P_{max} \wedge t_0 \leq t \leq t_1 \end{cases} \tag{7}$$

Thus, the first metric M_1 is defined as:

$$M_1 = \frac{\sum_{t=t_0}^{t_1} \delta(t)}{Np} \tag{8}$$

where Np is the total number of data points contained in the period of analysis. This metric ranges from 0 to $+\infty$, the higher it is, the lower the system ability is to absorb the disruption.

Concerning the recovery measure, it is possible to define the points outside the thresholds as:

$$P_{\text{out}}(t) = \begin{cases} 1 \ \forall t : P(t) < P_{min} \wedge P(t) > P_{max} \wedge t_0 \leq t \leq t_1 \\ 0 \quad \forall t : P_{min} \leq P(t) \leq P_{max} \wedge t_0 \leq t \leq t_1 \end{cases} \tag{9}$$

M_2 can be associated with the time the system needs to return into an acceptable state. Under the assumption that from t_0 to t_1 the system is able to completely restore its acceptable performance, the second metric (recovery) M_2 is defined as:

$$M_2 = \frac{\sum_{t=t_0}^{t_1} P_{\text{out}}(t)}{Np} \tag{10}$$

M_2 represents the number of points (i.e., consequently portion of time) in which the system is experiencing a non-acceptable performance, over the total number of points (or total time of analysis). This metric ranges between 0 to 1, the higher the metrics, the lower the system capacity to recovery its functioning.

Since different performance of the system targeted by cyber attacks are available, specifically, one for each d_{ti}, the values μ_{Ci}, M_{Ci}, m_{Ci}, P_{Ci}^{x}, can be used as different thresholds values in the sense of P_{max}, P_{min}. The fields ts_j and te_j identify the different ranges in which the metrics have to be calculated for each j-th cyber attack. Equations (7)–(13) can be re-written and computed accordingly.

At this stage, for each couple of cyber attack (j) and system component (i), a value of M_1 and M_2 is computed. Accordingly, it is possible to define two matrices containing these results as:

$$M1 = m1_{ji} = \begin{pmatrix} m1_{11} & \cdots & m1_{1I} \\ \vdots & \ddots & \vdots \\ m1_{J1} & \cdots & m1_{JI} \end{pmatrix} \tag{11}$$

$$M2 = m2_{ji} = \begin{pmatrix} m2_{11} & \cdots & m2_{1I} \\ \vdots & \ddots & \vdots \\ m2_{J1} & \cdots & m2_{JI} \end{pmatrix} \tag{12}$$

where J is the total number of cyber attack considered ($J = 36$ in this case study), and I is the total number of components in the system ($I = 52$ in this case study).

For clarity purposes, matrices are translated into heatmaps, which show every value in $M1$ and $M2$ as colored bits following a severity scale ranging from 0 to 1. Notice that $M1$ values have been divided by the highest distance δ registered for each component, to normalize results on a common scale of criticality. Figure 3 and Fig. 4 report respectively heatmaps for *M1* and *M2* considering three different threshold levels (i.e., the mean value μ_{Ci}, the 25th percentiles P_{Ci}^{x1} and 75th percentile P_{Ci}^{x2}, and the maximum values M_{Ci} and minimum values m_{Ci}).

The heatmaps can be read from two points of view. By entering data from a value of *j* it is possible to look at a single attack effect on the system evaluating the values of the metrics on the horizontal line. This perspective stresses a comprehensive approach to evaluate the response to the cyber attack. On the other hand, by entering data from a value of *i* it is possible to assess the cyber resilience of every single component of the system under different cyber attacks at a more atomic level. Both perspectives can give information about most critical settings once cyber attacks occur. In the first case, different cyber attacks strategies can be compared by means of the effects they have on the resilient performance of the system. This action permits pointing at the most critical typologies of attacks that have to be considered when managing system operations. At component level, the cyber resilience assessment permit to point at most critical components under the occurrence of cyber attacks, highlighting points of improvements and investments. For example, it can be noted from Fig. 3 that component 6 (P-102) is particularly critical looking at its absorption capacity, since looking at the best scenario (Pmax $= M_{Ci}$, Pmin $= m_{Ci}$) it turns out to be the component with lower *M1* values overall. The results suggest paying higher attention to a specific pump of the system as it register major changes in its operations when several different cyber attacks occur.

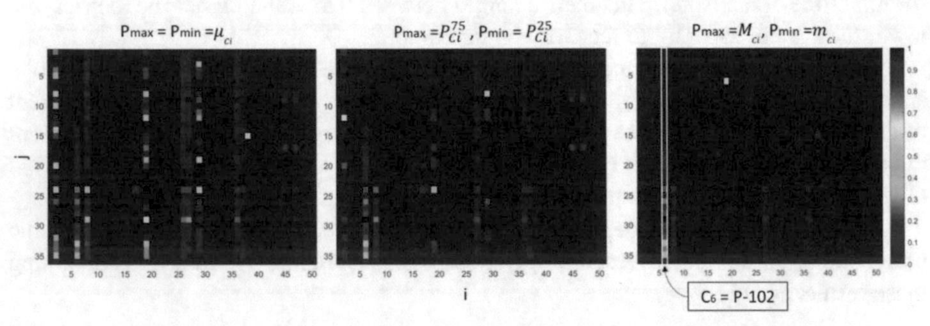

Fig. 3. Heatmaps presenting results for the absorption metric M_1 in the three different scenarios.

With reference to Fig. 4, it is clear how, by looking at the worst scenario (Pmax = Pmin $= \mu_{Ci}$) the last stages of the system are more prone to be affected by cyber attacks in terms of their recovery capacity. Component 8 (AIT-202), component 27 (AIT-401), and component 36 (AIT-501) are shown to be the most critical instead by looking at their recovery capacity since, even if considering the best scenario (Pmax $= M_{Ci}$, Pmin $= m_{Ci}$) they have among the highest impacts. Being these components mostly involved in water quality, it can be stated that the desalination plant has difficulties in the quick disposal of contaminant in water after a cyber attack occurrence. Particular response strategy might be implemented in this sense to overcome this limitation.

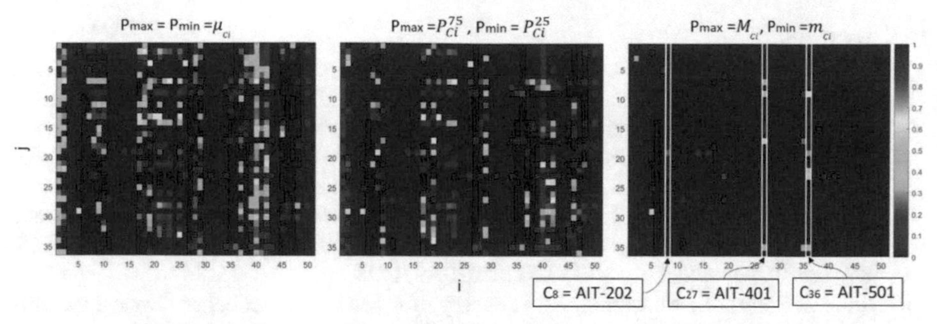

Fig. 4. Heatmaps presenting results for the recovery metric M_2 in the three different scenarios.

4 Conclusion

The complex structure of CPSs due to the combination of cyber, cyber-physical, and physical components, opens new challenges to assess overall system capacity to deal with cyber attacks. Operationalizing the notion of cyber resilience represents a solution to deal with this family of problems. For this purpose, a data-driven methodology to assess systems cyber resilience has been presented and instantiated through a case study related to a seawater desalination plant using SWaT testbed. The obtained results show the benefits of applying such methodology to calculate system cyber resilience and to support decision-making on how to deal with cyber attack. In this regard, water industry actors (management of purification treatment plant, or municipalities) might benefit of these results in a collaborative environment to ground decision prior to (e.g., prioritizing investments for critical components), during (e.g., drive risk communication for inhabitants), or following (e.g., define post-emergency procedures) a cyber attack. Considering the 2030 Sustainable Development Goals, this study contributes to (i) the "Industry, innovation and infrastructure" goal for the proposal of innovative operations strategies aimed at improving system resilience and prevention to cyber attacks, and to (ii) the "Sustainable cities and communities" goal for the significant technology transfer possibilities toward resilience and prevention of cyber attacks from industrial environments to societal services with impacts on communities. Even if the methodology has been presented in a general and reproducible way, its effectiveness in other applications must be proved. Further studies may address this verification. More resilience metrics can be defined and calculated to extend the resilience assessment. In this paper each point outside the thresholds has been considered to have the same weight as the others, without considering performance patterns and recovery trends during time [14]. Also, each cyber attack duration has been assumed to last from the beginning of the attack itself to the starting point of its following attack; a more precise strategy to evaluate the time range of analysis may be implemented to ensure wider application. Further studies may investigate these dimensions to enhance the methodology accordingly. Overall, the present work proves the potential of data-driven analysis to quantitatively assess complex systems cyber resilience, and suggests the benefit of further research in the area.

References

1. Sachdeva, N., Obheroi, R.K., Srivastava, A., Nehal, S.K.: Diffusion of industry 4.0 in manufacturing sector-an innovative framework. In: 2017 International Conference on Infocom Technologies and Unmanned Systems: Trends and Future Directions, ICTUS 2017, vol. 2018-January, pp. 1–5 (2018). https://doi.org/10.1109/ICTUS.2017.8286025
2. Zhang, B., Zhang, P., Vu, T., Chow, M.-Y.: Guest editorial: special section on resilience, reliability, and security in cyber-physical systems. IEEE Trans. Ind. Informatics **16**(7), 4865–4867 (2020). https://doi.org/10.1109/TII.2020.2971725
3. Duo, W., Zhou, M., Abusorrah, A.: A survey of cyber attacks on cyber physical systems: recent advances and challenges. IEEE/CAA J. Autom. Sin. **9**(5), 784–800 (2022). https://doi.org/10.1109/JAS.2022.105548
4. Björck, F., Henkel, M., Stirna, J., Zdravkovic, J.: Cyber resilience – fundamentals for a definition. In: Rocha, A., Correia, A.M., Costanzo, S., Reis, L.P. (eds.) New Contributions in Information Systems and Technologies. AISC, vol. 353, pp. 311–316. Springer, Cham (2015). https://doi.org/10.1007/978-3-319-16486-1_31
5. Simone, F., Patriarca, R.: A simulation-driven cyber resilience assessment for water treatment plants. In: Proceedings of the 32nd European Safety and Reliability Conference (2022)
6. Rashid, N., Wan, J., Quirós, G., Canedo, A., Faruque, M.A.A.: Modeling and simulation of cyberattacks for resilient cyber-physical systems. In: IEEE International Conference on Automation Science and Engineering, vol. 2017-Augus, pp. 988–993 (2017). https://doi.org/10.1109/COASE.2017.8256231
7. Md Haque, A., Shetty, S., Krishnappa, B.: Modeling cyber resilience for energy delivery systems using critical system functionality. In: Proceedings - 2019 Resilience Week, RWS 2019, pp. 33–41 (2019). https://doi.org/10.1109/RWS47064.2019.8971974
8. Patriarca, R., Simone, F., Di Gravio, G.: Modelling cyber resilience in a water treatment and distribution system. Reliab. Eng. Syst. Saf. (2022). https://doi.org/10.1016/j.ress.2022.108653
9. Nakhal, A.J., Patriarca, R., Di Gravio, G., Antonioni, G., Paltrinieri, N.: Business intelligence for the analysis of industrial accidents based on MHIDAS database. Chem. Eng. Trans. **86**(January), 229–234 (2021). https://doi.org/10.3303/CET2186039
10. Yarveisy, R., Gao, C., Khan, F.: A simple yet robust resilience assessment metrics. Reliab. Eng. Syst. Saf. **197**, 106810 (2020). https://doi.org/10.1016/j.ress.2020.106810
11. Goh, J., Adepu, S., Junejo, K.N., Mathur, A.: A dataset to support research in the design of secure water treatment systems. In: Havarneanu, G., Setola, R., Nassopoulos, H., Wolthusen, S. (eds.) CRITIS 2016. LNCS (LNAI and LNB), vol. 10242, pp. 88–99. Springer, Cham (2017). https://doi.org/10.1007/978-3-319-71368-7_8
12. Buros, O.K., Cox, R.B., Nusbaum, I., El-Nashar, A.M., Bakish, R.: The U.S.A.I.D. desalination manual : a planning tool for those considering the use of desalination to assist in the development of water resources. International Desalination and Environmental Association, Teaneck (1981)
13. Ayyub, B.M.: Practical resilience metrics for planning, design, and decision making. ASCE-ASME J. Risk Uncertain. Eng. Syst. Part A Civ. Eng. **1**(3) (2015). https://doi.org/10.1061/AJRUA6.0000826
14. Sung, D.C.L., Gauthama Raman, M.R., Mathur, A.P.: Design-knowledge in learning plant dynamics for detecting process anomalies in water treatment plants. Comput. Secur. **113** (2022). https://doi.org/10.1016/j.cose.2021.102532

Parallelization Algorithm for the Calculation of Typical Testors Based on YYC

Ariana Soria-Salgado, Julio Ibarra-Fiallo(✉), and Eduardo Alba-Cabrera

Colegio de Ciencias e Ingenierías, Universidad San Francisco de Quito, Cumbayá, Ecuador
{asoria,jibarra,ealba}@usfq.edu.ec
https://www.usfq.edu.ec

Abstract. In the present work, a new method is proposed to find typical testors, which helps reduce the number of features needed to carry out classification processes. Using the strategy of divide and conquer, we divide a basic matrix into blocks, then we find the typical testors of the defined blocks. To find the typical testors of the complete basic matrix, unions between the elements of these sets obtained from the blocks are tested. A criterion is developed to determine when the unions of typical testors of blocks form typical testors of the complete matrix. The performance of the method is evaluated using synthetic matrices. The execution time of the method in parallel and sequential versions was compared and contrasted with the YYC algorithm used for the complete basic matrix. Finally, its performance is analyzed in a real database obtained from the UCI Repository.

Keywords: pattern recognition · classification · typical testors · neural networks · accuracy · computational efficiency · time execution

1 Introduction

Testor theory is used to identify reduced sets of significant features to discriminate objects of different classes. Many algorithms have been developed to find the set of irreducible testors of a given basic matrix B [1–3]. The work of finding typical testors is a computationally intensive problem. Sometimes the search set becomes unmanageable; since in a set with N characteristics, the possible candidates to be typical testors are 2^N sets.

Although there are several algorithms, no one is considered the best for any type of basic matrix. The algorithm's efficiency usually depends on the characteristics of the basic matrix such as the density of ones, dimensions, number of testors, etc. In 2014, YYC Algorithm was introduced as a new computational strategy to find the complete set of typical testors of any basic matrix [3]. YYC is based on breaking down the calculation of typical testors up to some i row, instead of calculating the typical testors of the whole matrix at once.

A. Rocha et al. (Eds.): WorldCIST 2023, LNNS 799, pp. 477–489, 2024.
https://doi.org/10.1007/978-3-031-45642-8_47

This paper proposes a new computational method to find typical testors. Furthermore, our method involves parallel programming. The characteristic of the YYC algorithm to build typical testors row by row motivated the idea of using parallel computing. Since the calculation of typical testors can be done separately in a partition by rows of the basic matrix. The proposed method does not turn out to be the best in all cases. Experimentation allows us to conclude cases in which an improvement is visible in terms of computational efficiency. In addition, the method allows for future modifications that could increase its efficiency.

2 Testor Theory

Let U be the universe of objects for study, each object has p characteristics that define it. For each object of our universe, we have a set $P = \{x_1, x_2, x_3 \ldots, x_p\}$ of attributes, which is a p-dimensional variable. The objects of our universe can be classified into N disjoint classes, with $N > 1$.

Definition 1. *By comparing feature by feature each pair of objects belonging to different classes we can construct a Differentiation Matrix M_D.*

$$M_D = [m_{ij}]_{N \times p}, \qquad m_{ij} \in \{0, 1\} \tag{1}$$

If $m_{ij} = 0$, the pair of objects i are similar in the jth attribute. If $m_{ij} = 1$, the pair of objects i are different in the jth attribute.

Hence, we need to compare each pair of objects of different classes, feature by feature. If two objects in a feature j take the same value, in the differentiation matrix we put zero in this entry, otherwise we put one. We carry out this process until all possible pairs of objects of different classes have been compared.

Definition 2. *Let f and g be any two rows of M_D. We say that f is less than g if $\forall i$ $f_i \leq g_i$ and $\exists j$ such that $f_j \neq g_j$.*

Definition 3. *A row f is a basic row of M_D if there is no row $g \in M_D$ such that g is less than f.*

Definition 4. *The matrix M that only contains the basic rows of M_D and does not contain repeated rows is called the basic matrix of M_D.*

Definition 5. *A feature subset $T \subseteq P$, $T = \{x_{k_1}, x_{k_2}, x_{k_3} \ldots, x_{k_s}\}$ is a testor if and only if when all features are eliminated, except those in T, there is not any pair of similar sub-descriptions in different classes. Therefore, a testor is a feature subset, which allows complete differentiation of objects from different classes* [4].

If $T \subseteq P$, we define $M_{|T}$ as the matrix obtained from M by eliminating all the columns of M that do not belong to the set T.

In terms of M, we say that T is a testor if $M_{|T}$ does not have any rows of zeros. Now, we can define the concept of a typical testor [5]:

Definition 6. *A feature subset $T \subseteq P$ is a typical testor or irreducible testor if and only if T is a testor and there is no other testor τ such that $\tau \subset T$. This means that every feature in T is essential, so if we eliminated any, the resulting set is no longer a testor.*

Definition 7. *The attribute $j_{k_r} \in T$ is typical with respect to T and M if $\exists q$, $q \in \{1, 2, 3, \ldots, s\}$ such that $f_{i_q j_{k_r}} = 1$ and $\forall l \neq r$, $f_{i_q j_{k_l}} = 0$.*

Thus a set T has the typical property with respect to the matrix M if all its elements are typical attributes with respect to T and M.

Proposition 1. *A set $T = \{j_{k_1}, j_{k_2}, j_{k_3} \ldots, j_{k_s}\} \subseteq P$ has the typical property with respect to the matrix M if and only if we can obtain an identity matrix in $M_{|T}$ by exchanging and eliminating some rows* [6].

In terms of the basic matrix M, T is a typical testor of M if it is a testor and has the typical property with respect to M.

Proposition 2. *Let $\Psi(M_D)$ be the set of all typical testor of M_D and $\Psi(M)$ be the set of all typical testor of M, with M the basic matrix of M_D then*

$$\Psi(M_D) = \Psi(M). \tag{2}$$

By Proposition 2, we conclude that in terms of computational efficiency, it is better to work with M than with M_D because M has a less or equal number of rows than M_D [6].

Also, it is interesting to mention that there is a connection between testor theory and rough set theory. A testor is a super reduct and a typical testor is the equivalent of a reduct. So algorithms we could apply algorithms to find reducts and evaluate the performance of the proposed method with these variants [7,8].

3 Block's Algorithm Based on YYC

In this section, we present a new approach to obtaining typical testors of a basic matrix B. The strategy is inspired by the phrase "divide and conquer". Based on typical testors of blocks of B we will find typical testors of all of B. In addition to seeking a more efficient algorithm, we want to propose a computational method that can be parallelized.

Definition 8. *A block B of M is a submatrix of M consisting of all columns but only some rows.*

Proposition 3. *Let B be a basic matrix. Define B_1 as a block of B. Then B_1 is a smaller basic matrix.*

Proof. Because B is a basic matrix, it is composed only of basic (incomparable) rows. Thus if we choose just some rows of B but all of its columns, we end up with a smaller matrix that fulfills all the properties to be a basic matrix since it's formed only by basic rows.

3.1 Two Blocks of a Basic Matrix

Let's explore relationships between the testors of two different blocks of a basic matrix. Although we will only work with two blocks, in the case we have more, the ideas can be generalized.

Proposition 4. *Let B be a basic matrix. Suppose T_1 and T_2 are any typical testors of two different blocks of B, with $T_1 \neq T_2$ and $T_1 \cap T_2 \neq \emptyset$, then the intersections of typical testors of blocks will not form typical testors of B.*

Proof. Let $T = T_1 \cap T_2$. Without loss of generality assume there are one or more features that have been eliminated from T_1. Then since T_1 is a minimal set and by hypothesis ($T_1 \neq T_2$), if we eliminated any feature the resulting set is no longer a testor. Thus $B_{|T}$ will have one or more rows of zeros, i.e. T is not a testor of B.

Following a similar argument as above, the operation of difference of sets is not of our interest because it can eliminate essential features. From now on, we will focus only on typical testors of B that are the result of the union of typical testors of the blocks of B.

Remark: Given the properties of typical testors we can conclude that if B is a basic matrix, T_1 and T_2 be testors of B then $T_1 \cup T_2$ is a testor of B.

Proposition 5. *Let B be a basic matrix, B_1 and B_2 be blocks of B. If T is a typical testor of B_1 and B_2 then T is a typical testor of B.*

Proof. By the remark above, we know that T_1 is a testor of B since $T_1 \cup T_2 = T_1$. Because T_1 has the typical property in B_1 and B_2, then by Proposition 1 we can find an identity matrix in $B_{|T_1}$. Hence, T_1 has the typical property with respect to B and is a testor so T_1 must be a typical testor of B.

3.2 YYC in the Block's Approximation

The YYC algorithm has been designed to find the complete set of typical testors of a basic matrix, and since it builds typical testors up to a row i, it needs to reach the last row of the basic matrix to find its irreducible testors. This causes the algorithm to slow down if the number of rows is large. Also, since it uses the concept of finding compatible sets, the algorithm becomes slower as the number of compatible sets grows.

To deal with these disadvantages, we will apply the Block's Approximation. We will divide the basic matrix into two blocks. We will apply the YYC algorithm to each of its blocks to obtain the complete set of typical testors of the blocks and then using these two sets as inputs we will find the typical testors of the entire matrix.

3.3 Block's Approximation Algorithm

Once we divide the basic matrix into two blocks and we know the complete sets of typical testors of each block, we can test unions of elements of these sets to find irreducible testors.

We know that the union of typical testors of different blocks is a testor of the whole matrix. Thus, we only need to verify that the union has the typical property with respect to the entire basic matrix. To test whether the union results in a typical testor we can make use of Proposition 1 and formulate an extended version for our proposed method:

Proposition 6. *Let B be a basic matrix. Suppose T_1 and T_2 are any typical testors of two different blocks of B. The set $T = T_1 \cup T_2$ has the typical property with respect to the matrix B if and only if we can obtain an identity matrix in $B_{|T}$ by exchanging and eliminating some rows. Furthermore, T is a typical testor of B.*

Recall that mathematical unions of sets cannot guarantee the uniqueness results, we need to take this consideration into account to determine that each typical testor obtained by the union of typical testors of the blocks is unique in the final result of the blocks function.

3.4 Parallelization of Block's Algorithm

Technological innovation has advanced on a large scale in recent decades and one of the most recent areas of computational development is parallel programming. Currently, the existence of computers with multicore processors or hyper-threading makes parallel computing more economically feasible [10]. Although in everyday programming it is not so common to apply parallel programming for computationally intensive processes, its use is highly recommended. This is why one of the advantages of the way the proposed algorithm works is that it can be parallelized and improve its performance. For the Block approximation, the parallelization occurs in the phase in which the complete set of testors typical of the Blocks that have been defined is found. Instead of obtaining the complete set of typical testors sequentially, we can perform this task simultaneously by taking advantage of the available cores.

Due to the fact that this work is a first proposal of a more efficient algorithm, the proposed parallelization will serve for two blocks as the first stage with no order criterium in the rows of the blocks. The first step is to divide the basic matrix into Blocks. It is important to mention that the Blocks that are formed are disjoint and the number of rows that each block has can vary. For computational convenience, the rows are chosen in an orderly manner. The second step consists in obtaining the set of typical testors for each block. This part of the algorithm is going to be parallelized so that the calculations are done simultaneously. The last step consists of taking all the sets of typical testors obtained in Step 2 and testing unions of irreducible testors of these sets to obtain typical testors for the complete basic matrix.

Although it is true that this strategy guarantees that we will find the typical checkers of the basic matrix and in some cases, this process will be more efficient in terms of time, there is no guarantee that they will all be found. However, it is possible to carry out classification processes without having to know the complete set of typical testors.

4 Experiments and Application

It is necessary to mention that the codes used for this work are implemented in the *Matlab* 2022*a* platform and the experiments carried out were executed on a MacBook Pro (13-in., M1, 2020) personal computer with macOS Monterey operating system and memory of 8 GB.

4.1 Synthetic Basic Matrices

To test the proposed method, we construct a set of synthetic basic matrices that will help us to make the conclusions of this work. The advantage of using synthetic basic matrices and operators is that we can know prior the exact number of typical testors [6,9]. So is a good practice to test algorithms with a set of synthetics matrices and this practice has become more common in recent years.

The sequential version of the algorithm was applied to two different partitions in order to be able to conclude if the way in which we build the blocks is relevant or not. For the first partition, we took the top half of the rows for *Block* 1 and the bottom half of the rows for *Block* 2. For Partition 2, a proportion of 1/4 of the top rows of the complete basic matrix will be used for *Block* 1 and the rest of the rows for *Block* 2.

For each partition, we recorded the time it takes for the YYC algorithm to obtain the typical testors of *Block* 1 and then the time it takes to find the typical testors of *Block* 2. Then these sets will be the arguments for the block algorithm, it is also necessary to register the time it takes to execute. Experimentation showed that there is no definite pattern that makes one of the partitions better than the other. Experimental evidence is not included in this work due to length. We carried out experiments on around 80 basic synthetic matrices but in this work, we will present the results of representative matrices.

Operator θ Applied to I_5: The basic matrix I_5 has a dimension 5×5 with a density of ones of 0.2. This matrix has a single typical testor of cardinality 5. When we apply the θ-Operator recursively to the matrix I_5, the following experimental results are obtained (Table 1):

Table 1. Results parallelization for I_5 using the θ operator.

$\theta^N(I_5)$	Rows	Columns	Density	TT	YYC	Sequential	Parallel
1	5	5	0.200000	1	0.003416	0.031299	0.089988
2	25	10	0.080000	2	0.059171	0.076950	0.113450
3	125	15	0.024000	3	0.055460	0.094993	0.116216
4	625	20	0.006400	4	0.288344	0.291037	0.239069
5	3125	25	0.001600	5	4.557188	3.835996	2.545781
6	15625	30	0.000384	6	425.640116	143.165474	52.509040

A comparative graph of the efficiency in time of the algorithms is shown (Fig. 1):

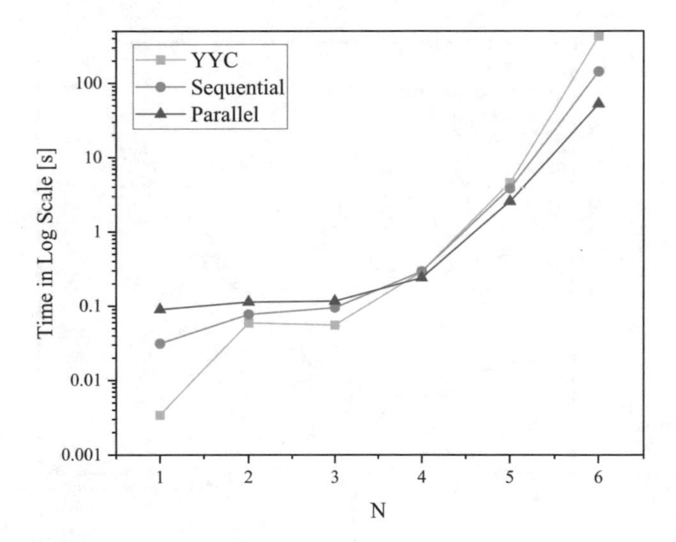

Fig. 1. Parallelization Results for I_5 using the θ operator.

Operator θ Applied to A: The basic matrix A has dimension 4×5 with a density of ones of $0,55$. This matrix has 5 typical testors, 4 of cardinality 2 and one of cardinality 3. When the operator θ is applied recursively to A and we tested the study algorithms, the following execution times dependent on N are obtained (Table 2):

Table 2. Results parallelization for A using the θ operator.

$\theta^N(A)$	Rows	Columns	Density	TT	YYC	Sequential	Parallel
1	4	5	0.5500	5	0.02955	0.16035	0.20046
2	16	10	0.3438	10	0.03722	0.08056	0.12627
3	64	15	0.1289	15	0.04003	0.06671	0.12347
4	256	20	0.0430	20	0.16886	0.22805	0.18971
5	1024	25	0.0134	25	1.98210	1.49374	0.99190
6	4096	30	0.0040	30	35.94585	22.66953	14.12854
7	16384	35	0.0012	35	762.99113	445.54098	276.71959

Next, a comparative graph of the performance of the algorithms (Fig. 2):

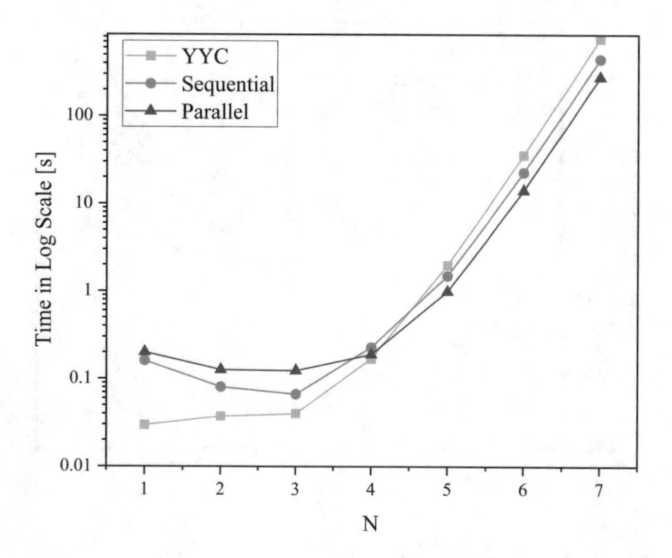

Fig. 2. Parallelization Results for A using the θ operator.

Operator θ Applied to B: The basic matrix B has dimension 4×5 with a density of ones of $0,50$. This matrix has 5 typical testors, three testors of cardinality 2 and two of cardinality 3. The experimental results obtained for B matrix when θ operator is applied (Table 3):

Table 3. Results parallelization for B using the θ operator.

$\theta^N(B)$	Rows	Columns	Density	TT	YYC	Sequential	Parallel
1	4	5	0.5000	5	0.004750	0.030756	0.078621
2	16	10	0.3125	10	0.008634	0.033356	0.070697
3	64	15	0.1172	15	0.066651	0.092834	0.108287
4	256	20	0.0391	20	0.552101	0.391459	0.293576
5	1024	25	0.0122	25	11.563568	5.175375	3.289233
6	4096	30	0.0037	30	303.740156	121.208505	77.532677

The success cases turn out to be when the matrix has a large number of rows, or is generally large in dimension, but has relatively few typical testors or the goal is not to find the complete set of irreducible testors. Regarding the parallelization of the proposed method, it can be concluded that it is successful when the dimension of the matrix is large. The proposed method manages to obtain the complete set of typical testors in all the experiments made.

In general, the efficiency of the Block's algorithm drops considerably when the number of unions of typical testors to be tested is very large. This slowdown of the method cannot be solved by using parallel programming but could be avoided by not using the full set of typical testors for the blocks.

The following is a comparative graph of the performance of the algorithms (Fig. 3):

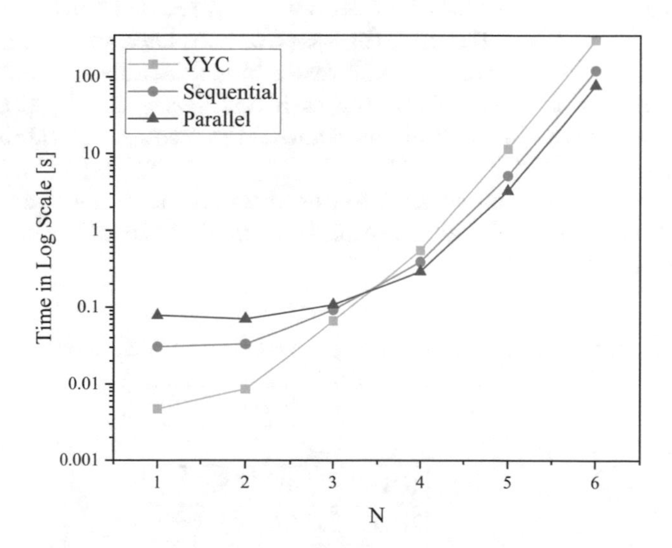

Fig. 3. Parallelization Results for B using the θ operator.

Also as part of future work, it would be interesting to use other types of algorithms to find typical testors in the Blocks. Depending on the characteristics of the basic matrices, the most efficient algorithm for that case could be chosen and used in Step 1 of the Block's algorithm. The rest of the method process would remain the same. With this, we want to note that this proposed method can be used with algorithms other than YYC and could be beneficial in computational terms.

4.2 Application to QSAR Biodegradation Data Set

To test the performance of the algorithm proposed in this work in application problems, a dataset from the UCI Repository [11] will be used. This data set was built in the Milano Chemometrics and QSAR Research Group. They study chemical stricture and biodegradation molecules [12].

First, the classification neural network was designed using the full number of base features. Of the 1055 instances, 500 were used for the training process and 555 for testing. Because the base is not balanced, 250 random items from each class were taken for the training process. The remaining elements of each class were part of the test set. The basic matrix obtained in this case had a dimension of 706×41.

In the training stage, the accuracy of the neural network was $90,00\%$. What turns out to be acceptable considering that the base is not equitable in the number of elements per class. For the test stage, the neural network had 85.77% accuracy. Below is the confusion matrix obtained for the neural network using all the features:

The percentage of elements of the RB class and NRB class that are misclassified as NRB is due to the fact that the base does not have an equal distribution of elements of classes. As they are 356 ready biodegradable molecules and 699 non-ready. But this is a risk of working with real bases, that a priori we don't know if it will be possible to obtain an equitable percentage of training for each class.

This preamble helps us to analyze the database and the neural network. Now the proposed method and the results obtained for classification using only selected features are applied.

Table 4. Confusion Matrix for the model trained with all attributes

		Predicted	
		RB	NRB
Actual	RB	87,74%	12,26%
	NRB	14,70%	85,30%

Below is a table that records the execution time (measured in seconds) to find the typical testors of the basic matrix for the biodegradable base QSAR [12] (Table 5):

Table 5. Executions times

Algorithm	Total Execution Time
YYC	18.8944
Sequential Block's algorithm	10.6227
Parallel Block's algorithm	7.9238

Once the typical testors for this base were obtained, the same neural network used above was trained using only the characteristics belonging to one single testor of length six. We preserved the neural network structure that allowed us to obtain the results shown in Table 4.

Out of a total of 41 characteristics, the new classification model only needs 6 to be able to predict. In other words, we reduce by approximately 85% the number of characteristics necessary to know if a molecule is ready or not ready biodegradable. The features selected for the neural network are the number of heavy atoms, frequency of C-N at topological distance 3, frequency of C-O at topological distance 3, number of CRX3, leading eigenvalue from adjacency matrix (Lovasz-Pelikan index), and second Mohar index from Laplace matrix.

It is worth mentioning that to obtain a detailed list of all the characteristics, you can go to the reference [12]. In general, the new model for the training set obtains an accuracy of 84,20%. For the test set, we obtained 81,57% of accuracy. Below is the confusion matrix obtained after training the neural network only with the selected features using Testor Theory (Table 6):

Table 6. Confusion Matrix for the model trained with only selected features

		Predicted	
		RB	NRB
Actual	RB	88,68%	11,32%
	NRB	19,60%	80,40%

Although we are still affected by the inequality of elements by class of the base, the percentage of well-classified for RB increases. This is favorable since it is the class with fewer members. On the other hand, the number of well-classified elements belonging to the NRB class has been reduced by about 5%. This is not entirely inefficient considering that we only used 6 of 41 variables.

5 Conclusions and Further Work

One of the advantages of the proposed method, but still to be explored, is that when is applied to real bases the number of typical testors that to find may be intentionally limited. In real application problems, it is not necessary to find the complete set of typical testors. For example, in classification problems, it is enough to find one typical testor that efficiently classifies the classes and thus a neural network can be trained more efficiently without using the full set of features.

In future work, the idea would be to try defining more than two blocks for large arrays. And then test the unions. In this case, applying parallelization could be even more profitable for the method. Also, will be interesting to use algorithms from rough set theory. We could apply algorithms to find minimal reducts and evaluate the performance of the proposed method with these variants. For future work, it could be interesting to investigate if some order criterium to the rows could help improve time efficiency.

References

1. Lias-Rodríguez, A., Pons-Porrata, A.: BR: a new method for computing all typical testors. In: Bayro-Corrochano, E., Eklundh, J.-O. (eds.) CIARP 2009. LNCS, vol. 5856, pp. 433–440. Springer, Heidelberg (2009). https://doi.org/10.1007/978-3-642-10268-4_50
2. Santiesteban-Algaza, Y., Pons-Porrata, A.: LEX: a new algorithm for the calculus of all typical testors, vol. 1, pp. 85–95 (2003)
3. Alba-Cabrera, E., Ibarra-Fiallo, J., Godoy-Calderon, S., Cervantes-Alonso, F.: YYC: a fast performance incremental algorithm for finding typical testors. In: Bayro-Corrochano, E., Hancock, E. (eds.) CIARP 2014. LNCS, vol. 8827, pp. 416–423. Springer, Cham (2014). https://doi.org/10.1007/978-3-319-12568-8_51
4. Martínez, J.F., Santos, J.A., Carrasco, A.: Feature selection using typical testors applied to estimation of stellar parameters. Comput. Sist. **8**(1), 15–23 (2004). https://www.redalyc.org/articulo.oa?id=61580103. ISSN 1405-5546
5. Muenala, K., Ibarra-Fiallo, J., Intriago-Pazmiño, M.: Study of the number recognition algorithms efficiency after a reduction of the characteristic space using typical testors. In: Rocha, Á., Adeli, H., Reis, L.P., Costanzo, S. (eds.) WorldCIST'19 2019. AISC, vol. 930, pp. 875–885. Springer, Cham (2019). https://doi.org/10.1007/978-3-030-16181-1_82
6. Alba-Cabrera, E., Godoy-Calderon, S., Lazo-Cortés, M.S., Martinez-Trinidad, J.F., Carrasco-Ochoa, J.A.: On the relation between the concepts of irreducible testor and minimal transversal. IEEE Access **7**, 82809–82816 (2019)
7. Lazo-Cortés, M., Martínez-Trinidad, J.F., Carrasco-Ochoa, J., Sanchez, G.: On the relation between rough set reducts and typical testors. Inf. Sci. **294**, 152–163 (2015)
8. Torres-Constante, E., Ibarra-Fiallo, J., Intriago-Pazmiño, M.: A new approach for optimal selection of features for classification based on rough sets, evolution and neural networks. In: Arai, K. (ed.) IntelliSys 2022. LNNS, vol. 542, pp. 211–225. Springer, Cham (2023). https://doi.org/10.1007/978-3-031-16072-1_16

9. Alba, E., Ibarra, J., Godoy, S.: Generating synthetic test matrices as a benchmark for the computational behavior of typical testor-finding algorithms. Pattern Recogn. Lett. **80**, 46–51 (2016)
10. Rauber, T., Rünger, G.: Parallel Programming, pp. 169–226. Springer, Berlin, Germany (2013). https://doi.org/10.1007/978-3-642-37801-0
11. Bache, K., Lichman, M.: UCI machine learning repository (2013). https://archive.ics.uci.edu/ml/index.php
12. Mansouri, K., Ringsted, T., Ballabio, D., Todeschini, R., Consonni, V.: Quantitative structure-activity relationship models for ready biodegradability of chemicals. J. Chem. Inf. Model. **53**, 867–878 (2013). https://archive.ics.uci.edu/ml/datasets/QSAR+biodegradation

AI Data Analysis and SOM for the Monitoring and Improvement of Quality in Rolled Steel Bars

Marco Vannucci[1](✉), Valentina Colla[1], and Alberto Giacomini[2]

[1] Scuola Superiore Sant'Anna, Istituto TeCIP, Pisa, Italy
{marco.vannucci,valentina.colla}@santannapisa.it
[2] Danieli Automation, Udine, Italy
a.giacomini@dca.it

Abstract. In the steel sector, within the production of high value hot rolled long products, assessment of components health and compliance with quality targets are fundamental aspects that need to be ensured. In this paper, an AI–based system for monitoring the conditions of the rolling process of bars with round section and estimating the final bars ovality is presented. The developed system, based on a Self Organising Map, is trained and evaluated by using data coming from a real plant and pre–processed by means of advanced AI techniques. The system allows monitoring a wide range of process variables affecting the ovality issue and actively supports plant operators in the task of avoiding quality–critical conditions and possible machine faults.

Keywords: steel bars · self organizing map · variables selection

1 Introduction

Steel bars manufacturing is an extremely challenging process due to tight quality standards to be met by producers to be competitive within the global market, also considering that the achievement of these requirements need to be pursued for a wide variety of products in terms of shapes, types of material and mechanical properties. On the other hand, productivity is a further fundamental factor, whose maximisation is targeted by steel companies together with achievement of high quality standards. Continuous monitoring of process conditions for assessing machinery health status is fundamental to ensure both productivity and compliance with quality standards [2,13]. Unlucky, bars rolling is a harsh task implying relevant stress conditions for many components of the machines that are part of the manufacturing chain. In fact, the continuous work on the rolls and the friction with the semi-finished bars wear the rolls and strain the engines. In the case of bars with round section, this factors can alter the shape of the section by deviating from the desired circular shape and giving rise to so–called *ovality* issue. In this context, prompt identification of deviations from quality requirements enables implementation of suitable countermeasures (e.g. replacement of worn rolls) to avoid quality deviations and minimize machine downtimes.

A. Rocha et al. (Eds.): WorldCIST 2023, LNNS 799, pp. 490–499, 2024.
https://doi.org/10.1007/978-3-031-45642-8_48

In this work, Artificial Intelligence (AI)-base data analytics and Self Organizing Maps (SOM) are adopted for monitoring and controlling the hot rolling process of round steel bars. The goal of the developed system is twofold: on one hand, it enables early identification of situations that might lead to quality downgrading and machinery faults by analysing process conditions and machinery ageing status. On the other hand, it supports plant managers in efficiently changing process parameters when critical situations arise. The system operates on an advanced hot rolling machine, fully equipped with sensors and an IT framework for data collection and management.

The paper is organised as follows: Sect. 2 presents the industrial set–up, with an overview of the data gathered for the modelling purpose of this work. The pre–processing techniques adopted to prepare the data to be fed to AI–models are described in Sect. 3. In Sect. 4 the system for bar quality monitoring and process operators support is described in detail, including its main achievement in the light of the improvement of product quality and process stability. Finally Sect. 5 provides some concluding consideration and ideas for future developments.

2 Industrial Context and Set–up

The work described in this paper refers to the production of round bars in a Special Bar Quality (SBQ) rolling mill. The equipment under consideration is the so–called *Draw Sizing Danieli (DSD)* produced by the Italian company Danieli. Differently from conventional SBQs, where the multiple roll-pass sequence feeding the sizing stands requires several mill changes decreasing the mill productivity, the DSD consists of a sturdy module with 4 consecutive rolling stands: two 2-rolls high-reduction modules and two 4-rolls low-reduction modules for precise sizing. The high-reduction stands make it possible to adopt a single pass sequence up to the sizing block for the whole product mix, while the 4-rolls stands keep the product within the required geometric tolerance.

Round bars manufacturing is subject to strict quality requirements in terms of *ovality*. An ovality index Ov is defined as the difference between the greatest diameter and the smallest diameter in the same bar section (orthogonal to the main axis). Ideally it should be $Ov = 0$, in practice Ov shall not exceed 75% of the tolerance range for the nominal diameter. Nowadays, the fulfilment of the base requirement of the international standard is considered insufficient: steel producers want to achieve dimension tolerances that are a fraction (from 1/2 down to 1/8) of the admissible ranges. According to this consideration and to the industrial practice, the ovality threshold is generally set to 0.375 mm for all the bars whose nominal diameter is in the range 36 mm–80 mm.

2.1 Data Collection and Management Framework

In order to implement Industry 4.0 practices, the machine is equipped with a broad set of updated sensors that can continuously collect information about the rolled bars and the main process parameters that are stored in a dedicated database. In addition, the DSD is integrated in a plant characterised by a modern design and automation solution, which integrates the above mentioned system with information coming from different processes or levels to implement an efficient and optimized maintenance approach. Due to the high productivity standard and the large amount of data produced during the manufacturing, an IT framework was designed to collect and manage data coming from different sources to be subsequently fed and processed by a centralized monitoring system.

The data are collected from two main sources: the Programmable Logic Controller (PLC) used for the Level–1 automation of the DSD and the Level–2. From the PLC a variety of data is collected, which are acquired from sensors and devices used to operate the DSD and check its working conditions. In particular, the data considered in the present analysis concern the bar ovality measured by the *HiProfile* gauge, a high-precision laser-based sensor used to measure bar diameters. Level–2 source, among all production data, stores information about the status of the machine, the type of the manufactured bar, starting and final dimensions, machine setup. Here information about the ageing of the rolls located in the different stands of the DSD is stored. Such information is in the form of the total manufactured tons of steel by each couple of rolls since their installation until the end–of–life due to significant wear of the roll itself.

The data acquisition system is part of a bigger framework for big data ingestion, ML-based software development and deployment and information retrieval through a Digital Twin, developed under the RFCS project named *CyberMan4.0*.

2.2 Available Data Description

This work is based on the data collected by the previously described IT framework throughout two months of standard production. These data were used for the development of a process monitoring system with functionality of decision support system (DSS) to help avoiding products ovality and machinery faults. The dataset includes PLC, Level–2 data and refers to about 13 thousand bars whose diameter lies in the range 52 mm–80 mm. The ovality of observed bars varies in the range 0 mm–2 mm, which is representative of the usual mill performance. The dataset in its original form includes more than 90 variables, most of them coming from the PLC. Such variables represent, among the others, inter–rolls gaps, engine speeds (in RPM), temperatures measured throughout the whole length of the manufactured bars. Within the dataset, for each bar several features were calculated for each individual bar starting from the original variables measures. This operation aimed at extracting meaningful information from the data without loss of informative content. Extracted features include: median (MED), standard deviation (STD) and variability range (SPAN) of the original measure for the bar. This latter operation rises the number of observed features up to more than 250. In addition to PLC information, there are the variables coming from the Level–2 database that include steel type, nominal bar diameter and, for each DSD gauge, amount of steel processed in tons.

3 AI–Based Data Pre–processing

The framework depicted in Sect. 2.2 provides a large amount of data for the subsequent modelling purposes in terms of both number of features and observations. The aim of models and modelling tools is to link process conditions and rolls wear status to products ovality. In this context, data pre–processing is fundamental to achieve satisfactory results since, for instance, not all the collected features are connected to ovality, and not all the information collected by sensors is reliable. In this work, data preparation is pursued by exploiting both the experience of plant personnel and AI–based techniques for data filtering and selection.

3.1 Outliers Detection and Removal Through AI Techniques

In the industrial field, when collecting data from sensors, it is always possible to handle *outliers*, namely measures that markedly deviate from other observations and are thus suspected to be generated by a different mechanism or, in many cases, are due to sensors malfunctions or changes in the measuring conditions (i.e. process conditions, product status, ..). This consideration is particularly true in the steel industry, which is characterized by harsh environmental conditions and high productivity rates. The detrimental effect of outliers when developing data–driven models is well known [7]: outliers introduce uncertainty and noise in the training dataset, reducing the predictive performance of the model and robustness. For this reason an outliers detection and removal step was pursued.

In this work the FUCOD [3] method was used. It mixes through a Fuzzy Inference System (FIS) different approaches to outliers identification to exploit their well known strengths and avoid their weaknesses in presence of different kinds of outliers. FUCOD works with multidimensional data so as to determine outliers by considering the interactions and joint distributions of all the dimensions of the data. These characteristics make FUCOD particularly suitable to filter industrial dataset, such as demonstrated in the numerous applications, even within the steel industry [8,10]. In addition to the FUCOD method, other approaches have been tested on the available data. The *z–score* for individual variables was calculated but this approach was not able to detect any outlier, likely due to its mono–dimensional nature. On the other hand, the Local–Outlier–Factor (LOF) algorithm - that belongs to the family of multi–dimensional methods - was applied: it led results similar to FUCOD but it required a substantially higher computational time. The application of the FUCOD method to the available data led to the identification of 6% of observations suspected to be outliers. Considering this low figure, no data imputation technique was applied to fix individual observations, which were removed from the dataset.

3.2 Variables Selection

Variables selection is performed to improve the quality of the training dataset. This step is efficient in presence of large datasets, formed by numerous variables that can even be not or very weakly related to the target of the modelling task. In this latter case uncorrelated variables do not bring any additional informative content to the model and can reduce its accuracy. In addition, in order to mitigate such effect, the complexity of models might unnecessarily grow, lowering its generalization capabilities [6]. This problem is common to both supervised and unsupervised methods and has already been analysed in [4] and especially in [5] for the SOM, which is the type of neural network adopted in this work. Here a two stage variables selection approach was adopted: the first one based on the suggestions of expert operators working on the plant, the subsequent based on AI techniques. Referring to the above described dataset, operators recommended to keep the information coming from Level–2 database related to product characteristics (i.e. steel grade) and the tons of product manufactured by each of the 4 gauges of the DSD (named TONS_PASS_13 to TONS_PASS_16 respectively in the following) which are expected to be strictly related to rolls ageing problems and ovality issues. Then, within the remaining PLC features, experts filtered out those that have no relation with ovality reducing the number of this set to 56 features that include stands gaps, temperatures, rotation speeds and torques for the 4 stands (DSD_1 to DSD_4) of the DSD.

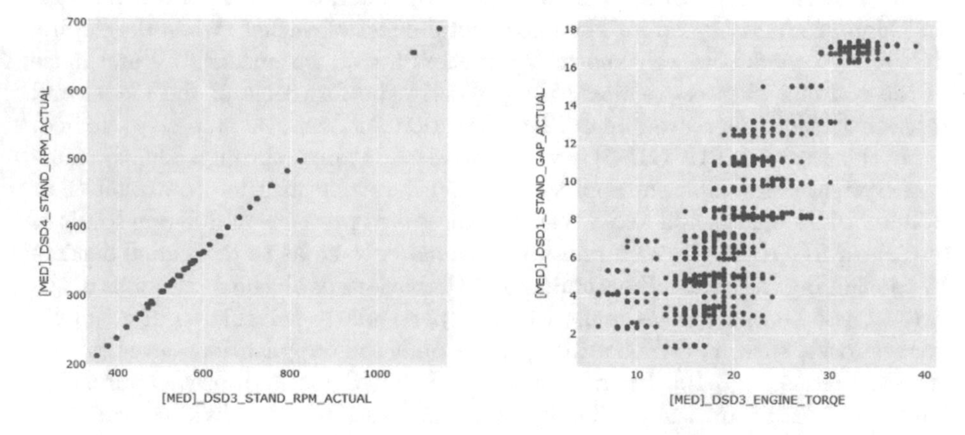

Fig. 1. Related variables from PLC data. On the left, the Pearson correlation is 0.97 ($p-value = 0.012$) thus the two variables are considered *redundant* whilst on the right the two variables, although related, are not (correlation is 0.79).

Many of these variables are strictly related to each other. In order to cut *redundant* variables a specific algorithm was adopted based on the dominating sets theory. In this method, features V are organized into a graph G, where vertex in V are connected each others through edges according to an arbitrary symmetric measure of similarity. Once this set of graphs is built, the algorithm looks for the *minimum dominating set* (MDS) of variables that *covers* V either by including an arbitrary node v or another node adjacent to it. More formally, for a graph G and a subset S of the vertex set $V(G)$, let us denote with $N_G[S]$ the set of vertices in G that are in S or adjacent to a vertex in S. If $N_G[S] = V(G)$, S is said to be a dominating set (of vertices in G). A dominating set of smallest size is called a *minimum dominating set* as sketched in Fig. 2. In this context, the identification of the MDS allows the selection of the minimum set of variables *related* to the whole set, minimizing the information loss.

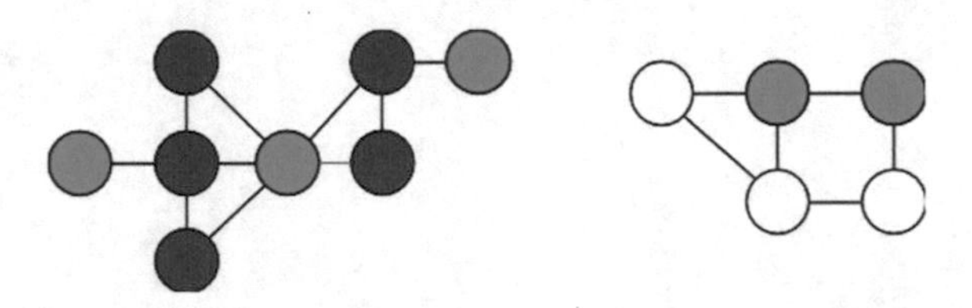

Fig. 2. Representation of minimum dominating set in green for a generic graph. The substitution of green vertices in the left side isolated sub–graph with the red ones leads to a dominating set which is not minimum.

In this application two variables are considered redundant on the basis of Pearson's linear correlation coefficient. More in detail, if such measure is higher in absolute value than a predetermined threshold set to 0.95 and the corresponding significance value (*p–value*) is lower than 0.05, then two variables are considered redundant. This condition is very strict and aims at removing only clearly redundant variables, such as highlighted in Fig. 1. Since the problem of the determination of the *minimum dominating set* is computationally NP-hard, an iterative algorithm leading to a sub–optimal solution was adopted to determine a (quasi) minimal set of non–redundant variables [9]. The application of this method led to the further reduction of the potential input variables to 35. The final step of the variables selection is supervised and involves the target of the model to better identify those variables mostly related to it. The *RReliefF* algorithm was employed [12]. The core idea behind this algorithms is the ranking of input variables on the basis of how well they can *distinguish* between instances that are near to each other. The main advantage of this method is that it does not make any independence assumption among the input variables and that it is able to catch dependencies between attributes. The outcome of RReliefF is a score associate to each variable that determines a ranking. The RReliefF algorithm was applied to the potential inputs dataset. The achieved scores are shown in Fig. 3. The associate ranking is used in the design phase of the SOM–based model described in next section.

4 SOM for Quality Monitornig and Process Control

For the monitoring and decision support purpose of this work, a SOM was used. SOM is a particular neural network that performs unsupervised learning and is widely used for clustering and visualization tasks. SOMs can map highly dimensional spaces into a lower dimensional space (two-dimensional, in the present

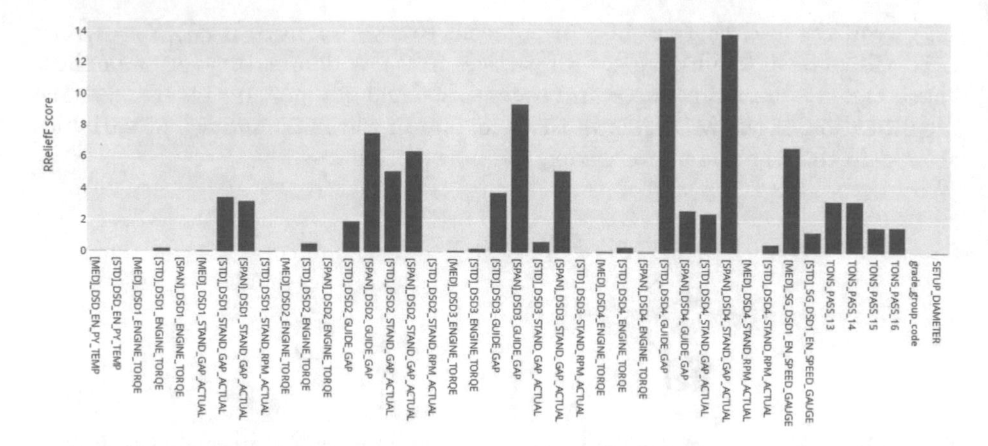

Fig. 3. RReliefF score obtained by the variables included in the input dataset resulting from the previous variables selection steps.

case), which can be easily interpreted by human operators [1]. The resulting clusters group observations with similar operating conditions and can be labelled by using an additional target indicator (ovality here). SOM clustering, differently from other unsupervised learning approaches, preserves the original distribution and topology of training data: it means that a higher number of clusters depicts the frequent operating conditions, resulting in a higher granularity for the monitoring system, where the process is actually more often operated (distribution preservation). Furthermore, topology preservation grants that observations with similar process conditions are associated to the same cluster or to a neighboring one, taking into account product category and steel grade.

4.1 Design and Training of the SOM Model

The SOM was trained by using the dataset described in Sect. 2.2. Data were standardized to avoid the SOM to be biased toward higher magnitude variables. Several sets of input variables with different number of variables were tested, picking (in order) from the ranking depicted in Fig. 3. The goodness of each set was assessed according to the trained SOM quantization error, the average distance of samples from closest cluster centroid (aka Best Matching Unit, BMU) to select the best performing one. According to the results shown in Tab. 1 the selected SOM has 16 inputs. The achieved low quantization error highlights the very good quality of the obtained clustering.

The optimal dimension of the SOM grid was selected by using the *Silhouette* method [11], which assesses the goodness of a clustering by simultaneously taking into account the cohesion of single clusters and their inter–clusters level of separation. The outcome of the silhouette test by varying the *side* dimension of the square SOM in a range 3–15 led to the selection of a 7×7 topology. For the training, the σ parameters that controls the neighborhood radius around

Table 1. Quantization error (QE) calculated for different input variables number.

Num. variables	Quantization error
8	0.203
10	0.095
12	0.095
16	0.089
20	0.122

the BMU was set to 0.5 and the learning rate was set to 0.3. Once the training procedure is complete and the clusters are determined, the associated ovality reference values are computed. More in detail, to each cluster the average ovality of the data samples associated to the cluster is assigned. The final outcome of the SOM training is depicted in Fig. 4 that shows the percent number of hits of the different clusters (*a*) and the ovality (*b*).

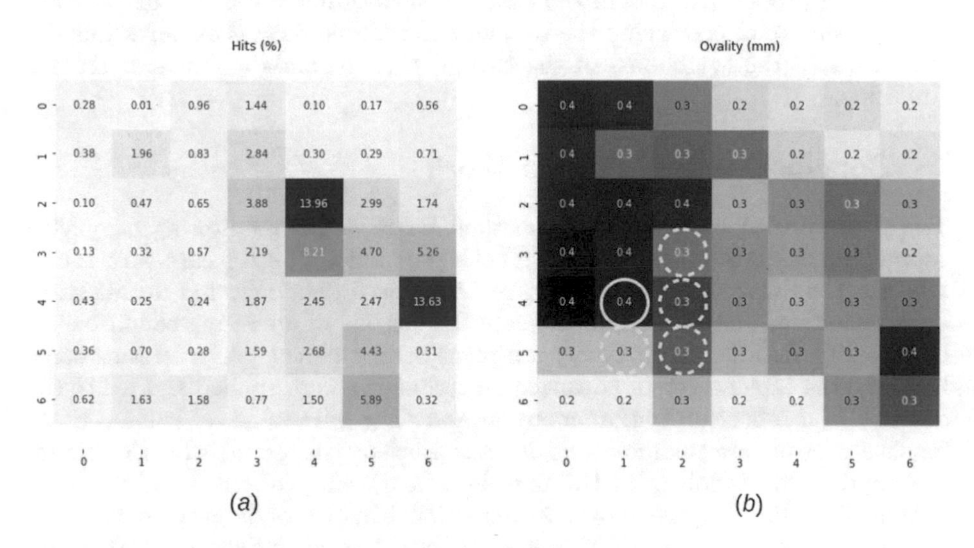

Fig. 4. Example of the main monitoring board that shows: (a) the SOM hit map; (b) the clusters with the associated estimated ovality values.

4.2 Exploitation of the SOM Model

The whole set of parameters that identify both the SOM and the ovality can be loaded by the dashboarding software that is currently used on plant human–machine–interfaces (HMI). The collected information include the weights of all neurons, the percent hits and, for each cluster, the associated ovality value. Once the HMI loaded the information stored in the above mentioned files, it connects to the plant database from which it continuously extracts actual operating conditions used as inputs by the SOM. These variables are used to determine in real–time the cluster corresponding to the current operating conditions. For monitoring purposes, the HMI shows the SOM grid where the active cluster and the ovality are put highlighted together with the SOM–hit map, such as shown in Fig. 4. All the details of the active clusters, including the corresponding process conditions, are shown. Moreover, by clicking on an arbitrary cluster the same information are provided. Figure 4 clearly highlights regions characterized by high ovality values (top–left of the map), while other ones are not affected by this issue. The tuned SOM can support operators during process control, when they need to change some process parameters to escape undesired process conditions that lead to ovality issues. In these cases, when in an ovality–critical situation (i.e. yellow continuous circle in Fig. 4), the system suggests to switch to a *close* operating condition that leads to ovality reduction. To this aim, among the neighbours of the actual cluster with lower associated ovality (i.e. yellow dashed circle in Fig. 4), the one is selected that requires the smallest changes in terms of process parameters, to ensure a smooth operation. In this way the SOM capability of preserving the topology of training data is exploited, as the search within the neighbours of the current cluster grants a similar operating condition.

5 Conclusions and Future Work

The paper presents a SOM-based method for real-time monitoring of quality problems and decision support during the production of steel bars with round section. The SOM is tuned by using real industrial data collected during standard operations and is used to correlate the clusters of operating conditions to bars ovality. In order to profitably use plant data, different pre-processing steps based on the exploitation of advanced AI techniques were applied and led to the creation of a robust and informative dataset. The advantages of using a SOM for this purpose are manifold and include efficient clustering, which preserves distribution and topology of training data, and ability of efficiently mapping a high dimensional space into a 2–dimensional map that is easy to visualize and interpret. The tuned SOM was embedded into an industrial HMI allowing real-time monitoring of process conditions and quality indicators through a user–friendly interface. In addition, the SOM is used within a decision support system to suggest the user a convenient strategy to recover from ovality-critical situations. In the future the system will be extended to allow simultaneous control of more than one quality indicators, and will be equipped with advanced optimization techniques to automatically set its hyper-parameters and improve the quality of suggestions provided by the decision support system implementing multi-objective strategies.

Acknowledgments. The work described in the present paper has been developed within the project entitled *Cyber-Physical System-based approach for intelligent data-*

driven maintenance operations applied to the rolling area (Ref. CyberMan4.0, Grant Agreement No. 800657) that has received funding from the Research Fund for Coal and Steel of the European Union, which is gratefully acknowledged. The sole responsibility of the issues treated in the present paper lies with the authors; the Commission is not responsible for any use that may be made of the information contained therein.

References

1. Alhoniemi, E., Hollmén, J., Simula, O., Vesanto, J.: Process monitoring and modeling using the self-organizing map. Integr. Comput.-Aided Eng. **6**(1), 3–14 (1999)
2. Brandenburger, J., Colla, V., Nastasi, G., Ferro, F., Schirm, C., Melcher, J.: Big data solution for quality monitoring and improvement on flat steel production. IFAC-PapersOnLine **49**(20), 55–60 (2016)
3. Cateni, S., Colla, V., Vannucci, M.: A fuzzy logic-based method for outliers detection. In: Proceedings of the 7th IASTED International Conference on Artificial Intelligence and Applications AIA 2007, pp. 605–610 (2007)
4. Cateni, S., Colla, V., Vannucci, M.: Variable selection through genetic algorithms for classification purposes. In: Proceedings of the 10th IASTED International Conference on Artificial Intelligence and Applications AIA 2010, pp. 605–610 (2007)
5. Cateni, S., Colla, V., Vannucci, M.: A genetic algorithm-based approach for selecting input variables and setting relevant network parameters of a SOM-based classifier. Int. J. Simul. Syst. Sci. Technol. **12**(2), 30–37 (2011)
6. Chandrashekar, G., Sahin, F.: A survey on feature selection methods. Comput. Electr. Eng. **40**(1), 16–28 (2014)
7. Di Bella, A., Fortuna, L., Graziani, S., Napoli, G., Xibilia, M.: A comparative analysis of the influence of methods for outliers detection on the performance of data driven models. In: 2007 IEEE Instrumentation & Measurement Technology Conference IMTC 2007, pp. 1–5. IEEE (2007)
8. Dimatteo, A., Vannucci, M., Colla, V.: Prediction of hot deformation resistance during processing of microalloyed steels in plate rolling process. Int. J. Adv. Manuf. Technol. **66**(9), 1511–1521 (2013)
9. Guha, S., Khuller, S.: Approximation algorithms for connected dominating sets. Algorithmica **20**(4), 374–387 (1998)
10. Matino, I., Dettori, S., Colla, V., Weben, V., Salame, S.: Forecasting blast furnace gas production and demand through echo state neural network-based models: pave the way to off-gas optimized management. Appl. Energy **253**, 113578 (2019)
11. Rousseeuw, P.J.: Silhouettes: a graphical aid to the interpretation and validation of cluster analysis. J. Comput. Appl. Math. **20**, 53–65 (1987)
12. Urbanowicz, R.J., Meeker, M., La Cava, W., Olson, R.S., Moore, J.H.: Relief-based feature selection: introduction and review. J. Biomed. Inform. **85**, 189–203 (2018)
13. Waters, C., Klemme, B., Talla, R., Jain, P., Mehta, N.: Transforming metal production by maximizing revenue generation with operational AI. In: AISTech - Iron and Steel Technology Conference Proceedings, pp. 1164–1174 (2021)

Performance Comparisons of Association Rule Learning Algorithms on a Supermarket Data

Monerah Mohammed Alawadh(✉) and Ahmed Mohammed Barnawi

King Abdulaziz University, Jeddah, Saudi Arabia
maalawadh@stu.kau.edu.sa, ambarnawi@kau.edu.sa

Abstract. In today's business environment, sustaining growth and profitability is never a guarantee. This concept of instability encourages business owners to seek new opportunities using technological and scientific advances to grasp the sustainability and maximizing profit. Nowadays, Machine Learning is helping the retail sector in many ways. Such improved life cycles of products, services, and business models. Association Rule Learning (ARL) is an unsupervised machine learning technique having great potential to extract interesting and useful associations and/or correlations among frequent item sets in transactional dataset. ARL techniques are widely employed in a variety of applications, including web click data, medical diagnostics, consumer behavior analysis, bioinformatics, and many more. In this paper, we conduct a comprehensive performance analysis of six distinct ARL algorithms, which was carried out to find the best fit algorithm for our data. The data is a large transactional dataset of one of the biggest supermarket chains in Saudi Arabia. Those algorithms are; Apriori, FP-Growth, GCD, Top-K Rules, TNR, and Closed AR. Note that these algorithms are uniquely different in terms of the number of datasets scanning which influences performance. The evaluation study shows that FP-growth algorithm based on FP-Tree outperformed other algorithms. FP-growth was more efficient and scalable than other algorithms for the given data in terms of memory usage and processing time.

Keywords: Association Rule Learning · Algorithm Comparison · Machine Learning · Apriori · pattern reconstruction · FP-Growth Algorithm

1 Introduction

Nowadays, under the recent developments about the information technologies, companies can store their data faster and easier with lower costs. All transactions (sales, web click, invoicing, …etc.) performed in companies during the day combine at the end of the day to form big datasets. It is possible to extract valuable information through these datasets with machine learning and data mining techniques [1, 2]. And this has become more important for companies in terms of today's conditions where the competition in the market is high.

Unsupervised Co-occurrences grouping is a machine learning method for finding associations between entities. This called association rule learning (ARL) which used

A. Rocha et al. (Eds.): WorldCIST 2023, LNNS 799, pp. 500–517, 2024.
https://doi.org/10.1007/978-3-031-45642-8_49

to enable market basket analysis. ARL seeks to discover affinities among traits and find rules for quantifying the relationship between these attributes [3–5].

Association Rule learning (ARL) has drawn the interest of data mining practitioners and researchers. ARL is one of the best signed and glowing researched methods of data mining [2]. Association Rules continues to prove instrumental in different fields including marketing, molding in biological databases, web personalization, clinical databases, risk management and inventory control, text mining, and statistics telecommunications. [6] introduces that one of the core functions of the association rule is to discover strong rules within databases by employing different attributes of "interestingness". [5, 7] underlines that association rules are instrumental in identifying hidden patterns which continue to form the basis of excellency for most commercial decision-making processes such as market analysis, cross marketing, and catalogue design [8].

This paper aims to evaluate several ARL algorithms toward finding the best fit algorithm for large transactional dataset of one of the biggest supermarket chains in Saudi Arabia. **The first** algorithm is Apriori and it should be distinguished that the Apriori algorithm [9, 10] is the classical and most important algorithm for association rule's extracting. However, [4] notes that it relies on multiple passes over the database. The performance of Apriori algorithm was improved by applying an ingenious compact data structure known as the frequent pattern tree (FP-Tree). So, **the second** Algorithm in this study is FP-Growth algorithm based on FP-Tree [1]. Both Apriori and FP-Growth algorithms required to set a minimum support (*minsup*) parameter. Instead of that, **the third** algorithm, GCD [11], uses the greatest common divisors value (GCD). Those values are hard to set and usually users set it by trial and error, which is time consuming. **The fourth** algorithm is Top-K Rules [8, 12, 13] which solves this problem by letting users directly indicate k, the number of top 'k' rules to be discovered instead of using any threshold values. Moreover, the result of any ARL algorithm usually contains a high level of redundancy (for example, hundreds of rules can be found that are variation of other rules having the same support and confidence). So, **the fifth** algorithm is a Top Non-Redundant (TNR) [14] algorithm, which provides a solution to both problems by letting users directly indicate k, the number of rules to be discovered, and by eliminating redundancy in results.

On the other hand, one of the challenges that is associate with ARL algorithms is that it can increasingly become unwieldy. This can be caused by an increased number of transactions while the confidence and support thresholds are uniquely small [7]. This follows the understanding that as the number of frequent item sets rises, the number of rules handled by the user also proportionately increases. [4] observes that many of such rules are likely to be redundant. So, **the sixth** algorithm is Closed association rule algorithm [15, 16], where the shortcoming from the "many frequent item sets" is usually eliminated by the adoption of the closed itemset concept. This creates an absence of superset which have similar support or frequency.

In this paper, we aimed to conduct a performance comparison of six association rule learning algorithms on a supermarket data to find the best fit algorithm for our big dataset. We tried to highlight the pros and cons of each algorithm and compare their performance with the dataset. This paper divided into five major sections. Section 2 constitutes a literature review study in the field of association rule learning algorithms.

Followed by Sect. 3 introduces and explain the six association rule algorithms involve in this study. Those algorithms are; Apriori, FP-Growth, GCD, Top-K Rules, TNR, and Closed AR. Noted that these algorithms are uniquely different in terms of the number of datasets scanning which influences performance. Sections 4, the Experiment section which is the main part of this paper. This section divided into four main parts: the *data used*, where we describe the dataset and the *preprocessing*. *Algorithms run,* where we show a results sample of each run. *Result & Discussion*: here we present the result of the performance comparison and the discussion and findings. Lastly, Sect. 5 is the conclusion of the paper.

2 Literature Review

Consumer Behavior Analysis (CBA) is one of the areas in which artificial intelligence (AI) and machine learning (ML) techniques are playing a major role. Recently, these techniques contributed to the creation of new opportunities that changed the shopping experience as well as delivered new methods of identifying and analyzing the market [17].

Many researchers [4, 5, 18–25] used ML techniques to find associated items toward analyzing consumers' behaviors. Association Rule Learning (ARL) is a popular ML method used for discovering relations between variables in extensive databases [18]. [18] suggested that having ARL's origin in the analysis of the marketing bucket. The exploration of association rules represents one of the essential activities and main applications of ML in data similarity analysis [4, 18]. ARL is the efficient discovery of valuable, non-obvious, previously un-known relationship between individual items in a large collection of data [18].

As mentioned in [4], the processes of generating association rule are based on two main functions: scanning the database to identify frequent item sets then generating association rules out of these frequent sets. So, it is fair to say: an algorithm over performs other algorithms for a given dataset, but this can't be widely generalized because changing the database may affects algorithm's performance. To illustrate, we see in [19] and [20] FP-Growth algorithm overperformed Apriori algorithm. While in a comparative survey on association rule mining algorithms, apriori algorithm outperforms other algorithms -including FP-Growth algorithm- in cases of closed item sets, i.e. fast, less candidate sets, generates candidate sets from only those items that were found large [18]. In [19], the author compared the effectiveness of two common association rules algorithms: Apriori and FP-Growth. The study aims to recognize consumers' buying preferences of an automotive service in order to improve the company's sales. In [20], the same algorithms were compared using some supermarket and voter datasets. The authors used different support and confidence values to compare. The comparison has been completed using WEKA tool.

Apriori, Charm, Eclat, Fp-Growth, and dEclat are applied to a dataset of a company which is selling car maintenance and repair products in Turkey in order to evaluate the performances by using execution times and memory usage metrics against varying data size and support value. It can be said that execution times generally increases inversely with the support values as most algorithms have higher execution time values for the lowest support value of 0.1 [4].

[2] gives the overview of six association rule learning algorithms. Namely AIS, SETM, AprioriTid, Apriori hybrid and Frequent Pattern-growth in which all algorithms are analyzed, and the merits and demerits are reported. Table 1 shows a comparison of these algorithms in different features: data support, speed in the initial and later phases, and accuracy. In terms of speed, the Apriori hybrid algorithm is on par with FP-Growth. However, in terms of Accuracy, the FP-Growth algorithm outperforms the Apriori hybrid [2].

Table 1. Comparison of ARL Algorithms [2]

	AIS	SETM	Apriori	AprioriTID	Apriori hybrid	FP-growth
Data support	Less	Less	medium	Often large	Very Large	Very Large
Speed in initial phase	Slow	Slow	High	Slow	High	High
Speed in later phase	Slow	Slow	Slow	High	High	High
Accuracy	Very less	Less	Less	Medium	More accurate than AprioriTID	Very fast,

[21] presents a comparison on three different association rule learning: FP Growth, Apriori, and Eclat. The comparision of this study was only based on execution time verses support and confidence. The highlighted outcome is Eclat algorithm noticed as the fastest algorithm. It was also identified that the execution time decreases with increasing confidence and support.

[5] also provides a study of the algorithms of association rule mining. LogEclat algorithm overperformed other algorithms involved in the comparison such as AIS, SETM, AprioriTid, AprioriHybrid .. etc. Elcat is similar to the recursive operation of the Apriori algorithm. It uses a tree such as the Tidset framework. The tidset starts with all risks in the database once in a while. The algorithm performs an in-depth search to assess the risk factor in all situations. This study describes ARL as technique has the ability of discovering and predicting future trends and behavior.

In 2018, a study uses two different techniques on a store' sales data. The authors used clustering for customers as well as association rule learning for the products. Based on all store visits, customers are divided into different categories. In contrast to comparable research, this study considers the products that individual consumer bought during multiple visits rather than only the things products he/she has purchased during a single visit [22].

Study [23] applies market basket analysis to sport equipment's database in order to increase the sales using FP growth. The study explains how to utilize the discovered information as to provide better recommendations and promotions for the customers. The study concluded with offering information about what spot items should be organized on the shelves together like towel, running shoes, socks etc. [23].

Similarly, a thesis study applied market basket analysis and Apriori on beauty products as a mean to offer competitive edge to the companies [24]. Furthermore, the study focused on the probability of buying products at different times of the day and in different seasons. an analysis like this will offer markets and stores managers with knowledge on when is the best season for promotion of some product categories and at what time of the day customers are more likely to buy items from a specific product category.

Moreover, according to [23] there are number of Stores widely used the market basket analyses to manage their storage areas more efficiently and re-organize goods in their store layout. Related products are placed together in such a manner that customers can logically find items he/she might buy which increases the customer satisfaction and enhancing their sales and profitability. Authors of [25] proposed IIMW algorithm. IIMW is an algorithm for Mining Weblog targeted the Infrequent Itemset, which was based on the Apriori-Rare and Apriori-Inverse algorithms. The data used for testing this algorithm was obtained from web traffic archives. Moreover, authors applied three algorithms on the dataset and compared the performance values [25].

It is apparent that employing association rules to discover the associations between products on different domains is so popular because it allows businesses to take better dissensions. Ultimately leading to better business performance with a competitive edge against competitors.

3 Association Rule Learning Algorithms

In this section six association rule learning algorithms are introduced and explained in detail:

3.1 Apriori Algorithm

Apriori algorithm has two key properties: (1) the "Downward closure" property which indicate that if an itemset appears frequently, then each of the existing subsets are also regarded as frequent. (2) the "Antimonotone" property of support value which is follows the rule that if an itemset S in a transaction T, then each subset J of S must exists within the transaction T, with support value greater or equal to the support of S. It should be underlined that the Apriori algorithm is distributed as level-based approach. This in essence means that each of the iteration itemsets are extracted equal to a given length [9, 10].

```
Algorithm: im_Apriori algorithm
Input: "D" a database of transaction
min_sup: the minimum support count threshold
Output: "L" frequent itemsets in D
L1=find_frequent_1_itemsets(A); //find L freq. itemsets, k-itemsets are used to explore (k+ 1)-
itemsets
for(k=2; Lk-10 ;k++) {
Ck= apriori_gen (Lk-1);
For each transaction a E A{//scan A for counts
Ca=subset(Ck,a); //get the subsets of t that are candidates For each candidate c ∈ Ca
c.count++;}
c.count =c.count*Wij
Lk={c E Ck|c.count>=min_sup}
}Return L=UkLk;
Phase2: Procedure rule_gen(Lk-1:frequent(k-1)-itemsets)
For each transaction a E A
if a[10]<k then
delete a[i];
else{
For each itemset l1ELk-1
For each itemset l2ELK-1
C=11 JOIN 12;// join step:generate candidates
If has_infrequent_subset(c, Lk-1)then
Delete c;//prune step:remove unfruitful candidate
Else add c to Ck;}}
return Ck;
```

3.2 FP-Growth Algorithm

The frequent pattern or FP-Growth is designed into 2 phases: First, locate frequent items after calculating item frequencies. This makes it different from other Apriori-like algorithms offering the same function [8]. The second phase, is the adoption of the suffix tree structure (FP-Tree) which is measured to encode transactions by circumventing the need to explicitly generate candidate sets. Such a step is remarkably essential because it allows frequent item sets to be extracted from the FP-Tree. So, in this implementation FP-Tree is used to generate frequent item sets instead of Apriori because FP-Tree in more efficient and claim to improve the processing time as well as memory usage [1].

Algorithm: FP-Growth algorithm
Input: Database, minimum support and confidence, FP- tree FT, Fpset.
Output: generated frequent rules
(1) If (FT has a single path as P) {
Foreach combination of nodes in this path (H)
HUFPset with support of minimum support of nodes in H
} Else foreach heather in FT as u {
H=u U FPset with the support = u.support
Create the conditional FP-tree of H as CTree
If CTree # 0; {
Go to the (1) and set the FT=CTree and FPset=H}
}
With the generated frequent list generate the associated rules.
Check that the confidence of generated rules be more that the minimum confidence value otherwise remove them.
Return the generated rules.

3.3 TOP-K Rules Algorithm

TOP-K Rules algorithm by default is designed to find TOP-K association rules. Here, K represents the number of association rules that can be discovered and determined by the user, unlike other Apriori-based algorithms. [13] notes that TOP-K algorithm is characterized by two key challenges. One is that TOP-K rules are not dependable on minimum support so as to effectively prune the search space. However, [8] observes that TOP-K rules possess the capacity to be adjusted to mine frequent itemsets having a minimum support = 1. This serves to ensure that each TOP-K rule can be produced correctly. Two, TOP-K association rules learning algorithm lacks the capacity to explore two-step process in order to mine the association rules. But, [7] presents that it is possible to modify Top-K rules through generation of rules. According to [7] efficient approach involved in the production of generating association rules is regarded as "rules expansion" which does not depend on the two-step process associate with mine association rules. It is important to indicate that TOP-K rules algorithm has the advantage of an alternative classical association rule in which users are keen on controlling the number of association rules generated.

Top-K Rules is an algorithm that can be fine-tuned to find the most (k) important association rules have been developed by a team of researchers [12]. The experimental results in [12] show that Top-K Rules has excellent performance and scalability, and that it is an advantageous alternative to classical association rule learning algorithms when the user wants to control the number of association rules generated. Top-K Rules algorithm takes as input a transaction database, a number k of rules that the user wants to discover, and the minconf threshold. Depending on the choice of parameters, Top-K Rules algorithm can generate varying number of rules [12].

```
Algorithm: Top-K Rules algorithm
TOPKRULES(T, k, minconf) R = ∅L = ∅ minsup = 0
Scan the database T once to record the tidset of each item.
FOR each pairs of items i, j such that |tids(i)|x|T|≥ minsup and |tids(j)|x|T| > minsup:
sup({i}→{j})= |tids(i) n tids(j)|/ |T |.
conf({i}→{j}): |tids(i) n tids(j)| / |tids(i)|.
IF sup({i}→{j}) >= minsup THEN
IF conf({i}→{j}) >= minconf THEN SAVE({i}→{j}, L, k, minsup).
Set flag expand LR of {i}→{j}to true.
Set flag expand LR of {j}→{i}to true.
R: RU{{i}→{j}, {j}→{i}}.
End If
End For
While ∃r E R AND sup(r) >= minsup DO
Select the rule rule having the highest support in R
IF rule.expand LR = true THEN
Expand-L(rule, L, R, k, minsup, minconf).
Expand-R(rule, L, R, k, minsup, minconf).
Else Expand-R(rule, L, R, k, minsup, minconf).
Remove rule from R.
Remove from R all rules r E R | sup(r) <minsup and End While
```

3.4 TNR Algorithm

TNR is an algorithm for discovering the top-k non-redundant association rules in any given transactional data. So, the difference between TNR and Top-K Rules lies in how to avoid generating redundant rules. TNR is an approximate algorithm in the sense that it always generates non-redundant rules. Although these rules may not always be the top-k non-redundant association rules but at least they are non-redundant rules. To illustrate, we could say that the below three points are the main different between TNR and Top-K rule algorithms [14]:

1. *TNR uses parameter named "delta (Δ)": a positive integer $> = 0$ that can be used to improve the chance that the result is exact (the higher the delta value, the more chances that the result will be exact).*
2. *During the search, TNR added only non-redundant rules to L. That means, for each rule ra that is generated such that sup(ra) $\geq$ minsup, if $\exists$ rb L | sup(rb) $=$ sup(ra) and ra is redundant with respect to rb, then ra is not added to L. Otherwise, ra is added to L.*
3. *If the number of redundant rules in L is never more than the delta (Δ) rules, then the algorithm result is exact and the k rules in L having the highest support will be the top-k non redundant rules.*

TNR algorithm has created an important resource in deep learning technique which has played an important function in pattern mining. [7] underlines that different model including Long Short-Term Memory (LSTM), Recurrent Neutral Network (RNN), and Convolutional Neutral Network (CNN) are designed to mine latent patterns. It should be highlighted that these techniques are founded on both flexibility and computational approaches which can be applied upon a distributed parallel environment.

3.5 Closed Association Rule Learning

Closed Association Rule Learning is an algorithm for mining "closed association rules", which are a concise subset of al association rules. Closed Association Rule Learning is an algorithm adopts a concept of "**closed itemset**". A closed itemset is an itemset that is strictly included in no itemset having the same support [4, 15].

[15] proposed a new framework for mining closed itemset without candidate generation and finding closed association rules. Closed itemset mining has emerged as an important research topic in data mining and has received a lot of attention in recent years [15, 16].

Algorithm: Closed algorithm
Phase1:
Input: D, min_support;
Output: The complete set of closed itemsets;
Scan D once to find promising items by calculating TWU of items. Scan D again to construct EU-List of itemset.
return(Ø, Ø, EU-List, min_support).
Phase2:
input: EU-List, PostSet(YC), PostSet(X);
output: Return Y's closure Yc. Update PostSet(Yc) and EU-List of Yc;
Yc← Y;
PostSet(Yc) ←Ø;
For each item ZEPostSet(X) do
If(TidSet(Yc) Є TidSet(X))
Yc ←Yc U Z;
For each Tid r Є TidSet(Yc) do
EL(Yc, Tr).EU ← EL(Yc, Tr).EU + EL(Z, T,).EU;
EL(Yc, Tr).PU← EL(Yc, Tr).PU — EL(Z, Tr).PU;
else PostSet(Yc) ← PostSet(Yc) U Z;
Return Yc;

This algorithm returns all closed association rules such that their support and confidence are respectively higher or equal to the minsup and minconf thresholds set by the user.

3.6 GCD (Greatest Common Divisors) Algorithm

GCD algorithm initially used for finding the association rules in a given sequence of events. But it is also applicable for the regular transactional data sets. GCD algorithm uses greatest common divisors "GCD" calculations for prime numbers. "maxcomb" parameter is used by the algorithm when finding the GCD between two transactions. For example, consider 385, which comes from the multiplication of (5, 7 and 11), this actually means that (5), (7), (11), (5, 7), (5, 11), (7, 11), (5, 7, 11) are all common combinations between these two transactions. For larger GCD's, calculating all combinations grows exponentially in both time and memory. Hence, we introduced this parameter, to limit the maximum combinations' length generated from a single GCD. Although increasing this number might seem to provide more accurate results, the experiments showed that larger association rules occur at lower support (less important to the user). Hence, setting "maxcomb" parameter to values from 1 to 4 produces reasonable results. Finally, the output is a set of association rules sorted by support in a descending order [11].

Algorithm: GCD algorithm
Input: Dataset D, min support, min confidence
pass items in D to calculate frequency
extract itemset
sort according to their frequency in an itemset
pass again to replace the most frequent itemset with the smallest not yet extracted
//The replacement stops at the minimum support percentage. The multiplication of the Repeated multiplications is eliminated while being represented by a single entry that holds the repetition count
Calculate GCD value
//The GCD is calculated between each entry's multiplication result and the other entries in the same set
The support of each found GCD is extracted
GCDs are sorted in a descending order according to support values
Confidence values are calculated while skipping some processing when confidence values drop below the specified minimum
Factorize each GCD to its original factors
return the original itemset along with the collected support and confidence values

4 Experiment

In this section, the whole experiment is described including data used, data preprocessing, run algorithms, result & discussion for the performance comparison of association rule algorithms.

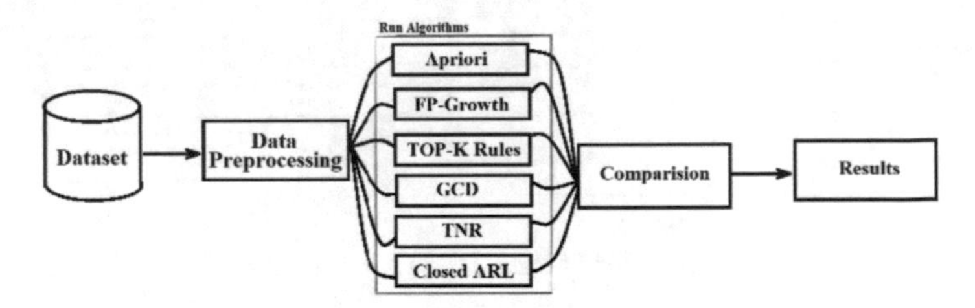

Fig. 1. Experiments Pipeline

Figure 1 shows the pipeline for the experiments as a block diagram for the work methodology.

4.1 Data Used

The data used in this performance evaluation study is a real supermarket data from one of the biggest supermarket chains in Saudi Arabia. It is a transactional data containing

1249 customer basket (receipts) with average size of 11 items in each basket. The dataset containing 8324 different items. Each transaction has a unique transaction ID and a list of items. The following Table 2 gives a summary of the data set.

Table 2. Data set Description Summary.

Criteria	Dataset
Number of transactions	1,249
Number of distinct items	8324
Maximum transaction size	29
Average transaction size	11.27
Total sealed items	23,922

4.2 Data Preprocessing

For the data preprocessing: 1) Data Cleaning, we removed any items that appears only once against all transactions. Then, we removed any transaction has only one item (even if this item is very frequent) because this will not make any sense for the association analysis. 2) Transformation, we transfer the transaction from a list to 0 and 1 data representation. Figure 2 shows the data after prepared. For data preprocessing we used Visual Basic for Applications (VBP) macro codes provided by Microsoft excel.

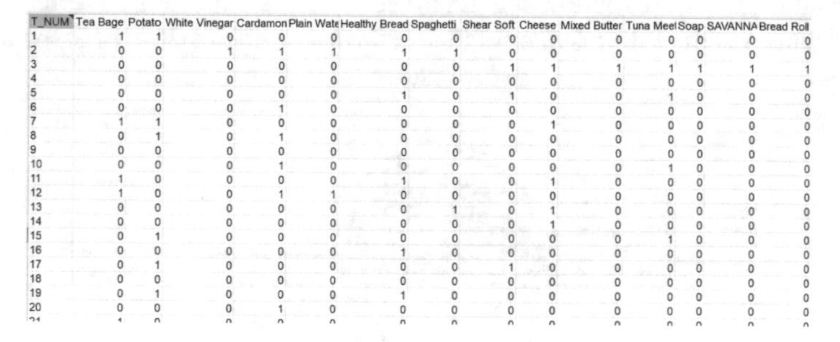

T_NUM	Tea Bage	Potato	White Vinegar	Cardamon	Plain Wate	Healthy Bread	Spaghetti	Shear Soft	Cheese	Mixed Butter	Tuna Meel	Soap	SAVANNA	Bread Roll
1	1	1	0	0	0	0	0	0	0	0	0	0	0	0
2	0	0	1	1	1	1	1	0	0	0	0	0	0	0
3	0	0	0	0	0	0	0	1	1	1	1	1	1	1
4	0	0	0	0	0	0	0	0	0	0	0	0	0	0
5	0	0	0	0	0	1	0	1	0	0	1	0	0	0
6	0	0	0	1	0	0	0	0	0	0	0	0	0	0
7	1	1	0	0	0	0	0	0	1	0	0	0	0	0
8	0	1	0	1	0	0	0	0	0	0	0	0	0	0
9	0	0	0	0	0	0	0	0	0	0	0	0	0	0
10	0	0	0	1	0	0	0	0	0	0	1	0	0	0
11	1	0	0	0	0	1	0	0	1	0	0	0	0	0
12	1	0	0	1	1	0	0	0	0	0	0	0	0	0
13	0	0	0	0	0	0	1	0	1	0	0	0	0	0
14	0	0	0	0	0	0	0	0	1	0	0	0	0	0
15	0	1	0	0	0	0	0	0	0	0	1	0	0	0
16	0	0	0	0	0	1	0	0	0	0	0	0	0	0
17	0	1	0	0	0	0	0	1	0	0	0	0	0	0
18	0	0	0	0	0	0	0	0	0	0	0	0	0	0
19	0	1	0	0	0	1	0	0	0	0	0	0	0	0
20	0	0	0	1	0	0	0	0	0	0	0	0	0	0

Fig. 2. Ready Data After Preprocessing

4.3 Run Algorithms

All algorithms were executed without any restrictions in the number of items in both lift and right-hand sides. The Support and confidence threshold values -if algorithm's required- are set to 0.3 and 0.45 respectively.

The first noticed result is: All algorithms generated almost the same result. Tables 3, 4, 5, 6, 7 and 8 summarized rules generated for each algorithm, each number represent an item ea.: 45 is a fresh cucumber, 48 is tomato, 5 is banana, 17 is peach, … etc.:

Table 3. Association Rules Generated by Apriori Algorithm

ASSOCIATION RULES	SUPPORT	CONFIDENCE
`If { 45 } AND { 48 } THEN { 5 }`	**0.335**	**0.755**
`If { 45 } THEN { 48 }`	0.444	0.803
`If { 45 } THEN { 5 }`	0.400	0.724
`If { 45 } THEN { 5 } AND {48 }`	0.335	0.606
`If { 48 } THEN { 45 }`	0.444	0.769
`If { 48 } THEN { 5 }`	0.411	0.712
`If { 48 } THEN { 5 } AND {45 }`	0.335	0.580
`If { 5 } AND { 45 } THEN {48 }`	0.335	0.838
`If { 5 } AND { 48 } THEN {45 }`	0.335	0.815
`If { 5 } THEN { 45 }`	0.400	0.666
`If { 5 } THEN { 45 } AND {48 }`	0.335	0.558
`If { 5 } THEN { 48 }`	0.411	0.684

Table 4. Association Rules Generated by Fb-Growth Algorithm

ASSOCIATION RULES	SUPPORT	CONFIDENCE
`If { 45 } THEN { 5 }`	0.400	0.724
`If { 5 } THEN { 45 }`	0.400	0.666
`If { 48 } THEN { 5 }`	0.411	0.712
`If { 5 } THEN { 48 }`	0.411	0.684
`If { 48 } THEN { 45 }`	0.444	0.769
`If { 45 } THEN { 48 }`	0.444	0.803
`If { 45 } AND { 48 } THEN {5 }`	0.335	0.755
`If { 5 } AND { 48 } THEN { 45 }`	0.335	0.815
`If { 5 } AND { 45 } THEN { 48 }`	0.335	0.838
`If { 48 } THEN { 5 } AND { 45 }`	0.335	0.580
`If { 45 } THEN { 5 } AND { 48 }`	0.335	0.606
`If { 5 } THEN { 45 } AND { 48 }`	0.335	0.558

Table 5. Association Rules Generated by Top-K Algorithm

ASSOCIATION RULES	SUPPORT	CONFIDENCE
`If { 5 } AND { 45 } THEN { 48 }`	0.335	0.838
`If { 5 } AND { 48 } THEN { 45 }`	0.335	0.815
`If { 45 } AND { 48 } THEN {5 }`	0.335	0.755
`If { 45 } THEN { 5 } AND { 48 }`	0.335	0.606
`If { 48 } THEN { 5 } AND { 45 }`	0.335	0.580
`If { 5 } THEN { 45 } AND { 48 }`	0.335	0.558
`If { 45 } THEN { 5 }`	0.400	0.724
`If { 5 } THEN { 45 }`	0.400	0.666
`If { 48 } THEN { 5 }`	0.411	0.712
`If { 5 } THEN { 48 }`	0.411	0.684
`If { 45 } THEN { 48 }`	0.444	0.803
`If { 48 } THEN { 45 }`	0.444	0.769

Table 6. Association Rules Generated by TNR Algorithm

ASSOCIATION RULES	SUPPORT	CONFIDENCE
If { 5 } THEN { 45 }	0.400	0.666
If { 5 } THEN { 48 }	0.411	0.684
If { 48 } THEN { 45 }	0.444	0.769
If { 45 } THEN { 48 }	0.444	0.803
If { 48 } THEN { 5 }	0.411	0.712
If { 45 } THEN { 5 }	0.400	0.724
If { 45 } AND { 48 } THEN {5 }	0.335	0.755
If { 5 } AND { 45 } THEN { 48 }	0.335	0.838
If { 5 } AND { 48 } THEN { 45 }	0.335	0.815
If { 17 } THEN { 5 }	0.293	0.743

Table 7. Association Rules Generated by Closed Algorithm

ASSOCIATION RULES	SUPPORT	CONFIDENCE
If { 5 } AND { 45 } THEN { 48 }	0.335	0.838
If { 5 } AND { 48 } THEN { 45 }	0.335	0.815
If { 45 } THEN { 48 }	0.444	0.803
If { 48 } THEN { 45 }	0.444	0.769
If { 45 } AND { 48 } THEN { 5 }	0.335	0.755
If { 45 } THEN { 5 }	0.400	0.724
If { 48 } THEN { 5 }	0.411	0.712
If { 5 } THEN { 48 }	0.411	0.684
If { 5 } THEN { 45 }	0.400	0.666
If { 45 } THEN { 5 } AND { 48 }	0.335	0.606
If { 48 } THEN { 5 } AND { 45 }	0.335	0.580

Table 8. Association Rules Generated by GCD Algorithm

ASSOCIATION RULES	SUPPORT	CONFIDENCE
If { 45 } THEN { 48 }	0.444	0.803
If { 48 } THEN { 45 }	0.444	0.769
If { 48 } THEN { 5 }	0.411	0.712
If { 5 } THEN { 48 }	0.411	0.684
If { 45 } THEN { 5 }	0.400	0.724
If { 5 } THEN { 45 }	0.400	0.666
If { 5 } AND { 45 } THEN { 48 }	0.335	0.838
If { 5 } AND { 48 } THEN { 45 }	0.335	0.815
If { 45 } AND { 48 } THEN {5 }	0.335	0.755
If { 45 } THEN { 5 } AND { 48 }	0.335	0.606
If { 48 } THEN { 5 } AND { 45 }	0.335	0.580
If { 5 } THEN { 45 } AND { 48 }	0.335	0.558

4.4 Result and Discussion

All algorithms: Apriori, FP-Growth, GCD, Top-K Rules, TNR, and Closed AR were implemented and executed in the Eclipse Layout Kernel under Java environment. The number of frequent itemset and association rule generated are vary from each algorithm and another. Figure 3 and Fig. 4 show a statistic of frequent itemset and association rule generated for each algorithm respectively.

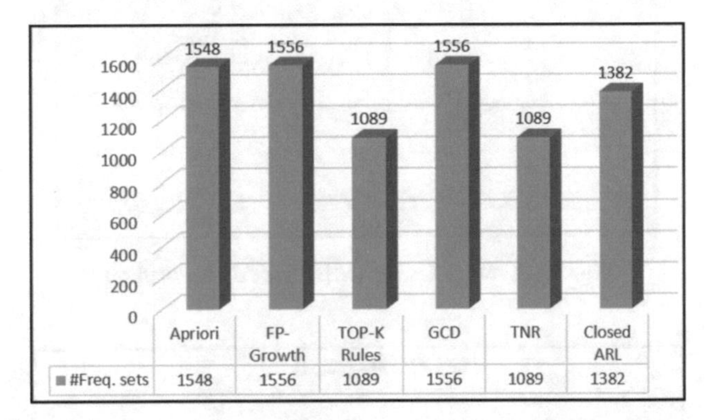

Fig. 3. Number of Frequent Itemset Generated for Each Algorithm

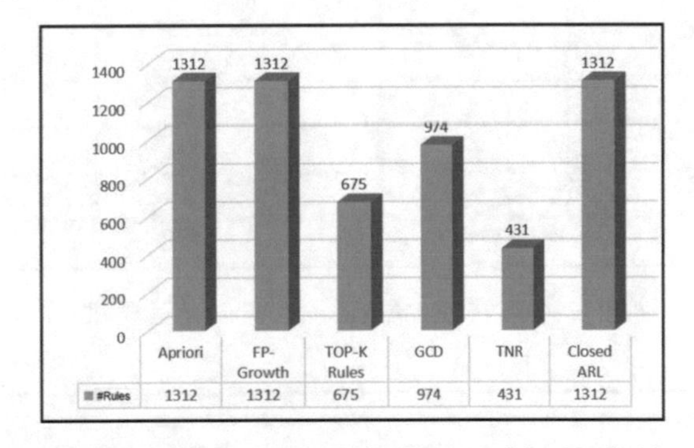

Fig. 4. Number of Association Rule Generated for Each Algorithm

Moreover, the performance of each algorithm is measured in terms of memory usage (in megabyte: 1024 kilobyte = 1 mb) and run time (in millisecond) as shown in Fig. 5 and Fig. 6 respectively.

To evaluate algorithms' performance, it is clear from Sect. 4 that these six algorithms are differ in the process of association rules generation. We could say that Apriori, FP-Growth, and GCD algorithm have the ability to generate all rules without any restrictions. While TNR, Top-K, and Closed ARL algorithm generate rules with different restrictions such as number of rules generated, the redundancy, and closed itemset concept. Thus, they can be categories into 2 categories and we could say FP-Growth algorithm is the best algorithm for the first category in terms of both run time and memory usage. And Closed ARL algorithm has the best performance over two other algorithms in the second category. Figure 7 shows the memory usage of all algorithms against different file sizes:

Also, it is clear from Sect. 4 that GCD algorithm has the higher cost in the scan of large data sets, due to the required storage space, and the execution time. On the

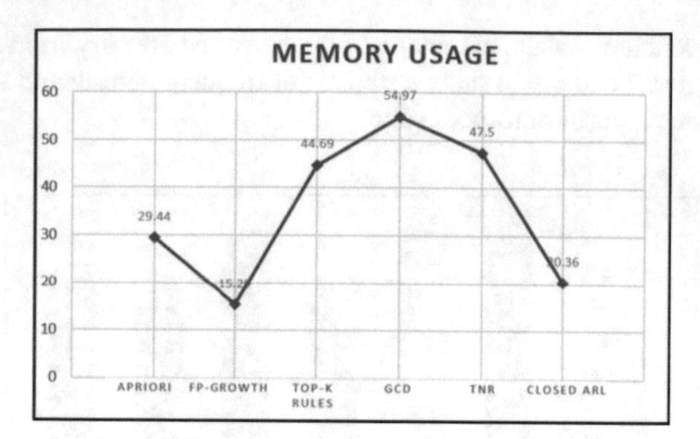

Fig. 5. Memory Usage (MB) For All Algorithms

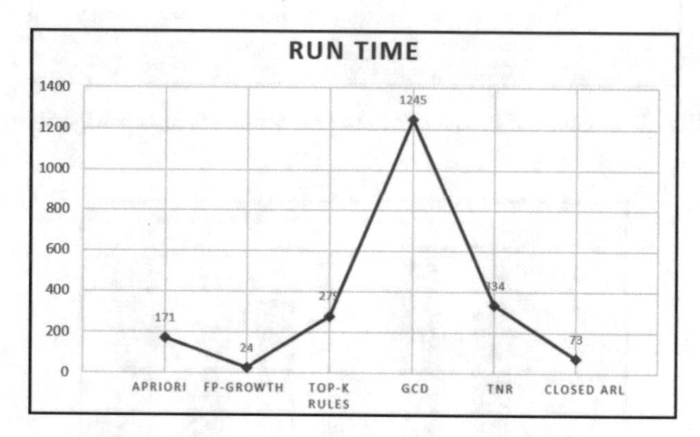

Fig. 6. Execution Time (MS) For All Algorithms

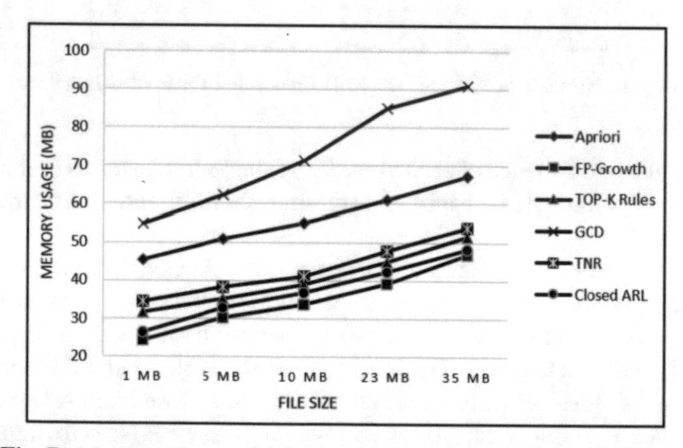

Fig. 7. Memory Usage of All Algorithms Against Different File Sizes

other hand, Apriori algorithm apply multiple scans to the data sets for each candidate set within the datasets, it may not be effective to be used on large data sets due to the time taken in each round of scans. Lastly, FP-growth algorithm which is implemented based on FP-Tree has good performance against other algorithm in this experiment.

TNR Algorithm is explained as a Top-K non-redundant association rules algorithm which combines the ideas of Top-K association rules with those of mining a set of non-redundant rules. [7] offers that TNR algorithm is structured on a rule expansion approach and adopts two strategies aimed at preventing the generation of redundant rules. While differentiating Apriori, [8] noted that as the minimum support diminishes, there is the generation of a large number of partial frequent item sets within the different datasets. Consequently, more time for execution is required. It is important to underline that in Apriori, the global infrequent itemset is significantly pruned in the first stage, which make it more reliable and efficient with regard to execution time.

5 Conclusion

The aim of this paper was to identify the best Association Rule Learning (ARL) algorithm for the given dataset. Which is a real supermarket data from one of the biggest supermarket chains in Saudi Arabia. There are various algorithms classified under the ARL algorithms. In this paper we covered Apriori, FP-Growth, GCD, Top-K Rules, TNR, and Closed AR. Each algorithm has its advantages and disadvantages and they are uniquely different in terms of the number of datasets scanning which influences performance. Therefore, in this study a comparison is done to find the best fit for our data.

Table 9 summarized algorithms' performance in terms of different factors such as: memory usage (in megabyte: 1024 kilobyte = 1 mb), frequent itemset count, run time (in millisecond), and number of association rule generated.

Table 9. ARL Algorithms Comparison

Algorithm	Comparison Factors			
	Max memory usage	Frequent item sets count	Total time	Number of association rules
Apriori	29.44 mb	1548	171 ms	1312
FP-Growth	15.25 mb	1556	24 ms	1312
TOP-K Rules	44.69 mb	1089	279 ms	675
GCD	54.97 mb	1556	1245 ms	974
TNR	47.5 mb	1089	334 ms	431
Closed ARL	20.36 mb	1382	73 ms	1312
Threshold Values		Minimum Support:	0.30	
		Minimum Confidence:	0.45	

From the algorithm comparison, it is clear that despite the effectiveness of each algorithm on a specified field, there is still major differences on the characteristics, which makes it much easier to decide which one is more efficient and effective than the other. The comparison of the algorithms outlines the FP-growth algorithm (based on FP-Tree) as the most effective algorithm to analyze the given data and generate association rules. FP-growth algorithm showcases considerably good results in the consideration of the characteristics of the algorithms among six algorithms as it has the ability to generate all rules in an acceptable processing time and memory usage.

Conflicts of Interest. The authors declare no conflicts of interest regarding the publication of this paper.

References

1. Ghafari, S.M., Tjortjis, C.: A survey on association rules mining using heuristics. WIREs Data Mining and Knowledge Discovery (2019). https://doi.org/10.1002/widm.1307
2. Prithiviraj, P., Porkodi, R.: A comparative analysis of association rule mining algorithms in data mining: a study (2015)
3. Luma-Osmani, S., Ismaili, F., Zenuni, X., Raufi, B.: A systematic literature review in causal association rules mining. In: 2020 11th IEEE Annual Information Technology, Electronics and Mobile Communication Conference (IEMCON), pp. 48–54 (2020). https://doi.org/10.1109/IEMCON51383.2020.92849
4. Nair, M., Kayaalp, F.: Performance comparison of association rule algorithms with SPMF on automotive industry data. J. Sci. Technol. **7**, 1985–2000 (2019)
5. Saxena, A., Rajpoot, V.: A comparative analysis of association rule mining algorithms. In: IOP Conference Series: Materials Science and Engineering, vol. 1099, no.1, p. 012032 (2021)
6. Belcastro, L., Marozzo, F., Talia, D., Trunfio, P.: ParSoDA: high-level parallel programming for social data mining. Soc. Netw. Anal. Min. **9**(4), 1–10 (2019)
7. Ren, R., Li, J., Yin, Y., Tian, S.: Failure prediction for large-scale clusters logs via mining frequent patterns. In: Gao, W., et al. (eds.) Intelligent Computing and Block Chain. FICC 2020. CCIS, vol. 1385, pp. 147–165. Springer, Singapore (2021). https://doi.org/10.1007/978-981-16-1160-5_13
8. Devi, S.G., Sabrigiriraj, M.: Swarm intelligent based online feature selection (OFS) and weighted entropy frequent pattern mining (WEFPM) algorithm for big data analysis. Clust. Comput. **22**, 11791–11803 (2019)
9. Liao, B.: An improved algorithm of Apriori. In: Computational Intelligence and Intelligent Systems, pp. 427–432 (2009)
10. Belcastro, L., Marozzo, F., Talia, D.: High-level parallel programming for social data mining. Soc. Netw. Anal. Min. **9**(4), 1–10 (2019)
11. SPMF, An Open-Source Data Mining Library, "GCD Association Rules Learning Algorithm", 2015, v.1.2. Copyright © 2008-2022 Philippe Fournier-Viger All rights reserved
12. Fournier-Viger, P., Wu, C.-W., Vincent, S.T.: Mining Top-K Association Rules, Canada (2022)
13. Kumar, S., Mohbey, K.K.: A review on big data parallel and distributed approaches of pattern mining. J. King Saud Univ. Comput. Inf. 1–10 (2019)
14. Fournier-Viger, P., Tseng, V.S.: Mining Top-K non-redundant association rules. In: Chen, L., Felfernig, A., Liu, J., Raś, Z.W. (eds.) Foundations of Intelligent Systems. ISMIS 2012. LNCS, vol. 7661, pp. 31–40. Springer, Berlin, Heidelberg (2012). https://doi.org/10.1007/978-3-642-34624-8_4
15. Wu, C.-W., Fournier-Viger, P., Gu, J.-Y., Tseng, V.S.: Mining closed+ high utility itemsets without candidate generation. In: 2015 Conference on Technologies and Applications of Artificial Intelligence (TAAI), pp. 187–194 (2015)

16. Fournier-Viger, P., Zida, S., Lin, J.C., Wu, C., Tseng, V.S.: EFIM-Closed: fast and memory efficient discovery of closed high-utility itemsets. In: Perner, P. (eds.) Machine Learning and Data Mining in Pattern Recognition. MLDM 2016. LNCS, vol. 9729, pp. 199–213. Springer, Cham (2016). https://doi.org/10.1007/978-3-319-41920-6_15
17. Alawadh, M.M., Barnawi, A.M.: A survey on methods and applications of intelligent market basket analysis based on association rule. J. Big Data **4**(1), 1–25 (2022)
18. Girotra, M., Nagpal, K., Minocha, S., Sharma, N.: Comparative survey on association rule mining. Int. J. Comput. Appl. **84**(10), 18–22 (2013)
19. Semra, T.: Comparison Of Apriori And Fp-Growth Algorithms On Determination Of Association Rules In Authorized Automobile Service Centres (2011)
20. Bala, A., et al.: Performance Analysis of Apriori and FP-Growth Algorithms (Association Rule Mining), vol. 7, no. 2, pp. 279–293 (2016)
21. Sinha, G., Ghosh, S.M.: Identification of best algorithm in association rule mining based on performance. Int. J. Comput. Sci. Mob. Comput. **3**(11), 38–45 (2014)
22. Griva, A., et al.: Retail business analytics: customer visit segmentation using market basket data. Expert Syst. Appl. **100**, 1–16 (2018)
23. Abbas, W.F., Ahmad, N.D., Zaini, N.B.: Discovering purchasing pattern of sport items using market basket analysis. In: 2013 International Conference on Advanced Computer Science Applications and Technologies, Kuching, pp. 120–125 (2013)
24. Gancheva, V.J.E.U.R.: Market basket analysis of beauty products, Master Thesis, Erasmus University Rotterdam, 2013
25. Bakariya, B., Thakur, G.S., Chaturvedi, K.: An efficient algorithm for extracting infrequent itemsets from weblog. Int. Arab. J. Inf. Technol. **16**(2), 275–280 (2019)
26. Mueller, A.: Fast sequential and parallel algorithms for association rule mining: a comparison (1998)

An Ontology-Based Approach for Risk Evaluation in Human-Machine Interaction

Marcílio F. O. Neto, Renato L. Cagnin, Ivan R. Guilherme(✉), and Jonas Queiroz

São Paulo State University (UNESP), IGCE/DEMAC, Rio Claro, Brazil
{marcilio.oliveira,renato.cagnin90,ivan.guilherme,jonas.queiroz}@unesp.br

Abstract. The Occupational Safety and Health areas have been changing to address the human-machine interaction (HMI) in collaborative workplaces. As regard, the Preliminary Hazard Analysis sub-area must identify, monitor, and evaluate new risks which emerge from such interaction. Although the classical HAZOP method covers this matter, there are many ambiguities in the hazardous scenario descriptions. Thus, cyber-physical systems come into play to help using computer technologies including ontologies, but the approaches do not have an expressive ontology capable of evaluating the risks covering the hazardous and HMI contexts. Therefore, this work aims to propose an application ontology to evaluate risk situations in a collaborative environment, capable of describing all the entities and their relationships. The proposed ontology was evaluated in a simulated scenario, being able to model different hazardous situations in HMI collaborative environments, and supporting the detection and characterization of hazardous events based on the distance between machines and operators.

Keywords: Risk Evaluation · SSN Ontology · Human-Machine Interaction · Hazard Ontology · Occupational Safety and Health

1 Introduction

The Occupational Safety and Health (OSH) demand has been changing along the last years since robots and humans have been working together at the same place seeking for a task to be completed in a more efficient way [1]. In this sense, the context of industrial manufacturing has also used robots to replace human efforts to prevent any damage and risks during repetitive tasks or potential risk/hazardous scenarios [2]. Therefore, these workplaces became collaborative environments allowing the occurrence of new risks [2], which forced the current OSH process to adopt different strategies to identify, analyze and monitor these risk situations [3].

In this context, a new area called Preliminary Hazard Analysis (PHA) [4] has emerged to attend these safety requirements and responsibilities. The PHA process has been applied in several scenarios and suggests a set of safety requirements [4] and strategies that aim to assess and mitigate the different levels of

A. Rocha et al. (Eds.): WorldCIST 2023, LNNS 799, pp. 518–527, 2024.
https://doi.org/10.1007/978-3-031-45642-8_50

risk, and to prevent any collaborator's harm [5]. A common method used in this process is called HAZOP [6].

Recently, Semantic Web and Ontology have been widely applied specifically in the PHA area [4,7], providing the tools that are needed to describe the risk scenarios and ensure unambiguous contexts that machines can understand and process. In [4], an ontological approach to safety analysis is presented to describe the risk concepts as well as the hazard identification, causes and consequences. Likewise, there are many approaches aiming to describe the robotic context to provide a safety collaborative environment, as detailed in [3] and [8]. However, there is no ontological approach that merges the concepts of risk situations and the human-robot interaction, and also provides details to address all the entities and the relationship present in such collaborative environments. Regarding the PHA area, where the main goal is to describe the risk representation itself, there are no detailed entities for the machines, activities, and relationships. Otherwise, the ontologies for robotics' context aim to describe only the physical entities and not how to identify, evaluate or mitigate the risks. In addition, considering safety approaches for this context, most of them just consider a kinematic point of view.

Thus, the main objective of this paper is to propose an ontology-based approach, which uses the Semantic Sensor Network (SSN) ontology to describe the robot context, combined with the Hazardous Situation ontology (presented in [9]), in order to describe the risk context. Moreover, in order to the integration of these ontologies, as well as the risk evaluation in human-machine interaction scenarios In addition, to support the risk evaluation in human-machine interaction scenarios, some other classes and properties were defined.

A simulated scenario, composed of two robots equipped with proximity sensors and human collaborators, has been developed to validate the proposed ontology. It comprises a simulation implemented using the Coppelia Simulator that defines a collaborative situation subject to possible hazardous situation, and a Java application that uses the Apache Jena framework to manage the ontology in order to perform the risk evaluation along the different scenario conditions. The proposed ontology-based application was able to detect different hazardous situations, based on the distance between the operation and machines in a collaborative environment, as well as describe the risks of the hardardous event, given both, the equipment type and operators body parts.

The remaining of this paper is organized as follows. Section 2 discusses the related works, while Sect. 3 describes the proposed ontology-based approach. Section 4 presents the experimental case study, discussing the evaluation and results, and Sect. 5 presents the conclusions and future work.

2 Background and Related Works

2.1 Preliminary Hazard Analysis (PHA)

The Industry 4.0 became a very-known term, where several technologies and research areas have contributed to increase the usage of robots, as well as their interaction with humans in collaborative environments. A review of this topic is presented by [2], that also discusses updates in industrial regulations involving human-robot collaboration, showing that new safety systems must be capable of identifying collisions and mitigating human-collaborator harms. Considering that, [10] proposes tech-categories which relates both the Industry 4.0 and the Occupational Safety and Healthy (OSH) context. As observed, risk management became an important area responsible for ensuring proactive actions to prevent harms and potential hazards in the workplace [11].

In this context, the PHA process is an important role in risk management once it is responsible for consolidating and analyzing all the risk information to decide actions about it, which means this area has the overall understanding of scenarios which can lead to a possible accident [4]. A common method used in this process [6] is the HAZOP (Hazard and Operability Analysis), a specialist's brainstorm of all different areas inside the workspace aiming to identify and mitigate hazardous scenarios [4]. This method can be also applied in the human-machine interaction, however, the HAZOP method demands a large amount of information and produces several others that can interfere on the management chain, leading to a huge human-effort [6].

Due to the human-dependency, the HAZOP method can provide a lot of general information based on text documents and different contexts for the same hazardous scenarios which can generate ambiguity and lack of semantic meaning [6]. This is an obstacle to the application of computational systems with reasoners capable of extracting real meaningful information.

2.2 Safety in Human-Machine Interaction

Human-Machine Interaction is not a simple task, because it needs several changes in the processes to make this interaction flexible, productive, and acceptable for the human-beings [12]. The increase of the usage of robots in the industrial environment has as consequence the emergence of new risks [2]. Thus, the machines must be capable of perceiving the environment and have some automated reasoning to identify the risk and make decisions to avoid them [12] easily and automatically.

Therefore, to achieve the capability of collecting and sharing data, the machines present in the industrial environment must be able to perceive the changes around and exchange information along the network. To allow that, the robots need to use sensors, such as vision sensor, distance sensor, and others to make the task of risk identification feasible. Basically, there are two possibilities to do that, the first one is the usage of fixed sensors along the environment. The second option is the usage of sensors coupled to the robots [3]. Nevertheless, many combinations between these approaches have been used to address that, but some technological limitations of sensors or even in the implementation aspects were found, making the sensor usage a point of interest to be analyzed, once these limitations impact systems in real time applications [2].

Furthermore, [2] detailed three main topics involving collision in a collaborative environment, which can cause human harm. To mitigate the risks, the three points of attention are: Level of Injury by Collision; Minimization of Injury during collision; and Prevent Collision.

2.3 Ontology for Risk

The Hazard Ontology proposed in [4] aims to define an interpretation about the risk concept as well as the hazard representation. The approach uses the fundamental ontology to represent some important concepts related to a risk in the OSH area, such as the situation, event, disposition, and role. Thus, it covers what triggers the hazardous event, the victims, the elements present in the scene and injuries.

The Hazardous Situation Ontology (HSO) proposed in [9] is based on the risk definition norm OSHAS 18001:2007. This application ontology covers the main entities involved in the risk situation, as the whole event, the hazardous situation, the hazardous event, the hazard itself, their causes and consequences, the present exposure and the object exposed to that hazardous event.

3 Ontology for Human-Machine Interaction - OHMI

The Ontology for Risk Evaluation in Human-Machine Interaction (OHMI), proposed in this work, uses the Semantic Sensor Network (SSN) ontology and the Hazardous Situation Ontology (HSO) proposed on [9]. Together they can be used to describe the entities in the industrial environment (e.g., machines, robots and sensors), as well as the hazardous and related situations that humans are subject to in an Human-Machine Interaction (HMI) collaboration scenario. Thus, enabling the description of industrial scenarios, based on their entities and relationships, as well as and development of systems to automate the monitoring and evaluation of hazardous situations and events in such HMI environments.

The SSN ontology defines concepts such as: System, Deployment, Platform, Procedure, Feature and Action - the actions have Result. Based on them, e.g., a robot can be described as a platform hosted on a system with sensors - that can sense (action) some feature - like a distance from the collaborator to the robot, and provide some result. The HSO defines concepts such as: Object, Hazardous Situation, Hazardous Event, Cause, and Consequence. Based on them, e.g., in a collaborative environment there are objects, as collaborators and robots, and a hazardous situation can have some hazardous event, like a collision, e.g., given by the short distance between these entities (cause) that can harm the collaborator (consequence).

The SSN ontology was used to complement the HSO, since the last only focuses on hazardous situation entities and relations, providing no fine grained classes to describe industrial entities, e.g., machines, robots and sensors. In this sense, in order to address the requirements for hazardous evaluation in HMI scenarios, some other relationships and classes were defined by the proposed ontology, such those represented in black in Fig. 1 that overviews the proposed ontology structure. Table 1 describes the main classes of the ontology, grouped by the concepts related to hazardous and industrial equipment. The link between the classes of these two ontologies was made by defining relationships between the class Range to the classes Event and HazardousEvent, where a Range classifies an Event and triggers a HazardousEvent.

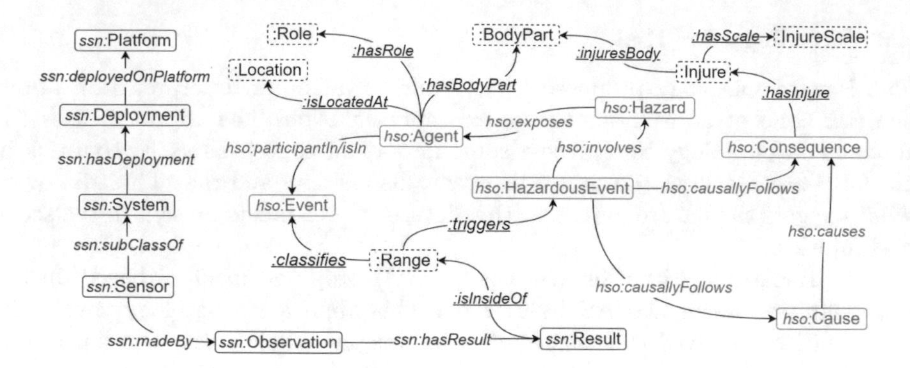

Fig. 1. Ontology structure: main classes and their relationships.

This way, the Range class is used to classify and define the distance between human and machines, enabling to use the ontology to evaluate the hazardous situations, e.g., a collision (described using the HSO concepts) that can be caused by a machine (described using the SSN concepts). In this sense, the proposed ontology has the goal to cover the main entities and relationships of a HMI collaborative environment, enabling queries to obtain the kind of hazardous in a given situation of a collaborative scenario, such as collisions.

In order to enable the use of the ontology to describe a HMI collaborative scenario and use it in a system to automate the monitoring and evaluation of hazardous situations and events, the approach illustrated in Fig. 2 is proposed.

This approach is divided in two phases as illustrated in Fig. 2. The first is part of the system development, comprising the setup of the collaborative environment. During this phase, the environment model should be described based on the proposed ontology. Modeling the environment includes the definition of the entities, their relationships and properties, as well as the several hazardous scenarios and events. Queries can be performed in SPARQL to verify the existence of some specific conditions (in this case those that characterize a hazardous event) based on the individuals, their relationships and properties.

Table 1. Main classes of the proposed hazardous evaluation ontology.

Domain	Class	Description
Hazardous	Agent	Human-beings; Collaborators
	Cause	Reason that leads a risk
	Consequence	Result of a risk
	Hazard	Definition of a risk
	HazardousEvent	Event contains the risk event
Industrial equipment	Platform	Machines
	System	Systems hosted by machines
	Sensor	Devices
	Observation	Devices actions
	Result	Actions results
	Range	The result can be classified by some range

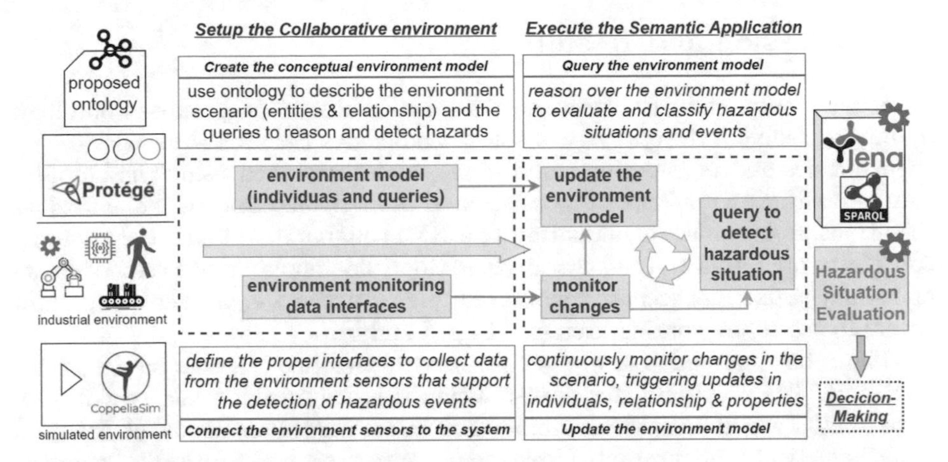

Fig. 2. Ontology-based approach to monitor and evaluate hazardous situations in HMI.

The continuous monitoring and evaluation of hazardous situation in HMI scenarios require information about the environment and its entities. In this sense, also part of this initial phase, is the definition of interfaces to connect the several sensors of the environment in order to retrieve the data that are related and can support the detection of hazardous events. For instance, to retrieve data from the sensors that can provide measurements of the distance between machines and operators, that can include sensors embedded in machines, carried by operators, or installed in the environment. As illustrated in Fig. 2, such information can be obtained from the entities of the industrial environment, but also from those defined in a simulated environment.

The second phase comprises the execution of the semantic application that can be organized in three main modules. The first is responsible to continuously monitor changes in the scenario (through the data retrieved by the sensors) that can also use sensor fusion techniques for triggering the other two modules. Based on the environment changes, the second module is responsible to update the environment model, by adding, removing or modifying the instances of entities, their relationship and properties. Note that some entities can be instantiated when they enter in the environment (e.g., an human operator is detected by a presence sensor in the environment), while other entities can be instantiated previously (e.g., entities that are always part of the environment). Such changes may also trigger the third module, that uses the updated model of the environment to execute the SPARQL queries, in order to detect and classify hazardous situations and events (e.g., when an operator enters in the robot operating area). The output of this module can be used by this or another application to evaluate the hazardous situation and make decision (or even take actions in the system) to avoid or mitigate their consequences. Tools such as Protégé and Apache Jena can be used to support these tasks.

4 Discussion and Results

An experimental case study was developed to evaluate the proposed ontology, mainly regarding if the ontology is able to completely describe the scenarios and their entities and relationships, and if the model described using the ontology can be used by an application to detect and properly describe all the hazardous situations. It comprises a simulation of a HMI collaborative industrial scenario, containing operators and robot arms. The Coppelia Simulator was used to create and simulate different scenarios with hazardous situations, e.g., when an operator enters in the robot working area (see Figs. 4 and 5).

The robot manipulators are equipped with proximity sensors to detect the operators. The proximity sensor disposition is based on the sensitive skin proposed in [3], which aims to distribute the sensors tagged on the robot based on a mathematical configuration. Three scenarios were created for the experiments, where the robot arm can execute 3 tasks in a collaborative environment: pick-and-place, welding and painting. Thus, when the operator crosses the sensors, the semantic application is triggered to detect a hazardous event.

Based on the proposed approach (described in Fig. 2), as part of its first phase, several application interfaces were implemented in the Coppelia Simulator to retrieve the sensor data. Additionally, the environment model of the scenarios was modeled based on the proposed ontology. The modeling of the HMI scenario included the creation of individuals regarding several classes, such as, Consequence (Injured Worker), Injure (Arm, Torso, Hip), Hazard (collision with robot), InjureScale (critical, medium, minimal), Cause (worker very close to robot), Range (long, short, medium), HazardousEvent (near robot). Figure 3 illustrates a fragment of the environment model of a given hazardous scenario.

Also part of the modeling of the environment, some SPARQL queries were defined, aiming to determine if there is a hazardous situation in a given scenario, as well as its type, the event that leads to it, its cause and consequence, the body part of the operator, the harm injured and its scale. Figure 4 illustrates a SPARQL query example that is used to identify a hazardous event, its cause, consequence and type. The result depends on the distance range between the operator and the robot that is retrieved from the environment sensors. In this sense, these queries are performed every time the changes in the environment model involves the distance between operators and robots, aiming to identify the hazardous of the given situation.

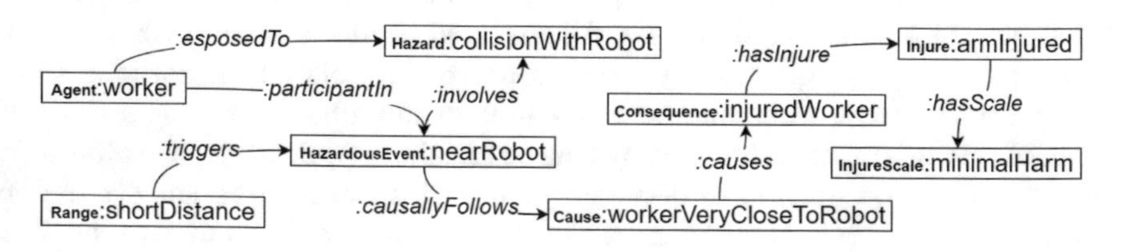

Fig. 3. A fragment of the environment model of a given hazardous scenario.

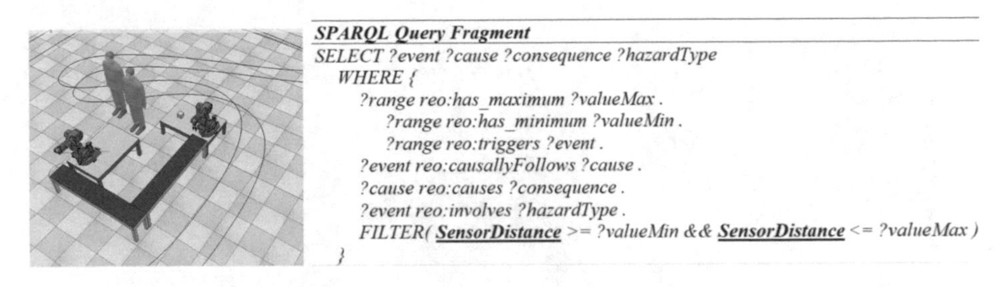

Fig. 4. Example of simulated scenario and SPARQL to identify a hazardous event.

The second phase of the proposed approach comprises an application developed in Java that uses the Apache Jena framework to manage the ontology and SPARQL queries. A module was implemented to interface with the Coppelia Simulator and retrieve the data from the sensors tagged on the robot arm, as well as the entities present on the scenario (e.g., the operators). The Coppelia Simulator can provide the location of the operators in the environment, detection of the operators in the sensors range, also providing the body part of the operator. Note that in a real environment some Computer Vision techniques could be used to obtain information regarding the body parts, since distance sensors cannot provide such kind of information.

Another module was implemented to update the environment model based on the information provided by the previous module. Thus, when the simulation starts, the application reads the operators in the scenario, their location and the measurements of the sensors embedded in the robot arm. This information is used to update the environment model, instantiating the ontology individuals and setting their relations and properties.

As described in the previous section, a third module was implemented to execute the SPARQL queries in order to detect the hazardous situations. Figure 5 illustrates two situations: (left) operator at a medium distance range (less or equal than 1.6 m); (right) operator at a short distance range (less or equal than 0.8 m), as defined by these scenarios. Both situations trigger the semantic application, that executes the SPARQL queries to detect a hazardous event. The short distance range situation (right) is identified as a hazardous situation with possibility of collision event, while the medium range (left) represents a walking event without any risk.

Table 2 presents the results of SPARQL queries for 3 different scenarios. In scenario A, the operator is walking closer to a robot arm that is performing a pick-and-place task. At 0.79 m, it is detected a hazardous situation where the robot can collide with the operator's hip leading to a medium harm. In the scenario B, the robot is performing a painting task, when the arm of the operator is detected in front of the tool, at 0.52 m. This is detected as a hazardous situation with a hazard of skin damage by paint with minimal harm. In the scenario C, the robot is performing a welding task, when the arm of the operator is detected in front of the tool, at 0.63 m. This is detected as a hazardous situation, and given

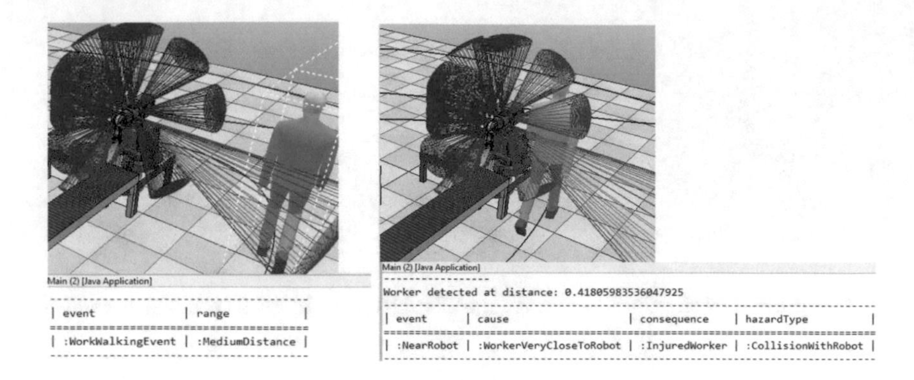

Fig. 5. SPARQL query result for a medium range distance.

Table 2. Hazardous situation detection for three scenarios.

Aspects	scenario A	scenario B	scenario C
Distance	0.7946	0.5216	0.6336
Event	NearRobot	MovingArm	MovingArm
Cause	WorkerVeryCloseToRobot	ArmInFrontPistol	ArmInFrontTorch
Consequence	InjuredWorker	InjuredWorkerByPaint	InjuredWorkerByTorch
HazardType	CollisionWithRobot	SkinDamagePaint	SkinDamageTorch
WorkerName	Worker2	Worker2	Worker2
BodyPart	WorkerHip	WorkerArm	WokerArm
Injure	HipInjured	ArmInjuredByPaint	ArmInjuredByBurn
InjureScale	MediumHarm	MinimalHarm	CriticalHarm

the kind of tool, it can lead to a critical harm by burn injure in the operator's arm.

The performed experiments illustrated the suitability of the proposed ontology to be used to model different hazardous situations and the related entities in HMI collaborative environments, as well as to support the detection and characterization of hazardous events. Besides that, the semantic application approach used to develop the experiments demonstrated how the ontology can be applied in a system to automate the monitoring and evaluation of hazardous situations.

5 Conclusions and Future Work

With the increasing adoption of Industry 4.0 and Cyber Physical Systems concepts, the industries are evolving towards more dynamic and human-machine collaborative environments. Among the several concerns and challenges to achieve and develop systems with such capabilities, is the monitoring and evaluation of hazardous situation during HMI.

In this sense, this paper proposed an ontology-based approach for risk evaluation that can model concepts of both hazards and machines in an HMI collaborative environment, being suitable to develop applications for automated monitoring and identification of hazardous situations. In the experiments the ontology allowed to evaluate risks along different scenario conditions, supporting the identification and characterization of hazardous events based on the distance between machines and operators.

As future work, the ontology can be extended for other risk situations, as well as levels of risks and actions to mitigate them. Additionally, the ontology-based approach can be used not only for standalone simulation applications, but also for those in Industry 4.0 scenarios that consider the use of Digital Twins, in this case represented by the interaction between the entities from physical and the cyber (simulated) environments.

References

1. Goetsch, D.L., Ozon, G.: Occupational Safety and Health for Technologists, Engineers, and Managers, 9 edn. Pearson, London (2019)
2. Robla-Gómez, S., Becerra, V.M., Llata, J.R., González-Sarabia, E., Torre-Ferrero, C., Pérez-Oria, J.: Working together: a review on safe human-robot collaboration in industrial environments. IEEE Access **5**, 26754–26773 (2017)
3. Buizza Avanzini, G., Ceriani, N.M., Zanchettin, A.M., Rocco, P., Bascetta, L.: Safety control of industrial robots based on a distributed distance sensor. IEEE Trans. Control Syst. Technol. **22**(6), 2127–2140 (2014)
4. Zhou, J.: An Ontological Approach to Safety Analysis of Safety-Critical Systems. Ph.D. thesis, Mälardalen University, January 2018. http://www.es.mdh.se/publications/5099-
5. Ericson, C.A.: Hazard Analysis Techniques for System Safety, 1 edn. John Wiley, Hoboken (2005)
6. Wu, C.G., Xu, X., Zhang, B.K., Na, Y.L.: Domain ontology for scenario-based hazard evaluation. Saf. Sci. **60**, 21–34 (2013)
7. Saeed, M., Chelmis, C., Prasanna, V., Thigpen, B., House, R., Blouin, J.: Semantic Web Technologies for External Corrosion Detection in Smart Oil Fields. SPE Western Regional Meeting, vol. All Days, April 2015. SPE-174042-MS
8. Dolganov, A., Letnev, K.: Securing interaction between human and robot using expert control system. In: 2020 International Conference on Industrial Engineering, Applications and Manufacturing (ICIEAM), pp. 1–5 (2020)
9. Lawrynowicz, A., Lawniczak, I.: The hazardous situation ontology design pattern. In: 6th Workshop on Ontology and Semantic Web Patterns (WOP) (2015)
10. Badri, A., Boudreau-Trudel, B., Souissi, A.S.: Occupational health and safety in the industry 4.0 era: a cause for major concern? Saf. Sci. **109**, 403–411 (2018)
11. Tan, X., Yew, K., Low, T.J.: Ontology design for process safety management. In: 2012 International Conference on Computer & Information Science (ICCIS), vol. 1, pp. 114–119 (2012)
12. Belouaer, L., Bouzid, M., Mouaddib, A.I.: Ontology based spatial planning for human-robot interaction. In: 17th International Symposium on Temporal Representation and Reasoning, pp. 103–110 (2010)

Speech Recognition Using HMM-CNN

Lyndainês Santos[1], Nícolas de Araújo Moreira[1,2](✉), Robson Sampaio[1], Raizielle Lima[1], and Francisco Carlos Mattos Brito Oliveira[1,3]

[1] Dell LEAD, Fortaleza, Brazil
{lyndaines.santos,fran}@dellead.com
[2] Teleinformatics Engineering Department, Federal University of Ceara, Building 725, Technology Center, Pici Campus, Fortaleza, Brazil
nicolas@ufc.br
[3] Ceara State University, Fortaleza, Brazil

Abstract. Nowadays, the largest part of speech recognition systems is based on statistical modeling, being Hidden Markov Model (HMM) the most popular among them, followed by Convolutional Neural Networks (CNN), whose learning is based directly on the raw signal. This work presents a study comparing HMM and CNN approaches and a hybrid approach based on HMM and CNN for speech recognition under noise, comparing metrics, such as accuracy, response time, and computational cost to generate models. The experimental results show that the integration between HMM and CNN increased the accuracy by 6% and 8% when compared to HMM and CNN isolated, respectively.

Keywords: Speech Recognition · Hidden Markov Model · Convolutional Neural Network

1 Introduction

Usually, machine learning-based speech recognition is divided into two steps: (i) database generation and training and; (ii) recognition process. The database generation step builds a speaker speech samples collection and the feature selection for selected words. The recognition is a process to identify a pronounced word, comparing the features of the current voice with the features of pre-stored voices. In real-time, the recognition defines, initially, the probability of unknown spoken words on a pre-stored known words database, and then the selection of word is made using maximum likelihood ([1]).

According to [2], the performance of ASR is one of the main challenges for its practical use, once the ideal performance of the system may be achieved using Gaussian Mixture Hidden Markov Model (HMMGMM) according to the arbitrary choice of gaussian mixture but that influences the results of system.

The authors in [3] propose an ASR system, whose different states of Gaussian mixture (GMM) were used to train the ASR system. This system is built to

A. Rocha et al. (Eds.): WorldCIST 2023, LNNS 799, pp. 528–537, 2024.
https://doi.org/10.1007/978-3-031-45642-8_51

recognize a vocabulary containing between 100 and 400 words. The best performance was achieved using 4 states of GMM. This system offers a 88% of accuracy for a vocabulary composed by 400 words [3].

The objective of this paper is to show a comparison of approaches using HMM, CNN and HMM integrated into CNN for speech recognition, comparing the metrics for accuracy, response time, size, and training time to generate the model. For this, the following methods were used: feature extraction, Mel-Frequency Cepstral Coefficients (MFCC) and HMMGMM.

The results show that Markov Model can reduce the recognition response time, and, in addition, reduce the allocated memory space, once it is 10 times smaller than the CNN model. So, HMM has a better processing time optimization and storage space beside the flexibility of the speech recognition system to be operated by users.

The remainder of this paper is structured as follows: Sect. 2 describes the methods and tools used to achieve the results, including information about libraries and requirements; in Sect. 3 we can find the results of accuracy for recognition of classes "one", "two" and "three" using HMM, CNN and HMM integrated to CNN. The conclusion is presented in Sect. 4.

2 Frameworks for Speech Recognition

The authors in [4] developed Julius framework. Julius is an open source high-performance large vocabulary continuous speech recognition (LVCSR) real-time decoder software for speech recognition based on HMM. Its architecture is composed of a dictionary (a set of words with their respective phonetic translation) and a grammar (a description of a structure of a sentence) [4].

A second framework for speech recognition is Sphinx Framework, which provides a state of art speech recognition algorithms, supporting several languages (e.g. English, French, German, etc.).

3 Methodology

The method for recognition is composed, in an overview, of two steps: Initially, feature extraction is done using MFCC. The second step is the input audio verification, represented by data given by extraction, which is analyzed by a Markov Model. This analysis consists of an accuracy comparison done over each previously estimated class on the model using a state probabilistic network. The output of this model represents the class corresponding to the characteristics a likelihood closer to the recorded audio from a speaker one.

3.1 Feature Extraction - Mel-Frequency Cepstral Coefficients

The feature extraction aims at the conversion of the speech waveform into a parametric representation. MFCC is used to extract exclusive resources from speech samples using the short-time energy spectrum of human speech. The MFCC technique uses two filters: linearly spaced filers and logarithmically spaced filters. To capture important phonetic features of speech, the signal is expressed in a Mel-frequency scale. The Mel scale is based mainly on the study of observation of tones or frequencies detected by human hearing. The scale is divided into Mel unities, usually with a linear mapping under 1 kHz and logarithmically spaced over 1 kHz.

The Eq. 1 is used to convert a frequency into a Mel scale:

$$Mel = 2595 \log 10 \left(1 + \frac{1}{700}\right). \tag{1}$$

MFCC is composed of the following steps ([5]): (1) Pre-emphasis - which emphasizes the magnitude of higher frequencies, improving the Signal-to-Noise Ratio (SNR); (2) Framing, which segments the voice sample into frames with a certain number of samples; (3) Hamming window, which divides each frame into windows, minimizing discontinuities - the coefficients of Hamming window is given by:

$$w(n) = 0.54 - 0.46 \cos(2\pi \frac{n}{N}), 0 \leq n \leq N \tag{2}$$

(4) Fast Fourier Transform (FFT), that converts from time to frequency domain:

$$X_k = \sum_{m=0}^{N/2-1} x_{2m} e^{-\frac{2\pi i}{N/2} mk} + e^{-\frac{2\pi i}{N} k} \sum_{m=0}^{N/2-1} x_{2m+1} e^{-\frac{2\pi i}{N/2} mk} \tag{3}$$

(5) Mel-scale filter bank and; (6) Discrete Cosine Transform, that converts the log Mel spectrum to time domain, generating Cepstrum Coefficients of Mel Frequencies (acoustic vectors).

$$X_k = \sum_{n=0}^{N-1} x_n \cos[\frac{\pi}{N} (n + 0.5)] \tag{4}$$

4 Markov Chains, Markov Processes, Markov Model, and Hidden Markov Model

Consider that a physical system is changing through time and it can assume a finite number of states. If some state can be predicted according to its previous state, this process is called Markov Process [6].

In a Markov Chain, the matrix corresponding to the transition state describes the probability of transition between two different states, except on the initial moment ($t = 0$) [7]. This way is necessary to insert a state distribution in $t = 0$ and consider that $\Pi = \{\pi_i = (\pi_1, \pi_2, ...\pi_N)\}$ represents:

$$\sum_{i=1}^{N} \pi_i = 1, \ 0 \leq \pi_i \leq 1 \tag{5}$$

Let's denote k the possible states of a Markov chain and p_{ij} the transition probability from the j-th state to the i-th state, and the stochastic matrix $P = [p_{ij}]$ is the transition matrix, so $\sum_i p_{ij} = 1$. $\mathbf{X}^n$ is the vector of states during the n-th observation, so $\mathbf{X}^{n+1} = P\mathbf{X}^n$ [6].

A transition matrix P is said to be regular if a positive power of this matrix has its elements positive, in other words, exists m such that all entries of P^m are positive. If P is a regular transition matrix and $\mathbf{x}$ is a probability vector so $P^n\mathbf{x} \rightarrow [q_i]_{k\times 1} = -\mathbf{q}$ when $k \rightarrow \infty$ [6].

So, according to [8], Markov can be described according to the following expression 6:

$$M = \{S, O, P, B, \Pi\}, \tag{6}$$

where S denotes the finite states of the model, O represents the observable sequence of outputs, P denotes the transition matrix, and B corresponds to the probability of outputs on the observable sequence of a state. and Π represents the initial state of probabilities of the model.

4.1 Hidden Markov Model

The Hidden Markov Model is a model based on the expansion of a Markov Chain, in other words, is a generalization of a model where the hidden variables which control the components to be selected on each observation are related by a Markov process instead of considering independent [6]. Thus, the model is based on the probability of a sequence of random variables, called states, each one assuming values of some set. The signal is processed in discrete states and time parameters. Thus, an HMM is causal (probabilities depend only upon previous states [9]). An HMM is said to be ergodic if every one of the states has a non-zero probability of occurring given some starting state [9]. There is no requirement that the transition probabilities be symmetric [9].

Now, given observations x_i generated according to the following a mixture model used parametric expression for modeling phenomena [10]:

$$g(x) \approx \hat{g}(x) = \sum_{i=1}^{k} p_i f(x|\theta_i) \tag{7}$$

where $f(x|\theta)$ denotes a standard density, $g(x)$ is a true density of a set of samples, with $p_1 + ... + p_k = 1$ and k large. The right side of the expression is called finite mixture distribution and the various distributions $f(x|\theta_i)$ are the components of the mixture. The usual kernel estimator is given by the Parzen window [10]:

$$\hat{g}(x) = \frac{1}{nh} \sum_{i=1}^{n} K\left(\frac{x - x_i}{h}\right) \tag{8}$$

but the missing data are states of a Markov chain $z_1, ..., z_n$ with state space $\{1, ..., k\}$, where z_i is an integer identifying the underlying state of observation

x_i. Thus, given z_i, $x_i\tilde{f}(x|\theta_{z_i})$ and $z_i = 1$ with probability $p_{z_{i-1}i}$ which depends on the underlying state of the previous observation z_i [10]. This model is called Hidden Markov Model (HMM).

Thus, HMM is a statistical model defined by a Markov model with hidden states (visible states are accessible to external measurements [9]). Because we have access only to the visible states, while ω_i are unobservable, such a full model is called a hidden Markov model [9]. It is widely used in pattern recognition. It is also possible to associate networks that are finite-state machines associated with transition probabilities (Markov networks) [9].

Consider a sequence of states at successive times, the state at any time t is denoted $\omega(t)$. A particular sequence of length T is denoted by $\omega^T = \{\omega(1), \omega(2), ..., \omega(T)\}$.

The transition probabilities are given by [9]: $P(\omega_j(t+1)|\omega(t)) = a_{ij}$, the time-independent probability of having state ω_j at step $t+1$ given that the state at time t is ω_i [9].

$$\begin{aligned} a_{ij} &= P(\omega_j(t+1)|\omega_i(t)) \\ b_{jk} &= P(v_k(t)|\omega_j(t)) \end{aligned} \tag{9}$$

Under normalization conditions:

$$\begin{aligned} \sum_j a_{ij} &= 1, \forall i \\ \sum_k b_{jk} &= 1, \forall j \end{aligned} \tag{10}$$

Assume now a visible symbol $v(t)$ and let's define a sequence of such visible states $\mathbf{V}^T = \{v(1), v(2), ..., v(T)\}$ [9].

In any state $\omega(t)$ the probability of emitting a particular visible state $v_k(t)$ is given by $P(v_k(t)|\omega_j(t)) = b_{jk}$ [9].

A Final or absorbing state ω_0 is one which, if entered, is never left ($a_{00} = 1$) [9].

The learning problem consists of determining the probabilities a_{ij} and b_{jk} given a set of training observations of visible symbols, number of states, and visible states.

5 Convolutional Neural Network

Convolutional Neural Network (ConvNet or CNN) is a deep learning algorithm, more precisely, a multi-level feed-forward network composed of hidden layers. Is a specialized type of neural network model designed for working with two-dimensional image data, although it can be used with one-dimensional and three-dimensional data, so is applied mainly for image and analysis of digital images. So, CNN has a good complex dataset processing capability [11] and solves two problems: feature extraction and pattern recognition/classification model.

A CNN has the ability to learn filters: the innovation of convolutional neural networks is the ability to automatically learn a large number of filters in parallel specific to a training dataset under the constraints of a specific predictive modeling problem, such as image classification.

CNN is widely used in speech analysis, image processing, and network sharing. CNN can classify high-volume and complex data. CNN has many advantages in image processing: it simplifies the process and reduces data extraction and reconstruction steps compared to traditional recognition techniques.

CNN demands less pre-processing than other classification algorithms. CNN reduces images into an easier-to-process format without losing important features for achieving a good prediction rate. It takes input images and assigns relevance (weights) to various aspects of the image.

The inspiration is based on the fact that individual neurons respond to stimuli only in a specific region of the visual field (Receptive Field).

The CNN structure is composed of multiple layers with nodes connected and each layer is composed of bidimensional planes. Each plane is composed of many different neurons. The input layer receives the initial data, the neurons extract local features of input data using the convolution kernel. So, each neuron on the following layer is associated with local resources from the previous layer.

A bidimensional convolution, commonly used in image processing, is the application of a filter/mask/kernel over a bidimensional signal and corresponds, in CNNs, to the multiplication of the weights to the input (dot product and then summed), extracting features such as edges. The training process modifies theses weights of the convolution kernel, so the learning process is based on the continuous weight modification of the convolution kernel [12]. Each kernel convolution will obtain the specific features of each position.

The network is also composed of one or more pairs of convolutional layers and a max pooling layer.

The Pooling layer reduces the size (dimension) of the Convolved Feature, decreasing the demanded computational effort, and can be compared to a diffusion filter. There are two types of Pooling: Max Pooling and Average Pooling. Max Pooling returns the maximum value from the portion of the image covered by the Kernel, in other words, the max pooling layer works as a maximum value filter and it samples the result of convolution and determines the maximum value within a given interval [13]. If the max pooling strategy is used, the robustness of the algorithm will be improved. On the other hand, Max Pooling also performs as a Noise Suppressant. It discards the noisy activation altogether and also performs de-noising along with reduction of dimension. Average Pooling returns the average of all the values from the portion of the image covered by the Kernel. The pooling layer increases the effect of translation invariance and tolerates a slight position deviation. The inferior input can deal with more complex high-level data through two transformation layers, and each layer is connected through connection layers, adequate for image processing [11]. The convolution and pooling layer extract only local features. ReLU stands for Rectified Linear Unit for a non-linear operation. The output is $f(x) = max(0, x)$. ReLU's purpose is to introduce non-linearity in our CNN.

The Fully-Connected Layer learns the non-linear combinations of high-level features and expands local features, so, this layer can be seen as a mapping of primary features. After converting the image into a suitable format for the Multi-Level Perceptron: the image is transformed into a vector and is used as input to a feed-forward neural network, being applied the error backpropagation algorithm for each iteration. In the end, the image is classified using Softmax. If supervised learning is used for classifying, the difference between the real output value and the expected output value is reduced using stochastic gradient descent.

5.1 Proposed Approach: CNN and HMM

In a situation where HMM is used as an acoustic model, the structure is demonstrated in Fig. 1, where b_{nm} indicates the transition probability from state n to the state m: As mentioned above, HMM has an incomplete capability of feature extraction for speech signals. The parameters feature state matrix obtained by HMM can be extracted, trained, and classified. Convolution, pooling, and fully connected layers can be good choices.

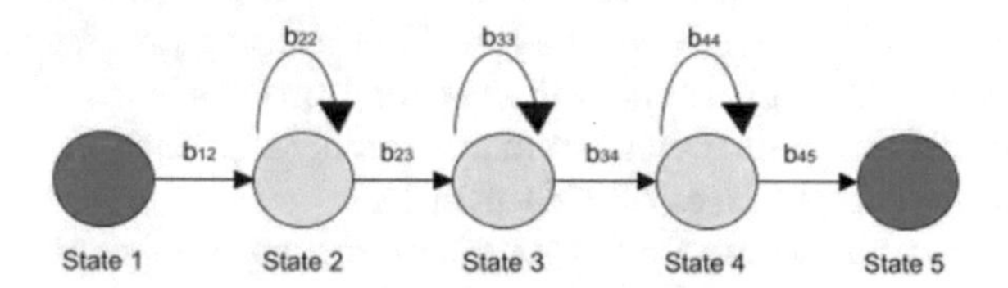

Fig. 1. Structures of acoustic model based on HMM.

Considering the speech signal processing aspect, data dimension can represent completely different meanings for input space, one is time-scale, and the other is frequency domain [14]. The effect of recognition is better on a determined frequency domain. On the convolutional layer, the input signal is a tridimensional matrix composed of a bi-dimensional matrix. The output also contains a bi-dimensional matrix. The input/output model uses the following expression:

$$y_j = \theta^{\left(\sum_i k_{ij} \otimes x_i + b_j\right)}, \tag{11}$$

where b_j denotes a polarization parameter that can be trained and optimized and is part of the convolution operator.

The proposed scheme is shown in Fig. 2: the raw speech data is the input of speech processing steps. The recognition step is then done by CNN followed by HMM. The combination of CNN and HMM demands particular attention to the following problems: (i) speech signals are signals characterized by variability

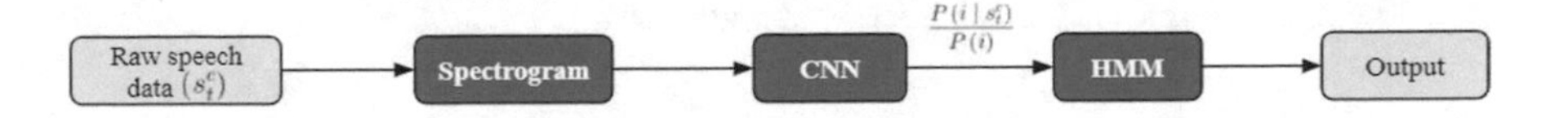

Fig. 2. Block diagram of CNN-HMM based method.

and uncertainty; (ii) the same pronunciation may lead to different results on feature extraction (iii) each data quantity on CNN must be the same, so the parameters matrix of HMM must be in accordance to the input of CNN.

6 Results, Discussions, and Conclusions

The Table 1 shows the precision, recall, and F1-score for the recognition of digits 0 to 9. The precision, in this case, varies from 71% (for number "8") up to 100% (for numbers "2" and "3"). The same accuracy measures are shown in Table 2. In this case, the precision varies from 87% (for number "9") up to 100% (for numbers "0", "2", "4", "5", "6", "8"). The plot for accuracy according to time of architectures using only HMM, only CNN, and using HMM with CNN is shown in Fig. 3. CNN integrated with HMM outperforms HMM or CNN-only based architectures almost every time, reaching accuracy above 90%, increasing the accuracy by 6% and 8%, respectively.

In addition, the results show that Markov Model can reduce the recognition response time, and, in addition, reduce the allocated memory space, once it is 10 times smaller than the CNN model. Thus, HMM presents a better processing time optimization and storage space beside the flexibility of the speech recognition system to be operated by users. Thus, a mixed architecture based on HMM and CNN proved to have a good performance for digit recognition.

Future works may explore the recognition of words and sentences under noise including a filtering step.

Table 1. Results for Precision, Recall and F1-Score using HMM.

Digit	Precision (%)	Recall (%)	F1-Score (%)
0	95	98	97
1	92	93	93
2	100	87	93
3	98	78	87
4	100	90	95
5	97	100	98
6	89	80	84
7	95	98	97
8	71	100	83
9	98	100	99

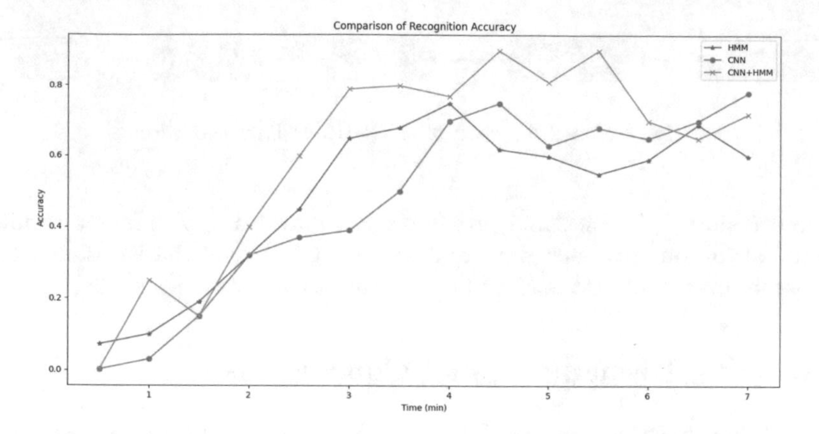

Fig. 3. Accuracy comparison between the three approaches according to time.

Table 2. Results for Precision, Recall and F1-Score using CNN.

Digit	Precision (%)	Recall (%)	F1-Score (%)
0	100	84	91
1	87	87	87
2	100	100	100
3	91	100	95
4	100	100	100
5	100	100	100
6	100	100	100
7	93	100	96
8	100	100	100
9	91	100	95

Acknowledgment. This work was supported by DELL, Inc., *Instituto Desenvolvimento, Estratégia e Conhecimento* (IDESCO) and *Instituto de Estudos, Pesquisas e Projetos da UECE* (IEPRO).

References

1. Shavan, R.S., Sable, G.: An Overview of Speech Recognition Using HMM, pp. 233–238, June 2013
2. Ankit, K., Mohit, D., Tripti, C.: Continuous Hindi speech recognition using Gaussian mixture HMM. In: 2014 IEEE Students' Conference on Electrical, Electronics and Computer Science, pp. 1–5 (2014). https://doi.org/10.1109/SCEECS.2014.6804519.

3. Aggarwal, R.K., Dave, M.: Discriminative techniques for Hindi speech recognition system. In: Singh, C., Singh Lehal, G., Sengupta, J., Sharma, D.V., Goyal, V. (eds.) Information Systems for Indian Languages. ICISIL 2011. CCIS, vol. 139, pp. 261–266. Springer, Berlin, Heidelberg (2011). https://doi.org/10.1007/978-3-642-19403-0_45
4. Lee, A., Kawahara, T., Shikano, K.: Julius - an open source real-time large vocabulary recognition engine. In: Proceedings of the European Conference on Speech Communication and Technology (EUROSPEECH), pp. 1691–1694 (2001)
5. Shafiq, A., et al.: Voice recongition system design aspects for robotic car control. Int J. Comput. Sci. Netw. Secur. **19**(1) (2019)
6. Anton, H., Rorres, C.: Algebra Linear Com Aplicacoes, 8th edn. Bookman, Porto Alegre (2001)
7. Karafiát, M., et al.: Analysis of multilingual sequence-to-sequence speech recognition systems. arXiv preprint arXiv:1811.03451 (2018)
8. Ting, W.: An acoustic recognition model for English speech based on improved HMM algorithm. In: 2019 11th International Conference on Measuring Technology and Mechatronics Automation (ICMTMA), pp. 729–732 (2019). https://doi.org/10.1109/ICMTMA.2019.00167.
9. Richard, O.D., Peter, E.H., David, G.S.: Pattern Classification, 2nd edn. John Wiley and Sons, Hoboken (2009)
10. Gilks, W.R., Richardson, S., Spiegelhalter, D.J.: Markov Chain Monte Carlo in Practice, 1st edn., 512 p. Springer Science (1996). eBook ISBN 9780429170232
11. Chorowski, J., et al.: Attention-based models for speech recognition. arXiv preprint arXiv:1506.07503 (2015)
12. Yin, W., et al.: ABCNN: attention-based convolutional neural network for modeling sentence pairs. Trans. Assoc. Comput. Linguist. **4**, 259–272 (2016)
13. Lo, S.-C.B., et al.: Artificial convolution neural network techniques and applications for lung nodule detection. IEEE Trans. Med. Imaging **14**(4), 711–718 (1995)
14. Ortega, D., et al.: Context-aware neural-based dialog act classification on automatically generated transcriptions. In: ICASSP 2019-2019 IEEE International Conference on Acoustics, Speech and Signal Processing (ICASSP), pp. 7265–7269. IEEE (2019)

AI Solutions for Inter-organisational Care: A Case Based Analysis

Jöran Lindeberg[1(✉)], Martin Henkel[1], Erik Perjons[1], Paul Johannesson[1], and Katarina Fast Lappalainen[2]

[1] Department of Computer and Systems Sciences, Stockholm University, Stockholm, Sweden
{joran,martinh,perjons,pajo}@dsv.su.se
[2] Department of Law, Stockholm University, Stockholm, Sweden
katarina.fast@juridicum.su.se

Abstract. Health care is a complex domain containing large amounts of data, including clinical and administrative data. Furthermore, the domain includes advanced decision-making utilising the collected data. Various IT systems based on AI technologies, such as machine learning, have been promoted as a way to improve both the quality and efficiency of health care. So far, the focus has been on supporting quite narrow and data-intensive activities carried out by a single actor, such as interpreting X-ray images and performing triage. However, providing health care for a single patient can involve a comprehensive process with numerous actors, ranging from home care and primary care to specialist care. In this paper, we examine how existing AI solutions can support a complex care process involving several collaborating actors. We base the examination on a health care case from Swedish elderly care. The case is used to identify multiple problem areas, which are then compared to existing AI solutions.

Keywords: AI · Inter-organisational Collaboration · Healthcare · Health Informatics · Elderly Care

1 Introduction

As recognized by Davenport & Kalakota [4], Artificial Intelligence (AI) and other data-driven solutions have great potential for improving health care. Furthermore, AI has already been used successfully to support single tasks in health care, for example in image processing. However, there are many other parts in healthcare that has the potential to be transformed by AI. From a system theory perspective, many healthcare systems can be characterised as Complex Adaptive Systems (CAS) [18], where the same patient often interacts with a plethora of more or less independent actors, public as well as private, which are governed by different governmental entities. Thus, taken as a whole, the health care of a patient becomes a complex inter-organisational process rather than a number of single isolated tasks.

A. Rocha et al. (Eds.): WorldCIST 2023, LNNS 799, pp. 538–549, 2024.
https://doi.org/10.1007/978-3-031-45642-8_52

If AI could be leveraged to improve the collaboration between healthcare actors that work in complex inter-organisational processes, it has the potential for strong outcomes at the system level. However, there is a lack of implementation of such system-supporting solutions in real healthcare processes. The aim of this paper is to examine how existing AI solutions could support complex care processes. This research is part of an ongoing case study examining how Region Stockholm and Stockholm Municipality care can improve collaboration in the health care area. Improved collaboration is important for all residents—but crucial for elderly people that are in more frequent need of special care (provided by the region) and home care (provided by the municipality). Tentative results from the case study focusing on issues with IT systems and hindering legislation have already been published in a report in Swedish [6]. Parts of the findings were presented as an archetype patient, "Alex", representing the health care contacts of an elderly person. The case was designed to illustrate how the different actors collaborate, and to what extent various IT systems support this process. In this paper, we are using the case of Alex to highlight the many problems we encountered and map them to possible AI solutions.

This paper is structured as follows. Section 2 presents AI and its use in health care, Sect. 3 explains the method employed in the case study, Sect. 4 describes the case study and the illustrative Alex case example, Sect. 5 presents the identified problems and AI solutions, Sect. 6 and Sect. 7 concludes the paper.

2 Related Research

There have been ambitious efforts to give general views of AI in health care. One systematic study [5] from 2019 of AI use in Swedish health care found that there has been much research but considerably less implementation. Furthermore, Mehta el al. [12] conducted a systematic mapping study of 2,421 research articles from between 2013 and 2019, presenting them according to a number of dimensions. Compared to these efforts, we have a different purpose: rather than trying to give a complete overview of all solutions, we particularly focus on solutions that help solve problems in real-world inter-organisational care process. Moreover, we concentrate on solutions that have already been incorporated into clinical practice, which limits the scope considerably. It relates primarily to what according to a classification scheme by Mehta el al. [12] is called "healthcare operations", one of 37 sub-categories.

Generally, AI is "the capability of a machine to imitate intelligent human behavior" [13]. Several different categories, and sub-categories, of AI has been implemented in health care. Drawing mostly from Davenport & Kalakota [4] and [10] we utilise the following high-level categories of AI in health care:

- *Supervised machine learning* uses labeled data as training data to be able to correctly categorize new, unlabeled data.
- *Unsupervised machine learning* finds hidden patterns in unlabeled data, e.g. finding clusters of data according to common characteristics.

- *Rule-based* AI utilizes decision rules based on human domain knowledge. It delivers predictable decisions based on pre-defined rules.
- *Robotic Process Automation* (RPA) is a type of AI that often acts as a human user in a system. It can be both rule based or learn from how a human performs certain simple tasks in e.g. a software interface.

Another way to classify health care AI is according to its data sources [10]. These include, e.g., patient records, communications (e.g. text messages), and surveys (e.g. questionnaires to patients).

The implementation of AI in clinical practice has been limited. According to Panch et al. [15], there are two primary reasons. First, healthcare systems are complex and fragmented, and will not easily change as a result of new technology. Second, most healthcare organisations lack the capacity to collect the necessary training data of sufficient quality while also respecting ethical principles and legal constraints. Thereby, health care AI still has a stronger presence in research than in practice. Existing AI solutions can be categorized according to their implementation level. Many are just concepts, others exist as prototypes, some have also been tested, or actually used in regular care practice.

3 Method

To study the existing care processes and collaborations, a case study was set up, and personnel at both the region and municipality was interviewed. In addition to this, information was also gathered from legislation (that sets rules for information exchange between parties) and documentation of the IT systems that were used. The interviews led to the identification of *problem areas* and also the documentation of an *illustrative example case* that was used to illustrate how the archetypical patient Alex is handled by the municipality and region and the problems encountered.

A total of 8 people were interviewed, out of these 3 were also involved in follow-up interviews to extend the data. Furthermore, 9 IT systems were examined. The problem areas were then synthesised into a thematic map using the Kumu.io tool. An overview of the method can be seen in Fig. 1.

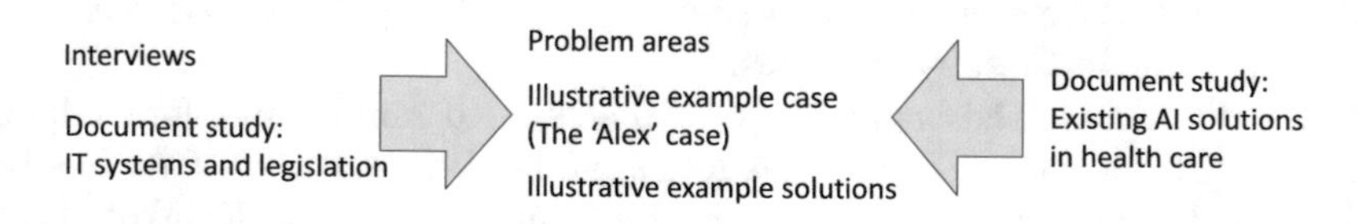

Fig. 1. Overview of the method

The existing AI-based solutions were found by a mapping review [8] of literature from 2015 and onward, including the two significant structured literature surveys [5,12] mentioned in Sect. 2.

Our main interest was in solutions already implemented in clinical practice, and such solutions may or may not have been covered by academic literature. Therefore, our search strategy included both articles from scholars and non-academic sources, such as reports from public organisations. Since the identified problem areas concern inter-organisational collaboration in healthcare, solutions without relevance for such aspects were excluded, e.g. for medical treatments.

To structure the solutions, they were first categorized according to three dimensions: AI type, data source type, and implementation level. Then, the solutions were mapped to the problem areas—each problem area was associated with those AI-based solutions that could help address the problems in that area.

4 The Stockholm Health Care Case

The healthcare in Stockholm is structured into two main responsible entities—Region Stockholm and Stockholm Municipality. While the region is responsible for primary care and special care, the municipality is responsible for home care and nursing homes. To illustrate how the actors get involved in the care we created the archetypical case of Alex based on the interviews (see Fig. 2).

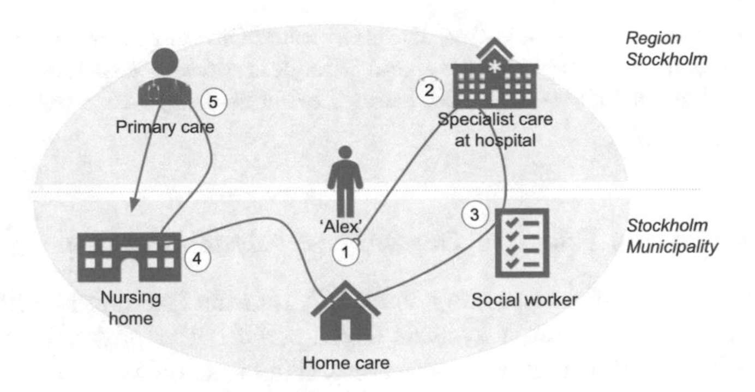

Fig. 2. Overview of the Alex case

The case starts with Alex living at home, receiving home care a few times per week (1). When Alex falls and gets injured, Alex needs to be transferred and treated at a hospital (2). When recovering, a municipality social worker needs to get notified and create a new home care plan (3). However, as there are complications Alex is transferred to a nursing home (4), and regularly needs to visit the primary care (5) for follow-ups.

While the case seems sequential and straightforward, there are a number of issues that makes the case more complex. To start with, the region and municipality are governed by different legislation, sometimes hindering information exchange. Furthermore, all actors use different IT systems that are only partially integrated. Therefore communication to some extent still relies on phone calls and fax.

5 Identified Problems and Solutions

This section presents the identified problems in the case study, and describes the found applicable AI solutions. Out of the 17 solutions, only five had been implemented in regular care.

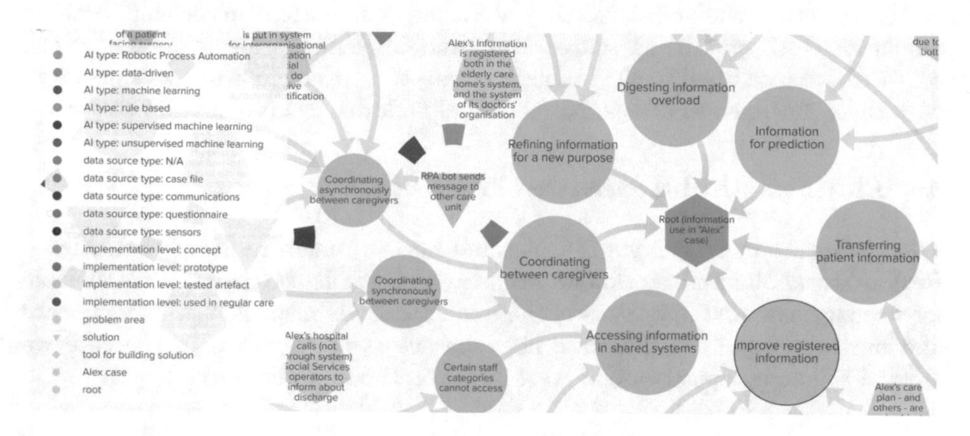

Fig. 3. Central view of the resulting Kumu.io map. Circles represent problem areas, pentagons are issues from the Alex case, and triangles represent solutions. The complete map is available at https://kumu.io/joran/stockholm-health-care-case-ai-solutions-map

5.1 Coordinating Between Caregivers: Asynchronously

When Alex is discharged from the hospital, it informs the municipality's elderly care unit through a common system developed for the particular purpose of exchanging information between caregivers. This is stipulated in specific provisions making the exchange of information possible in this area. This is important as the different organizations involved are generally bound by strict requirements of confidentiality in regard to patients.[1]

However, the system in use cannot send a notification to the elderly care unit. Instead, a social worker has to access the system several times a day to check manually. Obviously, if the communication between caregivers fail, Alex runs an increased risk of being left without attention.

From the point of view of the elderly care unit, a possible solution to address the absence of notifications could be Robotic Process Automation (RPA). One example has been used at a Swedish hospital [20]. An RPA digital agent was instructed to constantly check new arrivals to the Emergency Rooms (ER) unit

[1] Act regulating the cooperation in relation to discharge from inpatient health care *(Lag (2017:612) om samverkan vid utskrivning från sluten hälso- och sjukvård)* and Ch. 25 § 11 Public Access to Information and Secrecy Act (2009:400).

against cases referred by the elderly care administration in another system. When a match was found, the digital agent made a note and also alerted the ER staff. Thereby, no cases were missed and the administrative burden of the nurses was reduced.

As mentioned, an RPA digital agent can operate much like any human user via the application user interfaces. Therefore it can be implemented without the complexity of actual system integration. A disadvantage is that the performed tasks are relatively simple. Moreover, there can also be legal problems related to having a digital agent access a system originally designed for human users only. This was indeed the case of the Swedish region above [20].

Regarding the mentioned risk of follow-up failure, there are also other AI solutions in progress. Murphy et al. [14] has tested a solution that applies machine learning on patients' case file records to identify people who have received worrying results from X-ray scans but have not received follow-up. A similar solution could be used in inter-organisational contexts if enough parts of the patient case files are shared among different caregivers.

Coordination between caregivers can also be accomplished indirectly. In Sweden the different regions share the health service "1177", which people access by either a browser or by making a phone call. The advice that may be given via the service include referrals to caregivers. Hence, even if the actual communication is between 1177 and the patient, the result is a coordination between caregivers. One way to improve this coordination is the use of chatbots, of which one example is Operationskollen [9].

5.2 Coordinating Between Caregivers: Synchronously

In the case study, a problem with asynchronous communication (such as via email) was identified: in some cases the care givers simply needed a speaking partner to solve pressing issues. Therefore a need for the used systems to support synchronous communication between caregivers was identified. Presently, this is primarily solved by phone, but it could preferably be integrated into the systems. One advantage would be increased efficiency, e.g. if users could identify in the system who is available for communication. Moreover, productivity tools produce data about a socio-technical system's operations that in turn can be used for machine learning of other solutions based on machine learning. Consequently, if synchronous communication were logged by the system, this would pave the way for future AI solutions.

Systems with integrated support for synchronous communication between caregivers exist already. For example, such a communication platform can provide communication channels in the form of both chat, voice and video, in order for patients, caregivers, and family members to have a dialogue about the current health status [3].

5.3 Improve Registered Information

In modern health care, a single patient can generate much documented information. In spite of this, important aspects are sometimes omitted. For example, the hospital may miss documenting Alex's ability to handle everyday activities, i.e. Alex's Activities of Daily Living (ADL) status. ADL is important for the municipality's elderly care social workers to determine what help Alex needs at home.

Regardless of whether a piece of information has been registered by a human user or a robot, technology can be used to ensure its quality. Schneider-Kamp [19] has developed different concepts and prototypes of supervised machine learning solutions to ensure quality information in the context of Danish care. The training data was based on client case files, complemented by data about the users. This approach would allow creating solutions capable of understanding which pieces of information that can be expected to be in a report given the type of report and the other information therein. Thus, if Alex's ADL status was missing, this could be flagged for the hospital staff.

5.4 Managing Information Overload

When Alex is discharged from the hospital, there is a risk that a large amount of heterogeneous data is provided by different professionals in the hospital. This may result in information overload for the social worker receiving the discharge information, which may not be relevant to the task at hand [19]. It can also be problematic from both a patient security and data protection perspective.

Schneider-Kamp has, as mentioned above, developed different concepts and prototypes of supervised machine learning solutions [19]. Also for this problem as solution was developed with training data based on client case files, complemented by data about the users. This strategy allows creating solutions capable of: (1) highlighting the most relevant parts of a text mass in a particular situation, (2) showing the latest changes, which tend to contain the most relevant information, or (3) sorting sections according to the estimated needs of a user.

5.5 Refining Information for a New Purpose

During Alex's care process, on several occasions, partially overlapping information will be registered at different places. For example, when Alex is about to be discharged from the hospital, its staff will need to spend time to find information that is already in the case file, and register it in the common system for the exchange of patient information between caregivers.

We found no specific solution to this problem in the scope of this study. On a general level, however, Natural Language Processing (NLP), which often builds on supervised learning, can be used to analyse unstructured clinical notes [4]. Exports or printouts from other systems, as in Alex's case, could also be processed. Another AI solution would be to use RPA to copy and paste data from certain sections of the clinical notes, if the information in the source system was structured enough. Either way, the results would have to be checked and corrected by a human.

In addition, legal issues might arise regarding purpose limitation, which means that personal data, such as health data cannot be reused for another purpose than the original purpose, unless it can be legitimized by, for example, the public interest or research purposes. Repurposing thus has to be compatible with such provisions.[2] Within the EU, recently introduced legislation, the Data Governance Act and the proposed European Health Data Space might expand the possibilities to re-use health data and other personal data.[3]

5.6 Transferring Patient Information

Another recurring problem in Alex' care process was difficulties in transferring patient information due to lacking interoperability. Note that we mean interoperability in more than the technical sense, and also include e.g. legal and policy issues [16]. For example, if Alex, when living at a nursing home, decides to change doctor, there is a risk that the patient case file would have to be transferred by fax and partly re-digitized.

Just like the problem of refining information for a new purpose, no specific solution was encountered for this problem. For want of increased interoperability, a possible work-around could again include NLP and RPA. If paper has been used as medium, scanning with Optical Character Recognition (OCR) can return it to its digital origin. Legal and organizational changes should also be considered, to make way for more flexible solutions.

5.7 Accessing Information in Shared Systems

Yet another recurring problem during Alex's care process, is when relevant information is available in a system that is to some extent shared, but there still are certain organisations or staff categories that do not have access rights. For example, nursing homes as well as the hospital's nutritionists and work therapists are for different reasons excluded from the mentioned inter-organisational exchange system.

We found no solutions to this problem that are in the scope of this study. It is rather a question of regulation and organisation.

[2] See e.g. Art. 5.1 b Regulation (EU) 2016/679 of the European Parliament and of the Council of 27 April 2016 on the protection of natural persons with regard to the processing of personal data and on the free movement of such data, and repealing Directive 95/46/EC (General Data Protection Regulation), GDPR, Chap. 2 § 4 the Swedish Patient Data Act (2008:355).

[3] Regulation (EU) 2022/868 of the European Parliament and of the Council of 30 May 2022 on European data governance and amending Regulation (EU) 2018/1724 (Data Governance Act), European Commission, proposal for a regulation of the European Parliament and of the Council on the European health data space, COM(2022) 197 final.

5.8 Information for Prediction

Predictions can support inter-organisational collaboration with a shared view of a situation. In emergency care processes, patients like Alex risk getting stuck in bottlenecks, causing suffering and delayed interventions. Therefore, caregivers need to understand the patient flows.

Solutions for analysing patient flows include the work of Zlotnik et al. [21], who have tested support vector regression (a form of supervised machine learning) used for prediction and dynamic allocation of nurse staffing needs. The test was conducted with data from a 1,100-bed specialized care hospital. Furthermore, Alharbi et al. have tested an artefact that use unsupervised machine learning with data from patients case to identify patients flows, thereby reducing the time needed from domain experts to share their knowledge [1].

Another problem where accurate predictions are needed is on an individual level. For example: Is Alex likely to have to be readmitted in the near future if discharged? Chen el al. [2] have developed a method for applying machine learning to patient case files in order to predict the probability of hospital readmission. Furthermore, Glover et al. [7] have tested how to calculate how likely it is that a patient will not come to a scheduled consultation. Proactive actions were suggested, such as sending text messages or arranging transport. A related solution is provided by Resta el al. [17], who have tested how to use machine learning on case files to predict how long patients will stay in ER. A similar approach has been taken by Jiang et al. [11], but on a conceptual level.

As part of developing and deploying predictive models to prevent hospital readmission, legal and ethical aspects need to be included in both the design and the use of such solutions.[4]

6 Discussion

This study's focus on collaboration in healthcare affected the results in terms of the AI solutions that we found applicable. This focus on collaboration led to that some of the identified problem areas being left without matching AI solutions, to some extent this could be expected. AI is simply not a panacea for all problems. That said, we did expect it to be easier to find, in particular, fully implemented AI solutions, even if our scope was narrow.

Collaboration among actors means that several IT systems are involved—in the studied case, each actor had at least one system on their own. This pushes *RPA* as an approach with much potential. This is also what Davenport & Kalakota [4] have argued with regard to what they refer to as administrative applications. The lack of interoperability causes healthcare staff spending time copy/pasting, faxing and typing. While waiting for improved interoperability, RPA solutions can reduce parts of this workload.

Evident when performing the interviews in the case was that the sensitive nature of health data has so far had a cooling effect on information exchange.

[4] See e.g. art. 25 of the GDPR.

This effect is there for machine learning approaches as well—without training data it is difficult to research and build AI solutions. However, it must be highlighted that *rule-based* approaches do not need training data, which increases their applicability.

In spite of the limited access to health care data, *supervised machine learning* applications have been developed, for e.g. improving registered information and managing information overload. NLP of unstructured clinical data could be further developed for refining information for a new purpose.

The identified solutions relying on *unsupervised machine learning* were used for prediction, both on an aggregated level, such as patient flows, and on an individual level for predicting how patients would respond to different possible decisions.

The need for clear and transparent legal rules as well as ethical considerations in relation to the processing of sensitive data such as health data is crucial. Whether the legislative initiatives from, for example, the EU, such as the proposal for a European Health Data Space, are an answer to this seems likely, but to what extent remains an open question.

Regarding the value of this study, we would like to argue that grounding it in a case study lead to us to solutions that matter. We do not claim to present a complete picture of the existing solutions. We believe that this study has practical use for IT managers in the healthcare sector looking for possibilities, as well as for researchers searching for new territory in the rapidly expanding AI field.

7 Conclusions

In this paper, we utilised a case study to examine if AI solutions can support complex inter-organizational healthcare. Based on interviews with employees at Region Stockholm and Stockholm Municipality we derived seven problem areas—areas that currently hinder collaboration in healthcare. These problem areas ranged from the support of different communication forms to the transfer of information and coordination among care providers as well as the related legal hurdles.

Given the problem areas, we found several existing AI solutions that could help alleviate the problems. While this is encouraging, we also found that there is extensive research in AI for health care, but much less implemented solutions in practical use, particularly for supporting complex care processes.

The next step in the case study is to examine the IT systems closer to see if the quality and type of data are enough to implement some of the solutions within the existing legal and ethical framework.

Acknowledgements. This study is part of the project Data-driven Solutions in Complex Care Processes (DISAL) which is financed by Region Stockholm.

References

1. Alharbi, A., Bulpitt, A., Johnson, O.A.: Towards unsupervised detection of process models in healthcare. Stud. Health Technol. Inform. **247**, 381–385 (2018)
2. Chen, T., Madanian, S., Airehrour, D., Cherrington, M.: Machine learning methods for hospital readmission prediction: systematic analysis of literature. J. Reliab. Intell. Environ. **8**(1), 49–66 (2022). https://doi.org/10.1007/s40860-021-00165-y
3. Communication Platform. https://cuviva.com/en/offer/
4. Davenport, T., Kalakota, R.: The potential for artificial intelligence in healthcare. Futur. Healthc. J. **6**(2), 94–98 (2019). https://doi.org/10.7861/futurehosp.6-2-94
5. Digitala vårdtjänster och artificiell intelligens i hälso- och sjukvården. Technical report, National Board of Health and Welfare (Socialstyrelsen) (2019). https://www.socialstyrelsen.se/globalassets/sharepoint-dokument/artikelkatalog/ovrigt/2019-10-6431.pdf
6. Fast Lappalainen, K., Fors, U., Henkel, M., Magnusson Sjöberg, C., Perjons, E.: Digitalisering inom vård och omsorg : Ett projekt om samverkan inom och mellan Region Stockholm och Stockholms stad. Technical report. No. 21-003, Stockholm University, Stockholm (2021). http://urn.kb.se/resolve?urn=urn:nbn:se:su:diva-196640
7. Glover, M., et al.: Socioeconomic and demographic predictors of missed opportunities to provide advanced imaging services. J. Am. Coll. Radiol. **14**(11), 1403–1411 (2017). https://doi.org/10.1016/j.jacr.2017.05.015
8. Grant, M.J., Booth, A.: A typology of reviews: an analysis of 14 review types and associated methodologies. Health Inf. Libr. J. **26**(2), 91–108 (2009)
9. Hessius, J.: Håll koll på din operation. http://operationskollen.se
10. Iliashenko, O., Bikkulova, Z., Dubgorn, A.: Opportunities and challenges of artificial intelligence in healthcare. In: E3S Web of Conferences, vol. 110, p. 02028 (2019). https://doi.org/10.1051/e3sconf/201911002028
11. Jiang, S., Chin, K.S., Tsui, K.L.: A universal deep learning approach for modeling the flow of patients under different severities. Comput. Methods Programs Biomed. **154**, 191–203 (2018). https://doi.org/10.1016/j.cmpb.2017.11.003
12. Mehta, N., Pandit, A., Shukla, S.: Transforming healthcare with big data analytics and artificial intelligence: a systematic mapping study. J. Biomed. Inform. **100**, 103311 (2019). https://doi.org/10.1016/j.jbi.2019.103311
13. Dictionary by Merriam-Webster. https://www.merriam-webster.com/
14. Murphy, D.R., et al.: Computerized triggers of big data to detect delays in follow-up of chest imaging results. Chest **150**(3), 613–620 (2016). https://doi.org/10.1016/j.chest.2016.05.001
15. Panch, T., Mattie, H., Celi, L.A.: The inconvenient truth about AI in healthcare. NPJ Digit. Med. **2**(1), 1–3 (2019). https://doi.org/10.1038/s41746-019-0155-4, number: 1 Publisher: Nature Publishing Group
16. Refined eHealth European Interoperability Framework. Tech. Rep. Ref. Ares(2017)4754764, eHealth Network, Brussels (2015)
17. Resta, M., Sonnessa, M., Tànfani, E., Testi, A.: Unsupervised neural networks for clustering emergent patient flows. Oper. Res. Health Care **18**, 41–51 (2018). https://doi.org/10.1016/j.orhc.2017.08.002
18. Rouse, W.B.: Health care as a complex adaptive system: implications for design and management. Bridge-Washington-Natl. Acad. Eng. **38**(1), 17 (2008)
19. Schneider-Kamp, A.: The potential of AI in care optimization: insights from the user-driven co-development of a care integration system. INQUIRY J. Health Care Organ. Provis. Financ. **58** (2021). https://doi.org/10.1177/00469580211017992

20. Västra Götalandsregionen: Innovationsdagen 2019 - Administrativ förenkling med hjälp av datoriserade robotar, February 2019. https://www.youtube.com/watch?v=wan1O_tS4Qs
21. Zlotnik, A., Gallardo-Antolín, A., Alfaro, M.C., Pérez, M.C.P., Martínez, J.M.M.: Emergency department visit forecasting and dynamic nursing staff allocation using machine learning techniques with readily available open-source software. CIN Comput. Inform. Nurs. **33**(8), 368–377 (2015). https://doi.org/10.1097/CIN.0000000000000173

Breast Cancer Stage Determination Using Deep Learning

Elmehdi Aniq[1,2](✉), Mohamed Chakraoui[1], Naoual Mouhni[2,3], Abderrahim Aboulfalah[4], and Hanane Rais[5]

[1] LS3ME, Polydisciplinary Faculty of Khouribga, Sultan Moulay Slimane University, Béni Mellal, Morocco
elmehdi.aniq@gmail.com
[2] LAMIGEP, EMSI Marrakech, Marrakech, Morocco
[3] GL-ISI Team, Department of Informatics, Faculty of Sciences and Technics, UMI-Meknes, Errachidia, Morocco
[4] Obstetrics Gynecology Department, Faculty of Medicine and Pharmacy, Cadi Ayyad University, Marrakech, Morocco
[5] Pathological Anatomy Department, Faculty of Medicine and Pharmacy, Cadi Ayyad University, Marrakech, Morocco

Abstract. Ki-67 is a non-histone nuclear protein located in the nuclear cortex and is one of the essential biomarkers used to provide the proliferative status of cancer cells. Because of the variability in color, morphology and intensity of the cell nuclei Ki-67 is sensitive to chemotherapy and radiation therapy. The proliferation index is usually calculated visually by professional pathologists who assess the total percentage of positive (labeled) cells. This semi-quantitative counting can be the source of some inter- and intra-observer variability and is time consuming. These factors open up a new field of scientific and technological research and development. Artificial intelligence is attracting attention to solve these problems. Our solution is based on deep learning to calculate the percentage of cells labeled by ki-67 protein. The tumor area with x40 magnification is given by the pathologist to use to segment different types of positive, negative or TIL (tumor infiltrating lymphocytes) cells. The calculation of the percentage comes after the counting of the cells using classical image processing techniques. To give the model our satisfaction, we make a comparison with other datasets of the test and we compare it with the diagnosis of pathologists.

Keywords: Ki-67 · Proliferation index · Deep learning · Medical Artificial intelligence · Medical imaging · Digitization of Ki-67 · breast cancer

1 Introduction

Breast cancer is the most diagnosed female cancer in Morocco. The diagnosis of breast cancer is based on clinical examination, breast imaging (ultrasound, mammography, breast MRI). Diagnostic confirmation is based on the anatomo-pathological examination of the breast biopsy. There are several histopathological types of breast cancer according

A. Rocha et al. (Eds.): WorldCIST 2023, LNNS 799, pp. 550–558, 2024.
https://doi.org/10.1007/978-3-031-45642-8_53

to the 2019 WHO (World Health Organization) classification. Nonspecific infiltrating carcinoma is the most common histopathologic type worldwide, with a tendency to present in younger women. Ki-67 is a non-histone nuclear protein expressed only during the active phases of the cell cycle (G1, S, G2 and M), but not during the quiescent phases (G0 and G1 in early phase). It is strongly linked to cell proliferation and has therefore been indicated as an effective marker to assess the proliferation rate of tumors, including breast tumors. The anti Ki-67 antibody, together with the antiestrogen and progesterone antibodies and the HER-2 (Human Epidermal Growth Factor Receptor 2) antibodies are recognized as the main biological indicators to approach the molecular classification of breast cancer (luminal A, luminal B, triple negative, HER2 enriched and basal like), and therefore, it certainly helps to define the best therapeutic strategy especially chemotherapy as well as prognosis assessment. Ki-67 is an accurate marker to infer the proliferative status of carcinoma cells. The assessment of the percentage of labeled infiltrating carcinoma cells is performed on histopathological images analyzed by the pathologist. This semi-quantitative count may be the source of some inter-observer variability [1]. To increase the accuracy of its estimation, it has been suggested to count all tumor cells from different fields of a histopathological section of invasive breast carcinoma. If this is not possible, it is recommended that the pathologist count at least 500 to 1000 cells in representative areas of the entire section. The limitations mentioned above make it necessary to calculate this marker accurately and continuously, which could be evaluated using artificial intelligence (AI).

AI and specifically deep neural networks and convolutional algorithms have a main role in speeding up and improving the accuracy of clinical diagnoses after the great performance and excellent results in different problems both in medicine and in another field, but all this depends on the design of a generalized and robust method of experts in the field to extract handcrafted features. After the birth of convolutional neural networks (CNN) as a class of deep neural networks almost this problem is solved, since CNN have the ability to automate the extraction of features from input data [2], and which shows their effectiveness, especially in the processing of images in different areas such as robotics, bioinformatics, computer vision, etc.

2 Related Works

Currently, with the development of technologies and the large number of digital scans of the tissues, they have opened up new methodologies for diagnosis using major axes such as cell detection, image preprocessing and also semantic segmentation. Which allows for a challenge between researchers, as long as the error exists, despite its smallness.

Recently, research focused on detection and segmentation topics more than traditional methods. Schmidt et al. [3] in their article Cell Detection with Star-convex Polygons suggested a technique include per-pixel cell segmentation with subsequent pixel grouping, or localization of bounding boxes with subsequent shape refinement. In situations of crowded cells, these can be prone to segmentation errors, such as falsely merging bordering cells or suppressing valid cell instances due to the poor approximation with bounding boxes. To overcome these issues, they propose to localize cell nuclei via star convex polygons, which are a much better shape repre- sentation as compared to bounding boxes and thus do not need shape refinement. Negahbani et al. [4], their study is based on the semantic segmentation of different cell types of IHC Ki-67 (labeled by Ki-67, unlabeled and TILS) using their PathoNet model which is based on the autoencoder architecture. Fulawka et al. [5], propose a solution for calculating PI Ki-67 by deep learning and fuzzy interpretations for the detection of hot spots, all this to segment the relevant cells via conventional image processing methods. Sornapudi et al. [6] They apply a CNN, which allows the detection of nuclei on the features extracted from the clustering algorithms which extract the features located in the superpixels. Xie et al. [7] propose an approach they take is to use convolutional neural networks (CNNs) to regress a spatial density map of cells in the image. This approach is applicable to situations where traditional methods based on single cell segmentation do not work well due to clumping or overlapping of cells.

For the traditional methods we found a study based on the methods of mathematical morphologies in order to treat the distribution of the colors of different type, brown or yellow for positive cell and blue for negative cell of IHC Ki-67, that he proposes by Shi et al. [8], and they calculated the correlation between the pixels in order to classify each pixel using the neighboring color space.

On the other hand, Benaggoune et al. [9] propose a pipeline which divides by three parts, the first centered to segment the cells and the background using U-NET with RESNET, then they apply a classification in the second part to see nested cells. A watershed algorithm which separates the nuclei. Finally the application of the random forest to classify the positive or negative cells in order to calculate the PI.

3 Methods

Machine translation (auto-encoding) is a major problem area for semantic segmentation models, whose input and output are both variable length sequences (images). To handle this type of input and output, one can design an architecture with two major components. The first is an encoder: it takes a variable length sequence as input and transforms it into a fixed form state. The second component is a decoder: it maps the coded state of a fixed form to a variable length sequence. This is called an encoder-decoder architecture [10]. Auto-encoder is based most often on the CNN for that we presented in this section the methods and the materials that used in this model.

3.1 Dataset

The image base is generated from SHIDC (Shiraz Histopathological Imaging Data Center) which provides free access to histopathological datasets collected and labeled by experts from Shiraz University and Shiraz University of Medical Sciences. In order to provide a better infrastructure for collaborations between computer scientists and pathologists. This dataset contains microscopic biopsy images of malignant breast tumors, exclusively of invasive ductal carcinoma type. The database contains 2357 labeled images, 1656 images for training and 701 images for testing [4]. The patients who participated in this study were patients with pathology-confirmed diagnosis of breast cancer and breast truncation biopsies performed in the pathology laboratories of hospitals affiliated with Shiraz University of Medical Sciences, Iran. Each image is 1228 × 1228 as dimension and RGB color space (Fig. 1).

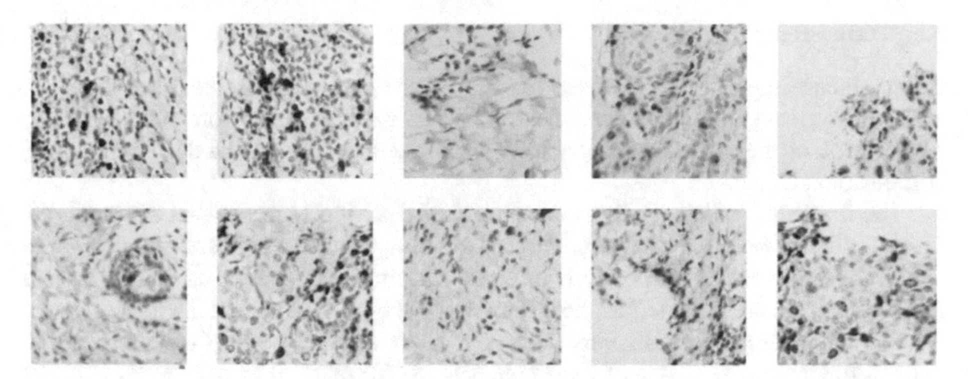

Fig. 1. Dataset samples

Labeling is a difficult, delicate and laborious step and plays a crucial role in the process of correct learning and accuracy of the methods. In the SHIDC-B-Ki-67 set each image contains 69 cells on average and a total of 162,998 cells, it requires the efforts and accuracy of experts, which can be tedious and error-prone for a large number of samples. The choice of the number of classes and the type of labels is another challenge for the process of our work (semantic segmentation), this phase requires to determine the class of each cell, to have small cells and nested cells that imposes the impossibility of the simple labeling (mask) that represents the cells in the image. To overcome this problem, the center of the cell and the cell type are selected as annotations in JSON files.

3.2 Methodology

The first step to implement the two general architectures U-Net [11] and Deep Labv3 [12] in order to see the impact in our dataset and also to control as much as possible what we do and to be able to measure the effects of the choices we will make afterwards (effect on the execution time, the relevance of the results etc. (Fig. 2).

[{"x": 88, "y": 105, "label_id": 1}, {"x": 390, "y":
614, "label_id": 1}, {"x": 542, "y": 623, "label_id":
{"x": 712, "y": 1060, "label_id": 1}, {"x": 611, "y":
"label_id": 2}, {"x": 141, "y": 143, "label_id": 2},
{"x": 1098, "y": 283, "label_id": 2}, {"x": 155, "y":
398, "label_id": 2}, {"x": 40, "y": 400, "label_id":
{"x": 508, "y": 534, "label_id": 2}, {"x": 769, "y":
661, "label_id": 2}, {"x": 894, "y": 678, "label_id":
2}, {"x": 69, "y": 769, "label_id": 2}, {"x": 189, "y
846, "label_id": 2}, {"x": 585, "y": 849, "label_id":
2}, {"x": 268, "y": 928, "label_id": 2}, {"x": 134, "
"y": 1016, "label_id": 2}, {"x": 438, "y": 1019, "lab
1088, "label_id": 2}, {"x": 887, "y": 1093, "label_id
"label_id": 2}, {"x": 748, "y": 1184, "label_id": 2},

Fig. 2. Example of annotation file

3.3 Data Preprocessing

In data science, the data never arrives in a form directly exploitable by the "data scientist", and for the great conditions of deep learning "better results come from better data". In reality 80% of the work is concentrated on the preprocessing of the data: form, type, composition, … etc.

The first thing to do is to always visualize the data on which we want to work later. This could allow us to make better choices about the nature and characteristics of the dataset we are presenting to the network. After having a general vision on the data we notice that there are three terms: the dimension, the type of color and the noise.

Change of dimension is a very important phase for the speed of our network. To make a training on an image of three channels (RGB) of 1228 × 1228 × 3 (4523952 pixel) it takes a lot of time that 256 × 256 × 3 (196608 pixels) (Table 1).

Table 1. Resources consumption and training time

image size	RAM	training time
1228 × 1228 × 3	3 Go	3.75 min
6 × 256 × 3	0.6 Go	0.78 min

And to avoid the problem of degradation between cells and the background. To facilitate the detection of counters we added a denoising step.

4 Results and Discussion

As said before, the networks that we have chosen initially are the general implementation of the u-net and deeplabv3 architecture because of their performance on semantic segmentation. For U-net and Deep Labv3, we notice that the result is bad in spite of the increase of the depth and the number of iterations (epochs), because of two factors, first of all the balancing of the classes since we have each image with a number of cells varying with an average of 69 cells, there are images with a percentage of 100% of the marked cells and vice versa. In order to train a model with u-net and deeplabv3 it is necessary to give for each image a labeling that shows the objects, but for our case we have except the positions of the cells are cancerous or not. The two previous factors are a problematic in the training part.

To solve this problematic we proceed as follow:

4.1 Cropping 256 x 256

The dataset contains 1656 training images, 1228 × 1228 × 3 as a dimension for each one and to make a cropping of 256 × 256 × 3 we will lose some part of the image, so before this step we will resize from 1228 × 1228 × 3 to 1024 × 1024 × 3 and we will end up making a cropping. Finally, for each image we get 16 parts with 5 cells as

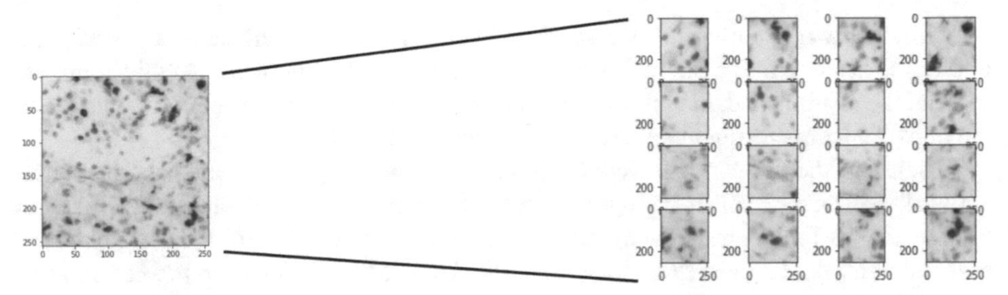

Fig. 3. Cropping the image into 16 patches

average, in order to solve the balancing problem as much as possible. And it also helps us to detect the background as well as possible (Fig. 3).

To solve the last labeling problem we represent each cell with a circle of diameter 9 (the average diameter of the cells) to cover at least 60% of the cell (Fig. 4).

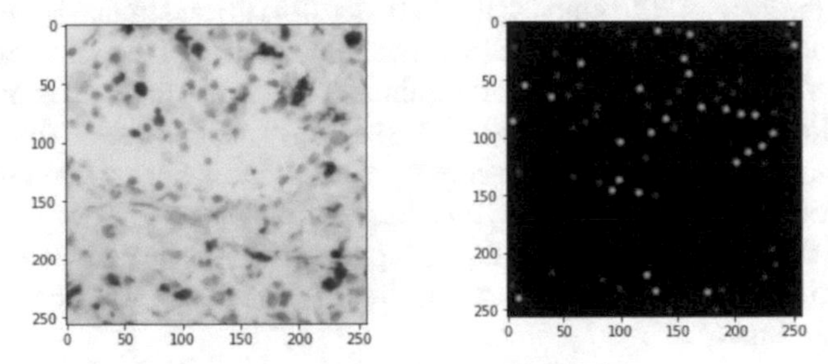

Fig. 4. Labeling

The first line corresponds to the input image, the second to its mask (the expected true response), the third to the U-Net architecture network and the last line to the prediction made by the deep (Fig. 5).

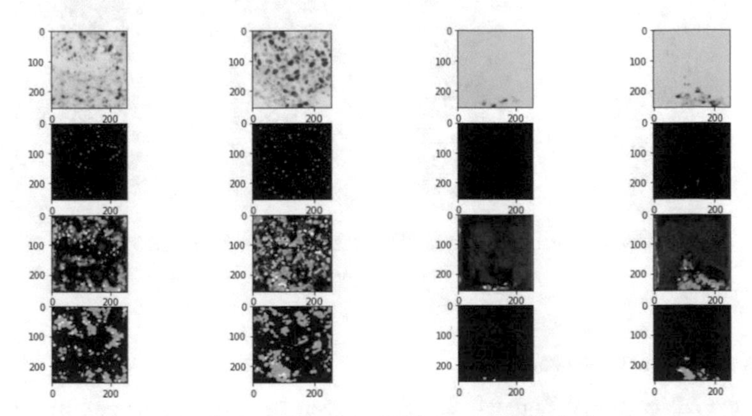

Fig. 5. Results of the U-Net and DeepLabv3

Labv3 network. The third one corresponds to the U-Net architecture network, and finally the last one to the prediction made by the deep Labv3 network. At first sight, we see that the result is bad, in fact, we still do not escape the case we wanted to avoid when the network learns to say that there is no impact all the time.

Therefore the network showed its limits, which was expected. We made the neural network deeper. The following figure shows the progression of the prediction linked to the increase of the neural network depth. The first two lines concern the image and the associated mask. The last three lines concern, from top to bottom, the prediction from the shallowest to the deepest network. We go from a network with 2,142,019 trainable parameters to 31,472,001.

Concretely, to get an idea of how to make the network deeper, we concatenate the first layers of deep Labv3 with the U-Net architecture. For the training we worked with the notion of "batch size" which allows the definition of the number of samples that will be propagated on the network. For example, let's suppose that you have 1050 training samples and that you want to configure a "batch size" equal to 10. The algorithm takes the first 1050/10 = 105 samples (1st to 105th) of the training dataset and trains the network. Then, it takes the second 105 samples (from 106th to 211th) and trains the network again. We can keep doing this procedure until we have propagated all the samples through the network [13]. And this gives a speed of training. Then we are taken "Adam Optimazer" (to converge to the solution (the minimum)) since it gives great tools (momentum and adaptive learning rate) to improve the convergence, the first have a main role of giving the best direction and the second gives a better step of change.

These satisfactory qualitative results are supported by a broader quantitative test that runs through the entire test set (not just four images as in the figure) For each of these three networks (Fig. 6).

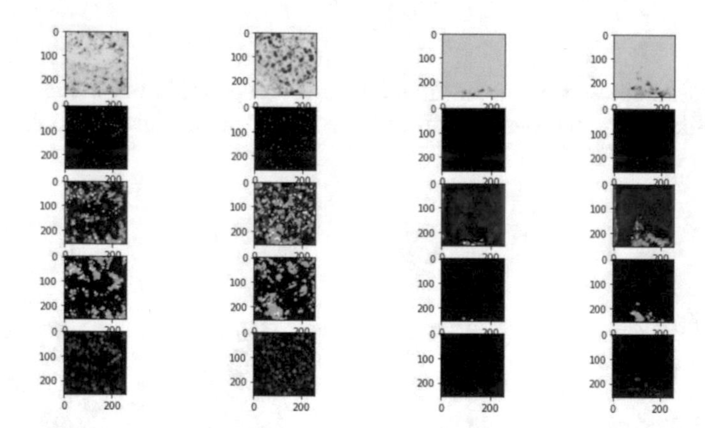

Fig. 6. Results of the U-Net, DeepLabv3 and Our Model

4.1.1 Mean Squared Error

The average MSE (Mean Squared Error) between the predictions and the associated masks was calculated. Here are the results (Table 2).

Table 2. Mean Squared Error

U-Net	deepLabv3	Our
4.117	4.679	0.104

4.1.2 Intersection-Over-Union (IOU)

The IoU is the intersection area between the predicted segmentation and the associated mask divided by the union area between the predicted segmentation and the associated mask [14] (Table 3).

Table 3. Mean Intersection-Over-Union

U-Net	deepLabv3	Our
13.72%	14.26%	50.00%

4.1.3 Dice Coefficient (F1 Score)

The Dice coefficient is equal to 2 * the intersection area divided by the total number of pixels in the two images [15] (Table 4).

Table 4. Dice

U-Net	deepLabv3	Our
68.90%	80.93%	97.01%

5 Conclusion

This paper digitalize of ki-67 proliferation index calculation in breast carcinoma to reduce as much as possible the intra and inter observer variability. We have proposed an approach to detect and classify ki-67 stained cells in order to calculate PI from histopathological images of tumor areas in breast biopsy specimens. This approach uses feature extraction techniques automatically by CNN. Our proposed approach showed high performance by evaluating between PI generated by our pipeline and PI calculated by expert pathologists in this field. Finally, we can increase the accuracy of our approach by adding traditional techniques like data augmentation, etc. Now we have worked on the detection of cancerous or normal cells to standardize a system that allows the extraction of the tumor area using histopathological images based on hematoxylin and eosin (HE) staining and the calculation of PI.

References

1. Fasanella, S., et al.: Proliferative activity in human breast cancer: ki-67 automated evaluation and the influence of different ki-67 equivalent antibodie. Diagn. Pathol. **6**, 1–6. Springer (2011)
2. Jogin, M., Madhulika, M.S., Divya, G.D., Meghana, R.K., Apoorva, S.: Feature extraction using convolution neural networks (CNN) and deep learning. In: 2018 3rd IEEE International Conference on Recent Trends in Electronics, Information Communication Technology (RTEICT), pp. 2319–2323 (2018). https://doi.org/10.1109/RTEICT42901.2018.9012507
3. Schmidt, U., Weigert, M., Broaddus, C., Myers, G.: Cell detection with star-convex polygons. In: Frangi, A., Schnabel, J., Davatzikos, C., Alberola-López, C., Fichtinger, G. (eds.) Medical Image Computing and Computer Assisted Intervention – MICCAI 2018. MICCAI 2018. LNCS, vol. 11071, pp. 265–273. Springer, Cham (2018). https://doi.org/10.1007/978-3-030-00934-2_30
4. Negahbani, F., et al.: Pathonet introduced as a deep neural network backend for evaluation of ki-67 and tumor-infiltrating lymphocytes in breast cancer. Sci. Rep. **11**(1), 1–13 (2021)
5. Fulawka, L., Blaszczyk, J., Tabakov, M., Halon, A.: Assessment of Ki-67 proliferation index with deep learning in DCIS (ductal carcinoma in situ). Sci. Rep. **12**(1), 1–12 (2022)
6. Sornapudi, S., et al.: Deep learning nuclei detection in digitized histology images by superpixels. J. Pathol. Inform. **9**(1), 5 (2018)
7. Xie, W., Noble, J.A., Zisserman, A.: Microscopy cell counting and detection with fully convolutional regression networks. Comput. Methods Biomech. Biomed. Eng. Imaging Vis. **6**(3), 283–292 (2018)
8. Shi, P., Zhong, J., Hong, J., Huang, R., Wang, K., Chen, Y.: Automated ki-67 quantification of immunohistochemical staining image of human nasopharyngeal carcinoma xenografts. Sci. Rep. **6**(1), 1–9 (2016)
9. Benaggoune, K., Masry, Z.A., Ma, J., Devalland, C., Mouss, L.H., Zerhouni, N.: A deep learning pipeline for breast cancer ki-67 proliferation index scoring (2022). arXiv preprint arXiv:2203.07452
10. Yu, E.M., Iglesias, J.E., Dalca, A.V., Sabuncu, M.R.: An auto-encoder strategy for adaptive image segmentation (2020). arXiv preprint arXiv:2004.13903
11. Ronneberger, O., Fischer, P., Brox, T.: U-net: convolutional networks for biomedical image segmentation (2015). https://doi.org/10.48550/ARXIV.1505.04597. https://arxiv.org/abs/1505.04597
12. Yurtkulu, S.C., Sahin, Y.H., Unal, G.: Semantic segmentation with extended deeplabv3 architecture. In: 2019 27th Signal Processing and Communications Applications Conference (SIU), pp. 1–4. IEEE (2019)
13. Mukhoti, J., Kulharia, V., Sanyal, A., Golodetz, S., Torr, P., Dokania, P.: Calibrating deep neural networks using focal loss. Adv. Neural. Inf. Process. Syst. **33**, 15288–15299 (2020)
14. Padilla, R., Netto, S.L., Da Silva, E.A.: A survey on performance metrics for objectdetection algorithms. In: 2020 International Conference on Systems, Signals and Image Processing (IWSSIP), pp. 237–242. IEEE (2020)
15. Eelbode, T., et al.: Optimization for medical image segmentation: theory and practice when evaluating with dice score or Jaccard index. IEEE Trans. Med. Imaging **39**(11), 3679–3690 (2020)

Optimal Cut-Off Points for Pancreatic Cancer Detection Using Deep Learning Techniques

Gintautas Dzemyda[1(✉)], Olga Kurasova[1], Viktor Medvedev[1], Aušra Šubonienė[1], Aistė Gulla[2], Artūras Samuilis[3], Džiugas Jagminas[3], and Kęstutis Strupas[2]

[1] Institute of Data Science and Digital Technologies, Vilnius University, Akademijos Str. 4, 08412 Vilnius, Lithuania
gintautas.dzemyda@mif.vu.lt

[2] Institute of Clinical Medicine, Faculty of Medicine, Vilnius University, M. K. Čiurlionio Str. 21/27, 03101 Vilnius, Lithuania

[3] Institute of Biomedical Sciences, Department of Radiology, Nuclear Medicine and Medical Physics, Faculty of Medicine, Vilnius University, M. K. Čiurlionio Str. 21/27, 03101 Vilnius, Lithuania

Abstract. Deep learning-based approaches are attracting increasing attention in medicine. Applying deep learning models to specific tasks in the medical field is very useful for early disease detection. In this study, the problem of detecting pancreatic cancer by classifying CT images was solved using the provided deep learning-based framework. The choice of the optimal cut-off point is particularly important for an effective assessment of the results of the classification. In order to investigate the capabilities of the deep learning-based framework and to maximise pancreatic cancer diagnostic performance through the selection of optimal cut-off points, experimental studies were carried out using open-access data. Four classification accuracy metrics (Youden index, closest-to-(0,1) criterion, balanced accuracy, g-mean) were used to find the optimal cut-off point in order to balance sensitivity and specificity. This study compares different approaches for finding the optimal cut-off points and selects those that are most clinically relevant.

Keywords: Pancreatic cancer · deep learning · convolutional neural networks · cut-off point · CT scans · Youden index · classification accuracy

1 Introduction

Pancreatic cancer remains one of the leading causes of cancer-related deaths in both men and women. Despite advance in state-of-the-art imaging techniques, early detection remains a challenge. Deep learning (DL) techniques, in general, are gaining increasing attention in medicine [2]. The effectiveness of DL-based algorithms for specific tasks in some medical cases speeds up and simplifies the

A. Rocha et al. (Eds.): WorldCIST 2023, LNNS 799, pp. 559–569, 2024.
https://doi.org/10.1007/978-3-031-45642-8_54

process compared to human-based approaches and is very useful for early disease detection. There is potential for DL approaches to medical image analysis in radiology, and in particular, to detect cancer tumours [4,10]. Computed tomography (CT) and magnetic resonance imaging (MRI) scans are frequently used in the management of diseases of the pancreas and are very useful in diagnosing pancreatic cancer [9]. A CT scan is the preferred initial imaging modality for patients with suspected pancreatic cancer because it is quick and easy to perform and is well tolerated [3,16]. One of the most challenging tasks in interpreting images is of disease detection: rapid differentiation of abnormalities from normal background anatomy [11].

DL-based methods became a powerful tool for diagnostic CT image analysis. Convolutional neural networks (CNNs) are indispensable tools for addressing image analysis problems. In [10], the authors applied CNN in distinguishing CT images of pancreatic cancer tissue from non-cancerous pancreatic tissue. The performance of CNN was compared with radiology reports and showed that the proposed CNN was able to achieve similar performance compared with radiologist results. The authors of [4] developed a tool comprising a segmentation convolutional neural network and a classifier ensembling five CNNs.

Experimental results in the scientific literature prove that the use of CNN or other techniques based on DL is feasible and promising for clinical use, however, pancreatic cancer segmentation, classification tasks, and strategies to improve accuracy remain under the scrutiny of the scientific community. Collaboration between radiologists, computer scientists and cancer researchers, as well as the exchange of experience in this field, recently made an important contribution to the problem of diagnosis and subsequent treatment of cancer. When dealing with medical image analysis, data shortages are inevitable. Taking into account the specificities of DL, the amount of available training data must be increased. The images corresponding to the CT scan slices are cropped into square subregions (patches) that are smaller in size than the images, thus, many patches can be extracted from one image. Another challenging task is to evaluate the performance of the DL-based model to determine whether a patch classified by the model is cancerous or non-cancerous and when a patient should be considered to have pancreatic cancer or healthy. When given an input patch, the trained CNN outputs probabilities for the different classes. When analysing medical data, the most common way of assessing the model performance is to measure sensitivity and specificity and to strike a trade-off between these criteria. To make a prediction of which class the patch is assigned to, we need to select a threshold value (cut-off point). An overview of the scientific literature (see [7,10]) reveals that the most commonly used metrics for cut-off points selection are: Youden index, balanced accuracy, geometric mean, and shortest distance to upper left corner ($(0,1)$ criterion). This paper aims to identify which cut-off point approach is most suitable for solving the pancreatic cancer detection problem.

2 Methodology

This section provides a CNN-based methodology to address the problem of pancreatic cancer detection. The methodology includes an approach for selecting optimal cut-off points to improve the overall classification accuracy. The image pre-processing used to train and test the CNN model is also described in detail.

2.1 Classification Accuracy and Cut-Off Point Selection

After DL-based model training, the model performance must be evaluated. In disease diagnosis, the simplest classification rule is binary, i.e. diseased or not diseased. The probability that a model correctly classifies a non-diseased or diseased subject is defined as specificity or sensitivity, respectively.

In the context of the binary classification problem, the confusion matrix is one of the best and most intuitive representations for evaluating the results of a classification model. The confusion matrix is not a metric for estimating the model, but it does provide an indication of the results obtained and most of the metrics are calculated from it [13]. Typically, elements of the confusion matrix are computed: number of correctly classified positive examples (True Positives, TP), number of incorrectly classified positive examples (False Negatives, FN), number of correctly classified negative examples (True Negatives, TN), number of incorrectly classified negative examples (False Positives, FP). Then the evaluated measures (sensitivity and specificity) can be computed. The specificity is the number of correctly predicted negative cases divided by the number of all negative cases [1]: $\text{Specificity} = \text{TN}/(\text{TN} + \text{FP})$. The sensitivity is the number of correctly predicted positive cases divided by the number of all positive cases: $\text{Sensitivity} = \text{TP}/(\text{TP} + \text{FN})$.

When dealing with classification problems of a medical nature, it is necessary to find a trade-off between sensitivity and specificity. In addition, to predict which class an object belongs to, it is necessary to select a cut-off point also known as a threshold. The optimal cut-off point is of importance in the analysis of medical data to strike a balance between sensitivity and specificity. The choice of a proper cut-off point is particularly important for the efficient estimation of the classification results. Classification accuracy metrics are individually maximized or minimized to obtain the most appropriate optimal cut-off points [1]. All the metrics are based on Receiver Operating Characteristic (ROC) curve analysis. The ROC curve is a plot of sensitivity versus 1-specificity for all possible cut-off points.

One frequently used metric for selecting the optimal cut-off point is the Youden index, first introduced in the medical literature by Youden [15]. Youden index [8] is defined for all points of a ROC curve, and the maximum value may be used as a criterion for selecting the optimal cut-off point (Eq. (1)) and ranges between 0 and 1. It measures the model ability to avoid miss-classifications. In the closest-to-(0,1) criterion, the optimal cut-off point is defined as the point closest to the point on the ROC curve [6,12]. The implication of this approach is that the point on the curve closest to perfection should correspond to the optimal cut-off point chosen from all available cut-off points, which intuitively minimises mis-classification. This criterion can be thought of as finding the shortest radius originating at (0,1) and ending at the ROC curve (Eq. (2)). Although the Youden index and the closest-to-(0,1) criterion use the values associated with the ROC curve, however, they may not select the same optimal cut-off point.

Balanced accuracy (Eq. (3)), the arithmetic mean of sensitivity and specificity, can also be used to determine the optimal cut-off point [7]. The geometric mean (Eq. (4)) of sensitivity and specificity is also used for this purpose.

$$\text{Youden index} = \max_t (\text{Sensitivity}(t) + \text{Specificity}(t) - 1) \quad (1)$$

$$\text{(0,1)-criterion} = \min_t \sqrt{(1 - \text{Sensitivity}(t))^2 + (1 - \text{Specificity}(t))^2} \quad (2)$$

$$\text{balanced accuracy} = \max_t ((\text{Sensitivity}(t) + \text{Specificity}(t))/2) \quad (3)$$

$$\text{g-mean} = \max_t \sqrt{\text{Sensitivity}(t) \times \text{Specificity}(t)} \quad (4)$$

Here, t denotes the cut-off point for which the corresponding metric is optimal.

The aim of this study is to compare different optimal cut-off points using several metrics and to determine which is the most clinically relevant for assessing diagnostic outcomes.

2.2 Image Sets and Data Pre-processing

To find optimal cut-off points and assess their impact on classification accuracy, open-access CT databases were used: the Memorial Sloan Kettering Cancer Center dataset consisted of 281 CT images of pancreatic cancer patients, and the TCIA dataset [5] consisted of CT images of 80 people with normal pancreas from the US National Institutes of Health Clinical Center [14]. All CT scans are annotated in such a way: pancreas contours in all images; pancreatic cancer contours in images with pancreatic cancer.

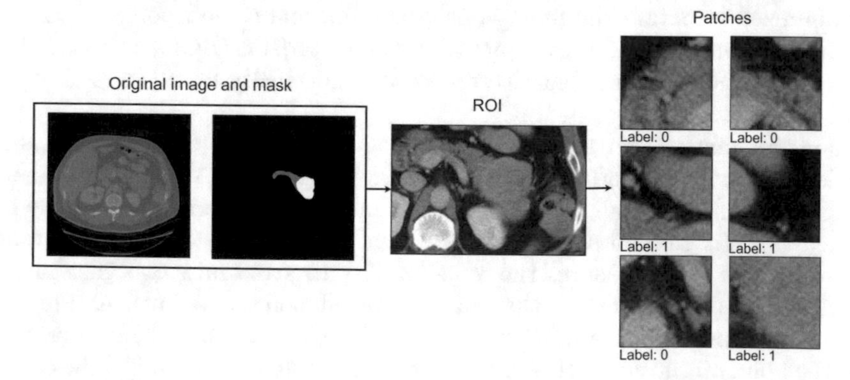

Fig. 1. The process of image pre-processing by cropping and labelling patches.

To achieve high classification performance and reliable diagnostic results, the number of images to train DL models should be as large as possible. One way of increasing the amount of data is to get many smaller images (patches) from one large CT image. In addition, the problem to be solved is confined to the detection of pancreatic cancer, so it is not appropriate to analyse all abdominal CT images but rather to limit the region of interest (ROI). In this case, ROI is the pancreas and the tumour. The process of image pre-processing is as follows (see Fig. 1): (1) Separate images (slices) are extracted from each patient's CT scan; (2) Each image is cropped according to the ROI (the ROI is a rectangle bounding the pancreas); (3) Pixel-level transformations are applied; (4) The resulting images are cropped into smaller patches (size of 160×160 px); (5) A label is assigned for each patch: 0 for a healthy pancreas, and 1 for a patch with a tumour. Data used for model training, validation, and testing are presented in Table 1.

2.3 Deep Learning-Based Framework

The main objective is to classify patients into healthy patients and those who have pancreatic cancer visible in CT. Since the number of patients is limited, patch-based classification is performed. A CNN model (see Fig. 2) is designed and trained using cancerous and non-cancerous patches. The architecture proposed by the authors of [10] was adopted as the basis since it demonstrated good classification results during the experiments. The model consists of six convolutional layers, followed by a ReLU activation function, three max-pooling layers, and two fully connected layers with a ReLU activation function and a sigmoid function at the output. The following hyper-parameter settings are used in the experimental investigation: Epochs = 50, BatchSize = 128, LearningRate = 0.001, Optimiser = Adam, and PatienceForEarlyStopping = 8.

The CNN model was trained using the data described in Table 1. Simultaneously, the model was validated to control the convergence process and to stop training early to avoid over-fitting. Finally, the model was tested by evaluating the classification performance and finding optimal cut-off points at which the classification accuracy would be highest. The results obtained are described in Sect. 3. The testing process is as follows:

1. Using the trained CNN model, the probability (neural network output) that a patch contains the cancer is determined. The patch is assigned to cancerous or non-cancerous based on the cut-off point (cut-off$_a$): if the probability is

Table 1. Data used for training, validation, and testing

	Train	Validation	Test	Total
Patients	253	54	54	361
Non-cancerous patches	43173	20800	24458	88431
Cancerous patches	11243	2303	2163	12303

```
==========================================================================================
Layer (type:depth-idx)          Input Shape            Output Shape           Kernel Shape
==========================================================================================
Net2                            [1, 1, 50, 50]         [1, 2]                 --
├─Conv2d: 1-1                   [1, 1, 50, 50]         [1, 16, 46, 46]        [5, 5]
├─ReLU: 1-2                     [1, 16, 46, 46]        [1, 16, 46, 46]        --
├─Conv2d: 1-3                   [1, 16, 46, 46]        [1, 32, 42, 42]        [5, 5]
├─ReLU: 1-4                     [1, 32, 42, 42]        [1, 32, 42, 42]        --
├─MaxPool2d: 1-5                [1, 32, 42, 42]        [1, 32, 21, 21]        2
├─Conv2d: 1-6                   [1, 32, 21, 21]        [1, 64, 19, 19]        [3, 3]
├─ReLU: 1-7                     [1, 64, 19, 19]        [1, 64, 19, 19]        --
├─Conv2d: 1-8                   [1, 64, 19, 19]        [1, 64, 17, 17]        [3, 3]
├─ReLU: 1-9                     [1, 64, 17, 17]        [1, 64, 17, 17]        --
├─MaxPool2d: 1-10               [1, 64, 17, 17]        [1, 64, 8, 8]          2
├─Conv2d: 1-11                  [1, 64, 8, 8]          [1, 128, 6, 6]         [3, 3]
├─ReLU: 1-12                    [1, 128, 6, 6]         [1, 128, 6, 6]         --
├─Linear: 1-22                  [1, 32]                [1, 2]                 --
├─Conv2d: 1-14                  [1, 128, 6, 6]         [1, 128, 4, 4]         [3, 3]
├─ReLU: 1-15                    [1, 128, 4, 4]         [1, 128, 4, 4]         --
├─MaxPool2d: 1-16               [1, 128, 4, 4]         [1, 128, 2, 2]         2
├─Linear: 1-17                  [1, 512]               [1, 32]                --
├─ReLU: 1-18                    [1, 32]                [1, 32]                --
├─Dropout2d: 1-19               [1, 32]                [1, 32]                --
├─Linear: 1-20                  [1, 32]                [1, 32]                --
├─ReLU: 1-21                    [1, 32]                [1, 32]                --
├─Linear: 1-22                  [1, 32]                [1, 2]                 --
==========================================================================================
```

Fig. 2. The summary of the CNN model.

less than cut-off$_a$, the patch is classified as non-cancerous; if the probability is higher than cut-off$_a$, the patch is classified as cancerous.

2. Patients are classified as having pancreatic cancer or not, depending on the proportion of patches predicted as cancerous by CNN using a cut-off point (cut-off$_b$): if the ratio of predicted cancerous and all cancerous patches for cancer patients is more than cut-off$_b$, then the patient is classified as having cancer; if the ratio of predicted cancerous and all patches with pancreas for healthy patients is less than cut-off$_b$, the patient is classified as healthy.

Thus, it is necessary to find the optimal values of two types of cut-off points: cut-off$_a$ and cut-off$_b$.

3 Evaluation

An experimental study was carried out to evaluate the CNN-based framework presented in Subsect. 2.3. Due to the fact that the patients are classified according to the patches from their CT images, not only the cut-off points for the patient classification (cut-off$_b$) but also cut-off points for the patch classification (cut-off$_a$) have to be evaluated.

The patches are classified by changing the cut-off$_a$ point in the interval $(0, 1)$ due to the interval of values of the activation function used in the CNN output layer. At each point, the four classification accuracy metrics (g-mean, (0,1)-criterion, Youden index, and balanced accuracy) are computed, and the results obtained are presented in Fig. 3. The dots on the curves and the dashed lines indicate the optimal cut-off points at which the values of classification metrics are minimum in the case of (0,1)-criterion and maximum otherwise. At the cut-off point of 0.5, the two metrics (Youden index and balanced accuracy) reach the highest values.

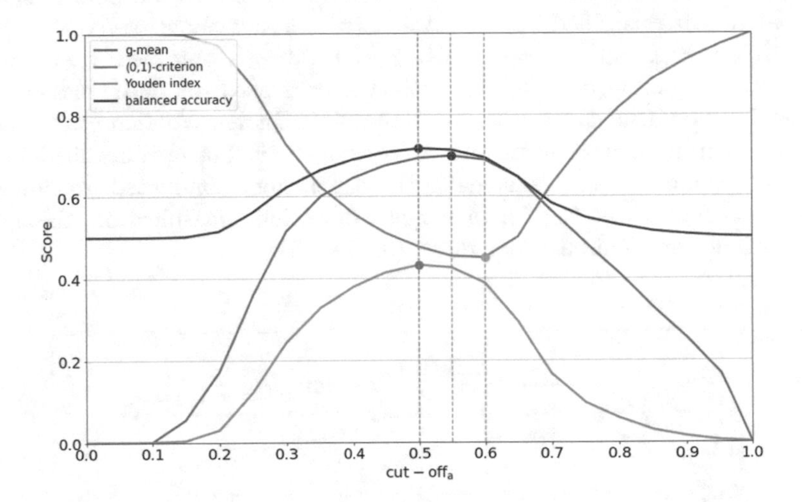

Fig. 3. The dependence of patch classification accuracy on the cut-off$_a$ points. The dots on the curves and the dashed lines represent the optimal cut-off points.

Table 2. The normalized confusion matrices: (a) patch classification, cut-off$_a$ = 0.5, (b) patch classification, cut-off$_a$ = 0.55, (c) patch classification, cut-off$_a$ = 0.6, (d) patient classification cut-off$_b$ = 0.5.

(a)

Actual \ Predicted	1	0
1	0.90	0.10
0	0.46	0.54

(b)

Actual \ Predicted	1	0
1	0.86	0.14
0	0.43	0.57

(c)

Actual \ Predicted	1	0
1	0.78	0.22
0	0.39	0.61

(d)

Actual \ Predicted	1	0
1	0.94	0.06
0	0.15	0.85

The highest g-mean value is obtained when the cut-off point is 0.55. The cut-off point of 0.6 is optimal by (0,1)-criterion. We can see that, in the case of patch classification, using these metrics yields quite similar optimal cut-off points that lie in the interval $(0.5, 0.6)$.

The normalized confusion matrices for each optimal cut-off point are presented in Table 2(a–c). It can be concluded that the confusion matrix given by Table 2(a) is the most appropriate to maximise the diagnostic performance, as the true positive (TP) value is the highest compared to the other cases, which implies a more accurate identification of cancerous patches. This confusion matrix is obtained when cutoff$_a$ = 0.5. Figure 3 indicates that this value of the cut-off point is optimal in terms of the Youden index and balanced accuracy. This indicates that these measures are the most appropriate for assessing the cut-off points for the data analyzed (see Table 1).

The dependence of the classification accuracy on the cut-off points (cut-off$_b$) when patients are classified into two classes (with and without pancreatic cancer) is depicted in Fig. 4. The classification performance is obtained when only the optimal cut-off point (cut-off$_a$) for each metric is used for patch classification. The results show that the optimal cut-off points for the Youden index and the Balanced accuracy are the same and are equal to 0.45. The optimal cut-off points for these two metrics also coincide in the experiment conducted previously for patch classification (see Fig. 3). In the case of patient classification, the optimal cut-off points are within a wider range (0.25, 0.45).

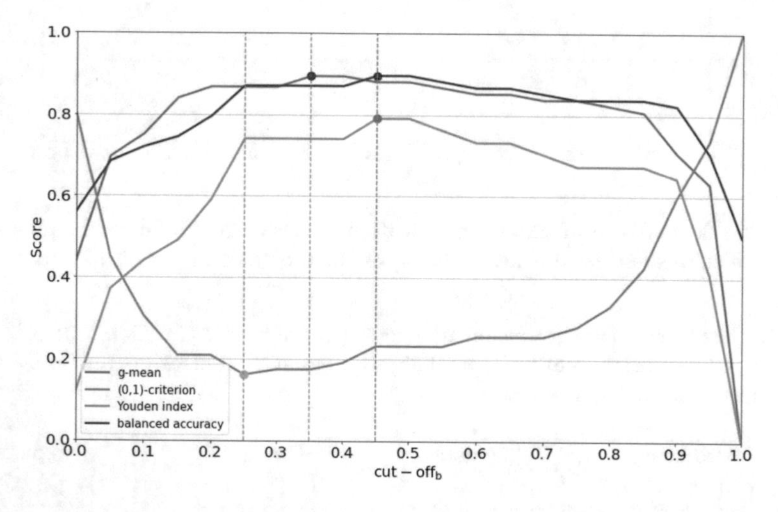

Fig. 4. The dependence of patient classification accuracy on the cut-off$_b$ points. Only the optimal cut-off$_a$ point for each metric was used for patch classification (see Fig. 3).

In order to investigate a patient classification approach where the same cut-off points (cut-off$_b$) are used for both patch and patient classification, another experiment was carried out. The results obtained indicate that the optimal cut-off point is the same for all four metrics (see Fig. 5). The normalized confusion matrix for patient classification is presented in Table 2(d). It can be concluded that the true positive (TP) rate is quite high, which means that patients with pancreatic cancer are accurately identified.

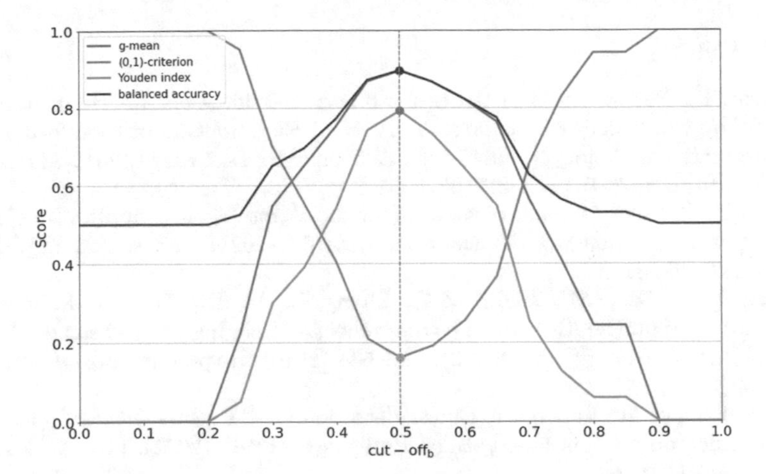

Fig. 5. The dependence of patient classification accuracy on the cut-off$_b$ points. The same cut-off points were used for both patch and patient classification.

4 Discussion and Conclusions

In this study, the problem of pancreatic cancer detection was addressed using the provided framework based on deep learning. CT images of patients with and without pancreatic cancer were analyzed. The images were cropped into patches and CNN was trained to determine the probability that the patch is cancerous and to classify the patch as cancerous or non-cancerous based on the optimal cut-off point. Patients were classified as pancreatic cancer or non-cancer based on the proportion of patches that the CNN identified as cancerous using the cut-off point. The choice of the optimal cut-off point is particularly important for an effective assessment of the results of the classification. Four classification accuracy metrics (Youden index, closest-to-(0,1) criterion, balanced accuracy, g-mean) were used to find the optimal cut-off point in order to balance sensitivity and specificity. In order to investigate the capabilities of the described framework and to maximise the diagnostic performance through the selection of optimal cut-off points, experimental studies were carried out using open-access data. It has been noted that the classification metrics used to determine the optimal cut-off point influence the classification results in order to maximise the number of cancer cases (true positives) that are actually cancerous. The study shows that the Youden index and balanced accuracy are the most appropriate for assessing the cut-off points for the data analyzed.

The preliminary results are potentially promising. The findings are expected to be useful for the diagnosis of pancreatic cancer by analysing CT images obtained at Vilnius University Hospital Santaros Klinikos.

This study was approved by the Institutional Ethics Committee of Vilnius University (protocol code Nr. 158200-17-941-455, date of approval October, 2017).

References

1. Berrar, D.: Performance measures for binary classification. In: Ranganathan, S., Gribskov, M., Nakai, K., Schönbach, C. (eds.) Encyclopedia of Bioinformatics and Computational Biology, pp. 546–560. Academic Press, Oxford (2019). https://doi.org/10.1016/B978-0-12-809633-8.20351-8
2. Cardobi, N., et al.: An overview of artificial intelligence applications in liver and pancreatic imaging. Cancers **13**(9), 2162 (2021). https://doi.org/10.3390/cancers13092162
3. Chen, F.M., Ni, J.M., Zhang, Z.Y., Zhang, L., Li, B., Jiang, C.J.: Presurgical evaluation of pancreatic cancer: a comprehensive imaging comparison of CT versus MRI. Am. J. Roentgenol. **206**(3), 526–535 (2016). https://doi.org/10.2214/AJR.15.15236
4. Chen, P.T., et al.: Pancreatic cancer detection on CT scans with deep learning: a nationwide population-based study. Radiology 220152 (2022). https://doi.org/10.1148/radiol.220152
5. Clark, K., et al.: The cancer imaging archive (TCIA): maintaining and operating a public information repository. J. Digit. Imaging **26**(6), 1045–1057 (2013). https://doi.org/10.1007/s10278-013-9622-7
6. Coffin, M., Sukhatme, S.: Receiver operating characteristic studies and measurement errors. Biometrics 823–837 (1997)
7. Cubuk, C., et al.: Clinical likelihood ratios and balanced accuracy for 44 in silico tools against multiple large-scale functional assays of cancer susceptibility genes. Genet. Med. **23**(11), 2096–2104 (2021). https://doi.org/10.1038/s41436-021-01265-z
8. Fluss, R., Faraggi, D., Reiser, B.: Estimation of the Youden index and its associated cutoff point. Biom. J. **47**(4), 458–472 (2005). https://doi.org/10.1002/bimj.200410135
9. Kumar, H., DeSouza, S.V., Petrov, M.S.: Automated pancreas segmentation from computed tomography and magnetic resonance images: a systematic review. Comput. Methods Programs Biomed. **178**, 319–328 (2019). https://doi.org/10.1016/j.cmpb.2019.07.002
10. Liu, K.L., et al.: Deep learning to distinguish pancreatic cancer tissue from non-cancerous pancreatic tissue: a retrospective study with cross-racial external validation. Lancet Digit. Health **2**(6), e303–e313 (2020). https://doi.org/10.1016/S2589-7500(20)30078-9
11. Mazurowski, M.A., Buda, M., Saha, A., Bashir, M.R.: Deep learning in radiology: an overview of the concepts and a survey of the state of the art with focus on MRI. J. Magn. Reson. Imaging **49**(4), 939–954 (2019). https://doi.org/10.1002/jmri.26534
12. Perkins, N.J., Schisterman, E.F.: The inconsistency of "optimal" cutpoints obtained using two criteria based on the receiver operating characteristic curve. Am. J. Epidemiol. **163**(7), 670–675 (2006). https://doi.org/10.1093/aje/kwj063
13. Redondo, A.R., Navarro, J., Fernández, R.R., de Diego, I.M., Moguerza, J.M., Fernández-Muñoz, J.J.: Unified performance measure for binary classification problems. In: Analide, C., Novais, P., Camacho, D., Yin, H. (eds.) IDEAL 2020. LNCS, vol. 12490, pp. 104–112. Springer, Cham (2020). https://doi.org/10.1007/978-3-030-62365-4_10
14. Roth, H.R., Farag, A., Turkbey, E., Lu, L., Liu, J., Summers, R.M.: Data from pancreas-CT. The cancer imaging archive (2016). https://doi.org/10.7937/K9/TCIA.2016.tNB1kqBU

15. Youden, W.J.: Index for rating diagnostic tests. Cancer **3**(1), 32–35 (1950). https://doi.org/10.1002/1097-0142(1950)3:1<32::aid-cncr2820030106>3.0.co;2-3
16. Zhang, Y., et al.: A deep learning framework for pancreas segmentation with multi-atlas registration and 3D level-set. Med. Image Anal. **68**, 101,884 (2021). https://doi.org/10.1016/j.media.2020.101884

Computational Approaches for the Automatic Quantification of Cells from Brain Images

Diogo Lopes[1(✉)], Ana Bela Campos[2,3], Patrícia Maciel[2,3], Paulo Novais[1], and Bruno Fernandes[1]

[1] ALGORITMI Research Centre/LASI, University of Minho, Braga, Portugal
diogoalexlopes@hotmail.com, pjon@di.uminho.pt, bruno.fernandes@algoritmi.uminho.pt

[2] Life and Health Sciences Research Institute (ICVS), School of Medicine, University of Minho, Braga, Portugal
pmaciel@med.uminho.pt

[3] ICVS/3B's, PT Government Associate Laboratory, Braga, Portugal

Abstract. Microglia are glial cells residing in the central nervous system (CNS). They represent the first line of immune defence within the CNS and are responsible for fundamental physiological and pathological processes. Given their importance, the quantification of these cells is fundamental in a clinical context. However, this process is a major challenge, as conventional cell counting involves a specific set of tools and devices, being extremely costly. Currently, most cell-counting processes are manual. Such processes are time-consuming, tedious, and imprecise, being heavily dependent on the operator. To address this, new approaches have been developed to improve the quantification process. Indeed, an automated solution can greatly increase the standardised process as it shows better accuracy and efficiency. In this work, we compare and demonstrate that, on the one hand, we have classical computational approaches, where software and assistants for automatic cell counting, such as ImageJ, are applied in images that contain scattered cells. On the other hand, deep learning approaches show similar accuracy to the manual counting process but present a significant enhancement in reproducibility and efficiency.

Keywords: Computer Vision · Deep Learning · ImageJ · Image Analysis

1 Introduction

Microglia represent a population of macrophage cells residing in the central nervous system. They participate in multiple important events in the brain including neurodevelopment, neuronal activity, regulation of synaptic architecture and neurogenesis, which are crucial to sustaining normal and healthy brain functions [2,8]. Given the importance that these cells have in maintaining a healthy

A. Rocha et al. (Eds.): WorldCIST 2023, LNNS 799, pp. 570–580, 2024.
https://doi.org/10.1007/978-3-031-45642-8_55

microenvironment, their identification for further quantification is fundamental in a medical context. However, this is one of the biggest tasks in image analyses and medical diagnoses. As the conventional cell counting process involves a specific set of tools and devices developed for that specific purpose, and because cell counting is an important procedure, multiple studies are focusing on the development of new approaches to automate cell counting and make it a more time-efficient process.

Related to the so-called classical approaches, image-processing software solutions emerge to automate cell counting. Such software requires specific settings on an image to obtain reasonable accuracy. Therefore, allied to these solutions, it is always necessary to resort to image processing and analysis techniques. Presented by the National Institute of Health, ImageJ proves to be a powerful solution to this problem [5]. ImageJ has a set of protocols that can be used to quantify multiple and single cells [5,8]. Novel segmentation routines implemented within the FIJI-ImageJ ecosystem to quantify retinal microglia [2] make ImageJ a suitable solution for the automatic quantification of these cells.

In recent years, deep learning-based approaches evidenced promising performance in various image analysis tasks, such as classification, detection, and segmentation [1]. Regarding the cell counting problem, this approach shows similar accuracy to manual counting but with a significant enhancement in reproducibility, throughput efficiency, and reduced error from human factors [3]. Through the use of convolutional neural networks (CNNs), one can produce an automated counting process of brain cells [7]. Within the deep learning approaches, one also has deep neural network-based methods that enable automated counts with high levels of accuracy compared to manual counting methods [3].

The study here presented sets the starting point for future studies in the computer vision domain, namely for the automatic quantification of cells from brain images using classical and/or deep learning approaches. Hence, considering that the goal of a review is to look for the most recent scientific manuscripts to obtain an overview of the current state of the art, to substantiate new approaches that rival current ones, the following research questions have been elicited:

- **RQ1.** What strategies and tools have already been developed and used to implement a solution to automatic quantify microglial cells regarding a more classical approach to computer vision?
- **RQ2.** What strategies and tools have already been developed and used to implement a solution to automatic quantify microglial cells regarding a computer vision deep learning-based approach?

This paper is structured as follows: Sect. 2 describes the methodology carried out to do the research and review process, taking into account its main steps, i.e., the selection of data sources, the search strategy, the selection criteria and the respective results. Section 3 provides an overview of the results obtained throughout the relevant literature search and review described in the previous section. Next, Sect. 4 describes our proposed approaches that rival the current state-of-the-art ones. Finally, Sect. 5 summarises and compiles the conclusions and lessons learned throughout this study.

2 Methodology

This review was conducted based on PRISMA[1] (*Preferred Reporting Items for Systematic Reviews and Meta-Analyses*) statement and respective checklist [6]. This choice was mainly due to PRISMA being widely accepted by the scientific community. Therefore, the following steps were taken into consideration:

1. Identification of the study's research questions and relevant keywords;
2. Creation of the research query;
3. Definition of the eligibility criteria to filter and reduce the article's sample;
4. Analysis of the resultant set of studies and papers;
5. Presentation and discussion of the results.

The preliminary research was carried out on January 18th, 2022. The used data source was SCOPUS due to its size, quality assurance, and wide coverage in terms of publication subjects. To carry out the relevant literature and documentation research, some keywords were defined as a starting point and were applied in the title, abstract, and keywords. To organize the search resources, the keywords were organized into two groups, which are combined in conjunction. The keywords in each group are combined with disjunctions. This choice fulfils the purpose of each group selecting all documents that include at least one of the keywords and then ensuring that only documents that contain at least one term from each of the groups are selected. The first group is related to the areas and technical subjects directly related to the research topic. The second group aims to filter by broader areas of the technological scope. Therefore, by applying the strategy described above, the following research query arose:

```
1 ( TITLE-ABS-KEY ( microglia )
2 OR  TITLE-ABS-KEY ( cell AND counting )
3 AND
4 ( TITLE-ABS-KEY ( automated AND counting )
5 OR  TITLE-ABS-KEY ( imagej )
6 OR  TITLE-ABS-KEY ( fiji )
7 OR  TITLE-ABS-KEY ( deep  AND learning ) ))
```

To screen the collected articles and studies, some eligibility criteria (in the form of exclusion criteria) were defined. As such, all documents that matched any of the following criteria were excluded:

EC1: Not accessible in *OPEN ACCESS* mode;
EC2: Were not produced in the last 4 years or have not yet been published;
EC3: Do not belong to the *Comp. Science*, *Engineering*, or *Neuroscience* fields;
EC4: Not articles or reviews/surveys and not written in English;
EC5: Do not have a relevant number of citations;
EC6: Do not focus on the studied variables or are out of context.

[1] http://www.prisma-statement.org.

3 Results

The initial database resulting from all the relevant bibliographic research identified 2 651 studies, followed by the application of the eligibility criteria. The first five criteria (EC1 to EC5) were applied using the SCOPUS direct filtering system. The results are a subset of computer science, engineering, and neuroscience articles or reviews that are freely accessible and were written in English in the last 4 years. As a result, 40 studies were left for a full reading. A final set of 7 studies were included and considered in the review. The other 33 studies were excluded due to not focusing on classical and deep learning-based approaches for automatic cell counting (EC6). Figure 1 shows a PRISMA flowchart with a general overview of the mentioned process.

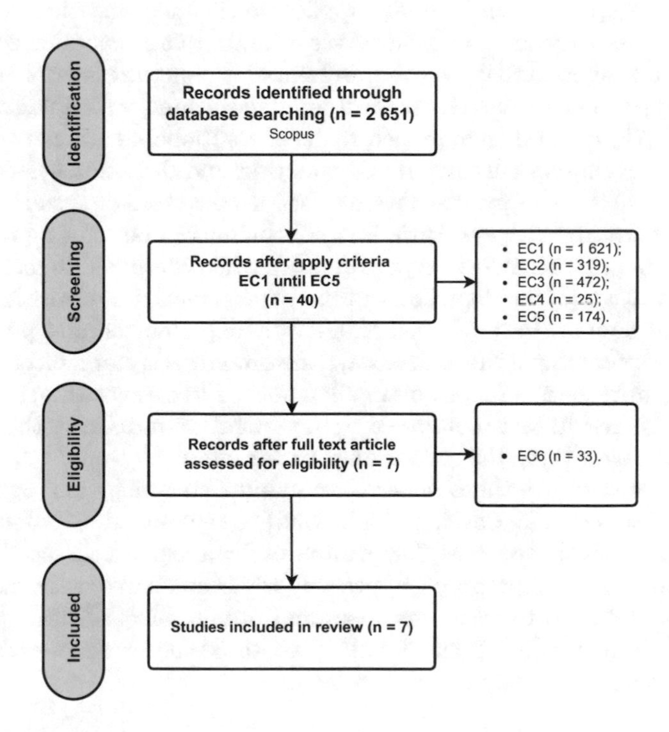

Fig. 1. Adaptation of the PRISMA Flowchart applied in this study.

Considering the final set of suitable studies, the one of Ash et al. [2] implemented, within the FIJI-ImageJ ecosystem, a new segmentation routine to perform automated segmentation and cell counting in retinal microglia. As the algorithms that automate microglia counts are increasing in popularity, few of those are adequate for segmentation and cell counting of retina microglia. Therefore, their algorithm built entirely with open-source software, addresses the presented problem. Throughout the results obtained, they showed that their routine can perform cell counts with similar accuracy to manual counting.

O'Brien et al. [5] developed two plugins within ImageJ for the task of automated hemacytometers and migration/invasion cell counting. These plugins are open-source, easy to use and optimized for cell counting. Through the combination of the core principles of the Cell Counter feature on ImageJ with a counting algorithm, they were able to increase the accuracy of the cell counting process.

Young et al. [8] developed a protocol with recommended ImageJ plugins for automatically accessing the morphology and quantifying microglial cells. The protocol converts fluorescence photomicrographs into binary skeletonized images. They concluded that the use of this protocol makes microglia quantification accessible to all laboratories, as the plugins used are open-source.

Dökümcüoğlu et al. [4] demonstrated the usefulness of image processing approaches for the quantification of cells, which in their case was a microalga culture. Their approach can be easily applied to the quantification of microglial cells problem, as they are so similar since a brain image contains several glial cells. This study showed that resorting to ImageJ allows the operator to complete the counting process 4 times faster than a manual count, with similar accuracy.

Xie et al. [7] proposed an approach with convolutional neural networks across an image to develop an automated cell counting and detection system through microscopy images. The results obtained set a new state-of-the-art methodology for cell counting with synthetic image benchmarks but can be used in real microscopy images with increased performance when compared to manual count.

Phoulady et al. [1] developed an automatic segmentation algorithm to apply to stereology counts. Part of their validation experiments was performed in microglial cells containing the ionized calcium-binding adapter molecule 1 marker (IBA1). The microglia were automatically counted in tissue sections of a mouse neocortex. The results were compared with manual counts. Using the automatic segmentation algorithm, they were able to automatically annotate images of microglia more than five times faster than manual stereology and counts.

Finally, Dave et al. [3] developed a method to separate stain colour channels on images from tissue sections. The samples of microglial cells used contained the IBA1 marker. The proposed approach, when compared with state-of-the-art methods, manages to overcome them and obtain more accurate results. To conclude, their methodology can function as an input for deep learning-based algorithms to automatically quantify cells.

4 Proposed Approaches

Having studied and analyzed relevant literature throughout the findings of the review against the research questions, this work propose new developments to implement a solution for the automatic quantification of microglial cells. Subsection 4.1 presents a protocol that helps to automate microglial cell counts based on a more classical computational approach. On the other hand, Subsect. 4.2 presents 2 classification procedures that aim to solve the raised problem. Based on the conclusions drawn from the review, the classic and deep learning-based methodology experiments rival the current state-of-the-art. To conduct these procedures the research team used 16 brain images of 4 different animals. Given the complexity of the microglial cells images, to obtain proper and more accurate results, the areas of interest for the quantification were divided into several images, thus composing a dataset set with 661 images.

4.1 Classical Computational Approach

Within the so-called classic approaches emerges ImageJ. Allied with image-processing and analysis tasks a protocol was developed to automate the microglial cell quantification. The division performed in the areas of interest reduces image noise and helps to obtain a clearer view of the cells for the required image-processing tasks inherent to our protocol. The phases that make up this protocol are the following: 1 - Channel Separation; 2 - Threshold Adjustment; 3 - Image Conversion; 4 - Cells Separation; 5 - Cells Quantification. The first stage is where we use the ImageJ functionality to separate channels, as the brain cells are captured in a fluorescent format, and the microglial cells are only visible in the red channel. The second stage is where we guarantee all areas of interest are included with or without overlap, as we decreased the overlap as much as possible using the sliders for manual adjustment of the threshold values. Following this, the third stage is where the image is firstly converted to binary, resulting in an 8-bit image. The fourth stage is the application of the watershed algorithm. Finally, to finalize the cell counting process comes the fifth phase of the developed protocol. With the Analyze Particles functionality of ImageJ, we defined the circularity and the size of objects, ensuring that we are not excluding any cell in the quantification process. Table 1 is an example of the quantification settings applied to the first slice of brain images of animal CN276 2FD. The area of interest to the quantification, in this case, was lobule 2 which was divided into 4 images. These procedures were applied individually to each of the 661 images that compose the dataset. This protocol address the problem raised by the first research question and help ground a classic methodology to computer vision for the automatic quantification of microglial cells.

Table 1. Quantification Settings for Lobule 2 - Animal CN276 2FD - Slice 1.

		1st	2nd	3rd	4th
Threshold	Min	0	0	0	0
	Max	1461	1750	1526	1574
		96.42%	96.26%	95.18%	95.04%
Analyze	Algorithm	Default	Default	Default	Default
	Size (micron^2)	17-Infinity	17-Infinity	17-Infinity	17-Infinity
	Circularity	0.05–1.00	0.05–1.00	0.05–1.00	0.05–1.00

Table 2 presents some of the quantification results obtained with the application of the developed protocol. On the right, the results of manual quantification are presented. When we compare the results with the automated quantification, we obtain adequate results, and above all, we turn a process that takes weeks to be finalized into a process of days. This protocol guarantee that within a more classical computational approach, we can automate the cell counting process.

Table 2. Quantification Settings Animal CN276 2FD - Slice 1.

	Number of Cells (Manual Quantification)	Number of Cells (Protocol Quantification)
Lobule 2	73	68
Lobule 3	80	74
Lobule 4	54	55
Lobule 5	38	48
Lobule 6	43	45

4.2 Deep Learning Approach

CNNs were chosen as the basis of our solution to automate cell counting based on a deep-learning approach. This methodology can deliver faster results with the same levels of accuracy as manual counting processes. When taking a classification approach to automate the counting process, CNNs prove to be a viable solution. In total, 2 classification approaches were developed, one based on the number of cells contained in the image and another based on the percentage of area that the cells occupy in the image. Both approaches address the problem raised by the second research question and help ground deep learning-based approaches for the automatic quantification of microglial cells.

In a classification approach, the goal is to classify one or several images using previously defined classes. That said, in our study, we defined 3 classes. The results are 339 images labelled as "Few", 203 as "Average", and 119 as "Many". After labelling, we trained the model assigning the most complex values to the model parameters. Given the complexity of the images, assigning complex values gives us a better perspective of how the model behaves in complicated scenarios.

The sequential API of Keras was used where the activation function of the microarchitecture of the model used was "relu". The input shape defined for the images was 256×256, with "same" for the padding value. The number of neurons was 128 and the activation function for the dropout layer was "softmax".

After properly adjusting the model, followed tuning the model by setting the ideal values for the model parameters. The metric defined for the model was "accuracy" since it is a classification approach. The number of splits for the model was 5. We cross-validate with 5 folds, and the error is the average of the 5 folds. In this way, we don't add too much difficulty to the model and manage to have an equally effective but faster process. That said, we are left with 529 images for training and 132 for testing. This is done for each fold, which makes training and testing a little more time-consuming, but will generate more reasonable results. Table 3 presents the Top 15 best experiment results.

Table 3. Deep Learning Classification Results per Number of Cells.

batch_size	epochs	learning_rate	data_augmentation	score	loss	run_time	str(score_list)
32	20	0.0005	false	0.9021	0.2479	326.4157	[0.5714, 0.9393, 1.0000, 1.0000, 1.0000]
32	20	0.0010	false	0.9006	0.2765	260.9422	[0.5789, 0.9318, 0.9924, 1.0000, 1.0000]
16	20	0.0005	false	0.8690	0.3291	333.8006	[0.4586, 0.8863, 1.0000, 1.0000, 1.0000]
32	15	0.0010	false	0.8672	0.3326	202.2335	[0.5864, 0.7500, 1.0000, 1.0000, 1.0000]
16	15	0.0005	false	0.8520	0.4301	293.1848	[0.6466, 0.6136, 1.0000, 1.0000, 1.0000]
32	10	0.0005	false	0.8489	0.3904	150.9446	[0.6691, 0.6363, 0.9469, 0.9924, 1.0000]
32	15	0.0005	false	0.8475	0.4282	206.0727	[0.5939, 0.6590, 0.9848, 1.0000, 1.0000]
32	10	0.0010	false	0.8475	0.3738	179.2750	[0.6240, 0.6363, 0.9848, 1.0000, 0.9924]
16	10	0.0005	false	0.7959	0.5477	156.7259	[0.6691, 0.5454, 0.7651, 1.0000, 1.0000]
8	15	0.0005	false	0.7944	0.5796	266.0632	[0.6691, 0.5530, 0.7575, 0.9924, 1.0000]
16	15	0.0010	false	0.7595	0.8250	209.3782	[0.6691, 0.8409, 0.3560, 0.9393, 0.9924]
32	5	0.0010	false	0.7462	0.6996	76.5467	[0.6090, 0.8181, 0.5151, 0.8409, 0.9469]
16	5	0.0005	false	0.6611	0.8956	93.8852	[0.6691, 0.8030, 0.3636, 0.6212, 0.8484]
32	5	0.0005	false	0.6520	0.8672	84.6684	[0.6691, 0.8333, 0.3484, 0.5378, 0.8712]
32	4	0.0005	false	0.5989	0.9619	86.9296	[0.6691, 0.8409, 0.3181, 0.4318, 0.7348]

Looking at the results, the first 8 experiments should be highlighted. In at least 1 of the folds, the 132 images were correctly classified, which attests to the solidity of this approach. Beyond the already mentioned difficulty caused by the complexity of the images, the values of the parameters that gave rise to the best result were the most complex possible, except for the data augmentation. With all this difficulty, the model was able to correctly classify more than 90% of the cases, which means it correctly classified 596 images out of 661. As if the score wasn't enough, the model was able to classify all 661 images in 5 min and 44 s.

Once again, in this classification approach, we defined 3 classes. An image that does not contain a percentage of cell area greater than 0.45 is labelled as "Few", an image that contains a percentage of cell area greater than 0.45 and less than 0.85 as "Average", and an image that contains a percentage of cell area greater than 6.9 as "Many". Similar to the previous procedure, we trained the model assigning the most complex values to the model parameters. The metric defined for the model was also "accuracy", and the number of splits for the model was again 5, so we can cross-validate with 5 folds. Table 4 presents the Top 15 best experiment results.

Table 4. Deep Learning Classification Results per Area of Cells.

batch_size	epochs	learning_rate	data_augmentation	score	loss	run_time	str(score_list)
16	25	0.0005	false	0.8312	0.5005	186.4194	[0.3533, 0.8030, 1.0000, 1.0000, 1.0000]
16	20	0.0005	false	0.7827	0.7110	145.2895	[0.3533, 0.5606, 1.0000, 1.0000, 1.0000]
32	25	0.0005	false	0.7721	0.6995	181.540	[0.3759, 0.5000, 0.9848, 1.0000, 1.0000]
8	25	0.0005	false	0.7525	0.7375	250.4927	[0.3383, 0.4318, 0.9924, 1.0000, 1.0000]
32	20	0.0005	false	0.7479	0.5524	146.7203	[0.3383 0.4848, 0.9166, 1.0000, 1.0000]
8	20	0.0005	false	0.7479	0.8725	206.0794	[0.3383, 0.4242, 0.9772, 1.0000, 1.0000]
16	15	0.0005	false	0.7464	0.6019	129.4396	[0.3383, 0.4848, 0.9166, 0.9924, 1.0000]
32	15	0.0005	false	0.6919	0.6063	100.8088	[0.3383, 0.4242, 0.7121, 0.9848, 1.0000]
16	15	0.0010	false	0.6706	0.7802	109.5244	[0.3383, 0.4469, 0.5757, 0.9924, 1.0000]
32	25	0.0010	false	0.6510	0.6611	172.7800	[0.3383, 0.4393, 0.5757, 0.9090, 0.9924]
16	10	0.0005	false	0.6449	0.7910	80.5878	[0.3383, 0.4469, 0.4924, 0.9469, 1.0000]
32	20	0.0010	false	0.6313	0.7594	162.2263	[0.3383, 0.4318, 0.4924, 0.9015, 0.9924]
8	15	0.0005	false	0.6222	0.8347	151.2869	[0.3383, 0.4469, 0.4545, 0.8712, 1.0000]
32	10	0.0005	false	0.5903	0.7720	67.7552	[0.3383, 0.4318, 0.5303, 0.7803, 0.8712]
16	10	0.0010	false	0.5706	0.8302	80.6591	[0.3383, 0.4469, 0.4545, 0.6287, 0.9848]

Once again, looking at the results obtained, the first 7 experiments should be highlighted as the score percentage obtained is higher than 74%. This means that in these experiments at least 493 images were correctly classified. In the best experience, the model was able to correctly classify more than 83% of the cases, which means it correctly classified 548 images out of 661. The classification time took 3 min and 10 s which attests to the benefits and applicability of a deep learning-based approach o automate the process of microglial cell count.

5 Conclusion

Computational approaches are the cornerstone to implement automatic quantification systems for cell counting. Most cell counting processes are manual and involve a specific set of tools and devices developed for that purpose. An automated solution can greatly increase the standardised process as it shows better accuracy and efficiency. Thus, the involvement of the scientific community is extremely relevant for the development of new strategies and tools.

In this paper, a review was conducted to analyse all the relevant literature produced related to different computational approaches for the automatic quantification of cells from brain images. The review was based on the PRISMA model using the SCOPUS database as source. Of the 2 651 identified studies, the application of the PRISMA methodology resulted in a final set of 7 studies that were relevant in regard to the previously defined research questions.

From the proposed research questions, the state-of-the-art was assessed regarding the role and purpose of computational approaches for cell counting. The collected studies suggest that automated systems obtain faster results in the counting process and show an increased value of accuracy when compared to manual counting processes. Finally, in the analysed studies, all authors agree that the implementation of automated systems and solutions can improve the tedious, time-consuming, and imprecise manual counting process.

From the proposed approaches, this study has proven that to automate the cell counting process emerge classic and deep learning methodologies. Within the so-called classic methodology, software solutions and assistants, like ImageJ, emerge to automate the quantification process. Needs to be pointed out that to develop a solution to automate the cell counting process, as was the case of our procedure, specific settings are required on an image to obtain reasonable accuracy. Deep learning-based approaches evidenced promising performance in various image analysis tasks, such as classification. Deep learning approaches showed similar accuracy to manual counting but a significant enhancement in reproducibility, throughput efficiency and reduced error from human factors. Through our experiments, we concluded that the models based on CNNs emerge since regression-based cell counting avoids the challenging task of detection or segmentation.

To conclude, given all the compiled information, and all the results acquired from the developed experiments, it can be inferred that deep learning-based methods are evolving to become a performant solution to automate the cell counting process. Such methods evidence a significant enhancement in reproducibility, as they don't need to be allied with image processing and analysis techniques. They also do not require specific settings on an image to obtain reasonable accuracy, unlike software solutions and assistants. Both the classical and deep-learning approaches helped to turn a process that normally is tedious, time-consuming, with high labour costs and imprecise because it is dependent on the operator into an automated process. This process is now less time-consuming, and more accurate, where the error associated with users is reduced since a large part of the process is done by machine. The classical and deep learning-based procedures produced are a competitive solution with state-of-the-art methods and help to optimize systems for the automatic quantification of cells from brain images.

Acknowledgements. This work has been supported by *FCT - Fundação para a Ciência e Tecnologia* within the R&D Units project scope: UIDB/00319/2020.

References

1. Ahmady Phoulady, H., Goldgof, D., Hall, L.O., Mouton, P.R.: Automatic ground truth for deep learning stereology of immunostained neurons and microglia in mouse neocortex. J. Chem. Neuroanat. **98**, 1–7 (2019). https://doi.org/10.1016/j.jchemneu.2019.02.006
2. Ash, N.F., Massengill, M.T., Harmer, L., Jafri, A., Lewin, A.S.: Automated segmentation and analysis of retinal microglia within ImageJ. Exp. Eye Res. **203**, 108416 (2021). https://doi.org/10.1016/j.exer.2020.108416
3. Dave, P., Alahmari, S., Goldgof, D., Hall, L.O., Morera, H., Mouton, P.R.: An adaptive digital stain separation method for deep learning-based automatic cell profile counts. J. Neurosci. Methods **354**, 109102 (2021). https://doi.org/10.1016/j.jneumeth.2021.109102
4. Dökümcüoğlu, V.E., Yılmaz, M.: Assessment of cell counting method based on image processing for a microalga culture. Mediterr. Fish. Aquac. Res. **3**(2), 75–81 (2020)

5. O'Brien, J., Hayder, H., Peng, C.: Automated quantification and analysis of cell counting procedures using ImageJ plugins. J. Visualized Exp. (117) (2016). https://doi.org/10.3791/54719
6. Page, M.J., McKenzie, J.E., Bossuyt, P.M., Boutron, I., Hoffmann, T.C., et al.: The Prisma 2020 statement: an updated guideline for reporting systematic reviews. BMJ **372** (2021). https://doi.org/10.1136/bmj.n71
7. Xie, W., Noble, J.A., Zisserman, A.: Microscopy cell counting and detection with fully convolutional regression networks. Comput. Methods Biomech. Biomed. Eng. Imaging Vis. **6**(3), 283–292 (2018). https://doi.org/10.1080/21681163.2016.1149104
8. Young, K., Morrison, H.: Quantifying microglia morphology from photomicrographs of immunohistochemistry prepared tissue using ImageJ. J. Vis. Exp. (136) (2018). https://doi.org/10.3791/57648

Design and Implementation of a Well-Being Index for Knowledge Assessment

Fábio Senra and Orlando Belo(✉)

ALGORITMI Research Centre/LASI, University of Minho, 4710-057 Braga, Portugal
obelo@di.uminho.pt

Abstract. Due to their simplicity, meaning and representation, indexes are excellent communication tools. They have the ability for simplifying complex measurements. The fact that composite indexes allow for the division into several dimensions and associate weights to them makes their understanding easier. Their versatility justifies the increase of their use in many application fields. In the enormous spectrum of index application, well-being is one of the most potential application area, where we can use them for predicting the impact of programs and measures built taking into account the needs presented by people, institutions or systems. Over time, indexes were applied to various other domains, such as health, social progression, economics or education. In this paper, we conceived and implemented a well-being index system for helping students and teachers in learning and knowledge assessment processes, providing them personalized means for training and evaluating multidisciplinary knowledge based on student historical interaction with a specialized tutor system.

Keywords: Data Analysis · Well-Being Indexes · Knowledge Assessment · Dashboarding · eLearning Platforms

1 Introduction

Evaluation mechanisms development and application are very regular activities in most organizations whose aim is to assess the quality and "well-being" of their staffs or services and, in general, of the organization as a whole [1, 2]. The use of these mechanisms is not exclusive to organizations. We can apply them to people or to operational systems for monitoring activities and assessing performance. Currently, a special type of indicator is in high demand: indexes. Like all others, indexes are measures that usually do not have a unit. Their value and categorization vary according to the domain in which they are applied. However, despite being only measuring instruments, indexes can also be used for characterizing, reporting, assessing or diagnosing the most varied situations, based on a given set of previously established criteria. The design and implementation of a well-being index requires a detailed study of the various characteristics of its application domain. Its usefulness depending on the effectiveness of its application.

A well-being index [3, 4] communicates some kind of assessment through a single numerical value, facilitating discussion of its practical value and usefulness, and simplifying the reading of complex or multidimensional measurements that may have been

A. Rocha et al. (Eds.): WorldCIST 2023, LNNS 799, pp. 581–590, 2024.
https://doi.org/10.1007/978-3-031-45642-8_56

taken. The current vast range of application areas for well-being indexes reveals their practical importance and the recognition of their practical utility [5, 6]. They make it possible to detect gaps in various segments and areas of knowledge and provide elements for their correction, allowing and encouraging the implementation of continuous improvement policies. Areas such as quality of life, health, culture, or education, for example, have hosted numerous initiatives for applying well-being indexes. In education, several authors defend the need to measure and monitor students' well-being, justifying that is crucial for their achievement and involvement with all subjects that are studying [7]. The use of well-being indices, composed of several dimensions, the school and family environment for example, would play a very important role in measuring emotional, social and academic skills of students, helping teachers to have a more correct understanding of everything that influence students' learning processes in their different lines of working.

In this paper, we present and discuss a system we developed to create a well-being index for an educational institution, especially conceived for the environment of an artificial tutor system, aiming to support teachers during their activities, providing means for assessing student skills and knowledge in specific subject domains. The well-being index system provides means to students and teachers using a tutor system to access to the state of knowledge students have in subjects that are studying. This allows them to adapt their working methods and provide individualized feedback focused on their greatest difficulties. The remaining part of this paper is organized as follow: Sect. 2, exposes and discusses some of the fundamentals regarding well-being indexes and their applications; Sect. 3, presents the well-being system we developed, describing the application case and discussing how the index is calculated and its values analyzed; and, finally, Sect. 4, presents some conclusions and future work.

2 Well-Being Indexes

The development of mechanisms for evaluating organizations is a very common activity. Its objective is to measure the quality and "well-being" of the elements and services of an organization and, in general, of the organization itself. Today, this type of mechanism is applied in a large range of areas, such as health, banking or telecommunications. Among the mechanisms most used in this type of measurement, there is a special type of indicator of great interest: indexes. According to [8], an index is a measure that does not have a specific unit. Its value is distinct, depending on the domain to which it is applied. To have an index with high value is not always an advantage. For example, if we look at a mortality rate it is expected to reach a low value. On the other hand, if it is an average hope rate of life, the ideal is that its value is as high as possible. The use of indices has been increasing. In [1] indexes are used to classify countries, states, cities, evaluate public or private institutions, companies or organizations, and to explore the characteristics that these elements have. The reasons that are usually behind the elaboration of an index are the fact that it wants to demonstrate progress or decline, promoted by measures or programs, or just to show the data collected. An index is for measuring, but it also helps to characterize the system in which it was defined for comparing particularities and characteristics. Once some comparison criteria has been established, indexes can still be a mechanism for helping to report and diagnose issues that may be raised by their value, ultimately leading to their resolution [9]. All these small tasks serve as a basis for decisions that can be taken in a certain context. A type of index that is widely used today are the well-being indexes. These indexes allow for monitoring the evolution of well-being or the progress of people or institutions. For example, using them, it is possible, in a given context, to assess economic behavior, emphasizing as far as possible the well-being of people, regions or countries. Ryan and Deci [10] stated that the way well-being is defined influences all governmental, teaching, therapy, parenting practices, since all these behaviors have as their main objective to change human beings to the best. Therefore, the definition of a well-being index imposes the existence of a vision of what is best. In the field of education and teaching, Noble et al. [7] referred that the study of well-being is an important tool to assess the emotional, social and academic competences of students, in order to help teachers, so that they have a more correct understanding of everything, which influences students at various levels. The study of well-being intends to clarify which are the obstacles to learning processes, so that, in the future, they can be dissipated and, consequently, improve the results of students [11].

In the next section, we will present and discuss a concrete example of a well-being index for assessing knowledge of students in one or more subject domains. Our main goal was to evaluate the application of a well-being index on an educational institution, designing and implementing a system for analyzing responses posted to students by an artificial tutor, and evaluating them using a single value, an index, which has the ability for reflecting the well-being of students when studying some application domain. Based on this index, we believe that it is possible to provide to teachers valuable information for adapting, improving or constructing new teaching methodologies, monitoring student knowledge, or giving more attention to subject and topics that proved to be less clear to students. This index will improve the well-being of students and teachers.

3 Assessing Knowledge Using an Index

The index system developed within the scope of this work is part of a broader project for the development of an artificial tutor [12], which aims to improve the learning process of students in an institution, by measuring their expertise and knowledge in various subject areas. To provide a more positive evolution to learning processes, the tutor is able to adjust his behavior according to the performance of the students interacting with it [13], producing, whenever possible, a personalized feedback referencing the most positive or negative aspects of the students' learning processes. This feedback makes the tutor more effective and useful, particularly when it addresses students' progress and helps them to advance in their study and improve their performance in specific subject areas [14]. The focus of this system is clearly on the students. As such, it provides a very diverse range of learning tools, so that they can learn faster, using personalized means. However, we cannot forget that helping students can be made by also helping teachers, preparing teaching activities and monitoring results of their students, collectively and individually [15]. Thus, the tutoring system intends to be a useful tool for these two main actors (students and teachers), allowing them to customize their training and assessing processes, in an accessible way, 24 h a day, from any place having an Web access.

In order to simplify the analysis process of students' expertise and knowledge when carrying out some assessment tasks, we designed and implemented a well-being index, especially oriented for assessing student performance. During the assessment sessions, the system records a set of metrics, quite diverse, extracted from the records of the answers students give to the different questions that the system presented to them in these sessions. Through these metrics, we can assess student performance, taking into account the knowledge revealed by students, the attendance or the interest in one or more fields of study, and thus establish a performance index for a specific student or for all students of one or more domains of study.

In a simple way, the index allows for revealing how each student studied the subjects of each knowledge domain, as well as highlight the domains in which students are weaker or stronger. Thus, the index provides teachers with a complementary view of how to approach topics more effectively, based on the performance of their students. In addition, the index will make it possible to understand which questions students reveal to have greater difficulties. An index of this nature, it is a very important tool for teachers, allowing them for solving knowledge acquisition problems or, in the long term, contributing to change teaching methods, as well as helping how to enhance the perception of the best ways for students assimilate knowledge of the various study domains implemented in the tutor system.

The calculus of the index started with a detailed analysis of a dataset we prepared for this work. Then, we identified the most relevant data elements for the calculation and support of the index, taking into consideration data related to the user (the student) – name, gender, degree and course), the domain of the questions that were asked to students – domain and subdomain –, and the answers given to each of the questions – correction, response time and points obtained. After identifying these elements, we designed and implemented a data mart specifically oriented to receive the data required by the process of calculation and maintenance of the well-being index. The data mart includes two fact tables – "FT-StudentsIdx" and "FT-QuestionsIdx" – and 5 dimension tables – "DimStudents", "DimQuestions", "DimDomains", "DimMethods", and "DimCalendar" –, being the last three shared dimension tables, which means that they are used simultaneously by the two fact tables.

Both fact tables have an identical structure. However, they host different data – the "FT-StudentsIdx" table stores data for calculating the performance index per student, while the "FT-QuestionsIdx" table maintains data for calculating the performance index per question presented. Dimension tables store essentially information about calculation methods ("DimMethods"), students ("DimStudents"), subject domains ("DimDomains") and index calculation date and time ("DimCalendar"). The data multidimensional space provided by the data mart allows us to analyze the index values according to each of the dimensions and hierarchies they host. For example, using the fact table "TF-Students" (Fig. 1) and the dimension table "DimDomains", we can perform a roll-up operation, following the hierarchy "SubDomain$\gg$Domain". The roll-up operation results provide the index values of a student (or several), first, for one or more subdomains and then, aggregating the index values for these subdomains, reach the corresponding index value for the overall domain in which subdomains are integrated. Similar operations can be done using other dimension tables included in the schema.

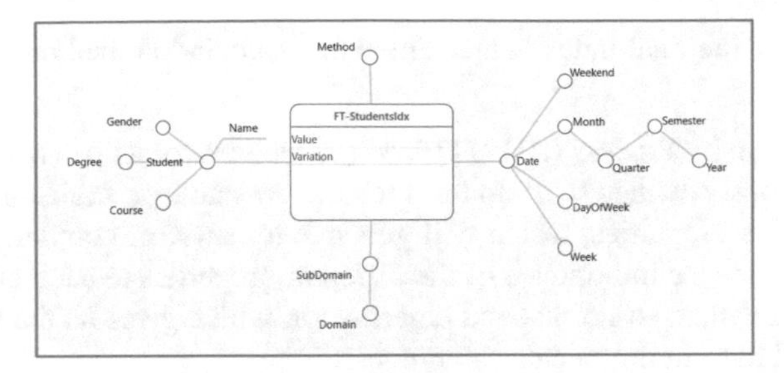

Fig. 1. A conceptual schema view of the well-being indexes data mart – students segment

Table 1. The properties of the index

Nr	Property	Description	Range	Example
1	Hit rate	Quotient between correct answers and the total of answers given	0..1	0,63
2	Average difficulty level of questions	Average of the difficulty levels of each of the questions answered	0..5	3,2
3	Subdomains rate	Ratio between the number of subdomains addressed and the total number of subdomains. Each knowledge domain can correspond to one or more subdomains	0..1	0,4
4	Hit rate in the last 10 days	Identical to the previous one, but which is temporally conditioned	0..1	0,28
5	Average response time	Average time taken by students to answer the questions	0..5	4,2
6	Consecutive right answer rate	Identical to the previous ones, but that verifies how many consecutive correct answers (maximum) a student managed to reach	0..1	0.21
7	Points rate	Ratio between the number of points obtained by the student (the number of points per question will never exceed 1) and the number of answers given	0..1	0.54

The index we defined is a composite index. When defining it, we took into account two fundamental aspects: the dimensions that characterize the index and its calculation method. This type of index allows for more detailed analyzes of students' performance and the questions asked to them. In Table 1, we can see the properties that supports the definition of the well-being index we defined. After defining the properties of the index, we established the way each one of the properties will be used by the calculation method chose in order to weight them, define their degree of influence (the weight) in the calculation of the final index value. The index calculation methods we chose were the following:

- Analytic Hierarchy Process (AHP) [16, 17], which performs pairwise comparisons, using a previously defined scale. In this method, we start by establishing comparisons between all the properties, which will generate a matrix of comparisons. With this table, we can see the importance of the different properties to each other. Using this matrix, we calculated a normalized eigenvector, which gives us the weight of each property will have in the calculation process.
- Individual Classification, which allows for classifying properties on a scale of 1 to 10, defining the influence we consider that a property should have in the calculation of the index. Having the properties' classifications, we sum them reaching a global value, and them we calculate each property's weight through the quotient of the classification of the property with the referred global value.
- Equalizer, a very simple, intuitive and easy calculation method, in which we have to indicate the percentage value that we want to assign to a certain property, taking into account that the sum of all the values defined must be equal to 100%.

The calculation process of the index takes place in two stages: configuration and analysis. In the first stage, we choose the weighing methods we want to apply, configuring each one according to its requirements. Next, we run the program related to the analysis method we chose before. The calculation program starts by collecting the information it needs from the system's data mart, according to the configuration (list of selected properties and required calculation method) and the selected filters, regarding the base characterization of a student or a question. Then, the system calculates the index, adding all the weighted values of the chosen properties. We established the weights for each property at the time we chose the calculation method (AHP, Individual Classification, or Equalizer). In turn, the values of the properties are calculated using the data about responses, that was collected and stored in its document store. With the final index value, the system stores it in the index data mart jointly with the data related to students, date and time on which the index was calculated, domain and subdomain for which the index was calculated, the weighing method used and, finally, the index value and its variation relative to the last calculated value. The process is similar for calculating the index for the questions area ("TF-QuestionsIdx").

Having the well-being index calculated, we may analyze and interpret it from its support values stored in the system's document store. The use of different weighing methods leads, notably, to obtaining a wide range of weights for the performance index. It is difficult to obtain the same weighing, since the forms of use differ. In order to carry out this process, we chose to analyze the index only according to the "Students" perspective, whose data structure ("TF-StudentsIdx"), as mentioned, was populated with the index values that were calculated based on the data set of responses prepared and on the index's properties.

To begin the analysis of this index, it is important to say that the same criterion was adopted when using each of the methods, namely:

Hit rate > Points rate > Hit rate in the last 10 days > Subdomains rate > Average difficulty level of questions > Average response time
= Consecutive right answer rate.

This hierarchy of properties will be used throughout the analysis of results about "Students". In this hierarchy, the hit rate is the most important property, having a slight superiority over the rate of points obtained. Next, there is the success rate in the last 10 days that allows for measuring attendance, the rate of subdomains covered, as it is important that the student explores all aspects of each domain and finally the average response time and the rate of correct answers consecutively, which will increase the index value as the student's knowledge increases.

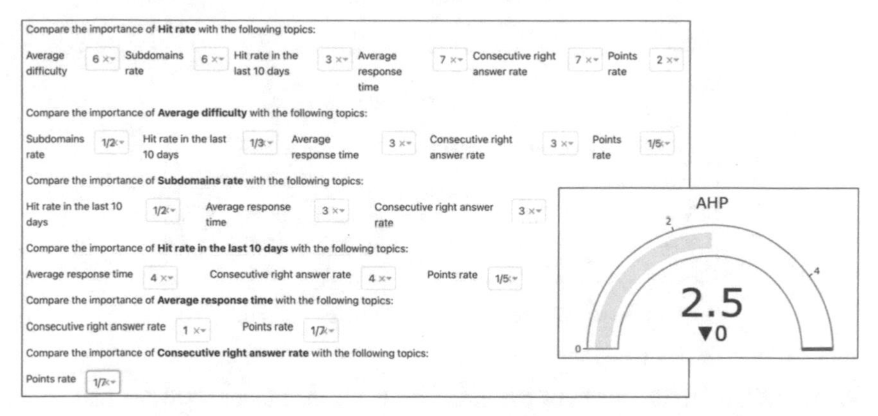

Fig. 2. The results of the AHP method

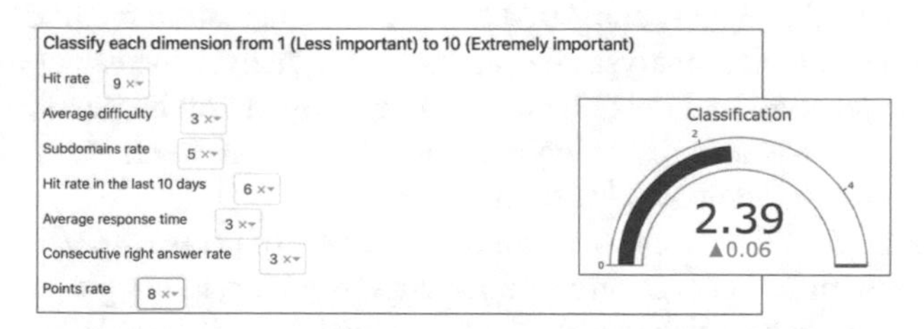

Fig. 3. The results of the Individual Classification method

Let us then see the results produced by each of the methods we have chosen, using the data generated for the well-being index for students ("TF-StudentsIdx"). In Fig. 2. we can see the results produced by the AHP method and compare the values obtained for each property, after selecting the data for a specific student. As can be seen in the figure, the order mentioned above is maintained, being the hit rate values the highest, and the average time rate of responses values and correct answers values being the lowest.

Taking into account the same data set and the same filters, when applying the "Individual Classification" method, we obtained the results shown in Fig. 3. The dashboard configuration presented in this figure includes the same properties analysis hierarchy, but it not contains the same values, as this method does not include comparisons between properties. This means that the values obtained are not exactly the same according the weights of the properties. As you can see, with the application of the Individual Classification method, the same student obtained now an index of 2.390. When compared to the previous method, this value is more penalizing, revealing a more negative situation for the student. Finally, using the "Equalizer" method, in which all weights are manipulated directly, but always using the same criterion per base, we got the results presented in Fig. 4.

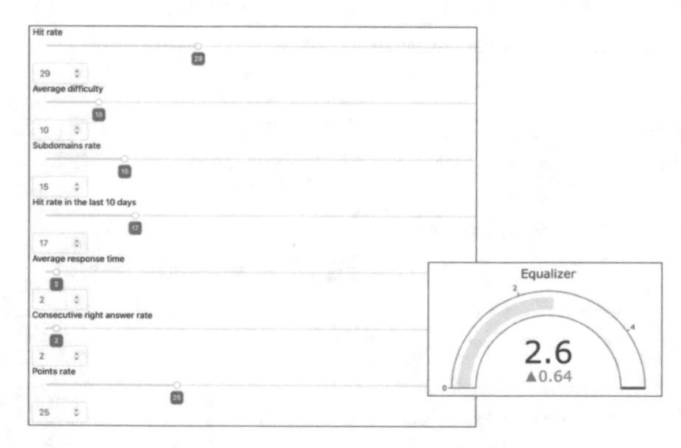

Fig. 4. The results of the Equalizer method

When we look to the value of the index for the same student, we see that with the Equalizer method the index presents a value of 2.600. Thus, this method proves to be more beneficial for students, since, having used the same data and the same criteria, the final value was the highest compared to the other calculation methods. For an even more detailed (and precise) analysis, it is interesting to analyze the values of the weights that were assigned to all methods (Table 2). In this way, it will be possible to conclude whether the criterion is actually being maintained, what are the main differences, and what is the impact on the final value of the index.

By analyzing Table 2, we can see that the hierarchy of properties was always maintained, with slight differences in terms of percentage, which, although small, have some impact on the final value of the index. As mentioned, the hit rate (HR) is the property that has the greatest weight, accompanied by the points rate (PR). Regarding the lowest values, these also continue to be the average response time (ART) and the rate of consecutively correct responses (CRAR). However, in this case, there is the greatest variation in weights in the size of the hit rate, which reaches 9%.

Table 2. Weights assigned by method for the student perspective

	HR	AD	SR	HR10	ART	CRAR	PR
AHP	0.315	0.081	0.097	0.143	0.032	0.032	0.300
Individual Classification	0.220	0.122	0.146	0.171	0.073	0.073	0.195
Equalizer	0.290	0.100	0.150	0.170	0.020	0.020	0.250

Obs.: HR: Hit rate; AD: Average difficulty level of questions; SR: Subdomains rate; HR10; Hit rate in the last 10 days; ART: Average response time; CRAR: Consecutive right answer rate; PR: Points rate.

4 Conclusions and Future Work

The process of designing an index requires a previous study on several factors, ranging from its target audience, through the way in which it can be divided into dimensions, and the way to assign their respective weights, to the way to proceed with its implementation, to develop something that is easily understood by all its users. In addition, it also implies the study of its various forms of development, whether being supported through scales and questionnaires or through the elaboration of calculation expressions, involving different weights and dimensions of analysis. This, it is one of the aspects will have the greatest impact on the final analysis of an index. The form of using the index has great influence in the interest and flexibility of then index. Having different calculation methods for an index also favors its use, as it allows for different analyzes using diverse means.

The definition of a well-being index for the area of education is still considered difficult, because there is not a global established definition. However, several authors agree about their application in the education domain. However, the application of well-being indices in the area of education proved to be an effective, complete and transparent form of evaluation, as can be seen from the case analyzed throughout this work. The integration of an index system, such as the one we have developed, in a tutor system will enable the autonomous development of the student, allowing him to monitor his development in the different fields of study. Although, the main beneficiaries of this system will be the teachers, since through its use they will be able to verify the impact of the various methods and forms of teaching adopted, provide personalized support to each student or receive elements that indicate that they should focus more on the some content in which students have greatest difficulties.

As future work, we intend to identify new calculation methods, as well as increase the number of properties for supporting the calculation of the index. Despite the system already offers three different methods and the index already has a varied and interesting range, it would be useful to integrate other alternative methods of property weighting. The emergence of new forms of analysis is always well seen, since the existence of different points of view on student performance will enable a more complete analysis and, consequently, an even more attentive feedback. The evolution of this type of systems is important, as it directly interferes with the training given to students and may contribute to improve the learning processes.

Acknowledgements. This work has been supported by FCT – Fundação para a Ciência e Tecnologia within the R&D Units Project Scope: UIDB/00319/2020.

References

1. Bandura, R., Del Campo, C.: Survey of Composite Indices Measuring Country Performance: 2006 Update (2009)
2. Cox, D., Frere, M., West, S., Wiseman, J.: Developing and using local community wellbeing indicators: learning from the experience of community indicators Victoria. Aust. J. Soc. Issues **45**(1), 71–88 (2010). https://doi.org/10.1002/j.1839-4655.2010.tb00164.x
3. Sharpe, A.: A Survey of Indicators of Economic and Social Well-Being an Overview of Social Indicators. Centre for the Study of Living Standards, Ottawa (1999)
4. Lowen, R.: Index Analysis: Approach Theory at Work (2015). https://doi.org/10.1007/978-1-4471-6485-2
5. Kamaletdinov, A., Ksenofontov, A.: Index method of evaluating the performance of economic activities. Financ. Theory Pract. **23**, 82–95 (2019). https://doi.org/10.26794/2587-5671-2019-23-3-82-95
6. Sackett, D., Chambers, L., MacPherson, A., Goldsmith, C., Mcauley, R.: The development and application of indices of health: general methods and a summary of results. Am. J. Public Health **67**, 423–428 (1977). https://doi.org/10.2105/AJPH.67.5.423
7. Noble, T., Wyatt, T., McGrath, H., Roffey, S., Rowling, L.: Scoping Study into Approaches to Student Wellbeing: Final (2008)
8. Bland, J., Altman, D.: Validating scales and indexes. BMJ **324**(7337), 606–607 (2002). https://doi.org/10.1136/bmj.324.7337.606
9. Foley, E., Mishook, J., Thompson, J., Kubiak, M., Supovitz, J., Rhude-Faust, M.: Beyond Test Scores: Leading Indicator for Education. Annenberg Institute for School Reform at Brown University (NJ1) (2008)
10. Ryan, R., Deci, E.: On happiness and human potentials: a review of research on hedonic and eudaimonic well-being. Annu. Rev. Psychol. **52**(1), 141–166 (2001). https://doi.org/10.1146/annurev.psych.52.1.141
11. Noble, T., Mcgrath, H.: The positive educational practices framework: a tool for facilitating the work of educational psychologists in promoting pupil wellbeing. Educ. Child Psychol. **25**, 119–134 (2008)
12. Belo, O., Coelho, J., Fernandes, L.: An evolutionary software tool for evaluating students on undergraduate courses. In: Proceedings of 13th Annual International Conference of Education, Research and Innovation (ICERI 2019), Seville, Spain, 11–13 November 2019 (2019). https://doi.org/10.21125/iceri.2019.0703
13. Graesser, A., Hu, X., Sottilare, R.: Intelligent tutoring systems. In: Fischer, F., Hmelo-Silver, C.E., Goldman, S.R., Reimann, P. (eds.) International Handbook of the Learning Sciences. Routledge Handbooks Online (2018)
14. Neto, J., Nascimento, E.: Intelligent tutoring system for distance education. J. Inf. Syst. Technol. Manag. **9**(1), 109–122 (2012). https://doi.org/10.4301/s1807-17752012000100006
15. Yacef, K.: Intelligent teaching assistant systems. In: International Conference on Computers in Education, pp. 136–40. IEEE (2002). https://doi.org/10.1109/CIE.2002.1185885
16. Saaty, T.: That is not the analytic hierarchy process: what the AHP is and what it is not. J. Multi-Criteria Decis. Anal. **6**(6), 324–335 (1997). 10.1002/(SICI)1099-1360(199711)6:6<324::AID-MCDA167>3.0.CO;2-Q
17. Vargas, L.: An overview of the analytic hierarchy process and its applications. Eur. J. Oper. Res. **48**(1), 2–8 (1990). https://doi.org/10.1016/0377-2217(90)90056-H

Information Technologies in Radiocommunications

Experimental Assessment of Fractal Miniaturization for Planar Inverted-F Antennas

Sandra Costanzo[1,2,3](✉) and Adil Masoud Qureshi[1]

[1] DIMES, University of Calabria, Rende, CS, Italy
costanzo@dimes.unical.it

[2] CNR - Institute for Electromagnetic Sensing of the Environment (IREA), Naples, Italy

[3] ICEmB, Inter-University National Research Center on Interactions Between Electromagnetic Fields and Biosystems, Genoa, Italy

Abstract. The comparison of two miniaturized Planar Inverted-F Antenna (PIFA) designs is presented in this work. In the first one, PIFA miniaturization is accomplished by adopting a pre-fractal-based geometry, while in the second design an improved version of the first configuration is considered, which uniquely maintains those geometry features contributing to the desired resonant frequency. Both designs are realized and tested to validate the simulations. Measurement features of both prototypes, including return loss, boresight gain, radiation pattern and efficiency are presented. The discussed results fully validate the utility of pre-fractal geometries for PIFAs miniaturization.

Keywords: PIFA · Efficiency · Measurements · Miniaturization

1 Introduction

During the past four to five decades, we have seen a dramatic decline in the size and weight of electronic devices, while at the same time the capabilities of these devices continued to improve. The final result has been an unprecedented proliferation of consumer technology into our daily lives. The bulk of the credit for this revolution goes to rapid advancement in semiconductor technology. Smaller more efficient transistors have fueled this relentless march towards smaller and smaller form-factors. But none of this would have materialized without corresponding advances in other fields such as energy storage, optical imaging and antenna design.

The majority of the research in the field of antennas is focused on techniques for realizing compactness [1–3]. Most miniaturization techniques mainly belong to one of three groups, namely cavity loading (via dielectrics), circuit loading (lumped impedances, shorting posts etc.), and geometry modification (including folding). Recently, miniaturization has been produced using metamaterial substrates and superstrates, however the cost and the complexity of these techniques remains an issue.

A. Rocha et al. (Eds.): WorldCIST 2023, LNNS 799, pp. 593–598, 2024.
https://doi.org/10.1007/978-3-031-45642-8_57

The use of pre-fractal configuration, belonging to one of the categories described above, has been shown to be useful for planar antennas such as microstrip patch antennas and reflectarrays [4, 5]. The miniaturization thus produced is explained by the apparent increase in the length of the path followed by the currents on the radiating element [6]. This means that the miniaturization technique is independent on the substrate material. Furthermore, it does not require the use of additional components or structures, allowing it to be cheaply manufactured. This paper presents a 'side-by-side' comparison of two miniaturized PIFA designs, that use a pre-fractal inspired radiating element. The illustrated-comparison provides some interesting clues about the mechanism that leads to miniaturization in pre-fractal-based designs.

2 Miniaturized PIFA Designs

2.1 Minkowski Pre-fractal Design

The initial design uses a radiating element in the shape of a first-generation Minkowski pre-fractal. The pre-fractal shaped radiating element has a larger perimeter than a square shaped radiating element of the same overall size. The pre-fractal based PIFA resonated at ~2.3 GHz, as compared to ~2.7 GHz for an identically sized square radiating element. The shift in the resonant frequency corresponds to an effective miniaturization of around ~15%. The design also employs a T-shaped ground plane [7], for improving the bandwidth. Figure 1 shows the simulated reflection coefficient of the Minkowski Pre-fractal design as compared with a typical PIFA having the same dimensions.

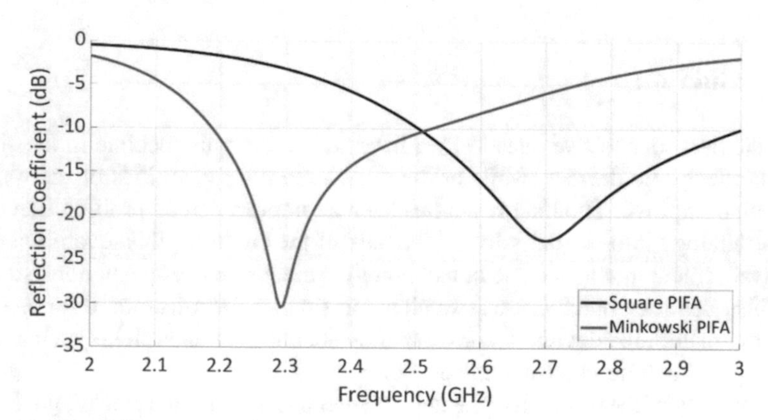

Fig. 1. Simulated Reflection Coefficient of the Minkowski Pre-fractal based PIFA, compared with an identically sized Square PIFA.

2.2 Evolution of the Minkowski Pre-fractal Design

Parametric analyses of the size and location of individual indentations shows that a miniaturization result similar to that reported in Fig. 1 can be achieved using only two indentations (see Fig. 2). Furthermore, the resonant frequency shows a strong correlation with the location of the indentations, while the size (and as a result, the perimeter) remains constant. The results in Fig. 2, along with the lack of correlation between the perimeter and the resonant frequency of the 1st iteration Minkowski design, suggest that the pre-fractal shape is not very effective at miniaturization.

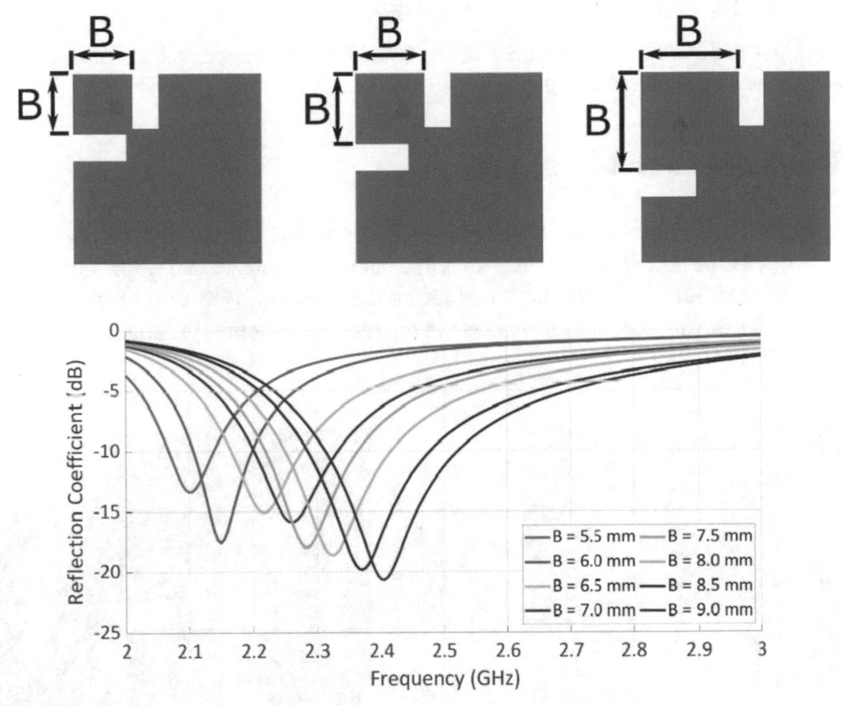

Fig. 2. Simulated Reflection Coefficient with only two indentations in the radiating element, at different positions.

Based on the previous analysis, a new simplified design is created using only two indentations. Simulations show that the simplified design achieves similar performance as that relative to the 1st iteration Minkowski pre-fractal geometry. In particular, Fig. 3 shows the simulated reflection coefficient of the simplified design, very similar as compared to the 1st iteration Minkowski pre-fractal design.

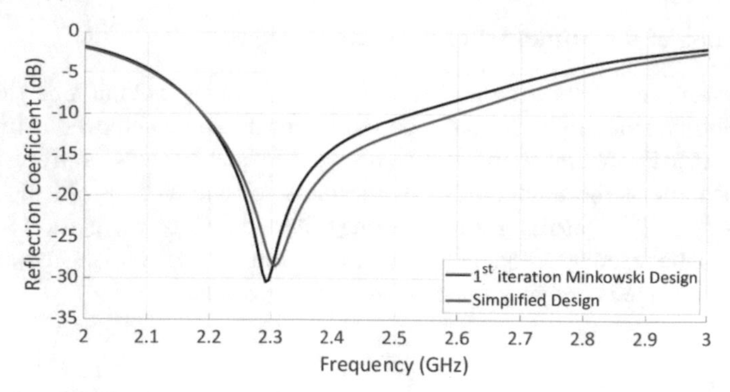

Fig. 3. Simulated Reflection Coefficient of the Simplified Design, compared with the original Minkowski pre-fractal-based design.

3 Realization and Measurement

Prototypes of both designs are fabricated in-house at the ERMIAS Laboratory of the University of Calabria. The antennas are CNC realized using 0.3mm thick copper sheet for the entire structure. Panel mount SMA connectors are used to realize the coaxial probe feed for the antennas. Figure 4 shows the realized antennas, along with a detail of the probe-feed.

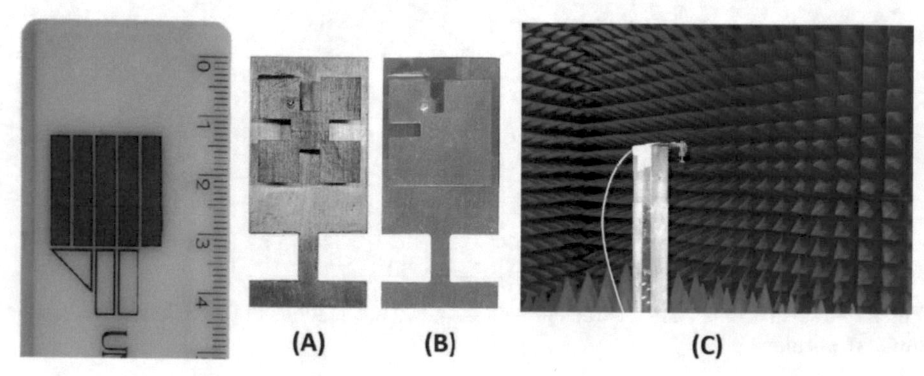

Fig. 4. Fabricated prototypes of the Minkowski pre-fractal design (A), simplified design (B), and the antenna undergoing testing in the anechoic chamber at Università della Calabria (C).

3.1 Impedance Bandwidth

Reflection coefficient is measured for both antennas in the frequency range from 2 to 3 GHz, using a vector network analyzer. Comparing the results (Fig. 5), it can be seen that both antennas have a nearly identical response. The simplified design has a slightly larger 10 dB impedance bandwidth at approximately 290 MHz, as compared to ~270 MHz for the pre-fractal design.

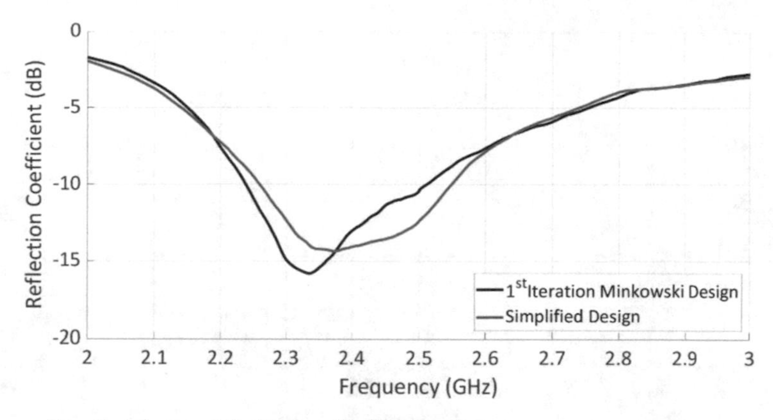

Fig. 5. Measured Reflection Coefficient of the two prototype antennas.

3.2 Boresight Gain

Far-field measurements are also carried out for both antennas in the anechoic chamber (Fig. 6). Again, a similar trend is observed as in the case of the reflection coefficient measurements. Both antennas perform nearly identically with the simplified design exhibiting slightly higher gain in the upper frequency range.

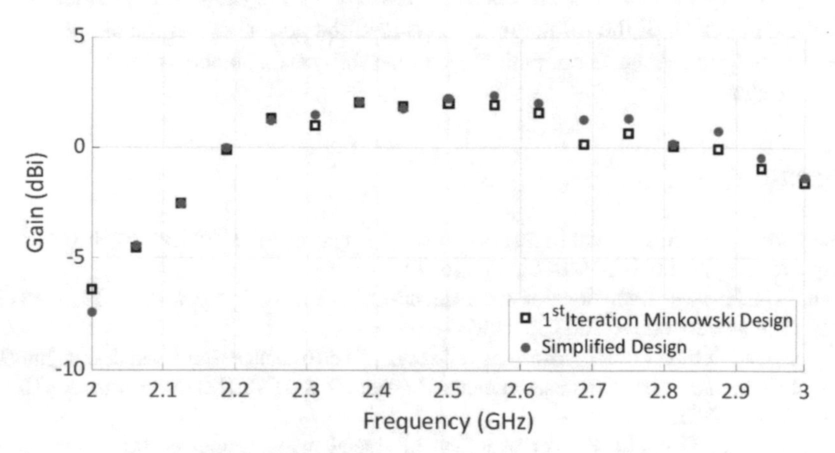

Fig. 6. Measured boresight Gain of the two prototype antennas.

3.3 Radiation Efficiency

Efficiency is calculated for both antennas based on the far-field measurements. Results show that the simplified antenna performs just as well as the 1st iteration Minkowski based design (Fig. 7).

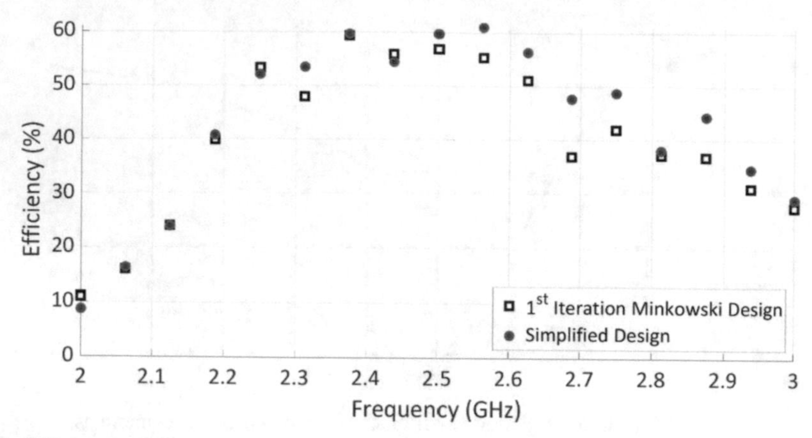

Fig. 7. Efficiency of the two prototype antennas evaluated on the basis of measured gain and radiation pattern.

4 Conclusion

Measured performance characteristics of two different miniaturized PIFA designs have been presented in this work. It has been shown that miniaturization produced by the fractal geometry in a PIFA design is not a function of the perimeter. Furthermore, by careful examination of the miniaturization effect, the geometry can be simplified, thus reducing complexity, while maintaining comparable performance as the full pre-fractal shaped design.

References

1. Wheeler, H.A.: Fundamental limitations of small antennas. Proc. IRE **35**, 1479–1484 (1947). https://doi.org/10.1109/JRPROC.1947.226199
2. Chu, L.J.: Physical limitations of omni-directional antennas. J. Appl. Phys. **19**, 1163–1175 (1948). https://doi.org/10.1063/1.1715038
3. Sievenpiper, D.F., et al.: Experimental validation of performance limits and design guidelines for small antennas. IEEE Trans. Antennas Propag. **60**, 8–19 (2012). https://doi.org/10.1109/TAP.2011.2167938
4. Werner, D.H., Ganguly, S.: An overview of fractal antenna engineering research. IEEE Antennas Propag. Mag. **45**, 38–57 (2003). https://doi.org/10.1109/MAP.2003.1189650
5. Costanzo, S., Venneri, F., Di Massa, G., Borgia, A., Costanzo, A., Raffo, A.: Fractal reflectarray antennas: state of art and new opportunities. Int. J. Antennas Propag. **2016** (2016). https://doi.org/10.1155/2016/7165143
6. Gianvittorio, J.P., Rahmat-Samii, Y.: Fractal antennas: a novel antenna miniaturization technique, and applications. IEEE Antennas Propag. Mag. **44**, 20–36 (2002). https://doi.org/10.1109/74.997888
7. Wang, F., Du, Z., Wang, Q., Gong, K.: Enhanced-bandwidth PIFA with T-shaped ground plane. Electron. Lett. **40**, 1504–1505 (2004). https://doi.org/10.1049/el:20046055

Author Index

A. Rocha et al. (Eds.): WorldCIST 2023, LNNS 799, pp. 599–601, 2024.
https://doi.org/10.1007/978-3-031-45642-8

Printed in the United States
by Baker & Taylor Publisher Services